THE CONCISE HANDBOOK OF ALGEBRA

The Concise Handbook of Algebra

by

Alexander V. Mikhalev
Department of Mechanics and Mathematics,
Moscow State University, Moscow, Russia

and

Günter F. Pilz
Institute of Algebra,,
Johannes Kepler University-Linz, Linz, Austria

Springer-Science+Business Media, B.V.

A C.I.P. Catalogue record for this book is available from the Library of Congress.

ISBN 978-94-017-3269-7   ISBN 978-94-017-3267-3 (eBook)
DOI 10.1007/978-94-017-3267-3

Printed on acid-free paper

Contents

Preface   xv

A SEMIGROUPS   1

A.1 Ideals and Green's Relations
by Lev N. Shevrin in Ekaterinburg, Russia 1

A.2 Bands of Semigroups
by Lev N. Shevrin in Ekaterinburg, Russia 5

A.3 Free Semigroups
by Lev N. Shevrin in Ekaterinburg, Russia 8

A.4 Presentations and Word Problems
*by Peter M. Higgins in Colchester and Nik Ruškuc in
St. Andrews, UK* 13

A.5 Simple Semigroups
by Lev N. Shevrin in Ekaterinburg, Russia 17

A.6 Epigroups
by Lev N. Shevrin in Ekaterinburg, Russia 23

A.7 Periodic Semigroups
by Lev N. Shevrin in Ekaterinburg, Russia 28

A.8 Finite Semigroups and Pseudovarieties
by Jorge Almeida in Porto, Portugal 31

A.9 Regular Semigroups
by Peter G. Trotter in Hobart, Australia 35

A.10 Completely Regular Semigroups
by John M. Howie in St. Andrews, U.K. 39

A.11 Inverse Semigroups
by Peter G. Trotter in Hobart, Australia 42

A.12 Separated Transformation Semigroups
by Kenneth D. Magill, Jr. in Buffalo, NY, USA 46

A.13 Matrix Semigroups
by Jan Okniński in Warsaw, Poland 49

A.14 Subsemigroup Lattices
*by A. J. Ovsyannikov and Lev N. Shevrin in Ekaterinburg,
Russia* 53

A.15 Varieties of Semigroups
*by Lev N. Shevrin and Mikhail V. Volkov in Ekaterinburg,
Russia* 58

A.16 Compact Semigroups
by Karl H. Hofmann in Darmstadt, Germany 62

A.17 Applications of Semigroups
*by Günter F. Pilz in Linz, Austria, Lev N. Shevrin in
Ekaterinburg, Russia, and Peter G. Trotter in Hobart,
Australia* 66

B GROUPS   **71**

B.1 Abelian Groups
by Alexander V. Mikhalev in Moscow, Russia 71

B.2 p-Groups: Basics and the Coclass Project
by Charles R. Leedham-Green in London, UK 75

B.3 p-Groups: Other Approaches
by Charles R. Leedham-Green in London, UK 79

B.4 Soluble Groups
by Hans Lausch in Melbourne, Australia 82

B.5 Permutation Groups
by Peter J. Cameron in London, UK 86

B.6 Lie Groups
by Joachim Hilgert in Clausthal, Germany 89

B.7 Lie Groups and Differential Equations
by Peter J. Olver in Minneapolis, MN, USA 92

B.8 Simple Groups
by Ronald Solomon in Columbus, OH, USA 97

B.9 Free and Relatively Free Groups
by Gilbert Baumslag in New York, NY, USA 102

B.10 Free Products, Amalgamated Products, and HNN Extensions
by Gilbert Baumslag in New York, NY, USA 104

B.11 Word Problems in Groups
by Derek Holt in Warwick, UK 107

B.12 Combinatorial Group Theory
by Gilbert Baumslag in New York, NY, USA 109

B.13 The Burnside Problems
by *Rostislav I. Grigorchuk and Igor G. Lysenok in
Moscow, Russia* 111
B.14 Automatic and Hyperbolic Groups
by *Derek Holt in Warwick, UK* 115
B.15 Actions of Finite Groups
by *Peter Paule in Linz, Austria* 119
B.16 Wreath Products
by *John D. P.Meldrum in Edinburgh, UK* 123
B.17 Automorphism Groups of Algebraic Systems
by *Boris I. Plotkin in Jerusalem, Israel* 127
B.18 Frobenius Groups
by *Peter Fleischmann in Kent, UK* 131
B.19 Covered and Fibered Groups
by *Carlton J. Maxson in College Station, TX, USA* 134
B.20 Subgroup Lattices of Finite Abelian Groups
by *Frank Vogt in Darmstadt, Germany* 137
B.21 Varieties of Groups
by *Awad A. Iskander in Lafayette, LA, USA* 140
B.22 Groups and Absolute Geometry
by *Helmut Karzel in Munich, Germany* 144
B.23 Groups for Particle Physics
by *Gudrun Kalmbach H. E. in Ulm, Germany* 148

C RINGS, MODULES, ALGEBRAS   153
C.1 Commutative Algebra I: Ideal Decompositions
by *Robert Gilmer in Tallahassee, FL, USA* 153
C.2 Commutative Algebra II: The Krull Dimension
by *Robert Gilmer in Tallahassee, FL, USA* 156
C.3 Factorization and Decomposition of Polynomials
by *Joachim von zur Gathen in Paderborn, Germany* . . . 159
C.4 Element Factorization in Integral Domains
by *David F. Anderson in Knoxville, TN, USA* 161
C.5 Dedekind and Prüfer Domains
by *Marco Fontana in Rome, Italy and Ira J. Papick in
Columbia, MO, USA* 165
C.6 Local Rings
by *T. Y. Lam in Berkeley, CA, USA* 169
C.7 Semilocal Rings
by *T. Y. Lam in Berkeley, CA, USA* 173

C.8 Cohen-Macaulay Rings
  by Jürgen Herzog in Essen, Germany 177
C.9 Rings of Formal Power Series
  by Askar A. Tuganbaev in Moscow, Russia 182
C.10 Automorphisms of Polynomial Algebras and the Jacobian
  Conjecture
  by Jie-Tai Yu in Hong Kong, China 185
C.11 Rings of Integer-Valued Polynomials
  *by Paul-Jean Cahen in Marseille and Jean-Luc Chabert
  in Amiens, France* 188
C.12 Commutativity Conditions
  by Howard E. Bell in St. Catharines, Canada 192
C.13 Finite Dimensional Division Algebras
  by David Saltman in Austin, TX, USA 193
C.14 Infinite Dimensional Division Rings
  by Paul M. Cohn in London, UK 198
C.15 Rings of Quotients
  by Wallace S. Martindale, 3rd, in Glenside, PA, USA . . . 202
C.16 Endomorphism and Matrix Rings
  *by Alexander V. Mikhalev and Askar A. Tuganbaev in
  Moscow, Russia* 205
C.17 Primitive Rings and Semisimplicity
  by Kostia Beidar in Tainan, Taiwan 209
C.18 The Jacobson Radical
  by Richard Wiegandt in Budapest, Hungary 213
C.19 General Radical Theory
  by Richard Wiegandt in Budapest, Hungary 215
C.20 Free and Projective Modules
  *by Alexander V. Mikhalev and Askar A. Tuganbaev in
  Moscow, Russia* 219
C.21 Injective Modules
  by Askar A. Tuganbaev in Moscow, Russia 223
C.22 Flat Modules
  by Askar A. Tuganbaev in Moscow, Russia 227
C.23 Free Ideal Rings and Free Algebras
  by Paul M. Cohn in London, UK 231
C.24 Simple Rings
  by Kostia Beidar in Tainan, Taiwan 235
C.25 Principal Ideal Domains
  by Paul M. Cohn in London, UK 239

C.26 Noetherian Rings
 by Lance W. Small in San Diego, CA, USA 242

C.27 Artinian Rings
 by Victor T. Markov in Moscow, Russia 244

C.28 Maximal, Prime, and Semiprime
 by Gary F. Birkenmeier in Lafayette, LA, USA 249

C.29 Regular Rings
 *by Alexander V. Mikhalev and Askar A. Tuganbaev in
 Moscow, Russia* 253

C.30 Quasi-Frobenius and Self-injective Rings
 *by Alexander V. Mikhalev and Askar A. Tuganbaev in
 Moscow, Russia* 258

C.31 Gröbner Bases: The Commutative Case
 by Bruno Buchberger in Linz, Austria 262

C.32 Gröbner Bases: Applications
 by Bruno Buchberger in Linz, Austria 265

C.33 Gröbner Bases: The Non-commutative Case
 by Leonid Bokut' in Novosibirsk, Russia 268

C.34 Orders and Maximal Orders
 by T. Y. Lam in Berkeley, CA, USA 273

C.35 Representations of Orders
 by Vladimir Kirichenko in Kiev, Ukraine 278

C.36 Distributive Modules and Rings
 by Askar A. Tuganbaev in Moscow, Russia 280

C.37 PI-Algebras
 *by Alexei J. Belov in Perth, Australia, and Louis H.
 Rowen in Ramat Gan, Israel* 284

C.38 Clifford Algebras
 by Jaques Helmstetter in Grenoble, France 289

C.39 Rings with Involution
 by Henry E. Heatherly in Lafayette, LA, USA 293

C.40 Rings of Operators
 by Henry E. Heatherly in Lafayette, LA, USA 296

C.41 The Gel'fand-Kirillov Dimension
 by Vesselin Drensky in Sofia, Bulgaria 300

C.42 Coalgebras
 by Vyacheslav A. Artamonov in Moscow, Russia 305

C.43 Hopf Algebras
 by Vyacheslav A. Artamonov in Moscow, Russia 308

C.44 Ordered Rings
    by Laszlo Fuchs in New Orleans, LA, USA 311

C.45 Topological Rings and Modules
    by S.T.Glavatsky and A.V.Mikhalev in Moscow, Russia . 314

C.46 Semirings and Semifields
    *by Udo Hebisch in Freiberg and Hanns J. Weinert in
    Clausthal, Germany* 318

C.47 Near-Rings and Near-Fields
    by Günter F. Pilz in Linz, Austria 322

C.48 Conformal Algebras
    by Efim Zelmanov in New Haven, CT, USA 326

C.49 Lie Algebras
    by Mikhael Zaicev in Moscow, Russia 330

C.50 Simple Lie Algebras
    by Andrej A. Zolotykh in Moscow, Russia 333

C.51 Varieties of Lie Algebras
    by Mikhael Zaicev in Moscow, Russia 337

C.52 Lie Superalgebras
    by Alexander A. Mikhalev in Moscow, Russia 339

C.53 Other Nonassociative Algebras
    by Luiz A. Peresi in São Paulo, Brazil 343

C.54 Computational Ring Theory
    by Franz Winkler in Linz, Austria 347

C.55 Applications of Rings
    by Günter F. Pilz in Linz, Austria 350

D FIELDS   **355**

D.1 Field Extensions
    by Shreeram S. Abhyankar in Lafayette, IN, USA 355

D.2 Finite Fields
    by Harald Niederreiter in Singapore 359

D.3 Galois Theory
    by Joseph J. Rotman in Urbana, IL, USA 362

D.4 Differential Galois Theory
    by Andy R. Magid in Norman, OK, USA 366

D.5 Differential Dimension Polynomials
    by A. V. Mikhalev and E. V. Pankratiev in Moscow, Russia 369

D.6 Ordered Fields
    by T. Y. Lam in Berkeley, CA, USA 373

D.7 Applications of Fields
     by Franz Binder in Linz, Austria and Rudi Lidl in Hobart,
     Australia 377

E REPRESENTATION THEORY   383
   E.1 Group Rings
        by Donald S. Passman in Madison, WI, USA 383
   E.2 Character Theory
        by I. Martin Isaacs in Madison, WI, USA 386
   E.3 Ordinary Representations I
        by Albert Fässler in Biel, Switzerland 390
   E.4 Ordinary Representations II
        by Albert Fässler in Biel, Switzerland 392
   E.5 Modular Representations: General theory
        by Peter Fleischmann in Kent, UK 398
   E.6 Modular Representations: Brauer's Theorems
        by Peter Fleischmann in Kent, UK 400
   E.7 Characters of the Symmetric and Alternating Groups
        by Albert Fässler in Biel, Switzerland 403
   E.8 Representation Theory of Lie Groups
        by Joachim Hilgert in Clausthal, Germany 405
   E.9 Groups in 2D/3D vision problems
        by Kenichi Kanatani in Okayama, Japan and Günter F.
        Pilz in Linz, Austria 409
   E.10 Representation Theory and Statistics
        by Albert Fässler in Biel, Switzerland 412

F LATTICES   417
   F.1 Congruences and Constructions
        by George Grätzer in Winnipeg, Canada and E. Tamás
        Schmidt in Budapest, Hungary 417
   F.2 Modular Lattices
        by Ralph Freese in Honolulu, HI, USA 420
   F.3 Distributive Lattices; Heyting and Post Algebras
        by Viacheslav N. Salii in Saratov (Russia) 423
   F.4 Complemented and Orthocomplemented Lattices
        by Gudrun Kalmbach H. E. in Ulm, Germany 426
   F.5 Orthomodular Lattices
        by Gudrun Kalmbach H. E. in Ulm, Germany 429

F.6 Boolean Algebras
by Carlton J. Maxson in College Station, TX, USA and Günter F. Pilz in Linz, Austria 432

F.7 Complete Lattices
by Viacheslav N. Saliĭ in Saratov, Russia 436

F.8 Free Lattices
by Ralph Freese in Honolulu, HI, USA 439

F.9 Varieties of Lattices
by George Grätzer in Winnipeg, Canada 442

F.10 Applications of Lattices: Formal Concept Analysis
by Rudolf Wille in Darmstadt, Germany 446

G UNIVERSAL ALGEBRA   **451**

G.1 Constructions in Universal Algebras
by Vyacheslav A. Artamonov in Moscow, Russia, and Günter F. Pilz in Linz, Austria 451

G.2 Automorphisms and Endomorphisms in Universal Algebra
by Boris I. Plotkin in Jerusalem, Israel 455

G.3 Polynomials and Polynomial Completeness
by Kalle Kaarli in Tartu, Estonia 457

G.4 Free Algebras
by Lev N. Shevrin and Evgeny V. Sukhanov in Ekaterinburg, Russia 461

G.5 Varieties and Quasi-varieties
by Lev N. Shevrin and Mikhail V. Volkov in Ekaterinburg, Russia 465

G.6 Congruence Modular Varieties
by A. G. Pinus in Novosibirsk, Russia and Yefim Katsov in Hanover, IN, USA 469

G.7 Word Problems and Rewriting Systems
by Leonid Bokut' in Novosibirsk, Russia 474

G.8 Relational Algebras
by Hajnal Andréka, Judit X. Madarász, and István Németi in Budapest, Hungary 478

G.9 Partial Algebras
by Peter Burmeister in Darmstadt, Germany 482

G.10 Abstract Data Types
by Hans-Dieter Ehrich in Braunschweig, Germany 486

H HOMOLOGICAL ALGEBRA   **491**

H.1 Foundations of Homological Algebra
     by Peter Hilton in Binghampton, NY, USA 491

H.2 Universal Constructions
     by Yefim Katsov in Hanover, IN, USA 493

H.3 Projective and Injective Resolutions
     by Yefim Katsov in Hanover, IN, USA 497

H.4 Homological Dimension
     by Peter Hilton in Binghampton, NY, USA 502

H.5 Homological Characterizations of Rings: The Commutative Case
     by Sarah Glaz in Storrs, CT, USA 505

H.6 Homological Characterizations of Rings: The General Case
     by Alexander V. Mikhalev and Askar A. Tuganbaev in Moscow, Russia 508

H.7 Algebraic K-Theory I: K_0 and K_1
     by Jonathan M. Rosenberg in College Park, MD, USA . . 513

H.8 Algebraic K-theory II: K_2, Milnor K-theory, and Symbols
     by Jonathan M. Rosenberg in College Park, MD, USA . . 517

H.9 Algebraic K-Theory III: Higher Algebraic K-Theory
     by Jonathan M. Rosenberg in College Park, MD, USA . . 520

I MISCELLANEOUS   **523**

I.1 Model Theory for Algebraists
     by H. Jerome Keisler in Madison, WI, USA 523

I.2 Linear Codes over Fields
     by Günter F. Pilz in Linz, Austria 526

I.3 Linear Codes over Rings
     by V. T. Markov, A. V. Mikhalev, and A. A. Nechaev in Moscow, Russia 530

I.4 History of Algebra before 1500
     by Hans Kaiser in Vienna, Austria 535

I.5 History of Algebra after 1500
     by John Stillwell in Melbourne, Australia 539

Bibliography   **543**

About the Authors   **585**

Index   **595**

Preface

It is by no means clear what comprises the "heart" or "core" of algebra, the part of algebra which every algebraist should know. Hence we feel that a book on "our heart" might be useful. We have tried to catch this heart in a collection of about 150 short sections, written by leading algebraists in these areas. These sections are organized in 9 chapters A, B, ..., I. Of course, the selection is partly based on personal preferences, and we ask you for your understanding if some selections do not meet your taste (for unknown reasons, we only had problems in the chapter "Groups" to get enough articles in time). We hope that this book sets up a standard of what all algebraists are supposed to know in "their" chapters; interested people from other areas should be able to get a quick idea about the area. So the target group consists of anyone interested in algebra, from graduate students to established researchers, including those who want to obtain a quick overview or a better understanding of our selected topics. The prerequisites are something like the contents of standard textbooks on higher algebra. This book should also enable the reader to read the "big" Handbook (Hazewinkel 1999–) and other handbooks.

In case of multiple authors, the authors are listed alphabetically; so their order has nothing to do with the amounts of their contributions. At the end of the book, you will find a list of all contributors with their addresses and (in most cases) photos.

We have tried to make the presentation fairly uniform; cross references link the sections. So this book is not just a collection of independent articles. Standard conventions are $\mathbb{N}, \mathbb{N}_0, \mathbb{Z}, \mathbb{Z}_n, \mathbb{Q}, \mathbb{R}, \mathbb{C}, \mathbb{H}, \mathbb{B}$ for the sets of natural numbers, natural numbers together with zero, integers, integers mod n, rationals, reals, quaternions, and for the 2-element Boolean algebra $\{0, 1\}$, respectively. $\mathrm{Mat}_{m \times n}(R)$ is the collection of all $m \times n$-matrices over some ring R, and $\mathrm{Mat}_n R$ are the $n \times n$-matrices over R. A^+ denotes the free semigroup over the alphabet A, while A^* is the free monoid over A. The symmetric group over n symbols is denoted by

S_n. $I \trianglelefteq A$ denotes that I is a normal subgroup or an ideal of A; if it is proper, we write $(I \triangleleft A)$. $R[x]$, $R(x)$, $R\langle x \rangle$, $R[[x]]$, and $R\langle\langle x \rangle\rangle$ denote the poynomials over R (commuting variable), the rational functions, the polynomials (non-commuting variable, i.e., the free R-algebra), the formal power series, and the Laurent series, respectively. In case of several variables $x_1, x_2, \ldots$, we write $R[X]$, and so on. Finally, $\mathbb{F}_{q^m}$ denotes the finite field of order q^m.

A "domain" is a non-zero ring without non-trivial divisors of zero; it is not necessarily commutative. Also, a principal ideal domain is not automatically assumed to be commutative. If G is a Lie group, the associated Lie algebra is denoted by $\mathfrak{g}$ or by $\mathrm{Lie}(G)$. The group algebra of G over R is denoted by $R[G]$ or simply by RG.

Unless stated otherwise, all algebraic structures are supposed to be non-empty.

A very big "Thank you" goes to Franz Binder. He is at the same time one of the authors and also the main person who helped enormously in the preparation of the manuscript. Thanks are also due to Mrs. Waltraud Eidljörg (Linz) for their help in the production process and to Dr. Liesbeth Mol and Mrs. Patricia de Vries (Kluwer) for their patience and most pleasant cooperation. And, of course, cordial thanks to those who made the book: the authors.

Please kindly report any improvements which come to your mind. We hope that the readers will enjoy this book. It should not be in the shelves of our colleagues, but on their working tables!

ALEXANDER V. MIKHALEV IN MOSCOW (RUSSIA)
GÜNTER F. PILZ IN LINZ (AUSTRIA)

Chapter A

SEMIGROUPS

A.1 Ideals and Green's Relations

by Lev N. Shevrin in Ekaterinburg, Russia

Generalities

Let S be a semigroup. A non-empty subset A of S is called a **left** [**right**] **ideal** if $SA \subseteq A$ [$AS \subseteq A$]. If A is both a left and a right ideal of S, then A is called a **two-sided ideal**, or simply an **ideal**. Clearly, S itself is an ideal of S; an ideal (of any kind) which differs from S is called **proper**.

Ideals play a substantial role in various semigroup-theoretic considerations. Many properties of semigroups are connected with the behavior of (all or certain of) their ideals, many concepts are based on the consideration of ideals. Sometimes the corresponding connections are direct: a number of definitions or characterizations of important types of semigroups can be given in terms of ideals. Such are, for example, the definition of an archimedean semigroup (see below), and a characterization of regular semigroups (Section A.9): a semigroup S is regular iff for any left ideal L and any right ideal R of S, the equality $RL = R \cap L$ holds. Restrictions on a semigroup to be "poor" of ideals of a certain kind provide rather rich in content considerations of different types of simple semigroups, see Section A.5.

In contrast to ring theory, where any congruence of a ring is determined by some ideal, this is not the case in semigroup theory. However, there is an important type of congruences associated with ideals. It is defined as follows. Let I be an ideal of a semigroup S. The binary relation ρ_I defined by the condition

$$a \; \rho_I \; b \iff \text{either } a = b \text{ or both } a \text{ and } b \text{ belong to } I$$

1

is a congruence. It is called the **Rees congruence modulo** I; this type
of congruences was first considered in (Rees 1940). The equivalence
classes of S mod ρ_I are I itself and (if I is proper) every one-element
set $\{x\}$ with $x \in S \backslash I$. The quotient semigroup S/ρ_I is commonly written
as S/I and is called the **Rees quotient semigroup** (modulo I). Clearly,
S/I has a zero, namely I. A semigroup S is called an **ideal extension**
of a semigroup A by a semigroup B if A is an ideal of S and the quotient
semigroup S/A is isomorphic with B. The study of different types of
ideal extensions of semigroups form a rather fruitful trend in semigroup
theory; a survey of many achievements in this trend is presented in
(Petrich 1970) and (Petrich 1973).

For any $a \in S$, the set $L(a) = \{a\} \cup Sa$ $[R(a) = \{a\} \cup aS, J(a) =$
$\{a\} \cup Sa \cup aS \cup SaS]$ is the least of the left [right, two-sided] ideals
containing a. The set $L(a)$ $[R(a), J(a)]$ is called the **principal left**
[right, two-sided] ideal of S generated by a.

A semigroup S is called **archimedean** if for any $a, b \in S$ there exists a
natural number n such that $a^n \in J(b)$. The structure of archimedean epigroups is presented in Section A.6. In general, archimedean semigroups,
even commutative ones, may have a rather complicated structure; see,
for example, (Clifford and Preston 1967a, p. 136–137). A certain general
construction of commutative archimedean semigroups without idempotents (Tamura's theorem) is presented in (Grillet 1995, Ch. IV). About
the role of archimedean semigroups in some decompositions of semigroups, see Section A.2 and Section A.6.

A survey of investigations on ideals of commutative semigroups is
given in (Anderson and Jonson 1984).

The set $\mathrm{Id}_l\, S$ $[\mathrm{Id}_r\, S]$ of all left [right] ideals of a semigroup S is an
upper semilattice under set-theoretic inclusion. The set $\mathrm{Id}\, S$ of all two-
sided ideals of S is a (distributive) lattice, not necessarily complete. If
S has a zero 0, then $\{0\}$ is the least element of $\mathrm{Id}_l\, S$, $\mathrm{Id}_r\, S$, and $\mathrm{Id}\, S$,
so these sets turn out to be complete lattices. A minimal element of
the set $\mathrm{Id}_l\, S$ $[\mathrm{Id}_r\, S, \mathrm{Id}\, S]$ is called a **minimal left [right, two-sided]**
ideal of a semigroup S. Not any semigroup possesses such ideals: the
simplest example is provided by $(\mathbb{N}, +)$. A minimal two-sided ideal, if it
exists, is unique. It is the least ideal and called the **kernel** of the corresponding semigroup. For example, the kernel of the full transformation
semigroup $\mathcal{T}_X$ is the right zero semigroup consisting of all constant transformations on X; the kernel of a finite cyclic semigroup is a cyclic group.
If K is the kernel of a semigroup S, then K does neither contain ideals
of S different from K, nor, moreover, proper ideals of K.

If a semigroup S has a minimal left ideal (m.l.i.), then every left ideal of S contains some m.l.i. and the union of all m.l.i. of S is the kernel of S. Naturally there are similar statements concerning minimal right ideals (m.r.i.). If a semigroup S has a m.l.i. L and a m.r.i. R, then $R \cap L = RL$ is a maximal subgroup of S, and $L = Se$, $R = eS$, where e is the identity of this subgroup; furthermore, LR is the kernel of S.

For a semigroup S with zero, it is natural to consider nonzero ideals, and a minimal element in $\mathrm{Id}_l\, S \setminus \{0\}$ [$\mathrm{Id}_r\, S \setminus \{0\}$, $\mathrm{Id}\, S \setminus \{0\}$] is called a **0-minimal** left [right, two-sided] ideal. Properties of 0-minimal ideals are partly parallel to those of minimal ones, but there are some differences. For instance, a 0-minimal two-sided ideal is not necessarily unique as well as may contain proper non-zero ideals of it.

Minimal and 0-minimal ideals play an appreciable role in the structure theory of semigroups. Some illustration of this role is presented in Section A.5. For more detailed information, consult (Clifford and Preston 1967a, § 2.5, 2.7), and (Clifford and Preston 1967b, Ch. 6, § 7.7, 8.2, 8.3).

Green's relations

The relations featured in this title are five equivalences $\mathscr{L}, \mathscr{R}, \mathscr{J}, \mathscr{D}, \mathscr{H}$ on an arbitrary semigroup defined in terms of principal ideals as follows:

$$x\mathscr{L}y \Leftrightarrow L(x) = L(y), \quad x\mathscr{R}y \Leftrightarrow R(x) = R(y),$$

$$x\mathscr{J}y \Leftrightarrow J(x) = J(y), \quad \mathscr{D} = \mathscr{L} \vee \mathscr{R}, \quad \mathscr{H} = \mathscr{L} \cap \mathscr{R}.$$

Here $\vee$ means the join in the lattice of equivalences on a given semigroup. In the semigroup of binary relations, $\mathscr{L}$ and $\mathscr{R}$ commute, so in fact $\mathscr{D} = \mathscr{L} \cdot \mathscr{R} = \mathscr{R} \cdot \mathscr{L}$. Clearly, for a commutative semigroup, $\mathscr{L} = \mathscr{R} = \mathscr{J} = \mathscr{D} = \mathscr{H}$. In the general case, $\mathscr{D} \subseteq \mathscr{J}$ while $\mathscr{L}$ and $\mathscr{R}$ can be incomparable with respect to inclusion. There are semigroups for which all Green's relations are distinct; such a semigroup is, for instance, the monoid given by generators a, b, c and a defining relation $abc = 1$. Rather often, the equality $\mathscr{J} = \mathscr{D}$ holds; it takes place, for instance, for epigroups (and so, in particular, for finite semigroups, see Section A.6). It is also possible that, for a semigroup S, an utmost difference between $\mathscr{D}$ and $\mathscr{J}$ takes place: namely $\mathscr{D}$ is the equality relation while $\mathscr{J}$ is the universal relation $S \times S$.

The name for the relations under consideration is due to their first study in (Green 1951). They give a useful tool for elucidating the structure of semigroups in general, and have played a fundamental role in the development of semigroup theory.

There is a standard notation for the equivalence classes of Green's relations: the $\mathscr{L}$-class [$\mathscr{R}$-class, $\mathscr{J}$-class, $\mathscr{D}$-class, $\mathscr{H}$-class] containing the element a is denoted by L_a [R_a, J_a, D_a, H_a]. The quotient sets $S/\mathscr{L}$, $S/\mathscr{R}$ and $S/\mathscr{J}$ can naturally be made into posets by the following rules: $L_a \leq L_b$ [$R_a \leq R_b$, $J_a \leq J_b$] means $L(a) \subseteq L(b)$ [$R(a) \subseteq R(b)$, $J(a) \subseteq J(b)$]. The poset $S/\mathscr{J}$ is called the *frame* of S; the posets $S/\mathscr{L}$ and $S/\mathscr{R}$ may be called the *left* and the *right frames*, respectively. For example, the frame of the full transformation semigroup T_X is a chain of the same cardinality as X. The frame of an arbitrary semigroup is obviously a down directed poset. It is quite non-trivial that the converse takes place as well: any down directed poset is isomorphic to the frame of some semigroup (even inverse with certain additional properties), see (Ash 1979) and (Meakin 1980). For a finite poset, the property of being down directed is the same as the property to have a least element, and any such a poset is isomorphic to the frame of some *finite* semigroup (again even with additional properties), see (Hall 1973b).

An $\mathscr{L}$-class and an $\mathscr{R}$-class have non-empty intersection iff they are contained in the same $\mathscr{D}$-class. All $\mathscr{L}$-classes [$\mathscr{R}$-classes, $\mathscr{H}$-classes] contained in the same $\mathscr{D}$-class are of equal cardinality. An arbitrary $\mathscr{D}$-class may be visualized as a rectangular pattern (which, following Clifford and Preston, is commonly called the *egg-box picture*),

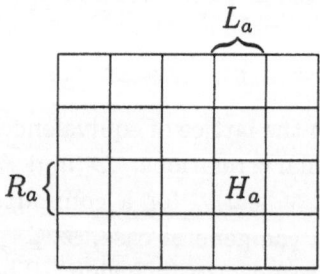

where each row [column, cell] represents an $\mathscr{R}$-class [$\mathscr{L}$-class, $\mathscr{H}$-class].

Those $\mathscr{H}$-classes which contain idempotents are precisely the maximal subgroups of a given semigroup; such classes are called *group $\mathscr{H}$-classes*. The group $\mathscr{H}$-classes contained in the same $\mathscr{D}$-class are isomorphic groups. The following conditions for a $\mathscr{D}$-class D are equivalent: (i) D contains an idempotent; (ii) D contains a regular element; (iii) all elements of D are regular. A $\mathscr{D}$-class with this property is called *regular*. The structure of a regular semigroup depends to a large extent both on the structure of its $\mathscr{D}$-classes and, if there are distinct $\mathscr{D}$-classes, on interrelations between them. Semigroups having a single $\mathscr{D}$-class will be considered in Section A.5.

For more detailed information on Green's relations and related topics, consult some general semigroup-theoretic sources such as (Clifford and Preston 1967a), (Lallement 1979), (Shevrin 1991), (Grillet 1995), (Howie 1995).

A.2 Bands of Semigroups

by Lev N. Shevrin in Ekaterinburg, Russia

A **band** is another name for a semigroup of idempotents. The following types of bands deserve to be distinguished: a **semilattice**, i.e., a commutative band (in such a band, the relation $\leq$ defined by the rule $x \leq y \Leftrightarrow xy = x$ is a partial order with respect to which ab is the infimum of a and b); a **rectangular band** defined by the identity $xyx = x$ (which implies the identity $x^2 = x$); and a **left** [**right**] **zero** (or **left** [**right**] **singular**) **semigroup** defined by the identity $xy = x$ [$xy = y$]. A semigroup which is either left or right singular will be called **singular**.

Let S be a semigroup, and ρ be a congruence on S such that the quotient semigroup S/ρ is a band. This means that each ρ-class is a subsemigroup. Let $\{S_\alpha\}_{\alpha \in I}$ be the family of these subsemigroups. Then S is said to be a **band of subsemigroups** S_α, $\alpha \in I$. In other words, S is a band of subsemigroups S_α if they form a partition of S, and for any S_α, S_β, there exists S_γ such that $S_\alpha S_\beta \subseteq S_\gamma$. The subsemigroups S_α are called the **components** of the band under consideration. If all components belong to a certain class of semigroups $\mathcal{K}$, then we say that S is a **band** (or **decomposable into a band**) **of semigroups** of $\mathcal{K}$. For instance, we may consider bands of groups, bands of commutative semigroups, and the like. (In these terms, semigroups of idempotents, i.e., bands proper, are just bands of one-element semigroups.) On the other hand, we may take an interest in a type of the quotient semigroup S/ρ for a given decomposition of a semigroup S into a band. If S/ρ is a semilattice [rectangular band], then S is said to be a **semilattice** [**rectangular band**] **of the corresponding semigroups**. Similarly, we may speak about a **left** [**right**] **zero** (or simply a **left** [**right**]) **band of semigroups**; in this case, for any two components S_α, S_β of a given band, the inclusion $S_\alpha S_\beta \subseteq S_\alpha$ [$S_\alpha S_\beta \subseteq S_\beta$] holds, that is, all components are right [left] ideals. For example, a simple semigroup possessing minimal left [right] ideals is a right [left] band of these ideals. If S/ρ is a semilattice being a chain, we say that S is a *chain of the corresponding semigroups*.

A decomposition of a semigroup into a semilattice of some subsemi-groups is called a *semilattice decomposition*. Similarly, we obtain the notions of *rectangular*, *left* and *right decompositions* as well. A rectangular decomposition is also called a *matrix decomposition*. Both of these terms are justified by the fact that rectangular bands can be characterized as direct products of a left zero semigroup and a right zero semigroup. So the components of a matrix decomposition can be equipped with double indices running some sets I and Λ such that the rule

$$S_{i\lambda}S_{j\mu} \subseteq S_{i\mu}, \; i,j \in I, \; \lambda, \mu \in \Lambda$$

is fulfilled.

The notion of band has been introduced explicitly in (Clifford 1954), although in effect it appeared in the basic work (Clifford 1941), where it has been established that, as we now formulate, a completely regular semigroup is a semilattice of completely simple semigroups. One may say, however, that this phenomenon appeared in a "latent" form already in the pioneer work (Suschkewitsch 1928), where the structure of a finite simple semigroup has been described. It is immediately seen from this description that, in today's terminology, such semigroups are precisely finite rectangular bands of groups; see Section A.5 for a more detailed exposition of the corresponding topic.

Different decompositions into bands play an important role in the structure theory of semigroups, since semigroups from many classes turn out to be decomposable into bands of more specific semigroups. It becomes thereby possible, to some extent, to reduce the examination of such semigroups to consideration of types to which the components of a band belong, as well as to using the peculiarities of interaction of these components.

For a class of semigroups $\mathcal{K}$, by a $\mathcal{K}$-*congruence* we mean a congru-ence of a given semigroup such that the corresponding quotient semi-group belongs to $\mathcal{K}$. If $\mathcal{K}$ is a variety, on any semigroup there exists the smallest $\mathcal{K}$-congruence. If this congruence on a semigroup S coin-cides with the universal relation, S is called $\mathcal{K}$-*indecomposable*. In particular, since the property of a semigroup to be a band is defined by identities, there exists the smallest band congruence on any semi-group. The corresponding partition is the *greatest band decomposi-tion*. Similarly, we may consider, among others, the *greatest semi-lattice [matrix, left, right] decompositions* of any semigroup. Such decompositions can elucidate the structure of semigroups belonging to many classes having been the object of attention. The two examples of a decomposition into a band mentioned above (for completely regular and completely simple semigroups) present just the greatest semilattice and matrix decompositions, respectively. We give now some more examples for the case of commutative semigroups (where "band" is clearly the same as "semilattice". A commutative semigroup is called *separative* if it obeys the condition $x^2 = y^2 = xy \Rightarrow x = y$. The greatest semilattice decomposition of an arbitrary commutative semigroup is its (single) decomposition into a semilattice of archimedean subsemigroups. The components of this decomposition are called the **archimedean components**. A commutative semigroup is **regular** [**separative**] iff its archimedean components are groups [cancellative]. The archimedean components of a commutative epigroup (in particular, finite semigroup) are unipotent, i.e., each contains a single idempotent. It is notable that the components of the greatest semilattice decomposition of any semigroup are semilattice-indecomposable. For the components of the greatest matrix decomposition, this is not the case.

Semilattice and matrix decompositions play a peculiar role not only because of their clearness (it is rather convenient to deal with certain visual attributes of a partial order as well as with a rectangular pattern) but also due to the following fact: any band of semigroups forming a family $\{S_\alpha\}$ is a semilattice of rectangular bands of semigroups from $\{S_\alpha\}$; that is, the components S_α can be grouped in subfamilies such that the union of the components belonging to each subfamily is a rectangular band of these components, while the initial semigroup is the semilattice of the unions indicated. In particular, any semigroup of idempotents is a semilattice of rectangular bands.

If not every semigroup of a type under examination is decomposed into a certain band with "nice" components, it is natural to describe those semigroups of a given type which possess such decompositions. Some of such examples are presented in Section A.6.

There are some specific types of decompositions into bands. One of them involves strong semilattices. They are defined in terms of a transitive system of homomorphisms; see Section A.10. A semilattice of groups is automatically strong. Another important particular case is given by the Rees matrix semigroup $\mathcal{M}[T; I, \Lambda; P]$ over an arbitrary semigroup T (see Section A.5); such a semigroup is a rectangular band of subsemigroups $M_{i\lambda} = \{(i, t, \lambda) \mid t \in T\}$. Not every rectangular band of semigroups can be realized in this way, but for rectangular bands of groups this is just the case. One more particular case is given by a so-called *U-band*; this is a band of semigroups such that for any elements a, b of distinct components, we have $ab \in \langle a \rangle \cup \langle b \rangle$, that is, ab is equal to some power of a or b. Such bands naturally occur in descriptions of semigroups with certain restrictions on the lattice of subsemigroups, see Section A.14. A special case of a U-band is an ordinal sum. A semigroup S is called an *ordinal sum* of subsemigroups S_α, $\alpha \in I$, if S is a chain of these subsemigroups (so the set I is regarded to be linearly ordered), and for any S_α, S_β such that $\alpha < \beta$, we have $ab = ba = a$ for any $a \in S_\alpha$ and $b \in S_\beta$. The maximally "splitting" U-band is that whose components contain only a single element; in other words, this is a semigroup (clearly band) in which the product of any two elements is equal to one of them. We call such a semigroup *crumbly*. A semigroup is crumbly iff it is an ordinal sum of singular semigroups.

For some basic facts concerning decompositions into bands, see (Clifford and Preston 1967a, § 1.8, 4.1–4.3). Various questions relating to semilattice and matrix decompositions of semigroups are extensively examined in (Petrich 1973) and (Petrich 1977). The topic "Bands of semigroups" is surveyed with due fullness in (Shevrin 1991, § 2.3). A comprehensive survey of investigations concerned various kinds of decompositions (including certain ones for semigroups with zero) is given in (Ćirić and Bogdanović 1995).

A.3 Free Semigroups

by Lev N. Shevrin in Ekaterinburg, Russia

Generalities

Let A be a non-empty set, called an *alphabet*, whose elements are called *letters*. Denote by A^+ the set of all finite sequences of letters from A. We define a binary operation of multiplication on A^+ obtained by concatenating two sequences:

$$(a_1, \ldots, a_n)(b_1, \ldots, b_m) = (a_1, \ldots, a_n, b_1, \ldots, b_m).$$

With respect to this operation, A^+ is a semigroup, called the *free semigroup* over A (or on A). It is customary to identify each letter of A with the sequence of length 1 consisting of this letter. This implies that a sequence $(a_1, \ldots, a_n)$ may be expressed as the product $a_1 \ldots a_n$. The elements of A^+ written in such a form are called *words* (over A). For a word $w = a_1 \ldots a_n$, the number n is called the **length** of w and is denoted by $|w|$ or by $l(w)$. Words v and w of A^+ are **equal** if they co-incide as sequences, i.e., $|v| = |w|$ and these words have the same letters at the same places; rephrasing it, one may say that every word of A^+ is uniquely expressible as a product of letters of A.

We will point out some elementary properties of free semigroups. From the observation above it follows that any free semigroup is cancellative, i.e., the implications $xy = xz \Rightarrow y = z$ and $yx = zx \Rightarrow y = z$ hold. Taking into account the definition of a free group (see Section B.9), one sees that A^+ is embeddable in the free group over A. If A contains more than one letter, then A^+ is not commutative. For words $u, v \in A^+$ the following conditions are equivalent: (1) $uv = vu$; (2) u and v are powers of the same word of A^+; (3) some powers of u and v are equal. A word is called **primitive** if it is not a power of another word. Every word of A^+ is equal to a power of some primitive word, and such a representation is unique.

It is easy to verify that the free semigroup A^+ possesses the universal mapping property with respect to A, in other words, A^+ is freely gen-erated by A (see Section G.4). Conversely, if a semigroup S is freely generated by a generating set X, then $S \cong X^+$. Thus, applying the term *free semigroup* not only to any semigroup of all words over some alphabet, but also to any semigroup isomorphic to such a semigroup, we conclude that, for the class of all semigroups, free semigroups are just the free algebras in the sense of universal algebra. Using the notation accepted in universal algebra, one denotes the free semigroup over A by $\mathcal{F}(A)$; sometimes the notation $\mathcal{F}_A$ is used as well. However, the nota-tion A^+ has become the most popular. According to universal-algebraic terminology, the set A is a **free base** of A^+. In general, a free base is a particular case of a base, that is, a minimal (with respect to set-theoretic inclusion) generating set. A specific feature of the semigroup case is that the set A is not only a minimal but the least and thereby the unique base of A^+. This, by the way, is contrasting with the situation in group theory: the free group $(\mathbb{Z}, +)$ has two bases: $\{1\}$ and $\{-1\}$, and, every non-cyclic free group has even infinitely many free bases. The cardinality of the free base of a free semigroup is called, in accordance with universal-algebraic terminology, the **rank** of this semigroup. Two free semigroups are isomorphic iff their ranks are equal.

From the fact that free semigroups are freely generated by their bases, it follows that any semigroup is a homomorphic image of an appropri-ate free semigroup. This explains a fundamental role of free semigroups in general semigroup theory. When examining semigroups, it is often convenient to regard them in fact as quotient semigroups of a free semigroup and to deal with certain congruence classes of it. This is especially typical for combinatorial semigroup theory and, in particular, in investigations of semigroups given by presentations (see Section A.4). Diverse problems stem from the fact that an arbitrary element of such a semigroup has, as a rule, different representations as words over a generating set.

If we adjoin an identity 1 to A^+, we obtain a monoid called the *free monoid* over A and denoted by A^*. In A^*, 1 is regarded as the *empty word* whose length is assigned to be equal to 0. (Some other symbols are used as well to denote the empty word, for instance, ϵ, λ.) Properties of A^+ and A^* are, in principle, parallel to each other, but in many situations it is more convenient to deal with A^*; in particular, it often permits to consider a certain single expression instead of several ones. A typical example: a word $u \in A^*$ is called a *factor* of a word $v \in A^*$ if there exist words $x, y \in A^*$ such that $v = xuy$; in A^+ the same definition would require to consider four equalities: $v = u$, $v = xu$, $v = uy$, $v = xuy$. If in the condition $v = xuy$ we have $x = 1$ [$y = 1$], then u is called a *left* [*right*] *factor* (or a *prefix* [*suffix*]) of v. A factor u of v is called *proper* if $u \neq v$. Many authors use the term *subword* as a synonym of *factor*, however sometimes by a subword of a word v is meant, more generally, a word obtained from v by deleting some letters (or equal to v). So, having encountered the term *subword* in some text, the reader should consult the corresponding definition to avoid possible misunderstanding.

Besides their fundamental role within semigroup theory, free semigroups play a significant role in applications. This originates from the fact that words over some alphabet provide an appropriate mathematical model for strings of symbols (letters, signs, signals etc.) being encountered in very diverse contexts. One should mention in this connection, first of all, the theories of automata, languages and codes. Some information on the semigroup-theoretic approach to these areas is given in Section A.17. Note that in many considerations, especially in applications, the alphabet of a free semigroup under consideration is assumed to be finite.

Some basic properties of free semigroups are considered in several sources; we mention, in particular, (Clifford and Preston 1967a, § 1.12), (Clifford and Preston 1967b, § 9.1), (Lallement 1979, Ch. 5), (Howie 1995, Ch. 7).

Characterizations

The first characterization of free semigroups has already been mentioned above: in terms of the universal mapping property. From this characterization, it can be deduced that the uniqueness of a representation of every word in a free semigroup as a product of letters is in fact a characteristic property, i.e., a semigroup S is free iff S has a subset such that every element of S is uniquely expressible as a product of elements of this subset.

We give two more, not so transparent, characterizations based on a property which is called equidivisibility. This property first appeared in (Levi 1944) and was used there as well as in (Dubreil-Jacotin 1947) for the characterizations we are going to give. The definition is a little shorter for monoids, so we shall consider here just monoids; one easily derives a characterization of free semigroups from that of free monoids. A monoid M is called **equidivisible** if, for all x_1, x_2, y_1, y_2 of M, $x_1 x_2 = y_1 y_2$ implies

either $\quad x_1 = y_1 u$ and $ux_2 = y_2$ for some $u \in M$,

or $\quad x_1 v = y_1$ and $x_2 = vy_2$ for some $v \in M$.

The definition of a left [right] factor of an arbitrary monoid is the same as for free monoids (see above). For a monoid M, the following conditions are equivalent: (1) M is free; (2) M is equidivisible, the set $S = M \setminus \{1\}$ is a subsemigroup (in other words, 1 is a single invertible element of M), and $\bigcap_{n \in \mathbb{N}} S^n = \varnothing$; (3) M is cancellative and equidivisible, 1 is a single invertible element, and every element of M has finitely many left factors. We remark that in (3), *left* may be replaced by *right*.

Subsemigroups

Not only a free semigroup itself, but each of its subsemigroups has a unique base, consisting of those elements which are indecomposable into a product within this subsemigroup. At the same time, not every subsemigroup of a free semigroup is also free. (This is again in contrast to the group case: according to the Nielsen-Schreier theorem, non-trivial subgroups of free groups are free.) The simplest counterexample is given by the subsemigroup $\{a\}^+ \setminus \{a\}$ of $\{a\}^+$; indeed, its base is $\{a^2, a^3\}$, and, for instance, the element a^6 of this subsemigroup has two expressions as a product of the generators, since $a^6 = a^2 a^2 a^2 = a^3 a^3$. A subsemigroup T of the free semigroup A^+ is itself free iff, for any $w \in A^+$,

$$wT \cap T \neq \varnothing \text{ and } Tw \cap T \neq \varnothing \text{ together imply } w \in T.$$

This criterion was first found in (Schützenberger 1956) and rediscovered in (Shevrin 1960). It is remarkable that the free semigroup of rank 2 has free subsemigroups of countable rank; hence it contains free subsemigroups of any finite rank. For example, in the semigroup $\{a, b\}^+$, the words ab^n, $n \in \mathbb{N}$, form the free base of the subsemigroup generated by them.

A finitely generated subsemigroup of a free semigroup need not be finitely presented (f.p.—see Section A.4). There is a 4-generated (i.e., with 4 generators) subsemigroup which is not f.p.; such is, for example, the subsemigroup of $\{a, b\}^+$ generated by a, ab, ba, bb: this can be deduced from an algorithmic criterion of finite presentability of a subsemigroup of a free semigroup given in (Markov 1971). The bound 4 is precise here. Indeed, as it has been established in (Budkina and Markov 1973), all 3-generated subsemigroups of a free semigroup are f.p. As to 2-generated subsemigroups, every such one is either commutative (and so is contained in some cyclic subsemigroup) or free, see (Blum 1965); in the first case, finite presentability is provided by Redei's theorem (see Section A.4), in the latter, we have simply the empty set of defining relations.

Combinatorics on words

A huge variety of situations where free semigroups occur includes a wide range of problems related to analysis of the structure of words: diverse configurations of letters and factors, factorizations, rearrangements and the like. The broad study of such problems of combinatorial character has led to forming a peculiar mathematical area having fruitful connections with a number of branches of mathematics and the great number of applications in various fields. This area received the name *combinatorics on words*. For familiarization with it, the reader is referred to (Lothaire 1983), (Lothaire 2002), and (Choffrut and Karhumäki 1997), see also (de Luca and Varricchio 1999, Ch. 1, 2). Here we shall touch only one subject, namely power-free words.

Let k be an integer > 1. A word w over an alphabet A is called *k-power-free* if for any $u \in A^+$, u^k does not occur among the factors of w. For $k = 2$ [$k = 3$] it is common to use the term *square-free* [*cube-free*]. There are only six square-free words over a two-letter alphabet, say, $\{a, b\}$, they are a, b, ab, ba, aba, bab. This observation provokes the following natural questions: how many square-free [cube-free] words are there over a three-letter alphabet [two-letter alphabet]? Both these questions have the same answer: infinitely many. Certain words with the properties required have been discovered in (Thue 1906) and (Thue 1912). In essence, combinatorics on words originates from these pioneer works. The classical results concerning the words mentioned belong to the most wonderful facts of combinatorics on words. The same words were rediscovered in (Morse 1921), so it is accepted to call them the **Thue-Morse words**. These famous words are defined as follows.

Let μ [τ] be the endomorphism of $\{a, b, c\}^+$ [$\{a, b\}^+$, respectively] given by the defining conditions

$$\mu(a) = abc, \ \mu(b) = ac, \ \mu(c) = b, \quad [\tau(a) = ab, \ \tau(b) = ba].$$

Consider the iterations

$$m_1 = \mu(a), m_2 = \mu^2(a), \ \ldots, \ m_n = \mu^n(a), \ \ldots$$
$$[t_1 = \tau(a), \ \ t_2 = \tau^2(a), \ \ldots, \ \ t_n = \tau^n(a), \ \ldots].$$

For instance, $m_2 = abcacb$, $m_3 = abcacbabcbac$, $t_2 = abba$, $t_3 = abbabaab$. The words m_n [t_n], $n \in \mathbb{N}$, are just the square-free [cube-free] Thue-Morse words. Since each word m_i [t_i] is a prefix of all the words m_j [t_j] with $j > i$, it is usual to consider, instead of the families $\{m_i\}$ and $\{t_i\}$, two infinite words (which are defined as the maps $m \colon \mathbb{N} \longrightarrow \{a, b, c\}$ and $t \colon \mathbb{N} \longrightarrow \{a, b\}$) with the prefixes m_i and t_i, respectively. They are called the **infinite Thue-Morse words**.

A special direction of combinatorics on words is devoted to study of equations in a free semigroup (or equations in words). This topic is reflected in (Lentin 1972), (Hmelevskii 1976) as well as in (Lothaire 1983, Ch. 9). We mention here only the remarkable result that the problem to determine whether a given equation with constants has a solution is decidable (Makanin 1977).

A.4 Presentations and Word Problems

by Peter M. Higgins in Colchester and Nik Ruškuc in St. Andrews, UK

Presentations

Presentations are means of defining semigroups as homomorphic images of free semigroups. Recall (Section A.3) that the free semigroup A^+ on a set A consists of all non-empty words over A multiplied by concatenation. The free monoid A^* is A^+ together with the empty word (denoted by 1).

A (*semigroup*) *presentation* with generators A is an ordered pair $P = \langle A \mid R \rangle$, where $R \subseteq A^+ \times A^+$. A typical element $(u, v) \in R$ is written as $u = v$ and referred to as a *defining relation*. The semigroup defined by P is the quotient $S = A^+/\rho$, where ρ is the smallest congruence on A^+ which contains R. Intuitively, S is the semigroup generated by A in which these generators satisfy all the relations from R and all their consequences, but no other relations. Any other semigroup in which generators satisfy the relations from R is a homomorphic image of S. Two words $w_1, w_2 \in A^+$ represent the same element of S (i.e., $w_1 \rho = w_2 \rho$) if and only if w_2 can be obtained from w_1 by applying relations from R. Every semigroup can be defined by a presentation; if it can be defined by a presentation $\langle A \mid R \rangle$ in which both A and R are finite, it is said to be *finitely presented* (or *fp*, for short).

For example, consider the semigroup S defined by the presentation

$$\langle a, b, c, d, e \mid ac = ca, \ ad = da, \ bc = cb, \ bd = db,$$
$$eca = ce, \ edb = de, \ cca = ccae \rangle. \tag{1}$$

The words cde and $eabcd$ represent the same element of S because one can be transformed into the other by using the relations in the following sequence of steps:

$$cde \rightarrow cedb \rightarrow ecadb \rightarrow ecabd \rightarrow eacbd.$$

On the other hand, the words $ccdaeedb$ and $acbdeae$ have different numbers of d's, and so are not equal in S because an application of any relation preserves the number of d's.

By working over the free monoid A^* instead of the free semigroup, one arrives at the notions of a monoid presentations and monoids defined by them. These allow defining relations of the form $u = 1$. For example, the presentation $\langle a, b \mid ab = 1 \rangle$ defines the so-called *bicyclic monoid* (see Section A.5).

There are two general aspects regarding presentations and semigroups defined by them. The first is to find suitable presentations for particular semigroups of interest. The second, which is the subject of this article, is to determine properties of a semigroup defined by a presentation.

The word problem and other decidability problems

The word problem (cf. Section G.7) for a given presentation $P = \langle A \mid R \rangle$ asks if there is an algorithm that decides whether or not two given words

$u, v \in A^+$ represent the same element in the semigroup defined by P. If such an algorithm exists, we say that the word problem is **decidable**, otherwise it is **undecidable**. For example, in the free semigroup A^+, which is defined by $\langle A \mid \rangle$, two words are equal if and only if they are identical; therefore the word problem is decidable. Also, if $\langle A \mid R \rangle$ is a presentation for a finite semigroup S, then the word problem is decidable: indeed, to check if two words $u, v \in A^+$ are equal it is sufficient to compute their values using the (finite) Cayley table.

However, not every finite presentation has a decidable word problem. The first examples were given by Markov (1947) and Post (1947). Ceĭtin (1958) proved that the presentation (1) has an undecidable word problem. Using this, Matijasevič (1967) devised a presentation with two generators and three relations, and showed that its word problem was undecidable as a consequence of undecidability of (1).

Markov (1951) used a presentation with an undecidable word problem in order to show that a host of natural properties (known as Markov properties) are undecidable for finitely presented semigroups. A property $\mathscr{P}$ of semigroups is called a **Markov property** if: (1) it is preserved under isomorphisms; (2) there is an fp semigroup with property $\mathscr{P}$; (3) not every fp semigroup can be embedded into one with property $\mathscr{P}$. For example, all the following properties are Markov: being trivial, finiteness, commutativity, cancellativity, being a group, being free. Markov proved that, given a Markov property $\mathscr{P}$, there is no algorithm which decides whether a given presentation defines a semigroup satisfying $\mathscr{P}$.

Special classes of semigroups and presentations

As a positive result, we mention

A.4.1 Theorem (M. V. Sapir, 1985) *Let $\mathscr{M}$ be a finitely based nonperiodic (i.e., satisfying no identity of the form $x^n = x^m$ for $n \neq m$) semigroup variety. Then the following conditions are equivalent:*
(1) *$\mathscr{M}$ has a solvable word problem for any f.p. semigroup;*
(2) *The elementary theory of any f.p. semigroup of $\mathscr{M}$ is solvable;*
(3) *Any f.p. semigroup of $\mathscr{M}$ is residually finite;*
(4) *Any f.p. semigroup of $\mathscr{M}$ is representable by matrices over a field.*

In order to obtain positive decidability results, and possible computational applications, one needs to restrict attention to special types of presentations of semigroups.

A classical result of Adian (1967) is: if $u, v \in A^+$ are two words such that they have different first letters and different last letters, then the word problem for $P = \langle A \mid u = v \rangle$ is decidable. Adjan proves this by associating two graphs with any semigroup presentation, and showing that, if they have no cycles, the semigroup defined by the presentation embeds naturally into the group defined by the same presentation. Applied to P, this yields that the semigroup defined by it embeds into the corresponding one-relator group, which has a decidable word problem by Magnus's theorem (Magnus et al. 1976, Theorem 4.14). Adjan also showed that the word problem for the monoid presentation of the form $\langle A \mid u = 1 \rangle$ is decidable. However, it is still an open problem whether every presentation with only one defining relation has decidable word problem, although many special cases have been resolved in the affirmative (see, for example, Lallement 1986, Watier 1997). A variety of techniques have been brought to bear on this difficult problem, including the geometric techniques of diagrams (Higgins 1992, Chapter 5) and pictures (Pride 1995).

A *convergent rewriting system* (also known as a *complete rewriting system*) is a special type of presentation designed with the aim of having an easy and natural solution to the word problem (Section G.7). Intuitively, these are presentations which guarantee that a unique normal form for every word can be obtained by a sequence of transitions, each replacing a left hand side of a relation by the right hand side (but not the other way round). For details see Book and Otto (1993).

The best known natural class of semigroups with good algorithmic properties is that of commutative semigroups. Redei in 1965 proved that every finitely generated commutative semigroup is finitely presented; for a short proof see Grillet (1995, Section 4.9). In addition, every commutative semigroup has a finite convergent rewriting system (Diekert 1986), and therefore a decidable word problem.

Standard algorithms

The two best known algorithms for computing with finitely presented semigroups are the *Todd-Coxeter* and the *Knuth-Bendix algorithm*. The former takes as its input a finite semigroup presentation P, and terminates if and only if the semigroup S defined by P is finite, giving the right regular representation of S. (This does not contradict the fact that the finiteness and triviality problems are undecidable; the time in which the algorithm terminates is not bounded by either the size of P or the order of S.) The algorithm was first formulated for groups (Todd and Coxeter 1936), and modified for semigroups (Neumann 1967a, Jura 1978). The algorithm can be modified to enumerate various types of ideals in possibly infinite semigroups (Campbell et al. 1995).

Knuth-Bendix takes as its input a finite presentation $P = \langle A \mid R \rangle$ and a well-ordering $\preceq$ on A^+. It terminates if and only if the semigroup defined by P has a finite convergent rewriting system $\langle A \mid Q \rangle$ which respects $\preceq$ (i.e., $u \succeq v$ for all $(u, v) \in Q$) and returns such a system. For the description of the algorithm and further references see Book and Otto (1993, Section 2.4) and Sims (1994, Chapter 2).

The computer algebra system GAP (GAP 2000) contains implementations of both the Todd-Coxeter and Knuth-Bendix algorithms.

A.5 Simple Semigroups

by Lev N. Shevrin in Ekaterinburg, Russia

In group theory and in ring theory, the notion of simplicity may be treated in two ways: as is customary, a group or ring is simple if it contains only the trivial normal subgroups or ideals, respectively, but this is equivalent to saying that it has only the trivial congruences, i.e., the equality relation Δ and the universal relation ∇. In semigroup theory, this is not the case: a semigroup without (non-zero or any) proper ideals may have non-trivial congruences. So one can suggest at least two variants to call a semigroup simple. In fact there are even quite a number of different and natural notions of simplicity for semigroups. The uniting sign for all members of the family of such notions is the absence of proper ideals or equivalences of a certain kind. Below, we consider the main members of this family and discuss interrelations between them.

Main types of simple semigroups

Let S be a semigroup. S is *simple* if it has no proper (two-sided) ideals. S is *left [right] simple* if it has no proper left [right] ideals. S is *completely simple* if it is simple and contains a primitive idempotent, or it is a singleton. (An idempotent $e \in S$ is called *primitive* if e is a minimal non-zero idempotent with respect to the partial order $\leq$ defined on the set of all idempotents of S as follows: $x \leq y$ means $xy = yx = x$.) Further, S is *bisimple* if Green's relation $\mathscr{D}$ coincides with ∇, i.e., S has a single $\mathscr{D}$-class.

For semigroups with zero, the notions just listed give a degenerate case to be a singleton, and here one has to deal with non-zero ideals. Let S be a semigroup with zero 0. Then S is called **0-*simple* [*left* 0-*simple*, *right* 0-*simple*]** if $S^2 \neq \{0\}$ and S has no proper non-zero two-sided [left, right] ideals. (We remark that, for some reason, it is natural to exclude the case of a *null semigroup*, when $S^2 = \{0\}$, in these definitions.) S is called *completely* **0-*simple*** if S is 0-simple and contains a primitive idempotent. S is **0-*bisimple*** if S has exactly two $\mathscr{D}$-classes, one of which is $\{0\}$.

And one more definition, in terms of congruences: a semigroup is called *congruence-simple* (or, most commonly, *congruence-free*) if its congruences are exhausted by Δ and ∇.

Every left [right] simple semigroup is bisimple. Every bisimple semigroup is simple, but there exists a simple semigroup which is not bisimple and, moreover, in which $\mathscr{D}$ coincides with Δ (such a semigroup is thereby as far from being bisimple as any semigroup could be); see a corresponding example in (Clifford and Preston 1967a, Exercises 8–10 for § 2.1). There is an alternative to definitions and a more transparent characterization of simple and 0-simple semigroups: a semigroup S [with zero] is [0-]simple iff $SaS = S$ for every [non-zero] element $a \in S$. However, one cannot hope to get any satisfactory structural characterizations of simple and 0-simple semigroups in general. The matter is that any semigroup can be embedded in a simple one and even in an (automatically regular) bisimple monoid; see, for instance, (Clifford and Preston 1967b, § 8.5, 8.6). Furthermore, any semigroup is embeddable in a bisimple semigroup, generated by idempotents, see (Meakin 1985); any semigroup is embeddable in a congruence-simple semigroup, see (Bokut' 1963), (Shutov 1963).

A semigroup is both left simple and right simple iff it is a group. It is noteworthy that groups often occur as certain "blocks" in constructions describing simple semigroups of a number of specific types. This can take place if semigroups under consideration contain idempotents; below we shall point out several examples of such occurrences. As regards *idempotent-free* semigroups, i.e., semigroups without idempotents, perhaps the most remarkable phenomenon here is that in the class of simple idempotent-free semigroups possessing minimal right ideals (and so being the union of such ideals), there exist fairly transparent *universal* semigroups. This means that any semigroup of the class indicated can be embedded in some semigroup of this class belonging to a certain family whose members are so-called *Croisot-Teissier semigroups*;

they are defined as a special type of transformation semigroups. When such semigroups are right simple, they are called **Teissier semigroups** and, in a certain particular case, **Baer-Levi semigroups**. The last ones are right cancellative, that is, are obeying the law $yx = zx \Rightarrow y = z$. Any right simple idempotent-free [and right cancellative] semigroup is embeddable in an appropriate Teissier [Baer-Levi] semigroup. Clearly, there are "left" versions of all these facts. A good exposition of topics concerning one-sided simple idempotent-free semigroups is given in (Clifford and Preston 1967b, § 8.1, 8.2).

Every left [right] 0-simple semigroup arises from a left [right] simple semigroup by the adjunction of a zero element. This shows that there is no need to examine left [right] 0-simple semigroups, and it is sufficient to restrict oneself with left [right] simple ones. In the two-sided case, the situation is partly different: the adjunction of a zero to a simple semigroup gives a 0-simple semigroup, however not every 0-simple semigroup can be obtained by this way. Indeed, in a 0-simple semigroup S, the set $S \setminus \{0\}$ need not be a subsemigroup, in other words, S may have (**proper**) **divisors of zero**. The same is true, in particular, for interrelations between completely simple and completely 0-simple semigroups, both being apparently the most important types of semigroups under discussion. We consider these types in a special subsection of this section.

Various kinds of simplicity may be revealed when studying arbitrary semigroups, in particular, at their **factors** (or **divisors**), i.e., homomorphic images of subsemigroups. A particular role is played by so-called principal factors. If a is an arbitrary element of a semigroup S, then the set $N(a) = J(a) \setminus J_a$ (see Section A.1), if it is nonempty, is the greatest ideal of S contained in $J(a)$. The quotient semigroup $J(a)/N(a)$, or merely $J(a)$ if $N(a) = \varnothing$, is called the **principal factor** corresponding to a. Each principal factor is either simple (if it is the kernel of a given semigroup), or 0-simple, or null. One may say that any semigroup is covered by the "web" of semigroups being its principal factors. This web naturally exerts influence on the structure of a semigroup, particularly if there are no null semigroups among the principal factors. Semigroups obeying the last condition are called **semisimple**. For instance, every regular semigroup is semisimple. A stronger condition is that each principal factor is either completely simple (if it is the kernel of a given semigroup) or completely 0-simple. Semigroups obeying this condition are called **completely semisimple**. A semigroup is completely semisimple iff it is regular and contains no bicyclic subsemigroup.

The **bicyclic semigroup** $\mathcal{B}$ just mentioned, which is a monoid given by the presentation $\langle a, b \mid ab = 1 \rangle$, occurs in diverse contexts of semigroup-theoretic investigations. In the context of this article, it is relevant to note that $\mathcal{B}$ is one of the important representatives of the class of bisimple but not completely simple semigroups. Another remarkable representative of the same class is the so-called **four-spiral semigroup** Sp_4 given by the presentation

$$\langle a, b, c, d \mid ab = b, \ ba = a, \ bc = b, \ cb = c, \ cd = d, \ dc = c, \ da = d \rangle.$$

From these defining relations, it follows that the generators a, b, c, d are idempotents. The semigroup Sp_4 first appeared at the semigroup-theoretic stage in (Byleen et al. 1978) as a "minimal" example of an idempotent-generated semigroup from the above-mentioned class. It possesses a number of peculiar properties and, by the way, has an interesting link with the bicyclic semigroup: Sp_4 may be represented as a Rees matrix semigroup (see the definition below) over $\mathcal{B}$. For fuller information about this semigroup, see (Meakin 1985, § 5).

Completely simple and completely 0-simple semigroups

We will use hereinafter the abbreviations c.s. and c.0-s. for semigroups featured in the title of this section. According to a remark given in the previous section, there are c.0-s. semigroups with divisors of zero. They thereby have no "companions" among c.s. semigroups, that is, such a semigroup cannot be obtained by the adjunction of a zero to some c.s. semigroup. As minimal (in quantity of elements) examples of c.0-s. semigroups with divisors of zero, the following two notable five-element semigroups A_2 and B_2 should be mentioned. The latter is particularly known and belongs, along with the bicyclic semigroup, to, so to say, the elite of famous individual semigroups; it occurs in diverse areas of semigroup theory (see, for instance, Section A.6 and Section A.15). Here are the representations of them as multiplicative semigroups of 2×2-matrices:

$$A_2 = \{ (\begin{smallmatrix} 1 & 0 \\ 0 & 0 \end{smallmatrix}), (\begin{smallmatrix} 0 & 1 \\ 0 & 0 \end{smallmatrix}), (\begin{smallmatrix} 1 & 0 \\ 1 & 0 \end{smallmatrix}), (\begin{smallmatrix} 0 & 1 \\ 0 & 1 \end{smallmatrix}), (\begin{smallmatrix} 0 & 0 \\ 0 & 0 \end{smallmatrix}) \};$$
$$B_2 = \{ (\begin{smallmatrix} 1 & 0 \\ 0 & 0 \end{smallmatrix}), (\begin{smallmatrix} 0 & 1 \\ 0 & 0 \end{smallmatrix}), (\begin{smallmatrix} 0 & 0 \\ 1 & 0 \end{smallmatrix}), (\begin{smallmatrix} 0 & 0 \\ 0 & 1 \end{smallmatrix}), (\begin{smallmatrix} 0 & 0 \\ 0 & 0 \end{smallmatrix}) \}.$$

For a semigroup S, the following conditions are equivalent: (1) S is c.[0-]s.; (2) S is a [0-]simple epigroup; (3) S is [0-]simple and possesses at least one [0-]minimal left ideal and at least one [0-]minimal right ideal;

(4) S is [0-]simple, possesses a [0-]minimal left or right ideal and contains a [non-zero] idempotent. In particular, any finite [0-]simple semigroup is c.[0-]s. Every c.[0-]s. semigroup is regular, [0-]bisimple and is the union of both its [0-]minimal left ideals and its [0-]minimal right ideals; all non-zero idempotents of such a semigroup are primitive. C.0-s. inverse semigroups are precisely the so-called **Brandt semigroups**; for the original definition of these semigroups and the origination of their name, see (Clifford and Preston 1967a, § 3.3). The property of a semigroup S of being c.s. is equivalent, besides the corresponding versions of the conditions (2)–(4), to each of the following conditions: (5) S is a rectangular band of (automatically isomorphic) groups (see Section A.2); (6) S is regular and, if S is not a singleton, all its idempotents are primitive. In view of (5), any c.s. semigroup is completely regular. The identity distinguishing c.s. semigroups among completely regular ones (regarded with the unary operation $x \mapsto x^{-1}$) is presented in Section A.10.

   Left [right] simple semigroups may also be treated as semigroups *with left [right] division* or as semigroups *with left [right] invertibility*. This is justified by an observation that a semigroup S is left [right] simple iff for any $a, b \in S$, the equation $xa = b$ [$ax = b$] has a solution. If in a semigroup S such a solution is always unique, S is called a *left [right] group*. This is a particular case of c.s. semigroups. For definiteness, we shall consider only the "right" case. For a semigroup S, the following conditions are equivalent: (a) S is a right group; (b) S is right simple and contains an idempotent; (c) S is right simple and left cancellative; (d) S is regular and left cancellative; (e) S is regular and the set of its idempotents is a right zero semigroup; (f) S is a right band of (automatically isomorphic) groups; (g) S is isomorphic to the direct product $G \times R$ of a group G and a right zero semigroup R.

   There is quite a constructive characterization of c.s. and c.0-s. semigroups in terms of so-called the Rees matrix semigroups. This classical fact of semigroup theory is the subject of the remainder of this section.

   Let T be a semigroup, I and Λ be nonempty sets, and $P = (p_{\lambda i})$ be a mapping of the Cartesian product $\Lambda \times I$ into T. Note that P may be regarded as a $\Lambda \times I$ matrix with entries in T. We define a multiplication on the set $I \times T \times \Lambda$ by the formula

$$(i, a, \lambda)(j, b, \mu) = (i, a p_{\lambda j} b, \mu). \tag{1}$$

Then M turns into a semigroup, which is denoted by $\mathcal{M}[T; I, \Lambda; P]$. This is the **Rees matrix semigroup over the semigroup T with sandwich matrix P**; see also Section A.13.

For our purpose, T is assumed to be either a group or a group with zero. If T is obtained from a group G by adjoining a zero 0, then it is denoted by G^0. The set $N = \{(i, 0, \lambda) \mid i \in I, \lambda \in \Lambda\}$ is an ideal of $\mathcal{M}[G^0; I, \Lambda; P]$. The quotient semigroup $\mathcal{M}[G^0; I, \Lambda; P]/N$ is commonly denoted by $\mathcal{M}^0[G; I, \Lambda; P]$ and called the **Rees matrix semigroup over the group with zero** G^0 with sandwich matrix P. The elements of it are, in essence, all triples (i, g, λ) with $g \neq 0$ and a zero element which may be written as $(i, 0, \lambda)$ for any i and λ.

Another treatment of $\mathcal{M}^0[G; I, \Lambda; P]$ may be given in terms of $I \times \Lambda$ matrices with entries from G^0, where each triple (i, a, λ) is replaced by the matrix $(a)_{i\lambda}$ having entry a in the (i, λ) position and zeros in the remaining ones. Then the right-hand side of the formula (1) turns into the matrix product $(a)_{i\lambda} P(b)_{j\mu}$ (well-defined, in spite of possible infinity), which visually justifies the term *sandwich matrix*. A sandwich matrix is called **regular** if each row and each column of it contains a non-zero entry. The semigroup $\mathcal{M}^0[G; I, \Lambda; P]$ is regular iff the matrix P is regular.

A semigroup S is completely 0-simple iff S is isomorphic to a regular Rees matrix semigroup over a group with zero; hence a semigroup S is completely simple iff S is isomorphic to a Rees matrix semigroup over a group.

It is justly customary to call the above statement the **Rees-Suschkewitsch theorem** (sometimes it is called the **Rees theorem**). In the general form it has been established in (Rees 1940); its version for completely simple semigroups is in effect due to Suschkewitsch who examined in (Suschkewitsch 1928) the case of finite semigroups. It should be noted that this pioneer paper of Suschkewitsch marks the beginning of semigroup theory. The Rees-Suschkewitsch theorem gives an effective tool for studying completely [0-]semigroups from different points of view: congruences, subsemigroups, identities, and so on. But its role goes far beyond examinations of only one type of semigroups. It has had a significant influence on the later development of many trends in semigroup theory, both through the use of it in a great variety of situations and as an expressive sample of fruitful constructions.

The topics discussed in this article have been surveyed more or less thoroughly in the corresponding chapters or sections of (Clifford and Preston 1967a), (Clifford and Preston 1967b), (Lallement 1979), (Shevrin 1991), (Grillet 1995), (Howie 1995).

A.6 Epigroups

by Lev N. Shevrin in Ekaterinburg, Russia

Introductory remarks

We begin with some preliminary notions and notation connected with idempotents of a semigroup. The set of all idempotents of a semigroup S will be denoted by E_S or simply by E. For each $e \in E$, the maximal subgroup of S with e as identity is unique and denoted by G_e. Let $\operatorname{Gr} S = \bigcup_{e \in E} G_e$; this set will be called the *group part* of S, any element of $\operatorname{Gr} S$ is a *group element* of S. We remark that S is completely regular if $\operatorname{Gr} S = S$. A semigroup S is *periodic* (or *torsion*) if each element of S is of finite order, i.e., generates a finite cyclic subsemigroup; this is equivalent to the condition that some power of any element is an idempotent.

Among semigroups containing idempotents, there are such ones in which the set of all idempotents plays a role of a certain framework. One of the most important types of such semigroups is represented by the regular semigroups. Indeed, in a regular semigroup every element is "surrounded" by the idempotents from the same $\mathscr{D}$-class. Another such type is represented by epigroups. A semigroup S is called an *epigroup* if for any $x \in S$ there exists a natural number n such that x^n lies in some subgroup of S. A subgroup in this definition may be taken as being maximal. If m is the minimal number such that $x^m \in G_e$ (this m is called the *index* of x), then $x^n \in G_e$ for any $n > m$. So one may imagine that elements of an epigroup S are pulled by idempotents to the group part of S, and each element is "attached" to a single idempotent. This induces a partition of S into the subsets K_e defined as follows:

$$K_e = \{x \in S \mid x^n \in G_e \text{ for some } n\}.$$

The set K_e is called the *unipotency class* corresponding to the idempotent e. If S is a periodic semigroup, the unipotency classes are usually called *torsion classes*.

The class of epigroups is very large. It contains all periodic semigroups and, in particular, all finite semigroups, all completely regular semigroups, all completely 0-simple semigroups. It also contains some important concrete semigroups; for example, the semigroup of all matrices over a division ring is an epigroup. In the course of semigroup-theoretic investigations during several decades, some properties of epigroups were revealed via the study of particular types of epigroups; for instance, the coincidence of Green's relations $\mathscr{D}$ and $\mathscr{J}$ was observed originally for periodic semigroups. The first work where epigroups were considered *per se* was the paper (Munn 1961), in which these semigroups were called **pseudoinvertible**. This paper was, in turn, influenced by the paper (Drazin 1958), in which the notion of a pseudoinverse of an element (essential for epigroups, see the last subsection below) had been introduced. Epigroups were considered later (as one of the objects or the main object of attention) by different authors and under different names: **quasi-completely regular**, **group-bound**, **quasiperiodic** and some others. The term *epigroup* was suggested by the present writer at the end of the 80s and has become fairly common by now. A rather coherent account of many primary facts about epigroups is contained in (Shevrin 1995, §1). The topic *Epigroups* is extensively surveyed in (Shevrin 1991, §6).

Certain decompositions

An element x of a semigroup with zero is called a **nil-element** if $x^n = 0$ for some n. A semigroup with zero is called a **nilsemigroup** if all its elements are nil-elements. A **nil-extension** of a semigroup T is a semigroup being an ideal extension of T by a nilsemigroup. A semigroup is called **unipotent** if it has a single idempotent. Two of the extreme types of unipotent epigroups are given by groups and nilsemigroups (the latter are even periodic). It turns out that an arbitrary unipotent epigroup can be built of these polar types. Namely, the following conditions for a semigroup S are equivalent: (i) S is a unipotent epigroup; (ii) S is a nil-extension of a group; (iii) S is a subdirect product of a group and a nil-semigroup.

A subsemigroup of an epigroup need not be an epigroup; a **subepigroup** is a subsemigroup being epigroup. Any epigroup S is covered by unipotent subepigroups. If S has a partition into unipotent subepigroups, this means that such a partition is unique and its components are precisely the unipotency classes; in this case an epigroup will be called **unipotently partitionable**. Not every epigroup possesses this property. A simplest counter-example is given by the famous five-element Brandt semigroup B_2 (see Section A.5). Indeed, in B_2, the unipotency class K_0 is not a subsemigroup. It is rather surprising that this minimal particular example is in reality determining the matter as a *forbidden* part for any unipotently partitionable epigroup. Namely, an epigroup is unipotently partitionable iff it contains no subsemigroup being an ideal extension of a unipotent epigroup by B_2.

For the forthcoming criteria we need some more definitions and notations. We will write $a \mid b$ (an element a **divides** an element b) if $b \in J(a)$. We will view E as a partially ordered set with respect to the natural order: $e \leq f$ means $ef = fe = e$; it is an **antichain** if its elements are pairwise incomparable. We will denote by $\text{Reg} \, S$ [Nil S] the set of all regular [nil-]elements of a semigroup S [with zero]. A **retract** of a semigroup S is a subsemigroup T for which there exists a homomorphism from S onto T that is identical on T. A homomorphic image [Rees quotient semigroup] of a subsemigroup of a semigroup S is called a [**Rees**] **divisor** (or **factor**) of S; if a divisor is obtained from a subepigroup of an epigroup, we will call it **epidivisor**.

The following conditions for a semigroup S are equivalent: (a) S is an archimedean epigroup; (b) S is an epigroup in which E_S is an antichain; (c) S is a nil-extension of a completely simple semigroup. A completely simple semigroup featured in (c) is nothing else than the group part $\text{Gr} \, S$, which is the kernel of S in this case. An epigroup S is a subdirect product of a completely simple semigroup and a nilsemigroup iff $\text{Gr} \, S$ is a retract ideal of S. Epigroups (in particular, finite semigroups) being decomposable into a band of archimedean epigroups occurred in semigroup-theoretic investigations of diverse directions. They can be characterized from various points of view. The following conditions for an epigroup S are equivalent: (1a) S is a band of archimedean epigroups; (1b) S is a semilattice of archimedean epigroups; (2) in every homomorphic image S' of S such that S' has a zero, the set Nil S' is an ideal; (3a) there are no semigroups A_2, B_2 among the epidivisors of S; (3b) the same as in (3a) but for the Rees epidivisors; (4a) $\text{Reg} \, S = \text{Gr} \, S$; (4b) each regular $\mathscr{D}$-class is (automatically a completely simple) subsemigroup; (5) for any $e, f, i \in E$, $i \mid e$ and $f \mid e$ together imply $if \mid e$. A fuller list of such equivalent conditions is presented in (Shevrin 1995). This work contains also quite a number of results giving characterizations of epigroups possessing decompositions into bands of certain more restrictive types, for instance, into bands of unipotent epigroups, etc. An allied material is contained in (Bogdanović 1985, Ch. IX–XI).

Finiteness conditions

Given a class of algebraic systems, by a **finiteness condition** is meant any property which is possessed by all finite systems of this class. Imposing finiteness conditions is a classical approach in investigations of algebraic systems of different kinds. For semigroups, one of the broad group of such conditions deals with conditions formulated in terms of ideals or congruences of certain types, in particular, in terms of frames or one-sided congruences. Some facts in this direction are presented in (Clifford and Preston 1967b, §6.6); see also later works (Hotzel 1975–1976), (Hall and Munn 1979), (Hotzel 1979), (Kozhukhov 1980) as well as (de Luca and Varricchio 1999, §3.6). Here and in Section A.7, we shall focus attention on another group of conditions, which are expressed basically in terms of subsemigroups; they just distinguish some important subclasses within the class of epigroups. Note that the property of being an epigroup is itself a finiteness condition.

A semigroup is called *finitely assembled* (f. a.) if it has finitely many non-group elements and idempotents. Any f. a. semigroup is evidently an epigroup. The term *finitely assembled* is justified by the observation that such a semigroup looks as if it is assembled from a finite family of groups and a finite set of (non-group) elements. If in such a semigroup S all its maximal subgroups belong to a fixed class $\mathcal{K}$, we say that S is f. a. *from $\mathcal{K}$-groups*. The structure of f. a. semigroup is fairly clarified by the following proposition: a semigroup S is f. a. iff S has a finite series of ideals in which each factor is either finite or is a rectangular band of finitely many infinite groups (this takes place for the kernel of S only) or is obtained from such a band by the adjunction of a zero. A key sufficient (and, for the case of periodic semigroups, also necessary) criterion can be given in terms of bases, that is, minimal generating sets. Namely, if an epigroup has no subsemigroups with a unique infinite basis, then it is f. a. This criterion leads to the following general scheme discovered in (Shevrin 1974). Let $\mathcal{K}$ be a finiteness condition satisfying the following requirements: A) $\mathcal{K}$ is hereditary for subsemigroups; B) no $\mathcal{K}$-epigroup has a unique infinite basis; C) any epigroup covered by a finite system of $\mathcal{K}$-subsemigroups is a $\mathcal{K}$-epigroup. Then an epigroup is a $\mathcal{K}$-semigroup iff it is f. a. from $\mathcal{K}$-groups. This theorem gives a description of $\mathcal{K}$-epigroups via a complete reduction to the group case. In most particular cases for various concrete finiteness conditions, semigroups under consideration turn out to be periodic, so the requirement for such a semigroup to be *a priori* an epigroup can be omitted. A typical example of such a situation is given by semigroups with the minimal condition for subsemigroups: a semigroup possesses this property iff it is f. a. from groups with the minimal condition for subgroups. A detailed exposition of this subject is given in (Shevrin and Ovsyannikov 1996, Ch. IV).

Epigroups as unary semigroups

A *unary semigroup* is a semigroup with an additional unary operation. Important representatives of such algebras are completely regular and inverse semigroups (see Section A.10 and Section A.11). Epigroups can be treated as unary semigroups as well. To introduce the corresponding unary operation we need few preliminary remarks. Let S be an epigroup, x be an arbitrary element of S. The idempotent of the unipotency class to which x belongs will be denoted by e_x. We have $xe_x \in G_{e_x}$, hence one may consider the element

$$\bar{x} = (xe_x)^{-1},$$

where the right-hand side is the inverse in the group G_{e_x}. This element is called the *pseudoinverse* of x, and the operation $x \mapsto \bar{x}$ is just the announced one. The following equalities hold:

$$x\bar{x} = \bar{x}x, \ x\bar{x}^2 = \bar{x}, \ x^{n+1}\bar{x} = x^n \text{ for some } n. \tag{1}$$

The equalities (1) have been used in (Drazin 1958) for the original definition of pseudoinverses. In a particular case when S is completely regular, the operation $x \mapsto \bar{x}$ turns into the usual operation $x \mapsto x^{-1}$ of such semigroups.

The idea of viewing epigroups as unary semigroups has been proposed in (Shevrin 1995), and this approach was manifested in three lines there: in formulations of statements solving some problems examined, in posing some problems, in a technique. In particular, it permits to develop the theory of varieties of epigroups as unary semigroups, and some basic premises of this theory are presented in the work mentioned. The class $\mathscr{E}$ of all epigroups is not a variety: it is not closed under (infinite) direct products. This means that the lattice $L(\mathscr{E})$ of all varieties of epigroups has no greatest element. In $L(\mathscr{E})$ there is a naturally distinguished chain

$$\mathscr{E}_1 \subset \mathscr{E}_2 \subset \cdots \subset \mathscr{E}_n \subset \ldots, \tag{2}$$

where $\mathscr{E}_n$ is the variety of all epigroups of index $\leq n$ (the index of an epigroup is the maximum of indices of its elements). If one regards the equalities (1) as identities, then $\mathscr{E}_n$ can be given just by them together with the identity of associativity. The chain (2) can be regarded as the "spine" of $L(\mathscr{E})$. The reason is the following observation: any variety of epigroups is contained in $\mathscr{E}_n$ for some n. The variety $\mathscr{E}_1$ is none other than the variety of all completely regular semigroups treated as unary semigroups.

Many natural classes of epigroups can be characterized in terms of identities; see various examples in (Shevrin 1995). In doing so, along with the *basic* unary operation $x \mapsto \bar{x}$, it is useful to consider a *derivated* operation $x \mapsto e_x$. The latter is linked with the former by the formula $e_x = x\bar{x}$. It is customary to denote e_x by x^ω in the theory of finite semigroups and by x^0 in the theory of completely regular semigroups. We remark in conclusion that interesting advances in the trend under discussion have been effected in (Auinger and Szendrei 1999) and (Zhil'tsov 2000).

A.7 Periodic Semigroups

by Lev N. Shevrin in Ekaterinburg, Russia

Periodic semigroups have been presented in Section A.6 as one of the most important types of epigroups. All properties of epigroups discussed in the section cited can clearly be applied, in particular, to periodic semigroups. In this section, we shall focus attention on two topics which are specific for periodic semigroups. About a role of periodic semigroups in the theory of semigroup varieties, see Section A.15.

Burnside type problems

An important subclass of the class of periodic semigroups is formed by locally finite semigroups. A semigroup S is **locally finite** (l. f.) if each of its finitely generated subsemigroups is finite. Not every periodic semigroup is l. f.: as is well known, there exist even periodic groups being not l. f. The corresponding facts are grouped around the famous Burnside problem for groups (see Section B.12). Therefore, when examining problems of both detecting periodic but not l. f. semigroups within certain classes of semigroups and, on the other hand, finding conditions under which periodic semigroups are l. f., it is common to call them **Burnside type** (or simply **Burnside**) **problems** (Section B.13). There exist examples of periodic but not l. f. semigroups among the types of semigroups which are far from groups, first of all, among nilsemigroups. The best known (and the "smallest") ones are the infinite 3-generated semigroup that satisfies the identity $x^2 = 0$ and the 2-generated semigroup with the identity $x^3 = 0$. Such semigroups can be easily constructed by using the Thue-Morse words (see Section A.3). A number of results in this direction can be expressed in terms of the free semigroup of rank k in the variety given by the identity $x^m = x^n$, $m < n$; this semigroup is usually denoted by $B(k, m, n)$. For instance, $B(k, m, n)$ with $k \geq 2$ and $m \geq 2$ is infinite. Free semigroups (not necessarily of finite rank) in varieties of the kind just indicated are sometimes called *free Burnside semigroups*. Certain details of the structure of such semigroups are considered in the survey (do Lago and Simon 2001).

A band of l. f. semigroups is l. f. (Shevrin 1965). This implies that if in a completely regular semigroup S all its maximal subgroups are l. f., then S (being a semilattice of rectangular bands, i.e., a band of bands of these subgroups) is itself l. f. Since semigroups satisfying the identity $x^n = x$ with $n > 1$ are completely regular, the last fact implies, in turn, a known theorem obtained originally in (Green and Rees 1952) and stating that the following two conditions are equivalent: (s) all semigroups satisfying the identity $x^n = x$ are l. f.; (g) all groups satisfying the identity $x^{n-1} = 1$ are l. f. Semigroups of idempotents can be treated as a particular case of the above situations, so any semigroup of idempotents is l. f. In turn, if one takes in account the last fact as known, the following statement generalizes the one about bands of l. f. semigroups. If ρ is a congruence on a semigroup S such that the quotient semigroup S/ρ is l. f. and all the ρ-classes being subsemigroups are l. f. semigroups, then S is l. f. (Brown 1968). A commutative periodic semigroup is obviously l. f.; some generalizations of this observation expressed in terms of so-called permutability are presented in (de Luca and Varricchio 1999, §3.4). A periodic semigroup of matrices over a field is l. f.; this result was obtained independently in (McNaughton and Zalstein 1975), (Ponizovskii 1975) and (Shneperman 1975). There is a more general result having established local finiteness of periodic semigroups of linear relations (Shneperman 1982).

The Burnside problems are considered in (Lallement 1979, Ch. 10), (de Luca and Varricchio 1999, Ch. 3), as well as in (Kharlampovich and Sapir 1995, §3.3).

Nilsemigroups

The definition of a nilsemigroup was given in Section A.6. Nilsemigroups being of independent interest as one of the main types of periodic semigroups may occur also as certain components of the structure of epigroups in general. Note that they are presented "latently" (as divisors) in any semigroup which is not periodic completely regular (nil-divisors of a periodic completely regular semigroup are only singletons), but in descriptions of some types of semigroups they arise explicitly, see examples in the section just cited.

In this subsection, we shall consider interrelations between several particular types of nilsemigroups distinguished in the course of investigations. All of these types are actually finiteness conditions, and as the smallest one the property of being a nilpotent semigroup will stand out. A semigroup S is *nilpotent* if there exists n such that for any $x_1, \ldots, x_n \in S$ we have $x_1 \ldots x_n = 0$. A semigroup is called *locally nilpotent* (l. n.) if each of its finitely generated subsemigroups is nilpotent. A semigroup all of whose proper subsemigroups are nilpotent will be called *almost nilpotent*. Of course, any nilpotent semigroup is almost nilpotent, but an almost nilpotent semigroup need not even be a nilsemigroup: the matter is that the zero of a subsemigroup of a semigroup S need not be the zero of S. (For l. n. semigroups, this is not the case, so every l. n. semigroup is a nilsemigroup. A finite nilsemigroup is nilpotent, a finitely generated nilpotent semigroup is finite, hence l. n. semigroups are precisely locally finite nilsemigroups.) Almost nilpotent semigroups being not nilsemigroups are exhausted by 2-element bands and cyclic groups of prime order. As to the nil case, see below the respective comments. The *idealizer* of a subsemigroup T in a semigroup S is the greatest subsemigroup in S containing T as an ideal. A semigroup S is said to satisfy the *idealizer condition*, or to be an *I-semigroup*, if any proper subsemigroup of S is distinct from its idealizer. A well ordered series $\{0\} = A_0 \subset A_1 \subset \cdots \subset A_\alpha \subset A_{\alpha+1} \subset \cdots \subset A_\beta = S$ of ideals of a semigroup S with zero (where, as usual, $A_\alpha = \bigcup_{\gamma < \alpha} A_\gamma$ for every limit ordinal α) is called an *ascending annihilator series* (a. a. s.) of S if $SA_{\alpha+1} \cup A_{\alpha+1}S \subseteq A_\alpha$ for every $\alpha < \beta$. A nilsemigroup has an a. a. s. [is nilpotent] iff its frame satisfies the minimal condition [is of finite length]. If all nilpotent subsemigroups of a nilsemigroup S are finite, then S is finite.

The following diagram of inclusions shows the interrelations between the above-mentioned classes of semigroups.

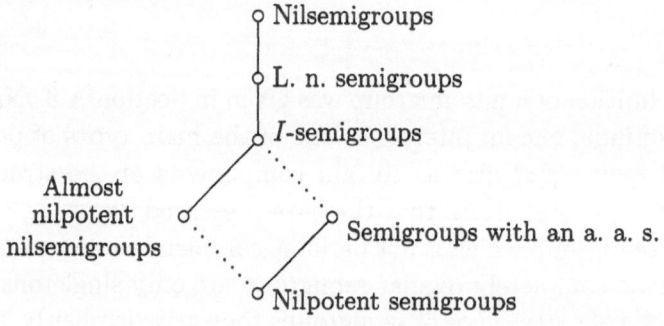

Continuous lines correspond to strict inclusions, dotted ones mean that we have two unsolved questions here. Both of them (posed by the author of the present section) arose in 1961. In the work (Shevrin 1992) written by the 30th anniversary of these problems, this matter is discussed in detail: history and connections, particular cases, properties of hypothetic counter-examples.

A.8 Finite Semigroups and Pseudovarieties

by Jorge Almeida in Porto, Portugal

Pseudovarieties

The most developed part of the theory of finite semigroups consists in their classification in classes closed under taking homomorphic images, subsemigroups and finite direct products, called ***pseudovarieties***. There are many familiar examples of pseudovarieties such as: $\mathscr{G}$ (finite groups), $\mathscr{G}_{sol}$ (finite solvable groups), $\mathscr{G}_{nil}$ (finite nilpotent groups), $\mathscr{C}$ (finite commutative semigroups), $\mathscr{N}$ (finite nil semigroups). The first systematic study of pseudovarieties of semigroups was made by Eilenberg (1976). A basic introduction to pseudovarieties can be found in (Pin 1986) while (Almeida 1995) contains a more elaborate study. The proceedings volume (Fountain 1995) contains several introductory articles on or relating to finite semigroups.

A strong motivation for the study of pseudovarieties comes from a correspondence due to Eilenberg (1976) with so-called ***varieties of languages***. These are mappings $\mathcal{V}: A \mapsto \mathcal{V}A$ associating with each finite alphabet A a set $\mathcal{V}A$ of rational languages over A which is a Boolean algebra closed under left and right quotients, and such that, for a homomorphism $\varphi: A^+ \to B^+$ between finitely generated free semigroups, if $L \in \mathcal{V}B$ then $\varphi^{-1}L \in \mathcal{V}A$. A language $L \subseteq A^+$ is said to be ***recognized*** by a semigroup S if there is a homomorphism $\varphi: A^+ \to S$ such that $L = \varphi^{-1}(\varphi(L))$. Eilenberg's correspondence associates with each pseudovariety $\mathcal{V}$ the mapping $\mathcal{V}$ whose image $\mathcal{V}A$ for a finite alphabet A consists of all rational languages over A recognized by semigroups from $\mathcal{V}$.

Instances of Eilenberg's correspondence which predate it and led to it are given by the characterizations of star-free languages, piecewise testable languages, and locally testable languages (see e.g., Eilenberg 1976). They correspond respectively to the pseudovarieties $\mathscr{A}$ of finite ***aperiodic*** (that is with no non-trivial subgroups) semigroups, $\mathscr{J}$

of finite $\mathcal{J}$-*trivial* semigroups (whose principal ideals admit only one generator), and $\mathscr{L}Sl$ of finite *local semilattices* (that is semigroups whose submonoids are semilattices). In all these examples, the algebraic characterization of combinatorial properties of languages provides an algorithm to decide whether these properties hold. A pseudovariety is said to be *decidable* if there is an algorithm to test membership of finite semigroups in it.

Pseudovarieties are often defined in terms of generators, with the generators being constructed from members of other pseudovarieties by applying some natural algebraic operator such as direct, semidirect or Schützenberger product, or the power semigroup construction. Each such algebraic operator induces a corresponding operator which usually carries the same name. An exception is the join in the lattice of pseudovarieties which is induced by the direct product. The semidirect product of pseudovarieties is also induced by the wreath product from which it follows that it is associative.

A general problem in finite semigroup theory is to determine conditions on the arguments of an operator on pseudovarieties which will ensure that the resulting pseudovariety is decidable. One might naïvely expect that decidability of the arguments would suffice but Albert et al. (1992) and Rhodes (1999) have shown that this is not the case for the join and for the semidirect and Mal'cev products.

Profinite semigroups

Looking at n-generated members of a pseudovariety $\mathcal{V}$, there is sometimes a most general one of which all others are homomorphic images, but this is usually not the case. In complete generality, to find a most general n-generated semigroup with respect to $\mathcal{V}$, it is necessary to leave the realm of finite semigroups and take the projective limit of all n-generated semigroups in $\mathcal{V}$, where here and elsewhere finite semigroups are viewed as discrete topological semigroups. One thus obtains a *profinite semigroup* (which may be axiomatized as a compact semigroup which is residually finite as a topological semigroup) or, more specifically, a pro-$\mathcal{V}$ semigroup (meaning a profinite semigroup which is residually in $\mathcal{V}$). This projective limit is the free pro-$\mathcal{V}$ semigroup on n generators, denoted $\overline{\Omega}_n \mathcal{V}$. An evaluation of the generators as elements $s_1, \ldots, s_n$ of a semigroup S from $\mathcal{V}$ yields a unique continuous homomorphism $\overline{\Omega}_n \mathcal{V} \to S$ and therefore to each element π of $\overline{\Omega}_n \mathcal{V}$ corresponds an element $\pi_S(s_1, \ldots, s_n)$ of S. This defines π as an n-ary operation on members of $\mathscr{V}$ which commutes with homomorphisms. Such an operation is said to be an *implicit operation*.

Denote by $\mathscr{S}$ the pseudovariety of all finite semigroups. An important example of an implicit operation on $\mathscr{S}$ is the unary operation associating with an element s of a finite semigroup the inverse $s^{\omega-1}$ of ss^{ω} in the maximal subgroup (with idempotent s^{ω}) of the subsemigroup generated by s. In other words, $s^{\omega-1}$ is the unique power t of s which is a *weak inverse* of s in the sense that $tst = t$.

Implicit operations may be composed in the natural way and may be multiplied pointwisely to obtain again implicit operations. The above correspondence from members of $\overline{\Omega}_n\mathscr{V}$ to n-ary implicit operations on $\mathscr{V}$ is in fact a semigroup isomorphism. The operation point of view turns out to be quite useful. If we fix a set σ of implicit operations on S containing the binary operation of multiplication, then every element of σ has a natural interpretation as an operation on every profinite semigroup. In particular, we may consider the σ-subsemigroup $\Omega_n^{\sigma}\mathscr{V}$ of $\overline{\Omega}_n\mathscr{V}$ generated by the component projections; this is precisely the free σ-semigroup in the variety of σ-semigroups generated by $\mathscr{V}$. The most often encountered example involves the set κ consisting of the binary operation of multiplication and the $\omega - 1$ power operation. For instance, denoting by $\mathscr{G}$ the pseudovariety of all finite groups, $\Omega_n^{\kappa}\mathscr{G}$ is the free group on n generators.

Tame pseudovarieties

To a finite (directed multi)graph Γ we associate a system of formal equations $x_{\alpha e}x_e = x_{we}$, where $\alpha e \xrightarrow{e} we$ runs over all edges of Γ, for which we seek solutions in the semigroup $\overline{\Omega}_n\mathscr{V}$. A labeling of Γ by elements of an n-generated finite semigroup S (with vertices possibly labeled by an adjoined identity element 1, if S is not a monoid) determines constraints for the solutions in the sense that we wish the value in $\overline{\Omega}_n\mathscr{V}$ of each unknown x_z to be chosen as the restriction to $\mathscr{V}$ of an implicit operation whose value on the generators of S is precisely the chosen label of the element z of Γ.

The pseudovariety $\mathscr{V}$ is said to be σ-*reducible* if, whenever such a system has a solution satisfying the constraints, it also has a solution in $\Omega_n^{\sigma}\mathscr{V}$ satisfying the constraints in the strong sense that there are implicit operations in $\Omega_n^{\sigma}\mathscr{S}$ whose restriction to $\mathscr{V}$ are the solutions and whose values in S are the corresponding given labels. We say that $\mathscr{V}$ is σ-*tame* if $\mathscr{V}$ is σ-reducible and the word problem is algorithmically solvable for each relatively free σ-semigroup $\Omega_n^\sigma \mathcal{V}$. Finally, a recursively enumerable pseudovariety is **tame** if it is σ-tame for some recursively enumerable set σ of computable implicit operations.

Examples of tame pseudovarieties are the pseudovariety $\mathcal{G}$ of all finite groups (Ash 1991), which is κ-tame, and the pseudovariety $\mathcal{G}_p$ of all finite p-groups (Almeida 2000), which is not κ-tame. Herwig and Lascar (2000) have proved a result in model theory with a formulation in terms of a property of free groups which turns out to be (formally) equivalent to κ-tameness of $\mathcal{G}$. A weak version of κ-tameness is intimately connected with properties of free groups with respect to profinite topologies. In particular, the special case of Ash's theorem corresponding to graphs with only one vertex, which was obtained independently by Ribes and Zalesskiĭ (1993), may be formulated by stating that the product of finitely generated subgroups of a free group is closed in the profinite topology.

J. Rhodes has announced that $\mathcal{A}$ is κ-tame.

The Krohn-Rhodes complexity

A semigroup S is said to **divide** a semigroup T if S is a homomorphic image of a subsemigroup of T. Just as finite groups may be reconstructed (up to embedding) from their composition factors using the wreath product, Krohn and Rhodes (1965) have shown that every finite semigroup S divides a wreath product whose factors are simple groups dividing S and copies of the 2-element right-zero semigroup with identity adjoined. Collecting together consecutive factors according to whether they are groups or aperiodic, one finds a decomposition of S (up to division) into a wreath product in which groups and aperiodic factors alternate. The minimal number of group factors in such a decomposition is called the **Krohn-Rhodes complexity** of S, denoted $c(S)$.

A central problem in the theory of finite semigroups has been to determine whether the complexity function c is computable. The problem reduces to finding a systematic algorithmic solution to the membership problem in semidirect product pseudovarieties of the form $\mathcal{A} * \mathcal{G} * \mathcal{A} * \cdots * \mathcal{G} * \mathcal{A}$. As a consequence, since we are then dealing with products of tame pseudovarieties, it will follow that c is computable.

It was announced by Almeida and Steinberg (2000) that the semidirect product of tame pseudovarieties is decidable. In view of the tameness of $\mathcal{G}$ and $\mathcal{A}$, this would entail the decidability of the Krohn-Rhodes complexity. Unfortunately, the result by Almeida and Steinberg depends on a characterization of Almeida and Weil (1998) of the semidirect product of two pseudovarieties in whose proof a gap has since been found. An alternative way of settling the decidability of complexity problem would be to show that the semidirect product of tame pseudovarieties is again tame, but this remains an open problem.

A.9 Regular Semigroups

by Peter G. Trotter in Hobart, Australia

An element x of a semigroup S is **regular** if and only if there exists $x' \in S$ such that $xx'x = x$ and $x'xx' = x'$; x' is then called an **inverse** of x and both xx' and $x'x$ are idempotents. A semigroup is regular if each of its elements is regular. Examples of regular semigroups include any band, inverse semigroup or completely regular semigroup (see sections A.2, A.11, and A.10); in particular, any group is a regular semigroup. In a semigroup S, the inverses of a regular element $x \in S$ all lie in the $\mathcal{D}$-class of x (see Section A.1) and every member of this $\mathcal{D}$-class is also regular. Any idempotent in S is the identity of a subgroup of S, so a semigroup with a regular element has a regular subsemigroup; it follows that any finite semigroup has a regular subsemigroup. The full transformation semigroup $\mathcal{T}_X$ on a set X (see Section A.12) is a regular semigroup. Each semigroup embeds in a regular semigroup; this is because by an elementary generalization of Cayley's Theorem for groups, each semigroup embeds in a full transformation semigroup. More remarkably, each semigroup embeds in an idempotent generated regular semigroup (see Howie (1995)). These comments indicate that the notion of regularity is at the center of the theory of semigroups.

Recall (from Section A.11) that the set of idempotents E of an inverse semigroup I plays a critical role in its structure theory. This set E is a subsemilattice of I; any semilattice is an inverse semigroup. There is an idempotent separating homomorphism from I into Munn's fundamental inverse semigroup T_E and by McAlister there is a surjective idempotent separating homomorphism from an E-unitary inverse semigroup onto I. A congruence on I is uniquely determined by its restriction to E and the union of its idempotent classes. These results have strongly influenced the structure theory of regular semigroups.

In the remainder of this article, S will denote a regular semigroup, $E(S)$ is its set of idempotents and for any $x \in S$, $V(x)$ is the set of all inverses of x in S.

Biordered sets

Generalizations to regular semigroup theory of the construction of Munn's fundamental inverse semigroups were independently obtained in the 1970's by Hall, Grillet, Nambooripad and Clifford. A regular semigroup is *fundamental* if and only if its greatest idempotent separating congruence is the equality relation. Given that $E(S)$ need not be a subsemigroup of the regular semigroup S it was not clear what algebraic structure should underlie a generalized Munn semigroup. Hall (1973a) constructed a fundamental regular semigroup $T_{<E(S)>}$ from the idempotent generated subsemigroup $< E(S) >$ of S such that $< E(T_{<E(S)>}) >$ is the maximal fundamental homomorphic image of $< E(S) >$. Clifford and Grillet concentrated on $E(S)$ as being a partial subgroupoid of S. Nambooripad (1979) extracted the essence of the relationship between S and $E(S)$ through his notion of a biordered set.

There are natural quasi-orderings $\leq_l$ and $\leq_r$ of $E(S)$ given by $e \leq_l f$ or $e \leq_r f$ if and only if $ef = e$ or $fe = e$ respectively. Nambooripad considered a partial binary operation on $E(S)$ such that for $e, f \in E(S)$, ef is defined if and only if e and f are comparable by $\leq_l$ or $\leq_r$. He took as axioms eight elementary properties of the partial operation and quasi-orders and called any partial algebra satisfying the axioms a *biordered set*. A *regular biordered set* satisfies the additional axiom that in S the condition $fV(ef)e \neq \varnothing$ holds; this non-empty set is the *sandwich set* of the ordered pair e, f of idempotents of S and is made up of some idempotent inverses of the product ef in S. So $E(S)$ is a regular biordered set, and Nambooripad proved that any biordered set is isomorphic to $E(S)$ for some regular semigroup S. He showed that a homomorphism $\psi : S \to S'$ of regular semigroups such that $\psi(E(S)) = E(S')$ is idempotent separating if and only if $\psi|_{E(S)}$ is a biorder isomorphism. From any regular biordered set E he constructed a fundamental regular semigroup T_E whose biordered set of idempotents is isomorphic to E. Conversely, any regular semigroup S is fundamental if and only if it embeds in $T_{E(S)}$.

Congruences

It is noted in Clifford and Preston (1967b) that any congruence ρ on a regular semigroup S is uniquely determined by its idempotent classes. In particular, let $\ker \rho = \cup\{\rho(e) : e \in E(S)\}$ be the *kernel* of ρ and

$\operatorname{tr} \rho = \rho \cap (E(S) \times E(S))$ be the *trace* of ρ, then by Feigenbaum

$$a\rho b \iff a\rho \mathcal{H} b\rho, \; ab' \in \ker \rho \text{ for some [all] } b' \in V(b).$$

In the last statement, both *some* and *all* are correct. By Hall, $a\rho\mathcal{H}b\rho = a(\mathcal{L} \operatorname{tr} \rho \mathcal{L} \operatorname{tr} \rho \mathcal{L} \cap \mathcal{R} \operatorname{tr} \rho \mathcal{R} \operatorname{tr} \rho \mathcal{R})b$ where $\mathcal{H}$, $\mathcal{L}$ and $\mathcal{R}$ are Green's relations (see Section A.1). There remains the problem of characterizing those partitions of $E(S)$ that are traces of congruences on S, of those subsets of S that are kernels of congruences on S, and of the connections between such partitions and subsets that together make them the trace and kernel of a congruence; Feigenbaum and (independently) Trotter have found (necessarily) technical descriptions of the partitions and subsets while Pastijn and Petrich (1986) have nice criteria for the connections.

Let $\mathcal{C}(S)$, $\mathcal{E}(S)$ and $\mathcal{K}(S)$ denote respectively the lattices of congruences on S, equivalence relations on S and kernels of congruences on S; all ordered by inclusion. By Trotter (1996), the map $\operatorname{tr}: \mathcal{C}(S) \to \mathcal{E}(S)$ is a complete lattice homomorphism. The congruence induced by the map tr has complete modular sublattices of $\mathcal{C}(S)$ for its classes while the congruence induced by the map $\ker: \mathcal{C}(S) \to \mathcal{K}(S)$ has complete sublattices for its classes. Interesting congruences include the greatest idempotent separating congruence μ on S due to Hall:

$a\mu b \iff a\mathcal{H}b$ and
$$\text{for some [any] } a' \in V(a), \; b' \in V(b) \text{ where } a'\mathcal{H}b',$$
$$a'ea = b'eb \text{ for each idempotent } e \le aa'.$$

By Feigenbaum, the least group congruence σ on S is given by

$$a\sigma b \iff au = vb \text{ for some } u, v \in C_\infty(S)$$

where $C_\infty(S)$ is the least subsemigroup of S that contains $E(S)$ and is self conjugate in that for any $x \in S$ and $x' \in V(x)$, $xC_\infty(S)x' \subseteq C_\infty(S)$. If S has the property that $< E(S) >$ is completely regular (that is, S is **E-solid**) then Trotter has shown that the least inverse congruence on S has $C_\infty(S)$ for its kernel.

e-varieties

The classes of inverse (Section A.11) and of completely regular semigroups (Section A.10) have been extensively researched as subvarieties of the variety of regular unary semigroups; that is of semigroups with a inverse unary operation $^{-1}$ (satisfying $xx^{-1}x = x, x^{-1}xx^{-1} = x^{-1}$).

By the axiom of choice, any regular semigroup comes endowed with (possibly many) inverse unary operations but usually these operations cannot be characterized in a natural way. As a consequence, the powerful concepts associated with varieties of algebras have, until recently, been applied only to the above mentioned subvarieties.

A new and fruitful way of studying classes of regular semigroups was independently introduced by Hall (1989) and by Kaďourek and Szendrei (1990) that partially utilizes the universal algebraic results of Birkhoff. A class of regular semigroups forms an *e-variety* of regular semigroups if it is closed under taking direct products, homomorphic images and *regular* subsemigroups.

Kadourek and Szendrei generalized the notion of a free object in a variety as follows. Let X be a denumerable set and $X^{-1} = \{x^{-1} : x \in X\}$ be a copy of X that is disjoint from X. Put $\overline{X} = X \cup X^{-1}$ and $(x^{-1})^{-1} = x$. For any regular semigroup S, a map $\theta: \overline{X} \to S$ is *matched* if $\theta(y^{-1}) \in V(\theta(y))$ for each $y \in \overline{X}$. In an e-variety $\mathcal{V}$ an object $F_{\mathcal{V}}(X) \in \mathcal{V}$ is *bi-free* in $\mathcal{V}$ on X if and only if there is a matched map $\iota: \overline{X} \to F_{\mathcal{V}}(X)$ such that for each matched map $\theta: \overline{X} \to T$, $T \in \mathcal{V}$, there is a unique homomorphism $\varphi: F_{\mathcal{V}}(X) \to T$ where $\iota\varphi = \theta$. In a remarkable result, Yeh proved that for $|X| \geq 2$, $F_{\mathcal{V}}(X)$ exists if and only if $\mathcal{V}$ is an e-variety of E-solid regular semigroups or of locally inverse semigroups (these are regular semigroups whose submonoids are all inverse subsemigroups).

The notion of a bi-free object on a set X in an e-variety has been extended by Kadourek to tri-free objects. They also have the analogous property to the universal property of free objects but occur in a strictly larger class of e-varieties than do bi-free objects. Their definition involves the extension of the double alphabet $\overline{X}$ in the above to include the sandwich sets of pairs xx^{-1}, yy^{-1} for $x, y \in \overline{X}$. It has been shown by Churchill and Trotter that tri-free objects on at least 3 elements exist precisely in e-varieties of locally E-solid regular semigroups (the submonoids of these are E-solid regular subsemigroups).

For details on e-varieties see the survey by Trotter (1996) and the article of Churchill and Trotter (2000).

A.10 Completely Regular Semigroups

by John M. Howie in St. Andrews, U.K.

Just as the study of non-abelian groups focuses on various conditions of *quasi-commutative* type, such as solvability and nilpotency, so semigroup theory has fixed attention on classes of semigroups that are in some sense *group-like*. The class Gp of groups can be considered as a variety with a binary operation $(a, b) \mapsto ab$ and a unary operation $a \mapsto a^{-1}$, satisfying the identities

$$(xy)z = x(yz), \quad xx^{-1}x = x, \quad (x^{-1})^{-1} = x, \tag{1}$$

and

$$xx^{-1} = yy^{-1},$$

and two important generalizations of groups can be characterized in a similar way. If we have the identities (1) and the extra identity

$$xx^{-1}yy^{-1} = yy^{-1}xx^{-1},$$

we obtain the class Inv of *inverse* semigroups (see Section A.11); if we have the identities (1) and the extra identity

$$xx^{-1} = x^{-1}x,$$

then we have the class CR of *completely regular* semigroups.

It is clear that Gp $\subseteq$ CR, and only slightly less clear that the containment is proper: any non-trivial semigroup consisting entirely of idempotents, in which we define $x^{-1} = x$, belongs to CR. It is clear also that if $S \in$ CR then every element of S lies in a subgroup of S; for we know that $ss^{-1} = s^{-1}s$ $(= e$, say), and it is easy to show that $\{t \in S \mid tt^{-1} = t^{-1}t = e\}$ is a subgroup of S with identity e. In fact the concept of a completely regular semigroup was first introduced by Clifford (1941) as a semigroup which is a union of groups.

The class CS of *completely simple* semigroups (Section A.5) is properly contained in CR, being characterized within CR by the identity $xx^{-1} = (xyx)(xyx)^{-1}$. All finite semigroups that are *simple*, in the ring-theoretic sense of having no proper ideals, are completely simple, and some (but not all) of those are simple in the group-theoretic sense of having no proper homomorphic images. The finite simple groups are of course examples of finite *congruence-free* semigroups, but there are others, classified by Tamura (1956); for example, the semigroup

	0	a	b	e	f
0	0	0	0	0	0
a	0	0	e	0	a
b	0	f	0	b	0
e	0	a	0	e	0
f	0	0	b	0	f

is a congruence-free inverse semigroup.

The structure of completely simple semigroups was clarified (modulo groups) at an early stage by Suschkewitsch (1928) and Rees (1940). (See Section A.3 for this description and historic remarks, and (Howie 1995) or (Petrich and Reilly 1999) for more on this work.) For completely regular semigroups the crucial theorem, due to Clifford (1941), is that every such semigroup S is a **semilattice of completely simple semigroups**. That is to say, S is a disjoint union $\bigcup_{\alpha \in Y} S_\alpha$, where Y is a (lower) semilattice, each S_α is completely simple, and, for all α, β in Y,

$$S_\alpha S_\beta \subseteq S_{\alpha\beta}.$$

(Here $\alpha\beta$ means $\min\{\alpha, \beta\}$.)

This theorem elucidates the 'gross' structure of a completely regular semigroup. The 'fine' structure is another matter entirely. Elements x_α and y_β in S_α and S_β, respectively, can be described in terms of the Rees-Suschkewitsch theorem, and so can their product $x_\alpha y_\beta$ within the completely simple semigroup $S_{\alpha\beta}$. It is, however, a very complicated matter to give the description of $x_\alpha y_\beta$ in terms of the descriptions of x_α and y_β. See (Lallement 1967), (Petrich 1974), and (Clifford and Petrich 1977).

More detailed information is available for the variety $\mathsf{Cliff} = \mathsf{CR} \cap \mathsf{Inv}$ of so-called **Clifford semigroups**. If $S \in \mathsf{Cliff}$, then the components S_α, being in the class $\mathsf{CS} \cap \mathsf{Inv}$, are all groups, and we get what is called a **strong semilattice structure**, in which, for all $\alpha \geq \beta$ in Y, there is a homomorphism $\varphi_{\alpha,\beta} \colon S_\alpha \to S_\beta$ such that

(i) $\varphi_{\alpha,\alpha}$ is the identity mapping of S_α for every α in Y;

(ii) $\varphi_{\alpha,\beta}\varphi_{\beta,\gamma} = \varphi_{\alpha,\gamma}$ for all $\alpha \geq \beta \geq \gamma$ in Y.

Then the multiplication in S is given by the rule that, for each x_α in S_α and y_β in S_β,

$$x_\alpha y_\beta = (x_\alpha \varphi_{\alpha,\alpha\beta})(y_\beta \varphi_{\beta,\alpha\beta}),$$

where the product on the right takes place within the group $S_{\alpha\beta}$.

The Clifford semigroup

	e	a	f	b
e	e	a	f	b
a	a	e	b	f
f	f	b	f	b
b	b	f	b	f

provides a remarkably simple example to show that even a semigroup very close to being a group does not have a modular lattice of congruences.

Other special cases can be described in a comparably transparent way. We have already observed that a **band** B (a semigroup consisting entirely of idempotents, Section A.2) is an example of a completely regular semigroup, and here the components S_α are **rectangular bands** (semigroups satisfying the identity $xyx = x$. A band is a strong semilattice of rectangular bands if and only if it satisfies the identity $xyzx = xzyx$. Such a band is called **normal**.

Other particular cases that have been considered are **orthogroups** (completely regular semigroups whose idempotents form a subsemigroup) and **cryptogroups** (completely regular semigroups in which Green's relation $\mathcal{H}$ (see Section A.1) is a congruence. Two typical results, both due to Petrich (1972) will give the flavor of this area, where more detailed information can be obtained by placing restrictions on the nature of the idempotents

First, the simplest kind of orthogroup is a **rectangular group**, defined as a direct product of a group and a rectangular band. It is a special case of what is called a **normal orthogroup**, an orthogroup whose idempotents form a normal band. Then a semigroup S is a normal orthogroup if and only if it is a strong semilattice of rectangular groups.

Similarly, the simplest kind of cryptogroup is a completely simple semigroup, and this is a special case of a **normal cryptogroup**, a cryptogroup S for which $S/\mathcal{H}$ is a normal band. Then a semigroup S is a normal cryptogroup if and only if it is a strong semilattice of completely simple semigroups.

The class CR, as we have defined it, is evidently a $(2,1)$-variety, and much attention has been devoted in recent years to subvarieties of CR. This study is made more difficult by the fact that the free object is quite hard to describe—see (Gerhard 1983, Trotter 1984). The lattice of all varieties of bands has been described (see Section A.15 for details) by Gerhard (1970), Biryukov (1970) and Fennemore (1971a,b), but a comparably complete classification of all varieties of completely regular semigroups is too much to hope for. An excellent account of this area is in (Petrich and Reilly 1999).

A.11 Inverse Semigroups

by Peter G. Trotter in Hobart, Australia

A substantial theory of inverse semigroups has been developed since their introduction by Wagner (1952) and Preston (1954) as regular semigroups with commuting idempotents. Inverse semigroups are also the regular semigroups in which each element has a unique inverse. That is, the class of inverse semigroups is the variety of algebras with an associative binary operation, and a unary operation $^{-1}$, that satisfies the identities

$$xx^{-1}x = x, \quad (x^{-1})^{-1} = x, \quad xx^{-1}y^{-1}y = y^{-1}yxx^{-1}.$$

The variety of groups is the subvariety that satisfies the extra identity $xx^{-1} = y^{-1}y$, while the subvariety of **semilattices** (that is, commutative semigroups of idempotents) satisfies the extra identity $x^2 = x$. It is therefore natural that groups and semilattices play important roles in the structural theory of inverse semigroups. Notice that a non-commutative semigroup of idempotents is not an inverse semigroup; for example, any right zero semigroup consists of idempotents and is such that every element is an inverse of each element. Apart from groups and semilattices, the class of inverse semigroups includes the bicyclic semigroup, and well known classes such as the Brandt semigroups (Section A.5) (these include any groups with an adjoined zero, as well as the 5-element congruence-free inverse semigroup), the Clifford semigroups, and the symmetric inverse semigroups.

In this article, S denotes an inverse semigroup, a^{-1} is the inverse of $a \in S$ and $E(S)$ is the subsemigroup of all idempotents of S (so $E(S)$ is a subsemilattice of S).

Symmetric inverse semigroups

Permutation groups are fundamental to group theory; they are a source of examples, are often involved in applications and most importantly, each group embeds in a symmetric group. Wagner and Preston investigated a generalization of the symmetric group on a non-empty set X, namely the **symmetric inverse semigroup** $\mathcal{I}_X$. The semigroup $\mathcal{I}_X$ consists of all one to one partial mappings of X into itself, with the binary operation being composition of partial mappings. Any inverse semigroup S is embeddable in $\mathcal{I}_S$; symmetric inverse semigroups are therefore a ready source of examples of inverse semigroups.

Congruences

The congruence theory is well developed and is "user friendly". In particular, any congruence ρ on S is completely determined by its idempotent classes. This was noticed by Wagner and Preston and in (Scheiblich 1974); the observation was developed into the *trace-kernel theory of congruences*. By this theory, ρ is uniquely determined by its **trace** $\operatorname{tr}\rho = \rho \cap (E(S) \times E(S))$ and **kernel** $\ker\rho = \bigcup\{\rho(e) \mid e \in E(S)\}$. In fact, $(a,b) \in \rho$ if and only if $(aa^{-1}, bb^{-1}) \in \operatorname{tr}\rho$ and $ab^{-1} \in \ker\rho$. Scheiblich characterized the relations on $E(S)$ and the subsemigroups of S that form the trace-kernel pairs constituting congruences. If S is a group, $|E(S)| = 1$ and $\ker\rho$ is a normal subgroup.

Let $\mathcal{C}(S)$, $\mathcal{N}(S)$ and $\mathcal{K}(S)$ respectively denote the lattices of congruences on S, traces of congruences on S, and kernels of congruences on S, all ordered by inclusion; these are complete lattices. By Reilly and Scheiblich, the map $\operatorname{tr}\colon \mathcal{C}(S) \to \mathcal{N}(S)$ is a complete lattice homomorphism and the congruence θ induced by tr has complete modular sublattices for its classes. The kernel map is not so nice; by Green, $\ker\colon \mathcal{C}(S) \to \mathcal{K}(S)$ is a complete $\cap$-homomorphism and the classes of the induced congruence are complete sublattices. The greatest and least congruences in any θ-class are known; the **greatest idempotent separating congruence** μ on S, by Howie, is given by

$$(a,b) \in \mu \Leftrightarrow aea^{-1} = beb^{-1} \quad \forall e \in E(S).$$

When ρ is the universal congruence on S, the least congruence in the θ-class of ρ is the **least group congruence** σ on S which, by Munn, is given by

$$(a,b) \in \sigma \Leftrightarrow ae = be \text{ for some } e \in E(S).$$

The semigroup S is called **E-unitary** if the least group congruence σ has ker $\sigma = E(S)$. Alternatively, S is **fundamental** if its greatest idempotent separating congruence μ is the equality relation (for details, see Petrich 1984).

The fundamental and the E-unitary inverse semigroups are key components in the structure theory of inverse semigroups. Any semilattice is fundamental and E-unitary, any non-trivial group is E-unitary but not fundamental, and a free inverse semigroup is both fundamental and E-unitary.

Fundamental inverse semigroups

Suppose that E is a semilattice. Munn (1970) constructed an inverse semigroup T_E from E; T_E is the semigroup consisting of all isomorphisms amongst the principal ideals of E (these are one to one partial mappings of E into itself) with composition of partial maps as the binary operation. This object has remarkable properties; $E = E(T_E)$, T_E embeds any fundamental inverse semigroup whose semilattice of idempotents is isomorphic to E and furthermore, any inverse subsemigroup S of T_E with $E(S) = E$ is fundamental. Hence for any inverse semigroup S there is an idempotent separating homomorphism of S into $T_{E(S)}$.

E-unitary inverse semigroups

McAlister (1974a) showed that for any S there exists an E-unitary inverse semigroup T and a surjective idempotent separating homomorphism from T onto S; furthermore, McAlister and Reilly showed that any such T is a subdirect product of S by the maximal group homomorphic image of T. This result is useful in the structure theory of inverse semigroups because of the relatively simple structure of E-unitary inverse semigroups and of idempotent separating congruences. McAlister (1974b) provided a structure theorem for a typical E-unitary inverse semigroup that was reinterpreted by O'Carroll (1976) to characterize the E-unitary inverse semigroups as the inverse semigroups that are embeddable in semidirect products of semilattices and groups. A **semidirect product** of a semilattice Y and a group G is the Cartesian product $Y \times G$ with multiplication given by $(\alpha, g)(\beta, h) = (\alpha g(\beta), gh)$ where G acts on Y by automorphisms.

Free inverse semigroups

The existence of the free inverse semigroup I_X on a non-empty set X
was noticed by Wagner. Scheiblich (1973) produced a very usable model
for I_X. From the model it is easy to see that I_X is both E-unitary and
fundamental, in fact it has only trivial subgroups. There is an alternative and also very usable model of I_X due to Munn (1974); in this model
the elements of I_X are represented by graphs that are trees with two
distinguished vertices and that are labelled by members of the double
alphabet $X \cup X^{-1}$. Various authors have since made use of graphical
methods as used by Munn, or based on graphical techniques of combinatorial group theory or automata theory to explore word problems,
free products of inverse semigroups and other problems (for a survey see
Meakin 1993). Although it is not true that all inverse subsemigroups of
I_X are free, Reilly characterized the subsets of I_X that freely generate
free inverse subsemigroups. Jones proved that any inverse subsemigroup
of I_X has a minimal generating set and all of its minimal generating sets
have the same cardinality.

Varieties

The variety $\mathbf{I}$ of inverse semigroups is not the join of a finite number
of its proper subvarieties, but it is generated by its finite members with
only trivial subgroups. Kleiman (1977) initiated the investigation of the
lattice $\mathcal{L}(\mathbf{I})$ of subvarieties of $\mathbf{I}$. He showed that the join of members
of $\mathcal{L}(\mathbf{I})$ with the variety $\mathbf{G}$ of all groups determines a complete lattice
homomorphism of $\mathcal{L}(\mathbf{I})$ onto the interval $[\mathbf{G}, \mathbf{I}]$ of $\mathcal{L}(\mathbf{I})$, and under the
induced lattice congruence, the class of $\mathbf{V} \in \mathcal{L}(\mathbf{I})$ is the interval bounded
below by the variety generated by the fundamental members of $\mathbf{V}$, and
above by $\mathbf{V} \vee \mathbf{G}$. Hence the lattice $\mathcal{L}(\mathbf{I})$ consists of layers of these intervals. The bottom layer is the lattice $\mathcal{L}(\mathbf{G})$ of all varieties of groups. The
next two layers are each isomorphic to $\mathcal{L}(\mathbf{G})$. The remainder of the lattice $\mathcal{L}(\mathbf{I})$ is largely unexplored and is more complex; for example, there
is a bottom layer that covers the first three layers, these four layers form
a non-modular sublattice of $\mathcal{L}(\mathbf{I})$ (see Reilly 1980), and the least variety
of the fourth layer has no irreducible basis of identities (Kaďourek 1992).
However, good information has been obtained about various segments
of $\mathcal{L}(\mathbf{I})$. The lattice has been explored via the anti-isomorphic lattice
of fully invariant congruences on I_X where X is denumerable, particularly with respect to intervals of fully invariant congruences that have a
common trace or have a common kernel (see Reilly and Trotter 1986).

This section has briefly summarized a small portion of the theory of inverse semigroups. For fuller details, see Petrich (1984) and Lawson (1998).

A.12 Separated Transformation Semigroups
by Kenneth D. Magill, Jr. in Buffalo, NY, USA

For information about semigroups in general, one can consult Clifford and Preston (1967a,b). We denote by T_X the **full transformation semigroup**, under composition, on a nonempty set X. That is, $f \circ g(x) = f(g(x))$ for all $x \in X$ and $f, g \in T_X$. It is well known that every semigroup S is isomorphic to a subsemigroup of some T_X. Furthermore, semigroups of transformations occur quite naturally. Examples include the semigroup of all linear transformations on a vector space, the semigroup of all continuous selfmaps of a topological space, and the semigroup of all closed selfmaps of a topological space. By a closed selfmap we mean any (not necessarily continuous) selfmap with the property that the image of each closed subset is closed. Throughout these pages, $L(V)$ will denote the semigroup of all linear transformations of a vector space V over a field F, $S(X)$ will denote the semigroup of all continuous selfmaps of a topological space X, and $\Gamma(X)$ will denote the semigroup of all closed selfmaps of a topological space X. Our goal here is to indicate how one can investigate rather general transformation semigroups in order to obtain results which then can be applied to various "natural semigroups of transformations". For this purpose, we have chosen to discuss some of the results in Magill (1993) and that paper may be consulted for proofs and any other further details. For any nonempty set X, we denote by id_X (or, more simply by id, when X is apparent) the identity function on X. Let $T(X)$ be any subsemigroup of T_X. The range of a function $f \in T(X)$ will be denoted by $Ran(f)$.

A.12.1 Definition The range of an idempotent element of $T(X)$ is referred to as a **T-retract** of X.

A.12.2 Definition Let $A, B \subseteq X$. We say that $f \in T(X)$ maps A **T-isomorphically** onto B if $f[A] \subseteq B$ and there exists a $g \in T(X)$ such that $g[B] \subseteq A$ and $g \circ f|A = \mathrm{id}_A$ and $f \circ g|B = \mathrm{id}_B$. When this is the case, we say that A and B are **T-isomorphic** and we refer to $f|A$ as a **T-isomorphism** from A onto B.

When the semigroup under consideration is $L(V)$, we will speak of L-retracts and L-isomorphisms rather than T-retracts and T-isomorphisms. Similarly, when the semigroup is $S(X)$, we will speak of S-retracts and S-isomorphisms and when it is $\Gamma(X)$, we will speak of Γ-retracts and Γ-isomorphisms. The L-retracts are precisely the linear subspaces of V and the L-isomorphisms between subspaces of V are just the linear isomorphisms. For any topological space X, the S-retracts of X are just the retracts in the usual sense. An S-isomorphism between two subspaces of a topological space X is a stronger condition than homeomorphism. If any two subspaces of X are S-isomorphic, then they are certainly homeomorphic, but two subspaces may be homeomorphic and not S-isomorphic. For example, the two subspaces $A = \{(x, y) \in \mathbb{R}^2 \mid y = 0\}$ and $B = \{(x, y) \in \mathbb{R}^2 \mid x^2 + y^2 = 1\} \setminus \{(1, 0)\}$ of $\mathbb{R}^2$ are homeomorphic but not S-isomorphic. The Γ-retracts of a T_1 space X are precisely the nonempty closed subsets of X. It is immediate that a Γ-retract of X must be a nonempty closed subset of X. To see that any such subset must be a Γ-retract, let H be any nonempty closed subset, choose $p \in H$ and define a selfmap f of X by $f(x) = x$ for $x \in H$ and $f(x) = p$ for $x \in X \setminus H$. For any closed subset K of X, we have $f[K] = (K \cap H) \cup \{p\}$ which is closed since X is T_1. Thus, $f \in \Gamma(X)$ and since f is idempotent, we conclude that H is a Γ-retract. It is easy to verify that for any topological space X, two closed subsets of X are Γ-isomorphic if and only if they are homeomorphic.

Now, suppose a semigroup S has an identity and a proper ideal. Then it has a largest proper ideal which we denote by $M(S)$. The group of units of S we denote by $G(S)$. It is immediate that $G(S) \cap M(S) = \varnothing$ for each semigroup S. However, $G(S) \cup M(S) \neq S$ for some semigroups S while $G(S) \cup M(S) = S$ for others.

A.12.3 Definition A semigroup S is said to be **separated** if it has an identity and a proper ideal and $G(S) \cup M(S) = S$.

In our main theorem, we give several characterizations of separated semigroups and we then apply this result to semigroups of linear transformations, semigroups of continuous selfmaps, and semigroups of closed selfmaps. But first, we need to discuss several concepts. Let J be any ideal of a semigroup S. The Rees factor semigroup of S modulo J is denoted by S/J and is the semigroup obtained from S by identifying all of J to a single point (Section A.1). The **bycyclic semigroup** B is the semigroup with identity generated by a two element set $\{x, y\}$ subject to the generating relation $xy = 1$ (see Section A.4). Finally, we need to discuss Green's equivalence relations $\mathcal{L}, \mathcal{R}, \mathcal{H}, \mathcal{D}$, and $\mathcal{J}$ on a semi-group S. We write $a\mathcal{L}b$ if a and b generate the same left ideal. We write $a\mathcal{R}b$ if a and b generate the same right ideal and we write $a\mathcal{J}b$ if a and b generate the same ideal. The relation $\mathcal{H}$ is defined by $\mathcal{H} = \mathcal{L} \cap \mathcal{R}$. The equivalence relations $\mathcal{L}$ and $\mathcal{R}$ commute so that the relation $\mathcal{D} = \mathcal{L} \circ R$ is also an equivalence relation. The corresponding equivalence classes containing an element $a \in S$ will be denoted by $\mathcal{L}_a, \mathcal{R}_a, \mathcal{H}_a, \mathcal{D}_a$, and $\mathcal{J}_a$. For any group G, we let $G^0 = G \cup \{0\}$ where $0 \notin G$ and we extend the multiplication in G over G^0 by defining $a0 = 0a = 0$ for all $a \in G$. Then G^0 is a semigroup and is referred to as a **group with zero**. If a semigroup S contains a proper ideal and an identity, then it contains a largest left ideal, a largest right ideal and a largest ideal. These will be denoted by $M_L(S), M_R(S)$, and $M(S)$ respectively. We are now in a position to state our main result. One may consult Magill (1993) for a proof of this result.

A.12.4 Theorem *Suppose $T(X)$ contains a proper ideal and an identity* id. *Then the following statements are equivalent.*

(1) *$T(X)$ is a separated semigroup.*

(2) *There is a Rees-factor semigroup of $T(X)$ that is a group with zero.*

(3) *If B is any bicyclic subsemigroup of $T(X)$, then the identity of B differs from that of $T(X)$.*

(4) *The equivalence classes $\mathcal{L}_{\mathrm{id}}, \mathcal{R}_{\mathrm{id}}, \mathcal{H}_{\mathrm{id}}, \mathcal{D}_{\mathrm{id}}$, and $\mathcal{J}_{\mathrm{id}}$ all coincide.*

(5) *$M_L(T(X)), M_R(T(X))$, and $M(T(X))$ all coincide.*

(6) *The only T-retract of X that is T-isomorphic to $\mathrm{Ran}(\mathrm{id})$ is $\mathrm{Ran}(\mathrm{id})$ itself.*

(7) *If the product of two elements in $T(X)$ is a unit, then each of the elements is a unit.*

We now determine when $L(V)$, $S(X)$, and $\Gamma(X)$ are separated and hence have all the properties listed in Theorem A.12.4. We denote by $\dim(V)$ the dimension of a vector space V. $L(V)$ will have a proper ideal if and only if $\dim(V) \geq 1$ and one easily verifies that the only L-retract of $L(V)$ which is isomorphic to V is V itself if and only if $\dim(V)$ is finite. Our next result follows from these observations and Theorem A.12.4.

A.12.5 Theorem *$L(V)$ is a separated semigroup if and only if $\dim(V) \geq 1$ and $\dim(V)$ is finite.*

We denote by $\mathbb{R}^n$ the Euclidean n-space. It follows from Brouwer's well known theorem on the invariance of domain that $\mathbb{R}^n$ is not homeomorphic to any proper closed subspace of $\mathbb{R}^n$. Since retracts of $\mathbb{R}^n$ are closed, Theorem A.12.4 implies

A.12.6 Theorem *Both $S(\mathbb{R}^n)$ and $\Gamma(\mathbb{R}^n)$ are separated semigroups.*

Since the Euclidean n-cell $\mathbb{I}^n$ is homeomorphic to many S-retracts and Γ-retracts, it readily follows from Theorem 4 that neither $S(\mathbb{I}^n)$ nor $\Gamma(\mathbb{I}^n)$ is separated. Since the Euclidean n-sphere $\mathbb{S}^n$ is not homeomorphic to any proper closed subspace, Theorem A.12.4 also implies

A.12.7 Theorem *Both $S(\mathbb{S}^n)$ and $\Gamma(\mathbb{S}^n)$ are separated semigroups.*

Suppose X is a Hausdorff space. Since retracts of Hausdorff spaces are closed, if X is not homeomorphic to any proper closed subspaces, it is certainly not homeomorphic to any proper retract and all this, together with Theorem A.12.4 implies

A.12.8 Theorem *Let X be a Hausdorff space. If $\Gamma(X)$ is separated, then so is $S(X)$.*

A.13 Matrix Semigroups

by Jan Okniński in Warsaw, Poland

A *matrix semigroup* is a subsemigroup of the multiplicative monoid $\mathrm{Mat}_{n\times n}(K)$ of $n \times n$ matrices over a field K. The approach to the structure of semigroups of this type is based on an attempt to imitate the ideal theory of finite semigroups, using the theory of linear groups (Wehrfritz 1973, Merzljakov 1986) and especially linear algebraic groups (Humphreys 1975).

We let $\mathrm{Mat}_{n\times n}(K)$ act on the space K^n of column vectors by left multiplication.

A.13.1 Lemma *Green's relations (cf. Section A.1) on $\mathrm{Mat}_{n\times n}(K)$ are given by: $a\mathcal{J}b$ if and only if $\mathrm{rank}(a) = \mathrm{rank}(b)$, $a\mathcal{R}b$ if and only if $\mathrm{im}(a) = \mathrm{im}(b)$, and $a\mathcal{L}b$ if and only if $\ker(a) = \ker(b)$.*

Let Y_j, $1 \le j \le n$, be the set of all matrices of rank j which are in the reduced row elementary form. Let X_j be the transpose of Y_j. If $e \in Y_j$ is the diagonal idempotent, let G_j be the group of units of the monoid $e\,\mathrm{Mat}_{n\times n}(K)e$. Then $G_j \simeq \mathrm{GL}(j; K)$ and every matrix of rank j has a unique presentation in the form xgy, where $x \in X_j, y \in Y_j, g \in G_j$.

Let $M_j = \{a \in \mathrm{Mat}_{n \times n}(K)|\ \mathrm{rank}(a) \leq j\}$ for $j = 0, 1, \ldots, n$. The semigroup structure of $\mathrm{Mat}_{n \times n}(K)$ can be described as follows.

A.13.2 Theorem *The sets* $0 = M_0 \subset M_1 \subset \cdots \subset M_n = \mathrm{Mat}_{n \times n}(K)$ *are the only ideals of the monoid* $\mathrm{Mat}_{n \times n}(K)$. *Each Rees factor* M_j/M_{j-1} *is isomorphic to the completely 0-simple semigroup* $\mathcal{M}(G_j, X_j, Y_j, Q_j)$ *(see Section A.3), where the matrix* $Q_j = (q_{yx})$ *is defined for* $x \in X_j, y \in Y_j$ *by* $q_{yx} = yx$ *if* yx *is of rank* j *and otherwise* $q_{yx} = \theta$, *the zero element.*

We introduce a class of semigroups that is crucial for the approach to arbitrary matrix semigroups.

A.13.3 Lemma *Let* U *be a subsemigroup of* M_j/M_{j-1}, $j \geq 1$, *which intersects nontrivially every* $\mathcal{H}$-*class of a completely 0-simple subsemigroup of* M_j/M_{j-1}. *Then there exists a unique smallest completely 0-simple subsemigroup* $\widehat{U}$ *of* M_j/M_{j-1} *containing* U. *Moreover, if* G *is any maximal subgroup of* $\widehat{U}$, *then* G *is the group generated by* $U \cap G$.

A subsemigroup U of M_j/M_{j-1} satisfying the condition of the lemma is called a **uniform semigroup**, while $\widehat{U}$ is called the **completely 0-simple closure** of U.

We say that a subset A of a semigroup T is a **0-disjoint union** of certain $A_\alpha, \alpha \in \mathcal{A}$, if $A = \bigcup_\alpha A_\alpha$ and $\bigcap_\alpha A_\alpha$ is the zero of T. We adopt the convention $S/I = S$ for $I = \varnothing$.

A.13.4 Theorem *For a matrix semigroup* $S \subseteq \mathrm{Mat}_{n \times n}(K)$ *define the sets* $S_j = \{a \in S \mid \mathrm{rank}(a) \leq j\}$ *and* $T_j = \{a \in S_j|\ SaS$ *does not intersect any maximal subgroup of* $\mathrm{Mat}_{n \times n}(K)$ *contained in* $M_j \setminus M_{j-1}\}$. *Then*

$$S_0 \subseteq T_1 \subseteq S_1 \subseteq T_2 \subseteq \cdots \subseteq S_{n-1} = T_n \subseteq S_n = S$$

are ideals of S *(if nonempty). Moreover, for every* j *we have*

(1) $T_j^{\binom{n}{j}} \subseteq S_{j-1}$, *and* $N_j = T_j/S_{j-1}$ *is a nilpotent ideal of* S/S_{j-1} *if* $T_j \neq \varnothing$;

(2) $(S_j \setminus T_j) \cup \{\theta\} \subseteq M_j/M_{j-1}$ *is a 0-disjoint union of uniform subsemigroups* $U_1^{(j)}, \ldots, U_{n_j}^{(j)}$, $n_j \leq \binom{n}{j}$, *of* M_j/M_{j-1} *that intersect different* $\mathcal{R}$- *and different* $\mathcal{L}$-*classes of* M_j/M_{j-1};

(3) $U_i^{(j)} U_k^{(j)} \subseteq N_j$ *for every* $k \neq i$; *moreover* $U_i^{(j)} N_j$ *and* $N_j U_i^{(j)}$ *are contained in* N_j, *so each* $U_i^{(j)}$ *can be considered as an ideal of* S/T_j.

So, the Rees factors $S_j/S_{j-1} \subseteq M_j/M_{j-1}$ represent the 'layers' of S and are approached via the egg-box picture (see Section A.1) on the completely 0-simple semigroup M_j/M_{j-1}. This description of S shows a strong analogy to Wedderburn's structure theorem for finite dimensional algebras. Namely, N_j is the maximal nil ideal of S_j/S_{j-1} and the U_i^j play the role of simple homomorphic images of the algebra.

The groups associated to S that come from the same component U_i^j are conjugate, because of the following observation.

A.13.5 Lemma *Let U be a uniform subsemigroup of some M_j/M_{j-1}. Assume that U intersects maximal subgroups D_1, D_2 of $M_j \setminus M_{j-1}$. If G_1, G_2 are the groups generated by $U \cap D_1, U \cap D_2$, then G_1, G_2 are conjugate in $\mathrm{Mat}_{n \times n}(K)$.*

The common (up to isomorphism) group H generated by the intersections $U \cap G$ with maximal subgroups G of $\hat{U}$ will be called the **group associated** to the uniform semigroup U.

If the minimal rank of matrices in S is $j \geq 1$, so that $S_{j-1} = \varnothing$, then we must have $n_j = 1$ and $N_j = \varnothing$. All $U_i^j \setminus \{\theta\}$ coming from all possible layers $S_{(j)} = S_j/S_{j-1}$ are called the **uniform components** of S. By the **nilpotent components** of S we mean all nonempty $T_j \setminus S_{j-1}$. The philosophy of studying $S \subseteq \mathrm{Mat}_{n \times n}(K)$ via uniform components, associated groups and the corresponding sandwich matrices is a natural extension of the structural approach to finite semigroups via completely 0-simple principal factors.

As an example of this philosophy, we state the following semigroup analogue of Tits alternative.

A.13.6 Theorem *Let $S \subseteq \mathrm{Mat}_{n \times n}(K)$ be a matrix semigroup, K a finitely generated field. Then S has no free noncommutative subsemigroups if and only if the associated linear groups are almost nilpotent, which is also equivalent to the fact that S satisfies a semigroup identity.*

Here, by an **almost nilpotent** group we mean a finite extension of a nilpotent group. If $S \subseteq \mathrm{Mat}_{n \times n}(D)$ for a division algebra D, then the assertions of the theorem hold with two qualitative differences: (1) there may be infinitely many uniform components consisting of matrices of a given rank, (2) the nilpotent components are just unions of nilpotent ideals of index two, but are not necessarily nilpotent.

Assume that K is algebraically closed. Then the Zariski closure $\overline{S}$ of a semigroup $S \subseteq \mathrm{Mat}_{n \times n}(K)$ in $\mathrm{Mat}_{n \times n}(K)$ is also a semigroup.

For example, $\overline{\mathrm{GL}(n;K)} = \mathrm{Mat}_{n \times n}(K)$ and $\overline{\mathrm{Mat}_{n \times n}(\mathbb{Z})} = \mathrm{Mat}_{n \times n}(\mathbb{C})$ if $K = \mathbb{C}$. The structure of $\overline{S}$ is related to that of S as follows.

A.13.7 Theorem *If $S \subseteq \mathrm{Mat}_{n \times n}(K)$, then any uniform component of $\overline{S}$ (with zero adjoined) is completely 0-simple. If X is a uniform (respectively, nilpotent) component of S, then $A = A' \cap S$ for some uniform (nilpotent) component A' of $\overline{S}$.*

An **algebraic linear semigroup** is a Zariski closed subsemigroup of some $\mathrm{Mat}_{n \times n}(K)$. A special class, with particularly nice properties, is the one consisting of semigroups which are irreducible as algebraic varieties. They are analogs of connected algebraic groups. In particular: if M is an irreducible algebraic monoid with zero, then M is (von Neumann) regular if and only if the group G of units of M is **reductive**. That is, G is a nontrivial connected group with no nontrivial connected unipotent normal subgroups, $\mathrm{GL}(n;K)$ being the key example. Moreover, in this case, $M = \bigcup_{r \in R} BrB$ where R is an inverse monoid (called the **Renner monoid** of M) and B is the Borel subgroup of G. R is the analogue of the Weyl group of G. As in reductive groups, the idea is to reduce problems on M to certain problems for R. For a survey on reductive monoids, and their finite analogs (monoids of Lie type), we refer to Putcha (1995) and Solomon (1995).

For example, let R be the set of all matrices in $\mathrm{Mat}_{n \times n}(K)$ with at most one nonzero entry in each row and in each column. R is called the **symmetric inverse monoid**. It is the semigroup analogue of the symmetric group. Let $B \subseteq \mathrm{GL}(n;K)$ be the group of upper triangular matrices. B is a Borel subgroup of $\mathrm{GL}(n;K)$. The corresponding Weyl group is W, the group of permutation matrices. Then the above analogue of the Bruhat decomposition takes the following form.

A.13.8 Theorem $\mathrm{Mat}_{n \times n}(K) = \bigcup_{\sigma \in R} B\sigma B$ *and if $B\sigma B = B\sigma'B$, then $\sigma = \sigma'$. If $a \in \mathrm{Mat}_{n \times n}(K)$, then we have*

$$\mathrm{GL}(n;K)\, a\, \mathrm{GL}(n;K) = \mathrm{GL}(n;K)\, e\, \mathrm{GL}(n;K) = \bigcup_{\sigma \in WeW} B\sigma B,$$

where $e = e^2 \in R$ is such that $\mathrm{rank}(a) = \mathrm{rank}(e)$.

The theory of matrix semigroups is presented in Okniński (1998). For linear algebraic semigroups we refer to Putcha (1988). Approaches to the theory of linear representations of semigroups are given by Clifford and Preston (1967b), McAlister (1971), and Okniński (1991), while more results on semigroups $S \subseteq \text{Mat}_{n \times n}(K)$ that act irreducibly on K^n can be found in Ponizovskii (1982). The structure of the semigroup of (real) nonnegative matrices, and its applications, are described in Berman and Plemmons (1979).

A.14 Subsemigroup Lattices

by A. J. Ovsyannikov and Lev N. Shevrin in Ekaterinburg, Russia

For any algebraic system S, the set $\text{Sub}\,S$ of all subsystems of S, partially ordered by inclusion, forms a lattice. If S is a semigroup, then there may be a situation when two subsemigroups have the empty intersection; so, in order to guarantee that $\text{Sub}\,S$ is a lattice, we have to treat the empty set as a subsemigroup. The study of various interrelations between algebraic systems and their subsystem lattices is a rather large area in algebra being developed for a long time. This trend was formed first in group theory more than seventy years ago. In semigroup theory, investigations devoted to subsemigroup lattices are in progress during more than five decades, so plentiful and diverse material has been accumulated here. One may distinguish three main aspects in such investigations, which are reflected in the next three subsections.

Restrictions on subsemigroup lattices

A general problem in this aspect is describing the structure of semigroups S for which the lattice $\text{Sub}\,S$ satisfies given conditions. Quite a number of well-known lattice-theoretic properties have been imposed as such conditions: modularity, distributivity, different kinds of complementability, various finiteness conditions, etc. In each case, as a rule, a clearly outlined class of semigroups arises, sometimes with a reduction to groups. For a periodic group, its subsemigroups are subgroups, so in the corresponding cases one deals merely with the subgroup lattice of a group. We give several typical examples of descriptions under discussion; for a number of notions used below, the reader is referred to Section A.2, Section A.5, Section A.6. Any semigroup in which the join of every two subsemigroups coincides with their set-theoretic union is called a *U-semigroup*.

Let $\mathcal{X}$ be an arbitrary lattice variety consisting of modular lattices. For a semigroup S, the lattice $\text{Sub}\,S$ belongs to $\mathcal{X}$ iff S is periodic and is decomposable into a U-band of semigroups, each of which is an ideal extension of a group whose subgroup lattice belong to $\mathcal{X}$ by a nilpotent U-semigroup. It is notable that, in general, if Sub S satisfies a non-trivial lattice identity, then S has to be periodic; this means clearly that the subsemigroup lattice of the infinite cyclic semigroup does not satisfy any non-trivial identity.

A semigroup S is called a **K-semigroup** if Sub S is a complemented lattice. Any K-semigroup is a chain of simple semigroups. It is easy to see that any periodic K-semigroup is idempotent-generated (i.e., generated by idempotents). For periodic completely simple semigroups, the converse is true as well. As to non-periodic completely simple semigroups, the condition of being idempotent-generated is probably neither necessary nor sufficient for the property of being a K-semigroup. But corresponding examples are still unknown. Moreover, no examples of K-semigroups which are not idempotent-generated are known. A *local* structure of an arbitrary K-semigroup can be arbitrarily complicated, as any semigroup is embeddable in a K-semigroup. The following conditions for a semigroup S are equivalent: (a) Sub S is relatively complemented; (b) Sub S is initially complemented, in other words, all subsemigroups of S are K-semigroups; (c) S is an ordinal sum of rectangular bands.

Many finiteness conditions are formulated just in terms of subsystem lattices; the most important examples are the minimal and the maximal conditions (which are equivalent to the descending and the ascending chain conditions, respectively). The overwhelming majority of such conditions for semigroups imply periodicity of a semigroup and can be covered by the general scheme described in the subsection *Finiteness conditions* in Section A.6. We give only one illustration here. A lattice L is said to be of $\wedge$-breadth [$\vee$-breadth] r if the meet [join] of any finite set of elements of L is equal to the meet [join] of at most r of them, and r is the least number with this property. It can be shown that the $\wedge$-breadth coincides with the $\vee$-breadth, so we may speak about the **breadth** of a lattice. We shall say that a semigroup S is *of finite breadth* if Sub S has a finite breadth. A group of finite breadth is periodic, hence, according to the remark given at the beginning of this section, such a group can be treated as a special type of a semigroup of finite breadth. A semigroup S is of finite breadth iff S is finitely assembled from groups of finite breadth. So, as usual in the scheme mentioned, we have a complete reduction to the group case. An arbitrary group of finite breadth may have a structure which is not transparent. However, for locally finite groups the situation is quite different: such a group is of finite breadth iff it is an extension of a direct product of finitely many quasicyclic groups by a finite group. It is interesting that this is the same structure as for a locally finite group with the minimal condition for subgroups. For groups which are not locally finite, these two conditions do not imply each other; the corresponding results have solved the question which has been open for more than 20 years. For details concerning these and some other questions in this topic, both solved and still open, see (Shevrin and Ovsyannikov 1996, Ch. IV).

Properties of subsemigroup lattices

A general problem here is to examine, for a given class of semigroups, the lattices Sub S with S belonging to this class, in particular, to characterize such lattices and to describe lattices embeddable in the subsemigroup lattices under consideration. We say that a class Sub K of semigroups is *lattice-characterized* if one can indicate conditions formulated in lattice-theoretic terms such that the lattice Sub S of a semigroup S satisfies these conditions if and only if $S \in$ Sub K; such conditions form a *lattice characteristic* of the class Sub K. Speaking about lattice-theoretic conditions, we mean *abstract* properties (i.e., inherited by isomorphic images) so each lattice-characterized class is clearly abstract. If Sub K consists, up to isomorphism, of a single semigroup, we say that this *semigroup* is lattice-characterized.

Among lattice-characterized classes and individual semigroups, we mention the following ones: the infinite cyclic semigroup, the infinite cyclic group, the classes of torsion-free groups, non-periodic groups, abelian non-periodic groups, orderable groups, the classes of periodic semigroups, idempotent semigroups, idempotent-free semigroups, free semigroups, free commutative semigroups. All of them are in fact *lattice-elementary*, that is, their lattice characteristics can be formulated by sentences of the first-order logic.

It should be noted that the lattice characteristics under discussion are *relative*, i.e., they distinguish the corresponding classes of lattices within the class of *subsemigroup lattices*. It is natural to set also the problem of finding *absolute* lattice characteristics, i.e., given in the class of *all lattices*. Very little is known on this subject. There is a problem (posed in (Jónsson 1972)) of finding conditions for an algebraic lattice under which it is isomorphic to the subsemigroup lattice of some semigroup. One can show that such necessary and sufficient conditions cannot be given by sentences of the first-order logic.

A class of semigroups Sub K is called **lattice-universal** if any lattice can be embedded in Sub S for some $S \in$ Sub K. Not only the class of all semigroups is lattice-universal but even some "small" classes possess this property. Such are, in particular, the class of all commutative nilsemigroups of index 2 and the class of all semilattices. Both of them are quite far from the class of groups whose lattice universality (cited sometimes as **Whitman's theorem**) is a well-known fact. For some specific types of semigroups, which are not lattice-universal, one can describe those finite lattices which are embeddable in subsemigroup lattices of semigroups of a given type. This has been effected, in particular, for the following types: free and free commutative semigroups, the infinite cyclic semigroup, finite nilpotent semigroups, free semilattices, finite semilattices. It was rather unexpected that the class of finite lattices embeddable in subsemigroup lattices of semigroups of each of these so different types is the same: this is the class of so-called **lower bounded** lattices, which occur in lattice theory when studying homomorphisms of free lattices and in connection with some other questions.

Lattice isomorphisms

Let S and S' be semigroups. An isomorphism of the lattice Sub S onto the lattice Sub S' is called a **lattice isomorphism** (sometimes a **projectivity**) of S upon S'. The relation of being lattice isomorphic establishes, so to say, certain kinship between semigroups. It is natural to know a "measure" of this kinship, that is, to find properties of semigroups which are retained by lattice isomorphisms or, as the most complete picture for an arbitrary semigroup S of a given class, to describe the whole family of semigroups lattice isomorphic to S. In doing so, one succeeds often in finding the mappings which induce lattice isomorphisms of semigroups under consideration. There are quite a number of results of such a character for cancellative semigroups (including groups), commutative archimedean semigroups, semilattices and other bands, completely simple semigroups and others.

A mapping φ of a semigroup S onto a semigroup S' is called an **anti-isomorphism** if φ is a bijection and $\varphi(xy) = \varphi(y)\varphi(x)$ for all $x, y \in S$. Isomorphic or anti-isomorphic semigroups are obviously lattice isomorphic. The converse does not hold in general, therefore an utmost case when it takes place, that is, the family of semigroups lattice isomorphic to a given semigroup S, is exhausted by isomorphic and maybe anti-isomorphic images of S, is of special interest. In this case, a semigroup S is said to be **lattice-determined**. Among lattice-determined semigroups, we mention the following ones: any free semigroup and any semigroup decomposable into a free product; any semigroup which is free in some overcommutative variety of semigroups; any commutative cancellative non-periodic semigroup; any locally nilpotent non-periodic group; certain types of completely 0-simple semigroups with divisors of zero; any semigroup admitting a presentation with n relations over an alphabet which contains at least $4n - 1$ letters; a free semilattice of rank ≥ 3; any rectangular non-singular semigroup; the bicyclic semigroup.

Additional remarks

As is known (see e.g., Section A.6, Section A.9, Section A.10, and Section A.11), semigroups of certain types can be treated as *unary* semigroups, i.e., semigroups with an additional unary operation. The subsystem lattice of a unary semigroup is the lattice of its unary subsemigroups. For an inverse semigroup S, this is the lattice $\mathrm{Subi}\,S$ of all inverse subsemigroups. Amongst unary semigroups, just the case of inverse semigroups has been to a large extent the object of investigations from the viewpoint in question, predominantly for the first and the third of the aspects noted at the beginning of this article. Sometimes imposing natural restrictions on the lattice $\mathrm{Subi}\,S$ turns out to be a rather strong requirement. One may say that, from the point of view of such restrictions, the lattice $\mathrm{Subi}\,S$ is "too rich". However, there exists, so to say, a less rich lattice which is naturally associated with an inverse semigroup S and influences substantially the structure of S (and in the group case transforming into the subgroup lattice); this is the lattice of all **full inverse subsemigroups**, that is, inverse subsemigroups, containing the set of all idempotents of S.

The book (Shevrin and Ovsyannikov 1996) gives a comprehensive presentation of the area discussed in this article; see also (Shevrin and Ovsyannikov 1983). The material pertaining to inverse semigroups is contained in three (out of fourteen) chapters of the book just cited; see also (Jones 1990). Quite a number of facts of this area relevant mostly to the first of the aspects under discussion are presented in (Petrich 1977, Ch. V). In the survey (Arshinov and Sadovskii 1972), two (out of five) sections are devoted to lattice isomorphisms of semigroups of certain types.

A.15 Varieties of Semigroups

by Lev N. Shevrin and Mikhail V. Volkov in Ekaterinburg, Russia

A ***semigroup variety*** is a class of semigroups closed under taking homomorphic images, subsemigroups and direct products. Equivalently, varieties are classes defined by *semigroup identities* like the commutativity law $xy = yx$ or the idempotency law $x^2 = x$. See Section G.5 for a discussion of varieties in the general universal-algebraic setting.

The study of semigroup varieties or, more generally, the varietal approach to studying semigroups has proved to be very successful and now underlies many directions of modern semigroup theory. The reason for this success appears to be twofold. First of all, it turns out that several important structural properties of semigroups are efficiently caught by the language of identities and varieties and admit a clear description in this language. On the other hand, the equational behavior of semigroups is rich and diverse, thus creating a convenient framework for classifying semigroups with respect to their equational properties.

As a simple illustration of how equational and structural properties of semigroups interact, we have chosen a fact which arises as a combination of results by Chrislock (1969) and by Saliĭ (1969). To formulate it, we need a definition concerning identities and a definition concerning semigroups. An identity $u = v$ is said to be ***homotypical*** [***heterotypical***] if the sets of variables occurring in the terms u and v coincide [are different]. A semigroup is called ***archimedean*** if, for any two of its elements, each divides a power of the other. We also say that a variety $\mathscr{V}$ *satisfies an identity* $u = v$ if every member of $\mathscr{V}$ does so. For a semigroup variety $\mathscr{V}$, the following are equivalent: (i) $\mathscr{V}$ satisfies a heterotypical identity; (ii) $\mathscr{V}$ satisfies the identity $(x^r y^r x^r)^r = x^r$ for some r; (iii) $\mathscr{V}$ consists of archimedean semigroups; (iv) $\mathscr{V}$ does not contain the two-element semilattice.

The reader can see that this indeed links purely syntactic properties of identities holding in $\mathscr{V}$ with structural properties of semigroups in $\mathscr{V}$. This interplay of syntax and semantics is extremely profitable for both sides: for instance, on the syntactic side it reveals that any heterotypical semigroup identity implies an identity of the form $(x^r y^r x^r)^r = x^r$; it seems difficult to extract this identity directly from the deduction rules of equational logic. On the structural side, since condition (ii) implies that all semigroups in $\mathscr{V}$ are periodic, we can make use of the rather transparent description of periodic archimedean semigroups presented in Section A.6. We also want to draw the reader's attention to condition (iv). It provides an additional characterization in terms of "forbidden" objects or *indicators*. An ***indicator characterization*** for varieties with a property Θ is a list of semigroups such that $\mathcal{V}$ possesses Θ if and only if $\mathcal{V}$ contains none of the listed semigroups. Characterizations of such kind are known to be very useful, especially when the list of "forbidden" objects consists of a few finite semigroups with simple structure.

There is a large body of deep research which has resulted in equational characterizations of many important structure properties of semigroup varieties; in many cases, indicator characterizations have also been provided. As a typical example of such a property we mention the existence of a band decomposition whose components belong to certain specific classes of semigroups—for a discussion see Section A.2. The comprehensive surveys (Shevrin and Sukhanov 1989) and (Kharlampovich and Sapir 1995) contain a detailed account of related results.

As mentioned, the other aspect of the theory consists in classifying semigroups with respect to their equational properties. The most important of them is the *finite basis property*. A semigroup is said to be ***finitely based*** if the variety it generates can be defined by a finite identity system. The problem of deciding whether or not a given semigroup is finitely based turns out to be very difficult and still remains open even if restricted to the case of finite semigroups. The first example of a non-finitely based finite semigroup is due to Perkins (1969): it is the 6-element monoid B_2^1 obtained by adjoining an identity element to the famous semigroup B_2 (see Section A.5). The miniature size of B_2^1 gave rise to an intensive study of the finite basis property for semigroups with ≤ 5 elements. This study which involved many researchers was completed by Trahtman (1983, 1991); it has led to the conclusion that B_2^1 is a minimal (with respect to the number of elements) non-finitely based semigroup. Yet another remarkable property of B_2^1 has been discovered by Sapir (1988): if the variety generated by a finite semigroup S contains B_2^1, then S is non-finitely based. Thus, non-finitely based finite semigroups are plentiful—but the same may be said about finitely based ones because, for instance, commutative or nilpotent semigroups, bands, finite groups are all finitely based. We refer the reader to the surveys (Shevrin and Volkov 1985) and (Volkov 2001) for a detailed discussion of this intriguing topic; the first survey also covers various other aspects of the study of semigroup identities.

The collection **S** of all semigroup varieties forms a complete lattice with respect to the class-theoretical inclusion. The structure of **S** has proved to be rather complex: as a striking example of its complexity, we mention that **S** has intervals which are dually isomorphic to the partition lattice of a countably infinite set (Burris and Nelson 1971, Ježek 1976). This immediately implies that the subvariety lattice of **every** variety of algebras of finite type embeds into **S**, that **S** satisfies no non-trivial lattice law, and that **S** has the cardinality of the continuum. Perhaps the only "positive" property of the lattice **S** in large is that every proper semigroup variety $\mathscr{V}$ has a cover in **S** ($\mathscr{W}$ is said to *cover* $\mathscr{V}$ if $\mathscr{W} \supset \mathscr{V}$ and there is no variety between $\mathscr{W}$ and $\mathscr{V}$)—see (Trahtman 1974).

The lattice **S** consists of two big parts with contrasting properties:

- the filter **OC** of all *overcommutative* varieties, that is, varieties containing the variety of all commutative semigroups;
- the ideal **Per** of all *periodic* varieties, that is, varieties consisting of periodic semigroups.

It turns out that the lattice **OC** admits a relatively easy description in terms of the congruence lattices of certain unary algebras (see Volkov 1994); thus, the complexity of **S** is concentrated within the lattice **Per**. A sketch map on Figure 1 shows the relative location of the main parts of this lattice.

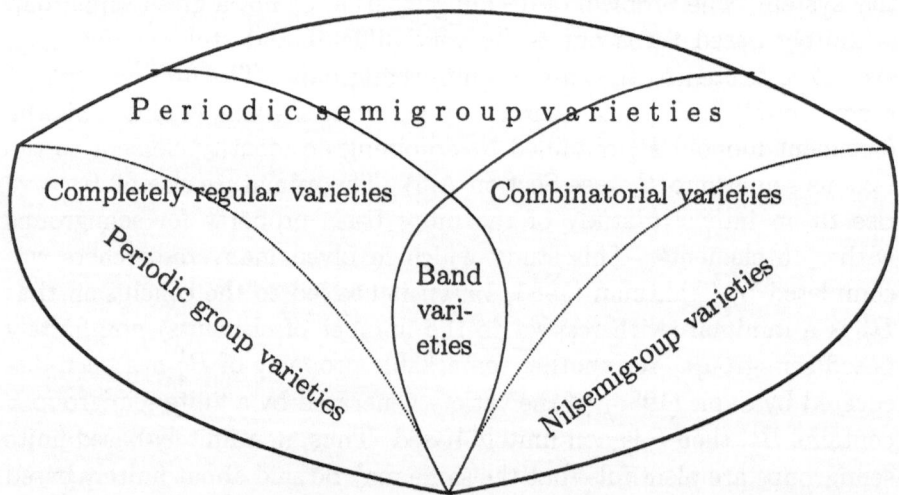

Figure 1: The "map" of the lattice of periodic semigroup varieties

The lattice ideals **CR** of all varieties of completely regular periodic semigroups and **Comb** of all varieties of combinatorial periodic semigroups are the two extremes with respect to groups—recall that a semi-group S is called **completely regular** [**combinatorial**] if S is a union of groups [if all subgroups of S are singletons]. Their intersection is the lattice **B** of varieties of **bands** (idempotent semigroups), which has been completely described by Gerhard (1970), Biryukov (1970) and Fennemore (1971a,b); **B** is a countably infinite distributive lattice of width 3, see Figure 2. Here $\mathscr{B}$ stands for the variety of all bands and $\mathscr{SL}$, $\mathscr{LZ}$, and $\mathscr{RZ}$ denote the varieties of semilattices, left zero, and right zero semigroups, respectively.

The lattice **CR** was the subject of intensive studies in the 80s, and now its structure is relatively well understood—of course, modulo the lattice of periodic group varieties. In particular, **CR** has been proved to be modular, see (Pastijn 1990).

In contrast, almost nothing is known about the structure of the lattice **Comb**. In fact, the aforementioned embedding results by (Burris and Nelson 1971, Ježek 1976) apply to this lattice, whence **Comb** and even its ideal **Nil** consisting of nilsemigroup varieties are in a sense as complex as the whole lattice **S**. For a certain approach to studying the lattice **Nil** see (Vernikov and Volkov 2000).

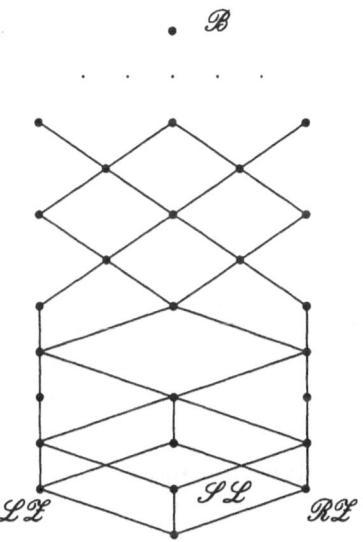

Figure 2: The lattice of varieties of bands

In many cases, it is natural to equip semigroups with additional unary operations of various kinds and to consider corresponding varieties of type $\langle 2, 1 \rangle$, see sections (A.6, A.11, A.10). Moreover, even in situations when one cannot establish varietal structure by simply adding new operations, it turns out to be reasonable to consider variety-like classes rather than individual semigroups. In this way one comes to the notions such as a *pseudovariety of finite semigroups*, see Section A.8, an *e-variety of regular semigroups*, see Section A.9 as well as the survey (Trotter 1996), etc.

Notice that equational, structural and lattice aspects of the theory of semigroup varieties are surveyed also in Section 7 of a book chapter (Shevrin 1991). Chapter XII in (Petrich 1984) is entirely devoted to va-

rieties of inverse semigroups; varieties of completely regular semigroups
are examined in several chapters of (Petrich and Reilly 1999).

A.16 Compact Semigroups

by Karl H. Hofmann in Darmstadt, Germany

Groups tend to occur as subgroups of the automorphism group $\text{Aut}(E)$
of some structure E and semigroups as subsemigroups of its endomorphism semigroup $\text{End}(E)$. If E is an euclidean vector space then $\text{Aut}(E)$
is the group $O(E)$ of linear orthogonal self-maps of E, and $\text{End}(E)$ contains the subsemigroup $O_{\leq}(E)$ of all linear self-maps which preserve
or decrease distance, i.e., of all *linear contractions*. The semigroup
$S = O_{\leq}(E)$ can be viewed as a compact subsemigroup of the *monoid*
(=semigroup with identity) of all real $n \times n$- matrices ($n = \dim E$), and
the multiplication $(x, y) \mapsto xy : S \times S \to S$ is continuous. A topological
space with an associative continuous multiplication is called a *topological semigroup*. A handy elementary example is the complex unit disc
under multiplication. The subgroup $G = O(E)$ is exactly the subset of
all *units* (=invertible elements), and in addition to the continuity of
its multiplication, also inversion $g \mapsto g^{-1} : G \to G$ is continuous: Thus
G is a compact *topological group*. If E denotes a (complex) Hilbert
space then the monoid $S = O_{\leq}(E)$ of all linear contractions is a compact space with respect to the weak operator topology (the coarsest one
making all functions $s \mapsto \langle s(v), w \rangle : S \to \mathbb{C}$, v, $w \in E$ continuous). The
group of units is the group $U(E)$ of all unitary operators on E. On S,
multiplication is not continuous if E is infinite dimensional; however, all
translations $x \mapsto sx$, $xs : S \to S$ are continuous; S is a *semitopological
semigroup*. Sometimes semigroups occur on a compact space S such
that only one-sided translations $x \mapsto sx : S \to S$, $s \in S$, are continuous;
such semigroups are called *one-sidedly topological semigroups*. An
example is the Stone-Čech compactification $\beta \mathbb{N}$ of the discrete additive
semigroup of integers.

All of these examples are *compact semigroups*, differing, however,
in the quality of their continuity properties. Consequently, compact
topological semigroups have a rich connections to topology (and compact transformation group theory), semitopological semigroups pertain
to harmonic and functional analysis, and one sidedly topological semigroups have yielded applications through semigroups like $\beta \mathbb{N}$.

For a formulation of the principal theorems on compact semigroups, one needs the following definition (Rees, Suschkewitsch):

A.16.1 Construction Let G be a group, and X and Y arbitrary sets. Assume that we are given any function $(y, x) \mapsto [y, x] : Y \times X \to G$. Then the set $X \times G \times Y$ becomes a semigroup with respect to the multiplication $(x_1, g_1, y_1)(x_2, g_2, y_2) = (x_1, g_1[y_1, x_2]g_2, y_2)$. Every $\{x\} \times G \times \{y\}$ is a group isomorphic to G whose identity $e = (x, g, y)$ must satisfy $(x, g, y) = e = e^2 = (x, g, y)(x, g, y) = (x, g[y, x]g, y)$, that is, $g = [y, x]^{-1}$.

The "rows" $X \times G \times \{y\}$ are (minimal) left ideals while the "columns" $\{x\} \times G \times Y$ are (minimal) right ideals which are isomorphic to $X \times G$ ($G \times Y$, respectively) with $x_1 x_2 = x_1$ on X and $y_1 y_2 = y_2$ on Y. For more details on this construction, see Section A.3.

Such semigroups are called **completely simple semigroups** or **paragroups**. Clifford and Preston (1967b) speak of the **egg-box picture** (cf. Section A.1) in order to visualize a paragroup:

$$
\begin{array}{cc}
\mathbf{x_1} & \mathbf{x_2} \\
\mathbf{y_2} \begin{pmatrix} \bullet\,(\mathbf{x_1}, g_1[y_1, x_2]g_2, \mathbf{y_2}) & \leftarrow & \bullet\,(x_2, g_2, \mathbf{y_2}) \\ \uparrow & & \\ \mathbf{y_1} \;\; \bullet\,(\mathbf{x_1}, g_1, \mathbf{y_1}) & & \end{pmatrix}
\end{array}
$$

A.16.2 Theorem (Suschkewitsch-Rees-Wallace; cf. Section A.3)
A compact one-sidedly topological semigroup S possesses a unique minimal ideal $M(S)$ which is a paragroup.

If S is semitopological, then the maximal subgroups of $M(S)$ are compact topological groups and the minimal left [right] ideals are compact topological semigroups; S supports a probability measure which is invariant under $x \mapsto sx$ for all s iff $M(S) \cong G \times Y$ and each such measure is supported on $M(S)$.

If S is topological, then $M(S)$ is a compact topological semigroup. If S is connected and has an identity, then the Čech cohomology is supported by any of the maximal groups of the minimal ideal; in particular, the Čech cohomology of a compact connected monoid is that of a compact connected group.

Consequences of this Theorem are, for example: Every compact one-sidedly topological semigroup S contains at least one idempotent. If it has only one idempotent e, its minimal ideal is a group, and if e is an identity, S is a group. Also, a topologically and multiplicatively closed subset of a compact group is a subgroup.

For applications of the Suschkewitsch-Rees-Wallace Theorem in compact one-sidedly topological semigroups such as $\beta\mathbb{N}$ to number theory (such as a proof of Van der Waerden's Theorem on arithmetic progressions) see (Hindman and Strauss 1998). For invariant measure theory on compact semitopological semigroups and applications to ergodic theory (see e.g., Berglund et al. 1989). For the Čech cohomology on compact connected groups and therefore compact connected monoids (see e.g., Hofmann and Morris 1998). If a semigroup has an identity 1, then the set $H(1)$ of invertible elements with respect to 1 is a subgroup, called *the group of units*.

A.16.3 Theorem (Hofmann-Mostert) *Every compact connected topological semigroup with identity contains a compact connected abelian subsemigroup A such that $A \cap H(1) = \{1\}$ and $A \cap M(S) \neq \varnothing$.*

An action of a a group G on a semigroup S is called **automorphic** if $s \mapsto g \cdot s$ is an automorphism of S for all $g \in G$.

A.16.4 Proposition (Borel-Conner-Floyd-Hofmann-Mostert) *If $G \times S \to S$ is an automorphic action of a compact connected* abelian *group G on a compact connected topological monoid S with zero, then the fixed point set F is connected.*

It is immediate that F is a compact submonoid containing zero and the identity. The result itself is far from elementary (see Hofmann and Mostert 1966, p. 62ff). It is an open problem to prove or disprove the following statement:

A.16.5 Conjecture *If $G \times S \to S$ is an automorphic action of a compact group G on a compact connected topological monoid S with zero then the fixed point set F is connected.*

In contrast to Proposition A.16.4, in Conjecture A.16.5 we merely assume that G is a compact group which need neither be connected nor abelian. Theorem A.16.3 has substantial consequences (Hofmann and Mostert 1966). One of them is Theorem A.16.6 below, which, however, requires additional apparatus.

In a purely topological fashion, one can define a point p to be *peripheral*, if it has arbitrarily small neighborhoods U such that the cohomology with compact supports $H_c^*(U) = 0$. On a compact manifold with boundary, the boundary points are the peripheral ones. A finite dimensional compact space always possesses nonperipheral points

(Hofmann-Mostert, Lawson-Madison). In particular, if such a space is, in addition, homogeneous, i.e., has a transitive homeomorphism group, then it has no peripheral points. "Peripherality" allows us to talk about "boundary" points of spaces even if they are not subspaces of larger ones.

A.16.6 Theorem (Hofmann-Mostert, Lawson-Madison) *In a compact semitopological semigroup S, the group $H(1)$ is a topological group; in fact, the action $(h, s) \mapsto sh : H(1) \times S \to S$ and the analogous action on the other side, as well as inversion on $H(1)$, are continuous (J. D. Lawson). The minimal ideal of the closure $\overline{H(1)}$ is a a compact topological group (Berglund-Hofmann-Troallic). If S is topological, then $H(1)$ is closed, and an element $s \in S$ belongs to $H(1)$ iff it has an inverse on the right (Koch). In a compact topological semigroup with identity which is not a group, all units are peripheral.*

So, "in a proper compact connected topological monoid, units are boundary points". For more information, see (Ruppert 1984) and (Berglund et al. 1989)

Some consequences of Theorem A.16.6: If G is a locally compact Hausdorff space which is a group and a semitopological semigroup, then $G \cup \{\infty\}$, the one-point compactification of G with ∞ acting as a zero is a semitopological compact semigroup S with $H(1) = G$. Hence:

- If G is a semitopological semigroup of an locally compact space and G is algebraically a group, then it is a topological group (Ellis).
- A compact semitopological semilattice (i.e., idempotent commutative semigroup) is topological.
- A homogeneous compact connected topological finite dimensional monoid is a group. (A.Hudson-Mostert)
- A compact topological semigroup cannot contain a copy of the bicyclic subsemigroup (i.e, the free monoid generated by p and q subject to the relation $pq = 1$; (see e.g., Clifford and Preston 1967b), cf. Section A.5. Green's Relations $\mathcal{D}$ and $\mathcal{J}$ coincide.

For an application of the absence of bicyclic semigroups to logic and theoretical computer science, see (Hofmann and Mislove 1996). The one point compactification of the discrete bicyclic semigroup with ∞ acting as zero is a compact semitopological monoid.

Compact abelian topological semigroups can be complicated. This remains true even for compact semilattices, a familiar example of which is the unit interval $\mathbb{I} = [0, 1]$ with multiplication $(x, y) \mapsto \min\{x, y\}$.

There are compact topological semilattices with no nonconstant homo-morphisms into $\mathbb{I}$ (J. D. Lawson, see e.g., (Gierz et al. 1980)). Compact semilattices have substantial applications to theoretical computer science. Let X be a topological space. Write $x \leq y$ if $x \in \overline{\{y\}}$. Then $\leq$ is a transitive relation. X is said to be a T_0 space if $x \leq y$ and $y \leq x$ implies $x = y$. A T_0 space L is called *injective* (according to D. S. Scott) whenever a continuous function $f \colon X \to L$ form a subspace X of a space Y extends to a continuous function $F \colon Y \to L$. Every injective space L is complete lattice. Scott called the complete lattices arising in this fashion *continuous lattices*. Each compact topological semilattice is a complete lattice with respect to the the partial order defined by $x \leq y$ iff $xy = x$.

A.16.7 Theorem (Lawson, Hofmann-Stralka) *For a compact topological semilattice S, the following statements are equivalent:*

(1) *The continuous homomorphisms $S \to \mathbb{I}$ separate the points.*
(2) *S is isomorphic to a closed subsemigroup of some cube $\mathbb{I}^X$.*
(3) *Every element of S has arbitrarily small multiplicatively closed neighborhoods.*

If these conditions are satisfied, then $(S, \leq)$ is a continuous lattice, and conversely, any continuous lattice carries a compact Hausdorff topology with respect to which it is a compact topological semilattice satisfying (1,2,3).

For more on compact semigroups, see (Gierz et al. 1980).

A.17 Applications of Semigroups

by Günter F. Pilz in Linz, Austria, Lev N. Shevrin in Ekaterinburg, Russia, and Peter G. Trotter in Hobart, Australia

A huge variety of structures studied by mathematicians are sets endowed with associative binary operations. Because of this we can say that different semigroup phenomena occur, explicitly or implicitly, in an immense field of applications of mathematics. But of course it is also the case that most applications make minimal use of the rich algebraic theory of semigroups that has been detailed in this chapter. However there are some important areas in which the semigroup-theoretic approach is quite substantial and semigroup theory is more fully utilized.

The most significant (and the best known) such areas are the theories of automata, formal languages and codes; one sometimes calls applications of semigroups to these areas *combinatorial applications*. We touch below on these applications and in addition mention some other situations where semigroups naturally appear.

Automata

Let $\mathcal{A}$ be a *complete deterministic finite state automaton*; that is, $\mathcal{A}$ consists of a finite set S of states, an input alphabet A and a next-state function $\delta\colon S \times A \to S$. Such an automaton models a machine whose action at any instant is uniquely determined by its mode and content. Each input $a \in A$ gives rise to a function $f_a\colon S \to S$ that maps $s \in S$ to the next state $\delta(s, a)$; the set $\{f_a \mid a \in A\}$ generates the *transformation monoid* $T(\mathcal{A})$ of $\mathcal{A}$ as a submonoid of the monoid of all mappings from S to S. Conversely, any finite monoid is isomorphic to the transformation monoid of some such automaton.

The Krohn-Rhodes Theorem is an important structural result for finite semigroups. By the theorem, any finite semigroup R is a homomorphic image of a subsemigroup of a finite semigroup T (that is, R divides T, see Section A.8) such that T is a wreath product whose components are finite simple groups or copies of the two element right zero semigroup with an adjoined identity. This theorem extends the well known corollary of a theorem by Krasner and Kaloujine by which any finite group embeds in a wreath product of finite simple groups. For an automaton $\mathcal{A}$, application of the Krohn-Rhodes Theorem to $T(\mathcal{A})$ results in a simulation of $\mathcal{A}$ by an automaton that is constructed using "parallel" and "series" compositions of "irreducible" automata.

Amongst the sources where the topics of this and the next subsection are extensively examined, we mention (Eilenberg 1974), (Lallement 1979), (Salomaa 1981), (Pin 1986), (Howie 1991).

Languages

A *formal language* is any subset of a free monoid A^* over some alphabet A. Formal languages have been studied from a grammatical point of view or via automata and semigroups. Some classes of languages are of special importance, for instance the rational languages. A language $L \subseteq A^*$ is *rational* if it can be obtained from one-element subsets of A^* by using a finite number of the constructions of unions, products and generation of submonoids.

Consider complete deterministic finite state automata as above, but with the additional property that they each include an initial state and a set of final states. By Kleene's Theorem, the rational languages of A^* coincide with those subsets L of A^* for which such an automaton exists that turns precisely the members of L into a designated set of final states.

Given a language $L \subseteq A^*$, the **syntactic congruence** for L is the greatest congruence ρ on A^* such that L is a union of ρ-classes. Then the **syntactic monoid** Syn L is A^*/ρ. A language L is rational if and only if Syn L is finite. Rational languages can be classified according to the pseudovarieties in which their syntactic monoids lie; recall (Section A.5) that a pseudovariety is a class of finite semigroups closed under taking homomorphic images, subsemigroups and finite direct products. For example, a "star-free" language is a rational language that is obtained from one-element sets by using a finite number of the constructions of unions, products and complementations; by Schutzenberger a language is star-free if and only if its syntactic monoid is finite and its subgroups are all trivial.

Some sources for the topic of this section are mentioned in the previous subsection. Note that certain connections between languages and pseudovarieties of finite semigroups are considered in Section A.8.

Codes

Let C be a subset of a free semigroup A^+ and let x be an arbitrary element of $\langle C \rangle$, the subsemigroup generated by C. If we consider the expression of x as a word over A, can we uniquely decode x as a product of words of C? The answer will be positive for every element of $\langle C \rangle$ if this subsemigroup is free and C is its base (see Section A.3). This explains the origin of the following concept. A **code** (or **variable length code**) over an alphabet A is a subset of A^+ which is the base of some free subsemigroup of A^+. Note that there are other variants of the concept of a code; see Section I.2 and Section I.3.

For example, the set A^n of all words of length n of A^+ is a code called the **uniform** (or **homogeneous**) code of degree n. Further examples include the following codes over the alphabet $\{a, b\}$: i) $\{a^2, ba^2, ba\}$; ii) $\{a^n b \mid n \in \mathbb{N}\}$; iii) $\{a^n b^n \mid n \in \mathbb{N}\}$. The set $\{a, ab, ba\}$ is not a code over $\{a, b\}$, since aba may be decoded both as $a \cdot ba$ and as $ab \cdot a$. Given a finite subset C of a free semigroup, it is natural to ask whether or not there is an algorithm to decide if C is a code. Such algorithms do exist; the first one (discovered in 1953) and apparently the best known is the so-called Sardinas-Patterson algorithm. A description of it as well as some references to other algorithms for the same purpose can be found, for instance, in (Lallement 1979, Ch. 5).

Sometimes we can easily determine whether a given subset is a code. We point out an important case of such a situation. A subset C of A^+ is called **prefix** [**suffix**] (or has the **prefix** [**suffix**] **property**) if $CA^+ \cap C = \emptyset$ [$A^+C \cap C = \emptyset$]. This may be rephrased as: no word in C is a left [right] factor of another word of C. It is easy to prove that any prefix [suffix] subset is a code. A code which is both prefix and suffix is called **biprefix**. For example, $\{ab, ba\}$ is a biprefix code over $\{a, b\}$, every uniform code is biprefix, and the code iii) indicated above is also a biprefix code. The code ii) is prefix but not suffix, while the code i) is neither prefix nor suffix.

In the course of the numerous investigations devoted to codes, diverse types of codes have been distinguished. Various properties, constructions and decompositions of codes have been studied and many connections have been established. The theory of variable length codes has become a large field of mathematics which may also be considered to be a part of theoretical computer science related not only to semigroup theory but also to combinatorics, automata theory, language theory, and to probability theory.

The reader is referred for further information on the topic of this subsection to (Lallement 1979, Ch. 5 and 8), (Salomaa 1981, Ch. 4), (Shyr 1991, Ch. 2, 5, 9–11), and particularly to the book (Berstel and Perrin 1985) which presents a comprehensive exposition of this topic.

Some other applications

There are areas of scientific endeavour where the application of semigroups can shed light on some phenomena but where, as yet, the applications are not as fundamental as in the areas discussed in the three previous subsections. Such areas include solid state physics, biology and even sociology. Some connections between semigroups and the last two of these areas are provided in Lidl and Pilz (1998). We begin with a brief outline of an application to sociology that stems from **relation semigroups**. The reader is referred to the cited book for further details.

The product $\diamond$ of two binary relations R and S on a set X is given by $x\,(R \diamond S)\,y \Leftrightarrow \exists z \in X \colon xRz \wedge zSy$. So if M is the relation "is mother of" and F is the relation "is father of" then $M \diamond F$ denotes "is grand-mother on father's side of". So questions like "is the friend of an enemy the enemy of a friend?" translate into a question of commutativity, and "is the enemy of an enemy a friend?" can be written as "enemy^2 = friend?". Since all societies are finite many relational products coincide, and the questions characterize societies in sociology. For instance, a society which does not distinguish between grandmothers from the father's and from the mother's sides fulfills $M \diamond F = M \diamond M$, with M and F as above. Frequently, inverse semigroups arise.

Another interesting area for applications is presented by tilings, and the remainder of the section is devoted to this topic. A solid (such as a crystal) can be modeled by a tiling where each tile represents an atom or molecule of the solid; a modeling like this is relevant to solid state physics. A common assumption is that the tiling is periodic and that its group of symmetries provides good structural detail of the tiling. However, the group of symmetries is concerned with global rather than local symmetries and, as was demonstrated by Weinstein (1996), it may provide negligible local detail even when the tiling is periodic. There are solids, namely quasi-crystals, that have non-periodic tilings. The comments on symmetric inverse semigroups in Section A.11 indicate that we should expect inverse semigroups to play a role in any theory that involves local symmetry. A theory of this type has recently been developed for application in solid state physics by J. Kellendonk, in which the initial step is the construction from a tiling of its tiling semigroup.

A *tile* in $\mathbb{R}^n$ is a connected bounded subset of $\mathbb{R}^n$ which is the closure of its interior. A *tiling* of $\mathbb{R}^n$ is an infinite set of tiles that cover $\mathbb{R}^n$ but overlap at most in their boundaries. A finite connected union of tiles in a tiling is a *pattern*. Given a tiling, the *tiling semigroup* is defined as follows. On the set $\mathcal{C}$ of all triples (p, P, q), where P is a pattern that includes tiles p and q, define a partial product by $(p, P, q)(r, R, s) = (p, P \cup R, s)$ if and only if $q = r$. Let G be the group of all translations of $\mathbb{R}^n$ and ρ be the equivalence relation on $\mathcal{C}$ given by $(p, P, q) \rho (r, R, s)$ if and only if $g(p) = r$, $g(P) = R$ and $g(q) = s$ for some $g \in G$. Then the quotient $\mathcal{C}/\rho$, along with an adjoined zero element, form the tiling semigroup in which products are inherited from the partial product on $\mathcal{C}$, or if not so defined, are zero. The tiling semigroup is an inverse semigroup.

For more details on tilings, see (Lawson 1998).

Chapter B

GROUPS

B.1 Abelian Groups

by Alexander V. Mikhalev in Moscow, Russia

We say that a group G is **abelian** if the group operation (usually written as addition) is commutative. The theory of abelian groups can be considered on the one hand as a part of the general theory of groups and also on the other hand as a part of module theory since every abelian group is a module over the ring $\mathbb{Z}$ of integers. But at the same time the theory of abelian groups now is an independent branch of algebra.

The beginning of the theory of abelian groups dates back to the 19-th century. We could find some finite abelian groups in the works of Gauss. N. Abel evaluated the role of abelian groups in the theory of algebraic equations. Frobenius and Stickelberger (1878) proved the first structure theorem of finite abelian groups.

The cyclic groups (in particular, the group $\mathbb{Z}$ of integers), the groups $\mathbb{Z}(p^\infty)$ of type p^∞ (so called **quasicyclic groups**, i.e., the union of the ascending chain of nonzero cyclic subgroups), p a prime number, the groups isomorphic to the group $\mathbb{Q}$ of rationals (which are locally cyclic), their direct sums and subgroups give us abelian groups with the simplest structure.

If the orders of all elements of G are finite then G is called a **torsion group**. In case when all nonzero elements of G have infinite order G is called **torsion-free**. All finite order elements of an abelian group G form a subgroup tG called the **torsion part** of G, $G/\text{t}G$ is torsion-free, and in general the canonical extension

$$0 \to \text{t}G \to G \to G/\text{t}G \to 0$$

can be nonsplittable.

A direct sum $F = \bigoplus_{i \in I} \langle x_i \rangle$ of infinite cyclic groups $\langle x_i \rangle$ is said to be a **free abelian group** ($X = \{x_i, \ i \in I\}$ is a **free set of generators** of F; cf. Section B.9). Each subgroup of a free abelian group is free; each subgroup of a finitely generated abelian group is finitely generated.

Every abelian group G is a homomorphic image of a free abelian group F. Moreover, if we fix a set $M = \{g_i, \ i \in I\}$ of generators of G, then we may take the free abelian group $F = \bigotimes_{i \in I} \langle x_i \rangle$ and $\eta \colon F \to G$,

$$\eta\left(\sum_{i=1}^{s} k_i x_i \right) = \sum k_i g_i, \ k_i \in \mathbb{Z}. \text{ Let } H = ker\,\eta \text{ be the kernel of } \eta. \text{ Then}$$

$G \cong F/H$, H is free and the relation $k_1 g_1 + \cdots + k_s g_s = 0$ holds in G if and only if $k_1 x_1 + \cdots + k_s x_s \in ker\,\eta = H$. Any generating subset L of H gives us a set of **defining relations** among generators $g_i, \ i \in I$.

Any finitely generated abelian group G is a direct sum of a finite number of indecomposable cyclic subgroups (some of them are cyclic groups of prime power orders, others are infinite cyclic). In particular, each finite abelian group is a direct sum of cyclic groups of prime power orders. Such a decomposition is not unique, but two such decompositions are isomorphic, and therefore the numbers of infinite and primary (for each prime) cyclic direct summands are invariants of G which define G up to an isomorphism.

An abelian group G is called **divisible**, if for all $a \in G$ and $n \in \mathbb{Z}$ the equation $nx = a$ has a solution in G. Every divisible group is a direct sum of groups isomorphic to $\mathbb{Q}$ and/or groups of type p^∞ for some p. Every group G contains the greatest divisible subgroup dG (the sum of all divisible subgroups, called the **divisible part** of G). If $dG = 0$, the group G is called **reduced**. Each divisible group is a direct summand in every group containing it as a subgroup. In particular, $G = dG \oplus R$, where R is a reduced group. Each abelian group can be embedded as a subgroup into a divisible group D; moreover, there exists a minimal divisible subgroup of D containing G (the so called **divisible hull** of G), uniquely determined up to an isomorphism over G. Divisible groups are precisely injective objects of the category of all abelian groups.

Kulikov's criterion describes p-groups which are decomposable into direct sum of cyclic groups. From that, we get the **Prüfer theorems** as important corollaries:

B.1.1 Theorem
(1) *Any bounded abelian group is a direct sum of cyclic groups;*
(2) *A countable p-group is a direct sum of cyclic groups if and only if it*

does not contain nonzero elements of infinite height.

The countable p-groups can be characterized by numerical invariants (so called Ulm invariants).

The **rank** $r(G)$ of a torsion-free group G is defined as cardinality of a maximal linearly independent (over $\mathbb{Z}$) system of elements of G. Each torsion-free group of rank 1 is isomorphic to some subgroup of $\mathbb{Q}$, and vice versa (torsion-free groups of rank 1 can be characterized up to an isomorphism by their types).

If G is a direct sum of torsion-free rank 1 groups then G is called **completely decomposable**. Every two such decompositions are isomorphic. Any direct summand of a completely decomposable group is also completely decomposable (but it is not true for subgroups).

There are some characterizations of pure subgroups of completely decomposable finite rank torsion-free groups. Some *separability conditions* for torsion-free groups G are under consideration, in particular, "if every finite system of elements of G is contained in a completely decomposable direct summand (respectively, in a free direct summand)".

For any infinite cardinal $\mathfrak{m}$ there exists $2^{\mathfrak{m}}$ non-isomorphic torsion-free indecomposable groups of cardinality $\mathfrak{m}$ (indecomposable torsion groups are only $\mathbb{Z}(p^k)$, $k \leq \infty$, p prime; there are no mixed indecomposable groups at all). Every torsion-free group G of finite rank is either indecomposable or the direct sum of a finite number of indecomposable groups (but different such decompositions may be non-isomorphic; there are only a finite number of non-isomorphic direct summands). Some "bad decompositions" of a torsion-free group into direct sum of indecomposable subgroups are usefull for constructing some fine examples in ring theory.

We say that two abelian groups A and B are **quasi-isomorphic** if there exist subgroups $A' \subseteq A$ and $B' \subseteq B$ such that $A' \cong B'$ and A/A', B/B' are bounded groups. Quasi-isomorphisms play now an essential role in the structure theory of abelian groups. For example, two torsion groups G and H are quasi-isomorphic if and only if their p-components G_p and H_p are quasi-isomorphic for all prime p and, moreover, $G_p \cong H_p$ for almost all p.

There are some sufficient conditions for splitting of a mixed abelian group G, i.e. $G = tG \oplus H$ for some subgroup H. Some classes of nonsplitting mixed groups can be described by means of invariants. Sometimes, it is usefull to investigate mixed groups as an extension of a torsion-free group by a torsion group (instead of the traditional extension of a torsion group by a torsion-free group).

In succeding years the theory of abelian groups was breaking down into three main parts: torsion groups; torsion-free groups; mixed groups. But many problems and methods considered now are not related to exactly one of these parts.

The *endomorphism ring* (see Section C.16) End(G), the additive group of End(G) (the *endomorphism group* of G), the *automorphism group* Aut G (see Section B.17) are important algebraic structures associated with an abelian group G.

B.1.2 Theorem

(1) (Baer-Kaplanski) *If A and C are torsion groups with isomorphic endomorphism rings* End(A) *and* End(C) *then the groups A and C are isomorphic, and any ring isomorphism ψ:* End(A) $\to$ End(C) *is induced by a certain group isomorphism φ: $A \to C$;*

(2) (Leptin-Liebert) *If G and H are abelian p-groups (p $\geq$ 3) and* Aut $G \cong$ Aut H, *then $G \cong H$.*

Let us consider the group Hom(A, B) of all homomorphisms from an abelian group A to an abelian group B. The group Hom($A, \mathbb{R}/\mathbb{Z}$) is called the *character group* of A. The classical character duality for finite abelian groups found many generalizations in the theory of abelian groups, in particular in the duality theory of locally compact groups.

The group Ext(B, A) of all classes of equivalent extensions of A by B is called the *group of extensions* of A by B (the derived functor of the functor Hom(B, A)). A complete description of Ext(B, A) is known only for some classes of groups A and B.

The *tensor product* $A \otimes B$ of two abelian groups A and B and the *torsion product* Tor(A, B) (the derived functor of $\otimes$) are also under consideration.

Many results of the theory of abelian groups were extended and generalized for modules over rings and for abelian categories. Some interesting results have been obtained for abelian groups regarded as *modules over their endomorphism rings*, for *additive groups of rings*, for *topological* and *partially ordered abelian groups*.

Logical aspects of abelian groups are also under consideration (in particular, the elementary theory of abelian groups is solvable).

Finally, many important mathematical invariants are, by their nature, abelian groups, for example, homology groups H$_n(X)$ of topological spaces X, and K-groups K$_n(R)$ of rings R.

B.2 p-Groups: Basics and the Coclass Project

by Charles R. Leedham-Green in London, UK

A finite p-***group*** is a group whose order divides p^n for some prime p and $n \geq 0$. What can be said about finite p-groups? For subgroups H and K of a group G, define $[H, K] = \langle \{h^{-1}k^{-1}hk | h \in H; k \in K\} \rangle$. Then defining $\gamma_1(G) = G$ and $\gamma_{i+1}(G) = [\gamma_i(G), G]$ for $i > 1$ gives the lower central series $G = \gamma_1(G) \geq \gamma_2(G) \geq \cdots$ of the group G. If $\gamma_c(G) > \gamma_{c+1}(G) = \langle 1 \rangle$ then G is said to be ***nilpotent of class*** c. The most basic result about finite p-groups is the observation that if G is of order p^n with $n > 1$ then G is nilpotent of class at most $n - 1$.

The most widely read basic introduction to the theory of finite p-groups is Chapter 4 of (Huppert 1983). The reader of that chapter will come away with the feeling that finite p-groups are intruiging objects, without any very unifying properties. Various interesting classes of p-groups can be defined; indeed seven sections of that chapter are devoted to special classes of p-groups. This reference is now 35 years old. If a new book were to be written in the same style, it would certainly include a section on powerful p-groups. For an odd prime p, a p-group P is ***powerful*** if P/P^p is abelian; that is to say, every commutator is a product of p-th powers. For $p = 2$ this condition is vacuous, and so is replaced by the requirement that every commutator is a product of 4-th powers. This does not look like a very promising concept, especially as it needs adjusting for $p = 2$. It was introduced by Mann, and developed by him and Lubotsky, and has played a central role in the modern theory of p-groups, as described in [(Dixon, du Sautoy, Mann, and Segal 1999) and (McKay 2000). There are other important strands missing from (Huppert 1983).

1. The most pervasive is the use of Lie algebras and Lie rings. These turn up in the theory of p-groups in all kinds of ways. The most obvious construction is to define

$$L(P) = \bigoplus_{i=1}^{c} \gamma_i(P)/\gamma_{i+1}(P)$$

where c is the nilpotency class of P, and the Lie product is defined by commutation. One of the most striking uses of Lie algebras is Shalev's proof of the coclass conjectures, as described in (McKay and Leedham-Green 2002). Although restricted to an exercise in (Huppert 1983), this concept is given due prominence in (Huppert and Blackburn 1982). For a modern account, see Chapter 1 of (Sautoy, Segal, and Shalev 2000).

2. There is some emphasis now on trying to describe the class of all finite p-groups. I shall return to this later.

3. The primary source of p-groups for many purposes is now taken to be the set of finite quotients of pro-p-groups, as pre-figured in (Huppert 1983, Chapter 4, Section 17). The principal reference for pro-p-groups is (Sautoy, Segal, and Shalev 2000). This is a very rich and expanding area of mathematics.

4. In a different direction, one should mention Zelmanov's celebrated positive solution to the restricted Burnside problem, following on Kostrikin's partial solution, and the earlier negative solution to the Burnside problem for large enough odd exponents due to Adian and Novikov (cf. Section B.9). Perhaps the most challenging question is to decide whether or not there is an infinite 2-generator group of exponent 5. Sims has devised and implemented many powerful techniques in a still unsuccessful attempt to prove that all such groups are finite. For a general account of the attack on restricted Burnside problems, see (Vaughan-Lee 1990).

5. Khukhro has lead an interesting attack on the attempt to describe p-groups in terms of automorphisms of p-groups having automorphisms with few fixed points, see (Khukhro 1993).

6. The mechanical enumeration of all isomorphism classes of p-groups of small order has reached new heights. The output of these programs is strictly for mechanical reading. The groups of order 2^6 were classified by Hall and Senior (1964). Extending this to obtain the groups of order 2^8 was regarded as one of the first challenges for computational algebra. This was a hard challenge, successfully met by O'Brien, who has constructed all groups of order dividing 2^9, and has counted (or rather has persuaded a computer to count) the isomorphism classes of groups of order 2^{10} (see Besche, Eick, and O'Brien 2001). The attempt to describe in meaningful ways the class of all finite p-groups is perhaps best summarised by the co-class project. This can be described in outline in a very elementary way.

The **dihedral group** D_{2^n} of order 2^n is the group of symmetries of a regular 2^{n-1}-ogon. Thus it has a cyclic subgroup of order 2^{n-1} and index 2. We shall also wish to consider two similar families of groups; Q_{2^n} is the **quaternion group** of order 2^n, defined for $n \geq 3$, and the **semi-dihedral group** SD_{2^n} of order 2^n, defined for $n \geq 4$. We need not concern ourselves too closely with these groups. They too contain cyclic subgroups of order 2^{n-1}. From our point of view, the most important property of these groups is the fact that they have nilpotency class $n - 1$.

A p-group of order p^n where $n \geq 3$ and nilpotency class $n-1$ is said to be of **maximal class** for obvious reasons. Now an elementary result states that the 2-groups of maximal class are precisely the diheral, quaternion and semi-dihedral groups. Here is a list of results about these groups that need to be emphasised.

B.2.1 Theorem *Let P be a 2-group of maximal class. Then:*

(1) *P contains a cyclic subgroup of index 2.*

(2) *There is a unique infinite group D_{2^∞} that is an inverse limit of 2-groups of maximal class. This group is an extension of the 2-adic completion $\mathbb{Z}_2$ of the additive group of the integers by a cyclic group of order 2 that acts by multiplication by -1.*

(3) *The central quotient of any 2-group of maximal class is isomorphic to a dihedral group, and hence to the unique quotient of D_{2^∞} of that order.*

(4) *The derived quotient of any 2-group of maximal class is isomorphic to the Klein four group V; that is, to the direct product of two cyclic groups of order 2; and conversely, if P is any 2-group with $P/\gamma_2(P)$ isomorphic to V then P is of maximal class.*

None of these results is hard. The surprising thing is that similar results were first proved for p-groups of maximal class for all p, and later for p-groups of order p^n and class $n - r$ for all p and n and r, thus providing a description of all finite p-groups. But it needs to be stated at once that the description is only useful when n is much larger than r.

In outline, let us see what these more general results look like. We shall content ourselves with a rough outline. A precise account of the results and their proofs will be found in (McKay and Leedham-Green 2002).

B.2.2 Theorem *Let P be a p-group of maximal class. Then:*

(1) *P contains a subgroup P_1 of index p such that $\gamma_3(P_1)$ has order bounded by a function of p. In particular, the derived length of P is bounded in terms of p.*

(2) *There is a unique infinite group D_{p^∞} that is an inverse limit (see Section G.1) of p-groups of maximal class. This group is an extension of the additive group of integers in the p-adic completion of the p-th cyclotomic field by a complex p-th root of 1.*

(3) *P has a normal subgroup M of order bounded by a function of p such that P/M can be obtained from a quotient of D_{p^∞} by a certain*

twisting process. Also, if P has order p^n then the normal subgroup N of P of order $p^{a+n/2}$, where a is a function of p, has the property that P/N is isomorphic to a quotient of D_{p^∞}.

(4) *Let Q be any finite p-group with $Q/\gamma_{p+1}(Q)$ of maximal class. Then Q is of maximal class.*

The last property is related to property (iv) above since it states that to determine whether or not a finite p-group Q is of maximal class it suffices to consider $Q/\gamma_{p+1}(Q)$ for odd primes, as opposed to having only to consider $Q/\gamma_2(Q)$ if $p = 2$. The question of when important structural information about a finite p-group (or pro-p-group) Q can be determined by the quotient of Q by some given term of the lower central series of Q is an important and recurrent theme.

These results in part are due to Blackburn and appear in Chapter III, §14 of (Huppert 1983), and in part are due to Shepherd and independently to Leedham-Green and McKay. The proofs depend on rather detailed computations, but require no sophisticated machinery.

Now let us turn to the more striking generalisation. If P is a p-group of order p^n and nilpotency class $n - r$ then P is said to have **coclass** r.

Let P be a p-group of coclass r. Then:

(i) P contains a normal subgroup N of index bounded by a function of p and r (and independent of n) such that N is of nilpotency class at most 2 (for $p = 2$ we may require N to be abelian). In particular, the derived length of P is bounded.

(ii) There are only finitely many infinite groups that are inverse limits of p-groups of coclass r. These groups are extensions of $\mathbb{Z}_p$-lattices of rank a multiple of $p - 1$, acted on by p-groups of order bounded as a function of p and r.

(iii) This generalises from the maximal class case very simply. D_{p^∞} must be replaced by any one of the infinite groups in (ii). For $p = 2$ the twisting process is not needed, so that P/M is isomorphic to a quotient of one of these infinite groups, whre the order of M is bounded in terms of r.

(iv) This also generalises, but with $p + 1$ being replaced by a function of p and r.

Explicit bounds are obtained for these results. They come in two kinds. There are absolute bounds, and smaller generic bounds that hold with finitely many exceptional groups. The proofs of these results are considerably more sophisticated than in the case of groups of maximal class. None the less, most of the machinery is of general interest, and the proof is described, almost from first principles, in (McKay and Leedham-Green 2002). The coclass project started in 1980 with the publication in (Leedham-Green and Newman 1980) of the coclass conjectures. These conjectures were labelled A to E (in order of decreasing strength). Conjecture C states that the inverse limit of an inverse system of p-groups of fixed coclass is soluble. A critical point was reached when Donkin proved this conjecture when the pro-p-group is p-adic analytic, at least for $p > 3$. Tits produced a proof along similar lines, using the theory of buildings, in conversation with Donkin, his proof covering all primes. However, Shalev's proof of Conjecture A (and hence of all the conjectures) removed the need for this theory. As a result, the theory of algebraic groups no longer plays a role in the proof of the conjectures, and it seems unlikely that Tits' argument will be published. However, Donkin's proof established the fact that there are p-adic analytic groups of great interest that are the inverse limits of systems of finite p-groups whose coclass only tends very slowly to infinity, and so the spirit of his work lives on.

To produce alternative descriptions of the class of all finite p-groups, it may be worth considering the coclass project in more general terms. The first step was to filter the class of all finite p-groups into quotient closed classes. Thus if $\Gamma_{p,r}$ is the set of isomorphism classes of finite p-groups of coclass at most r, then this set is quotient closed, and the union of all such sets, for fixed p, is the class of all p-groups. Now turn $\Gamma_{p,r}$ into a graph by joining any p-group P (strictly its isomorphism class) to $P/\gamma_c(P)$, where c is the class of P. The aim is then to describe the graph $\Gamma_{p,r}$; although an exact description is unattainable, we have produced an interesting description of this graph, the maximal infinite chains corresponding to the inverse limits described above; that is to say, the pro-p-groups of finite coclass.

For a very powerful and profound analysis of the fine structure of $\Gamma_{p,r}$ using model theory and zeta functions see (du Sautoy 2000).

B.3 p-Groups: Other Approaches

by Charles R. Leedham-Green in London, UK

The question that now arises is to decide what other filtrations on the class of all finite p-groups might lend themselves to obtaining interesting results. Clearly the **pro-p-groups** obtained by taking inverse limits of groups in any fixed class will play a dominant role; so consider the conditions that a good pro-p-group might be expected to satisfy. Here are some examples.

0. *Countably based.* A pro-p-group is **countably based** if and only if it is the inverse limit of a sequence

$$\cdots \to P_2 \to P_1 \to P_0$$

of finite p-groups, so all the pro-p-groups we are interested in are countably based. A pro-p-group is countably based if and only if it is a metric space with respect to the pro-p-topology.

1. *Finite width.* A pro-p-group G has **finite width** if the factors $\gamma_i(G)/\gamma_{i+1}(G)$ have uniformly bounded order.

2. *Finite rank.* A pro-p-group has **finite rank** if there is a uniform bound to the number of elements required to generate (topologically) any closed subgroup. These are precisely the finitely generated p-adic analytic pro-p-groups, and are of fundamental importance. To say that the group G is p-**adic analytic** simply means that G has an analytic structure over the p-adic completion $\mathbb{Q}_p$ of the rationals, with respect to which the group operations are locally analytic.

3. *Just infinite.* An infinite pro-p-group is **just infinite** if it has no non-trivial closed normal subgroups of infinite index. This condition is a natural analogue of the concept of 'simple' in finite groups, but may be the wrong analogue. It has the disadvantage in the present context that it does not in any way define a useful condition on finite p-groups, as they have normal subgroups of every order dividing the order of the group. This is in contrast to the first two conditions; one can define a quotient closed class of finite p-groups by bounding the width, or the rank, or both.

4. *Finite obliquity.* A pro-p-group G has **finite obliquity** if for some fixed k every normal subgroup of G is sandwiched between $\gamma_i(G)$ and $\gamma_{i+k}(G)$; equivalently if there is a uniform bound to the order of $MN/M \cap N$ as (M, N) varies over all pairs of normal subgroups of the same index in G. This condition clearly implies that G is just infinite, and is in fact a stronger condition. It has the advantage that obliquity can be quantified in some way or another so as to ensure that the class of finite p-groups of obliquity at most o is quotient closed, and then, by bounding the obliquity in a class of p-groups, we will define pro-p-groups that have finite obliquity, and hence are just infinite.

5. *Hereditarily just infinite.* An infinite pro-p-group is **hereditarily just infinite** if it has no non-trivial closed sub-normal subgroups of infinite index. This raises the question of whether a simple group would

be more naturally defined as one that has no non-trivial subnormal subgroups. The class of hereditarily just infinite pro-p-groups is seriously more restrictive than the class of just infinite pro-p-groups, but we are nowhere near understanding this class.

6. *Analytic groups.* We have seen that the finitely generated pro-p-groups that are analytic over $\mathbb{Q}_p$ are those of finite rank. But analytic groups over local fields of characteristic p, and over other rings, are also of great interest.

7. *Branch groups.* The definition of 'branch group' in Chapter 4 of (Sautoy, Segal, and Shalev 2000) is rather technical. The crucial point is that if G is a branch group then G contains, as an open subgroup of itself, a subgroup that is isomorphic to the direct product of p copies of an open subgroup of G. The global example is the iterated wreath product W of p copies of the cyclic group of order p. Clearly W contains a maximal subgroup isomorphic to the direct product of p copies of itself. Branch groups are studied as subgroups of W, and hence as acting on the infinite rooted tree in which every node has p children. The most famous branch group arises from the Grigorchuk group $\mathcal{G}$. This is an infinite finitely generated 2-group. From the current prospective one should consider its 2-adic completion $\overline{\mathcal{G}}$. It is clear that a branch group cannot be hereditarily just infinite, and apparently cannot have finite obliquity. However, $\overline{\mathcal{G}}$ has finite width and is just infinite. For the latest information on branch groups, see Chapter 4 of (Sautoy, Segal, and Shalev 2000), and also (Bartholdi and Grigorchuk 2000) and (Bartholdi 2001).

8. *The Nottingham group and Fersenko's groups.* The **Nottingham group** $\mathcal{N}_p$ is the pro-p-group of index $p-1$ in the group of ring automorphisms of $\mathbb{F}_p[[x]]$. R. Camina proved in (Camina 1997) the startling result that $\mathcal{N}_p$ contains a copy of every countably based pro-p-group. ($\overline{\mathcal{G}}$ has the same property for $p=2$). However, apart from the cyclic group of order p, it is generally hard to find copies of any specific p-group or pro-p-group in $\mathcal{N}_p$. Fersenko has proved that various naturally defined closed subgroups of $\mathcal{N}_p$ have similar properties to $\mathcal{N}_p$ itself. The basic properties of $\mathcal{N}_p$ include the following. The group is hereditarily just infinite, of finite width. It has finite obliquity for $p > 2$, but not for $p = 2$. For $p > 2$ it is related to the first Witt Lie algebra; and it appears to be finitely presented, requiring five relators on 2 generators (as pro-p-group), though this has not yet been proved, and appears to be a very difficult problem. See also Camina's account in Chapter 6 of (Sautoy, Segal, and Shalev 2000). The Nottingham group was so named in hon-

our of the pioneering work on this group by Johnson and his student York at the University of Nottingham.

One can take combinations of these properties. Consider for example the class of just infinite pro-p-groups of finite rank. Call such a group *maximal* if it is not a proper open subgroup of another such. Then all just infinite pro-p-groups of finite rank are open subgroups of maximal groups. To simplify the situation slightly, restrict attention now to the hereditarily just infinite pro-p-groups of finite rank. These correspond bijectively to the set of central simple Lie algebras over local fields of characteristic 0 (ignoring the group $D_{2\infty}$). For details of these groups see (Klaas, Leedham-Green, and Plesken 1997). They are sufficiently well understood that one can hope to classify finite p-groups in the above style, in terms of width, rank and obliquity. For a start in this direction see (McKay and Leedham-Green 2002). The difficult challenge is to remove the restriction on the rank. It is shown in (Klaas, Leedham-Green, and Plesken 1997) that there are uncountably many pro-p-groups to deal with in this case, but the uncountably many examples produced (for given p) are all virtually isomorphic; that is to say, they can all be constructed as finite extensions of one fixed group. The possibility remains therefore that there is a classification theorem for hereditarily just infinite pro-p-groups of finite width up to virtual isomorphism. If such a theorem exists it will be very hard to prove. Even the small progress that would be achieved by producing an internal characterisation of the Nottingham group seems very hard. It seems quite possible that for some value of n, such as $n = 2p+2$, the only infinite pro-p-group P with $P/\gamma_n(P) \cong \mathcal{N}_p/\gamma_n(\mathcal{N}_p)$ is $\mathcal{N}_p$ itself. There must also be further gold to be mined from the subgroup structure of the Nottingham group.

The properties and classes of pro-p-groups that I have considered here have been biased towards the study of finite p-groups. To produce a proper bibliography would be to produce an article that was more bibliography than matter. The reader is urged to study the bibliographies in the books and articles mentioned, and to take an active role in this exciting and rapidly changing area of algebra.

B.4 Soluble Groups

by Hans Lausch in Melbourne, Australia

A series $G = G_0 \unrhd G_1 \unrhd G_2 \unrhd \cdots \unrhd G_n = \{\,1\,\}$ in a group G is a *soluble series* if all factors G_i/G_{i+1}, $0 \le i \le n-1$ are abelian. G is *soluble* if it has a soluble series. If $N \trianglelefteq G$, then G is soluble iff this applies to both N and G/N. So, finite direct products of soluble groups are soluble, and each finite p-group is soluble. More generally, **Burnside's $p^a q^b$ Theorem** says that all groups of order $p^a q^b$ (p, q prime) are soluble. The famous **Feit-Thompson-Theorem** of 1963 says that every group of odd order is soluble. Nilpotent groups are soluble. S_n and A_n are not soluble for $n \geq 5$ (in fact, A_5 is the smallest non-soluble group). Of course, no non-abelian simple group can be soluble. **Hall's Theorem** says that a soluble group of order ab, where $\gcd(a, b) = 1$, has at least one subgroup of order a, and two such subgroups are conjugate.

Finite soluble groups

Finite soluble groups, introduced at the very beginning of group theory, have been explored to such an extent that a brief report can furnish only very few samples of an exceedingly rich theory. Luckily, the subject has been given *comprehensive coverage within a single monograph* which also provides an excellent bibliography, namely Doerk and Hawkes (1992).

If π is a set of primes, then a finite group is called a π-group if every prime dividing its order belongs to π. A π-subgroup H of a given group G is called a **Hall π-subgroup** of G if no prime in π divides $[G : H]$. In major investigations, Philip Hall exhibited the key role of these subgroups for the theory of soluble groups: a finite group G is soluble if and only if G possesses Hall π-subgroups for all sets π of primes, in which case, for each set π of primes, the Hall π-subgroups of G form a single conjugacy class in G.

For the remainder of this section, all groups are assumed to be finite and soluble, unless stated otherwise. One notes that Hall's Theorem extends the scope of Sylow's Theorem from p-groups to π-groups. Carter discovered another Sylow type theorem, where certain nilpotent subgroups of a group play the role of the Sylow subgroups (Doerk and Hawkes, 305–306). A paper by Gaschütz (1963) became crucial for further developments. In particular, classes of groups with certain closure properties have been the framework for investigating soluble groups ever since.

A collection $\mathscr{F}$ of groups is called a **class of groups** if $G \in \mathscr{F}$ and $H \simeq G$ implies $H \in \mathscr{F}$. Groups belonging to $\mathscr{F}$ are called $\mathscr{F}$-*groups*. In search of further Sylow type theorems, Gaschütz defined an $\mathscr{F}$-*projector* of a group G to be a subgroup E of G with the property that for every normal subgroup $N \trianglelefteq G$

(1) EN/N is an $\mathscr{F}$-group, and

(2) whenever K is a subgroup of G that contains EN as a proper subgroup, then K/N is not an $\mathscr{F}$-group.

For example, the Hall π-subgroups and the Carter subgroups of G are the $\mathscr{F}$-projectors of G if $\mathscr{F}$ is the class of π-groups and the class of nilpotent groups, respectively. A class $\mathscr{F}$ of groups is called a **Schunck class** if G is an $\mathscr{F}$-group whenever all primitive homomorphic images of G are $\mathscr{F}$-groups. Schunck and Gaschütz completely solved the problem of determining all classes $\mathscr{F}$ of groups with the property that $\mathscr{F}$-projectors satisfy a Sylow type theorem: Schunck classes are precisely those classes $\mathscr{F}$ of groups for which every group has $\mathscr{F}$-projectors; if $\mathscr{F}$ is a Schunck class, then, in each group G, the $\mathscr{F}$-projectors form a single conjugacy class (Doerk and Hawkes, 292 and 299).

Examples of Schunck classes are: the classes of π-groups, groups of nilpotency length at most n ($n = 1, 2, \ldots$), supersoluble groups, groups with a Sylow tower of given type, p-nilpotent groups, groups with nilpotent commutator subgroup. Note that the class of abelian groups is *not* a Schunck class.

For not necessarily soluble groups, $\mathscr{F}$-projectors and related concepts were studied thoroughly by Förster (Doerk and Hawkes, 290–298), whose work has exhibited the limits and limitations of Schunck class theory once the realm of soluble groups is gone beyond.

After initial work by Fischer, the Gaschütz-Schunck theory was, in a certain sense, dualized by Fischer, Gaschütz, and Hartley: if $\mathscr{F}$ is a class of groups, then a subgroup $V \leq G$ is called an $\mathscr{F}$-*injector* of G if, for every subnormal subgroup N of G,

(1) $V \cap N$ is an $\mathscr{F}$-group, and

(2) whenever K is a subgroup of N that contains $V \cap N$ as a proper subgroup, then K is not an $\mathscr{F}$-group.

The search for conditions on $\mathscr{F}$ that would produce Sylow type theorems led to the definition: $\mathscr{F}$ is a **Fitting class** if

(1) every normal subgroup of an $\mathscr{F}$-group is itself an $\mathscr{F}$-group, and

(2) the product of two normal $\mathscr{F}$-subgroups of a group is again an $\mathscr{F}$-subgroup of that group.

Indeed, Fitting classes are precisely those classes $\mathscr{F}$ of groups for which every group has $\mathscr{F}$-injectors; if $\mathscr{F}$ is a Fitting class then, in each group G, the $\mathscr{F}$-injectors form a single conjugacy class (Doerk and Hawkes, 564–565).

Examples of Fitting classes are the classes of π-groups, groups of nilpotency length at most n $(n = 1, 2, \ldots)$, p-nilpotent groups. The class of abelian groups is not a Fitting class, while the class of all groups with socle in their center is a Fitting class which is not a Schunck class. Note that the Hall π-subgroups are not only the $\mathscr{F}$-projectors but also the $\mathscr{F}$-injectors of a given group if $\mathscr{F}$ is the class of π-groups. On the other hand, if $\mathscr{F}$ is the class of nilpotent groups, then the $\mathscr{F}$-projectors (i.e., Carter subgroups) of the symmetric group S_3 are its Sylow-2 subgroups, whereas A_3 is its $\mathscr{F}$-injector. New Fitting classes have been obtained from given ones through a diversity of constructions (Doerk and Hawkes, 574–600).

Special, and highly useful, Fitting classes were discovered by Dark (Doerk and Hawkes, 630–647). For example, using Dark's idea, work by Bryce, Cossey, Ormerod, and McCann led to the problem of describing the smallest Fitting class containing S_4 being reduced to describing the smallest Fitting class containing A_4. Paradoxically, Fitting classes of groups of nilpotency length 2 appear to pose special problems (Doerk and Hawkes, 653–660).

In view of the conjugacy statements in the theorems above, the question arose: Which Schunck (Fitting) classes $\mathscr{F}$ are **normal**, i.e., have the property that every group has normal $\mathscr{F}$-projectors ($\mathscr{F}$-injectors)? Gaschütz and Blessenohl obtained a complete and straightforward answer for Schunck classes (Doerk and Hawkes, 303–305), while there is a much richer theory of normal Fitting classes (Doerk and Hawkes, 704–719). An example of a normal Fitting class is the class $\{G \mid$ for each $g \in G$, the map $x \mapsto g^{-1}xg$ is an even permutation on the largest odd-order normal subgroup of $G\}$. Lockett provided important results that generalize the concept of normal Fitting classes very effectively and helped streamline the study of Fitting classes (Doerk and Hawkes, 677–682). Blessenohl and Gaschütz also furnished a method for constructing normal Fitting classes, which Lausch proved to be universal and Bryce and Cossey, by way of generalization, related to Lockett's result (Doerk and Hawkes, 720–736).

Following a construction due to Berger, new normal Fitting classes, defined in terms of group theoretical transfer, were discovered by Laue, Lausch, and Pain; Berger then proved a deep theorem which has numerous applications (Doerk and Hawkes, 737–761), one of which provides algorithms that decide whether or not a given group G belongs to

- the smallest normal Fitting class,
- the smallest normal Fitting class containing a given group H.

Infinite soluble groups

Some ideas that play an important role for finite soluble groups have also been considered in the context of infinite ones, e.g., by Stonehewer (1966) and by Hartley, Gardiner, and Tomkinson (1971), while other developments went into quite different directions. Not unexpectedly, the structure of infinite soluble groups is considerably influenced by their abelian subgroups. The most famous theorems exhibiting this feature, due to Mal'cev and Schmidt, respectively, state that if each abelian subgroup of a soluble group G satisfies the maximum (minimum) condition, then G does, too. Finitely generated soluble groups form a rather complex class as a theorem of Neumann and Neumann (1959) shows: any countable soluble group of derived length d is a subgroup of some 2-generator soluble group of derived length at most $d + 2$. Philip Hall proved that a finitely generated group which is an extension of an abelian group by a nilpotent group is a subdirect product of finite groups (i.e., *residually finite*). A good reference to some central themes of infinite soluble groups is Robinson (1996), especially chapter 15 and bibliography.

B.5 Permutation Groups

by Peter J. Cameron in London, UK

In the early days of group theory, a 'group' was a set of transformations of a set closed under composition and inversion and containing the identity transformation. Now, groups are axiomatically defined, and the above concept is a *permutation group*, that is, a subgroup of the symmetric group. *Cayley's Theorem* establishes the relation between these concepts: *every group is isomorphic to a permutation group*. However, it is often convenient to deal with the more general concept of an *action* of a group G, a homomorphism from G to a symmetric group (see Section B.15); the advantage is that G can be represented in several different ways as a permutation group.

The first section of the paper gives a brief introduction to the notation and terminology. The study of finite permutation groups is one of the oldest parts of group theory, motivated initially by its connection with solvability of equations. Many of the problems which defeated early researchers such as Jordan have been resolved using the classification of finite simple groups. This is discussed in the second section of the paper. The study of infinite permutation groups is less well developed, and was recognized by the Mathematics Subject Classification only in the 1980s. Some of the threads are followed in the final section.

General references for permutation groups include Cameron (1999), Dixon and Mortimer (1996), Passman (1968), Wielandt (1964).

Introduction

Let G be a permutation group on the set Ω (a subgroup of the symmetric group on Ω). We denote the image of α under the permutation g by αg. We say that G is **transitive** if, for any $\alpha, \beta \in \Omega$, there exists $g \in G$ with $\alpha g = \beta$. In general, Ω is uniquely expressible as a disjoint union of subsets (called **orbits**) on each of which it is transitive, and G is a subcartesian product of transitive permutation groups. In the finite case, this is just a subdirect product.

We say that G is **primitive** if it is transitive and there is no G-invariant equivalence relation on Ω except the 'trivial' relations (equality and $\Omega \times \Omega$). An imprimitive group is a subgroup of the complete wreath product of simpler groups. In some cases (for example, finite groups), this reduction process terminates, and for many purposes it suffices to study primitive groups.

Let t be a positive integer with $t \le |\Omega|$. Then G is t-**transitive** if it is transitive in its induced action on t-tuples of distinct elements of Ω. For example, the symmetric group on Ω is t-transitive for all $t \le |\Omega|$.

Finite permutation groups

The watershed in the theory of finite permutation groups was the announcement of the O'Nan-Scott Theorem in 1979. This theorem, though elementary (and only a slight strengthening of a result of Jordan) opened the way for the application of the Classification of Finite Simple Groups (see Section B.8) to primitive permutation groups. The theorem divides primitive groups into several classes, depending on the action of the socle (the product of the minimal normal subgroups).

B.5.1 Theorem *One of the following holds for any finite primitive permutation group G:*

(1) *G is contained in a wreath product (in the product action);*
(2) *the socle of G is a product of isomorphic simple groups in a known action (regular or 'diagonal');*
(3) *G is 'almost simple', that is, $S \le G \le \operatorname{Aut}(S)$ for some nonabelian simple group S.*

In case (1) we have a reduction to smaller permutation groups similar to the reduction for imprimitive groups. In cases (2) and (3), information about the simple groups can be used.

In this way, many powerful results about primitive groups have been proved. Here are some examples.

(a) Classification theorems: all 2-transitive groups are known; other classifications include groups of rank 3 (that is, having three orbits on Ω^2), and primitive groups of odd degree.

(b) Asymptotic results: primitive groups are rare (there are at most $n^{c \log n}$ conjugacy classes in S_n, and there are none except S_n and A_n for almost all n, and of small order (at most $n^{c \log n}$ with known exceptions.

(c) Other results: there are no 6-transitive groups apart from symmetric and alternating groups; and *Sims' Conjecture* (that the order of the point stabilizer in a primitive group is bounded by a function of any one of its nontrivial orbit lengths) is true.

These results have many applications, in logic, number theory, Riemann surfaces, combinatorics, and theoretical computer science, as well as in other parts of algebra, including Galois theory, the area with the oldest applications (cf. Section I.5).

Infinite permutation groups

The study of infinite permutation groups is much more recent and less well developed. It is not possible in a short survey to do more than identify a few trends.

A permutation group G on Ω is *oligomorphic* if there are only finitely many orbits on Ω^n for all positive integers n. According to the theorem of Engeler, Ryll-Nardzewski, and Svenonius, these groups are essentially the same as automorphism groups of countable models of $\aleph_0$-categorical first-order theories. The amalgamation method, introduced by Fraïssé and refined by Hrushovski, is a powerful construction method. The growth rate of the orbit-counting sequence is interesting. All these matters are surveyed in Cameron (1990).

A permutation group is *finitary* if its elements move only finitely many points. The *Jordan-Wielandt Theorem* asserts that the only primitive finitary groups are the finitary symmetric and alternating groups. Neumann has developed this result into a general theory of finitary permutation groups. At the other extreme, a permutation group is *cofinitary* if its non-identity elements fix only finitely many points This class generalizes Frobenius groups (Section B.18). There are many examples but few theoretical results: see Cameron (1996).

An important recent piece of work is the classification of primitive *Jordan groups*, (those having a subgroup transitive on the points it moves) by Adeleke and Macpherson (1996).

Finally, there is a natural topology on the symmetric group, that of *pointwise convergence*: the closed subgroups are precisely the automorphism groups of first-order structures.

B.6 Lie Groups

by Joachim Hilgert in Clausthal, Germany

Definition and examples

A *Lie group* is a group that at the same time is a smooth manifold such that the group operations are differentiable. Usually one only considers real and complex manifolds, but there is also a p-adic version of Lie groups. A prime example of a (real) Lie group is the general linear group $GL(n; \mathbb{R})$ of invertible $n \times n$ matrices (over $\mathbb{R}$); it is open in the space of all $n \times n$ matrices. More generally, all closed subgroups of $GL(n; \mathbb{R})$ are Lie groups. Invented originally by Sophus Lie in the late 19th century to serve as symmetry groups (viz. Galois groups) of (partial) differential equations, Lie groups typically occur as symmetry groups of geometric objects such as isometry groups of quadratic and Hermitian forms (e.g., orthogonal and unitary groups) or groups of holomorphic automorphisms of bounded complex domains (e.g., fractional linear transformations or more general conformal automorphism groups).

More on this subject can be found in (Bourbaki 1960-1982, Chevalley 1946, Helgason 1978, Howie 1992, Knapp 1996, Serre 1992a)

Lie algebra and exponential map

A crucial tool in the study of a Lie group G is the *Lie algebra* $\mathfrak{g}$ associated with it. As a vector space, $\mathfrak{g}$ can be defined as the tangent space of G at the unit element e. In particular, we have $\dim \mathfrak{g} = \dim G$. The *Lie bracket* on $\mathfrak{g}$ is most easily visualized as the commutator $[X, Y] = XY - YX$ of vector fields (viewed as derivation operators on the space of smooth functions). Here, each tangent vector X at the identity e is identified with the (left invariant) vector field obtained by translating the vector via the left multiplication on the group. There is a natural **exponential map** exp: $\mathfrak{g} \to G$ which associates with $X \in \mathfrak{g}$ the value $\varphi(1)$ of the integral curve φ of X with $\varphi(0) = e$. In the case of the general linear group, the corresponding Lie algebra consists of all real $n \times n$ matrices with the matrix commutator bracket $[X, Y] = XY - YX$ and the exponential map is the matrix exponential map given by the exponential series $\exp(X) = \sum_{k=0}^{\infty} \frac{1}{k!} X^k$.

The exponential map is a local diffeomorphism and connects the algebraic structures of G and $\mathfrak{g}$ via the **Baker-Campbell-Hausdorff identity**

$$(\exp X)(\exp Y) = \exp(X * Y)$$

for small $X, Y \in \mathfrak{g}$, where $X * Y$ is a universal power series of iterated bracket products of X and Y which reduces to the sum $X + Y$ if X and Y commute (i.e., if $[X, Y] = 0$). One can use this to show that there is an equivalence between the category of real finite dimensional Lie algebras and *connected and simply connected Lie groups*. That explains why the exponential map can be used not only as a device to translate problems (typically non-linear) on the Lie group into (typically linear, hence more easily tractable) problems on the Lie algebra, but also to translate the solutions of the Lie algebra problems back to solutions of the original Lie group problems.

A typical example where this general principal can be put to work is the determination of structural information like central or commutator series for which one has analogous concepts on the Lie algebra level. In fact, concepts like *(semi)simplicity, solvability, nilpotency*, or *product decompositions* for Lie groups often are defined via the corresponding Lie algebra properties and differ slightly from the purely group theoretic versions. So for instance a Lie group is called **semisimple** if its Lie algebra is semisimple, i.e., if the Lie algebra has no nonzero solvable ideal.

Structure theory

Structural results of prime importance are various decomposition and conjugacy results. Any Lie group has a maximal compact subgroup which completely determines the homotopy type of the Lie group and all maximal compact subgroups are conjugate. Similarly, one always finds a decomposition (direct on the Lie algebra level and almost direct, i.e., up to finite intersections, on the Lie group level) into a semisimple subgroup and solvable normal subgroup. The solvable part, called the *radical*, is uniquely determined, whereas the semisimple part, called the *Levi part*, is unique only up to conjugacy. Semisimple Lie groups admit several important decompositions: The *Iwasawa decomposition* $G = KAN$, where K is a maximal compact subgroup, A is a maximal closed torsion free abelian subgroup, and N is a nilpotent subgroup, generalizes the Gauß decomposition of matrices, whereas the *Cartan decomposition* $G = KAK$ can be viewed as a generalization of the Gram-Schmidt decomposition of symmetric matrices.

Classification

There is no hope to classify all Lie groups. Even nilpotent and solvable Lie groups do not show enough rigidity to admit a classification beyond small dimensions. On the other hand, (semi)simple Lie groups have a rich but rigid structure and can be classified completely. The classification scheme starts out with complex simple Lie algebras. The are four infinite series of simple complex Lie algebras, as well as five exceptional ones. Once complex semisimple Lie algebras are classified, one studies involutions whose fixed point sets are real forms and in this way obtains a classification of the real simple Lie algebras. To classify all the connected real Lie groups, one has to study the corresponding fundamental groups, which can be identified with certain central subgroups.

Applications

As mentioned above, Lie groups often occur as symmetry groups and can thus be used to study the geometric and analytic properties of objects with sufficiently large groups of symmetries. The examples come from various fields of mathematics and physics: Geometry and complex analysis, but also number theory (e.g., the symmetry group of the moduli space of the lattices studied in the *geometry of numbers* is a symplectic group) or classical mechanics (e.g., in the context of Noether's Theorem on conserved quantities).

Of particular importance are spaces on which a Lie symmetry group acts transitively (e.g., symmetric spaces). Not only do such spaces come up naturally, but also they serve as model spaces on which one has additional (group theoretic) methods to study various problems (like solving differential equations, which gives rise to *harmonic analysis*, a generalization of Fourier analysis).

Related concepts

Obviously Lie groups are intimately related with Lie algebras, but there are also close connections to the theory of algebraic groups. Real and complex algebraic groups are at the same time Lie groups. On the other hand, *Ado's Theorem* says that any finite dimensional Lie algebra can be realized as a Lie algebra of matrices so that, up to covering maps, any Lie group can be realized as a group of matrices. Thus Lie groups are not so far from being algebraic (but evidently not every Lie group is an algebraic group). The relation is particularly close in the case of simple groups. Here the classification theory of algebraic groups parallels its Lie theoretic counterpart to quite some extent. The correspondence between Lie groups and Lie algebras, however, cannot be effected by an exponential map which is a transcendental object. On the other hand, whenever non-connected Lie groups enter the picture, the exponential map does not help in the study of the group of components, so very often one borrows methods and results from the theory of algebraic groups.

The fundamental importance of root systems in Lie theory explains why their symmetry groups, i.e., *reflection groups* play a crucial role in the theory as well: They occur in various forms as *Weyl groups*.

Directions of recent research

The structure theory and geometry of Lie groups is no longer a very active research area as such. New results are usually conceived and obtained in contexts where the Lie groups themselves only serve as tools (e.g., in Riemannian or symplectic geometry) or are at least are not the prime objects of interest (e.g., in harmonic analysis or representation theory). Generalizations like quantum groups or infinite dimensional groups (e.g., of diffeomorphisms or loops) have been studied extensively in the last twenty years. Motivation here very often comes from theoretical physics.

B.7 Lie Groups and Differential Equations
by Peter J. Olver in Minneapolis, MN, USA

The applications of Lie groups to solve differential equations dates back to the original work of Sophus Lie, who invented Lie groups for this purpose. The modern era begins with Birkhoff (1950), and was forged into a key tool of applied mathematics by Ovsyannikov (1982). Basic references are (Hydon 2000, Olver 1993, 1995).

First we review the geometric approach to systems of differential equations. We begin with a smooth m-dimensional manifold M; the reader will not experience any significant loss of generality by taking $M = \mathbb{R}^m$. Solutions will be identified as p-dimensional (smooth) submanifolds $S \subset M$. Local coordinates on M include a choice of independent variables $x = (x^1, \ldots, x^p)$, and dependent variables $u = (u^1, \ldots, u^q)$, where $p + q = m$, and so a (transverse) submanifold is given as the graph of a function $u = f(x)$. The derivatives of the dependent variables are represented by $u_J^\alpha = \partial_J f^\alpha(x)$, where $\alpha = 1, \ldots, q$ and J is a multi-index of order $0 \leq |J| \leq n$. These form a system of local coordinates, collectively denoted by $(x, u^{(n)})$, on the n-th order (extended) jet bundle $J^n \to M$. The jet bundle can be defined as the set of equivalence classes of p-dimensional submanifolds $S \subset M$, where $S \sim \widetilde{S}$ define the same equivalence class or *jet* $j_n S \in J^n$ at common point $z \in S \cap \widetilde{S}$ whenever the two submanifolds have n-th order contact at z. A system of differential equations $\Delta_\nu(x, u^{(n)}) = 0$, $\nu = 1, \ldots, s$ is *regular* if the Jacobian matrix of the Δ_ν has maximal rank s at all $(x, u^{(n)})$ that satisfy the system. A regular system can be viewed as a submanifold $S_\Delta = \{\Delta_\nu(x, u^{(n)}) = 0\} \subset J^n$. A classical (smooth) solution is, thus, a submanifold $S \subset M$ whose jet $j_n S \subset S_\Delta$. The system is *locally solvable* if there exists a smooth solution passing through each point $(x, u^{(n)}) \in S_\Delta$.

Let G be an r-dimensional Lie group acting smoothly on M. Since G preserves contact between submanifolds, there is an induced action, denoted $G^{(n)}$, on J^n, called the n-th order *prolonged action*, which tells us how G acts on the derivatives of functions. The action defines a *symmetry group* of a system of differential equations S_Δ if it maps solutions to solutions. Assuming local solvability this occurs if and only if S_Δ is a $G^{(n)}$-invariant subset of J^n.

A connected Lie group action is entirely determined by its *infinitesimal generators*, which are vector fields on the manifold M and can be identified with the Lie algebra $\mathfrak{g}$ of G. Each vector field

$$\mathbf{v} = \sum_{i=1}^p \xi^i(x, u) \frac{\partial}{\partial x^i} + \sum_{\alpha=1}^q \varphi^\alpha(x, u) \frac{\partial}{\partial u_J^\alpha} \in \mathfrak{g},$$

generates a one-parameter subgroup. The infinitesimal generator of the corresponding n-th prolonged one-parameter subgroup is a vector field

$$
\text{pr } \mathbf{v} = \sum_{i=1}^{p} \xi^{i}(x,u)\frac{\partial}{\partial x^{i}} + \sum_{\alpha=1}^{q}\sum_{j=\#J} \varphi_{J}^{\alpha}(x,u^{(j)})\frac{\partial}{\partial u_{J}^{\alpha}},
$$

on J^{n}. There is an explicit formula for the coefficients φ_{J}^{α} of the prolongation pr $\mathbf{v}$ in terms of the derivatives of the coefficients $\xi^{i}, \varphi^{\alpha}$ of $\mathbf{v}$. This *prolongation formula*, coupled with the following *infinitesimal symmetry criterion* allows us to explicitly compute the symmetry groups of almost any systems of differential equations. Indeed, there now exist a wide range of computer algebra packages for performing this computation, (Hereman 1994).

B.7.1 Theorem *A connected group of transformations G is a symmetry group of the regular system of differential equations S_{Δ} if and only if pr $\mathbf{v}(\Delta_{\nu})=0$, $\nu=1,\ldots,s$, on S_{Δ} for every $\mathbf{v} \in \mathfrak{g}$.*

Example

Consider the linear heat equation $u_{t} = u_{xx}$. Applying Theorem B.7.1, an infinitesimal symmetry $\mathbf{v} = \xi(x,t,u)\partial_{x} + \tau(x,t,u)\partial_{t} + \varphi(x,t,u)\partial_{u}$ must satisfy $\tau_{u} = \tau_{x} = \xi_{uu} = 0$, $-\xi_{u} = -2\tau_{xu} - 3\xi_{u}$, $\varphi_{uu} = 2\xi_{xu}$, $\varphi_{u} - \tau_{t} = -\tau_{xx} + \varphi_{u} - 2\xi_{x}$, $-\xi_{t} = 2\varphi_{xu} - \xi_{xx}$, $\varphi_{t} = \varphi_{xx}$. The solution space to this overdetermined linear system of partial differential equations yields the symmetry algebra of the heat equation, with basis $\mathbf{v}_{1} = \partial_{x}$, $\mathbf{v}_{2} = \partial_{t}$, $\mathbf{v}_{3} = u\partial_{u}$, $\mathbf{v}_{4} = x\partial_{x} + 2t\partial_{t}$, $\mathbf{v}_{5} = 2t\partial_{x} - xu\partial_{u}$, $\mathbf{v}_{6} = 4xt\partial_{x} + 4t^{2}\partial_{t} - (x^{2} + 2t)u\partial_{u}$, and $\mathbf{v}_{\alpha} = \alpha(x,t)\partial_{u}$, where $\alpha_{t} = \alpha_{xx}$. The corresponding one-parameter groups are, respectively, x and t translations, scaling in u, the scaling $(x,t) \mapsto (\lambda x, \lambda^{2}t)$, Galilean boosts, an "inversional symmetry", and the addition of solutions stemming from the linearity of the equation. Each of these groups maps solutions to solutions, e.g., the inversional group tells us that if $u = f(x,t)$ is any solution, so is $u = \frac{1}{1+4\varepsilon t} \exp\left\{\frac{-\varepsilon x^{2}}{1+4\varepsilon t}\right\} f\left(\frac{x}{1+4\varepsilon t}, \frac{t}{1+4\varepsilon t}\right)$, for any $\varepsilon \in \mathbb{R}$. The constant solution $u = 1/\sqrt{2\pi}$ produces the fundamental solution at $(0, -(4\varepsilon)^{-1})$. Thus, the symmetry group provides an effective mechanism for computing a wide variety of new solutions from known solutions. Further applications—to finding explicit group-invariant solutions, to determining conservation laws, to solution, to classification of differential equations with given symmetry groups, and so on—are described below.

Generalized symmetries

For ordinary or *point* symmetries, the coefficients ξ^i, φ^α of $\mathbf{v}$ depend only on x, u. Generalized or higher order symmetries, including contact transformations, allow dependence on the derivatives u_j^α as well. Higher order symmetries play an essential role in the study of integrable soliton equations, (Fokas 1980, Mikhalev et al. 1991, pp. 115–184, Sanders and Wang 1998). *Recursion operators* and *master symmetries* map symmetries to symmetries and thereby generate infinite hierarchies of generalized symmetries. The biHamiltonian structure theory of Magri (1978), provides an method for constructing recursion operators.

Linearization of partial differential equations

Any linear partial differential equation has an infinite-dimensional symmetry group: addition of solutions. A system of partial differential equations can be linearized if and only if it has an infinite-dimensional symmetry group of the proper form.

Noether's theorems

A variational problem admitting a symmetry group G leads to a G-invariant system of Euler-Lagrange equations. Noether's first theorem, (Noether 1918), associates a conservation law for the Euler-Lagrange equations with every one-parameter symmetry group of the variational problem. For instance, translation invariance leads to conservation of linear momentum, rotation invariance leads to conservation of angular momentum, and time translation invariance leads to conservation of energy. Noether's second theorem, of application in relativity and gauge theories, produces dependencies among the Euler-Lagrange equations arising from infinite-dimensional variational symmetry groups.

Integration of ordinary differential equations

Lie observed that virtually all the classical methods for solving ordinary differential equations (separable, homogeneous, exact, etc.) are instances of a general method for integrating ordinary differential equations that admit a symmetry group. An n-th order scalar ordinary differential equation admitting an n-dimensional solvable symmetry group can be integrated by quadrature. The associated conservation laws of variational problems and Hamiltonian systems doubles the order of symmetry reduction.

Symmetry reduction of partial differential equations

If the orbits of G are s-dimensional and transverse to the vertical fibers $\{x=c\}$, then the G-invariant solutions to a G-invariant system of differential equations are found by reducing to a system in $p-s$ variables. See (Bluman and Cole 1969, Olver and Rosenau 1987) for the nonclassical generalization, and (Anderson and Fels 1997) for the nontransverse case, of importance in many physical systems, e.g., relativity, fluid mechanics.

Differential invariants

A function $I\colon \mathrm{J}^n \to \mathbb{R}$ which is invariant under the action of $G^{(n)}$ is known as a differential invariant. Basic examples include curvature and torsion of curves, and the Gaussian and mean curvature of surfaces in three-dimensional Euclidean geometry. The differential invariants are the fundamental building blocks for constructing G-invariant differential equations, variational problems, etc., as well as solving the basic problems of equivalence and symmetry of submanifolds. For example, every Euclidean-invariant differential equation for space curves involves just the curvature, torsion and their arc-length derivatives: $F_\nu(\kappa, \kappa_s, \ldots, \tau, \tau_s, \ldots) = 0$.

Differential invariants are characterized by the infinitesimal invariance criterion pr $\mathbf{v}(I) = 0$ for all $\mathbf{v} \in \mathfrak{g}$. Cartan's method of moving frames, (Cartan 1935, Fels and Olver 1999), forms an effective tool for producing complete systems of differential invariants. An n-th order *moving frame* is a smooth, G-equivariant map $\rho\colon \mathrm{J}^n \to G$, where G acts on itself by left multiplication. The most familiar case is the moving frame for a curve in $\mathbb{R}^3$, consisting of a point z on the curve together with the unit tangent $\vec{t}$, normal $\vec{n}$ and binormal $\vec{b}$ at z. These form a left-equivariant map $\rho\colon \mathrm{J}^2 \to \mathrm{E}(3)$ from the second jet space to the Euclidean group, where we interpret $z \in \mathbb{R}^3$ as the translation component and the 3×3 matrix $[\vec{t}, \vec{n}, \vec{b}] \in \mathrm{O}(3)$ as the rotation component of the group element. In general, a moving frame exists if and only if $G^{(n)}$ acts freely and regularly, which holds in all practical examples for $n \gg 0$. Normalization amounts to setting $r = \dim G$ components of the prolonged group transformations $(g^{(n)})^{-1} \cdot (x, u^{(n)})$ to be suitably chosen constants. Solving for the group parameters and substituting into the remaining components produces a complete system of differential invariants. In the case of space curves, these are the curvature, torsion and their successive derivatives with respect to arc length.

Symmetry classification of ordinary differential equations

Lie's classification of all finite-dimensional Lie groups acting on the plane, (Lie 1924, Olver 1995), along with their differential invariants and Lie determinants leads to a complete symmetry classification of scalar ordinary differential equations, and possible symmetry reductions.

Discrete symmetries

Discrete symmetry groups also play an important role in differential equations, including Schwarz's theory of hypergeometric functions, Fuchsian and Kleinian groups, etc., (Hille 1976). Discrete symmetries can often be determined from the continuous symmetry group, (Hydon 2000).

B.8 Simple Groups
by Ronald Solomon in Columbus, OH, USA

The groups

Simple groups first entered the mathematical consciousness through the work of Galois, who recognized that nonabelian simple groups were the fundamental obstruction to the solution of polynomial equations *by radicals*(Section D.3). Initially the known examples of simple groups were the cyclic groups of prime order, the alternating groups of degree at least 5 and the fractional linear groups $PSL(2; p)$ for p a prime, $p \geq 5$. The first systematic cataloging of simple groups was begun by C. Jordan and extended by Dickson. Their work gives a complete description of the classical linear groups over finite fields and identifies their nonabelian simple composition factors. If G is a classical linear group acting as linear operators on a vector space V, then its center $Z(G)$ is a group of scalar matrices and the quotient group $PG = G/Z(G)$ acts naturally on the projective space $P(V)$ whose objects are the nonzero subspaces of V. If $\dim(V) > 1$, and in the case of orthogonal spaces if $\dim(V) > 4$, then the derived group $[PG, PG]$ is almost always a nonabelian simple group.

In 1861, Mathieu announced the discovery of two remarkable simple groups, M_{12} and M_{24}, which act as 5-fold transitive permutation groups on sets of size 12 and 24, respectively. They and their subgroups, M_{11}, M_{22}, and M_{23}, were the first five sporadic simple groups, as they were dubbed by Burnside.

The next major step was the construction in 1955 by Chevalley of finite analogues of all of the complex simple Lie groups. This was soon followed by the construction by Steinberg of certain *twisted* Chevalley groups. From a different perspective, Suzuki discovered a new family $SZ(2^{2n+1})$ of 2-transitive permutation groups; and shortly thereafter, Suzuki's groups and two further families were constructed by Ree, using a similar method to that of Steinberg. This completed the discovery of infinite families of finite simple groups. The groups constructed by Chevalley and Steinberg (which include all of the classical linear groups) as well as those constructed by Suzuki and Ree are collectively denoted the finite groups of Lie type.

Around this time, Tits discovered an elegant set of axioms for a class of groups called BN-pairs, which included all finite simple groups known in 1964.

B.8.1 Definition A group $G = BNB$ is a *BN-pair* if B and N are subgroups of G with $B \cap N \lhd N$ and with $W = N/B \cap N$ generated by a finite set S of involutions such that $sBs \neq B$ for all $s \in S$ and

$$BsB \cdot BwB \leq BwB \cup BswB \quad \text{for all } s \in S, w \in W.$$

We call $|S|$ the **rank** of the BN-pair G. A simple BN-pair of rank 1 is just a simple 2-transitive permutation group, whence every simple alternating group and the five Mathieu groups are simple BN-pairs of rank 1. Tits proved that every finite simple BN-pair of rank at least 3 is a finite simple group of Lie type with Weyl group W of rank at least 3, i.e., not a cyclic or dihedral group. In particular, every finite simple BN-pair of rank at least 3 is split, in the sense that $B = U(B \cap N)$, where U is a normal nilpotent subgroup of B. (A BN-pair has an associated geometry called its building. The split condition has the geometrical interpretation that the building is *Moufang*.) Fong and Seitz proved that every finite simple split BN-pair of rank 2 is a finite simple group of Lie type whose Weyl group is a dihedral group.

In 1965, Janko discovered a new sporadic simple group, dubbed J_1. In the ensuing decade twenty more sporadic simple groups were discovered to yield a total of 26. The Monster M is the largest of the sporadic simple groups, and twenty of the sporadic simple groups arise as quotients of subgroups of M. Although several interesting avenues have been pursued (e.g., diagram geometries, vertex operator algebras), no unifying principle has been discovered to explain all of the finite simple groups. Nevertheless the following **classification theorem** has been proved.

B.8.2 Theorem *Let G be a finite simple group. Then G is isomorphic to a member of one of the following families:*

(1) *the cyclic groups of prime cardinality;*
(2) *the alternating groups of degree at least 5;*
(3) *the finite simple groups of Lie type; or*
(4) *the 26 sporadic simple groups.*

The original classification proof occupies about 15,000 journal pages. The "second-generation proof", being constructed by Bender, Glauberman, Peterfalvi, Aschbacher, Smith, Gorenstein, Lyons, Solomon, and others, has an estimated length of approximately 5,000 pages.

The Classification Theorem

It was known to Sylow that a simple group of prime power order is cyclic of prime order. A dramatic extension of this result is the following theorem of Burnside.

B.8.3 Theorem (***Burnside's $p^a q^b$ Theorem***) *If G is a nonabelian finite simple group, then $|G|$ is divisible by at least three distinct primes.*

Burnside's theorem was one of the early triumphs of the theory of group characters. One way of establishing that a finite group satisfying certain hypotheses cannot be simple is to establish the existence of a non-trivial character of G whose kernel is not the identity subgroup, since this kernel is then a proper normal subgroup of G. In addition to Burnside's $p^a q^b$ Theorem, this method underlies "transfer" theorems and the following two non-simplicity results.

B.8.4 Theorem (Frobenius) *Let G be a finite group with a proper subgroup M such that $N_G(X) \leq M$ for every nonidentity subgroup X of M. Then G is not a simple group.*

B.8.5 Theorem (***Glauberman's Z^*-Theorem***) *Let G be a finite simple group of even order in which some involution (element of order 2) z does not commute with any distinct conjugate gzg^{-1}. Then G is cyclic of order 2.*

Most other applications of the theory of group characters to finite simple groups are of the nature of sophisticated counting arguments. Often these are coupled with character-free counting arguments which rely on the elementary but important observation that two involutions generate a dihedral group. A notable exception to this last statement is the use of character theory in the proof of the Odd Order Theorem of Feit and Thompson.

B.8.6 Theorem (*Odd Order Theorem*, Feit-Thompson)
A finite simple group of odd order is a cyclic group of odd prime order.

Character theory plays a prominent role in the classification of simple groups of 2-rank at most 3, i.e., having no abelian 2-subgroup of rank 3. Beyond that, character theory is only invoked via Glauberman's Z^*-Theorem.

The principal methodology for the remainder of the classification proof is called local group theory, i.e., the analysis of normalizers of nonidentity solvable subgroups. A p-local subgroup of a group G is the normalizer of a nonidentity p-subgroup of G. The point of departure for the analysis of p-local subgroups is Sylow's Theorem, which guarantees their existence.

B.8.7 Theorem (*Sylow's Theorem*) *Let G be a finite group with $|G| = p^a \cdot m$, where p is a prime and $(p, m) = 1$. Then G contains subgroups of order p^b for all $b \leq a$; all subgroups of order p^a are G-conjugate; and every subgroup of order p^b is contained in some subgroup of order p^a.*

The maximal p-subgroups of G are called **Sylow p-subgroups of G**.

The **Fitting subgroup** $F(G)$ of a finite group G is the unique maximal normal nilpotent subgroup of G. A subgroup E of G is called **semisimple** if $E = E_1 \ldots E_r$ with $E_i = [E_i, E_i]$, $E_i/Z(E_i)$ a nonabelian simple group and $[E_i, E_j] = 1$ for all $i \neq j$. G has a unique maximal normal semisimple subgroup, denoted $E(G)$. Then $[E(G), F(G)] = 1$ and we define the characteristic subgroup $F^*(G) = E(G)F(G)$. The following theorem is fundamental to the analysis of finite groups.

B.8.8 Theorem (Fitting, Bender) *Let G be a finite group. Then $C_G(F^*(G)) = Z(F(G))$. Thus $G/Z(F(G))$ is isomorphic to a subgroup of $\operatorname{Aut}(F^*(G))$.*

Thus the structure of a finite group G is "controlled" by the structure of $F^*(G)$. For π a set of primes and H a finite group, we define $O_\pi(H)$ to be the largest normal π-subgroup of H; and $O_{\pi',\pi}(H)$, $F_\pi^*(H)$ to be the full inverse image in H of $O_\pi(H/O_{\pi'}(H))$, $F^*(H/O_\pi(H))$, respectively. Much of the proof of the Classification Theorem is devoted to the analysis of the possible structures of $F_{p'}^*(H)$ or $F_p^*(H)$ for H a p-local subgroup of the simple group G.

Primarily the p-local analysis focuses on the case $p = 2$. The fact that 2-local analysis suffices to determine the nonabelian simple groups rests on the Odd Order Theorem and the following generalization of Frobenius' Theorem.

B.8.9 Theorem (*Strongly Embedded Theorem*, Suzuki, Bender)
Let G be a finite simple group having a proper subgroup M of even order such that $N_G(X) \leq M$ for all non-identity 2-subgroups of M. Then $G \cong \mathrm{SL}(2; 2^n)$, $\mathrm{SZ}(2^{2n+1})$ or $\mathrm{PSU}(3; 2^n)$.

We say that a group G is of **even type** if $F_{2'}^*(H) = O_{2',2}(H)$ for all 2-local subgroups H of G. Otherwise we say that G is of **odd type**. For nonabelian simple groups of odd type, a key result is the B-Theorem.

B.8.10 Theorem (*B-Theorem*) *Let G be a finite simple group of even order and H a 2-local subgroup of G. Then $F_{2'}^*(H) = E(H)O_{2',2}(H)$.*

The proof of this result relies heavily on the Signalizer Method pioneered by Thompson, Gorenstein, and Walter. Corollaries such as Aschbacher's Component Theorem give detailed information about the structure of $E(H)$ for H a 2-local subgroup of G. When G is of odd type, this permits the identification of G, generally by presentations due to Steinberg, Curtis and Tits.

When G is of even type, the key method is the analysis of the embedding in a 2-local H of the Thompson subgroup $J(T_H)$ of a Sylow 2-subgroup T_H of H. (If T is a p-group and d is the maximum rank of an elementary abelian subgroup of T, then $J(T)$ is the subgroup of T generated by all elementary abelian subgroups of rank d.) If $J(T_H) \leq O_{2',2}(H)$ for all 2-local subgroups H of the simple group G of even type, then it is possible (if $d \geq 3$) to reach the hypotheses of the Strongly Embedded Theorem. Otherwise the theory of failure of factorization modules (F-modules) gives sufficient control over the structure of $F_2^*(H)$ to permit the identification of G as a split BN-pair or as a group of automorphisms of some diagram geometry.

Aschbacher, Gorenstein and Lyons were leaders of the final phase of the classification project. The most important general references in this area are (Aschbacher 1986, Gorenstein 1982, Gorenstein et al. 1994, 1996, 1998, 1999, Suzuki 1982, 1986).

B.9 Free and Relatively Free Groups
by Gilbert Baumslag in New York, NY, USA

A *variety* is a class $\mathscr{V}$ of groups closed under subgroups, epimorphic images and unrestricted direct products (Section B.21 and Section G.5). Recall that epimorphisms are simply onto homomorphisms (see Neumann 1967b, 1937). The groups in $\mathscr{V}$ are referred to as $\mathscr{V}$-groups. A $\mathscr{V}$-group F is termed *free* (in $\mathscr{V}$) if it comes equipped with a set of generators X such that for every $\mathscr{V}$-group G and every map $\theta : X \longrightarrow G$, there exists a homomorphism from F into G which agrees with θ on X. X is sometimes called a *free generating set* and F is said to be *freely generated* by X. The free groups in the various varieties are collectively termed *relatively free* groups and the free groups in the variety of all groups are termed *absolutely free* groups. Every variety can be defined by a set of laws or identical relations. In order to explain, let E be an absolutely free group, freely generated by a countably infinite set $X = \{x_1, \ldots, x_n, \ldots\}$ and let $\mathscr{V}$ be a variety of groups. An element $v = v(x_1, \ldots, x_n) \in E$ is termed a *law* or an *identical relation* of $\mathscr{V}$ if for every $\mathscr{V}$-group G and every choice of elements $a_1, \ldots, a_n$ in G, $v(a_1, \ldots, a_n) = 1$ in G. The set of all laws of $\mathscr{V}$ forms a fully invariant subgroup V of E and the mapping $\mathscr{V} \to V$ is one-to-one and onto from the class of all varieties of groups to the class of all fully invariant subgroups of E. It follows that there are at most continuously many varieties. In fact, there are continuously many varieties of groups, most of which have not yet been studied (cf. Ol'shanskiĭ 1995).

Given any group G and any variety $\mathscr{V}$, the subgroup $V(G)$ of G consisting of all elements of the form $v(a_1, \ldots, a_n)$ where $v(x_1, \ldots, x_n)$ ranges over all of V and the a_i range over all elements of G, form a fully invariant subgroup of G. $G/V(G)$ is the largest factor group of G in $\mathscr{V}$. The free groups in V then all take the form $H/V(H)$, where H is a suitably chosen absolutely free group.

In many instances, a given variety $\mathscr{V}$ can be described by a finite set L of laws. This means, by definition, that the smallest fully invariant subgroup of E containing L, i.e., invariant under all endomorphisms of E, is V. In such a case, one terms V a *finitely based* variety. It turns out that the smallest variety containing any given finite group is always finitely based. In view of the fact that there are continuously many varieties and only a countably infinite number of finite subsets of E, it follows that there are continuously many varieties of groups that are not finitely based. Besides the varieties generated by finite groups, there are many other finitely based varieties. These include the variety $\mathcal{N}_c$ of all nilpotent groups of class at most c, for every choice of c. Perhaps the most well-known finitely based varieties are the so-called *Burnside varieties*. More precisely, for each positive integer n, let $\mathbb{B}_n$ be the variety defined by the law x_1^n. One of the most celebrated problems in this area is the so-called *Burnside Problem*—is every finitely generated free group in $\mathbb{B}_n$ finite? This problem is known to have a positive answer only for 2,3,4 and 6. However, if n is sufficiently large, the answer is has been proved to be negative. There is another related problem, called the *restricted Burnside problem*: given any pair of positive integers m and n is there a bound on the orders of the the finite groups with m generators in $\mathbb{B}_n$? Here the answer is positive (Zelmanov 1990a); cf. Section B.13.

Since every group in a given variety $\mathscr{V}$ is a factor group of a suitably chosen free group in $\mathscr{V}$, these free groups are the ones that have been studied the most. In particular, a great deal is known about the free $\mathcal{N}_c$-groups; here, there is a link between such groups and certain Lie algebras, which is very important in the study of nilpotent groups as a whole (Hall 1969). Other important relatively free groups are the free grops in the class $\mathscr{S}_d$ of all solvable groups of derived length at most d, the relatively free groups in the variety of so-called Engel groups and various other varieties defined by commutator identities. Many questions remain unanswered about these relatively free groups, questions such as whether or not these groups have any torsion, questions about their lower central series, and so on.

Varieties of groups admit to a multiplication. Given two varieties $\mathbb{U}$ and $\mathbb{V}$, the class of all groups with a normal subgroup in $\mathbb{U}$ with factor group in $\mathbb{V}$, is again a variety, denoted $\mathbb{U}.\mathbb{V}$. This product of varieties turns the class of all varieties into a free monoid; in other words every variety can be expressed as a product of varieties in one and only one way (see Neumann 1967b).

Among the relatively free groups, the absolutely free ones are the ones that are best understood. In particular, subgroups of absolutely free groups are free, the intersection of two finitely generated subgroups is again finitely generated and the automorphism groups of finitely generated absolutely free groups are finitely presented. Limited results of this kind hold also for certain varieties.

The variety of groups where all commutators commute, i.e., the so-called **metabelian groups** are reasonably well understood. In particular, the finitely generated relatively free groups in this variety satisfy the maximal condition for normal subgroups. This implies that there are, up to isomorphism, a countably infinite number of finitely generated metabelian groups and that these groups all have solvable word problems (Section B.11 and cf. (Hall 1954)). This is not true even for finitely generated solvable groups which are finitely presented in the class of all groups (Kharlampovich 1954). Perhaps the one outstanding problem for finitely generated metabelian groups is the isomorphism problem.

B.10 Free Products, Amalgamated Products, and HNN Extensions

by Gilbert Baumslag in New York, NY, USA

There is a construction in group theory, called an amalgamated product, which enables one to build new groups from known ones (Schreier 1927, Lyndon and Schupp 1977, Magnus et al. 1966). This construction is such that it allows for an analysis of the new groups, given sufficient information about the groups from which they have been built. The same remark holds if a given group turns out to be an amalgamated product. Many groups, in fact, turn out to be amalgamated products. These include, among others, the fundamental groups of surfaces of genus at least two, many subgroups of fundamental groups of three dimensional manifolds, various groups of matrices, Fuchsian groups and certain groups given in terms of generators and defining relators. These remarks indicate that amalgamated products are an important tool in group theory and that their study provides valuable information about a diverse collection of groups.

A group G is said to be the **product** of its subgroups A and B if it is generated by the union of A and B. This implies that if $g \in G$ then g can be expressed in the form

$$g = x_1 x_2 \ldots x_n \ (x_i \in A \cup B, \ i = 1, \ldots, n). \qquad (*)$$

One of the most important products is the so-called amalgamated product discussed above. It is defined as follows.

B.10.1 Definition Suppose that the group G is a product of its subgroups A and B and that $A \cap B = H$. If every product of the form $(*)$

satisfying the conditions

(1) $n > 0$,
(2) $x_i \notin H$, for $i = 1, \ldots, n$,
(3) for every $i = 1, \ldots, n - 1$, if either $x_i \in A$, then $x_{i+1} \in B$ or if
   $x_i \in B$, then $x_{i+1} \in A$,

is different from 1 in G, then G is called an **amalgamated product** of
A and B with H amalgamated.

In the event that $H = 1$, such an amalgamated product G is called
a **free product** of A and B. There is a more general version of this
definition in which A and B are replaced by more than two subgroups
of G, allowing also for infinitely many such subgroups (cf. Neumann
1954).

Amalgamated products were invented in 1926. The first major use
of these products took place in 1933 when it was proved that a group
defined by a single relator can be built up from cyclic groups using amalgamated products and HNN-extensions (see below for the definition).
This decomposition of a group with a single defining relator was then
used to show that every group defined by one relator has a solvable word
problemthe solution of the word problem for one relator groups). This
process of decomposing a group is one of the most ways in which amalgamated products are utilized. The main point here is that given a pair
ofgroups A and B which intersect set-theoretically in a common subgroup H, it is possible to explicitly construct a bigger group G, which
contains both A and B and is the amalgamated product of A and B
with H amalgamated. It is the explicit nature of this construction that
is so useful. It allows also for the construction a host of new groups with
interesting properties. A special case of such a construction is that of
an HNN-extension.

It is easy to describe what an HNN-extension is. To this end, suppose
that B is a given group, that H and K are a given pair of subgroups
of B and that $\varphi \colon H \to K$ is a given isomorphism from H to K. Then
the group E defined by adjoining a new generator, say t, to E, together
with the relations

$$t^{-1}ht = h\varphi, \ h \in H,$$

is termed an **HNN-extension** of B. The point here is that E is obtained from B by adding a new generator t which makes H conjugate
to K in accordance with the given isomorphism φ. Since E can be defined by means of amalgamated products, it follows that B is actually

embedded in E and that all of the relations that hold in E are easily understood because of the way in which it is constructed.

The knowledge that a group G is an amalgamated product makes it possible, as indicated already a number of times, to obtain important properties of G. For example, if G is the free product of its subgroups A and B (so here $H = 1$), then the mimimal number of generators of G is the sum of the minimal number of generators of A and the minimal number of generators of B.

Uses of amalgamated products and HNN-extensions abound. The first proof, around 1954, of the existence of a finitely presented group with an unsolvable word problem was subsequently simplified considerably about 5 years later by making use of amalgamated products and HNN extensions (see Novikov 1955).

The same two constructions were again critical in the characterization in 1961, of the finitely generated subgroups of finitely presented groups: a finitely generated group is a subgroup of a finitely presented group if and only if it can be defined by a recursively enumerable set of relators (cf. Higman 1961, Boone 1959, Britton 1963).

A clearer understanding of the nature of these constructions was eventually obtained from an unexpected quarter, namely by a consideration of groups acting without inversion on a tree (see Serre 1992b). Here a group G is said to *act without inversion* on a tree if it comes equipped with a homomorphism into the group of automorphisms of a tree such that none of the images of the elements of G interchanges the vertices of an edge. It turns out that amalgamated products (and hence also HNN-extensions) can be made to act without inversion on suitably chosen trees. The relevance of this fact is that it can be proved that every group that acts without inversion on a tree can be built up in a specific manner from the stabilisers of the vertices and edges of this tree, by unions, amalgamated products and HNN extensions. Since the subgroups of a group which acts without inversion on a tree act without inversion on the same tree, they too can be built up from unions, amalgamated products and HNN-extensions. It follows that the subgroups of amalgamated products and HNN-extensions can also be built up in the same way (Karrass and Solitar 1970). Free products of infinite cyclic groups, so-called free groups (see Section B.9), also act on trees, without inversion and without fixed points. One of the consequences of this remark is that it then follows that subgroups of free groups act without inversion and without fixed points on a tree. Every group with this property turns out to be free. Hence one sees that subgroups of free groups are free, an observation of considerable importance in much of group theory (see Schreier 1927).

B.11 Word Problems in Groups

by Derek Holt in Warwick, UK

In this section we shall be talking about groups G defined by a presentation $\langle X \mid R \rangle$ where X and R are both finite. Let A be the union $X \cup X^{-1}$, where X^{-1} is a set disjoint from and in one-one correspondence with X, and the element of X^{-1} corresponding to $x \in X$ is denoted by x^{-1}. Let A^* be the set of words in A, including the empty word ε. Then an element $w \in A^*$ maps onto an element $\overline{w} \in G$, where $\overline{x^{-1}} = \overline{x}^{-1}$ and ε maps onto the identity element 1_G of G. We write $w_1 =_G w_2$ if $\overline{w}_1 = \overline{w}_2$. For $w \in A^*$, $l(w)$ will denote the length of w and, for $g \in G$, $l_X(g)$ will denote $\min\{l(w) \mid w \in A^*, \overline{w} = g\}$. A word w is called **reduced** if it does not contain adjacent mutually inverse elements of A.

The **word problem** in G can be defined to be the the the set W of words in A^* that map onto 1_G or, equivalently, the set of words in A^* that are equal in the free group on X to a product of some conjugates of some defining relators in R (Section B.9).

The celebrated Novikov-Boone Theorem says that there exist groups G for which testing the membership in W is undecidable. It was proved independently at about the same time in the mid 1950's by Novikov and Boone. There is a very accessible proof, together with proofs of other undecidability results, such as the conjugacy problem, in the final chapter of Rotman (1995).

If the reduced word $w \in W$ can be expressed as a product of r conjugates of defining relators, then we can construct a **van Kampen diagram** for w with r bounded regions. This is a connected diagram drawn in the Euclidean plane with finitely many vertices and labelled directed edges, where the labels are elements of A. A directed edge with label x is regarded as being equivalent to an edge between the same two vertices in the reverse direction labelled x^{-1}. In the drawing, two edges intersect only at their endpoints, so the plane is divided by the diagram into regions, where exactly one of these regions is unbounded.

For each region, we can define boundary labels w by starting at some vertex on the boundary of that region, and proceeding around the boundary in one of the two possible directions, concatenating the labels on the edges of the boundary as we go, to form a word $w \in A^*$. The defining property of a van Kampen diagram for $w \in W$ is that w should be one of the labels for the unbounded region of the diagram, whereas each of the bounded (internal) regions must have an element of R as one of its labels. (Note that the different labels for a given region are all conjugates or inverses of each other, so if one of them is in W then so are all of them.)

See Chapter V, Section 1 of Lyndon and Schupp (1977) for a more detailed treatment of van Kampen diagrams and a proof of their existence for all reduced $w \in W$. As an example, Figure 1 shows a van Kampen diagram for the word a^4 in the group $\langle a, b \mid a^{-1}bab, a^2b^{-2}\rangle$.

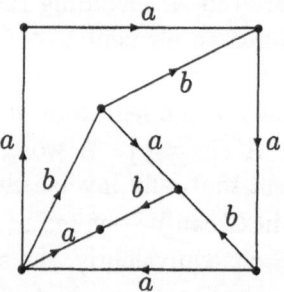

Figure 1: A van Kampen Diagram

For a given reduced word $w \in W$, let $\min_D(w)$ denote the smallest possible number of internal regions in a van Kampen diagram for w. The **Dehn function** of the presentation $\langle X \mid R\rangle$ is a function $D: \mathbb{N}_0 \to \mathbb{N}_0$ defined by putting $D(n)$ equal to the largest value of $\min_D(w)$ for reduced words $w \in W$ with $l(w) = n$. It can be shown that D is independent of the generating set X of G, at least up to linear equivalence of functions.

Consider, for example, the free abelian group of rank two defined by the presentation $\langle a, b \mid aba^{-1}b^{-1}\rangle$. For any $n > 0$, it is easy to construct a van Kampen diagram with n^2 bounded regions for the word $a^nb^na^{-n}b^{-n}$, and it is not hard to see that this gives rise to the largest possible value of $\min_D(w)$ for words w of length $4n$. Hence we have $D(4n) = n^2$ and similarly we find that $D(4n + 2) = n(n + 1)$, whereas $D(n) = 0$ for odd n, because all words in W have even length. Thus we can say that this group has quadratic Dehn function, and in fact this is the case for abelian groups in general.

In some sense, the Dehn function measures the complexity of W. Indeed, it is not difficult to see that membership in W is decidable if and only if D is a recursive function. However, this measure of complexity is misleading because the fact that the Dehn function is a quadratic function of n, for example, does not immediately imply that membership in W can be tested in quadratic time. In the worst conceivable case, we might need to enumerate all van Kampen diagrams having that number of regions. On the other hand, finitely generated infinite abelian groups do have quadratic Dehn functions, but membership in W can be tested in linear time.

During the mid 1980's there was increasing demand coming from geometers and algebraic topologists for practical algorithms for solving the word problem that could be implemented on a computer. Such algorithms exist for the classes of automatic and word-hyperbolic groups, which are to be discussed in Section B.14.

B.12 Combinatorial Group Theory

by Gilbert Baumslag in New York, NY, USA

Groups arise from many different sources and take many different forms. In particular, they arise in the study of algebraic equations as groups of permutations of the roots of such equations, in analysis as Fuchsian groups, in topology and geometry as fundamental groups and purely in abstract form as finitely presented groups. Many problems in diverse disciplines can be translated into questions about allied groups. The objective of combinatorial group theory is to develop techniques for answering these questions as well as to assemble a body of knowledge about many different kinds of groups (see also the remark below).

Three kinds of examples will help to explain these remarks:

The first goes back to Galois, who proved that the roots of a polynomial equation of the form

$$a_0 + a_1 x + \cdots + a_n x^n = 0,$$

where the coefficients a_i are integers, can be expressed in terms of radicals if and only if a certain group G of permutations of the roots of this equation is polycyclic. This means that G takes the form

$$1 = G_0 \trianglelefteq G_1 \trianglelefteq \cdots \trianglelefteq G_m = G,$$

where each of the subgroups G_i is a normal subgroup of the succeeding subgroup G_{i+1} and each of the factor groups G_{i+1}/G_i is cyclic.

The second kind of example has a number of different origins, in particular algebraic topology in the guise of fundamental group (Section H.1), the group of homotopy classes of loops in a space, with a given base point. Such a fundamental group, say G, in many instances, turns out to be finitely presented in the sense that it can be described by a finite set X of generators and a finite set R of so-called defining relators. This means first of all that every element of G can be expressed as a product of the elements of X and their inverses. Secondly, the elements of R, which are all given by expressions of the form $a_1 \ldots a_m$, where the a_i are either elements of X or inverses of elements of X, on computing their actual values in G, reduce to the identity element. Thirdly, everything about G can then be deduced from this information and the group laws. A typical question that arises is whether there is an algorithm which determines whether or not any loop in the given space is homotopic to the identity. This translates to a problem about the fundamental group G: is there an algorithm which determines whether or not any product of the form $a_1 \ldots a_m$ reduces to the identity? This problem is called the *word problem* for G (cf. Section B.11) and introduces a logical ingredient into group theory.

The third example, that of a group acting on a tree, serves to illustrate many diverse aspects of combinatorial group theory. It is worth recalling that a group *acts on a graph* if it comes equipped with a homomorphism into the group of all *automorphisms* of that graph. Here a one-to-one mapping of the set of vertices of a graph onto itself which preserves the graph structure is termed an automorphism. The particular case where the graph is a tree gives rise to many interesting kinds of groups, in particular free groups, amalgamated products, HNN extensions and various combinations of them.

Each of the examples above provides insights into what combinatorial group theory is all about. It is worth-while to reflect on them in turn.

A special case of a polycyclic group is that of a finitely generated abelian group. Finitely generated abelian groups can be decomposed into direct products of finitely many infinite cyclic groups and cyclic groups of prime-power order. The number and type of the cyclic groups involved completely determine these finitely generated abelian groups, i.e., provide one with a set of computable invariants that completely describe such groups. One part of combinatorial group is concerned with finding similar sets of invariants for other classes of groups, including, in particular the polycyclic ones.

The second example above leads to attempts not only to determine algorithmically whether certain classes of group have certain properties, but also to answer such questions in general. Among the kinds of questions that arise, the following are typical. Are all groups in a given class finite (e.g., is every finitely generated group all of whose elements have boundedly finite order finite; this is known as the **Burnside Problem** and the answer is in the negative)? Finitely generated? Finitely presented? What are the conjugates of a given element in a given group? What can one say about the subgroups of that group? Is there an algorithm for deciding for every pair of groups in a given class whether or not they are isomorphic? And so on.

Combinatorial group theory is the result of attempts to develop a body of algebraic techniques to settle such questions. In view of the scope of the subject and the extraordinary variety of groups involved, it is not surprising that no really general theory exists. Moreover, since almost every question about finitely presented groups has now been shown to be algorithmically undecidable, such attempts can never be completely successful.

The third example described above overlaps with the second. In particular, it involves various methods of decomposing groups into a combination of some of its subgroups, which provide some insight into the group itself. It also serves to bring to the fore other aspects of group theory which involve groups acting on special kinds of spaces, leading to such notions as hyperbolic and automatic groups (Section B.14). These hyperbolic groups have a close connection to hyperbolic geometry and their study has led to a deep and beautiful theory. The automatic groups are closely connected to computer science and their structure can only be described with aid of finite state automata. They give rise to a new and fascinating class of groups which has undergone much examination.

The foregoing remarks are designed to provide the reader with a glimpse into a remarkable world, filled with wonderfully complex objects, the domain of all of which is the realm of combinatorial group theory.

B.13 The Burnside Problems

by Rostislav I. Grigorchuk and Igor G. Lysenok in Moscow, Russia

A group G is said to be *periodic* (or *torsion*) if every element of G has finite order and *of exponent n* if G satisfies the identity $x^n = 1$. In 1902,

W. Burnside raised the question whether or not a finitely generated periodic group is finite. He emphasized the special case of the question when the group has exponent n and a fixed number m of generators, for given $m, n > 1$. Both questions are called *Burnside problems* and, to distinguish them, we use the terms the *unbounded Burnside problem* and the *bounded Burnside problem*, respectively.

In 1950, W. Magnus formulated the following problem: Given numbers m and n, does there exist a maximal (i.e. having the largest possible order) finite m-generated group of exponent n. He called it the *restricted Burnside problem*.

The unbounded Burnside problem was solved in negative by Golod in 1964 based on the construction of Golod–Shavarevich. The bounded Burnside problem turned out to be much more difficult to solve. The first examples of infinite finitely generated groups of exponent n were constructed in a series of papers by P. S. Novikov and S. I. Adian in 1968. This gave a negative solution to the bounded Burnside problem in general. An equivalent formulation of the problem is whether the free m-generated Burnside group $B(m, n) = \langle a_1, a_2, \ldots, a_m | X^n = 1 \rangle$ is finite. The following theorem summarizes all known cases of the positive solution of the bounded Burnside problem

B.13.1 Theorem (Burnside 1902, Sanov 1940, M. Hall 1957)
$B(m, n)$ is finite for $n = 2, 3, 4, 6$ and any m.

The book Adian (1979) contains an improved version of the Novikov–Adian method elaborated for the negative solution of the bounded Burnside problem. The result is known as the Novikov–Adian theorem:

B.13.2 Theorem *$B(m, n)$ is infinite for any odd $n \geq 665$ and $m \geq 2$.*

Observe that if $B(m, n)$ is infinite then $B(m, r)$ is also infinite for any r divisible by n, since $B(m, n)$ is a homomorphic image of $B(m, r)$ in this case. In particular, Theorem 2 implies the negative solution to the Burnside problem for any exponent r divisible by an odd number $n \geq 665$. The principal remaining case of 2-power exponent stayed open for a long time and was solved in papers by S. Ivanov Ivanov (1994) and I. Lysenok Lysënok (1996). The latter paper contains the best known bound for the case of even exponent:

B.13.3 Theorem *$B(m, n)$ is infinite for even $n \geq 8000$ and $m \geq 2$.*

The method developed by Novikov and Adian allows not only to prove the infiniteness of Burnside groups of sufficiently large exponents but also to study their other properties. For example, a description of all finite subgroups of groups $B(m,n)$ is known:

B.13.4 Theorem *Let G be a finite subgroup of $B(m,n)$.*

(1) *If n is odd and $n \geq 665$ then G is a cyclic group of order dividing n.*
(2) *If $n \geq 8000$ and n is divisible by 16 then G is a subgroup of a finite direct product $D_n \times D_r \times D_r \times \cdots \times D_r$ where D_k is the dihedral group of order $2k$ and r is the largest 2-power divisor of n.*

We can mention also that the groups $B(m,n)$ for $m \geq 2$ and n as in Theorem 4 have an exponential growth Adian (1979), Lysënok (1996) and, moreover, for odd $n \geq 665$ they are known to be non-amenable, see Adyan (1984).

The methods used to study Burnside groups turned out to be successful to construct numerous examples of groups with unusual properties which allowed to solve several difficult problems of the group theory. In 1980, A. Yu. Ol'shanskiĭ demonstrated that the geometric technique of Lyndon–van Kampen diagrams can be successfully applied to the study of the bounded Burnside problem and related questions. One of the strongest forms of the negative solution of the bounded Burnside problem is the following result which states existence of so called "Tarski monsters". It was first obtained by Ol'shanskiĭ for prime $n > 10^{70}$ (see Ol'shanskiĭ 1991).

B.13.5 Theorem (Ol'shanskiĭ, Atabekyan–Ivanov, Adian–Lysenok) *For any odd $n \geq 1003$, there exists an infinite 2-generated group all of whose proper subgroups are cyclic of order dividing n.*

An extensive study of the restricted Burnside problem by W. Magnus, O. Grün, H. Zassenhaus, R. Baer, I. N. Sanov, G. Higman, Ph.Hall and others showed that the problem for a primary exponent $n = p^k$ can be reduced to the question of local nilpotency of the Lie algebra over $\mathbb{F}_p$ with the Engel identity of degree $n - 1$

$$[\ldots [[x, y]\underbrace{y, \ldots, y}_{n-1}] = 0.$$

The affirmative answer to this question was given by A. I. Kostrikin (see Kostrikin (1986)) for $k = 1$, $p > 2$ and in the general case by E. I. Zelmanov in 1991 Zelmanov (1989, 1990b). Zelmanov's approach used an essential improvement of Kostrikin's sandwich method and the technique of Jordan algebras.

B.13.6 Theorem (Zelmanov) *For any n of the form p^k, p prime, and any m, there does exist a maximal m-generated finite group $R(m, n)$ of exponent n.*

In the case of non-primary exponent n, the positive solution of the restricted Burnside problem for primary exponents implies existence of the maximal *solvable* m-generated finite group of exponent n by the P. Hall–Higman theorem. Since a finite group of an odd exponent has an odd order (e.g. by the Sylow theorem) and any group of an odd order is solvable by the Faith–Thompson theorem, we have the following result.

B.13.7 Theorem (Zelmanov) *For any m and odd n, the group $R(m, n)$ exists. For any $m \geq 2$ and even non-primary n, the group $R(m, n)$ exists, provided the known list of finite simple groups is complete.*

Theorem B.13.6 and the Hall–Higman theorem imply also the positive solution of the restricted Burnside problem for *any* exponent n modulo the classification of finite simple groups.

After Golod's examples, series of finitely generated infinite periodic p-groups have been constructed by S. V. Aleshin, V. I. Sushchanskiĭ, R. I. Grigorchuk, Gupta–Sidki and other mathematicians, see Bartholdi et al. (2001), Gupta (1989). The simplest known examples of infinite finitely generated periodic groups are the Grigorchuk 3-generated 2-group and Gupta–Sidki infinite p-groups, $p \geq 3$. They can be defined as certain groups of automorphisms of the infinite rooted p-regular tree. All these groups are related to the so-called groups of finite automata, are residually finite and belong to one class of groups which is called *branch groups*. Constructing new types of finitely generated periodic groups is still an interesting and difficult problem of the group theory.

One of the most important open problems on periodic groups is the question of existence of *finitely presented* infinite periodic groups. The best known approximation to such a groups was given by examples by Grigorchuk (1998) and by Ol'shanskiĭ–Sapir (to appear). In the latter example, a finitely presented group G has an infinite normal subgroup H which is a finitely generated group of sufficiently large odd exponent n, with the infinite cyclic quotient G/H.

Due to the large popularity of the Burnside problems and the significance of periodic groups in algebra and applications, many other problems of the Burnside type have been appeared in different branches of mathematics. For example, there are analogues of the Burnside problems for monoids, rings and Lie algebras. For more information, the reader is reffered to books and surveys on the Burnside problem and related topics Adyan (1984), Ivanov and Ol'shanskiĭ (1991), Gupta (1989), Vaughan-Lee (1990)

B.14 Automatic and Hyperbolic Groups

by Derek Holt in Warwick, UK

Automatic groups

The definition of **automatic** groups arose from an observation by Thurston that some results of J. W. Cannon on discrete groups of isometries of hyperbolic groups (Cannon 1984) could be expressed in terms of finite state automata.

A groups is called **automatic** if there is a finite state automaton W (called the **word-acceptor**—see Section A.17) with alphabet A such that the language $L(W)$ of W maps onto G, and enjoys the following additional property. There is a finite state automaton M (called the **multiplier**) with alphabet $A \times A$ that accepts a pair of words $(w_1, w_2) \in A^* \times A^*$ if and only if $w_1, w_2 \in L(W)$ and either $w_1 =_G w_2$ or $w_1 =_G w_2 x$ for some $x \in A$.

A property equivalent to the existence of M is that there exists a constant K such that, for any pair of paths p_1, p_2 in the Cayley graph $\Gamma = \Gamma_X(G)$ of the presentation, we have $d_\Gamma(p_1(t), p_2(t)) \leq K$ for all t whenever p_1 and p_2 label words in L that start at the same vertex and finish at distance at most one apart in Γ (see Figure B.14). This is known as the **fellow-traveler** property, and plays a crucial rôle in the development of the theory.

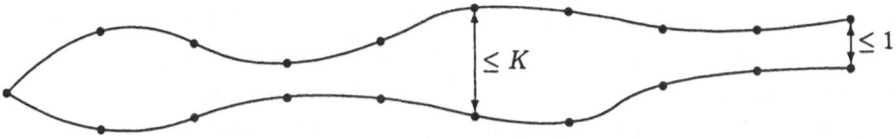

Figure 2: The fellow-traveler property

The book Epstein et al. (1992) is devoted to the theory of automatic groups. One important result (Theorem 2.4.1) is that the automaticity of a group does not depend on the choice of generating set X, so we are dealing with an algebraic as well as a geometric property of groups.

For many automatic groups, the automata W and M can be explicitly computed by using the author's software package KBMAG (Holt 1995). The algorithms involved are discussed in (Epstein et al. 1991). Automatic groups have quadratic (or sometimes linear) Dehn functions, but once W and M are known, the word problem can actually be solved in at worst quadratic time. It is then easy to decide finiteness of G, to enumerate unique words for each group element, and to compute the growth function of the group.

The picture in Figure 3, which is a frame from the video "Not Knot" (Gunn and Maxwell 1992), was produced with the help of the word-acceptor of the group

$$G_4 = \langle a, b, c, d, e, f \mid a^4 = b^4 = c^4 = d^4 = e^4 = f^4 =$$
$$aba^{-1}e = bcb^{-1}f = cdc^{-1}a = ded^{-1}b = efe^{-1}c = faf^{-1}d = 1 \rangle,$$

which is the symmetry group of the tessellation in the picture. Generating unique words for each group element enables the picture to be generated much more quickly and also more accurately.

An example of a group for which the finiteness had long been an open question and which was proved infinite by constructing the word-acceptor W is the Heineken group

$$G = \langle x, y, z \mid [x, [x, y]] = z, [y, [y, z]] = x, [z, [z, x]] = y \rangle,$$

where the commutator $[u, v]$ is defined to be $u^{-1}v^{-1}uv$. In fact this group turned out to be word-hyperbolic; see (Holt and Hurt 1999).

Many well-known types of finitely presented groups have been proved to be automatic. For example, groups satisfying the small cancellation hypotheses $C(6)$, $C(4) - T(4)$ or $C(3) - T(6)$ (see Section 2, Chapter 5 of Lyndon and Schupp (1977) for definitions) are proved automatic in (Gersten and Short 1991a). Other examples are word-hyperbolic groups (see below), finitely generated abelian groups, all Coxeter groups, Artin groups of finite type, Artin groups of extra large type, most (and probably all) knot groups, geometrically finite groups and mapping class groups. However, nilpotent groups are not automatic unless they are virtually abelian. Furthermore direct and free products of automatic groups are automatic, as are sub- and supergroups of finite index.

If we require the multiplier automaton M also to accept pairs of words (w_1, w_2) for which $w_1 =_G xw_2$ for some $x \in A$, then we get the more restricted class of **biautomatic** groups.

Figure 3: A tessellation of hyperbolic 3-space by dodecahedra

The conjugacy problem is solvable in biautomatic groups, but it is not known whether this is the case in all automatic groups. The theory of biautomatic groups is developed in Gersten and Short (1991b). There is currently no example of an automatic group that is known not to be biautomatic, and finding such an example or proving that there is none is probably the most challenging open problem in the area.

Several interesting generalizations of automatic groups have been studied. If the multiplier automaton M is allowed to read its two input words at different speeds, then we get the class of *asynchronous automatic* groups. The *Baumslag-Solitar group* groups $\langle x, y \mid y^{-1}x^m y = x^n \rangle$ are asynchronously automatic but only automatic when $m = \pm n$ (see Chapter 7 of Epstein et al. (1992)). If we relax the requirement for W to be the language of a finite state automaton, but keep the (synchronous or asynchronous) fellow-traveler property, then we get the class of *combable* groups. See Rees (1998) for a survey of combable groups; examples of non-automatic synchronously combable groups have been discovered recently by Bestvina and Brady and by Bridson. Groups that are automatic relative to a subgroup H of G have also been studied, and this sometimes allows finite presentations of subgroups with $|G : H|$ infinite to be computed (Holt and Hurt 1999). It is possible to define automatic monoids and semigroups, although we lose some of the nicest properties, such as the fellow-traveler property; see Duncan et al. (1999).

Hyperbolic groups

Historically, there have been several meanings attached to the term *hyperbolic* group, which would usually involve some kind of action on hyperbolic space or a negatively curved manifold. The current standard definition which follows is due to Gromov, and is stated in terms of the Cayley graph $\Gamma = \Gamma_X(G)$ of the group, although it turns out that hyperbolicity of a group is independent of the generating set chosen.

We call Γ *hyperbolic* if it has uniformly *slim* geodesic triangles. This means that there exists a constant δ such that, for any triangle in Γ whose sides a, b, c are geodesic paths, any point x on the side a satisfies $d_\Gamma(x, b \cup c) \le \delta$. Groups satisfying this condition are sometimes called *word-hyperbolic* to distinguish this meaning from earlier ones.

The best references for the general theory of hyperbolic groups are Alonso et al. (1991) and Ghys and de la Harpe (1990). They are automatic (indeed biautomatic), and the set of geodesic words $w \in A^*$ (that is, w such that $l(w) = l_G(w)$) can be chosen as the language of the word-acceptor. Examples include the *von Dyck groups* $\langle x, y \mid x^2, y^3, (xy)^n \rangle$ for $n \ge 7$.

A group presentation $G = \langle X \mid R \rangle$ is said to have a *Dehn algorithm* (see also Section B.11) if any word w in the word problem W can be reduced to the empty word by making a sequence of length-reducing substitutions $w_1 \to w_2$, for some subword w_1 of w, where $l(w_2) < l(w_1)$ and $w_1^{-1} w_2 \in R$. Dehn algorithms are studied in Section 4, Chapter 5 of Lyndon and Schupp (1977), where it is shown that presentations satisfying various small-cancellation hypotheses have Dehn algorithms. It turns out that any group having a Dehn algorithm is word-hyperbolic, and conversely, for any generating set X of a word-hyperbolic group G, there exists a finite set of defining relators R for G such that the resulting presentation has a Dehn algorithm. Furthermore, a group is hyperbolic if and only if its Dehn function is linear, and membership of the word problem W can be tested in linear time; indeed the Dehn algorithm provides such a test.

A significant result, which is proved in Papasoglu (1995), is that a group is hyperbolic under the weaker condition that geodesic bigons (which can be thought of as geodesic triangles in which two of the vertices are the same) are uniformly slim. This implies that a group is hyperbolic if and only if it is automatic with word acceptor W accepting the set of all geodesic words. This, in turn, makes it possible to prove computationally that a group is hyperbolic using the automatic groups programs. The Heineken group mentioned in Section B.14 was proved hyperbolic in this manner.

B.15 Actions of Finite Groups

by Peter Paule in Linz, Austria

Introduction

The development of powerful computers and computer algebra packages has stimulated renewed interest in definition, enumeration, and construction of certain unlabeled structures in mathematics and natural sciences. The structures in question are those which can be defined as equivalence classes on finite sets, and in particular on finite sets of mappings. Prominent examples are graphs, switching functions, physical states, and chemical isomers. The key to solving most of these problems is to replace the equivalence relation by a finite group action and to apply algebraic tools like the Cauchy-Frobenius Lemma and its refinements.

The whole theory is very rich and still expanding rapidly. It is linked to classical topics such as Pólya theory of enumeration, linear representations of the groups in question and in particular of symmetric groups (e.g., James and Kerber 1981), the theory of symmetric functions and generalizations like Schubert polynomials (e.g., Macdonald 1995), and the theory of partitions and q-hypergeometric series (e.g., Andrews 1976).

This list is by far not complete. But the book by (Kerber 1999), besides providing an elementary introduction, gives an excellent account on many of these relations and on recent developments in the field. Most of the material presented below is taken from this standard reference.

Basic facts and definitions

Let G denote a multiplicative group and X a non-empty set. An **action** of G on X (from the left) is described by a mapping

$$G \times X \to X, \ (g, x) \mapsto gx,$$

such that, for each $x \in X$ and any $g, g' \in G$, $g(g'x) = (gg')x$ and $1x = x$ hold. One says that G **acts** on X, or that X is a **G-set** (in short: $_G X$). In mathematics and physics, many objects of interest are defined as those which are invariant under certain group actions.

The mapping $\delta \colon G \to S_X$, $g \mapsto \overline{g}$, where $\overline{g} \colon x \mapsto gx$, is a homomorphism. It is called the **induced permutation representation** of the action G on X. We define $\overline{G} := \delta(G)$, and $G_X := \ker(\delta) := \{\, g \mid \forall x \in X \colon gx = x \,\}$ which is a normal subgroup of G. A trivial example is the natural action of S_X on X itself; here the induced permutation representation is the identity mapping. Further examples are given below.

An action G on X induces an equivalence relation on X, namely $x \sim x' \colon \iff \exists g \in G \colon x' = gx$. The equivalence classes $G(x) := \{\, gx \mid g \in G \,\}$ are called **orbits**. We denote by $G \backslash\backslash X := \{\, G(x) \mid x \in X \,\}$ the set of all orbits. Conversely, each equivalence relation on X is induced by a suitable group action G on X. Namely, take as G the subgroup of all elements from S_X that only permute elements which lie in the same equivalence class. In other words, *all the structures in mathematics and sciences that can be defined as equivalences classes on sets can be described as orbits of groups.*

There is a fundamental bijection between any fixed orbit $G(x)$ and the set of left cosets G/G_x of its **stabilizer** $G_x := \{\, g \in G \mid gx = x \,\}$ via $gx \leftrightarrow gG_x$. In particular, if G is finite, then $|G(x)| = |G|/|G_x|$. This implies the **Cauchy-Frobenius Lemma** (often called **Burnside Lemma**) for the number of orbits of a finite group G acting on a finite set X. There are two standard versions where the second one uses the fact that G can be partitioned into **conjugacy classes** $C^G(g) := \{\, hgh^{-1} \mid h \in G \,\}$:

$$|G \backslash\backslash X| = \frac{1}{|G|} \sum_{g \in G} |X_g| = \frac{1}{|G|} \sum_{g \in C} |C^G(g)| \, |X_g|, \qquad (1)$$

with **fixed point sets** $X_g := \{\, x \in X \mid gx = x \,\}$ and where C denotes a transversal of the conjugacy classes of G. Given a subgroup U of G, for fixed $x \in X$ the following bijection $U(gx) \leftrightarrow UgG_x$ between the orbits of U on $G(x)$ and the set $U \backslash G / G_x$ of (U, G_x)-double cosets is fundamental with respect to the construction of finite unlabeled structures.

Based on this bijection, and combining finite group action methods with the implementation of an improved version of the Lenstra-Lenstra-Lovasz (LLL) basis reduction algorithm, which transforms an arbitrary basis of a lattice in $\mathbb{R}^n$ into a reduced basis (see Winkler 1996)), the first 7- and 8-designs with small parameters were constructed. See Kerber (1999) for more details on this outstanding achievement and for construction techniques in general.

Symmetry classes of mappings

Many unlabeled structures such as graphs, incidence structures, and linear codes are **symmetry classes** of mappings, i.e., orbits of a group acting on a set Y^X, the set of all mappings from X to Y. For the sake of simplicity, we restrict ourselves to the case of **Pólya action**, namely: If G acts on X then we define an action of G on Y^X, Y being a set of indeterminates, by

$$G \times Y^X \to Y^X : (g, f) \mapsto f \circ \overline{g}^{-1}.$$

Example. Necklaces can be considered as colorings of the vertices of a regular n-gon in m colors, say. Thus labeled necklaces with n beads in at most m different colors taken from $Y := \{y_1, \ldots, y_m\}$ form the set of mappings $Y^{\underline{n}}$ where $\underline{n} = \{1, \ldots, n\}$. If we consider two labeled necklaces as equivalent if and only if one arises from the other by a cyclic shift, the set of unlabeled necklaces is nothing but $C_n \backslash\backslash Y^{\underline{n}}$, the set of orbits of the corresponding Pólya action with C_n being the cyclic group of order n.

If $\binom{n}{2}$ denotes the 2-element subsets of $\underline{n}$, then the set of mappings $\{y_0, y_1\}^{\binom{n}{2}}$ can be viewed as the set of simple labeled graphs over the vertex set $\underline{n}$. (One assigns y_1 or y_0 to $\{i, j\}$ depending on whether i and j are connected by an edge or not.) The symmetric group S_n acts on $\binom{n}{2}$ by $(\pi, \{i, j\}) \mapsto \{\pi i, \pi j\}$, therefore unlabeled graphs over the vertex set $\underline{n}$ are described by $S_n \backslash\backslash \{y_0, y_1\}^{\binom{n}{2}}$, the orbit set of the corresponding Pólya action.

For applying the Cauchy-Frobenius Lemma it is crucial to compute the cardinalities of the fixed point sets, a task which is straightforward in the case of Pólya actions. Namely, one has that $f \in \left(Y^X\right)_g$ if and only if f is constant on the cycles of $\overline{g}$. In general, if $\pi \in S_X$ let $a_k(\pi)$

be the number of cyclic factors of π of length k, and let $c(\pi)$ be their total number, i.e., $c(\pi) = a_1(\pi) + \cdots + a_{|X|}(\pi)$. With this notation we have for Pólya actions that $\left|(Y^X)_g\right| = |Y|^{a_1(\bar{g})+\cdots+a_{|X|}(\bar{g})} = |Y|^{c(\bar{g})}$, and thus

$$|G \setminus\setminus Y^X| = \frac{1}{|G|} \sum_{g \in G} |Y|^{c(\bar{g})} = \frac{1}{|G|} \sum_{g \in C} |C^G(g)| \, |Y|^{c(\bar{g})}.$$

Example. In the case of necklaces with n beads in at most m colors this implies

$$|C_n \setminus\setminus Y^{\underline{n}}| = \frac{1}{n} \sum_{g \in C} \left| C^{C_n}(g) \right| m^{c(\bar{g})} = \frac{1}{n} \sum_{d|n} \varphi(d) m^{n/d},$$

where φ stands for Euler's totient function. — For the number of simple unlabeled graphs on n vertices one obtains $\frac{1}{n!} \sum_{\pi \in S_n} 2^{c(\bar{\pi})}$, where

$$c(\bar{\pi}) = \frac{1}{2} \sum_k k \, a_k(\pi)^2 - \frac{1}{2} \sum_{k \text{ odd}} a_k(\pi) + \sum_{i<j} a_i(\pi) a_j(\pi) \gcd(i,j).$$

We define the **content** of $f \in Y^X$ as the mapping $\gamma(f,-): Y \to \mathbb{N}$, $y \mapsto |f^{-1}(y)|$, i.e., the multiplicity with which f takes on the value y. Since in the case of Pólya actions the content stays invariant on each orbit, i.e., $\gamma(f,-) = \gamma(gf,-)$ for all $g \in G$ and all $f \in Y^X$, we can enumerate orbits by given content. Namely, if $|G \setminus\setminus Y^X|_f$ denotes the number of orbits in $G \setminus\setminus Y^X$ which have the same content as a given $f \in Y^X$, then **Pólya's theorem**, a generating function version of (1), says that

$$|G \setminus\setminus Y^X|_f = \text{coefficient of } \prod_{y \in Y} y^{\gamma(f,y)} \text{ in } \frac{1}{|G|} \sum_{g \in G} \prod_{k=1}^{|X|} \left(\sum_{y \in Y} y^k \right)^{a_k(\bar{g})}. \quad (2)$$

Example. In the case of necklaces with $m = 2$ and $n = 5$ we obtain the generating function

$$\frac{1}{5} \left((y_1 + y_2)^5 + 4(y_1^5 + y_2^5) \right) = y_1^5 + y_1^4 y_2 + 2y_1^3 y_2^2 + 2y_1^2 y_2^3 + y_1 y_2^4 + y_2^5;$$

e.g., the summand $2y_1^3 y_2^2$ corresponds to the fact that there are exactly two different necklaces of 5 beads three of which with color y_1 and two of which with color y_2.

The polynomial $C(G,X) := \frac{1}{|G|} \sum_{g \in G} \prod_{k=1}^{|X|} z_k^{a_k(\bar{g})}$ is called **cycle indicator polynomial** or **cycle index** of $_G X$; e.g., $C(C_n, \underline{n}) =$

$\frac{1}{n}\sum_{d|n}\varphi(d)z_d^{n/d}$. The generating function in (2) is obtained from it by replacing each z_k by $\sum_{y\in Y}y^k$. More generally, this process of **Pólya substitution** is defined for any $p(y_1,\ldots,y_m)\in\mathbb{Q}[y_1,\ldots,y_m]$ by replacing each z_k in $C(G,X)$ by $p(y_1^k,\ldots,y_m^k)$; the resulting polynomial is denoted by $C(G,X\mid p(y_1,\ldots,y_m))$.

Pólya substitution establishes a connection between the algebraic information (cycle structure) and the combinatorial information (generating function). In books as (Kerber 1999) one finds tables of cycle indicator polynomials for specific groups together with formulae that describe how more complicated cycle indices can be built from more elementary ones; e.g., for finite actions $_GX$ and $_HY$, X and Y disjoint, we have $C(G\times H,X\cup Y)=C(G,X)\cdot C(H,Y)$. Many applications, in particular to chemistry, involve recursive methods; e.g., the generating function $a(x)$ for the number of alcohols satisfies the relation $a(x)=x\sum_{n=0}^{3}C(S_n,\underline{n}\mid a(x))$.

The whole theory has been generalized significantly. One important refinement is due to Burnside, namely the enumeration of orbits by stabilizer class. This is based on the fact that the elements of an orbit have as their stabilizers a full conjugacy class of subgroups of G, i.e., $\{\,G_{x'}\mid x'\in G(x)\,\}=\{\,gG_xg^{-1}\mid g\in G\,\}$ for any $x\in X$.

Finally, $C(G,X\mid\sum_{y\in Y}y)$ can be written as a linear combination of so-called Schur polynomials, which establishes a connection to representation theory.

B.16 Wreath Products

by John D. P.Meldrum in Edinburgh, UK

Two natural ways in which wreath products arise are in the context of imprimitive permutation groups and as the Sylow subgroups of permutation groups.

B.16.1 Definition Let (G,X) and (H,Y) be permutation groups. Write $Z=X\times Y$, $B=G^Y$, and define an action of B and H on Z by

$$(x,y)f=(xf(y),y),\quad(x,y)h=(x,yh),\quad\forall f\in B,h\in H.$$

Then $(B.H,Z)=:(G\operatorname{Wr}_YH,Z)$ is the *(complete) permutational wreath product* of (G,X) by (H,Y). If we replace B by the restricted power $G^{(Y)}$ of copies of G, we have the *restricted permutational wreath product* $(G\operatorname{wr}_YH,Z)$. Given two abstract groups K, L, we obtain the **standard** (**restricted** or **complete**) **wreath product** by using the right regular representation. In all cases, the group corresponding to B is called the **base group** and H (or L) the **top group**.

The most obvious natural context for these groups is as the Sylow subgroups of the finite symmetric groups S_n (Weir 1955).

B.16.2 Example If $n = p^m$, then the Sylow p-subgroups of Σ_n are iterated wreath products of m cyclic groups of order p. The explicit recursive construction uses the Sylow p-subgroup of $\Sigma_{n/p}$ and the element of order p given by $\left(1, p^{m-1} + 1, \ldots, (p-1)p^{m-1} + 1\right) \cdots \left(p^{m-1}, 2p^{m-1}, \ldots, p^m\right)$. If $n = a_k \ldots a_0$ to base p, $0 \leq a_i < p$, $0 \leq i \leq k$, then the Sylow p-subgroup of Σ_n is the direct product of the set of groups $\{P_{i,j} \mid 1 \leq j \leq a_i, 1 \leq i \leq k, a_i \neq 0\}$, where $P_{i,j}$ is isomorphic to a Sylow p-subgroup of Σ_{p^i}, hence is an iterated wreath product.

Some simple properties of wreath products relevant to the example follow.

B.16.3 Lemma
(1) *Permutational wreath products are associative, but standard wreath products are not.*
(2) *All wreath products of G by H are the split extension of the base group by the top group with the action given by $f^h(y) = f(yh^{-1})$.*

The fundamental link with imprimitivity is brought to light in the following result.

B.16.4 Theorem *Let G be an imprimitive permutation group on X with blocks of imprimitivity $\{X_\lambda\}_{\lambda \in \Lambda}$. Denote by $\theta \colon G \to K$ the representation of G as a permutation group on Λ and let H be the stabilizer of a fixed block X_ι. Then G can be embedded in a canonical way in $H \operatorname{Wr}_\Lambda K$ by a monomorphism φ such that $\varphi \pi_K = \theta$, where π_K is the projection $H \operatorname{Wr}_\Lambda K \to K$.*

The map φ is a permutation monomorphism, and can be defined fairly easily in the only way which can possibly work, starting from the correspondence between X and $X_\iota \times \Lambda$. Note that $\operatorname{Ker} \theta / \operatorname{Cor} H = \bigcap_{g \in G} H^g$. This leads to the Krasner-Kaluzhnin Theorem (Krasner and Kaloujnine 1951).

B.16.5 Theorem *Let G be a group, $N \lhd G$, $G/N \cong H$. Then there is a monomorphism $\theta\colon G \to N\operatorname{Wr}_H H = B.H$, where $G\theta\pi_H = G\nu_N$, and ν_N is the canonical homomorphism $G \to H$.*

One of the first general works on standard wreath products was by Neumann (1964). We give one result in this area, though not in its most advanced form.

B.16.6 Theorem *Let $G = A\operatorname{wr}_B B$ be a non-trivial wreath product. Then G is directly decomposable if and only if B is finite of order n and A has a non-trivial abelian direct factor C, say, which may be A itself, and C has unique nth roots for all its elements.*

In the process, a close examination of conjugacy in a wreath product and of normal subgroups, particularly those which project onto the top group, is made. In this same paper, the question of when the base group is characteristic was solved for standard wreath products.

B.16.7 Theorem *The base group in a standard wreath product, restricted or unrestricted, of a group A by a group B is characteristic except when B is a cyclic group of order 2 and A is a special dihedral group. In this case, the base group admits a subgroup of index 2 in the automorphism group of the wreath product.*

From here we can get results on the automorphism group of a wreath product.

B.16.8 Theorem *The wreath product with characteristic base group described in the previous theorem has an automorphism group which is the product of three subgroups: a subgroup isomorphic to $\operatorname{Aut} B$ acting naturally on B, a subgroup of automorphisms which act trivially on B, and which includes a copy of $\operatorname{Aut} A$ acting naturally on the copies of A, and a subgroup of automorphisms induced by conjugation by elements of the base group.*

A great deal of work has been done on the automorphisms of permutational wreath products. We next mention nilpotency: the question of which wreath products are nilpotent, and their nilpotency class, if nilpotent, has an interesting story of which we present the current state. For $A\operatorname{wr}_Y B$ to be nilpotent it is easy to see that the orbits of B must be finite, and not too difficult to reduce the case to B being transitive, and hence B finite (see Shield 1977).

B.16.9 Theorem *Let (B, Y) be finite and transitive. Then $A \operatorname{wr}_Y B$ is nilpotent if and only if for some prime p, A and B are p-groups and A is nilpotent of finite exponent.*

B.16.10 Definition For a group G, a prime p, define the K_p-*series* of G by

$$K_{i,p}(G) = \prod_{up^j \geq i} \gamma_u(G)^{p^j}$$

for all $i \geq 1$, where $\gamma_u(G)$ is the lower central series of G.

B.16.11 Theorem *Let p be a prime, A a p-group, nilpotent of class c, of finite exponent, B a finite p-group; let d be the maximal integer such that $K_{d,p}(B) \neq \{e\}$, for each v, $1 \leq v \leq d$, let $e(v) = \log_p |K_{v,p}(B)/K_{v+1,p}(B)|$, $a = 1 + (p-1)\sum_{v=1}^{d} v e(v)$, $b = (p-1)d$, $s(w) = \log_p(\text{exponent of } \gamma_w(A))$ for $1 \leq w \leq c$. Then the nilpotency class of $A \operatorname{wr}_B B$ is $\max_{1 \leq w \leq c}\{aw + (s(w) - 1)b\}$.*

Determining the nilpotency class of permutational wreath products is still in its early stages, and should provide a lot of fun.

Another early use of wreath products was in embedding theorems. Starting with Neumann and Neumann (1959) and going on with Hall (1962), one of the earliest is:

B.16.12 Theorem *Every countable group can be embedded in a two generator group.*

Wreath products offer an opportunity to combine groups with widely differing properties, or with unexpected properties.

B.16.13 Theorem *The wreath product $C_p \operatorname{wr} C_p$, C_p a cyclic group of order a prime p, is a minimal example of a non-regular p-group, and is a p-group of maximal nilpotency class (class p for a group of order p^{p+1}).*

One interesting connection is with Galois groups (Section D.3). If A and B are the Galois groups of the polynomials $f(x)$ and $g(x)$ respectively, then, in general, $A \operatorname{wr} B$ is the Galois group of $g(f(x))$. There has also been a number of papers written on the relationship between classes of groups and wreath products. A large number of papers referring to wreath products are in non-mathematical journals, where the key to the application is in the way wreath products arise as symmetry groups of certain structures.

There have been a number of generalizations of wreath products. The **verbal wreath product** (Shmel'kin 1964) is based on the restricted standard wreath product, when the base group is a verbal product instead of a direct product. It is of use in questions concerning varieties of groups. The **crown product** similarly replaces the base group by an amalgamated direct product. Neumann (1956, 1963) uses it for ascending derived series in a group. **Twisted wreath products** start from two groups A and B, a restricted direct power $A^{(Y)}$ of A, and an action of B on $A^{(Y)}$ which mixes an action of part of B as automorphisms of A, and an action of B on Y. The connections are mainly with representation theory and with finite soluble groups (Hawkes 1975)). There is also an interesting link between certain wreath products and dimension subgroups of group rings (Passi 1968).

The final generalizations we mention are those which involve a partially ordered set Λ and a permutation group (B_λ, X_λ) for each $\lambda \in \Lambda$, the normal case using a 2-element totally ordered set. The ensuing structure is somewhat complicated. It has been used to obtain certain simple groups which are only just simple (Hall 1974), a very general version of Theorem B.16.5 (Holland 1969), groups with prescribed lattices of normal subgroups (Silcock 1977) and much else. For a more complete account see (Meldrum 1995).

B.17 Automorphism Groups of Algebraic Systems

by Boris I. Plotkin in Jerusalem, Israel

The notion of the automorphism is one of the basic mathematical notions that can be applied to arbitrary algebraic, geometric and other structures.

In any category $\mathcal{C}$, for every object H there corresponds the group of automorphisms $\mathrm{Aut}(H)$ and the semigroup of endomorphisms $\mathrm{End}(H)$. The group $\mathrm{Aut}(H)$ consists of all invertible elements in $\mathrm{End}(H)$.

The notion of automorphism naturally formalizes the idea of symmetry. In this sense, the group $\mathrm{Aut}(H)$ determines to a great extent the structure of the given H. Galois theory brightly illuminates this matter. The same idea is illustrated by the theorem of Thompson below.

For the given object H, one can consider the structure of individual automorphisms and the structure of the whole group $\mathrm{Aut}(H)$. In the next sections we provide a glimpse on this topic.

Let us fix a variety Θ (see Section G.5) of algebras. The automorphisms of free Θ-algebras $W = W(X)$ are of special interest. Endomorphisms of such an algebra W are in one-to-one correspondence to arbitrary maps $X \to W$. The problem is to recognize those maps which lead to invertible endomorphisms, i.e., to automorphisms.

For free groups, this problem is solved in terms of Fox derivatives by Birman.

If Θ is the variety of semigroups then the problem to describe Aut W is solved trivially. In any free semigroup W there is a unique set of free generators and therefore all automorphisms of W are induced by permutations of X.

The picture for free groups is far from being so simple. Let $W = W(X)$ be a free group with a finite set X, $x \in X$, $Y = X \setminus x$. Consider automorphisms α_x, β_{xy}, $y \in Y$, defined by the rule: $\alpha_x(x) = x^{-1}$, $\beta_{xy}(x) = xy$, $\alpha_x(y_1) = \beta_{xy}(y_1) = y_1$ for every $y_1 \in Y$.

B.17.1 Theorem (Nielsen) *The group of all automorphisms* Aut(W) *is generated by all elementary automorphisms of the form α_x and β_{xy}.*

If $|X| = n$, then there is a canonical surjective homomorphism Aut$(W) \to \mathrm{GL}(n; \mathbb{Z})$ and the group $\mathrm{GL}(n; \mathbb{Z})$ is considered as the group of all automorphisms of the free abelian group of rank n. Nielsen, B. Neumann, Lyndon, Bestvina, Feighn, Halder, Sela, and others obtained results which look similar to some results for the general linear groups over a field. Most of these results have a geometric nature.

Let Θ be a variety of associative algebras with unity over a field P and $X = \{x_1, \ldots, x_n\}$ a finite set. Free Θ-algebras (Section G.4) with a finite number of generators have the form $P\langle X \rangle$, i.e., they are polynomial algebras with non-commuting variables. Maps σ of the type

$$x_i \to a_{i1}x_1 + \cdots + a_{in}x_n + b_i$$

where the invertible matrix of coefficients determines a linear automorphism σ (cf. Section C.10) of the algebra $P\langle X \rangle$.

There are also triangular automorphisms σ having the form

$$\sigma(x_1) = x_1 + f(x_2, \ldots, x_n), \qquad \sigma(x_2) = x_2, \quad \cdots, \quad \sigma(x_n) = x_n.$$

B.17.2 Definition An automorphism σ of $P\langle X \rangle$ is called **tame** if σ is either linear or triangular or a composition of such maps (Section C.10).

B.17.3 Conjecture (Cohn) *All automorphisms of $P\langle X \rangle$ are tame.*

This conjecture has been proved for $n = 1, 2$ by Makar-Limanov and Czerniakiewicz. The corresponding result for the algebra of commutative polynomials has been known (Jung, Nagata, Van der Kulk) long ago. Cohn proved this conjecture for free Lie algebras for arbitrary n (for details, see Cohn 1985).

One can consider various types of automorphisms. In semigroups, groups, and associative algebras, inner automorphisms play a special role. Let, for example, H be a semigroup and g an invertible element from H. To g we have an inner automorphism $\hat{g} \in \text{Aut}(H)$ defined by the rule: $\hat{g}(h) = ghg^{-1}$, $h \in H$. For every $\varphi \in \text{Aut}(H)$ we have $\varphi \hat{g} \varphi^{-1} = \widehat{\varphi(g)}$. This means that the group of inner automorphisms $\text{Inn}(H)$ is an invariant subgroup in the group $\text{Aut}(H)$. The factor group $\text{Aut}(H)/\text{Inn}(H) = \text{Out}(H)$ is called the group of **outer automorphisms** of H.

B.17.4 Conjecture *For every finite simple group, the group of outer automorphisms* $\text{Out}(H)$ *is solvable.*

B.17.5 Definition An automorphism σ of the group H is called **regular** if the only fixed point under the action of σ is 1.

B.17.6 Theorem (Thompson) *Every finite group having a regular automorphism of prime order is nilpotent.*

Consider a tower of groups of automorphisms. Its base consists of an algebra H. Then we take the groups

$$\text{Aut}(H), \text{Aut}(\text{Aut}(H)) = \text{Aut}^2(H), \ldots, \text{Aut}^n(H), \ldots \quad (1)$$

and so on, where $\text{Aut}^n(H) = \text{Aut}(\text{Aut}^{n-1}(H))$. The minimal n such that every automorphism of $\text{Aut}^n(H)$ is inner determines the height of the tower. It is clear that such an n does not necessarily exist.

B.17.7 Problem What can be said about a tower of groups of automorphisms if its base is a free associative algebra or free module over a ring, respectively?

For free semigroups $W = W(X)$, the problem is solved easily. The group $\text{Aut}(W)$ is isomorphic to the symmetric group S_X. All automorphisms of S_X are inner if $|X| \neq 6$ and $|X| \geq 3$ (Helder).

This problem for free groups is also solved.

B.17.8 Theorem (Dyer, Formanek) *The group* Aut(W) *is perfect if* $|X| > 1$, *(i.e., without center and all automorphisms of* Aut(W) *are inner).*

B.17.9 Theorem (Wielandt) *Let H be a finite group without center,* ord(H) $= b$. *Then the tower with the base H has a finite height. Moreover, if a is the order of the upper floor, then*

$$\log a < 3\frac{(\log b)^3}{(\log 2)^3}$$

For details, see (Cohn 1985), (Lyndon and Schupp 1977), and (Plotkin 1966).

Let Θ be the variety of modules over a ring R, $X = \{x_1, \dots, x_n\}$, and RX the free module. The group Aut(RX) is isomorphic to the general matrix group $GL_n(R)$. Automorphisms of this group and other linear groups over a different R are subjects of numerous papers. If R is a field, and in many other cases, the automorphisms of GL($n; R$) and other linear groups are described. See (O'Meara 1974) and (Merzljakov 1976).

A significant role in this theory is played by the notion of the semiinner automorphism. Like inner automorphisms of Aut(RX) are determined by automorphisms of the module RX, the semiinner automorphisms are determined by semiautomorphisms of RX. A *semiautomorphism* of any R-module H is a pair (σ, μ), where σ is an automorphism of the ring R, μ is an automorphism of the abelian group H and $\mu(\lambda a) = \sigma(\lambda)\mu(a)$, where $\lambda \in R$ and $a \in H$.

For $\sigma \in$ Aut(R), there is a semiautomorphism $(\sigma, \bar{\sigma})$ of $H = RX$ via

$$\bar{\sigma}(\lambda_1 x_1 + \cdots + \lambda_n x_n) = \sigma(\lambda_1)x_1 + \cdots + \sigma(\lambda_n)x_n.$$

Every semiautomorphism (σ, μ) of RX has the form $(\sigma, \mu) = (\sigma, \bar{\sigma})(1, \bar{\sigma}^{-1}\mu)$, where $(1, \bar{\sigma}^{-1}\mu)$ is an automorphism of RX. For every R-module H, the subgroup of all semiinner automorphisms is contained in the group Aut(H). This subgroup is not necessarily invariant in *Aut(H)*. Details and other information can be found in (O'Meara 1974) and (Merzljakov 1976).

B.18 Frobenius Groups

by Peter Fleischmann in Kent, UK

A finite transitive permutation group $1 < G \leq S_\Omega$ is called a **Frobenius group**, if it is not regular but no element different from the identity fixes more than one point (see Section B.5). These groups are named after Ferdinand Georg Frobenius (1849–1917), a German mathematician, who first defined them and investigated their structure.

The smallest example of a Frobenius group is the group S_3 of all permutations of three letters, where clearly any permutation fixing two letters also fixes the third. Moreover, in this group there are exactly two elements with no fixed points; together with the identity, these form a normal subgroup of order three. The group S_2 is not a Frobenius group, because it is regular and S_n for $n \geq 4$ is not a Frobenius group, because it contains nontrivial elements (e.g., the transposition (12)) fixing more than one point.

Identifying the transitive G-action on Ω with the permutation action given by multiplication on the set of cosets of a point stabilizer $H := G_\alpha$ with $\alpha \in \Omega$, one obtains the following purely group theoretic definition of Frobenius groups:

B.18.1 Definition The abstract finite group G is a **Frobenius group**, if it has a subgroup $1 < H < G$ satisfying $H \cap g^{-1}Hg = 1$ for all $g \in G \setminus H$.

In this case the group H is called a **Frobenius complement** and the set $K := G \setminus \biguplus_{g \in G} (g^{-1}Hg \setminus \{1\})$ is called a **Frobenius kernel** of G. In the setting of permutation groups, K is just the set of fixed point free elements in G, together with the identity. Immediately from the definitions one can see that $|K| = |\Omega| = |G : H| \equiv 1 \bmod |H|$. Moreover K is easily seen to be a normal set, i.e., $K^g := g^{-1}Kg = K$ for each $g \in G$. The surprising fact is, that K is indeed a group, and then of course a normal subgroup of G. This is the content of a famous theorem of Frobenius, whose only known proofs in full generality require methods from the theory of group characters (see Section E.2):

B.18.2 Theorem (Frobenius) *Let* $G \leq S_\Omega$ *be transitive and* $G_{\alpha,\beta} = 1$ *for all* $\alpha \neq \beta \in \Omega$. *Then* $K := G \setminus \bigcup_{g \in G} g^{-1}(G_\alpha \setminus \{1\})g$ *is the unique regular normal subgroup of* G.

This is in fact a special case of a more general result:

B.18.3 Theorem (Wielandt) *Let X be a finite group, $H^* \lhd H < X$, such that $H \cap x^{-1}Hx \le H^*$ for each $x \in X \setminus H$. Let*

$$X^* := X \setminus \bigcup_{x \in X} x^{-1}(H \setminus H^*)x.$$

Then X^* is a normal subgroup of X such that $X^* \cap H = H^*$ and $X^*H = X$.

Frobenius' theorem implies that G is a semidirect product of K and H. Moreover the set $G^\sharp$ of nontrivial elements can be written as a disjoint union

$$G^\sharp = K^\sharp \biguplus_{g \in G} (g^{-1}Hg)^\sharp$$

which is called a **Frobenius partition** of G and is unique for given G. In fact:

B.18.4 Theorem *Any two Frobenius complements are conjugate.*

B.18.5 Example Let $\mathbb{F}_q$ be a finite field of order q; the multiplicative group $\mathbb{F}_q^*$ acts on the additive group $\mathbb{F}_q^+$ by multiplication. Then the corresponding semidirect product $G_q := \mathbb{F}_q^+.\mathbb{F}_q^*$ is a Frobenius group with kernel $K := \mathbb{F}_q^+$ and complement $H := \mathbb{F}_q^*$.

The structure of Frobenius kernels is quite restricted: conjugation by any nontrivial element $h \in H$ induces a 'fixed point free' automorphism on K, i.e., an automorphism without fixed points other than the identity. Taking $h \in H$ to be of prime order, one can apply a theorem of Thompson, stating that any finite group admitting a fixed point free automorphism of prime order is nilpotent. This shows:

B.18.6 Theorem (Thompson) *Every Frobenius kernel is nilpotent.*

Frobenius complements are also far from arbitrary: let E be a minimal characteristic subgroup of K, then E is elementary abelian and can be viewed as a finite dimensional vector space V over a finite prime field, on which H acts linearly without nontrivial fixed points. All such linear groups have been described explicitly by Zassenhaus in the context of his classification of finite near fields (see Section C.47). In particular, any such perfect group is isomorphic to the special linear group $SL_2(5)$ (see (Zassenhaus 1939) or (Huppert and Blackburn 1982)).

The main structural result on Frobenius groups can be summarized as follows (see (Huppert 1983) V 8.18; see also (Passman 1968), (Gorenstein 1980)):

B.18.7 Theorem *Let G be a Frobenius group with kernel K and complement H. Then K is a nilpotent normal subgroup and G is a semidirect product of K and H. Moreover, the following hold:*

(1) *If H has even order, it contains a unique element s of order two. The kernel K is abelian and $k^s = k^{-1}$ for all $k \in K$.*

(2) *If H has odd order, it is metacyclic, i.e., it has a cyclic normal subgroup with cyclic factor group. Every $h \in H$ of prime order generates a normal subgroup of H, which is contained in the center $Z(H)$ or in the derived subgroup H'.*

(3) *In any case all Sylow groups of H are cyclic or generalized quaternion groups and the center $Z(H)$ of H is nontrivial.*

Related structures and generalizations

Frobenius groups are an important tool in the classification of sharply doubly transitive permutation groups. Let $Z \leq S_\Omega$ be a doubly transitive permutation group, such that every non-identity element has at most two fixed points. If moreover Z has a regular normal subgroup, then, due to a theorem of W. Feit, Z is either a Frobenius group or there is a bijection between Ω and the finite field $\mathbb{F}_{2^p}$ and Z is the full group of semilinear mappings $\lambda \mapsto a(\lambda^{2^i}) + b$ with $0 \neq a, b \in \mathbb{F}_{2^p}$. In the latter case, Z is solvable of order $2^p(2^p - 1)p$.

If Z has no regular normal subgroup, it is called a **Zassenhaus group**, named after Hans Zassenhaus, who initiated their investigation. Assume that Z is such a Zassenhaus group; in order to avoid repeating the case of Frobenius groups, we also assume that Z contains a nontrivial element with two fixed points. Then the one point stabilizers are (conjugate) Frobenius groups. Examples of Zassenhaus groups are $\mathrm{PGL}(2, p^f)$ for $p^f > 3$, $\mathrm{PSL}(2, p^f)$ for $p > 2$ and $p^f > 3$ and certain extensions $M(p^f) := \mathrm{PSL}(2, p^f).C_2$ in case $p > 2$ and f even. For a long time these were all the known Zassenhaus groups, until in 1960 Suzuki found a new infinite series of simple groups in this class, now called **Suzuki groups** and denoted by $\mathrm{Sz}(2^{2n+1})$ with $n = 1, 2, \ldots$. Together with the earlier examples, this completes the list of all Zassenhaus groups (see (Huppert and Blackburn 1982)).

A different generalization of Frobenius groups leads to so called *p-local*

Frobenius groups for a fixed prime p. They are defined as groups G containing a subgroup X such that $X \cap X^g$ is a p-group for each $g \in G \backslash X$. Unlike the class of Frobenius groups, the p-local class contains finite simple groups like $GL_3(2) \cong SL_2(7)$ with $X \cong S_4$ and $p = 2$. The list of all *primitive* p-local Frobenius groups has recently been completed with the help of the classification of finite simple groups (Lempken et al. 1998).

B.19 Covered and Fibered Groups

by Carlton J. Maxson in College Station, TX, USA

Basic concepts and results

A family $\mathscr{F} = \{ G_i \mid i \in I \}$ of nontrivial proper subgroups of a group $G = (G, +)$ is called *cover* of G if $\bigcup_{i \in I} G_i = G$. If no G_i is contained in another G_j, $\mathscr{F}$ is called a *geometric cover*. If, moreover, $G_i \cap G_j = \{0\}$ for $i \neq j$, then $\mathscr{F}$ is called a *fibration* or *partition* of G. A partition with $|G_i| = |G_j|$, for all $i, j \in I$, is called an *equal partition* and if $G_i + G_j = G$, for all $i \neq j$, then $\mathscr{F}$ is called a *congruence partition*.

The symmetric group S_3 is partitioned by its proper subgroups. It is easy to see that no group can be covered by two subgroups. Klein's 4-group, however, has an equal partition by its three subgroups. More generally, if G is the vector space F^n, F a field, then the collection of all lines through the origin forms an equal, congruence partition. On the other hand, the set of planes through $(0, 0, 0)$ in F^3 is a geometric cover but not a fibration.

Perhaps the first investigation on group partitions was that of Young (1927), who found that if a finite abelian group A has a nontrivial fibration, then A must be an elementary abelian p-group. Related to this is the fact that any p-group G of exponent p and with $|G| > p$ is equally partitioned by its cyclic subgroups. In fact, Isaacs (1973b) has determined that these are the only finite groups which can be equally partitioned.

B.19.1 Theorem (Isaacs 1973b) *Let G be a finite, equally partitioned group. Then G is a p-group of exponent p.*

There is a nice relationship between partitions of elementary abelian p-groups and certain codes.

B.19.2 Theorem (Herzog and Schønheim 1971) *Let $V_n(q)$ be an n-dimensional vector space over a finite field with q elements. Let $\{V_1, V_2, \ldots, V_k\}$ be a partition of $V_n(q)$ and let $W := V_1 \times V_2 \times \cdots \times V_k$. Then the kernel K of the linear transformation $T: W \to V_n(q)$, $T(v_1, \ldots, v_k) = v_1 + \cdots + v_k$, $v_i \in V_i$, is a single-error correcting perfect mixed linear code in W.*

Starting around 1940 and continuing for about 25 years, there were several investigations into the structure of partitioned groups. Kontorovitch, in a series of papers (Kontorovich 1939, 1946), initiated these investigations, which were continued by Baer, Kegel, and Suzuki. The objective here was to characterize (mainly finite) groups having a certain property and having a nontrivial partition. We refer the reader to the references in (Iwahori and Kondo 1965) for several of these results.

In (Iwahori and Kondo 1965), a criterion is given for the existence of a nontrivial partition of a finite group in terms of the existence of a certain permutation representation. They define a finite group G to be of **positive type** if there exists a positive integer ℓ and a set M on which G acts as a transformation group such that (i) every element $\sigma \in G \setminus \{0\}$ has exactly ℓ fixed points in M, and (ii) no point in M is fixed by all elements in G.

B.19.3 Theorem (Iwahori and Kondo 1965) *A finite group G has a nontrivial partition if and only if G is of positive type.*

B.19.4 Corollary *Let G be a finite group with no nontrivial abelian normal subgroups. Then G has a nontrivial partition if and only if G is isomorphic to $\mathrm{PGL}(2; q)$ or $\mathrm{PSL}(2; q)$ or $\mathrm{SZ}(q)$ (cf. B.8).*

Applications to geometry

We previously gave some examples of covers and partitions using vector spaces. These examples are typical since we now show that, in a certain sense, covers and fibrations come from geometry.

B.19.5 Definition A triple $(P, \mathscr{L}, \|)$ where P is a set of points, $\mathscr{L}$ a set of subsets (called *lines*) of P and $\|$ an equivalence relation (parallel) on $\mathscr{L}$ is called a **generalized translation structure** if

- Every 2 points in P are contained in at least one line of $\mathscr{L}$.
- There are at least 2 different lines and each line has at least 2 points.
- Euclid's parallel axiom holds.

- There exists an injective map Φ from P into the collineations of $(P, \mathscr{L}, \|)$ such that $\Phi(P)$ is a group acting transitively on P.

Let $(P, \mathscr{L}, \|)$ be a generalized translation structure. If (1) in the above definition is replaced by (1'): Every 2 different points are contained in exactly one line of $\mathscr{L}$, then $(P, \mathscr{L}, \|)$ is called a **translation structure**. If in addition, all lines have the same cardinality, then $(P, \mathscr{L}, \|)$ is called a **Sperner Space**.

We now relate group covers and geometry. This idea was introduced by André (1961).

B.19.6 Theorem (Maxson 1985) *If $(P, \mathscr{L}, \|)$ is a generalized translation structure then an addition, $+$, can be defined on P so that $(P, +)$ is a group with a geometric cover. Conversely, if $(G, +)$ is a group with a geometric cover $\mathscr{C} = \{ G_i \mid i \in I \}$, then by taking $\mathscr{L} := \{ x + G_i \mid x \in G, G_i \in \mathscr{C} \}$ and defining $(x + G_i) \parallel (y + G_j)$ if and only if $i = j$, one finds that $(G, \mathscr{L}, \|)$ is a generalized translation structure.*

From this one gets the following *dictionary* between geometry and covered groups:

Generalized translation structure	$\longleftrightarrow$	Geometric cover
Translation Structure	$\longleftrightarrow$	Fibration
Sperner Space	$\longleftrightarrow$	Equal fibration
Translation Plane	$\longleftrightarrow$	Equal congruence fibration

We next point out how near-rings can be associated with group coverings. Suppose $\mathscr{F}$ is a geometric cover of G. Let $\mathrm{End}(G, \mathscr{F}) := \{ \sigma \in \mathrm{End}\, G \mid \sigma(G_i) \subseteq G_i, \forall i \in I \}$. If G is abelian, $\mathrm{End}(G, \mathscr{F})$ is a subring of the ring $\mathrm{End}\, G$ of endomorphisms of G. If $\mathscr{F}$ is a partition of G, $\mathrm{End}(G, \mathscr{F})$ is an integral domain, and if G is finite, $\mathrm{End}(G, \mathscr{F})$ is a field. Geometers have used these fields to coordinate translation structures (Dembowski 1968).

Suppose G is not abelian. We consider the distributively generated near-ring $\mathrm{E}(G, \mathscr{F}) := \{ \sum \pm h_j \mid h_j \in \mathrm{End}(G, \mathscr{F}) \}$. The next result is rather surprising.

B.19.7 Theorem (Maxson and Pilz 1985) *Let $\mathscr{F}$ be a fibration of the finite group G. Then $\mathrm{E}(G, \mathscr{F})$ is always a ring. Either $\mathrm{E}(G, \mathscr{F}) \simeq \mathbb{Z}_n$ for some $n \in \mathbb{N}$, or $\mathrm{E}(G, \mathscr{F})$ is a finite field. In the latter case, if $p = \mathrm{char}(\mathrm{E}(G, \mathscr{F}))$, then $pG = \{0\}$ and G is nilpotent of class ≤ 2.*

The geometric significance of this result has yet to be explored.

We conclude with a result for arbitrary covers.

B.19.8 Theorem (Maxson and Pilz 1989) *Let* $(G, \mathscr{F})$ *be a finite covered group. Then* $\mathrm{E}(G, \mathscr{F})$ *is a field if and only if the lattice*

$$\mathrm{Lat}(G, \mathscr{F}) := \{ H \leq G \mid f(H) \subseteq H, \, \forall f \in \mathrm{E}(G, \mathscr{F}) \}$$

contains a fibration and G *is of prime exponent.*

B.20 Subgroup Lattices of Finite Abelian Groups

by Frank Vogt in Darmstadt, Germany

The subgroup lattices of finite abelian groups have been studied from various points of view (cf. Vogt 1995). For subgroup lattices of general groups, see (Schmidt 1994). Further interest in the abelian case gave rise to several counting formulas for the number of subgroups of a finite abelian group, some of which involve the structure of subgroup lattices and some of which do not. See also Section A.14.

Subgroup lattices of finite abelian groups may be understood as glued together by subgroup lattices of smaller abelian groups. This report focuses on this recursive structure of subgroup lattices and on the derivation of recursive counting formulas from that structure. Proofs and details are omitted, but they can be found in (Vogt 1995), where also an overview is given about properties of subgroup lattices and counting formulas for the number of subgroups of finite abelian groups.

The recursive structure of a subgroup lattice

The analysis of the recursive structure of the subgroup lattice is based on tolerance relations, which are generalizations of congruence relations: A binary relation on a finite lattice is called a **tolerance relation** if it is reflexive, symmetric, and compatible with the lattice operations. The notion of tolerance relations is not only interesting for finite lattices but also for infinite (complete) lattices (see Wille 1985). A tolerance relation may be viewed as a congruence relation which lacks transitivity. Hence, the *classes* of a tolerance relation are not necessarily disjoint but may overlap. Nevertheless, they still form a lattice.

B.20.1 Theorem *Let Θ be a tolerance relation on a finite lattice $\mathcal{L}$. Then the maximal subsets S of L with $x \Theta y$ for all $x, y \in S$ are intervals of $\mathcal{L}$, which are called the **blocks** of Θ. Ordered by their least elements, the blocks of Θ form a lattice, the **quotient lattice** $\mathcal{L}/\Theta$.*

*For every finite lattice $\mathcal{L}$, there is a smallest tolerance relation $\Sigma(\mathcal{L})$ that contains all covering pairs $x \prec y$ of $\mathcal{L}$. The quotient lattice $\mathcal{L}/\Sigma(\mathcal{L})$ is called the **skeleton** of $\mathcal{L}$, denoted by $\mathbf{S}(\mathcal{L})$.*

A remark on the usefulness of overlapping blocks of tolerance relations seems appropriate: Assume that you have to reconstruct a lattice $\mathcal{L}$ from the knowledge of the classes and the quotient lattice of some congruence Ψ of $\mathcal{L}$. Your problem would be that without additional information you will not be able to determine how the elements of two classes of Ψ compare within the lattice $\mathcal{L}$ itself. The situation becomes easier if you start with a tolerance relation Θ of $\mathcal{L}$ which has overlapping blocks. These overlappings provide precisely the information you need to glue the blocks together to obtain $\mathcal{L}$ (see the example in Figure 4). Compare this situation to a road atlas: Usually, the different maps in the atlas overlap, making it much easier to *glue* them together to a large map. The skeleton tolerance $\Sigma(\mathcal{L})$ is the best tolerance for this purpose because the overlappings are minimal. Hence, all necessary but no unnecessary information is provided.

Fundamental for the structure of the subgroup lattice is the structure of the group itself: The well-known basis theorem for finite abelian groups states that every finite abelian group is isomorphic to a direct product of cyclic groups of prime-power order (see Cohn 1982). It is easy to see that for different primes, the direct product structure of the group extends to the subgroup lattice. Hence, it is enough to consider the primary components of a finite abelian group.

B.20.2 Theorem *Let $\mathcal{G}$ be a finite abelian p-group isomorphic to the direct product $C_{p^{k_1}} \times \cdots \times C_{p^{k_m}} \times C_p^{n-m}$, p prime, $0 \leq m \leq n$ and $k_i \geq 2$ for $i = 1, \ldots, m$.*

(1) *The skeleton $\mathbf{S}(\mathrm{Sub}(\mathcal{G}))$ is isomorphic to $\mathrm{Sub}(C_{p^{k_1}-1} \times \cdots \times C_{p^{k_n}-1})$.*
(2) *All blocks of $\Sigma(\mathrm{Sub}(\mathcal{G}))$ are isomorphic to $\mathrm{Sub}(C_p^n)$.*
(3) *For two blocks $B_1, B_2 \in \Sigma(\mathrm{Sub}(\mathcal{G}))$ with $B_1 \leq B_2$, let l be the length of the interval $[B_1, B_2]$ in $\mathbf{S}(\mathrm{Sub}(\mathcal{G}))$. Then $B_1 \cap B_2$ is isomorphic to $\mathrm{Sub}(C_p^{n-l})$ if $(B_1, B_2) \in \Sigma(\mathbf{S}(\mathrm{Sub}(\mathcal{G})))$, and empty otherwise.*

Essentially, this Theorem says that the skeleton as well as the blocks of the skeleton tolerance of the subgroup lattice of the finite abelian

p-group $\mathcal{G}$ are subgroup lattices of smaller abelian p-groups. The dimension n of $\mathcal{G}$ determines the blocks (2), and the nontrivial exponents k_i determine the skeleton (1). Observe that the dimension of the *skeleton group* $C_{p^{k_1}-1} \times \cdots \times C_{p^{k_m}-1}$ may be less than n. Part (3) gives detailed information about the structure of the overlappings of blocks.

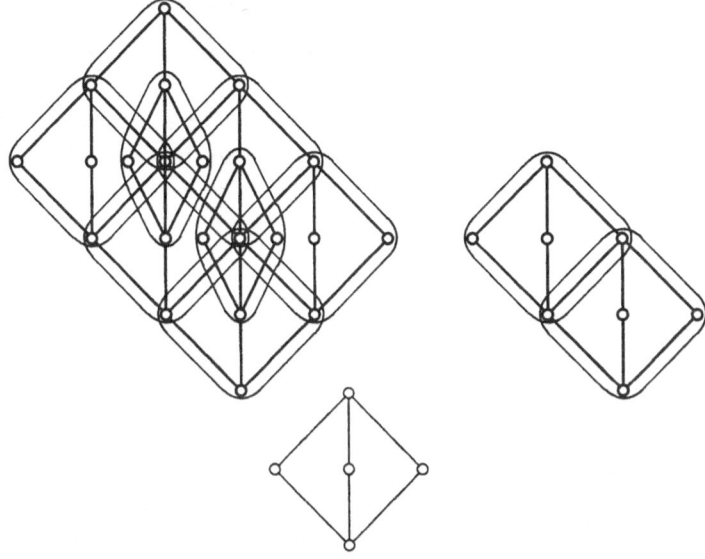

Figure 4: The subgroup lattices of $C_8 \times C_4$, $C_4 \times C_2$, and $C_2 \times C_2$

The example in Figure 4 shows the subgroup lattice of $C_8 \times C_4$ (left). The blocks of its skeleton tolerance are framed: They are obviously isomorphic to the subgroup lattice of $C_2 \times C_2$ (right). The skeleton, i.e., the set of blocks with the induced order, is isomorphic to $\mathrm{Sub}(C_4 \times C_2)$ (center). The skeleton itself can be decomposed similarly: The blocks of its skeleton tolerance are again isomorphic to $\mathrm{Sub}(C_2 \times C_2)$, and its skeleton is isomorphic to $\mathrm{Sub}(C_2)$, the two-element chain.

The number of subgroups

In (Wille 1985) and (Vogt 1995), counting formulas for elements and subsets of a lattice, respectively, are introduced which are based on tolerance relations and Möbius functions. The formula from (Vogt 1995), applied to Theorem B.20.2, yields a recursive formula for the number of subgroups of a finite abelian group. Again, it is sufficient to consider p-groups.

In order to state the formula, we have to introduce the **Gaussian polynomials** (cf. Stanley 1986): For $j \geq 0$, let $(j)_p := \frac{p^j - 1}{p - 1}$ and $(j)!_p := (1)_p \cdots (j)_p$. Then the Gaussian polynomials are defined by

$$\binom{n}{k}_p := \frac{(n)!_p}{(k)!_p \cdot (n - k)!_p}$$

for all $0 \leq k \leq n$. Now, the counting formula can be stated as follows:

B.20.3 Theorem *Let $\mathcal{G}$ be as in Theorem B.20.2. For $0 \leq l \leq n$, let $a_l(\mathcal{G})$ denote the number of atomistic intervals in $\mathrm{Sub}(\mathcal{G})$ of length l. Then*

1. $a_l(\mathcal{G}) = \sum_{j=l}^{n} \binom{n}{j}_p \cdot \binom{j}{l}_p$ if $m = 0$, and

2. $a_l(\mathcal{G}) = \sum_{j=0}^{\min\{m, n-l\}} (-1)^j \cdot p^{\frac{j \cdot (j-1)}{2}} \cdot a_j(C_{p^{k_1} - 1} \times \cdots \times C_{p^{k_m} - 1}) \cdot a_l(C_p^{n-j})$ if $m > 0$.

Obviously, for $l = 0$, the formula yields the number of subgroups of $\mathcal{G}$. The recursion is rather complicated but involves only fundamental operations and the integers m, n, l, and k_i as parameters. Therefore, it can be easily programmed on a computer, although it is rather hard to calculate even small examples by hand.

B.21 Varieties of Groups

by Awad A. Iskander in Lafayette, LA, USA

A **variety** of groups is the class of all groups satisfying a given set of *laws*, also called *identities*. The books by Neumann (1967b) and Adian (1979) give excellent accounts of the theory of group varieties. We denote by X_ω the *free group* (Section B.9), *freely generated* by the countable alphabet $x_1, x_2, \ldots$. For each $n \geq 1$, the free group X_n generated by $x_1, x_2, \ldots, x_n$ is considered as a subgroup of X_ω. The **rank** of X_n is n. Members of X_ω are called **words**. Every word $w \neq 1$ can be written uniquely in the form $x_{i_1}^{n_1} \ldots x_{i_k}^{n_k}$, $i_j \neq i_{j+1}$, $1 \leq j < k$ and n_i is a nonzero integer, $1 \leq i \leq k$. For any group A and $a \in A^\omega$, $w(a) = (a_{i_1}^{n_1} \ldots a_{i_k}^{n_k})$. The main result on varieties of abelian groups is the following. Let $\mathscr{A}_m$ be the class of all abelian groups which satisfy the law $x^m = 1$.

B.21.1 Theorem *All varieties of abelian groups are given by the classes $\mathscr{A}_m$, with $m \in \mathbb{N}_0$.*

The **verbal subgroup** $W(A)$ corresponding to $W \subseteq X_\omega$ is the subgroup of A generated by $\{ w(a) \mid w \in W, a \in A^\omega \}$. W is a set of laws in a group A iff $W(A) = \{1\}$. A subgroup B of a group A is **fully invariant** if $\alpha(B) \subseteq B$ for every endomorphism α of A. Every verbal subgroup of A is fully invariant. The converse is not true. Indeed, let $A = \mathbb{Z}_2 \times \mathbb{Z}_4$. A is an abelian group of order 8 and exponent 4. The subgroup B of all elements of A of order at most 2 is a Klein 4-group. It is clear that B is fully invariant in A. But B is not verbal in A since, for the given group A, the verbal subgroups are $\{(0,0)\}$, $\{(0,0),(0,2)\}$—the squares and A. However, every fully invariant subgroup U of a free group F is verbal ($U = U(F)$). The variety of groups defined by W coincides with the variety of groups defined by $W(X_\omega)$. There is a one-to-one correspondence between varieties of groups and verbal subgroups of X_ω. If U, V are verbal subgroups of X_ω and $\mathscr{U}$, $\mathscr{V}$ are the varieties satisfying the laws U, V respectively, then $U \subseteq V$ iff $\mathscr{U} \supseteq \mathscr{V}$. The verbal subgroups of X_ω form an algebraic modular lattice under set inclusion. The group varieties form a complete modular lattice under class inclusion that is anti-isomorphic to the lattice of fully invariant subgroups of X_ω. The lattice of group varieties is not distributive Higman (1967).

For any class $\mathscr{K}$ of groups, **Var** $\mathscr{K}$ is the smallest variety containing $\mathscr{K}$. We also say that $\mathscr{K}$ generates **Var** $\mathscr{K}$. Also $\mathbf{S}\,\mathscr{K}$ is the class of all subgroups of members of $\mathscr{K}$, $\mathbf{H}\,\mathscr{K}$ is the class of all homomorphic images of groups in $\mathscr{K}$, and $\mathbf{P}\,\mathscr{K}$ is the class of all Cartesian products of families of groups in $\mathscr{K}$. The classical theorem of Birkhoff (1935) (valid for general types of algebras) establishes that varieties can be described by their closure properties:

B.21.2 Theorem (Birkhoff) *A class $\mathscr{K}$ of groups is a variety if and only if $\mathbf{S}\,\mathscr{K}$, $\mathbf{H}\,\mathscr{K}$, $\mathbf{P}\,\mathscr{K} \subseteq \mathscr{K}$. Moreover, if $\mathscr{L}$ is a class of groups, then* **Var** $\mathscr{L} = \mathbf{H}\,\mathbf{S}\,\mathbf{P}\,\mathscr{L}$.

The variety $\mathscr{G}$ of all groups is defined by the law 1. The variety $\mathscr{E}$ of all trivial groups is defined by the law x_1. The left normed commutator of weight $c + 1$ is defined inductively by $[x_1, x_2] = x_1^{-1} x_2^{-1} x_1 x_2$, $[x_1, \ldots, x_c, x_{c+1}] = [[x_1, \ldots, x_c], x_{c+1}]$. The variety $\mathscr{N}_c$ of all nilpotent groups of class at most c is defined by the law $[x_1, \ldots, x_{c+1}]$. The variety $\mathscr{S}_l$ of all soluble groups of class at most l is defined by the law s_l where $s_1 = [x_1, x_2]$, $s_2 = [[x_1, x_2], [x_3, x_4]]$ and $s_{k+1} = [s_k, t_k]$; t_k is obtained from s_k by substituting x_{i+2^k} for x_i, $1 \leq i \leq 2^k$. The variety $\mathscr{S}_2$ is also called **metabelian**. Every variety $\mathscr{V}$ has its own free groups ($\mathscr{V}$-free) of any given rank. The free member of $\mathscr{V}$ of rank t is denoted by

$F_t(\mathcal{V})$. $F_t(\mathcal{V}) \cong X_t/V(X_t)$. The **Burnside variety** $\mathcal{B}_m$ of all groups of exponent dividing m is defined by the law $x_1{}^m$. For sufficiently large m, $F_2(\mathcal{B}_m)$ is infinite Adian (1979), Ivanov (1994). A group A is called **locally finite** if every finite set of A generates a finite subgroup. Kostrikin (1959) proved that the locally finite groups of a given prime exponent p form a variety denoted by $\mathcal{K}_p$. For sufficiently large primes p, $\mathcal{K}_p$ is a proper subvariety of the Burnside variety $\mathcal{B}_p$. This result is sharpened by Zelmanov (1991) for the case $\mathcal{K}_{p^k}$ for sufficiently large primes p and for any positive integer k.

Every variety $\mathcal{V}$ is generated by its free members $F_n(\mathcal{V})$, $n \geq 1$. Of course, $F_\omega(\mathcal{V})$ generates $\mathcal{V}$. The n-variable words of the verbal subgroup V of $\mathcal{V}$ are the members of $V_n = V \cap X_n$. An n-generator group G (generated by n or less elements) belongs to $\mathcal{V}$ iff V_n are laws in G. The variety $\mathcal{V}^{(n)}$ is the variety of all groups whose n-generator subgroups belong to $\mathcal{V}$. Thus $\mathcal{V}^{(1)} \supseteq \mathcal{V}^{(2)} \supseteq \ldots \supseteq \mathcal{V} = \bigcap \{\mathcal{V}^{(n)} \mid n \geq 1\}$. It is clear that $\mathcal{A}^{(1)} = \mathcal{G}$, $\mathcal{A}^{(2)} = \mathcal{A}$, and $\mathcal{S}_2^{(4)} = \mathcal{S}_2$. If $\mathcal{U}$ is a nilpotent variety of class c, then $\mathcal{U}^{(c+1)} = \mathcal{U}$. The inclusions, in general, are strict. On the other hand, **Var** $F_1(\mathcal{V}) \subseteq$ **Var** $F_2(\mathcal{V}) \subseteq \cdots \subseteq \mathcal{V} = \bigvee \{$**Var** $F_n(\mathcal{V}) \mid n \geq 1\}$ Again, the inclusions here can be proper.

B.21.3 Definition [Philip Hall] The product $\mathcal{K}\mathcal{L}$ of classes of groups $\mathcal{K}$, $\mathcal{L}$ is the class of all groups A with a normal subgroup $B \in \mathcal{K}$ whose quotient $A/B \in \mathcal{L}$.

Thus $\mathcal{A}^l = \mathcal{S}_l$. The **polynilpotent variety** of class $(c_1, \ldots, c_k)$ is $\mathcal{N}_{c_1}\mathcal{N}_{c_2}\ldots\mathcal{N}_{c_k}$. If $\mathcal{U}$, $\mathcal{V}$ are varieties whose verbal subgroups are U, V respectively, then $\mathcal{U}\mathcal{V}$ is a variety whose verbal subgroup is $U(V)$. Since fully invariant subgroups of fully invariant subgroups are automatically fully invariant, the product is associative on group varieties and so the group varieties form a monoid with zero under variety product. The unit in this monoid is the variety $\mathcal{E}$ of all trivial groups and the zero is the variety $\mathcal{G}$ of all groups. $F_\omega(\mathcal{U}\mathcal{V})$ is an extension of $F_\omega(\mathcal{U})$ by $F_\omega(\mathcal{V})$. Also, $\mathcal{U}\mathcal{V}$ is generated by the Wreath product of $F_\omega(\mathcal{U})$ with $F_\omega(\mathcal{V})$. The monoid of varieties is partially ordered under class inclusion. $(\bigvee \{\mathcal{U}_i \mid i \in I\})\mathcal{V} = \bigvee \{\mathcal{U}_i \mathcal{V} \mid i \in I\}$ and $(\bigwedge \{\mathcal{U}_i \mid i \in I\})\mathcal{V} = \bigwedge \{\mathcal{U}_i \mathcal{V} \mid i \in I\}$. However, the inclusions $\mathcal{U}(\mathcal{V}_1 \bigvee \mathcal{V}_2) \supseteq \mathcal{U}\mathcal{V}_1 \bigvee \mathcal{U}\mathcal{V}_2$ and $\mathcal{U}(\mathcal{V}_1 \bigwedge \mathcal{V}_2) \subseteq \mathcal{U}\mathcal{V}_1 \bigwedge \mathcal{U}\mathcal{V}_2$ may be proper.

The structure of the lattice of group varieties is still far from being completely described. However, the structure of the monoid of varieties is completely characterized independently by Neumann et al. (1962) and Shmel'kin (1963).

B.21.4 Theorem *The group varieties under variety product form a free monoid with zero.*

In other words, if a variety $\mathscr{V} \neq \mathscr{G}, \mathscr{E}$, then $\mathscr{V} = \mathscr{U}_1\mathscr{U}_2\ldots\mathscr{U}_k$ where $\mathscr{U}_i$ is an irreducible variety, $1 \leq i \leq k$. Moreover, if $\mathscr{V} = \mathscr{W}_1\ldots\mathscr{W}_l$ and $\mathscr{W}_j$ is an irreducible variety, $1 \leq j \leq l$, then $k = l$ and $\mathscr{U}_i = \mathscr{W}_i$ for every $1 \leq i \leq k$. There are uncountably many irreducible varieties. Every simple group generates an irreducible variety. Every nilpotent variety is irreducible.

A law v is a **consequence** of a set of laws W if $v \in W(X_\omega)$. A set of laws W is **independent** if no $v \in W$ is a consequence of $\{w \mid w \in W, w \neq v\}$; equivalently, W is independent if for any $v \in W$, there is a group A such that w is a law in A for every $w \in W$, $w \neq v$, but v is not a law in A. Klejman (1984) found a variety of soluble groups which cannot be defined by any independent set of laws. A variety $\mathscr{V}$ is **finitely based** if there is a finite set of laws that defines $\mathscr{V}$. Adian (1970), Ol'shanskiĭ (1970), and Vaughan-Lee (1970) independently proved that there are uncountably many group varieties. We quote here:

B.21.5 Theorem (Adian) *For every $n \geq 1003$, the set of laws $[x^{pn}, y^{pn}]^n$, where p runs over all primes, is independent.*

Thus there are group varieties that are not finitely based. From Ol'shanskiĭ and Vaughan-Lee there are locally finite varieties that are not finitely based. On the positive side,

B.21.6 Theorem (Lyndon 1952) *Every nilpotent group variety is finitely based.*

This was generalized by Higman (1967) to the variety product $\mathscr{U}\mathscr{V}$ where $\mathscr{U}$ is a nilpotent variety and $\mathscr{V}$ is a finitely based group variety. Krasil'nikov (1991) showed that every variety generated by a group with nilpotent commutator subgroup is finitely based.

B.21.7 Theorem (Oates and Powell 1964) *The variety generated by a finite group is finitely based.*

B.21.8 Theorem (Schreier) *Every subgroup H of a free group F is free. Moreover, if $H \subseteq X_n$ and j is the index of H in X_n and m is the rank of H, then $m + j = nj + 1$.*

If $\mathscr{V}$ is the abelian variety $\mathscr{A}$ or $\mathscr{A}_p$, where p is a prime, then every subgroup of a $\mathscr{V}$-free group is $\mathscr{V}$-free. Such varieties are called **Schreier varieties**.

B.21.9 Theorem (Neumann and Wiegold 1964) *If $\mathscr{V}$ is a Schreier group variety, then $\mathscr{V} = \mathscr{G}$, $\mathscr{V} = \mathscr{A}$ or $\mathscr{V} = \mathscr{A}_p$ for some prime p.*

The set of all locally finite varieties is a sublattice of the lattice of all varieties that is also closed under the variety product. A finite group A is **critical** if $A \notin \mathbf{Var}(\mathbf{QS\text{-}I})A$, the variety generated by all the homomorphic images of subgroups of A that are not isomorphic to A. Every locally finite variety is generated by its critical groups. However, Higman (1967) showed that there are three join-irreducible metabelian varieties of prime exponent p whose joins in pairs coincide.

The author thanks Mikhail V. Volkov for several valuable comments.

B.22 Groups and Absolute Geometry

by Helmut Karzel in Munich, Germany

The theory of groups originated from two sources, the problem of solving equations of higher degree and especially from symmetry considerations in geometry (cf. Karzel and Kroll 1988, §8 and p. 101f). The interplay between geometry and groups has created many interesting insights and was the source for new research areas, for which we will give some examples. If one takes Euclid's system of axioms and cancels the parallel axiom then one calls the theory, which is based on this reduced system, **absolute geometry**. In an absolute geometry $(P, \mathscr{L}, \equiv, \beta)$ (P denotes the set of points, $\mathscr{L}$ the set of lines, $\equiv$ the congruence and β the betweenness), one can still define:

Reflections in points: For $p \in P$, let $\widetilde{p} \colon P \to P$ be the map fixing p such that for $x \neq p$, $\widetilde{p}(x) \neq x$ is the point of the line $\overline{p, x}$ joining p and x with $(p, \widetilde{p}(x)) \equiv (p, x)$.

Reflections in lines: For $L \in \mathscr{L}$, let $\widetilde{L} \colon P \to P$ be the map fixing the points of L such that for $x \notin L$, $\widetilde{L}(x) \neq x$ is the point of the plane spanned by x and L with $(\widetilde{L}(x), l) \equiv (x, l)$ for all $l \in L$.

The **motion group** $\mathrm{M} := \mathrm{M}(P, \mathscr{L}, \equiv)$ consists of all permutations $\mu \in S_P$ conserving lines (i.e., $\forall L \in \mathscr{L} \colon \mu(L) \in \mathscr{L}$) and such that $\forall a, b \in P \colon (\mu(a), \mu(b)) \equiv (a, b)$.

One can show:

(1) If $S = P$ or $S = \mathscr{L}$ then $\tilde{S} := \{\tilde{s} | s \in S\} \subset$ M and $\forall a, b \in S: \tilde{a} \circ \tilde{a} =$ id and $\widetilde{\tilde{a}(b)} = \tilde{a} \circ \tilde{b} \circ \tilde{a}$ (cf. Karzel, Sörensen, and Windelberg 1973, 18.3).

(2) $\forall a, b, c, x, y, z \in S: a \neq b$ and $\tilde{a} \circ \tilde{b} \circ \tilde{x}, \tilde{a} \circ \tilde{b} \circ \tilde{y}, \tilde{a} \circ \tilde{b} \circ \tilde{z} \in \tilde{S} \implies \tilde{x} \circ \tilde{y} \circ \tilde{z} \in \tilde{S}$ (cf. Karzel, Sörensen, and Windelberg 1973, 18.8).

(3) $\forall a, b \in P \ \exists_1 c \in P: \tilde{c}(a) = b$ and $\operatorname{Fix} \tilde{a} := \{ x \in P \mid \tilde{a}(x) = x \} = \{a\}$ (cf. Karzel, Sörensen, and Windelberg 1973, 16.11, 17.8).

On the basis of Hilbert's axiom system with continuity axiom—in this case, $(P, \mathscr{L}, \equiv, \beta)$ is called a **classical absolute geometry**—one has a bifurcation in Euclidean geometry by $\tilde{P} \circ \tilde{P} \circ \tilde{P} = \tilde{P}$ and in hyperbolic geometry by $\tilde{P} \circ \tilde{P} \circ \tilde{P} \neq \tilde{P}$ (cf. Karzel, Sörensen, and Windelberg 1973, 23.3, 21.3).

Now we start from the following abstract situation:

B.22.1 Theorem *Let P be a set, let $\sim: P \to S_P, p \mapsto \tilde{p}$ be a map such that the properties (1) and (3) are satisfied. Then $(P, \sim)$ is called an* **invariant reflection structure**. *Let $o \in P$ be fixed and for $a \in P$, let $a' \in P$ be such that $\tilde{a}'(o) = a$. Then $(P, +)$ with $a + b := \tilde{a}' \circ \tilde{o}(b)$ is a* **Bruck loop** *(or what is the same a K-loop), i.e., the following conditions are satisfied:*

(L) *$\forall a, b \in P$ there exist unique $x, y \in P$ such that $a + x = b$, $y + a = b$, and $a + o = o + a = a$. Let $a^+: P \to P; x \mapsto a + x$ and $P^+ := \{a^+ | a \in P\}$.*

(B) *(Bol condition) $\forall a, b \in P: a^+ \circ b^+ \circ a^+ = (a + (b + a))^+$.*

(I) *If $-a$ is defined by $a + (-a) = o$ then $\nu: P \to P; x \mapsto -x$ is an automorphism of $(P, +)$, i.e., $-(a+b) = -a + (-b)$, with $\operatorname{Fix} \nu = \{o\}$.*

Moreover, $(P, +)$ is **uniquely divisible by 2**, *i.e., $\forall a \in P, \exists_1 b \in P: b + b = a$, and $(P, +)$ is a group (and then a commutative one) if and only if $\tilde{P} \circ \tilde{P} \circ \tilde{P} = \tilde{P}$.*

The converse is also true:

B.22.2 Theorem *Let $(P, +)$ be a Bruck loop uniquely divisible by 2 and, for $a \in P$, let $\tilde{a} := (a + a)^+ \circ \nu$. Then $(P, \sim)$ is an invariant reflection structure.*

In a loop $(P, +)$, one can associate to any $a, b \in P$ the permutation $\delta_{a,b} := ((a+b)^+)^{-1} \circ a^+ \circ b^+$ fixing the neutral element o. By A. Kreuzer, we have the following result:

B.22.3 Theorem *If $(P, +)$ is a Bruck loop then the following conditions, which, together with* **(I)** *characterize* **K-loops***, are satisfied:*

(A) $\forall a, b \in P$: $\delta_{a,b} \in \mathrm{Aut}(P, +)$;

(K) $\forall a, b \in P$: $\delta_{a,b+a} = \delta_{a,b}$.

Any loop $(P, +)$ satisfying **(A)** can be embedded via an *outer quasidirect product* into a group $P \rtimes_Q A$, where A is any subgroup of the automorphism group $\mathrm{Aut}(P, +)$ containing the set $\{\delta_{a,b} \mid a, b \in P\}$. The set $P \times A$ becomes a group $P \rtimes_Q A$ via

$$(a, \alpha) \circ (b, \beta) := (a + \alpha(b), \delta_{a,\alpha(b)} \alpha\beta).$$

Moreover, the map $\psi \colon P \rtimes_Q A \to P^+ \circ A; (a, \alpha) \mapsto a^+ \circ \alpha$ is a monomorphism. Hence $\Gamma := P^+ \circ A$ is a subgroup of the permutation group S_P and one can prove that $(\Gamma, \circ)$ is the *inner quasidirect product* of P^+ and A which is defined by: (i) $\mathrm{id} \in P^+, A \leq \Gamma$, (ii) $\forall \gamma \in \Gamma \exists_1 (a^+, \alpha) \in P^+ \times A: \gamma = a^+ \circ \alpha$. (iii) $\forall \alpha \in \Gamma: \alpha \circ P^+ \circ \alpha^{-1} = P^+$. (iv) $\forall a^+, b^+ \in P^+: |P^+ \circ a^+ \cap b^+ \circ A| = 1$.

On the other hand, let a group $(\Gamma, \cdot)$ be an inner quasidirect product of P and A and for $a, b \in P$ let $a + b \in P$ and $\zeta \in A$ such that $a \cdot b = (a+b) \cdot \zeta$ (cf. (ii)). Then $(P, +)$ is a loop satisfying **(A)**.

Now we apply these results to an important example, which is the link between various geometrical, algebraical, and physical structures.

Let $P := \{A = \left(\begin{smallmatrix} \alpha & a \\ \bar{a} & \alpha' \end{smallmatrix}\right) \mid a \in \mathbb{C}, \alpha \in \mathbb{R}, \alpha > 0, \alpha' = \alpha^{-1}(1 + a\bar{a})\}$ be the set of all Hermitian 2×2-matrices with determinant 1 and positive trace $\mathrm{tr}\, A$. P contains the identity matrix I and can be provided with the following operations and structures:

For a 2×2-matrix $M = (m_{ij})$ let $\widehat{M} := (\widehat{m}_{ij})$ with $\widehat{m}_{11} := m_{22}, \widehat{m}_{12} := -m_{12}, \widehat{m}_{22} := m_{11}$ and $\widehat{m}_{21} := -m_{21}$ and for $A, X \in P$ let $\sqrt{A} := (2 + \mathrm{tr}\, A)^{-\frac{1}{2}}(I + A)$ and $\tilde{A}(X) := A\widehat{X}A$. Then

(1) $\widehat{A}, \sqrt{A}, \tilde{A}(X) \in P$ and the maps $\widehat{} \colon P \to P; X \mapsto \widehat{X}$, $\sqrt{} \colon P \to P; X \mapsto \sqrt{X}$, $\tilde{A} \colon P \to P; X \mapsto \tilde{A}(X)$ are permutations of S_P and $\tilde{I} = \widehat{}$.

(2) $(P, \widetilde{})$ is an invariant reflection structure.

(3) $(P, \oplus)$ with $A \oplus B := \sqrt{A} B \sqrt{A}$ is a Bruck loop with neutral element I.

(4) *Collinearity* κ: For $A, B, C \in P$ let $(A, B, C) \in \kappa :\iff A\widehat{B}C = C\widehat{B}A$. We call a subset T of P *collinear* if for all $X, Y, Z \in T, (X, Y, Z) \in \kappa$; the maximal collinear subsets of P are called *lines*. If $\mathscr{L}$ denotes the set of all lines then $(P, \mathscr{L})$ is an *incidence space* (i.e., any two distinct points can be joined by exactly one line).

For each $A \in P$, the *translation (= Lorentz boost)* $A^{\oplus} : P \rightarrow P; X \mapsto A \oplus X$ is a collineation of $(P, \mathscr{L})$ and therefore the structure $(P, \mathscr{L}, \oplus)$ is called an *incidence loop*.

(5) *Betweenness* β: For $A, B, C \in P$, let $(A, B, C) \in \beta :\iff (A, B, C) \in \kappa$ and $(B|A, C) := \text{sgn}(A\widehat{B}C\widehat{B} + B\widehat{C}B\widehat{A} - A\widehat{C} - C\widehat{A}) = -1$.

(6) *Congruence* $\equiv$: For $A, B, C, D \in P$, let $(A, B) \equiv (C, D) :\iff A\widehat{B} + B\widehat{A} = C\widehat{D} + D\widehat{C}$.

Now we can state the:

B.22.4 Foundation theorem $\mathcal{H} := (P, \mathscr{L}, \equiv, \beta)$ *is a classical 3-dimensional hyperbolic geometry, and each classical 3-dimensional hyperbolic geometry is isomorphic to this model* $\mathcal{H}$.

For each $A \in P$, the map $\widetilde{A}$ is a reflection in the point A and the sets $\widetilde{P}$ and $P^{\oplus}$ are subsets of the motion group $M := M(\mathcal{H})$. Other motions can be obtained from the special linear group $SL(2; \mathbb{C})$. For $S \in SL(2; \mathbb{C})$, let $\bar{S} := (\bar{s}_{ij})$, $S^{T} = (s_{ij}^{T})$ with $s_{ij}^{T} := s_{ji}$, and $S^{*} := \bar{S}^{T}$. Then $\bar{S}, S^{T}, S^{*} \in SL(2; \mathbb{C})$, and for $X \in P$ the matrix $S^{\square}(X) := SXS^{*}$ is again an element of P, and moreover $S^{\square}$ is a motion. Since $S^{\square} \circ R^{\square} = (SR)^{\square}$, the pair $(SL(2; \mathbb{C})^{\square}, \circ)$ with $SL(2; \mathbb{C})^{\square} := \{S^{\square} | S \in SL(2; \mathbb{C})\}$ is a group, and since $S^{\square} = \text{id} \iff S \in \{I, -I\}$, $SL(2; \mathbb{C})^{\square}$ is isomorphic to the projective special linear group $PSL(2; \mathbb{C}) := SL(2; \mathbb{C})/\{I, -I\}$.

B.22.5 Fundamental theorem of hyperbolic geometry
$M = SL(2; \mathbb{C})^{\square} \times \{\text{id}, \widehat{}\}$, *i.e.*, $SL(2; \mathbb{C})^{\square}$ *is the set of all proper motions of the 3-dimensional hyperbolic space* $\mathcal{H}$, *and* $(SL(2; \mathbb{C})^{\square} \circ \widehat{})$ *of all improper ones.*

If $S^{\square}$ is an automorphism of the loop $(P, \oplus)$ then $I = S^{\square}(I) = SS^{*}$, thus $S^{*} = \widehat{S}$. All complex 2×2-matrices X satisfying the functional equation $X^{*} = \widehat{X}$ form a field $\mathbb{H}$, the field of quaternions, and for $\mathbb{H}_{1} := \mathbb{H} \cap SL(2; \mathbb{C})$, we have $\mathbb{H}_{1}^{\square} \leq \text{Aut}(P, \oplus)$. Furthermore, $SL(2; \mathbb{C})^{\square} = P^{\oplus} \rtimes_{Q} \mathbb{H}_{1}^{\square}$ is the inner quasidirect product of $P^{\oplus}$ and $\mathbb{H}_{1}^{\square}$.

On the other hand, $SL(2; \mathbb{C})^{\square}$ can be considered as the homogeneous proper orthochronous Lorentz group of the Minkowski world, where the set of events is given by the set $\mathfrak{H}$ of all Hermitian matrices and the structure by the symmetric bilinear form $f(X, Y) := X\widehat{Y} + Y\widehat{X}$. Here, a map α is called *orthochronous*, if it preserves time, i.e., $A \rightarrow B$ implies $\alpha(A) \rightarrow \alpha(B)$; the map $x \mapsto -x$ is not orthochronous. In particular, causality $\rightarrow$ is described by: $A \rightarrow B :\Leftrightarrow B - A \in \mathfrak{H}^{++} := \{X \in \mathfrak{H} \mid \det X > 0, \operatorname{tr} X > 0\}$.

In both cases (the hyperbolic geometry and the Minkowski world), the quasidirect product gives us insight into the structure of the motion group (Lorentz group, respectively). Each $S^{\square}$ can be written as a product $S^{\square} = A^{\oplus} \circ H^{\square}$, where $A^{\oplus}$ is a hyperbolic translation (= Lorentz boost) and $H^{\square}$ is a rotation. The 3-dimensional subspace $I^{\perp} := \{X \in \mathfrak{H} \mid f(I, X) = X + \widehat{X} = 0\}$ of $\mathfrak{H}$ is fixed by $H^{\square}$, and f induces a Euclidean metric of $I^{\perp}$; the restriction $H^{\square}/I^{\perp}$ is a Euclidean rotation.

There is a great variety of Bruck loops. The Bruck loops $(P, +)$ derived from the point reflections of a hyperbolic geometry according to Theorem B.22.4 are called *hyperbolic loops*. They satisfy the following *centralizer condition*:

(**C**) For $a \in P$, let $[a] := \{x \in P \mid x + a = a + x\}$. If $a, b \in P \setminus \{o\}$ and $b \in [a]$ then $[a] = [b]$.

In the same way as an affine space is derived from a vector space, one can associate a Bruck loop $(P, +)$ with (**C**) to an incidence structure $\mathscr{L}$: We replace the one dimensional linear subspaces by the centralizers $[a]$ with $a \in P \setminus \{o\}$, and define the cosets as lines, thus $\mathscr{L} = \{b + [a] \mid b, a \in P, a \neq o\}$. In the case of a hyperbolic loop, $(P, \mathscr{L})$ is the incidence structure of the hyperbolic geometry. It remains to determine all other Bruck loops satisfying (**C**).

B.23 Groups for Particle Physics

by Gudrun Kalmbach H. E. in Ulm, Germany

Symmetries in laws, patterns or objects, such as crystals, play an important role in nature. They may occur in physics as discrete or continuous *transformation groups* G corresponding to conservation laws, in which the magnitudes of some (physical) quantities F (*observables, operators*) are kept constant when members of G are applied. Beside global symmetries, there also exist local ones, which, more generally, keep the relevant physical laws, variation principles, or field equations invariant. Local symmetries may produce forces between space points or structures.

B.23.1 Example The following list contains first a quantity and then the associated transformation group:

(1) *energy* ($F = const.$) $\rightarrow$ *group of time translations*, $t \rightarrow t + a$ ($\frac{dF}{dt} = 0$),

(2) *momentum* $\rightarrow$ *group of space translations*, $u \rightarrow u + a$,

(3) *angular momentum* (also spin) $\rightarrow$ *group of spatial rotations R*, $(u - c) \rightarrow R(u - c)$,

(4) *charge conjugation* $\rightarrow$ $\mathbb{Z}_2$ with two elements: *id* (identity map) and *charge parity C*, (which replaces particles by their antiparticles, changes all internal quantum numbers and signs of all nonzero generalized charges,)

(5) *phase* $\rightarrow$ *group(s)* of complex unities $e^{iN\varphi}$ for relative phase angles,

(6) *barycenter* $\rightarrow$ *group of Lorentz rotations*,

(7) *Minkowski metric* $ds^2 = c^2 dt^2 - dr^2$ (t time, $r = (r_x, r_y, r_z)$ radius, c speed of light) in space-time $\rightarrow$ *group $\mathcal{L}$ of Lorentz transformations*. $\mathcal{L}$ has as continuous Lie group 10 parameters: 4 for translations, 3 for rotations and 3 for Lorentz transformations. All individual *inertial (coordinate) systems* in $\mathbb{R}^4$ are equivalent (special relativity).

(8) The *electromagnetic* interaction EMI, observed as electrical and chemical forces, has as carrier *photons* γ. The *unitary group* $U(1)$ belongs to EMI, its geometry is given by the complex unit circle, it is isomorphic to the 2-dimensional rotation group SO(2).

There are *four basic interactions* (forces or accelerations/second derivatives) in physics. Recall that *quantum field theories* explain the forces acting on or between several systems or particles in terms of other exchanged (anti-)particles which can be emitted or absorbed: Photons for EMI; nuclear forces have W-bosons for WI (*weak interaction*), gluons for SI (*strong interaction*); *gravity* may have *gravitons*, which are not experimentally found today. All electrical charged particles have electromagnetic interactions. *Leptons* (electrons, positrons, neutrinos, ...) have weak, but no strong interactions. *Hadrons* (mesons or baryons) have weak and strong interactions.

We concentrate now on *unitary symmetries* for WI and SI used in particle physics. For a unitary (complex) matrix A with $AA^* = I$, A^* Hermitian conjugate of A, *special* means $\det A = 1$. In a complex *Lie* group with an analytic Hausdorff structure, the operations are analytic and the underlying space is an analytic manifold. If operations or analytic infinitesimal transformations (for instance in matrix form $A, B, \ldots$) on such a space are studied, *Lie algebras* are associated with them which carry the additional structure of commutators or Lie products $[A, B] = AB - BA$, satisfying the **Jacobi identity**

$[[X,Y],Z] + [[Y,Z],X] + [[Z,X],Y] = 0$. The noncommutativity of such matrices (i.e., $[A,B] \neq 0$) lead to the *Heisenberg uncertainty relations*. For instance, the position and the momentum of a (physical) system cannot be measured simultaneously.

The *special unitary Lie groups* SU(2) and SU(3) of two- and three-dimensional special unitary matrices are used in particle physics in connection with the two *interactions* WI, SI. They have as their geometry the 3-dimensional unit sphere $S^3 \subset \mathbb{R}^4$ or $\mathbb{C}^2$ and the generalized torus $S^3 \times S^5$, where $S^5 \subset \mathbb{C}^3$ is the 5-dimensional unit sphere. The generators A of the Lie groups SU(2) and SU(3) can be chosen to have *trace* 0.

B.23.2 Example WI: is observed in radioactivity, decays, has short range, breaks charge conjugation and parity (the space transformation $u \rightarrow -u$), has coupling strength 10^{-5} and as carriers three W-*bosons* $W^{\pm}, Z^0$ with mass. Their operator-matrix representations are related to spin theory and the three Pauli generators of SU(2), for instance $W^{\pm} = S_x \pm iS_y$ holds. Z^0 is neutral, $W^{\pm}$ carry the indicated electrical charge, all three have spin 1. They are produced in collisions, for instance of an electron and either an antineutrino or a positron. The colliding particles are annihilated; in the first (second) case a W^- (Z^0) is produced, which decays into the same kind of particles, or in the second case also two photons can be generated from Z^0. Atomic kernels emit in radioactivity W-bosons.

The WI-group SU(2) is a subalgebra of SU(3). In a 4D-space-time coordinate representation, the SU(2) elements A are complex 2×2 matrices, which are obtained from the general SU(3)-matrix below by setting $w = 0$ and deleting the third row and column. SU(2) is generated by the three Pauli (scaled 3D-spin, isospin $S = (S_x, S_y, S_z)$) matrices:

$$\sigma_1 = \begin{pmatrix} 0 & 1 \\ 1 & 0 \end{pmatrix}, \quad \sigma_2 = \begin{pmatrix} 0 & -i \\ i & 0 \end{pmatrix}, \quad \sigma_3 = \begin{pmatrix} 1 & 0 \\ 0 & -1 \end{pmatrix}$$

B.23.3 Example SI: is observed as nuclear forces, has short range, coupling strength 1 and as carriers 8 (or 9) *gluons*: $G_{x\bar{y}}$, $x \neq y$, $x, y \in \{red = r, green = g, blue = b\}$ plus three additional found superpositions of these 6 color/anticolor charges $x, \bar{x}$, (carried for instance by quarks), where the bar denotes antimatter (anticolor, ...). Gluons have spin 1 and mass 0. They are related to 8 (or 9) generating matrices of SU(3). The theory is called *quantum chromodynamics*. The gluon-exchange between 2 or 3 quarks/antiquarks keeps these particles together in a *meson* or *baryon* (*confinement* of quarks in a bag). There are no single quarks. Nucleons (protons or neutrons) consist of 3 quarks. The SU(3) *classification scheme* of octcts/multiplets, the eightfold way, is used for mesons and baryons.

The SI-group SU(3) coordinates are represented in the following combined (very unusually added!) matrix A, constructed from $\begin{pmatrix} 1 & 0 & 0 \\ 0 & 1 & 0 \\ 0 & 0 & 0 \end{pmatrix}$ and eight SU(3)-generating Gellmann matrices λ_i with trace 0, which are suitably multiplied by independent real or pure imaginary variables, and the matrix entries of A are listed in vector form, due to a diagonal Gellmann matrix $\lambda_8 = \frac{1}{\sqrt{3}}\mathrm{diag}[1, 1, -2]$.

$$
A = \begin{pmatrix} (z_1, w) & z_2 & z_3 \\ -\overline{z_2} & (\overline{z_1}, w) & z_4 \\ -\overline{z_3} & -\overline{z_4} & -2w \end{pmatrix}, \ w \in \mathbb{R}, z_i \in \mathbb{C},
$$

is a $\mathbb{R}^9$-space representation with the generators λ_i. The commutation rules are $\left[\frac{\lambda_i}{2}, \frac{\lambda_j}{2}\right] = if_{ijk}\frac{\lambda_k}{2}$, $f_{ijk} = -\frac{i}{4} \cdot \mathrm{tr}([\lambda_i, \lambda_j]\lambda_k)$.

B.23.4 Remark Gravity does not fit until now into these theories. A new group D_3 and a unifying geometrical (9-dimensional) 9D/SU(3)-model for the four interactions can be found in Kalmbach (1998). Mass is there a 3D- or 6D-*Gleason operator*, gravitational effects are explained by a 5D-graviton-exchange between systems. Some new diffusion-differential equations are added for quark-bags potentials and gravitational potentials. This model is not related to string theory.

For more information see (Cracknell 1968), (Ho-Kim and Pham 1998), and (Kalmbach 1998).

Chapter C

RINGS, MODULES, ALGEBRAS

C.1 Commutative Algebra I: Ideal Decompositions

by Robert Gilmer in Tallahassee, FL, USA

Many problems in commutative algebra treat various ways that a fixed ideal (or each ideal of a given class of ideals) of a commutative ring can be decomposed. Generally speaking, early problems of this type that arose from algebraic geometry concerned representations of ideals as intersections, while those arising from algebraic number theory involved representations in terms of products.

We first discuss primary decomposition of ideals. An ideal Q of a commutative ring R is said to be **primary** if the conditions $ab \in Q$ and $a \notin Q$ imply that Q contains a power of b, for all $a, b \in R$. Equivalently, Q is primary if each zero-divisor of the ring R/Q is nilpotent. If Q is primary, then the radical P of Q is a prime ideal; Q is said to be **primary for** P or **P-primary**, and P is called the **associated prime ideal** of Q. If Q is P-primary and $P^k \subseteq Q$ for some positive integer k, then Q is **strongly primary**; in order for a P-primary ideal to be strongly primary, it is sufficient that P is finitely generated.

In the ring $\mathbb{Z}$, a nonzero proper ideal $n\mathbb{Z}$ is primary iff $n\mathbb{Z} = p^k\mathbb{Z}$ for some prime integer p. Thus the prime factorization $n = \pm \prod_{i=1}^{r} p_i^{e_i}$ gives rise to the representation $n\mathbb{Z} = \bigcap_{i=1}^{r} p_i^{e_i}\mathbb{Z}$ of $n\mathbb{Z}$ as a finite intersection of primary ideals of $\mathbb{Z}$, and the same result is true in any principal ideal domain. Lasker (1905) showed that each ideal of $\mathbb{Z}[x_1, \ldots, x_n]$ or of $K[x_1, \ldots, x_n]$, for a field K, can be expressed as a finite intersection of primary ideals. One of the strokes of genius of Emmy Noether was her recognition of the central role played by the ascending chain condition (a.c.c.) for ideals in general commutative rings, and in her fundamental work (Noether 1921), she proved that, in any ring satisfying the a.c.c. for ideals (that is, a Noetherian ring), each ideal is a finite intersection of primary ideals. This theorem is now standard, having appeared, for example, in all editions of van der Waerden's Modern Algebra (van der Waerden 1950).

In considering a primary decomposition $I = \bigcap_{i=1}^{r} Q_i$ of I, where $P_i = \sqrt{Q_i}$, there is no loss of generality in assuming that $P_i \neq P_j$ for $i \neq j$ and that $Q_i \not\supseteq \bigcap_{j \neq i} Q_j$ for each i. We refer to such a decomposition as **reduced**. While the ideals Q_i in a reduced primary decomposition of I may not be uniquely determined, the set $\{P_i\}_{i=1}^{r}$ is uniquely determined, and the prime ideals of this set are referred to as the **associated primes**[1] of I. Moreover, if the ideal P_{i_0} is minimal in the set $\{P_i\}_{i=1}^{r}$, then Q_{i_0} is independent of the reduced primary decomposition of I chosen. It follows, for example, that in a Noetherian domain in which nonzero proper prime ideals are maximal, each ideal admits a unique reduced primary decomposition.

A ring R with identity is **Laskerian** if each ideal of R admits a finite primary decomposition; R is **strongly Laskerian** if each ideal of R is a finite intersection of strongly primary ideals. Krull (1929) determined equivalent conditions for a ring to be Laskerian or strongly Laskerian (see also Atiyah and MacDonald 1969, chap. 4); his conditions involve the following notation. If I is an ideal and S is a multiplicative system in a ring R, the contraction of the extension of I with respect to the quotient ring R_S is denoted by $S(I)$. Thus $S(I) = \{x \in R \mid xs \in I \text{ for some } s \in S\}$; Krull called $S(I)$ an **isolated component ideal (i.K.i.)** of I.

C.1.1 Theorem *A commutative ring R with identity is Laskerian if and only if the following conditions (1) and (2) are satisfied.*

(1) *For each ideal I and each nonempty multiplicative system S in R, there exists $s \in S$ such that $S(I) = (I : s)$.*

(2) *Each decreasing sequence $S_1(I) \supseteq S_2(I) \supseteq \ldots$ of isolated component ideals of a fixed ideal I becomes stable, for each ideal I of R.*

---

[1]The notion of an associated prime of I has been extended, in several different ways, to an arbitrary ideal I, whether or not I admits a primary decomposition (see (Bourbaki 1972a) or (Heinzer and Ohm 1971)). The most useful of these seems to be that of an associated prime in the **weak-Bourbaki** sense: P is an associated prime of I if P is a minimal prime of $I : x$ for some $x \in R - I$.

Moreover, R is strongly Laskerian iff (2) and the following condition (3) are satisfied.

(3) *For all ideals* $I, J_1, J_2, \ldots$ *of R, the ascending sequence* $(I : J_1) \subseteq (I : J_1 J_2) \subseteq (I : J_1 J_2 J_3) \subseteq \ldots$ *becomes stable.*

Krull (1929, p. 12) asked for equivalent conditions for a fixed ideal I to admit a finite primary decomposition. This question is addressed by Fu (1999).

A Laskerian ring need not be Noetherian, but results of Heinzer and Ohm (1972) and Gilmer and Heinzer (1972) show that if either the polynomial ring $R[x]$ or the power series ring $R[[x]]$ is Laskerian, then R (hence also $R[x]$ and $R[[x]]$) are Noetherian.

The class of commutative rings in which each ideal admits a unique reduced primary representation has been determined by S. Mori. A short proof of Mori's result for rings with identity appears in (Gilmer 1963), which also contains an extension to the case where each primary ideal in the unique reduced representation is strongly primary.

The main question about ideal decomposition arising from algebraic number theory asked for a characterization of integral domains in which each nonzero proper ideal is a finite product of prime ideals. Such domains are now referred to as **Dedekind domains** (see Section C.4 and Section C.5, in honor of Richard Dedekind, who showed that the ring of all algebraic integers in a finite algebraic number field has the required property (Dedekind 1969). While many characterizations of Dedekind domains are known (see Gilmer 1992, chap. VI)), probably the most familiar of these states that an integral domain D is a Dedekind domain if and only if (1) D is Noetherian, (2) each nonzero proper prime ideal of D is maximal, and (3) D is integrally closed. While origins of this characterization can be traced to the seminal paper by Noether (1927) in this area, Noether actually worked with five axioms in (Noether 1927) and she considered unique factorization as a product of maximal ideals, rather than as products of prime ideals. For a brief account of the historical development of the characterization cited, see (Cohen 1950, Sect. 3).

If L is an infinite algebraic number field and $\mathcal{O}_L$ is the ring of algebraic integers in L, then each nonzero proper prime ideal of $\mathcal{O}_L$ is maximal and $\mathcal{O}_L$ is integrally closed, but $\mathcal{O}_L$ need not be Noetherian. Equivalent conditions for $\mathcal{O}_L$ to be Noetherian are given in (MacLane and Schilling 1939).

Factorization into products of prime ideals has also been considered in commutative rings with zero-divisors. Thus a commutative ring R is a **general ZPI-ring**[2] if each proper ideal of R is a finite product of proper prime ideals; R is a **ZPI-ring** if this factorization is unique. Asano (1951) gave the following characterization of general ZPI-rings. Recall that a special principal ideal ring is a local principal ideal ring in which the maximal ideal is nilpotent.

C.1.2 Theorem *A commutative ring R is a general ZPI-ring iff R is a finite direct sum of Dedekind domains and special principal ideal rings. Hence, R is a ZPI-ring iff R is either a Dedekind domain or a special principal ideal ring.*

As in the case of primary decomposition, the problem of determining conditions under which a fixed ideal I can be expressed as a product of prime ideals (and uniqueness of such a decomposition) has been considered. See, for example, (Gilmer 1972b).

C.2 Commutative Algebra II: The Krull Dimension

by Robert Gilmer in Tallahassee, FL, USA

The Krull dimension of a commutative ring is an important concept in much of commutative algebra. The definition, which is due to Krull (1951), is as follows. Suppose R is a commutative ring and m is a nonnegative integer. The (**Krull**) **dimension** of R is m (written $\dim(R) = m$) if there exists a chain $P_0 < P_1 < \cdots < P_m$ of proper prime ideals of R, but no longer such chain; if there is no upper bound on the lengths of chains of proper prime ideals of R, then R is **infinite-dimensional**, written as $\dim(R) = \infty$ (cf. Section C.6).

By definition, a ring R is zero-dimensional if each proper prime ideal of R is maximal. Equivalently, $\dim(R) = 0$ iff a power of each principal ideal of R is idempotent (Huckaba 1995). The class of zero-dimensional rings contains the fields, finite rings, Artinian rings, Boolean rings, von Neumann regular rings, and proper homomorphic images of one-dimensional integral domains. The monograph edited by Anderson and Dobbs (1995) contains an account of some of the more recent work in the theory of zero-dimensional rings.

---

[2]ZPI arose as an abbreviation for *Zerlegung in Primideale*.

Many of the integral domains that arise in algebraic number theory are one-dimensional. The ring $\mathbb{Z}$ is, of course, one-dimensional, as is any principal ideal domain that is not a field. In general, if R is a subring of the ring S, an element $s \in S$ is said to be *integral* over R if s is the root of a monic polynomial with coefficients in R, and S is *integral* over R if each element of S is integral over R. Krull (1937) showed that dimension is stable under integral extensions, that is, if S is integral over its subring R, then S and R have the same Krull dimension. Thus, if $\mathcal{O}$ is the ring of all algebraic integers in $\mathbb{C}$, then each subring of $\mathcal{O}$ is one-dimensional.

If K is a field, the polynomial ring $K[x_1, \ldots, x_n]$ is n-dimensional. Moreover, different *maximal* chains of prime ideals in $K[x_1, \ldots, x_n]$ are well-behaved according to the following definition: If $P < Q$ are proper prime ideals of the ring R, a chain $P = P_0 < P_1 < \cdots < P_s = Q$ of prime ideals is said to be a *maximal chain of primes between P and Q* if there is no prime of R lying strictly between P_{i-1} and P_i for $1 \leq i \leq s$; this chain is a *maximal chain of primes in R* if Q is a maximal ideal of R and P contains no prime ideal of R properly. The ring R satisfies the *first chain condition (f.c.c.)* if any two maximal chains of primes in R have the same length. Finitely generated affine domains $K[a_1, a_2, \ldots, a_n]$ over a field satisfy f.c.c; a proof of this result is usually based on the fact that for each proper prime ideal P of $K[a_1, a_2, \ldots, a_n]$, the Krull dimension of the domain $T = K[a_1, a_2, \ldots, a_n]/P$ is equal to the transcendence degree of T over K. This last statement is based, in turn, on a result known as Noether's Normalization Lemma.

C.2.1 Theorem *Let D be an integral domain that is finitely generated over the field K. If the transcendence degree of D over K is r, then there exist elements $x_1, \ldots, x_r \in D$ such that $\{x_1, \ldots x_r\}$ is algebraically independent over K and D is integral over $K[x_1, \ldots, x_r]$.*

Chains of prime ideals are not nearly as well behaved in arbitrary rings as the situation for $K[x_1, \ldots, x_n]$ might suggest. Some sufficient conditions for the equality $\dim R[x_1, \ldots, x_n] = \dim R + n$ to be valid are that R is Noetherian, zero-dimensional, or a Prüfer domain (see Section C.5). In general, Seidenberg (1953) showed that if $\dim R = n$ then $n+1 \leq \dim R[x_1] \leq 2n+1$; moreover, given nonnegative integers n and m such that $n + 1 \leq m \leq 2n + 1$, there exists an integral domain D such that $\dim D = n$ and $\dim D[x_1] = m$. The monograph by Jaffard (1960) is devoted to a study of the dimension theory of $R[x_1, \ldots, x_n]$ and of the behavior of chains of prime ideals in such rings. Arnold and Gilmer

(1974) determined all sequences $\{n_0, n_1, \dots\}$ of nonnegative integers which can be realized in the form $\{\dim R, \dim R[x], \dim R[x_1, x_2], \dots\}$ for some commutative ring R.

A powerful tool in the work on the dimension theory of Noetherian rings is **Krull's principal ideal theorem** (part (1) below) and its generalization (part (2) below). The statement of this theorem uses the following terminology: if P is a proper prime ideal of a ring R, the **height** of P is defined to be the dimension of the localization R_P of R at P.

C.2.2 Theorem *Let $I = (a_1, \dots, a_n)$ be a proper ideal of a Noetherian ring R and let P be a prime ideal minimal in the set of primes of R that contain I.*

(1) *If $n = 1$, the height of P is at most 1.*
(2) *In general, the height of P is at most n.*

This theorem implies that a Noetherian local ring (R, M) has finite dimension, which is at most the number of elements in any set of generators for an M-primary ideal of R. It is also used to show that $\dim \widehat{R} = \dim R$, where $\widehat{R}$ is the M-adic completion of R, and that for any Noetherian ring S, the dimension of the formal power series $S[[x_1, \dots, x_n]]$ in n analytic indeterminates over S is $\dim S + n$. While each maximal ideal of a Noetherian ring S has finite height, the set of these heights need not be bounded, and hence S need not be finite-dimensional (see Nagata 1962, App. 1)).

The dimension theory of formal power series rings is much less tractable than its polynomial ring counterpart. Arnold (1973a,b) has done pioneering and difficult work in the area. In particular, he has shown that $R[[x]]$ may be infinite-dimensional, even when R is zero-dimensional. A key concept in Arnold's work is that of an SFT-ring. The definition is as follows. An ideal I of a ring R is an **SFT-ideal** (SFT stand for **strong finite type**) if there exists a finitely generated ideal J contained in I and a positive integer k such that $a^k \in J$ for each $a \in I$, and R is an **SFT-ring** if each ideal of R is an SFT-ideal. Arnold showed that $R[[x]]$ is infinite-dimensional if R is not an SFT-ring. The status of the converse of this statement is an open question. A rank-one non-discrete valuation domain V does not satisfy the SFT-condition, and hence $V[[x]]$ is infinite-dimensional. On the other hand, if D is an n-dimensional Prüfer domain satisfying SFT, then $\dim D[[x]] = n + 1$. In general it is possible for $R[[x]]$ to be finite-dimensional, but for its dimension to be greater than $\dim R + 1$.

General references for the material in this section are (Gilmer 1992, Chap. II, Sect. 30), (Zariski and Samuel 1960, Chaps. VII and VIII), (Kunz 1985, Chap. II), (Kaplansky 1970, Chaps. 1–3), and (Jaffard 1960).

C.3 Factorization and Decomposition of Polynomials

by Joachim von zur Gathen in Paderborn, Germany

Carl Friedrich Gauß proved that (multivariate) polynomials over a field or over the integers form a Unique Factorization Domain. The computational version of this fundamental result asks for an algorithm which, given a polynomial as input, finds its irreducible factors. And the complexity-theoretic version asks to do this as efficiently as possible. This question and its answers form one of the successful areas of *computer algebra*.

An easy kind of factorization is to calculate the greatest common divisor (gcd) of two polynomials. Euclid's algorithm—"the granddaddy of all algorithms" (Knuth 1998, p. 318)—does this for univariate polynomials over a field. It has numerous applications, for example in the Chinese Remainder Algorithm, for modular arithmetic and computation in algebraic extensions, in coding theory (Berlekamp-Massey algorithm), and for fast linear algebra on *sparse* matrices. The theory of subresultants gives important structural insights and leads to efficient algorithms, for example the modular methods for integer and for multivariate polynomials.

If $f_1, \ldots, f_r$ are the distinct irreducible monic factors of $f = f_1^{e_1} \cdots f_r^{e_r} \in F[x]$ over a field F, then $f_1 \cdots f_r$ is the **squarefree part** of f. It can be computed as $f / \gcd(f, f')$ in characteristic zero, while in characteristic $p > 0$, also pth roots have to be extracted. This is easy over all fields of practical interest, such as finite fields or function fields over them, but there are sufficiently bizarre (but still *computable*) fields over which squarefreeness is undecidable, in the sense of Turing. The squarefree part is useful in the symbolic integration of rational functions.

Algorithms for the factorization of polynomials are built in a hierarchical way: One starts with univariate polynomials over finite fields, then over $\mathbb{Q}$, and then over algebraic extensions and of multivariate polynomials.

The first modern factorization methods for $f \in \mathbb{F}_q[x]$ of degree n, where $\mathbb{F}_q$ is a finite field with q elements, are due to Berlekamp. His motivation came from coding theory, and he found an algorithm based on linear algebra that uses $O^\sim(n^3 + nq)$ operations in $\mathbb{F}_q$. Here, the "soft Oh" notation $O^\sim$ means that we ignore factors of $\log n$. This is ok for small q, such as $q = 2$, which is particularly important in coding theory. But for large q, this is not polynomial in the "input length" $n \log q$ of f. A milestone was Berlekamp's invention of a polynomial time algorithm, using $O^\sim(n^3 + n \log q)$ operations. This is a probabilistic algorithm of Las Vegas type, whose output is always correct but whose running time is a random variable whose mean is given above.

A decade later, Cantor and Zassenhaus (1981) presented an algorithm which proceeds in two stages. In the ***distinct-degree factorization*** stage, the (squarefree) input is factored into a product of polynomials each of whose irreducible factors has the same degree. This is achieved by taking the gcd with $g_i = x^{q^i} - x$ for $i = 1, 2, \ldots$, and based on the fact that g_i is the product of all monic irreducible polynomials in $\mathbb{F}_q[x]$ whose degree divides i. This fact, and the squarefree and the distinct-degree factorization are in Gauß' *Nachlaß* (posthumous works).

The second ***equal-degree factorization*** stage factors a polynomial all of whose irreducible factors have the same degree. This is done by a probabilistic algorithm, the rudiments of which can already be found in Legendre's work. It remains an open question for the theory whether this can be achieved deterministically in polynomial time.

These algorithms have been improved in the 1990's and there is now a variety of algorithms which are optimized for a specific range of the proportion of degree n to field length $\log q$. The corresponding software can attack enormously large problems; in 2000, polynomials of degree one million (over $\mathbb{F}_2$) can be factored.

The next task is to factor integer polynomials. Gauß' Lemma reduces this to factoring in $\mathbb{Q}[x]$ and in $\mathbb{Z}$. The latter seems computationally hard (at our state of knowledge). Zassenhaus' algorithm for factoring in $\mathbb{Q}[x]$ works by first factoring modulo a (small) prime p, then applying ***Hensel lifting*** to get a factorization modulo a sufficiently large power of p, and finally trying out all combinations of the modulo factors to find the true factors. This works well for small inputs but uses exponential time. Lenstra, Lenstra and Lovász gave an efficient algorithm to find *short* vectors in integer lattices. Among many other computational applications, this also provides a polynomial-time algorithm for factoring in $\mathbb{Q}[x]$.

The next task are bivariate polynomials. Again, a judicious application of modular factorization, Hensel lifting, and an (efficient) factor combination yields an efficient factorization algorithm. Factorization over algebraic extensions can be handled in a similar way.

For polynomials in more than two variables, one may apply the same technology. However, the input length of the *dense representation* of polynomials, where the coefficient of each term up to the degree has to be specified, grows too quickly in size. It is more desirable to use concise forms such as the *sparse representation*, where only the nonzero coefficients are given. It is a remarkable achievement, mainly due to Kaltofen, to factor polynomials probabilistically in time polynomial in the input length for even more concise representations, namely by *arithmetic circuits* (a.k.a. straight-line programs) or by *black boxes*. The main ingredient are efficient versions of Hilbert's irreducibility theorem, as proved by Kaltofen and this author.

Factorization in $\mathbb{Z}_m[x]$, for $m \in \mathbb{N}$, exhibits a number of unusual properties. For example, there are polynomials with exponentially many factorizations into irreducible polynomials, but these can all be represented by a polynomial-sized data structure.

Detailed discussions, references, and reports on implementations can be found in Gathen von zur and Gerhard (1999).

The **composition** of two univariate polynomials $g, h \in F[x]$ is $g \circ h = g(h) \in F[x]$. In the **decomposition problem**, we are given $f \in F[x]$ and ask whether there exist $g, h \in F[x]$ so that $f = g \circ h$ and $2 \leq \deg g < \deg f$. The first polynomial-time algorithm was presented by Kozen and Landau in 1986, and the fastest one, by this author, uses $O(n \log^2 n \log\log n)$ operations in F, where $n = \deg f$ and $r = \deg g$ is prescribed, with $\operatorname{char} F \nmid r$ (the so-called **tame case**). Ritt showed in 1922 that such decompositions are essentially unique. Many generalizations have been studied: rational functions by Zippel, algebraic functions by Kozen, Landau, and Zippel, and multivariate decompositions by several authors. The problem has important applications in the simplification of parameterizations of rational algebraic curves and of the inverse kinematic equations in robotics (see also Section D.7).

C.4 Element Factorization in Integral Domains
by David F. Anderson in Knoxville, TN, USA

The ring $\mathbb{Z}$ of integers is a unique factorization domain (UFD), that is, each nonzero nonunit is a product of irreducible elements, and this factorization is unique up to order and associates. Beginning in the nineteenth century, the "ring of integers" of subfields of $\mathbb{C}$ other than $\mathbb{Q}$ started to play an important role in number theory. Precisely, let F be a subfield of $\mathbb{C}$ with $[F : \mathbb{Q}]$ finite. Then the *ring of integers of F* is the subring of all elements of F which satisfy a monic polynomial with coefficients in $\mathbb{Z}$. In a ring of integers, each nonzero nonunit is a product of irreducible elements. Some of these rings of integers, such as $\mathbb{Z}$ and $\mathbb{Z}[i]$, have unique factorization; others, such as $\mathbb{Z}[\sqrt{-5}]$ and $\mathbb{Z}[\frac{1+\sqrt{-31}}{2}]$, do not have unique factorization since $6 = 2 \cdot 3 = (1 + \sqrt{-5})(1 - \sqrt{-5})$ and $8 = 2^3 = (\frac{1+\sqrt{-31}}{2})(\frac{1-\sqrt{-31}}{2})$ are each distinct factorizations into irreducible elements.

This lack of unique factorization of elements lead to E. Kummer's ideal numbers, and later to R. Dedekind's theory of unique factorization of ideals for rings of integers. Today such an integral domain is called a *Dedekind domain* if it is Noetherian, integrally closed, and each nonzero prime ideal of R is maximal. Dedekind domains are precisely the integral domains in which each nonzero proper ideal is (uniquely) a product of prime ideals (cf. Section C.1 and Section C.5). For example, any PID is a Dedekind domain, but $\mathbb{Z}[\sqrt{-5}]$ and $\mathbb{Z}[\frac{1+\sqrt{-31}}{2}]$ are Dedekind domains which are not PIDs. The *ideal class group* of a Dedekind domain R, denoted by $\mathrm{Cl}(R)$, is the abelian group of invertible fractional ideals of R under the usual ideal multiplication modulo its subgroup of principal fractional ideals. A Dedekind domain R is a UFD (in fact, a PID) if and only if $\mathrm{Cl}(R) = 0$. A ring of integers has a finite ideal class group, but in general, $\mathrm{Cl}(R)$ may be any abelian group.

An integral domain R is a *Krull domain* if $R = \bigcap_{P \in Y} R_P$, where $Y = X^{(1)}(R)$ is the set of *height-one* (i.e., nonzero minimal) *prime ideals* of R, each nonzero nonunit of R is contained in only a finite number of $P \in X^{(1)}(R)$, and each R_P is a local PID(i.e., has a unique prime, up to associates). The class of Krull domains includes UFDs and integrally closed Noetherian domains (and hence Dedekind domains). For example, the polynomial ring $\mathbb{Z}[\sqrt{-5}][X]$ is a Krull domain which is neither a UFD nor a Dedekind domain. The divisor class group of a Krull domain R, $\mathrm{Cl}(R)$, is the abelian group of fractional divisorial ideals of R under v-multiplication modulo its subgroup of principal fractional ideals. Dedekind domains are then just the Krull domains in which each nonzero prime ideal is maximal, and a Krull domain R is a UFD if and only if $\mathrm{Cl}(R) = 0$. More generally, $\mathrm{Cl}(R)$ is torsion if and only if each nonzero nonunit of R is a product of primary elements, if and only if for each nonzero $x, y \in R$, there is an integer $n = n(x, y) \geq 1$ such that $x^n R \cap y^n R$ is principal (such Krull domains are called **almost factorial**). Thus, in some sense, $\mathrm{Cl}(R)$ measures how far R is from being a UFD. For more on Krull and Dedekind domains, see (Fossum 1973, Gilmer 1992).

The study of nonunique factorization has its roots in algebraic number theory, and it has recently received considerable renewed interest, particularly in the context of arbitrary integral domains. We say that an integral domain R is **atomic** if each nonzero nonunit of R is a product of irreducible elements (atoms). A Noetherian domain, Krull domain, or any domain satisfying the ascending chain condition on principal ideals is atomic (but not conversely). Unlike Krull domains or UFDs, the class of atomic domains is not very stable since R atomic does not imply that $R[X], R[[X]]$, or R_S is atomic. Although we are interested in factorization in arbitrary atomic integral domains, the best results are for Krull domains, in particular, Dedekind domains, with nice divisor class groups, or for one-dimensional Noetherian domains.

For a nonzero nonunit $x \in R$, we say that $x = x_1 \cdots x_n$, where each $x_i \in R$ is irreducible, is a factorization of length n. In an atomic integral domain, an element may have factorizations of different lengths, or even factorizations of arbitrarily long length. For example, in $\mathbb{Q}[X^2, X^3]$, for each integer $n \geq 0$, $X^6(1 - X^2)^n = [X^3(1 - X)^n][X^3(1 + X)^n] = (X^2)^3(1 - X^2)^n$ are factorizations of length 2 and $n + 3$, respectively. For a nonzero nonunit $x \in R$, define $L(x) = \{n \mid x \text{ has a factorization of length } n\}$ and $\mathcal{L}(R) = \{L(x) \mid x \in R \text{ is a nonzero nonunit}\}$. Basic problems in factorization theory concern lengths of factorizations, the number of nonassociate irreducible elements in a factorization, just how *good* or *bad* such factorizations can be, and the extent to which $\mathcal{L}(R)$ determines the arithmetic of R. See (Anderson 1997) and (Chapman and Glaz 2000) for more details and references on factorization in integral domains and its generalizations.

An atomic integral domain R is said to be a **half-factorial domain (HFD)** if any two factorizations of a given nonzero nonunit of R into irreducible elements have the same length. The starting point for studying HFDs was L. Carlitz's observation in 1960 that the ring of integers R in a number field is an HFD if and only if $|\mathrm{Cl}(R)| \leq 2$. For example, $\mathbb{Z}[\sqrt{-5}]$ is a Dedekind HFD since $\mathrm{Cl}(\mathbb{Z}[\sqrt{-5}]) = \mathbb{Z}/2\mathbb{Z}$, and $\mathbb{R} + X\mathbb{C}[X]$ is a non-Dedekind HFD, but neither is a UFD. In fact, in $\mathbb{R} + X\mathbb{C}[X]$, $X^2 = (aX)(a^{-1}X)$ for any nonzero $a \in \mathbb{C}$; so X^2 has infinitely many nonassociate factorizations. We have already seen that $\mathbb{Q}[X^2, X^3]$ and

$\mathbb{Z}[\frac{1+\sqrt{-31}}{2}]$ are not HFDs.

We are interested in invariants which measure the lack of unique factorization in an atomic integral domain. One such invariant is the **elasticity** of R, $\rho(R) = \sup\{m/n \mid x_1 \cdots x_m = y_1 \cdots y_n$ with each $x_i, y_j \in R$ irreducible$\}$. Then $1 \leq \rho(R) \leq \infty$, with $\rho(R) = 1$ if and only if R is an HFD. Thus $\rho(R)$ measures how far an atomic integral domain R is from being an HFD. Even for Dedekind domains, $\rho(R)$ may be any real number $r \geq 1$ or infinity. However, if R is a Krull domain with $\mathrm{Cl}(R)$ finite, then $\rho(R) = m/n$ is rational, and moreover, $x_1 \cdots x_m = y_1 \cdots y_n$ for some irreducible $x_i, y_j \in R$. If R is a Krull domain which is not a UFD, then $1 \leq \rho(R) \leq \frac{D(\mathrm{Cl}(R))}{2} \leq \frac{|\mathrm{Cl}(R)|}{2}$, where $D(G)$ is the **Davenport constant** of G. (For a finite abelian group G, $D(G)$ is the least positive integer d such that for any sequence $S \subseteq G$ with $|S| = d$, some nonempty subsequence of S has sum 0; $D(G) = \infty$ if G is infinite. We always have $D(G) \leq |G|$ and $D(\mathbb{Z}/n\mathbb{Z}) = n$.) But, we may have $\rho(R) < \frac{D(\mathrm{Cl}(R))}{2}$. For example, for any finite abelian group G, there is a Dedekind HFD with class group G. However, if each divisor class contains a height-one prime ideal (as is the case for algebraic number rings), then $\rho(R) = max\{\frac{D(\mathrm{Cl}(R))}{2}, 1\}$. For example, $\rho(\mathbb{Z}[\frac{1+\sqrt{-31}}{2}]) = \frac{3}{2}$ since $\mathrm{Cl}(\mathbb{Z}[\frac{1+\sqrt{-31}}{2}]) = \mathbb{Z}/3\mathbb{Z}$. Thus in a Krull domain R, factorization properties depend more on how the height-one prime ideals are distributed in the divisor classes in $\mathrm{Cl}(R)$, i.e., in the combinatorial properties of $\mathrm{Cl}(R)$, rather than in the group-theoretic properties of $\mathrm{Cl}(R)$ itself.

For a less trivial example, let K be a field. Then $\rho(K[X^2, X^3]) < \infty$ if and only if K is finite; and in this case, $\rho(K[X^2, X^3]) = \frac{D(K)+2}{2}$ ($= \frac{n(p-1)+3}{2}$ when K has p^n elements). However, $\rho(K[[X^2, X^3]]) = \frac{3}{2}$ for any field K. As another example, for (R, M) a local one-dimensional Noetherian domain, $\rho(R) < \infty$ if and only if R is analytically irreducible (i.e., the M-adic completion of R is an integral domain), and in this case, $\rho(R)$ is rational.

We have already seen that the divisor class group plays an important role in factorization theory in the Krull domain setting. For any integral domain R, define its t-class group $\mathrm{Cl}_t(R)$ to be the abelian group of t-invertible fractional t-ideals of R under t-multiplication modulo its subgroup of principal fractional ideals. If R is a Krull domain, then $\mathrm{Cl}_t(R) = \mathrm{Cl}(R)$, and $\mathrm{Cl}_t(R) = \mathrm{Pic}(R)$ if R is a one-dimensional integral domain. Instead of factorizations using prime or irreducible elements, one can also study factorizations using primary elements. An integral domain R is called weakly factorial (resp., almost weakly factorial) if each nonzero nonunit $x \in R$ (resp., some power of x) is a product of primary elements. Then, in analogy with Krull domains, R is weakly factorial (resp., almost weakly factorial) if and only if R is **weakly Krull** and $\mathrm{Cl}_t(R) = 0$ (resp., $\mathrm{Cl}_t(R)$ is torsion) (here, R is weakly Krull if $R = \bigcap R_P$, where this intersection is over $P \in X^{(1)}(R)$, and each nonzero nonunit of R is contained in only a finite number of $P \in X^{(1)}(R)$). Then for R an atomic weakly Krull domain, $\rho(R) \leq D(\mathrm{Cl}_t(R)) \sup \{\rho(R_P) | P \in X^{(1)}(R)\}$, and $\rho(R) = \sup \{\rho(R_P) | P \in X^{(1)}(R)\}$ if R is an atomic weakly factorial domain.

There are also many other invariants which relate to the length of factorizations, the structure of $\mathcal{L}(R)$, or the asymptotic behavior of factorizations. One can also consider factorization concepts in the contexts of commutative cancellative monoids or commutative rings with nonzero zerodivisors.

C.5 Dedekind and Prüfer Domains

by Marco Fontana in Rome, Italy and Ira J. Papick in Columbia, MO, USA

All rings in this Section are assumed to be commutative.

Dedekind domains

C.5.1 Definition An integral domain R is called a **Dedekind domain** if each nonzero proper ideal of R can be represented uniquely (apart from order) as a finite product of prime ideals of R (cf. Section C.1 and Section C.4).

Any principal ideal domain (e.g., $\mathbb{Z}$ or $K[X]$, where K is a field) is a Dedekind domain, and since the integral closure of a Dedekind domain in a finite field extension of its quotient field is also a Dedekind domain, it follows that the ring of algebraic integers in an algebraic number field and the ring of integral functions in a field of algebraic functions of one variable are also Dedekind domains (Zariski and Samuel 1958, Vol. 1, Section 6, pp. 270–271).

The theory of Dedekind domains was established by Emmy Noether, in the mid twenties, as a generalization of results concerning factorization properties of algebraic integers obtained primarily by Richard Dedekind in 1871. E. Noether provided a fundamental characterization for Dedekind domains which prominently exhibited the role of chain conditions for ideals. Namely,

C.5.2 Theorem *An integral domain R is a Dedekind domain if and only if R satisfies the following properties (called **Noether Axioms**):*
(1) *R is Noetherian (ascending chain condition for ideals);*
(2) *R is **integrally closed** (R coincides with the set of the roots of its monic polynomials inside the quotient field);*
(3) *R/I is Artinian (descending chain condition for ideals), for each nonzero ideal I of R.*

Note that condition (3) can be replaced by
(3') each nonzero prime ideal of R is maximal.

As a consequence, it follows that *Dedekind domains are Noetherian integral domains such that each localization at maximal ideals is a discrete valuation domain.* In particular, as we have mentioned already, each principal ideal domain is a Dedekind domain, but not conversely (e.g., $\mathbb{Z}[i\sqrt{5}]$).

The structural diversity of Dedekind domains can be observed by the following sampling of some classical characterizations:

C.5.3 Theorem *An integral domain R is a Dedekind domain if and only if one of the following statement holds:*
(1) *(W. Krull, 1935) the nonzero fractional ideals of R form a group under multiplication;*
(2) *(I. Kaplansky, 1952, the necessity, and S. U. Chase, 1960, the sufficiency) each extension of a torsion module of bounded order by a torsion-free module splits;*
(3) *(C. U. Jensen, 1963) R/I is a principal ideal ring, for each nonzero ideal I of R.*

The proofs or appropriate references for Theorem C.5.2 and Theorem C.5.3 can be found in (Kaplansky 1970, Gilmer 1972a, Narkiewicz 1990).

Dedekind domains have played a crucial role in the development of Algebraic Number Theory and Algebraic Geometry. In fact, it was observed by R. Dedekind and H. Weber in 1882 that several properties concerning rings of algebraic integers also apply to rings of integral elements in function fields. This realization contained the germ of the fact that the coordinate ring of a nonsingular irreducible curve is a Dedekind domain.

Prüfer domains

In view of Krull's characterization of Dedekind domains (see Theorem C.5.3 (a)), it is natural to consider the following:

C.5.4 Definition An integral domain R having its set of nonzero finitely generated fractional ideals form a group under multiplication is called ***Prüfer domain***.

Dedekind domains are Prüfer domains, but not conversely (see below). Papers of H. Prüfer in 1932 and of W. Krull in 1936 introduced the study of Prüfer domains, although the first place the name *Prüfer ring* appears in the literature seems to be in the classic book by Cartan and Eilenberg (1956). Prüfer and Krull were interested in the ideal and overring theory of Prüfer domains, whereas Cartan and Eilenberg examined the homological nature of these domains (by an ***overring*** of an integral domain R we mean a subring of its quotient field containing R as a subring). Many researchers during the second half of the 20th century investigated the structure of Prüfer domains and a reasonable sampling of this work has been chronicled in (Kaplansky 1970, Gilmer 1972a, Fontana, Huckaba, and Papick 1997).

A brief overview of some fundamental results concerning Prüfer domains will help to illustrate important research topics from the late 30's to the early 70's. The previously mentioned sources contain precise acknowledgement of the listed results.

C.5.5 Theorem *Let R be an integral domain. The following statements are equivalent to R being a Prüfer domain:*

(1) *Each valuation overring of R is a ring of fractions.*

(2) *Each finitely generated torsion-free R-module is projective.*

(3) *Each overring of R is integrally closed.*

(4) *Each overring S of R is R-flat (i.e., for each monomorphism of R-modules $N \to M$, $N \otimes_R S \to M \otimes_R S$ is also a monomorphism).*

(5) *R is integrally closed and there exists a positive integer $n \geq 2$ such that $(a, b)^n = (a^n, b^n)$, for each a, b in R.*

(6) *For all ideals I, J, L of R, $I + (J \cap L) = (I + J) \cap (I + L)$ (or, equivalently, the Chinese Remainder Theorem for ideals holds in R).*

From Theorem C.5.5 and Theorem C.5.3, it follows that: **(a)** *the integral closure of a valuation domain (and, more generally, of a Prüfer*

*domain) in any algebraic extension field of its field of quotients is a
Prüfer domain;* **(b)** *Prüfer domains coincide with the integral domains
such that each localization at maximal ideals is a valuation domain;* **(c)**
Noetherian Prüfer domains coincide with Dedekind domains.

Note that the ring of all algebraic integers (i.e., the integral closure of $\mathbb{Z}$
in the field of complex numbers) is a Prüfer *non* Dedekind domain, since
it is not Noetherian. Further interesting examples of Prüfer domains
include: **(1)** *Bezout domains* (i.e., integral domains in which every
finitely generated ideal is principal) and hence, in particular, Kronecker
function rings (W. Krull, 1936); **(2)** the ring of entire functions (O.
Helmer, 1940); **(3)** the intersection of any finite family of valuation
domains on a given field (M. Nagata, 1953); **(4)** the ring Int(R) of all
integer valued polynomials, where R is a Dedekind domain with finite
residue fields (D. Brizolis, 1979; D. L. McQuillan, 1985; J.-L. Chabert,
1987) hence, in particular, the classical ring of all integer-valued rational
polynomials Int($\mathbb{Z}$) := $\{f(X) \in \mathbb{Q}[X] \mid f(\mathbb{Z}) \subseteq \mathbb{Z}\}$ is a Prüfer (but *not* a
Dedekind) domain.

Extensions and ramifications of these ideas have been pursued in great
detail since the 70's, and more recent work focuses on specific properties
of ideals and modules over Prüfer domains. For example, an ideal in
a Dedekind domain requires no more than two generators and early
evidence suggested that finitely generated ideals in a Prüfer domain also
might require only two generators. In fact, H. Prüfer in his 1932 paper
showed that if each two-generated ideal of a domain R is invertible,
then R is a Prüfer domain. However, in 1979 H. Schülting gave an
example of a Prüfer domain with a finitely generated ideal requiring
three generators, which in turn, led to several new studies investigating
the number of generators needed for finitely generated ideals in Prüfer
domains.

Another notable example of recent directions in the theory of Prüfer
domains is the Y. Lequain and A. Simis extension of the D. Quillen and
A. Suslin solutions to the Serre Conjecture. They proved in 1980 that
if R is a Prüfer domain then, for all positive integers n, all finitely generated projective modules over $R[X_1, X_2, \ldots, X_n]$ are extended from R
(recall that, if S is an R-algebra, an S-module N is *extended* from R
if there exists an R-module M such that N is isomorphic to $M \otimes_R S$).

Research activity involving Prüfer domains has remained strong since
the initial work of H. Prüfer and W. Krull, and will continue to flour-
ish for many years to come (Chapman and Glaz 2000). The rich and
interesting structure of these rings, as well as their natural presence,

make them important and useful objects in the context of commutative algebra.

C.6 Local Rings

by T. Y. Lam in Berkeley, CA, USA

Examples, history, and conventions

Consider the ring $A = \{ a/b \mid a, b \in \mathbb{Z}, b \text{ odd} \}$. In A, all odd primes are units, and $2A$ is the only maximal ideal. The ring $B = k[[x, y]]$ of formal power series in x, y with coefficients in a field k has a similar feature: any power series outside of $xB + yB$ can be formally inverted, so $xB + yB$ is the only maximal ideal of B. Yet another "similar" ring is $C = \mathbb{Z}_{p^n}$, for any prime p, except that C is no longer a domain when $n > 1$.

A commutative ring R is called a *local ring* if it has exactly one maximal ideal, or equivalently, $R \neq 0$ and the nonunits of R form an ideal $\mathfrak{m}$. For such a local ring $(R, \mathfrak{m})$, $R/\mathfrak{m}$ is a field, called the *residue field* of R. The rings A, B, C in the last paragraph are good prototype examples.

Commutative local rings were introduced by Krull (1938), who called them "Stellenringe". Chevalley (1943) coined the English translation "local rings", and further developed their theory. In both (Krull 1938) and (Chevalley 1943), local rings were taken to be *noetherian*: a cause later championed by Bourbaki. Here, however, we follow Kaplansky, Atiyah-Macdonald, Matsumura (1989), and Bruns and Herzog (1998), and define local rings *without* the noetherian assumption; in Bourbaki's terminology, these are the "quasi-local" rings.

In the following, we give a quick overview of local rings (referring the reader to (Matsumura 1989) and (Bruns and Herzog 1998) for detailed expositions). Until further notice, *all rings are assumed to be commutative*.

Sources of local rings and algebraic geometry

Local rings are important because they are ubiquitous. First, all fields are local rings, although some of us would say that these shouldn't count. Any indecomposable artinian ring is (noetherian) local, and if R is a local ring, then so is any nonzero factor ring of R. Also, if $\mathfrak{m}$ is a maximal ideal of any ring R, then $R/\mathfrak{m}^n$ is a local ring for any $n \geq 1$. Lastly, a ring R contained in a field K is called a *valuation ring* of K if every $a \in K \setminus R$

has $a^{-1} \in R$. These rings arise from the theory of valuations and places, and are always local (though they are usually not noetherian). Krull was certainly deeply interested in these non-noetherian local rings as well.

If $\mathfrak{p}$ is a prime ideal in any ring R, then equivalence classes of the fractions $\{\, r/s \mid r, s \in R,\ s \notin \mathfrak{p} \,\}$ form a local ring with maximal ideal $\{\, p/s \mid p \in \mathfrak{p},\ s \notin \mathfrak{p} \,\}$. This ring, denoted by $R_{\mathfrak{p}}$, is called the *localization* of R at the prime $\mathfrak{p}$. By studying the local ring $R_{\mathfrak{p}}$, we hope to get insight on the prime $\mathfrak{p}$. We can localize R-modules too by tensoring them with $R_{\mathfrak{p}}$; this is known to preserve exact sequences. By localization with respect to *all* primes $\mathfrak{p}$, we get a powerful "local-global method" for studying R and its modules.

Local rings are very much "geometric" objects; they arise naturally in algebraic (resp. analytic) geometry as rings of "germs" of functions at a point of an algebraic (resp. analytic) variety. For instance, the local ring at the origin for the twisted cubic $x^2 = y^3$ is the localization of its coordinate ring $\mathbb{C}[x, y]/(x^2 - y^3)$ at the prime ideal $(\bar{x}, \bar{y})$. Classically, purely algebraic results on local rings were obtained with the goal of studying the local behavior of algebraic varieties. The theory of local rings developed by Krull, Chevalley, Zariski, Cohen, and others went a long way towards the rewriting of the algebraic foundations of classical algebraic geometry in the 1940s. Later, in the hands of Grothendieck, varieties were further generalized into "schemes". Each scheme X comes with a structural sheaf $\mathcal{O}_X$, whose "stalk" at each point $x \in X$ is a local ring $\mathcal{O}_{X,x}$.

Toolkit for local algebra

Some principal theorems on local rings $(R, \mathfrak{m})$ are as follows:

C.6.1 Krull's Intersection Theorem *If R is noetherian, then* $\bigcap_{i=1}^{\infty} \mathfrak{m}^i = 0$.

C.6.2 Krull's Dimension Theorem *If R is noetherian, it has finite* **Krull dimension** n *(the length of a longest chain of prime ideals—cf. Section C.2 and Section C.9). There exist $x_1, \ldots, x_n \in \mathfrak{m}$ which generate an ideal with radical $\mathfrak{m}$: this is called a* **system of parameters** *for R.*

C.6.3 Nakayama's Lemma *For any finitely generated R-module M, $\mathfrak{m}M = M$ implies that $M = 0$.*

C.6.4 Kaplansky's Theorem *Any projective module over R is free.*

Any noetherian local ring $(R, \mathfrak{m})$ comes with a Hausdorff topology whose system of neighborhoods at 0 is given by the powers of $\mathfrak{m}$. With respect to this topology, R has a completion $(\hat{R}, \hat{\mathfrak{m}})$, which is also a noetherian local ring (of the same Krull dimension). Going from R to $\hat{R}$ (and vice versa) is often a very effective tool in local algebra.

Every finitely generated module $M \neq 0$ over a noetherian local ring $(R, \mathfrak{m})$ comes with many invariants, such as $\dim(M)$ (Krull dimension of $R/\operatorname{ann}(M)$), the **projective dimension** $\operatorname{pd}_R(M)$ and the **injective dimension** $\operatorname{id}_R(M)$ (the lengths of the shortest projective and injective resolutions of M), and $\operatorname{depth}(\mathfrak{q}, M)$ (length of a maximal M-regular sequence in an ideal $\mathfrak{q} \subseteq \mathfrak{m}$). In case $\operatorname{pd}_R(M) < \infty$, one has the following depth formula of Auslander and Buchsbaum:

$$\operatorname{pd}_R(M) = \operatorname{depth}(\mathfrak{m}, R) - \operatorname{depth}(\mathfrak{m}, M).$$

In case $M/\mathfrak{q}M$ has finite length over R, there exists a **Hilbert-Samuel polynomial** $P_\mathfrak{q}(t, M) \in \mathbb{Q}[t]$ of degree $d = \dim(M)$ such that, for large n, $P_\mathfrak{q}(n, M) = \ell_R(M/\mathfrak{q}^{n+1}M)$. (For an explicit example of this, see the next section.) The leading term of $P_\mathfrak{q}(t, M)$ has the form $e_\mathfrak{q}(M)t^d/d!$, where $e_\mathfrak{q}(M)$ is a positive integer called the **multiplicity** of M with respect to $\mathfrak{q}$. Under suitable assumptions, $e_\mathfrak{q}(M)$ is given by the Euler-Poincaré characteristic of the homology of a Koszul complex. In the case $M = R/\mathfrak{p}$ where $\mathfrak{p} + \mathfrak{q}$ is $\mathfrak{m}$-primary, this enabled Serre to express the intersection multiplicity of two subvarieties at a point of a given variety in terms of lengths of certain Tor-modules.

Regular local rings and their generalizations

A noetherian local ring $(R, \mathfrak{m})$ of dimension n is said to be **regular** if the maximal ideal $\mathfrak{m}$ can be generated by n elements. Two further characterizations are as follows:

(1) $\dim_{R/\mathfrak{m}} \mathfrak{m}/\mathfrak{m}^2 = n$;
(2) the "associated graded ring" $\bigoplus_{i=0}^\infty \mathfrak{m}^i/\mathfrak{m}^{i+1}$ is isomorphic to the graded polynomial ring $k[x_1, \ldots, x_n]$, where $k = R/\mathfrak{m}$.

The characterization (2) is extremely powerful. For instance, it implies that R must be an integral domain, and that the Hilbert-Samuel polynomial $P_\mathfrak{m}(t, R)$ (for the free module R with respect to the maximal ideal $\mathfrak{m}$) has the very simple form $\binom{t+n}{n}$. Regular local rings were utilized effectively by Zariski to give a rigorous theory of simple points on an algebraic variety.

In general, if $(R, \mathfrak{m})$ is any regular local ring of dimension n, then so is its completion $(\hat{R}, \hat{\mathfrak{m}})$. For instance, the formal power series ring $k[[x_1, \ldots, x_n]]$ over any field k is a complete regular local ring of dimension n. The structure of complete regular local rings was determined by Cohen (1946). In that paper, Cohen showed how a general complete regular local ring is related to formal power series rings over fields and discrete valuation rings.

Two spectacular results on regular local rings were obtained in the 1950s. First, Serre and Auslander-Buchsbaum showed that regular local rings are precisely the noetherian local rings of finite global dimension (and indeed global dimension agrees with Krull dimension). This implies, for instance, Krull's conjecture that prime localizations of regular local rings remain regular. Second, Auslander and Buchsbaum proved that regular local rings are unique factorization domains, which is in part motivated by Weierstrass' Theorem that power series rings $k[[x_1, \ldots, x_n]]$ (over a field k) are UFDs.

One has the following hierarchy for noetherian local rings:

regular $\implies$ complete intersection $\implies$ Gorenstein $\implies$ Cohen-Macaulay,

which gives three interesting and increasingly broader classes of generalizations of regular local rings. "Complete intersection" means a factor ring of a regular local ring by an ideal generated by a regular sequence. "Gorenstein" means a noetherian local ring R with finite injective dimension over itself ($\mathrm{id}_R(R) < \infty$). "Cohen-Macaulay" means a noetherian local ring $(R, \mathfrak{m})$ of dimension n such that $\mathfrak{m}$ has a regular sequence of length n (see Section C.8). Prompted by Macaulay's classical work on ideals in polynomial rings, Cohen-Macaulay rings have now found deep applications in invariant theory, algebraic geometry, polytope geometry, and algebraic combinatorics. Most remarkably, Mark Haiman's recent solution (Haiman 2001) of the $n!$ Conjecture and Macdonald's Positivity Conjecture in symmetric function theory rests on proving that a certain isospectral Hilbert scheme is Gorenstein and Cohen-Macaulay.

Noncommutative local rings

In this subsection, we waive the commutativity assumption on our rings. Local rings are important in noncommutative ring theory as well. An arbitrary ring R is called **local** if it has a unique maximal left ideal $\mathfrak{m}$. We see easily that $\mathfrak{m}$ is also the unique maximal right ideal of R, and that $R \setminus \mathfrak{m}$ is exactly the group of units of R. Here we get a *residue* *division ring* $R/\mathfrak{m}$. Some examples of noncommutative local rings are: (1) division rings; (2) group rings kG for k a field of characteristic p and G any p-primary group; (3) $n \times n$ upper triangular matrices with constant diagonals over any division ring; and (4) the endomorphism ring of any indecomposable module of finite length (over any ring).

Various properties of commutative local rings still hold: for instance, Lemma C.6.3 and Theorem C.6.4 do not depend on commutativity. One of the most important theorems making use of noncommutative local rings is the following

C.6.5 Krull-Schmidt-Azumaya Theorem *Let $\{M_i, N_j\}$ be indecomposable modules over any ring A such that $\bigoplus_{i \in I} M_i \cong \bigoplus_{j \in J} N_j$. If the endomorphism rings $\mathrm{End}_A(M_i)$ are all local, then there is a bijection $\pi \colon I \to J$ such that $M_i \cong N_{\pi(i)}$ for all $i \in I$.*

Some attempts have been made toward developing a theory of regular local rings in the noncommutative setting, but the situation here is more complicated. It seems reasonable to define "regularity" of a local ring R by stipulating that R be 2-sided noetherian, with finite left and right global dimension (which must then be equal). *Is R then a domain?* This is known to be true if the global dimension is ≤ 3 (results of M. Ramras and R. Snider), but has remained unknown in general. On the other hand, there are other notions of noncommutative regularity (e.g., prompted by the consideration of graded rings), which we will not have space to explore here.

C.7 Semilocal Rings

by T. Y. Lam in Berkeley, CA, USA

From commutative to noncommutative

In commutative algebra, a ring R is said to be *local* if it has a unique maximal ideal, and *semilocal* if it has finitely many maximal ideals. If $\mathfrak{p}$ is a prime ideal in a commutative ring, we can "invert" the elements in $S = R \setminus \mathfrak{p}$ to form a local ring R_S, with the unique maximal ideal $\mathfrak{p}R_S$. More generally, if $\mathfrak{p}_1, \ldots, \mathfrak{p}_n$ are prime ideals in R, then we can "invert" the elements in $S = R \setminus (\mathfrak{p}_1 \cup \cdots \cup \mathfrak{p}_n)$ to form a semilocal ring R_S, with maximal ideals among $\mathfrak{p}_1 R_S, \ldots, \mathfrak{p}_n R_S$. Thus, for instance, all rational numbers r/s with s prime to 30 form a semilocal ring, with maximal ideals generated respectively by $2, 3$, and 5. This is a PID; as a matter of fact, any semilocal Dedekind domain is a PID. As a second instance of "semilocalization", note that if R is a commutative noetherian ring, then the set of zero-divisors is a union of finitely many prime ideals (namely, the "associated primes" of (0)). Thus, we obtain:

C.7.1 Theorem *If R is a commutative noetherian ring, then its classical ring of quotients (obtained from R by inverting all of its non zero-divisors) is a semilocal ring.*

How about noncommutative rings? We have seen in the last part of Section C.6 that local rings make sense in general: they are rings with a unique maximal left (or right) ideal. Some facts on commutative local rings generalize nicely to general local rings. The case of semilocal rings is, however, more tricky. The correct definition was first given by Bass (1964): *a ring R is **semilocal** if $R/\mathrm{Rad}(R)$ is a left (or right) artinian ring* (where $\mathrm{Rad}(R)$ denotes the Jacobson radical of R: see Section C.18). This condition turns out to be *weaker* (in the noncommutative case) than R having only finitely many maximal left (right) ideals. The case of *local rings* corresponds to the case where $R/\mathrm{Rad}(R)$ is a division ring.

Bass's definition was perhaps partly motivated by the following more classical definition: R is a **semiprimary ring** if $R/\mathrm{Rad}(R)$ is artinian and $\mathrm{Rad}(R)$ is nilpotent. Take away the latter condition and you get the definition of a semilocal ring. In particular, left (right) artinian rings are semilocal, which generalizes nicely the fact that commutative artinian rings have a finite number of maximal ideals. Other useful classes of semilocal rings include, for instance, finite direct products of local rings, matrix rings over local rings, and module-finite algebras over commutative semilocal rings.

Key results on semilocal rings

In studying the stable structure of general linear groups in algebraic K-theory, Bass proved the following basic result (ca. 1964) on the unit structure of semilocal rings.

C.7.2 Theorem *If R is a semilocal ring, then R has stable range 1, in the sense that, whenever $Ra + Rb = R$, there exists $r \in R$ such that $a + rb \in U(R)$.*

This theorem acquired added significance after E. G. Evans proved (ca. 1973) the following remarkable result on the cancellation of modules from direct sums.

C.7.3 Theorem *Let A be a ring, and M, P, Q be right A-modules. If $E = \mathrm{End}_A(M)$ has stable range 1 (e.g., E is semilocal), then $M \oplus P \cong M \oplus Q \Longrightarrow P \cong Q$.*

Camps and Dicks (1993) obtained a number of surprising new characterizations of semilocal rings. We shall state here only two of their many characterizations.

C.7.4 Theorem *The following conditions are each equivalent to R being semilocal:*

(1) *There exists an (artinian) semisimple ring S and a ring homomorphism $f\colon R \to S$ such that $f^{-1}(U(S)) \subseteq U(R)$;*
(2) *There exists a non-negative integer n and a function $d\colon R \to \{0, 1, \dots, n\}$ such that, for all $a, b \in R$, $d(1 - ab) + d(a) = d(a - aba)$, and $d(a) = 0$ only if $a \in U(R)$.*

Three remarkable consequences of the Camps-Dicks Theorem are recorded below. The first of these confirmed a conjecture of P. Menal, and was largely motivated by the classical fact that the endomorphism ring of a module of finite length is semiprimary.

C.7.5 Corollary *If M is an artinian right module over a ring A, then $\mathrm{End}_A(M)$ is semilocal. In particular, M cancels from direct sums in the category of right A-modules.*

C.7.6 Corollary *If R is a subring of a left artinian ring R' such that R'/R is artinian as a left R-module, then R is a semilocal ring.*

The third consequence of Theorem C.7.4 below is a generalization of Theorem C.7.1 to the noncommutative case. It was first proved by J. T. Stafford for left and right noetherian rings, and proved in the following form by F. Cedó as a consequence of the main results in (Camps and Dicks 1993).

C.7.7 Corollary *Let R be a "classical quotient ring" in the sense that every non zero-divisor in R is a unit. If R satisfies ACC on left as well as on right annihilators, then R is a semilocal ring.*

For modules with semilocal endomorphism rings, Facchini, Herbera, Levy, and Vámos (1995) proved the following "*n*-Cancellation" result for any positive integer n.

C.7.8 Theorem *Let M, N be right A-modules, with $\operatorname{End}_A(M)$ semilocal. If $M \oplus \cdots \oplus M \cong N \oplus \cdots \oplus N$ (n copies on each side), then $M \cong N$.*

The results in Corollary C.7.5 and Theorem C.7.8 generated a whole lot of new interest on the class $S(A)$ of right A-modules whose endomorphism rings are semilocal. A general result in this direction is that of Herbera and Shamsuddin (1995), which states that *right A-modules of finite uniform and co-uniform dimensions belong to $S(A)$.* (A module is said to have finite uniform dimension if it contains no infinite independent set of nonzero submodules, and finite co-uniform dimension if it contains no infinite co-independent set of proper submodules.) This implies, for instance, that any linearly compact A-module is in $S(A)$, as is any finite direct sum of uniserial A-modules. For the terminology used here, as well as an excellent presentation of the results surveyed in this section, see Facchini's recent monograph (Facchini 1998).

Special classes of semilocal rings

For more structural information and module-theoretic results to be available for semilocal rings, we often specialize to suitable subclasses of such rings. Foremost among these are the perfect and semiperfect rings introduced by Bass (ca. 1960) as "homological generalizations" of the semiprimary rings.

Recall that a *projective cover* for a module M is a projective module P with an epimorphism $f \colon P \to M$ such that $\ker(f)$ is "small" (that is, for any submodule $S \subseteq P$, $S + \ker(f) = P$ implies $S = P$). Such a projective cover may not exist for a given M, but if it exists, then it is unique (over M). (The definition of a projective cover was largely inspired by that of an injective hull.) According to Bass, a ring R is *left perfect* if every left R-module has a projective cover, and *left semiperfect* if every finitely generated (or just cyclic) left R-module has a projective cover.

It turns out that "semiperfect" is a left-right symmetric notion, so we may just speak of semiperfect rings. Bass proved that these are precisely those semilocal rings R for which idempotents in $R/\operatorname{Rad}(R)$ can be lifted to idempotents in R. [In the commutative category, semiperfect rings are precisely finite direct products of (commutative) local rings.] For a right module M over any ring A, there is a simple criterion for $\mathrm{End}_A(M)$ to be semiperfect: this is the case iff M has a decomposition $M_1 \oplus \cdots \oplus M_n$ where each M_i is **strongly indecomposable**, that is, $\mathrm{End}_A(M_i)$ is local. In particular, such a direct sum always belongs to $\mathcal{S}(A)$.

Contrary to the case of semiperfect rings, left perfect rings need not be right perfect. Bass proved that left perfect rings are precisely those semilocal rings R for which $\mathrm{Rad}(R)$ is "left T-nilpotent", that is, for any sequence $a_1, a_2, \cdots \in \mathrm{Rad}(R)$, $a_1 a_2 \cdots a_n = 0$ for some n; or equivalently, those semilocal rings R for which every nonzero left R-module has a maximal submodule. Amazingly, left perfect rings can also be characterized by right-sided conditions! Indeed, one of Bass's most beautiful results is that a ring is *left* perfect iff it satisfies DCC on principal *right* ideals. Detailed proofs of all of the above results can be found in my introductory text (Lam 2001, Chapter 8).

Since their inception in the early 1960s, perfect and semiperfect rings have permeated noncommutative ring theory, and have been used extensively in the research on Krull-Schmidt theory, Morita duality theory, representation theory, model-theoretic algebra, as well as on self-injective rings, exchange rings, cogenerator rings, CS rings, *inter alia*. For a more detailed survey on these applications of perfect and semiperfect rings, see (Lam 1999a).

C.8 Cohen-Macaulay Rings

by Jürgen Herzog in Essen, Germany

All rings in this section are supposed to be commutative.

Introduction

The origin of Cohen-Macaulay rings goes back to the unmixed theorems proved by Macaulay in the case of polynomial rings, and by Cohen for regular local rings. The theory in its present shape was created by Auslander and Buchsbaum.

Cohen-Macaulay rings occur naturally in geometric, homological and combinatorial contexts. Basic references for this material include the books by Bruns and Herzog (1998), Bruns and Vetter (1988), Eisenbud (1995), Matsumura (1989), Serre (1965), and Stanley (1996).

The concept of Buchsbaum rings is a natural generalization of Cohen-Macaulay rings. An account of this theory can be found in the books of Stückrad and Vogel (1986) and Schenzel (1982).

Regular sequences and depth

Let R be a Noetherian ring, and M a finitely generated R-module. A sequence $x_1, \ldots, x_n$ in R is called an *M-regular sequence*, if $(x_1, \ldots, x_n)M \neq M$, and x_{i+1} is not a zero-divisor on $M/(x_1, \ldots, x_i)M$ for $i = 0, \ldots, n-1$. For an ideal $I \subset R$ with $IM \neq M$, all maximal M-sequences in I have the same length, given by $\min\{i \mid \operatorname{Ext}^i(R/I, M) \neq 0\}$ and denoted by $\operatorname{depth}(I, M)$. If R is local with maximal ideal $\mathfrak{m}$, one sets $\operatorname{depth} M = \operatorname{depth}(\mathfrak{m}, M)$. The depth is bounded above by the Krull dimension of M, and M is called **Cohen-Macaulay** (**CM** for short), if the upper bound for the depth is reached, in other words, if $\operatorname{depth} M = \dim M$. In particular, R is CM if $\operatorname{depth} R = \dim R$. A Noetherian ring R (not necessarily local) is called CM if all localizations of R at its maximal ideals are local CM rings. Equivalently, R is CM if $\operatorname{depth}(I, R) = \operatorname{codim} I$ for every ideal $I \subset R$.

Let K be a field. A graded K-algebra R which is finitely generated as an algebra over K by elements of degree 1 is called standard graded, and the ideal $\mathfrak{m} = \bigoplus_{i>0} R_i$ is called the graded maximal ideal of R. For a standard standard graded K-algebra R, one has the following convenient criterion: R is CM if and only if $\operatorname{depth}(\mathfrak{m}, R) = \dim R$.

For example, for any integer $d > 0$, let $R_n^{(d)} \subset K[X_1, \ldots, X_n]$ be the subring of the polynomial ring which is generated over K by all monomials of degree d in the variables $X_1, \ldots, X_n$. This K-algebra is standard graded and of Krull dimension n. It is called the d-th **Veronese subring** of R. It can be shown that $X_1^d, \ldots, X_n^d$ is an $R_n^{(d)}$-regular sequence, so that $R_n^{(d)}$ is CM.

On the other hand, the standard graded K-algebra $S = K[X_1^4, X_1^3 X_2, X_1 X_2^3, X_2^4]$ which is obtained from $R_2^{(4)}$ by removing the generator $X_1^2 X_2^2$ has $\dim S = 2$ and $\operatorname{depth}(\mathfrak{m}, S) = 1$, so that S is not CM. In fact, $X_1^4 \in \mathfrak{m}$ is not a zero-divisor on S since S is a domain. However, the residue class of the element $X_1^6 X_2^2 \in S$ in $S/X_1^4 S$ is annihilated by $\mathfrak{m}$. Therefore X_1^4 is a maximal regular S-sequence in $\mathfrak{m}$.

The unmixedness theorem and permanence properties

A different characterization for CM rings is provided by the unmixedness theorem. An ideal is said to be *unmixed* if the associated prime ideals of R/I are the minimal prime ideals of I.

C.8.1 Theorem *A Noetherian ring R is CM if and only if every ideal $I \subset R$ generated by* $\operatorname{height}(I)$ *elements is unmixed.*

In the polynomial ring $R = K[X_1, \ldots, X_n]$ over a field K, any polynomial $f \in R$, $f \neq 0$, generates a principal ideal of height 1, and hence is unmixed. In fact, the associated prime ideals of (f) are the principal ideals (g) corresponding to the prime factors g of f. The next example shows that in the polynomial ring R (which is CM), an ideal I which is generated by more than height(I) elements needs not to be unmixed. Indeed, the ideal $I = (X_1^2, X_1 X_2, \ldots, X_1 X_n)$ is contained in (X_1), and hence is of height 1. The associated prime ideals of I are (X_1) and $(X_1, \ldots, X_n)$. Thus I is not unmixed if $n > 1$.

The unmixedness theorem also implies that if R and R/I are CM, then I is unmixed. Thus for $n > 1$, $K[X_1, \ldots, X_n]/(X_1^2, X_1 X_2, \ldots, X_1 X_n)$ is a K-algebra of Krull dimension $n - 1$ which is not CM.

The CM property behaves well with respect to polynomial extensions: the polynomial ring $R[X]$ over R is CM if and only if R is CM.

Cohen-Macaulay rings localize, that is, if R is CM and $S \subset R$ is a multiplicatively closed subset of R, then the ring of fractions $S^{-1}R$ is again CM. It follows that the polynomial ring over a field in a finite number of indeterminates and all its localizations are CM. This is a special case of the more general fact that any regular ring is CM.

If R is CM and height$(x_1, \ldots, x_n) = n$, then $x_1, \ldots, x_n$ is an R-regular sequence and $R/(x_1, \ldots, x_n)$ is CM. Thus for example, if K is a field, then the hypersurface ring $K[X_1, \ldots, X_n]/(f)$ defined by a homogeneous polynomial $f \neq 0$ of degree > 1 is a non-regular CM ring.

Let R be a Noetherian local ring with maximal ideal $\mathfrak{m}$, and suppose that $R \to T$ is a local flat extension of Noetherian local rings with CM fiber $T/\mathfrak{m}T$. Then R is CM if and only if T is CM. In particular, the $\mathfrak{m}$-adic completion of a local CM ring is CM.

Local cohomology and duality

The depth and the dimension of a module, and consequently Cohen-Macaulayness, can be expressed in terms of local cohomology. Let R be a Noetherian local ring with maximal ideal $\mathfrak{m}$, and M a finitely generated R-module. Then $H_{\mathfrak{m}}^i(M) = \varinjlim \operatorname{Ext}^i(R/\mathfrak{m}^n, M)$ is called the ith **local cohomology** of M.

C.8.2 Theorem (Grothendieck) *Let* $t = \operatorname{depth} M$ *and* $d = \operatorname{dim} M$. *Then*

(1) $H_{\mathfrak{m}}^i(M) = 0$ *for* $i < t$ *and* $i > d$.
(2) $H_{\mathfrak{m}}^t(M) \neq 0$ *and* $H_{\mathfrak{m}}^d(M) \neq 0$.

The theorem implies that R is CM if and only if $H_\mathfrak{m}^i(R) = 0$ for $i < \dim R$.

Let E denote the injective hull of $R/\mathfrak{m}$, and let $d = \dim R$. A finitely generated R-module ω_R whose $\mathfrak{m}$-adic completion is isomorphic to $\mathrm{Hom}_R(H_\mathfrak{m}^d(R), E)$ is called a *canonical module of* R. A canonical module, if it exists, is unique up to isomorphism. Complete local rings, and more generally rings which are factor rings of Gorenstein rings (see the next section) admit a canonical module. If R is CM, then ω_R is a CM-module and has finite injective dimension. The next theorem is known as the local duality theorem.

C.8.3 Theorem (Grothendieck) *Let R be a complete local CM ring of dimension d. Then for all finitely generated R-modules M and all i there exist natural isomorphisms*

$$\mathrm{Ext}_R^i(M, \omega_R) \simeq \mathrm{Hom}_R(H_\mathfrak{m}^{d-i}(M), E)$$

Gorenstein rings

Bass (1963) introduced Gorenstein rings. Daniel Gorenstein's name is attached to this class of rings since he studied in his thesis plane curves whose local rings are particular examples of rings which now are called Gorenstein rings.

C.8.4 Theorem (Bass) *Let R be a Noetherian local ring of dimension d with maximal ideal $\mathfrak{m}$. The following conditions are equivalent:*

(1) *R is CM and $\mathrm{Ext}_R^d(R/\mathfrak{m}, R) \simeq R/\mathfrak{m}$;*
(2) *R has finite injective dimension;*
(3) *the injective dimension of R equals its Krull dimension.*

Gorenstein rings are distinguished by the property that $\omega_R \simeq R$. One has the following implications:

Regular $\implies$ Complete intersections $\implies$ Gorenstein $\implies$ CM rings

Stanley Reisner rings and semigroup rings

The theory of CM rings has remarkable applications in combinatorics and the study of affine semigroups and integer polytopes. Let Δ be a simplicial complex of dimension $d-1$ on the vertex set $[n] = \{1, \ldots, n\}$, and let K be a field and $I_\Delta \subset K[X_1, \ldots, X_n]$ the ideal generated by all monomials $\prod_{i \in F} X_i$ where $F \notin \Delta$. Then $K[\Delta] = K[X_1, \ldots, X_n]/I_\Delta$

is called the **Stanley-Reisner ring of** Δ. The Krull dimension of $K[\Delta]$ is d, and by the Reisner criterion, $K[\Delta]$ is CM, if and only if $\tilde{H}_i(\operatorname{lk} F; K) = 0$ for all $F \in \Delta$ and all $i < \dim \operatorname{lk} F$. Here, for a simplicial complex Γ, $\tilde{H}(\Gamma; K)$ denotes reduced simplicial homology and $\operatorname{lk} F = \{ G \in \Delta \mid F \cup G \in \Delta, F \cap G = \varnothing \}$ the **link of** F.

The Stanley-Reisner ring of a triangulation of a sphere is Gorenstein, as shown by Stanley. As a consequence of this algebraic fact, Stanley proved the upper bound theorem for simplicial spheres, which was conjectured by Klee in 1964.

An **affine semigroup** is a finitely generated semigroup C which for some n is isomorphic to a subsemigroup of $\mathbb{Z}^n$ containing 0. Let $\mathbb{Z}C$ be the $\mathbb{Z}$-submodule of $\mathbb{Z}^n$ generated by C. C is called **normal**, if whenever $mz \in C$ for some $z \in \mathbb{Z}C$ and $m \in \mathbb{N}$, $m > 0$, then $z \in C$.

C.8.5 Theorem (Hochster) *Let C be a normal semigroup, and K a field. Then $K[C]$ is Cohen-Macaulay.*

Danilov and Stanley showed that the canonical module of $K[C]$ is isomorphic to the ideal generated by all monomials X^c where c belongs to the relative interior $\operatorname{relint} C$ of C. In particular, $K[C]$ is Gorenstein if and only if there exists $c \in \operatorname{relint} C$ with $c + C = \operatorname{relint} C$. These results have combinatorial applications in the theory of Ehrhart functions.

Rings of invariants

Let R be a CM ring, and G a finite group of automorphisms of R whose order is invertible in R. Then the ring R^G of invariants is CM. For infinite groups acting on a ring one has the following: Let K be an algebraically closed field. A linear algebraic group G over K is called **linearly reductive** if for every finite dimensional representation $G \to \mathrm{GL}(V)$, the K-vector space V splits into a direct sum of irreducible G-subspaces.

C.8.6 Theorem (Hochster-Roberts) *Let K be an algebraically closed field and G a linearly reductive group over K acting linearly on a polynomial ring R defined over K. Then R^G is Cohen-Macaulay.*

More generally, Hochster and Huneke (1995) proved that a pure subring of a regular ring containing a field, is CM.

C.9 Rings of Formal Power Series

by Askar A. Tuganbaev in Moscow, Russia

All rings are assumed to be associative and with nonzero identity element. Let A be a ring, and let φ be an injective endomorphism of A. We denote by $A_r[[x, \varphi]]$ or R the **right skew power series ring** consisting of formal series $\sum_{i=0}^{\infty} x^i a_i$ ($a_i \in A$), where $ax^i = x^i \varphi^i(a)$ for any $a \in A$ and each positive integer i. We denote the Jacobson radical of R by $J(R)$.

For any two series f and g in R, the series $1 + \sum_{i=1}^{\infty} (fxg)^i$ is well defined and this series is the inverse element for $1 - fxg$ in R. We have that $Rx \subseteq xR \subseteq J(R) = J(A) + xR$ and the factor ring R/xR is isomorphic to A. Therefore, the ring R is semilocal (resp., local) if and only if the ring A is semilocal (resp., local). (A ring is said to be **semilocal** (resp., **local**, cf. Section C.6 and Section C.7) if its factor ring with respect to its Jacobson radical is Artinian (resp., a division ring).)

Since $x \in J(R)$ and x is not a divisor of zero in R, the ring R does not have minimal right or left ideals; in particular, R is not left or right Artinian. A series $f \in R$ is right (resp., left) invertible in R if and only if the constant term f_0 is right (resp., left) invertible in A. Since a series $f \in R$ is a right (resp., left) divisor of zero in R if and only if the constant term f_0 is right (resp., left) divisor of zero in A, the ring $A_r[[x, \varphi]]$ is a domain if and only if the ring A is a domain. (A ring A is called **reduced** if A does not have nonzero divisors of zero.) The ring $A_r[[x, \varphi]]$ is reduced if and only if the ring A is reduced and $a\varphi^n(a) \neq 0$ for any nonzero $a \in A$ and each $n \geq 0$. For example, if F and G are two fields and α is the automorphism of the direct product $F \times G$ defined by $\alpha((f, g)) = (g, f)$, then $(x(f, g))^2 = 0$; therefore, skew power series rings with reduced coefficient rings are not necessarily reduced. If f is an idempotent of R, then the constant term of f is an idempotent of A. Every idempotent of the factor ring R/xR can be lifted to an idempotent of R; all idempotents of the factor ring $R/J(R)$ can be lifted to idempotents of R if and only if all idempotents of the factor ring $A/J(A)$ can be lifted to idempotents of A. Therefore, the ring $A_r[[x, \varphi]]$ is semiperfect if and only if A is semiperfect.

If A is a semisimple Artinian ring possessing n minimal two-sided ideals, then $A_r[[x, \varphi]]$ is a right Noetherian right hereditary ring and each right ideal of $A_r[[x, \varphi]]$ is generated by at most n elements. A ring is called a **right Bezout** ring if all its finitely generated right ideals are principal. The ring $A_r[[x, \varphi]]$ is a right Bezout domain if and only if A is a right Bezout domain and $\varphi(a)$ is invertible in A for any nonzero $a \in A$; in this case, φ is an automorphism of A if and only if $A_r[[x, \varphi]]$ is a left Bezout ring if and only if $A_r[[x, \varphi]]$ is a left Ore domain. If all right annihilator ideals of A are ideals, then $A_r[[x, \varphi]]$ is a right Bezout ring if and only if A is a right Bezout ring, all principal right ideals of A are projective, $\varphi(e) = e$ for every central idempotent $e \in A$, and the element $\varphi(a)$ is invertible in A for every element $a \in A$ that is not a divisor of zero.

A ring is said to be **strongly regular** if each its principal right or left ideal is generated by a central idempotent. A ring is said to be **right distributive** if the lattice of its right ideals is distributive. If A is strongly regular and $n \geq 2$, then $R/x^n R$ is a right Bezout ring if and only if $R/x^n R$ is a right distributive ring if and only if $R/x^n R$ is a right invariant ring if and only if $\varphi(e) = e$ for every central idempotent $e \in A$. A right A-module M is said to be **countably injective** if for every countably generated right ideal B of A, each homomorphism $B_A \to A_A$ can be extended to a homomorphism $A_A \to M$. If all idempotents of A are central and φ is an automorphism of A such that $\varphi(e) = e$ for any idempotent $e \in A$, then all submodules of flat $A_\ell[[x, \varphi]]$-modules are flat (Section C.22) if and only if all 2-generated right ideals of $A_\ell[[x, \varphi]]$ are flat if and only if A is strongly regular and right countably injective.

A ring is said to be **right uniserial** if any two of its right ideals are comparable with respect to inclusion. The ring $A_r[[x, \varphi]]$ is right uniserial if and only if A is right uniserial and $\varphi(a)$ is invertible in A for any nonzero $a \in A$; in this case, φ is an automorphism of A if and only if $A_r[[x, \varphi]]$ is left uniserial. The ring $A_r[[x, \varphi]]$ is an indecomposable right distributive ring if and only if A is a right distributive domain and $\varphi(a)$ is invertible in A for any nonzero $a \in A$. A ring is said to be **right invariant** if all its right ideals are ideals. If A is a division ring, then $A_r[[x, \varphi]]$ is a right uniserial right invariant right Noetherian domain; in this case, φ is an automorphism of A if and only if $A_r[[x, \varphi]]$ is left Noetherian if and only if $A_r[[x, \varphi]]$ is left invariant.

Let k be a field, F be the field of rational functions over k in countable number of variables $\{t_i\}_{i=0}^\infty$, and let $t \equiv t_0$. There exists an injective endomorphism α of F such that $\alpha(t_i) = t_{i+1}$ for all i and t is transcendental over $\alpha(F)$. Let B be the ring which is formed by all the fractions $f/(1+yg)$ such that f and g are arbitrary polynomials from the polynomial ring $F[y]$ in one variable y. By $\beta(y) = t$, we extend α to an injective endomorphism β of B. Then B is a commutative uniserial Noetherian domain such that B is not a field and $\beta(b) \in U(B)$ for any nonzero $b \in B$. Then $B[[x, \beta]]$ is a right uniserial right Noetherian domain that is not left uniserial or left Noetherian; in this case, $B[[x, \beta]]$ is a right distributive right Bezout domain that is not is not a left distributive ring or a left Bezout ring.

The ring $A_r[[x, \varphi]]$ is left distributive if and only if A is strongly regular, φ is an automorphism, $\varphi(e) = e$ for each idempotent $e \in A$, and for any countably generated left ideal B of A, each homomorphism ${}_A B \to {}_A A$ can be extended to an endomorphism of ${}_A A$; in this case, the ring $A_r[[x, \varphi]]$ is right distributive. If F is a field and B be the ring formed by all eventually constant sequences of elements of F, then B is a commutative regular ring such that the commutative ring $B[[x]]$ is not distributive and $B[[x]]$ has 2-generated ideals that are not flat.

If A is a right Noetherian right order in a right Artinian ring, then $A[[x]]$ is a right order in a right Artinian ring. Let F be a field, and let S be the commutative F-algebra with generators y and $\{t_i\}_{i \in \mathbb{Z}}$ and with relations $t_i t_j = 0$ $(\forall i, j)$ and $y t_i = t_{i+1}$ $(\forall i)$. Then S is a commutative order (Section C.34) in an Artinian ring and the commutative ring $S[[x]]$ is not an order in an Artinian ring.

Let D_0 be a skew field with center F and define D_{i+1} recursively as the field coproduct over F of D_i and $F[x_i, y_i, z_i]$. Take any F-subalgebra H_0 of D_0 and let H_{i+1} be the subring of D_{i+1} generated by $H_i, x_i, y_i, h^{-1} z_i$ for all nonzero $h \in H_i$. Then $H = \bigcup H_i$ is a right Ore domain (Section C.15), but the power series ring $H[[x]]$ is not: $\sum x_i x^i$ and $\sum y_i x^i$ have no common nonzero right multiple.

If A is a ring and φ is an automorphism of A, then we denote by $A_r((x, \varphi))$ the **right skew Laurent series ring** consisting of formal series $f = \sum_{i=m}^{\infty} x^i a_i$, where $m = m(f)$ is an integer and $a x^i = x^i \varphi^i(a)$. The ring $A_r((x, \varphi))$ is a division ring (resp., semisimple Artinian) if and only if A is a division ring (resp., semisimple Artinian). The ring $A_r((x, \varphi))$ is right Artinian (resp., right Noetherian) if and only if A is a right Artinian (resp., right Noetherian) ring. If A is a principal right ideal ring, then $A_r((x, \varphi))$ is a principal right ideal ring; $A((x, \varphi))$ is a principal right ideal domain if and only if A is a principal right ideal domain. A ring is said to be **regular** (Section C.29) if each of its principal right or left ideal is generated by an idempotent. The ring $A_r((x, \varphi))$ is reduced if and only if $AaA \bigcap A\varphi(a)A \neq 0$ for every nonzero element $a \in A$. A ring is said to be **biregular** if each of its principal two-sided ideals is generated by a central idempotent. The ring $A_r((x, \varphi))$ is biregular if and only if $A_r((x, \varphi))$ is a finite direct product of the simple rings if and only if A is a finite direct product of the rings $A_1, \ldots, A_n$ with identity elements $e_1, \ldots, e_n$, $\varphi(e_i) = e_i$ for all i, and every ring A_i coincides with any its nonzero ideal B such that $\varphi(B) = B$. The ring $A_r((x, \varphi))$ is strongly regular ring if and only if $A_r((x, \varphi))$ is a finite direct product of division rings if and only if A is a finite direct product of division rings and $\varphi(e) = e$ for every central idempotent e of A. If $(\varphi)^n \equiv 1$ for some positive integer n, then the ring $A((x, \varphi))$ is regular if and only if A is semisimple Artinian. The ring $A_r((x, \varphi))$ is right uniserial if and only if $A_r((x, \varphi))$ is a right Artinian right uniserial ring if and only if A is a right Artinian right uniserial ring. The ring $A_r((x, \varphi))$ is a semilocal right distributive ring if and only if $A_r((x, \varphi))$ is a finite direct product of right Artinian right uniserial rings if and only if A is a finite direct product of right Artinian right uniserial rings $A_1, \ldots, A_n$ and $\varphi(A_i) = A_i$ for all i. If B is the ring of all rational numbers with odd denominators, then B is a local principal ideal domain and the ring $B((x))$ is not local.

For every ordinal number γ, we define modules of Krull dimension γ (cf. Section C.2 and Section C.6). Modules of Krull dimension 0 coincide with Artinian modules, by definition. We say that a module M is a *module of Krull dimension* γ if for any ordinal $\delta < \gamma$, M is not a module of Krull dimension δ and for each countable descending chain $M_1 \supseteq M_2 \supseteq \ldots$ of submodules of M, we have that for all but finitely many subscripts i, the factor modules M_i/M_{i+1} are modules of Krull dimension $< \delta$; the Krull dimension of M is denoted by K.dim(A). Each Noetherian module has Krull dimension. The right global dimension of A, is denoted by A. The right $A((x))$-module $A((x))$ has Krull dimension if and only if $A((x))$ is a right Noetherian ring if and only if A is a right Noetherian ring; in this case, r.gl.dim($A((x))$) = r.gl.dim(A) and K.dim($A((x))_{A((x))}$) = K.dim(A_A).

For more information, see (Brewer 1981, Cohn 1985, Ruiz 1993, Tuganbaev 1999).

C.10 Automorphisms of Polynomial Algebras and the Jacobian Conjecture

by Jie-Tai Yu in Hong Kong, China

Let $\mathbb{K}$ be a field. Let $\mathbb{K}[\mathbf{x}] := \mathbb{K}[x_1, \ldots, x_n]$ be the polynomial algebra in n variables over $\mathbb{K}$. Let $\mathbf{f} := (f_1, \ldots, f_n) \in (\mathbb{K}[\mathbf{x}])^n$ be an n-tuple. Obviously, $\varphi \colon p(\mathbf{x}) \mapsto p(\mathbf{f})$ is an endomorphism of $\mathbb{K}[\mathbf{x}]$. On the other hand, every endomorphism of $\mathbb{K}[\mathbf{x}]$ may be defined in that way. To slightly abuse the language, sometimes we say that $\mathbf{f} := (f_1, \ldots, f_n)$ is an endomorphism of $\mathbb{K}[\mathbf{x}]$. In the sequel we sometimes denote $(x_1, \ldots, x_n)$ by $\mathbf{x}$, $(y_1, \ldots, y_n)$ by $\mathbf{y}$, $(f_1, \ldots, f_n)$ by $\mathbf{f}$, and so on.

If $\varphi \colon \mathbf{x} \mapsto \mathbf{f}$ is an automorphism of $\mathbb{K}[\mathbf{x}]$ and $\varphi^{-1} \colon \mathbf{x} \mapsto \mathbf{g}$ is the inverse, then $\mathbf{g} \circ \mathbf{f} = \mathbf{x}$. Hence $J(\mathbf{g} \circ \mathbf{f}) = J(\mathbf{x})$ where J denotes the usual Jacobian (matrix) operator. By the chain rule, $J(\mathbf{g})(\mathbf{f})J(\mathbf{f}) = \mathbf{1}$ where $\mathbf{1}$ is the identity matrix of order n. Hence $J(\mathbf{f}) \in \mathrm{GL}(n; \mathbb{K}[\mathbf{x}])$. The Jacobian conjecture states that the converse of the above statement is true.

C.10.1 Jacobian Conjecture *Let* $\varphi \colon \mathbf{x} \mapsto \mathbf{f}$ *be an endomorphism of* $\mathbb{K}[\mathbf{x}]$ *where* $\mathrm{char}\,\mathbb{K} = 0$. *If* $J(\mathbf{f}) \in \mathrm{GL}(n; \mathbb{K}[\mathbf{x}])$, *then* φ *is an automorphism.*

Formulated by Keller (1939), the conjecture is still open for $n \geq 2$ (the $n = 1$ conjecture is obviously true), to the best of our knowledge. For a history and background of the Jacobian conjecture, see Bass et al. (1982) and van den Essen (1997). Recently, Smale (1998) listed the Jacobian conjecture as one of the eighteenth important problems for the 21st century. It is almost certain that a solution to the Jacobian conjecture deserves a Fields Medal.

For arbitrary n, Keller (1939) himself proved the birational case (i.e., with an additional condition that $\mathbb{K}(\mathbf{x}) = \mathbb{K}(\mathbf{f})$) in 1939. Campbell (1973) proved the Galois case of the conjecture (i.e., with an additional condition that $\mathbb{K}(\mathbf{x})/\mathbb{K}(\mathbf{f})$ is a Galois extension). Wang (1980) proved the quadratic case of the Jacobian conjecture. In 1980s, Bass et al. (1982) and Jagžev (1980) reduced the Jacobian conjecture to the cubic homogeneous case. Namely, to solve the conjecture, one only needs to consider the case $\mathbf{f} = \mathbf{x} + H$ where every monomial in H is cubic (that implies that $J(H)$ is a nilpotent matrix). For $n = 2$, Moh (1983) proved the conjecture for the case $\max\{\deg(f_1), \deg(f_2)\} \leq 100$ in 1975. Yu (1995) reduced the Jacobian conjecture to the so-called 'positive case': to solve the Jacobian conjecture, one only needs to consider the case $f_i = x_i + H_i^{(2)} + H_i^{(3)} + H_i^{(4)} \in \mathbb{R}[x_1, \ldots, x_n]$ where $H_i^{(j)}$ are homogeneous of degree j and all coefficients in f_i are nonnegative. Recently the following reformulation of the Jacobian conjecture for two variables has been given by Shpilrain and Yu (2000): if φ is an endomorphism of $\mathbb{K}[x, y]$ with invertible Jacobian matrix, then for some automorphism α, the mapping $\alpha \cdot \varphi$ fixes a non-constant polynomial.

Now we consider the structure of the automorphism group of a polynomial algebra. The group $\mathrm{Aut}\,\mathbb{K}[\mathbf{x}]$ has two important subgroups.

The first one is the group $A(n, \mathbb{K})$ consisting of all **affine automorphisms**, $\mathbf{x} \mapsto A\mathbf{x}$. The second one $J(n, \mathbb{K})$ is generated by all triangular automorphisms defined as follows. Given invertible elements $a_1, \ldots, a_n$ in $\mathbb{K}$ and polynomials $f_i \in \mathbb{K}[x_{i+1}, \ldots, x_n]$, $i = 1, \ldots, n$, (f_n is a polynomial in zero variables, i.e., $f_n \in \mathbb{K}$) then the map $(a_1 x_1 + f_1, a_2 x_2 + f_2, \ldots, a_n x_n + f_n)$ is called a **triangular automorphism** of $\mathbb{K}[\mathbf{x}]$. The group generated by $A(n, \mathbb{K})$ and $J(n, \mathbb{K})$ is the group of **tame automorphisms** (cf. Section B.17) and is denoted by $T(n, \mathbb{K})$. The automorphisms of $\mathbb{K}[\mathbf{x}]$ which are not in $T(n, \mathbb{K})$ are called nontame or **wild**. Similarly, one can define tame and wild automorphisms of $\mathbb{K}[\mathbf{x}]$ when $\mathbb{K}$ is any commutative ring with identity.

C.10.2 Problem Is it true that $\operatorname{Aut} \mathbb{K}[x_1, \ldots, x_n] = T(n, \mathbb{K})$, i.e., is every automorphism in $\operatorname{Aut} \mathbb{K}[x_1, \ldots, x_n]$ tame?

The case $n = 2$ of the Problem has an affirmative answer. There are several known proofs of this result originally obtained by Jung (1942) for $\mathbb{K} = \mathbb{C}$ and by van der Kulk (1953) for any field $\mathbb{K}$.

Consider the Nagata automorphism $\nu = N = (n_1, n_2, n_3)$ of $\mathbb{K}[x, y, z]$: $n_1 = \nu(x) = x - 2y(y^2 + xz) - z(y^2 + xz)^2$, $n_2 = \nu(y) = y + z(y^2 + xz)$, $n_3 = \nu(z) = z$.

Nagata (1972) discovered that (n_1, n_2) is not tame considered as an automorphism in $\operatorname{Aut} \mathbb{K}[z][x, y]$. Based on that, he formulated in the same paper the following

C.10.3 Nagata's Conjecture *The Nagata automorphism* $N = (n_1, n_2, n_3)$ *is not tame.*

Nagata's conjecture is still open to the best of our knowledge. For related topics in both polynomial and free associative algebras, see Cohn (1985).

Moreover, it is still unknown whether there exist wild automorphisms, when $\mathbb{K}$ is an arbitrary field and $n > 2$. On the other hand, Smith (1989) discovered another important property of the Nagata automorphism. An automorphism φ in $\operatorname{Aut} \mathbb{K}[x_1, \ldots, x_n]$ is called **stably tame** if there exists $m > n$ such that φ can be considered as a tame automorphism in $\operatorname{Aut} \mathbb{K}[x_1, \ldots, x_m]$ fixing $x_{n+1}, \ldots, x_m$. Smith showed that N is stably tame and becomes tame if we extend it to an automorphism of $\mathbb{K}[x, y, z, t]$ fixing t. Recently, Drensky et al. (1999), and Le Bruyn (1997) found some new evidences that the Nagata automorphism is wild, considered as an automorphism of $\mathbb{K}[x, y, z]$. Drensky and Yu (2001b) classified all tame and nontame automorphisms algorithmically in $\operatorname{Aut} \mathbb{K}[z][x, y]$ in case char $\mathbb{K} = 0$. Hence, in view of Nagata, they gave many candidates for wild automorphisms in $\operatorname{Aut} \mathbb{K}[x, y, z]$.

The mathematical community believes that the following conjecture is true.

C.10.4 Stably Tame Conjecture

Every automorphism within $\operatorname{Aut} \mathbb{K}[x_1, \ldots, x_n]$ *is stably tame for any field* $\mathbb{K}$ *and any positive integer* n.

However, the above conjecture is again open. In fact, we do not even know whether every automorphism in $\operatorname{Aut} \mathbb{K}[x, y, z]$ fixing z is tame. In this direction, Drensky and Yu, and David Wright have done some positive work recently in case char $\mathbb{K} = 0$: they proved that such an automorphism is stably tame if it is a composition of up to three elementary automorphisms in $\operatorname{Aut} \mathbb{K}(z)[x, y]$. Here, φ is an **elementary automorphism** if it is either linear or triangular (for instance, $x \mapsto x$, $y \mapsto y + x^2$ would be a non-linear automorphism). See (Drensky and Yu 2001a,b) and references therein.

C.11 Rings of Integer-Valued Polynomials

by Paul-Jean Cahen in Marseille and Jean-Luc Chabert in Amiens, France

Binomial polynomials

The binomial polynomials $\binom{X}{n} = \frac{X(X-1)\ldots(X-n+1)}{n!}$ clearly take integral values on the integers, although their coefficients are not integers. In fact, every integer-valued polynomial is a linear combination with coefficients in $\mathbb{Z}$ of these polynomials, as shown by the Gregory-Newton interpolation formula:

$$f(X) = \sum_{n=0}^{\deg(f)} \Delta^n f(0) \binom{X}{n} \quad \text{where } \Delta^n f(0) = \sum_{k=0}^{n} (-1)^{n-k} \binom{n}{k} f(k).$$

Pólya (1919) considered similar expansions for integer-valued polynomials in a number field. But seemingly, never before the seventies was it noticed that integer-valued polynomials also form a ring, with interesting properties (providing for instance, the most 'natural' algebraic examples of non-Noetherian rings). But why restrict to number fields?

Considering polynomials as functions, is it not more natural to study integer-valued polynomials on a subset (as in the very recent literature)? Thus:

C.11.1 Definition Let D be an (integral) domain and let E be a subset of its quotient field K. The ring of *integer-valued polynomials* on E with respect to D is:

$$\text{Int}(E, D) = \{f \in K[X] \mid f(E) \subseteq D\}.$$

We simply write $\text{Int}(D)$ for $\text{Int}(D, D)$. To avoid trivialities, D is assumed not to be a field (and thus, to be infinite) and E to be a *fractional subset* of D (that is, $dE \subseteq D$ for $d \neq 0$ in D). Note that, for $E \neq F$, it may be that $\text{Int}(E, D) = \text{Int}(F, D)$ (E and F are then said to be *polynomially equivalent*, for instance, each cofinite subset of D is equivalent to D). Finally, if D is Noetherian or a Krull domain, good localization properties, for instance, $S^{-1}\text{Int}(D) = \text{Int}(S^{-1}D)$, allow to focus on quasi-local domains (cf. Section C.6).

The D-module structure

As in the case of the polynomials $\binom{X}{n}$ for $\text{Int}(\mathbb{Z})$, one looks for a *regular* basis $\{f_n\}$ of the D-module $\text{Int}(E, D)$ (that is, a basis such that $\deg(f_n) = n$ for each n). Clearly, if $\mathfrak{I}_n$ is the subset of D formed by the leading coefficients of the integer-valued polynomials with degree at most n, there exists such a basis if and only if, for each n, $\mathfrak{I}_n$ is a fractional principal ideal of D. In particular, Ostrowski obtained the following nice characterization, for Galoisian number fields (Pólya 1919):

C.11.2 Theorem *Let D be the ring of integers of a number field K, and suppose the extension $K/\mathbb{Q}$ to be Galoisian. Then $\text{Int}(D)$ has a regular basis if and only if, for each prime number p, the product of the maximal ideals of D lying over p is principal.*

As $K/\mathbb{Q}$ is Galoisian, note that pD decomposes in D as a product $pD = (\mathfrak{m}_1 \cdots \mathfrak{m}_g)^e$, and hence, the product $\mathfrak{m}_1 \cdots \mathfrak{m}_g$ of the maximal ideals lying over p is principal as soon as $e = 1$. Thus, we just have to consider the (finitely many) primes p which are ramified, that is, such that $e > 1$.

The inverse of $\mathfrak{I}_n$, that is, the ideal $\mathfrak{I}_n^{-1} = \{x \in D \mid x\mathfrak{I}_n \subseteq D\}$, has many properties of the classical factorials. It may be denoted by $(n!)_E$. For instance, in a Dedekind domain D, for every $a_0, a_1, \ldots, a_n \in E$, the product $\prod_{0 \leq i < j \leq n}(a_j - a_i)$ belongs to the ideal $\prod_{1 \leq k \leq n}(k!)_E$ (Bhargava 1998).

The Stone-Weierstrass theorem

Integer-valued polynomials on $\mathbb{Z}$ are continuous in the p-adic topology. As the p-adic completion $\widehat{\mathbb{Z}}_p$ of $\mathbb{Z}$ is compact, it follows from a p-adic version of the Stone-Weierstrass theorem (Dieudonné 1944) that $\text{Int}(\mathbb{Z})$ is dense in the ring $\mathcal{C}(\widehat{\mathbb{Z}}_p, \widehat{\mathbb{Z}}_p)$ of continuous functions. A similar result holds for rings of integers of number fields. For a quasi-local domain D, with maximal ideal $\mathfrak{m}$, note that compactness, in the $\mathfrak{m}$-adic topology, imposes D to be one-dimensional and $D/\mathfrak{m}$ to be finite. However, considering a *subset* of a valuation domain, we may relax the conditions on the Krull dimension and the residue field:

C.11.3 Theorem *Let E be a fractional subset of a valuation domain D. If the completion $\widehat{E}$ of E is compact, then $\text{Int}(E, D)$ is dense in $\mathcal{C}(\widehat{E}, \widehat{D})$ for the uniform convergence topology.*

On the contrary, note that $D[X]$ is never dense in $\mathcal{C}(\widehat{D}, \widehat{D})$. For a rank-one valuation domain, Theorem C.11.3 can be traced back to Kaplansky (1950).

Many interesting properties derive from the Stone-Weierstrass theorem. For instance, the fact that the prime ideals lying over the maximal ideal $\mathfrak{m}$ of D are in one-to-one correspondence with the elements α of $\widehat{E}$: to α corresponds the maximal ideal $\mathfrak{M}_\alpha = \{f \in \text{Int}(E, D) \mid f(\alpha) \in \widehat{\mathfrak{m}}\}$. Moreover $\mathfrak{M}_\alpha$ is finitely generated if and only if α is isolated in $\widehat{E}$.

Skolem properties

To what extent is an ideal $\mathfrak{A}$ of $\text{Int}(E, D)$ characterized by its **ideals of values** $\mathfrak{A}(a) = \{f(a) \mid f \in \mathfrak{A}\}$? One says that $\text{Int}(E, D)$ satisfies the **strong Skolem property** if, for two finitely generated ideals $\mathfrak{A}$ and $\mathfrak{B}$, the condition $\mathfrak{A}(a) = \mathfrak{B}(a)$ for each $a \in E$, implies $\mathfrak{A} = \mathfrak{B}$ (the classical *Skolem property* compares only an ideal $\mathfrak{A}$ with $\text{Int}(E, D)$ itself). These properties do not localize, in fact, they never hold in the quasi-local case: for t in the maximal ideal of D, the principal ideal $\mathfrak{J} = (1 + tX)$ differs from $\text{Int}(E, D)$ while $\mathfrak{J}(a) = D$ for each $a \in E$. In this case, one considers the **almost strong Skolem property**, restricting to **unitary ideals** (that is, ideals containing nonzero constants):

C.11.4 Proposition *Let E be a fractional subset of a valuation domain D. If the completion $\widehat{E}$ of E is compact, then $\mathrm{Int}(E, D)$ satisfies the almost strong Skolem property.*

The Skolem properties hold in the global case if, in addition to the almost Skolem properties, the integer-valued rational functions are polynomials. This is the case, for $E = D$ itself, when D is the ring of integers of a number field. In particular every integer-valued rational function on the ring of integers is an integer-valued polynomial.

In fact, and along the line of the original results by Skolem (1936) himself, these properties generalize to the case of several indeterminates. For a subset $\underline{E} \subseteq K^n$, we let $\mathrm{Int}(\underline{E}, D) = \{f \in K[X_1, \ldots, X_n] \mid f(\underline{E}) \subseteq D\}$. In particular, for $\underline{E} = D^n$:

C.11.5 Proposition *If D is the ring of integers of a number field, then $\mathrm{Int}(D^n, D)$ satisfies the* strong Skolem property.

On the contrary, note that in $\mathbb{Z}[X]$, the ideal generated by 3 and $X^2 + 1$ is not the whole ring, although 3 and $n^2 + 1$ are coprime for each n.

The Prüfer property

From the almost strong Skolem property [Proposition C.11.4], it follows that, if E is a fractional subset of a valuation domain D, and if $\widehat{E}$ is compact, then every finitely generated ideal $\mathfrak{A}$ of $\mathrm{Int}(E, D)$ is **strongly two generated**, that is, for each $g \in \mathfrak{A}$, $g \neq 0$, there is $h \in \mathfrak{A}$ such that $\mathfrak{A} = (g, h)$: assuming $\mathfrak{A}$ to be unitary, it is enough to find h such that $\mathfrak{A}$ and (g, h) have the same ideals of values. This property allows to conclude that $\mathrm{Int}(E, D)$ is a Prüfer domain (Section C.5):

C.11.6 Proposition *Let E be a fractional subset of a valuation domain D. If the completion $\widehat{E}$ of E is compact, then $\mathrm{Int}(E, D)$ is a Prüfer domain.*

By globalization, if D is a Dedekind domain with finite residue fields (and hence, for each maximal ideal $\mathfrak{m}$, the completion of D with respect to the $\mathfrak{m}$-adic topology is compact), then $\mathrm{Int}(D)$ is a Prüfer domain.

On the contrary, note that, as D is not a field, $D[X]$ is never a Prüfer domain.

Many more rings

One may also consider polynomials that are integer-valued together with their derivatives or finite differences (of various orders), integer-valued rational functions .... Some aspects are developed in (Cahen and Chabert 1997).

C.12 Commutativity Conditions

by Howard E. Bell in St. Catharines, Canada

Although the foundation theorem on commutativity of rings was Wedderburn's theorem of 1905 asserting that every finite division ring is commutative, intensive study of commutativity really began with Jacobson's famous "$a^n = a$ theorem" of 1945—an application of the structure theory based on the Jacobson radical.

C.12.1 Theorem (Jacobson) *If R is any ring such that for each $a \in R$ there exists an integer $n = n(a) > 1$ for which $a^n = a$, then R is commutative.*

During the 1950's, Herstein proved several deep theorems giving sufficient conditions for commutativity in arbitrary rings, the culminating theorem being

C.12.2 Theorem (Herstein 1953) *If for every $x \in R$ there exists $p(x) \in \mathbb{Z}[x]$ such that $x - x^2 p(x)$ is central, then R is commutative.*

An exposition of some of these results is in (Herstein 1968, Chapter 3). Herstein also investigated conditions which imply near-commutativity in the sense that the commutator ideal $C(R)$, i.e., the ideal generated by all commutators $xy - yx$, is a nil ideal. Various authors have explored this property as well as other notions of near-commutativity—e.g., $C(R)$ is periodic, $C(R)$ is finite, R is either finite or commutative, R contains a central ideal (see Bell, Klein, and Kappe 1997).

A large number of papers have dealt with commutativity of rings, often rings with 1, satisfying polynomial identities such as $(xy)^n = x^n y^n$ or $x^n y^n = y^n x^n$, or constraints involving commutators. A pretty, though uncharacteristically elementary, result is

C.12.3 Theorem (Ligh and Richoux) *If R is a ring with 1 satisfying the identities $(xy)^n = x^n y^n$ for three consecutive positive integers n, then R is commutative.*

In recent years, there has been considerable interest in conditions involving constrained mappings, typically derivations, usually in the setting of prime rings. This interest was sparked by a theorem of Posner (Brešar 1993 and Lanski 1997 contain many references.)

C.12.4 Definition A *derivation* on a ring R is an additive mapping $d\colon R \to R$ such that $d(xy) = xd(y) + d(x)y$ for all $x, y \in R$.

C.12.5 Definition For $k \geq 1$, we define the **extended commutators** $[x_1, x_2, \dots, x_k]$ of elements of R inductively by $[x_1, x_2] = x_1 x_2 - x_2 x_1$ and $[x_1, x_2, \dots, x_k] = [[x_1, x_2, \dots, x_{k-1}], x_k]$.

C.12.6 Theorem (Posner) *If R is a prime ring admitting a nonzero derivation d such that $xd(x) - d(x)x$ is central for all $x \in R$, then R is commutative.*

C.12.7 Theorem (Lanski 1997) *Let R be a semiprime ring, L a nonzero left ideal, and d a derivation on R such that $d(L) \neq \{0\}$. If there exist positive integers $t_0, t_1, \dots, t_n$ such that $[d(x^{t_0}), x^{t_1}, x^{t_2}, \dots, x^{t_n}] = 0$ for all $x \in L$, then R contains a nonzero central ideal. Hence, if R is prime, then R is commutative.*

The latest commutativity conditions to be studied are setwise conditions. For example, noting that in a commutative ring every set K of k elements has $|K^2| \leq \binom{k+1}{2}$, we are led to

C.12.8 Theorem (Bell and Klein) *Let R be a semiprime ring with 1, and let $k \geq 2$. If $|K^2| \leq \binom{k+1}{2}$ for every k-subset K of R, then R is either commutative or finite.*

C.13 Finite Dimensional Division Algebras
by David Saltman in Austin, TX, USA

The subject of finite dimensional division algebras has long fascinated mathematicians and specifically algebraists. A **division ring** is an associative unital algebra where every nonzero element is invertible. The **center** of a division ring is the set of elements which commute with all elements. Of course commutative division rings are fields. The subject is interesting because of the great wealth of noncommutative division rings. The center is necessarily a field, and so the division ring can be thought of as a vector space over its center. A **division algebra** is a division ring which is finite dimensional over its center. Since the center often must be displayed in the notation, we write a division algebra as D/F, F being the center. Note that a division algebra D/F must have dimension of the form n^2 over F (e.g., Saltman 1999, p. 10). We call n the degree of D and write is as $n(D)$.

This subject began, of course, when Hamilton defined the classical algebra of quaternions as $\mathbb{H} = \mathbb{R} + \mathbb{R}i + \mathbb{R}j + \mathbb{R}k$ where $i^2 = j^2 = -1$ and $ij = k = -ji$. It turns out that this is the only nontrivial (i.e., noncommutative) such algebra with center the real field $\mathbb{R}$. Furthermore, it is easy to see that there are no division algebras with center the complex field, or indeed with center any algebraically closed field. A justifiably famous theorem of Wedderburn says that there is no division algebra with center a finite field.

Despite these facts, there is a great wealth of division algebras over other fields. In fact, much of theory of these objects can be subsumed by the question: what do division algebras look like? In studying this question one can proceed by construction, creating new examples, or by elimination, showing some kinds of examples are impossible. Of course, the phrase "look like" is purposely ambiguous, so that we can consider all sorts of descriptions.

To begin with, however, let us think about describing a division algebra in the usual way, by giving the space and describing the multiplication. The example of Hamilton has within it generalizations that provide ways of constructing division algebras. To begin with, the quaternionic construction can be generalized to define so called *cyclic* algebras, as follows. Let L/F be a cyclic Galois extension of fields, of degree n, with Galois group generated by σ, and $a \in L^*$ a nonzero element. As a space, let $\Delta(L/F, \sigma, b)$ be $L \oplus Lz \oplus \cdots \oplus Lz^{n-1}$ with multiplication defined using the relations $zx = \sigma(x)z$ for all $x \in L$ and $z^n = b$. Of course, the quaternions are the cyclic algebra $\Delta(\mathbb{C}/\mathbb{R}, \sigma, -1)$ where σ is complex conjugation. Under suitable circumstances, cyclic algebras are division algebras. Under any circumstances, they are so called *central simple algebras*, meaning they are matrix algebras $M_r(D)$ over a division algebra D/F.

It turns out that division algebras can be constructed with more general Galois extensions, it is just that the description gets more complicated. Let L/F be a Galois extension of fields with Galois group G, and $c\colon G \times G \to K^*$ a so called 2– *cocycle map*, meaning c satisfies $c(\sigma, \tau)c(\sigma\tau, \eta) = \sigma(c(\tau, \eta))c(\sigma, \tau\eta)$ for all $\sigma, \tau, \eta \in G$. Given an L/F and c as above, one can form $A = \Delta(L, G, c) = \oplus_{\sigma \in G} Lu_\sigma$ with multiplication defined by $u_\sigma x = \sigma(x)u_\sigma$ and $u_\sigma u_\tau = c(\sigma, \tau)u_{\sigma\tau}$. One can show that if G is a cyclic group, then a ***crossed product*** $\Delta(L/F, G, c)$ is a cyclic algebra (e.g., Saltman 1999, p. 49). Once again a crossed product is a central simple algebra which is sometimes a division algebra.

Thus the first question might whether all division algebras are cyclic or crossed product algebras. Albert (see e.g., Saltman 1999, p. 85, for a generalization) constructed the first noncyclic algebra while much later Amitsur (e.g., Saltman 1999, p. 15) proved the existence of a noncrossed product division algebra. Brussel (cited in Saltman 1999) constructed more concrete noncrossed product division algebras D/K, where F has the form $F(t)$ for F a global field.

However, Amitsur's and Brussel's proof does not work for division algebras of prime degree, and so an extremely important open question is whether all division algebras of such degrees are cyclic (of course crossed products of prime degree must involve cyclic groups). An approach to this question is to consider crossed products $A = \Delta(L/F, G, c)$ where G is a nonabelian semidirect product $(\mathbb{Z}/p\mathbb{Z}) \rtimes (\mathbb{Z}/r\mathbb{Z})$. If one could show that one of these algebras is not a crossed product with respect to an abelian group, then one could show a noncyclic algebra of degree p has been constructed. However, Rowen and Saltman have shown (cited in Saltman 1999) that if F contains a p-th root of one, $(r, p) = 1$, and $r = 2, 3, 4, 6$, then such A are cyclic algebras.

On the other hand, there are some important positive results to note. Wedderburn observed that any division algebra of degree 2, and Albert showed that any division algebra of degree 3, were cyclic. Albert showed that any division algebra of degree 4 was a crossed product with group $\mathbb{Z}/2\mathbb{Z} \oplus \mathbb{Z}/2\mathbb{Z}$ (Albert 1961, p. 177, 179). Finally, a division algebra D/F is cyclic if F is a local or global field, or the field of fractions of the henselization of a normal surface over $\mathbb{C}$ (quoted in Saltman 1999, p. 71). Of course the above includes the famous ***Brauer-Hasse-Noether theorem***.

There are different ways the term "description" can be understood. For example, Brauer showed that the F-isomorphism classes of division algebras D/F (for fixed F) form a group now called the ***Brauer group*** and written $\mathrm{Br}(F)$. The crossed product construction, despite the noncrossed product result, can be used to identify the Brauer group with the (profinite) Galois cohomology group $H^2(G_F, F_s^*)$ where F_s is the separable closure of F and G_F is the Galois group of F_s/F. Thus all division algebras can in some sense be described as the elements of cohomology groups, and this point of view has enriched their study immeasurably.

In particular, one can use etale cohomology and Galois cohomology to study the Brauer group in many situations (Milne 1980, Grothendieck 1968, e.g.,).

Since division algebras D/F are elements of a group, they have an order usually said to be the their *exponent* which we will write as $e(D)$. This is an invariant of a division algebra that has a subtle connection to the other structural questions we have need discussing. For example, Amitsur showed that there are noncrossed product algebras of degree 8, but Rowen (cited in Saltman 1999) showed that a division algebra D/F with $n(D) = 8$, $e(D) = 2$ is always a crossed product. Vishne (preprint) showed that dihedral crossed products D/F, with F containing enough roots of unity, are abelian crossed products if $e(D) = 2$.

Combining the Brauer group and algebra points of view raises questions that are still not understood. For example how are the exponents of division algebras related to their degrees? One knows that the exponent divides the degree, and that the exponent and degree have the same prime divisors. The Hasse-Brauer-Noether theorem includes the result that for D/F a division algebra and F a global field, $n(D) = e(D)$. Artin showed this was also true if F is the field of fractions of the henselization of a surface (over $\mathbb{C}$) at a point. Ford and Saltman (cited in Saltman 1999) showed that for these same F, all division algebras are cyclic. Albert constructed algebras with $n(D) \neq e(D)$. Artin (cited in Saltman 1999) and Bloch showed that for F a C_2 field, say the function field of a surface over $\mathbb{C}$, a division algebra D/F has $n(D) = e(D)$ if $2, 3$ are the only primes dividing $n(D)$. Saltman showed that if F is the function field of a curve over a p-adic field, and D/F has degree prime to p, then n divides the square of the exponent of D. One might conjecture the following. Suppose that F is finitely generated over its prime subfield. Then there must be an f such that $n(D)$ divides $e(D)^f$ for all division algebras D/F. The works of Saltman and Brussel suggest a productive place to consider these questions is over fields that are in some sense two dimensional.

Amitsur's noncrossed product construction involved an important object, the so called *generic division algebra* $UD(F,n)/Z(F,n)$. It was this division algebra that Amitsur showed was a noncrossed product for certain n and F. Perhaps more importantly, he showed that all division algebras D/K of degree n, with $K \supset F$, are "specializations" of $UD(F,n)/Z(F,n)$. For this reason one can be said to have described all division algebras of degree n if one understands the structure of $Z(F,n)$. Much interest has focused on the rationality, stable rationality, or "retract rationality" of $Z(F, n)$, which would have consequences for division algebras (e.g., Saltman 1999, p. 116). Procesi showed that $Z(F, n)/F$ is rational for $n = 2$, while Formanek showed the same thing when $n = 3, 4$. Bessenrodt and Le Bruyn showed the stable rationality of $Z(F, n)$ for $n = 5, 7$, while Saltman showed the retract rationality of $Z(F, n)$ for n prime (Saltman 1999, p. 116). These questions have added interest because $Z(F, n)$ can also be described as an invariant field under the projective linear groups $PGL_n(F)$ (see e.g., Saltman 1999, p. 109).

The modern theory of Brauer groups and division algebras was revolutionized by the **Merkur'ev-Suslin Theorem** (Merkur'ev and Suslin 1982), which showed that in the presence of n roots of one, the n torsion part of the Brauer group $\mathrm{Br}(F)_n$ is naturally isomorphic to $K_2(F)/nK_2(F)$, where $K_2(F)$ is the algebraic K-group. A consequence of this is that for every division algebra D/F of exponent dividing n (for such F) there is a matrix algebra $M_r(D)$ which is an abelian crossed product and is in fact a tensor product of cyclic algebras. However, besides a feeling r can be quite large, little is known about it.

The theory of division algebras of exponent 2 is quite special because Albert showed such division algebras have a so called involution. Division algebras with involution also arise naturally in the classification of algebraic groups. We cannot say more about this large subject, except to refer the reader to the magnificent "Book of Involutions" (Knus, Merkurjev, Rost, and Tignol 1998).

In the above discussion we mentioned etale cohomology, and the specializations of the generic division algebras. To understand both of these directions, one must generalize Brauer groups from fields to commutative rings and even more generally to schemes. In this theory, corresponding to central algebras over fields, are the so called *Azumaya algebras* over schemes (including affine schemes i.e., commutative rings). An old but good reference for the beginning theory is (DeMeyer and Ingraham 1971), where these algebras are called central separable. The cohomological side is developed in (Milne 1980).

Brauer groups in this generality are a subtle invariant. For example, they are used to study the Hasse principle for varieties over global fields (e.g., Skorobogatov 1999, for recent progress). They have also been used to show unirational fields are nonrational (e.g., Saltman 1999, p. 88).

C.14 Infinite Dimensional Division Rings

by Paul M. Cohn in London, UK

In this section, all rings contain a unit element, which is preserved by homomorphisms and inherited by subrings.

A ring is called a **division ring** if its non-zero elements form a multiplicative group. A commutative division ring is called a field, whereas a non-commutative one is called a **skew field**.

A commutative ring can be embedded into a field precisely when it is an integral domain. This condition is also necessary in the general case, but not sufficient. For example, let R be the F-algebra (F a field) generated by a, b, c, d forming a 2×2 matrix $\mathbf{A}$ with relations expressing that in $\mathbf{A}^2$ only the element on position $(1, 2)$ is not zero. Then $\mathbf{A}^4 = 0$; for $\mathbf{A}$ in a subring of a field, this implies that $\mathbf{A}^2 = 0$, but this can be shown not to be so; one can also show that R is an integral domain.

The first skew field was the ring of quaternions, discovered by W. R. Hamilton in 1843. This was finite-dimensional over its centre, i.e., a **division algebra**, and such algebras have been much investigated. The first division ring infinite-dimensional over its centre was the power series field constructed by Hilbert around 1900. Take any field F with an automorphism α of infinite order, e.g., a rational function field $F = \mathbb{R}(t)$ with $\alpha \colon f(t) \mapsto f(2t)$, and form the ring of skew Laurent series in x with coefficients in F and commutation rule $cx = xc^\alpha$.

The problem of embedding a given ring in a skew field has been studied extensively. An important sufficient condition was found by O. Ore in 1931: An integral domain R can be embedded in a skew field in which every element f has the form $f = ab^{-1}$, for $a \in R$, $b \in R^*$, if and only if it is an **Ore domain**, i.e., it satisfies the **Ore right multiple condition**: $aR \cap bR \neq 0$, for any $a, b \in R^*$. Moreover, the skew field obtained in this way is unique up to isomorphism over R.

Various methods for embedding certain classes of non-Ore domains in skew fields were developed, until in 1971 a general criterion for embeddability was established (cf. Cohn 1985, Ch. 7, Cohn 1995, Ch. 4) To describe it, let us define, for any ring R, an **R-field** as a skew field K with a homomorphism φ from R to K. If φ is an epimorphism, which happens if and only if K is generated by $\mathrm{im}\, \varphi$, as skew field, we speak of an **epic R-field**. Clearly we have an embedding of R in K precisely when φ is injective; in that case the least R-field generated by φ is called a **field of fractions** of R. Henceforth, we shall often drop the prefix 'skew' and simply speak of a 'field'.

We shall want to turn the R-fields, for a given ring R, into a category. Of course a homomorphism between R-fields is necessarily injective; so we need a more general notion. We define a *local homomorphism* between two rings A, B as a homomorphism $f \colon A_0 \to B$, where the domain A_0 of f is a subring of A, and f maps non-units to non-units. If B is a field, this means that A_0 is a local ring with the unique maximal ideal Ker f. We now fix a ring R and consider only local homomorphisms $f \colon K \to L$ whose domains include the image of R in K. Two such local homomorphisms are said to be *equivalent* if there is a subring of K (containing the image of R) on which both are defined, and on which they agree and again define a local homomorphism. This is indeed an equivalence and the equivalence-class of local homomorphisms is called a *specialisation* from K to L over R. In this way, we obtain for each ring R, a category F_R of R-fields and specialisations; to avoid irrelevancies, one usually restricts oneself to epic R-fields, and we shall do so here. Then F_R is a small category and between any epic R-fields there is at most one specialisation, so F_R may be thought of as a partially ordered set. If there is a greatest element, it is called a *universal R-field*; clearly it is unique up to isomorphism, if it exists. Thus a universal R-field has the property that from it there exists, for any epic R-field L, a unique specialisation to L.

To give a simple illustration, take a commutative field K and form the polynomial ring $R = K[x]$. An epic R-field is either a simple field extension of K of finite degree or the rational function field $K(x)$, and the latter is the universal R-field. Here, each epic R-field F was formed by taking a prime ideal $\mathfrak{p}$ of R and forming the field of fractions of the integral domain $R/\mathfrak{p}$. There is another way of forming F; instead of putting the elements in $\mathfrak{p}$ to 0, we make the elements outside $\mathfrak{p}$ invertible, and so obtain a local ring $R_\mathfrak{p}$. Now, F is the residue-class field by its unique maximal ideal. It is this second method that we shall use in the non-commutative case. However, the prime ideal $\mathfrak{p}$ will need to be replaced by a set of matrices.

The main problem in the non-commutative case is the lack of a determinant; it can be overcome by generalising the notion of a singular matrix. An exact analogue is not to be expected, but the following notion will be of use: An $n \times n$ matrix $\mathbf{A}$ is said to be *full* if in any factorisation $\mathbf{A} = \mathbf{PQ}$, where $\mathbf{P}$ is $n \times r$ and $\mathbf{Q}$ is $r \times n$, we have $r \geq n$. Over a field, even skew, a square matrix is full if and only if it is regular, i.e., a non-zero divisor. Over a commutative field this is equivalent to the non-vanishing of the determinant, but the analogy must not be pushed too far: in the polynomial ring $\mathbb{Z}[x,y]$, the matrix $\begin{pmatrix} 0 & 2 & -y \\ -2 & 0 & x \\ y & -x & 0 \end{pmatrix}$ can be shown to be full, but it has zero determinant. We shall also need a form of addition for matrices. We recall that over a commutative ring, the determinant of the sum of two $n \times n$ matrices $\mathbf{A}$ and $\mathbf{B}$ can be written as a sum of 2^n determinants. What we need is an analogue of the sum of determinants, even though we do not have determinants. Thus let $\mathbf{A}, \mathbf{B}$ be two $n \times n$ matrices which differ only in the first column, say $\mathbf{A} = (\mathbf{a}_1, \mathbf{a}_2, \ldots, \mathbf{a}_n)$, $\mathbf{B} = (\mathbf{b}_1, \mathbf{a}_2, \ldots, \mathbf{a}_n)$. Then we define the *determinantal sum* of A and B with respect to the first column as the matrix $\mathbf{C} = (\mathbf{a}_1 + \mathbf{b}_1, \mathbf{a}_2, \ldots, \mathbf{a}_n)$. For rows, the definition is similar; we shall usually write this sum as $\mathbf{A} \triangledown \mathbf{B}$ without specifying the row or column to be added. Over a commutative ring we have $\det(\mathbf{A} \triangledown \mathbf{B}) = \det \mathbf{A} + \det \mathbf{B}$, while over any field, even skew, if $\mathbf{A}$ and $\mathbf{B}$ are singular, then $\mathbf{A} \triangledown \mathbf{B}$ is singular whenever it is defined. We shall also need an analogue of matrix products; this is given by the *diagonal sum* $\mathbf{A} \oplus \mathbf{B} = \begin{pmatrix} \mathbf{A} & \mathbf{O} \\ \mathbf{O} & \mathbf{B} \end{pmatrix}$. We now introduce the following analogue of an ideal.

C.14.1 Definition A *matrix ideal* in a ring R is a collection $\mathfrak{a}$ of square matrices such that

M.1 $\mathfrak{a}$ contains all non-full matrices;

M.2 If $\mathbf{A} \in \mathfrak{a}$, then $\mathbf{A} \oplus \mathbf{B} \in \mathfrak{a}$ for all square matrices $\mathbf{B}$ over R;

M.3 If $\mathbf{A}, \mathbf{B} \in \mathfrak{a}$ and $\mathbf{C} = \mathbf{A} \triangledown \mathbf{B}$ is defined with respect to some column, then $\mathbf{C} \in \mathfrak{a}$;

M.4 If $\mathbf{A} \oplus \mathbf{I} \in \mathfrak{a}$, then $\mathbf{A} \in \mathfrak{a}$.

$\mathfrak{a}$ is called a *prime matrix ideal* if, moreover,

M.5 $\mathfrak{a}$ is proper, i.e., there is a square matrix over R not in $\mathfrak{a}$;

M.6 $\mathbf{A}, \mathbf{B} \notin \mathfrak{a}$ implies $\mathbf{A} \oplus \mathbf{B} \notin \mathfrak{a}$.

It is easy to verify that for any R-field K with map $f: R \to K$, the singular kernel of f, i.e., the set of all square matrices over R mapped to singular matrices over K, is a prime matrix ideal. Conversely, it can be shown that for any prime matrix ideal $\mathfrak{p}$ of R there exists an epic R-field $K_{\mathfrak{p}}$ whose singular kernel is precisely $\mathfrak{p}$. This field is uniquely determined by $\mathfrak{p}$ up to isomorphism and is called the *localization* of R at $\mathfrak{p}$ (cf. Section C.6 and Section C.15).

The partially ordered set of prime matrix ideals of R is called the *field spectrum* of R and is denoted by $\operatorname{Spec}(R)$. For a commutative ring, it reduces to the familiar prime spectrum. For a general ring, we now have the following criterion for embeddability in a field (cf. Cohn 1985, Cor. 7.5.5, Cohn 1995, Theorem 4.4.5)

C.14.2 Theorem *A ring R has a skew field of fractions if and only if $R \neq 0$ and no diagonal matrix with non-zero diagonal entries can be written as a determinantal sum of non-full matrices.*

To find conditions for the existence of a universal R-field, we recall that in the commutative case each ideal $\mathfrak{a}$ has a radical $\sqrt{\mathfrak{a}} = \{\, x \in R \mid x^n \in \mathfrak{a}$ for some $n \geq 1 \,\}$, which equals the intersection of all prime ideals containing $\mathfrak{a}$. Similarly, one can for each matrix ideal $\mathfrak{a}$ define a radical $\sqrt{\mathfrak{a}} = \{\, \mathbf{A} \mid \mathbf{A} \oplus \cdots \oplus \mathbf{A} \in \mathfrak{a} \,\}$, and this turns out to be the intersection of all prime matrix ideals of R which contain $\mathfrak{a}$. It is easily seen from M.4 that the least matrix ideal (generated by $\varnothing$) $\mathfrak{n}$ is proper if and only if $\sqrt{\mathfrak{n}}$ is proper. Hence we obtain

C.14.3 Theorem *Let R be a ring and $\mathfrak{n}$ its least matrix ideal. Then there exists an R-field if and only if $\mathfrak{n}$ is proper. There is a universal R-field if and only if $\sqrt{\mathfrak{n}}$ is a prime matrix ideal, and it is a universal field of fractions if and only if $\sqrt{\mathfrak{n}}$ is prime and contains no non-zero 1×1 matrix.*

The best known examples are firs and semifirs (see Section C.23).

Given a family of fields K_λ with a common subfield F, we can form their ring coproduct $P = \coprod_F K_\lambda$ by taking the ring generated by these fields with the subfields F all identified and as defining relations the relations in each K_λ. This coproduct can be shown to be a fir (Cohn 1985, 1995); hence it has a universal field of fractions, called the **field coproduct** of the K_λ over F. Just as the free product of groups led to the HNN-construction, so one has an analogue for fields, using the field coproduct.

C.14.4 Theorem *Every skew field K (over a central subfield F) can be embedded in a skew field L over F in which any two elements with the same minimal equation over F or transcendental over F are conjugate.*

The same result can be proved with elements replaced by square matrices, where the matrices either satisfy the same minimal equation or are both **totally transcendental**, i.e., no polynomial other than 0 makes them singular.

There is also a form of Galois theory for skew field extensions; here the main difficulty is that in a polynomial equation the variable may be non-central. If K/E is a skew field extension, K can be regarded as left or right E-space, and so one has a left degree $[K : E]_L$ and a right degree, which in general are not related, though they are equal for **Galois** *extensions*, i.e., if E is the fixed field of a group of automorphisms of K. For any skew field K, using the Jacobson-Bourbaki correspondence between subfields of K and subrings of $\mathrm{End}(K)$ containing the right multiplications of K and closed in a finite topology, one proves the following analogue to the commutative theory.

C.14.5 Theorem *Let K/E be a Galois extension of finite left degree, with group G. Then the intermediate subfields D correspond to certain subgroups H of G, where the degree $[K : D]_L$ corresponds to an invariant of H (the 'reduced order').*

C.15 Rings of Quotients

by Wallace S. Martindale, 3rd, in Glenside, PA, USA

The term *ring* shall mean a ring with 1 which is not necessarily commutative. An element a of a ring R is called **regular** if it is not a zero-divisor in R. For (right) Artinian rings (i.e., every descending chain of right ideals "levels of" at some point) it is easy to show that every regular element is invertible. Taking the point of view that being able to invert regular elements is a desirable goal, given a ring R one defines a **classical ring of right quotients** of R to be a ring Q_{cl} satisfying:

(1) $R \subseteq Q_{\mathrm{cl}}$;
(2) x regular in R implies $x^{-1} \in Q_{\mathrm{cl}}$;
(3) $q \in Q_{\mathrm{cl}}$ implies $q = ab^{-1}$, $a, b \in R$, b regular in R.

In this situation, R is said to be a **right order** in Q_{cl} (cf. Section C.34. This Q_{cl} is unique when it exists. If it exists, let $a, b \in R$, b regular. Then $b^{-1}a = xy^{-1}$ for some $x, y \in R$, y regular. Thus a necessary condition for the existence of Q_{cl} is the **right Ore condition** (Ore 1933):

(∗) Given $a, b \in R$, b regular, there exists $x, y \in R$, y regular, such that $ay = bx$.

To show that (∗) is also a sufficient condition for the existence of Q_{cl} one simply constructs Q_{cl} as equivalence classes of ordered pairs (a, b), b regular, by making appropriate adjustments to the well-known method of constructing the field of fractions of a commutative integral domain.

Clearly any commutative ring satisfies (∗). On the other hand, we have the example of the free noncommutative algebra $F\langle x, y \rangle$ over a field F; this is a **domain** (i.e., all nonzero elements are regular) but is far from satisfying (∗). However, certain important classes of rings are right Ore rings, notably prime right Noetherian rings and prime PI rings.

We again assume that R is a right Ore ring, but this time we form Q_{cl} by a different method, using a construction which lends itself much more readily to generalization. We shall call a right ideal U **regular** if it contains a regular element, and we let $\mathscr{S}$ be the collection of all regular right ideals. We consider all right module maps $f \colon U \to R$, $U \in \mathscr{S}$, which we denote more concisely as (f, U). (Of course, if $b \in U$ is regular and $f(b) = a$, we still want to think of (f, U) as $f = ab^{-1}$.) Let (f, U) and (g, V) be two such "partial homomorphisms", with $b \in U$, $d \in V$ regular elements, and let $g^{-1}U = \{v \in V \mid g(v) \in U\}$. From $m = bx = dy$ (one shows that both x and y are regular) we see that $U \cap V$ is regular. Setting $c = g(d)$, writing $cy = bx$ with y regular, and noting that $g(dy) = g(d)y = cy = bx$, we see that $g^{-1}U$ is again a regular right ideal. We define $(f, U) \sim (g, V)$ if there exists $W \in \mathscr{S}$, $W \subseteq U \cap V$, such that $f = g$ on W. We then take Q_{cl} to be the set of all equivalence classes $[f, U]$ under this equivalence relation, and define $[f, U] + [g, V] = [f + g, U \cap V]$ and $[f, U][g, V] = [f(g), g^{-1}U]$. One proceeds to verify that Q_{cl} is indeed a ring and that it contains R via the embedding $a \mapsto [a_l, R]$, where a_l is the left multiplication determined by a.

The assumption of the right Ore condition is rather restrictive from the viewpoint of general noncommutative ring theory, and so we now search for a more general ring of "quotients" Q which can be defined for any ring. The idea of using "partial homomorphisms" from the preceding construction is encouraging, and what remains to be done is to replace the collection of regular right ideals with another collection $\mathscr{D}$ which clearly should share some of the attributes of $\mathscr{S}$, in particular:

(i) $aU = 0$, $a \in R$, $U \in \mathscr{D}$ implies $a = 0$;

(ii) $a \in R$, $U \in \mathscr{D}$ implies $a^{-1}U = \{x \in R \mid ax \in U\} \in \mathscr{D}$.

Assuming $\mathscr{D}$ satisfies (i) and (ii) we fix any $U \in \mathscr{D}$ and let $a, b \in R$, $a \neq 0$. By (ii), $b^{-1}U \in \mathscr{D}$ and so by (i), there exists $r \in b^{-1}U$ (hence $br \in U$) such that $ar \neq 0$. Accordingly we define a right ideal U of R to be **dense** if:

$(**)$ given $a, b \in R$, $a \neq 0$, there exists $r \in R$ such that $ar \neq 0$, $br \in U$.

We then let $\mathscr{D}$ denote the collection of all dense right ideals of R. $\mathscr{D}$ is nonempty since $R \in \mathscr{D}$. If $U, V \in \mathscr{D}$ a repeated application of $(**)$ shows that $U \cap V \in \mathscr{D}$. One now proceeds exactly as in the second construction of Q_{cl} (with regular right ideals replaced by dense right ideals) to construct a ring Q consisting of equivalence classes of "partial homomorphisms" $f: U_R \rightarrow R_R$, $U \in \mathscr{D}$. If (f, U) and (g, V) denote two such elements, we need to know that $g^{-1}U \in \mathscr{D}$: given $a \neq 0$, b by $(**)$ we get $ar \neq 0$ and $br \in V$, by $(**)$ again we get $ars \neq 0$ and $g(br)s = g(brs) \in U$, and so $a(rs) \neq 0$ and $b(rs) \in g^{-1}U$. Addition and multiplication are again determined by usual addition and composition of functions applied to respective domains $U \cap V$ and $g^{-1}U$.

The ring Q just constructed (due to Utumi 1956) is called the ***maximal right ring of quotients*** of R and is characterized by the following properties:

(a) $R \subseteq Q$;

(b) Given $f: U_R \rightarrow R_R$, $U \in \mathscr{D}$, there exists $q \in Q$ such that $qu = f(u)$ for all $u \in U$;

(c) If $qU = 0$, $q \in Q$, $U \in \mathscr{D}$, then $q = 0$;

(d) Given $q \in Q$ there exists $U \in \mathscr{D}$ such that $qU \subseteq R$.

We remark that (b) and (c), taken together, generalize (2), and that (d) generalizes (3).

It is tempting to replace the notion of dense right ideal with the more natural notion of *essential* right ideal (U is ***essential*** if $U \cap J \neq 0$ for every nonzero right ideal J), and often this can be done. However, although every dense right ideal is clearly essential, there are examples, even in primitive rings (Lawrence 1974), of essential right ideals U for which $aU = 0$ for some $a \neq 0$ (and so U can't be dense).

For the remainder of this note we assume R is a ***prime ring*** (i.e., the product of any two nonzero ideals is nonzero). Although it is sometimes natural to make use of Q, very often Q is too far removed from R. However, motivated by the observation that $\mathscr{D}$ clearly contains the set $\mathscr{F}$ consisting of all the nonzero two-sided ideals of R, we proceed to describe certain subrings of Q which have proved not only useful but necessary in a wide variety of problems. We first have $Q_r = \{q \in Q \mid qU \subseteq R \text{ for some } U \in \mathscr{F}\}$. In case R is simple then $Q_r = R$. Inside of Q_r we have the ***symmetric ring of quotients*** Q_s consisting of all $q \in Q$ such that $qU \subseteq R$ and $Uq \subseteq R$ for some $U \in \mathscr{F}$. We note that if R is a domain, then so is Q_s whereas Q_r may not be (e.g., take $R = F\langle x, y \rangle$). Most important of all perhaps is the center C of Q, called the ***extended center*** of R, which is a field containing the ordinary center Z of R. Accordingly one forms the useful ***central closure*** RC ($\subseteq Q_s$).

We close with a simple example to illustrate these notions. Let $F \subseteq E$ be two fields, let T be the ring of all countably infinite sized matrices over E which are column-finite, and let R be the subring of T consisting of all matrices of the form $A + cI$, where A is $n \times n$ over E, n varies, and $c \in F$. The center of R is F whereas the extended center of R is E. Thus the central closure is the same as R except that $c \in E$. It can be shown that $Q = Q_r = T$ and that Q_s consists of all matrices in T which are row-finite. The transpose map in R (an example of an involution) can be extended to R_s but not to R_r. Certain derivations and automorphisms (termed "X-inner") arise from elements outside of R but lying in R_s; in this example any $q \in R_s$ induces a derivation in R via $x \mapsto qx - xq$ and an element such as $s = \text{diag}(1, 1/2, 1/3, \dots)$ induces an automorphism of R via $x \mapsto s^{-1}xs$.

Among many general references we list (Beidar et al. 1996, Lambek 1976, Faith 1967, Lawrence 1974).

C.16 Endomorphism and Matrix Rings

by Alexander V. Mikhalev and Askar A. Tuganbaev in

Moscow, Russia

All rings are assumed to be associative and with nonzero identity element.

For a ring and a positive integer n, we denote by A_n the $n \times n$ matrix ring $\text{Mat}_{m \times n}(R)$ over A; the ring A is isomorphic to the unitary subring of all scalar matrices in A_n and the category $\text{Mod}(A)$ of all right A-modules is equivalent to the category $\text{Mod}(A_n)$ of all right A_n-modules. For every ideal B of A, the set B_n formed by all matrices whose entries are contained in B is an ideal of the matrix ring A_n and the mapping $\alpha : B \to B_n$ is an isomorphism of the ideal lattice of A onto the ideal lattice of A_n such that $\alpha(BC) = \alpha(B)\alpha(C)$ for any two ideals B and C of A.

If A and A' are two isomorphic rings, then $A_n \cong A'_n$ for all n. If A and A' are two division rings and $A_m \cong A'_n$ for some positive integers m and n, then $m = n$ and $A \cong A'$. Let D be the rational division algebra of generalized quaternions with basis $\{1, i, j, ij\}$ such that $i^2 = -1$ and $j^2 = -23$ $(ij = -ji)$, A be the subring of D generated by i and $(1+j)/2$, $B \equiv 3A + ((1+j)/2)A$, and let $A' \equiv BB^*$. Then $A_2 \cong A'_2$ and $A \not\cong A'$.

For a right A-module M, we denote by $\text{End}(M)$ the **endomorphism ring** of M with operations defined by $(f + g)(m) = f(m) + g(m)$ and $fg(m) = f(g(m))$. Any ring A can be identified with the ring $\text{End}(A_A)$, since there is the natural ring isomorphism $\alpha : A \to \text{End}(A_A)$ with

$\alpha(a)(x) = ax$. For any module M and each positive integer n, the endomorphism ring of the direct sum M^n of n copies of M is naturally isomorphic to the matrix ring $(\text{End}(M))_n$. If M is a module with endomorphism ring R and N is a direct summand of M with natural projection $e \colon M \to N$, then the ring $\text{End}(N)$ is isomorphic to the ring eRe. If M is a module with endomorphism ring R and $\{\varepsilon_i \mid i = 1, \ldots, n\}$ be a complete orthogonal system of idempotents of the ring R, then the correspondence

$$M = \varepsilon_1(M) \oplus \cdots \oplus \varepsilon_n(M) \longrightarrow R = \varepsilon_1 R \oplus \cdots \oplus \varepsilon_n R$$

between finite direct decompositions of the module M and decompositions of the ring R into finite direct sums of right ideals is one-to-one.

For a ring A, the matrix ring A_n is naturally isomorphic to the endomorphism ring of n copies of the module A_A; consequently, the endomorphism ring of any finitely generated free module is isomorphic to some matrix ring over A. For any positive integer n and each n-generated projective A-module M, there is an idempotent e of the matrix ring A_n such that $\text{End}(M) \cong eA_ne$.

If M is a simple right A-module, then the ring $\text{End}(M)$ is a division ring, the ring $\text{End}(M^n)$ is a simple Artinian ring isomorphic to the matrix ring $(\text{End}(M))_n$ for each positive integer n, and for arbitrary elements $y_1, \ldots, y_k$ in M and any elements $x_1, \ldots, x_k$ linearly independent in M over $\text{End}(M)$, there is an element $a \in A$ such that $x_i a = y_i$ for $i = 1, \ldots, k$. For a finitely generated semisimple module M, the ring $\text{End}(M)$ is isomorphic to a finite direct product of matrix rings over division rings. For a semisimple module M, the ring $\text{End}(M)$ is isomorphic to a direct product of rings of linear transformations of vector spaces over division rings. If M is a module such that all its submodules and all its factor modules are indecomposable, then the factor ring $\text{End}(M)/J(\text{End}(M))$ is either a division ring or a direct product of two division rings.

If M and N are two torsion abelian groups with isomorphic endomorphism rings, then every isomorphism $\text{End}(M_{\mathbb{Z}}) \to \text{End}(N_{\mathbb{Z}})$ is induced by some isomorphism $M \to N$ (cf. Section B.1). Let p be a prime integer, $\mathbb{Z}_{p^\infty}$ be the quasi-cyclic p-group, and let J_p be the additive group of the ring of p-adic integers. Then $\mathbb{Z}_{p^\infty} \not\cong J_p$ and the endomorphism rings of the $\mathbb{Z}$-modules $\mathbb{Z}_{p^\infty}$ and J_p are isomorphic to each other. We note that for every positive integer n, there exists a torsion-free abelian group M of rank n such that the ring $\text{End}(M)$ is isomorphic to $\mathbb{Z}$.

A module M is called a module with the ***finite exchange property***
(resp. the ***exchange property***) if for every module X and each finite
direct decomposition (resp. each direct decomposition) $X = M \oplus Y =$
$\oplus_{i \in I} N_i$, there are submodules $N_i' \subseteq N_i$ $(i \in I)$ with $X = M' \oplus (\oplus_{i \in I} N_i')$.
A ring A is called an ***exchange ring*** if A satisfies the following two
equivalent conditions: (1) for every element $a \in A$, there is an idempotent $e \in aA$ with $1 - e \in (1 - a)A$; (2) for every element $a \in A$, there is
an idempotent $e \in Aa$ with $1 - e \in A(1 - a)$. A module M is a module
with the finite exchange property if and only if $\mathrm{End}(M)$ is an exchange
ring.

For a module M, a submodule N of M is said to be ***closed*** (in M)
if N does not have proper essential extensions in M. A module M
is said to be ***continuous*** if every submodule of M isomorphic to a
closed submodule of M is a direct summand of M. If M is a continous
module and $J(\mathrm{End}(M))$ is the Jacobson radical of $\mathrm{End}(M)$, the factor
ring $\mathrm{End}(M)/J(\mathrm{End}(M))$ is regular, $J(\mathrm{End}(M))$ coincides with the set
of all the endomorphisms of M with essential kernels, and each countable
set of mutually orthogonal idempotents of the ring $\mathrm{End}(M)/J(\mathrm{End}(M))$
can be lifted to a countable set of mutually orthogonal idempotents of
the ring $\mathrm{End}(M)$.

For a right A-module X, a submodule Y of X is said to be ***pure***
(in X) if for every left A-module L, the natural group homomorphism
$Y \otimes_A L \to X \otimes_A L$ is a monomorphism. A right A-module M is said
to be ***pure-injective*** if for every right A-module X and each pure submodule Y of X, every homomorphism $Y \to M$ can be extended to a
homomorphism $X \to M$. If M is a pure-injective right A-module, then
for every right A-module X, the right $\mathrm{End}(X)$-module $\mathrm{Hom}(X, M)$ is
pure-injective. If M is a pure-injective right module, then $\mathrm{End}(M)$ is a
right pure-injective ring, the factor ring $\mathrm{End}(M)/J(\mathrm{End}(M))$ is regular,
and M a module with the exchange property. If M is an indecomposable
module that is either pure-injective or continuous, then the factor ring
$\mathrm{End}(M)/J(\mathrm{End}(M))$ is a division ring.

C.16.1 Theorem

(1) *An abelian group M is a simple $\mathrm{End}(M)$-module if either M is an*
    elementary p-group for some p or M is a divisible torsion-free group
    (i.e., a vector space over $\mathbb{Q}$).
(2) *An abelian group M is an Artinian and Noetherian $\mathrm{End}(M)$-module*
    if and only if $M = X \oplus Y$, where X is a bounded group and Y is a
    divisible torsion-free group.

(3) *An abelian torsion group M is a free $\text{End}(M)$-module if and only if M is a cyclic group.*

(4) *An abelian group M is an injective $\text{End}(M)$-module if and only if $M = \prod_p M_p \oplus D$, where M_p is a finite direct sum of cyclic p-groups, D is a divisible group of finite rank, and either D is a mixed group, or D is a nonmixed group and $M_p = 0$ for almost all p.*

If M is an abelian p-group and C is the center of the ring of $\text{End}(M)$, then M is a projective C-module if and only if M is a flat C-module if and only if M is a direct sum of cyclic groups of the same order.

Every injective right A-module containing the copy of A_A is a cyclic left $\text{End}(M)$-module. If M is a right A-module and Q is an injective cogenerator in the category of all right A-modules, then the M is a flat left module over its endomorphism ring if and only if the right $\text{End}(M)$-module $\text{Hom}(M_A, Q_A)$ is injective. Every faithful quasi-injective right A-module is a finitely generated flat left module over its endomorphism ring if and only if the module A_A is injective and has a finitely generated essential socle.

C.16.2 Theorem *For a ring A, the following are equivalent:*

(1) *A is an Artinian ring whose right or left principal ideals are principal;*

(2) *Every quasi-injective right A-module is a finitely generated flat left module over its endomorphism ring;*

(3) *A is right or left Noetherian and every right A-module is a projective left module over its endomorphism ring;*

If A is a ring satisfying a polynomial identity (Section C.37), then A is an Artinian ring whose right or left principal ideals are principal if and only if every right A-module is a flat left module over its endomorphism ring if and only if every right A-module is a projective left module over its endomorphism ring. If M is the subgroup of the additive group of rational numbers $\mathbb{Q}$ formed by all the fractions with denominators that are not divided by the square of any prime integer, then $\text{End}\,M \cong \mathbb{Z}$ and M is a flat $\text{End}\,M$-module that is not projective.

A module is said to be *distributive* (Section C.36) if the lattice of its submodules is distributive. A ring A is right distributive if and only if every projective left A-module is a distributive right module over its endomorphism ring if and only if every quasi-injective right A-module is a distributive left module over its endomorphism ring if and only if for every simple right A-module M, the injective hull of M is a distributive left module over its endomorphism ring. If A is a commutative ring and M is a distributive A-module, then $\text{End}(M)$ is a commutative ring that is a projective limit of commutative distributive rings; in addition, if the module M is finitely generated, then the ring $\text{End}(M)$ is isomorphic to the commutative distributive ring $A/r(M)$.

For more information on the above classes of modules and rings, see (Anderson and Fuller 1992), (Cartan and Eilenberg 1956), (Facchini 1998), (Faith 1973), (Jacobson 1964), (Kasch 1982), (Tuganbaev 1998), and (Wisbauer 1991).

C.17 Primitive Rings and Semisimplicity

by Kostia Beidar in Tainan, Taiwan

Given a ring R, a left R-module M is said to be **simple** (or **irreducible**) if $RM \neq 0$ and M has no proper nonzero submodules. The importance of simple modules is underlined by the fact that every irreducible representation of a group G (Lie algebra L over a field F) in a vector space V over a field F is exactly a simple module over the group algebra $F[G]$ (or the universal enveloping algebra $U(L)$, respectively). If M is a simple left R-module, then Schur's lemma asserts that the endomorphism ring $\text{End}_R(M)$ of M is a division ring and so M is a right vector space over D.

A module M is called **faithful** if $rM = 0$ implies $r = 0$ for all $r \in R$. A ring R is said to be **left primitive** if it has a faithful simple module. The concept of a right primitive ring is defined analogously. Every simple ring with unity is both left and right primitive. Bergman's example shows that there exists a left primitive ring which is not right primitive and so the concept is not left-right symmetric (see Herstein 1969). There exists a left primitive ring with nonzero left singular ideal, i.e., with nonzero elements whose left annihilators are essential left ideals (Lawrence). Furthermore, the class of primitive rings does not form an elementary class (Lawrence), and, moreover, Mekler proved that primitive rings are not definable in $L_{\infty,\infty}$ (see Jensen and Lenzing 1989), cf. Section I.1.

C.17.1 Theorem (Chevalley-Jacobson) *Let R be a left primitive ring with faithful simple left module M, let $D = \text{End}_R(M)$, let $x_1, x_2, \ldots, x_n \in M$ be linearly independent over D and let $y_1, y_2, \ldots, y_n \in M$. Then there exists $r \in R$ such that $rx_i = y_i$ for all $i = 1, 2, \ldots, n$.*

This celebrated Density Theorem has been generalized to larger classes of modules by a number of algebraists (see Wisbauer 1991, Zelmanowitz 1976 for further results and references). Being confined by context of primitive rings and simple modules, we state only the following generalization which is due to Beidar and Brešar (2001). A derivation d of R is said to be *M-inner* if there exists an additive transformation T of M such that $d(r)m = rT(m) - T(rm)$ for all $r \in R$ and $m \in M$; otherwise, d is called *M-outer*.

C.17.2 Theorem *Let R be a left primitive ring with faithful simple left module M, let $D = \mathrm{End}_R(M)$, let d be an M-outer derivation of R and let $p = \mathrm{char}(D)$ if $\mathrm{char}(D) > 0$ and $p = \infty$ if $\mathrm{char}(D) = 0$. Then for any positive integer $m < p$, for all elements $x_1, x_2, \ldots, x_n \in M$ linearly independent over D and for all $y_{ij} \in M$ there exists $r \in R$ with $d^i(r)x_j = y_{ij}$ for all $0 \le i \le m$, $1 \le j \le n$ (it is understood that $d^0(r) = r$).*

Besides the Density Theorem, the structure of primitive rings with nonzero socle, the description of their (anti)automorphisms and derivations, and Kaplansky's theorem on primitive PI-rings (see Section C.37) constitute the core of the classical theory of primitive rings. Allowing coefficients from a primitive ring to appear between variables of a polynomial in noncommutative variables, one gets an idea of a *generalized polynomial identity* (GPI). For example, if V is a vector space over a field F, $R = \mathrm{End}_F(V)$ and $e \in R$ is a projection of V onto a one-dimensional subspace, then $eRe = eF$ and so $exeye - eyexe$ is GPI because it vanishes under any substitution $x = a \in R$ and $y = b \in R$. Amitsur's theorem (Beidar et al. 1996) on primitive rings with GPI describes them as primitive rings with nonzero socle whose associated division ring is finite dimensional over its center. Generalizing the concept of the polynomial identity in different direction, Drazin introduced pivotal monomials. Let n be a positive integer and let S_n be the permutation group of degree n. If R is a ring and

$$r_1 r_2 \ldots r_n \in \sum_{\substack{\sigma \in S_n, \\ \sigma \ne 1}} R r_{\sigma(1)} r_{\sigma(2)} \ldots r_{\sigma(n)} \quad \text{for all } r_1, \ldots, r_n \in R,$$

then $x_1 x_2 \ldots x_n$ is (a particular example of) a pivotal monomial. Drazin characterized left primitive rings with pivotal monomial as matrix rings over division rings (see Beidar et al. 1996). The concept of a generalized pivotal monomial is obtained from that of the pivotal monomial in a similar way as the concept of GPI from that of PI. Amitsur characterized primitive rings with nonzero socle as left primitive rings with generalized pivotal monomial (see Beidar et al. 1996 for further details and results in this vein).

One of the main direction in the theory of primitive rings is the study of conditions which imply the primitivity of the ring in question. Kaplansky conjectured that a prime von Neumann regular ring (see Section C.29) is primitive. Fisher and Snider confirmed the conjecture for countable rings (see Faith 1999). Goodearl verified the conjecture for right self-injective rings (see Section C.21). Domanov constructed a counterexample (see Faith 1999).

We now describe Passman's primitivity machine, which allows one to reduce certain questions on prime rings to that on primitive ones. Let R be a prime algebra, let X be an infinite set of noncommutative indeterminates with $|X| \geq |R|$, let $T = R\langle\langle X \rangle\rangle$ be the ring of power series, let Y be a set with $|Y| = |T|$ and let $S = T\langle Y \rangle$ be the ring of polynomials in noncommutative indeterminates from Y. Then S is primitive.

The following result is due to Amitsur and Small (see McConnell and Robson 1987).

C.17.3 Theorem *Let D be a division ring with center F, let n be a positive integer and let $S = D[x_1, \ldots, x_n]$. Then S is primitive if and only if, for some m, the F-algebra $M_m(D)$ contains an isomorphic copy of the field $F(x_1, \ldots, x_n)$.*

If R is a domain and $G = A \star B$ is the free product of two nontrivial groups, not both of order 2, such that $|G| \geq |R|$, then the group ring $R[G]$ is primitive (Formanek).

Let R be an algebra over an infinite field F. A subset $S \subseteq R \setminus \{0\}$ is called a ***separating set*** if it intersects nontrivially every nonzero prime ideal of R. Consider the following conditions, known as ***Dixmier's conditions***:

(1) R is left primitive;
(2) If R is a (left) Goldie ring, then the center of the classical ring of quotients of R is algebraic over F;
(3) The intersection of all nonzero prime ideals of R is nonzero;
(3') R has a countable separating set.

The equivalence of these conditions in various classes of algebras has been studied by a number of algebraists. Dixmier proved that (1), (2) and (3') are equivalent when R is the enveloping algebra of a finite dimensional Lie algebra over $\mathbb{C}$ and conjectured that in this case (3) is also equivalent to these conditions (see Rowen 1988b). His conjecture was verified by Moeglin (see Rowen 1988b). We conclude our discussion of primitive rings with the following result due to Rowen.

C.17.4 Theorem *Let R be a prime algebra over an uncountable field F such that $\dim_F(R) < |F|$. Suppose that there exists $r \in R$ such that $\{r, r^2, \ldots, r^n, \ldots\}$ is a separating set. Then R is both left and right primitive.*

A ring R is called **semisimple** if it is a subdirect product of left primitive rings. It is known that every semisimple rings is also a subdirect product of right primitive rings. The famous Wedderburn-Artin theorem describes semisimple left Artinian rings as finite direct sums of matrix rings over division rings.

C.17.5 Theorem (Amitsur) *Let R be a ring having no nonzero nil ideals. Then $R[x]$ is semisimple.*

One of central problems on semisimple rings is: if F is a field of characteristic 0 and G is a group, is the group algebra $F[G]$ then semisimple?

In the connection with this problem, we note the following three results (see Passman 1971).

C.17.6 Theorem (Amitsur) *Let F be a field of characteristic 0 which is not algebraic over a prime subfield and let G be a group. Then $F[G]$ is semisimple.*

C.17.7 Theorem (Passman) *Let F be a field of characteristic $p > 0$ which is not algebraic over a prime subfield and let G be a group with no elements of order p. Then $F[G]$ is semisimple.*

The third result is due to Villiamayor, Wallace, Passman and Zalessky.

C.17.8 Theorem *Let G be a solvable group and let F be a field. Suppose that G has no elements of order p in case $\mathrm{char}(F) = p$. Then $F[G]$ is semisimple.*

C.18 The Jacobson Radical

by Richard Wiegandt in Budapest, Hungary

Among the various radicals the Jacobson radical turned out to be the most successful, in particular, it provides the best description of semisimple (i.e., radical free) rings.

The next definition of the Jacobson radical is perhaps the easiest to remember. In a ring R the circle operation $\circ$ is defined by $x \circ y = x + y - xy$. A ring R is called **quasi-regular** (or **quasi-invertible**), if $(R, \circ)$ is a group. Every ring R has a unique largest ideal $\mathscr{J}(R)$ which is a quasi-regular ring.

C.18.1 Definition The ideal $\mathscr{J}(R)$ is called the **Jacobson radical** of the ring R. The class

$$\mathscr{J} = \{R \mid \mathscr{J}(R) = R\}$$

is the **Jacobson radical class**, and rings of $\mathscr{J}$ are referred to as **Jacobson radical rings**.

In view of this definition the Jacobson radical can be represented as a sum

$$\mathscr{J}(R) = \sum \{I \triangleleft R \mid I \in \mathscr{J}\}.$$

The **semisimple class** belonging to the Jacobson radical is

$$\mathscr{S}\mathscr{J} = \{R \mid \mathscr{J}(R) = 0\}.$$

For describing the structure of Jacobson semisimple rings, an intersection representation of the Jacobson radical is needed. Call an *ideal I* of a ring R **primitive**, if $R/I \neq 0$ is a primitive ring (see Section C.17).

C.18.2 Theorem (Jacobson) *The Jacobson radical $\mathscr{J}(R)$ of a ring R is the intersection of all primitive ideals of R. A ring R is a Jacobson semisimple if and only if R is a subdirect sum of primitive rings. R is a Jacobson radical ring if and only if R has no primitive ideals.*

Jacobson semisimple rings are also called **semiprimitive** rings. Primitive rings are described by the Chevalley-Jacobson Density Theorem(see Section C.17), so Theorem C.18.2 is apparently the best possible decomposition theorem for rings.

A left ideal $L \lhd_l R$ is said to be **modular**, if there exists a right unity $e \in R$ modulo L, that is, $a - ae \in L$ for all $a \in R$. The Jacobson radical has also an intersection representation by left ideals:

$$\mathscr{J}(R) = \bigcap \{\, L \lhd_l R \mid L \text{ is a modular maximal left ideal}\,\}.$$

Noticing that for any modular maximal left ideal L the R-module R/L is faithful and simple, one can prove an external characterization of the Jacobson radical in terms of R-modules.

C.18.3 Theorem *R is a Jacobson radical ring if and only if there are no faithful simple R-modules. Moreover,*

$$\mathscr{J}(R) = \bigcap \{\, a \in R \mid aM = 0 \text{ for every faithful simple } R\text{-module } M \,\}.$$

Some important properties of the Jacobson radical:

- $\mathscr{J}$ is **hereditary**: $I \lhd R \in \mathscr{J}$ implies $I \in \mathscr{J}$, or equivalently, $\mathscr{J}(I) = I \cap \mathscr{J}(R)$ for all $I \lhd R$.
- $\mathscr{J}$ is **left hereditary**: $L \lhd_l R \in \mathscr{J}$ implies $L \in \mathscr{J}$.
- $\mathscr{J}$ is **left strong**: $\mathscr{J}(L) = L \lhd_l R$ implies $L \subseteq \mathscr{J}(R)$, or equivalently,

$$\mathscr{J}(R) = \sum \{\, L \lhd_l R \mid L \in \mathscr{J})\,\}.$$

- The class of all primitive rings is a special class, and so $\mathscr{J}$ is a **special radical** (see Section C.19).
- A Jacobson radical ring R does not contain non-zero idempotents.
- If $x \in R$ is a nilpotent element, then there exists an element $y \in R$ such that $x \circ y = 0$. Hence *every nil ring is a Jacobson radical ring.*
- The unique largest nil ideal $\mathcal{N}(R)$ of any ring R is contained in the Jacobson radical $\mathscr{J}(R)$. $\mathcal{N}(R)$ is called the **nil** or **Köthe radical** of R.
- $\mathcal{N}(R) \neq \mathscr{J}(R)$ *may happen.* For instance, the ring J consisting of rational numbers with even numerator and odd denominator is quasi-regular, whence $\mathscr{J}(I) = I$, but obviously $\mathcal{N}(I) = 0$.
- The Jacobson radical of an artinian ring coincides with its prime, nil and Brown-McCoy radical (see Section C.19).
- There exist simple Jacobson radical rings (Sąsiada).
- There exist Jacobson semisimple rings which cannot be mapped homomorphically onto subdirectly irreducible prime rings (Sąsiada and Suliński).

- The Jacobson radical is **matric extensible**: denoting by $M_n(R)$ the $n \times n$ matrix ring over R, the equality $\mathscr{J}(M_n(R)) = M_n(\mathscr{J}(R))$ holds.

One of the oldest but still open problems of ring theory is **Köthe's Problem** (1930): is the nil radical left (or right) strong, that is,

$$\mathscr{N}(R) = \sum \{ L \lhd_l R \mid L \text{ is a nil ring} \}?$$

An equivalent formulation asks as whether the matrix ring $M_2(R)$ over a nil ring R is nil.

The Jacobson radical plays a very subtle role in Köthe's Problem. Amitsur proved (see Section C.17) that if the polynomial ring $R[x]$ is Jacobson radical, then R is nil, or equivalently, $\mathscr{N}(R) = 0$ implies $\mathscr{J}(R[x]) = 0$. As shown by Krempa (1972), Köthe's Problem is equivalent to any of the following implications:

- Is the polynomial ring $R[x]$ Jacobson radical for every nil ring R?
- Does $\mathscr{J}(R[x]) = 0$ always imply $\mathscr{N}(R) = 0$?

Köthe's Problem has been verified for many different classes of rings, e.g., for algebras over uncountable fields.

An important issue is (e.g., concerning the weak Nullstellensatz or in the theory of group rings), when is the Jacobson radical nilpotent or nil.

- If R is an artinian ring then $\mathscr{J}(R)$ is nilpotent (though $\mathscr{J}(R)$ need not be artinian).
- If R is an algebraic algebra over a field then $\mathscr{J}(R)$ is nil.

Putting $J = \mathscr{J}(R)$, **Jacobson's Conjecture** asks as whether $\bigcap_{n=1}^{\infty} J^n = 0$ for every noetherian ring R. This has been solved for various classes of rings, but for left and right noetherian rings it is still open.

C.19 General Radical Theory

by Richard Wiegandt in Budapest, Hungary

Besides the Jacobson radical, several other radicals have been introduced, for instance, the **prime** or **Baer radical** $\beta(R)$ as the intersections of all prime ideals of R, the **Brown-McCoy radical** $\mathscr{G}(R)$ as the intersection of all ideals I of R such that R/I is a simple ring with identity, the **nil** or **Köthe radical** $\mathscr{N}(R)$ as the largest nil ideal of R, and the **von Neumann regular radical** $\nu(R)$ as the largest von Neumann regular ideal of R.

In the early fifties, Amitsur and Kurosh observed independently that all these radicals show some common properties.

C.19.1 Definition A class (that is, a property) γ of rings is said to be a *radical class* (in the sense of Kurosh and Amitsur), if

(a) γ is *homomorphically closed*: $f: R \to f(R)$ and $R \in \gamma$ imply $f(R) \in \gamma$,
(b) $\gamma(R) = \Sigma(I \lhd A \mid I \in \gamma) \in \gamma$ for all rings R,
(c) $\gamma(R/\gamma(R)) = 0$ for all rings R.

In this definition conditions (b) and (c) can be replaced by

(d) γ has the *inductive property*: if $\{I_\lambda \mid \lambda \in \Lambda\}$ is an ascending chain of ideals of a ring $R = \cup I_\lambda$ and $I_\lambda \in \gamma$ for each $\lambda \in \Lambda$, then also $R \in \gamma$,
(e) γ is *closed under extensions*: $I \lhd R$, $I \in \gamma$ and $R/I \in \gamma$ imply $R \in \gamma$.

The ideal $\gamma(R)$ of R is called the γ-*radical* of R of R. Obviously $\gamma = \{R \mid \gamma(R) = R\}$, further, the radical class γ defines the *semisimple class*

$$\mathscr{S}\gamma = \{R \mid \gamma(R) = 0\}.$$

The *radical* $\gamma(R)$ of R, which is defined as the sum of all γ-ideals of R, has also an intersection representation:

$$\gamma(R) = \cap(I \lhd A \mid \gamma(A/I) = 0).$$

A class σ of rings is said to be *regular*, if every non-zero ideal I of any ring $R \in \sigma$ has a nonzero homomorphic image in σ. The class

$$\mathscr{U}\sigma = \{R \mid R \text{ has no non-zero homomorphic image in } \sigma\}$$

is a radical class for every regular class σ. In particular, $\gamma = \mathscr{U}\mathscr{S}\gamma$ for every radical class, and so radical classes and semisimple classes determine each other.

$\mathscr{U}\sigma$ is the largest radical class having intersection $\{0\}$ with σ, and $\mathscr{U}\sigma$ is called the *upper radical* of the class σ.

C.19.2 Theorem (Anderson, Divinsky, Suliński) *Let γ be any radical. If $I \lhd R$, then $\gamma(I) \lhd R$.*

This Theorem is of fundamental importance in the general radical theory.

C.19.3 Corollary *Every semisimple class is hereditary:* $I \lhd R \in \mathscr{S}\gamma$ *implies* $I \in \mathscr{S}\gamma$. *A radical class* γ *is hereditary if and only if* $\gamma(I) = I \cap \gamma(R)$ *for every* $I \lhd R$.

Due to Theorem C.19.2, it is possible to characterize semisimple classes by conditions dual to those defining radical classes.

C.19.4 Theorem (Sands and van Leeuwen, Roos, Wiegandt)
A class σ *of rings is the semisimple class of the radical class* $\mathscr{U}\sigma$ *if and only if*

(a*) σ *is hereditary,*

(d*) σ *has the coinductive property: if* $\{I_\lambda \mid \lambda \in \Lambda\}$ *is a descending chain of ideals of a ring* R *such that* $\cap I_\lambda = 0$ *and each factor ring* $R/I \in \sigma$, *then* $R \in \sigma$.

(e*) = (e) σ *is closed under extensions.*

In this characterization condition (d*) can be replaced by

(b*) σ *is closed under forming subdirect products.*

The *good* radical classes (for which the semisimple rings are expected to be subdirect products of *nice* rings) are hereditary and contain all nilpotent rings. Such radicals are called **supernilpotent radicals**. An ideal I of a ring R is **essential** in R, if $I \cap K \neq 0$ for every non-zero ideal K of R. A hereditary class ϱ of semiprime rings is said to be a **weakly special class**, if ϱ is **closed under essential extensions**: if I is an essential ideal in R and $I \in \varrho$ then $R \in \varrho$.

C.19.5 Theorem *If* ϱ *is a weakly special class, then* $\gamma = \mathscr{U}\varrho$ *is a supernilpotent radical. If* γ *is a supernilpotent radical, then the class* $\mathscr{S}\gamma$ *is a weakly special class. A class* σ *is the semisimple class of a supernilpotent radical if and only if* σ *is hereditary, closed under subdirect products and essential extensions, is weakly homomorphically closed (that is, $I \lhd R \in \sigma$ and $I^2 = 0$ imply $R/I \in \sigma$) and σ is not the class of all rings.*

As far as structure theorems for semisimple rings are concerned, the *good* radicals are the special radicals. A hereditary class ϱ of prime rings which is closed under essential extensions, is called a **special class**, and the radical $\mathscr{U}\varrho$ is said to be a **special radical**.

C.19.6 Theorem (Andrunakievich) *If* ϱ *is a special class, then* $\gamma = \mathscr{U}\varrho$ *is a supernilpotent radical and every* γ *semisimple ring* R *is a subdirect product of* ϱ*-rings.*

Note that a supernilpotent radical need not be a special one. The prime, nil, Jacobson, and Brown-McCoy radicals are examples of special radicals.

The prime radical β, the Jacobson radical $\mathscr{J}$, and many other supernilpotent radicals are

- **left hereditary** if L is a left ideal of a radical ring R, then also L is radical,
- **left strong** if L is a radical left ideal of a ring R, then L is in the radical of R.

Left hereditary and left strong supernilpotent radicals can be characterized by a condition on Morita contexts. Köthe's Problem (is the nil radical left strong?, cf. Section C.18) is equivalent to the question as whether the nil radical satisfies this condition on Morita contexts.

Interesting questions have been settled concerning the lower radical constructions (that is, constructing the smallest radical class containing a given class of rings).

General radical theory provides a framework to introduce more concrete radicals with emphasis on (sub)direct decomposition theorems for semisimple rings, and to construct rings which distinguish properties of rings. In this latter aspect two recent decisive results should be mentioned. Agata Smoktunowicz solved two longstanding and difficult problems of Amitsur and Levitzki, respectively: *there exists a nil ring R such that the polynomial ring $R[x]$ is not nil* (Smoktunowicz 2000), and *there exists a simple prime nil ring,* (Preprint 1999), cf. Section C.24.

Further important benefits of the general radical theory are

  i) in studying and describing classes of rings closed under certain closure operations,
 ii) in interpreting and developing radical theory in other categories.

ad i) One may ask which are the homomorphically closed semisimple classes or the radical classes closed under subdirect products. The answer is

C.19.7 Theorem *For a class ϱ of rings, the following are equivalent:*

(1) ϱ *is a homomorphically closed semisimple class,*

(2) ϱ *is a radical class closed under subdirect products,*

(3) ϱ *is a radical class closed under essential extensions,*

(4) ϱ *is a radical and a semisimple class,*

(5) ϱ *is a subvariety of rings closed under (essential) extensions,*

(6) ϱ is the upper radical of a finite set $\mathscr{F}$ of finite fields such that every subfield of a field in $\mathscr{F}$ belongs to $\mathscr{F}$.

A more general question asks when is the essential closure of a radical class a semisimple class. Such non-trivial radical classes are exactly those which have a unique complement in the lattice of all hereditary radical classes, and they are explicitly described in terms of matrix rings over finite fields.

ad ii) The theory of (hereditary) radicals of modules is called **torsion theory**. Radical theory can be interpreted and can be developed for rings with involution, subvarieties of non-associative rings, near-rings, incidence algebras, Petri nets, etc., as well as in non-algebraic categories: the connectedness and disconnectedness theory of topological spaces and graphs, respectively.

C.20 Free and Projective Modules

by Alexander V. Mikhalev and Askar A. Tuganbaev in Moscow, Russia

In this section, all rings are assumed to be associative and with nonzero identity element.

Let M be a right module over a ring A. A subset X of M is said to be **linearly independent** if for any $x_1, \ldots, x_n \in X$ and any $a_1, \ldots, a_n \in A$, the relation $\sum_{i=1}^{n} x_i a_i = 0$ implies $a_1 = \ldots = a_n = 0$. The module M is said to be **free** if M has a linearly independent generator system X; in this case, the set X is called a **basis** of M. The module M is free if and only if there is a set J such that M is isomorphic to a direct sum of J isomorphic copies of the module A_A; the cardinality of J is not necessarily unique (e.g., it is unique if A has a nonzero ring homomorphism into a division ring). Each right (resp. left) A-module is free if and only if A is a division ring. Each nonzero finitely generated torsion-free abelian group is a free module over the ring of integers $\mathbb{Z}$. For a free algebra over a field, any of its right ideals is free.

A module isomorphic to a direct summand of a free module is called a **projective** module. All direct sums or direct summands of projective modules are projective. A nonzero subgroup of a free abelian group is free; in particular, all projective abelian groups are free. A direct product of infinitely many copies of the free cyclic $\mathbb{Z}$-module $\mathbb{Z}$ is not free. For a field F and each positive integer n, any nonzero projective module over the polynomial ring $F[x_1, \ldots, x_n]$ is free and the $F \times F$-module $F \times 0$ is a nonzero projective module that is not free. The additive group of rational numbers $\mathbb{Q}$ is a countable torsion-free abelian group that is not a projective $\mathbb{Z}$-module. Every projective module is a direct sum of countably generated projective modules.

For a module N, a module M is said to be **projective with respect to** N (or N-**projective**) if for every epimorphism $h : N \to \overline{N}$ and each homomorphism $\overline{f} : M \to \overline{N}$, there is a homomorphism $f : M \to N$ with $\overline{f} = hf$. A module M is said to be **injective** (Section C.21) if for every module N and any submodule N' of N, each homomorphism $N' \to M$ is extended to a homomorphism $N \to M$.

C.20.1 Theorem *For a module M over A, the following statements are equivalent:*

(1) *M_A is projective;*
(2) *M is projective with respect to every right A-module;*
(3) *M is projective with respect to each injective right A-module;*
(4) *There exist a set $\{m_i\}_{i \in I}$ of elements of M and a set $\{f_i\}_{i \in I}$ of homomorphisms $f_i \colon M \to A_A$ such that $m = \sum_{i \in I} m_i f_i(m)$ for any $m \in M$, where $f_i(m) = 0$ for almost all subscripts i.*

Assume that P_1 and P_2 are two projective modules, $h_1 \colon P_1 \to M$ and $h_2 \colon P_2 \to M$ are module epimorphisms with kernels Q_1 and Q_2, respectively. Then $P_1 \oplus Q_2 \cong P_2 \oplus Q_1$. Every right (resp. left) A-module is projective if and only if each simple right A-module is projective if and only if A is a classically semisimple ring (i.e., A is isomorphic to a finite direct product of full matrix ring over division rings).

For a module P, a submodule X of P is said to be **small** (in M) if $X + Y \neq P$ for every proper submodule Y of P. If M is a module and there is an epimorphism $f \colon P \to M$ such that the module P is projective and $\mathrm{Ker}(f)$ is a small submodule of P, then P is called a **projective cover** of M. If a module M has a projective cover P, then P is unique up to isomorphism. The $\mathbb{Z}$-module $\mathbb{Q}$ does not have a projective cover. All right modules over a ring A have projective covers if and only if A is a ring with the minimum condition on principal left ideals (such a ring is called a **right perfect** ring). If M is a projective module and each factor module of M has a projective cover, then M has the **finite exchange property** (this means that for every module X and each finite direct decomposition $X = \bigoplus_{i=1}^{n} N_i = M \oplus Y$, there are submodules $N_i' \subseteq N_i$ ($i = 1, \ldots, n$) with $X = M' \oplus (\bigoplus_{i=1}^{n} N_i')$).

A right A-module M is said to be **flat** if for each left A-module N and any submodule N' of N, the natural group homomorphism $M \otimes_A N' \to M \otimes_A N$ is a monomorphism. A projective module is flat and a finitely presented flat module is projective. The additive group $\mathbb{Q}$ is a flat $\mathbb{Z}$-module that is not projective. A right A-module M is flat if and only if the left A-module $\mathrm{Hom}(M_{\mathbb{Z}}, \mathbb{Q}/\mathbb{Z})_{\mathbb{Z}}$ is injective if and only if for each finitely generated left ideal B of A, the natural group homomorphism $M \otimes_A B \to MB$ is an isomorphism. All flat right A-modules are projective if and only if A is a right perfect ring. All right A-modules are flat if and only if for every element $a \in A$, there is an element $b \in A$ with $a = aba$ (such a ring is called a **regular ring**).

A ring A is called an **exchange** ring if for any element $a \in A$, there is an idempotent $e \in aA$ with $1 - e \in (1 - a)A$ (this is equivalent to the condition that for any element $a \in A$, there is an idempotent $f \in Aa$ with $1 - f \in A(1 - a)$). Every projective right module over an exchange ring A is isomorphic to $\bigoplus_{i \in I} e_A$, where all the e_i are idempotents of A. We note that any semiregular ring is an exchange ring (a ring A is said to be **semiregular** if the factor ring $A/J(A)$ is regular and any idempotent of $A/J(A)$ is lifted to an idempotent of A). Let B be the ring of all rational numbers with odd denominators, $\{Q_i\}_{i=1}^{\infty}$ be a countable set of copies of the field of rational numbers, D be the direct product of all the fields Q_i, and let A be the subring in D generated by the ideal $\bigoplus_{i=1}^{\infty} Q_i$ and by the subring $B' \equiv \{(b, b, b, \ldots) \mid b \in B\}$. Then A is a commutative exchange ring that is not semiregular.

For a ring A, all submodules of projective right A-modules are projective if and only if the ring A is right hereditary (i.e., every right ideal of A is projective). A principal right ideal domain is right hereditary. The residue ring $\mathbb{Z}_4$ and the polynomial ring $D[x, y]$ over a field D are not hereditary. A right hereditary semiprimary ring is left hereditary (a ring A is said to be **semiprimary** if its Jacobson radical $J(A)$ is nilpotent and the factor ring $A/J(A)$ is Artinian). A ring is said to be **right invariant** if all its right ideals are ideals. A right hereditary ring is right invariant if and only if it is right distributive (a module is said to be **distributive**—see Section C.36—if the lattice of its submodules is distributive).

A ring is said to be **right semihereditary** (resp. **right Rickartian**) if all its finitely generated (resp. principal) right ideals are projective. If A is an infinite direct product of fields, then the rings A and $A[x]$ are semihereditary rings that are not hereditary. A projective right module over a right semihereditary ring is isomorphic to a direct sum of finitely generated right ideals of the ring. A ring A is right semihereditary if and only if each finitely generated submodule of any projective right A-module is projective if and only if for every positive integer n, the matrix ring R_n is right Rickartian. A right semihereditary left or right Noetherian ring is left semihereditary. A ring A is right hereditary if and only if the endomorphism ring of every free right A-module is right Rickartian. If A is a right semihereditary ring, then all submodules of any flat right or left A-module is flat.

A ring A is said to be **right self-injective** if the module A_A is injective. Every injective right (resp. left) A-module is projective if and only if every projective right (resp. left) A-module is injective if and only if A is a right and left self-injective Artinian ring (i.e., A is a quasi-Frobenius ring).

A module is said to be **quasi-projective** if it is injective with respect to itself. A direct sum of simple modules is called a *semisimple* module. Every projective or semisimple module is quasi-projective and all cyclic right modules over a right invariant ring are quasi-projective. The $\mathbb{Z}$-module $\mathbb{Z}_4$ is a quasi-projective module that is not projective or semisimple. An abelian group M is quasi-projective if and only if either M is free or M is a torsion abelian group such that any primary component of M is a direct sum of isomorphic cyclic groups. An abelian group M is quasi-projective if and only if either M is free or M is a torsion group and any primary component of M is a direct sum of isomorphic cyclic groups. If P is a projective module X is a small submodule of P, then P/X is a quasi-projective module if and only if X is a fully invariant submodule of P. Any finite direct sum of isomorphic quasi-projective modules is a quasi-projective module. The direct sum $\mathbb{Z}_4 \oplus \mathbb{Z}_2$ of two quasi-projective abelian groups is not quasi-projective. A quasi-projective module M has the finite exchange property if and only if for every decomposition $M = N_1 + \cdots + N_n$ into a sum of submodules N_i of M, there is a direct decomposition $M = M_1 \oplus \cdots \oplus M_n$ such that $M_i \subseteq N_i$ for all i if and only if for every decomposition $M = L + N$ into a sum of submodules L and N of M, there is a direct summand M_1 of M such that $M_1 \subseteq L$ and $M = M_1 + N$.

A module M is said to be **skew-projective** if each endomorphism of any its factor module can be lifted to an endomorphism of M. Every quasi-projective module is skew-projective. A quasi-cyclic group is a skew-projective $\mathbb{Z}$-module that is not quasi-projective. An abelian group M is skew-projective if and only if one of the following conditions hold: (1) M is free; (2) $M = F \oplus T$, where F is free and T is a direct sum of pairwise nonisomorphic quasi-cyclic groups; (3) M is a torsion group and any primary component of M is either a quasi-cyclic group or a direct sum of isomorphic cyclic groups.

A module M is said to be π-**projective** if for any two its submodules X and Y with $X + Y = M$, there is an endomorphism f of M such that $f(M) \subseteq X$ and $(1 - f)(M) \subseteq Y$. Every skew-projective or uniserial module is π-projective (a module is said to be **uniserial** if any two of its submodules are comparable with respect to inclusion). An abelian group is π-projective if and only if it is skew-projective. If A is the subring of $\mathbb{Q}$ consisting of rational numbers with odd denominators, then $\mathbb{Q}$ is a π-projective A-module that is not skew-projective. If all 2-generated right A-modules are π-projective, then A is a classically semisimple ring.

For more information on the above classes of modules, see (Anderson and Fuller 1992), (Bass 1968), (Cartan and Eilenberg 1956), (Cohn 1985), (Facchini 1998), (Tuganbaev 1998), and (Wisbauer 1991).

C.21 Injective Modules

by Askar A. Tuganbaev in Moscow, Russia

Let A be a ring, and let M be a right A-module. Given a module N_A, the module M is said to be *injective with respect to* N (or N-*injective*@N-*injective*indexN-injective@N-injective) if for any submodule N' of N, each homomorphism $N' \to M$ can be extended to a homomorphism $N \to M$. The module M is said to be *injective* if M is injective with respect to every right A-module.

The module M is injective if and only if M is injective with respect to the module A_A. All direct products or direct summands of injective modules are injective. If M_A is an injective module, then given any $x \in M$ and each $a \in A$ such that the right annihilator $r_A(a)$ is contained in $r_A(m)$, there is an element $m \in M$ with $x = ma$. In particular, M is a *divisible module* (i.e., $Ma = M$ for every element a of the ring A that is not a right or left divisor of zero).

The additive group of rational numbers $\mathbb{Q}$ and each quasi-cyclic group $\mathbb{Z}(p^\infty)$ are examples of injective modules over the ring of integers $\mathbb{Z}$. An abelian group M is an injective $\mathbb{Z}$-module if and only if M is divisible if and only if M is a direct sum of quasi-cyclic groups and direct sums of copies of the group $\mathbb{Q}$.

For a ring A, the right A-module $\mathrm{Hom}(A_\mathbb{Z}, (\mathbb{Q}/\mathbb{Z})_\mathbb{Z})$ is injective, and every A-module is isomorphic to a submodule of a direct product of isomorphic copies of the module $\text{Hom}(A_{\mathbb{Z}}, (\mathbb{Q}/\mathbb{Z})_{\mathbb{Z}})$ (this module is injective).

For a module M, a submodule M' of M is said to be **essential** (in M) if $M' \cap X \neq 0$ for every nonzero submodule X of M; in this case, we say that M is an **essential extension** of M'. If M is an essential extension of M' and the module M is injective, then M is called an **injective hull** of M'. Every module has an injective hull that is unique up to isomorphism. The $\mathbb{Z}$-module $\mathbb{Q}$ is the injective hull of the module $\mathbb{Z}_{\mathbb{Z}}$. If D is a division ring and A is the ring consisting of all upper triangular matrices $\left(\begin{smallmatrix} x & y \\ 0 & z \end{smallmatrix}\right)$ with $x, y, z \in D$, then the injective hull of the module A_A is the A-module $\text{Mat}_{2\times 2}(D)$.

A direct sum of simple modules is called a **semisimple module**. Every right (resp. left) A-module is injective if and only if the ring A is right (resp. left) semisimple if and only if A is a classically semisimple ring (i.e., A is isomorphic to a finite direct product of full matrix ring over division rings). If all cyclic right A-modules are injective, then A is a classically semisimple ring.

A direct sum of injective modules is not necessarily injective. For a ring A, all direct sums of injective right A-modules are injective if and only if all countable direct sums of injective right A-modules are injective if and only if A is a right Noetherian ring (i.e., all right ideals of A are finitely generated). If A is a polynomial ring in an infinite set of variables over any ring, then there are direct sums of injective A-modules that are not injective. For a ring A, every injective right A-module is a direct sum of indecomposable modules if and only if the ring A is right Noetherian. Every injective right A-module is a direct sum of injective hulls of simple modules if and only if A is a right Artinian ring (i.e., A is a ring with the minimum condition on right ideals—see Section C.27).

A homomorphic image of an injective module is not necessarily injective. For a ring A, all homomorphic image of injective right A-modules are injective if and only if the ring A is right hereditary (i.e., every right ideal of A is isomorphic to a direct summand of a direct sum of copies of the module A_A). Therefore, all homomorphic image of injective abelian groups are injective. The residue ring $\mathbb{Z}_4$ is an injective module over itself and its factor module with respect to $2\mathbb{Z}_4$ is not injective. In addition, if $D[x, y]$ is the polynomial ring over a field D, then there are homomorphic images of injective $D[x, y]$-modules that are not injective. For a ring A, every right module has a largest injective submodule if and only if A is a right hereditary right Noetherian ring. Therefore, every abelian group has a largest injective subgroup. A ring A is said to be *right self-injective* if the module A_A is injective. Every injective right (resp. left) A-module is projective if and only if every projective right (resp. left) A-module is injective if and only if A is a right and left self-injective Artinian ring (i.e., A is a quasi-Frobenius ring). A group ring $A[G]$ (see Section E.1) is right self-injective if and only if A is a right self-injective ring and G is a finite group.

A right A-module is said to be *nonsingular* if the right annihilator of any nonzero element of the module is not an essential right ideal of A. A right self-injective right nonsingular ring is a regular ring (i.e., every principal right or left ideal of the ring is generated by an idempotent). For any right nonsingular ring A with the right injective hull Q_A, there is a natural ring monomorphism from A into the regular right right self-injective ring $\mathrm{End}(Q_A)$.

A commutative ring A is regular if and only if all simple A-modules are injective. Let F be the field consisting of 2 elements, B be an algebraic closure of F, φ be the automorphism of the algebraically closed field B defined by $\varphi(b) = b^2$, and let A be the left skew Laurent polynomial ring $B_\ell[x, x^{-1}, \varphi]$ consisting of the formal polynomials $\sum_{i=t}^n b_i x^i$ in an indeterminate x with canonical coefficients $b_i \in B$, where t is an integer and $x^i b = \varphi^i(b)x^i$ for every element b of B and all integers i. Then A is a simple nonregular domain and all simple right A-modules are injective modules that are isomorphic to each other.

A module is said to be *quasi-injective* if it is injective with respect to itself. Every injective or semisimple module is quasi-injective. The $\mathbb{Z}$-module $\mathbb{Z}_4$ is a quasi-injective module that is not injective or semisimple. A module is quasi-injective if and only if it is a fully invariant submodule of its injective hull. An abelian group M is quasi-injective if and only if either M is divisible or M is a torsion abelian group such that any primary component of M is either a direct sum of quasi-cyclic groups or a direct sum of isomorphic cyclic groups.

If M is a quasi-injective right module, then the Jacobson radical $J(\mathrm{End}(M))$ (see Section C.18) of the endomorphism ring $\mathrm{End}(M)$ coincides with the set of all the endomorphisms of M with essential kernels, the factor ring $\mathrm{End}(M)/J(\mathrm{End}(M))$ is a regular right self-injective ring, and every countable set of orthogonal idempotents of the factor ring $\mathrm{End}(M)/J(\mathrm{End}(M))$ is lifted to a set of orthogonal idempotents of the ring $\mathrm{End}(M)$. A module is said to be *uniform* any two of its nonzero submodules have nonzero intersection. Any indecomposable quasi-injective module is uniform with local endomorphism ring. The endomorphism ring of a quasi-injective nonsingular right module is a regular right self-injective ring.

A quasi-injective module M is a module with the exchange property; this means that for every module X and each direct decomposition $X = M \oplus Y = \oplus_{i \in I} N_i$, there are submodules $N_i' \subseteq N_i$ ($i \in I$) with $X = M' \oplus (\oplus_{i \in I} N_i')$.

A ring is said to be *right invariant* if all its right ideals are ideals. A module M is said to be *linearly compact* if every family of cosets of submodules of M that has the finite intersection property has nonempty intersection. For a ring A, all cyclic right (resp. left) A-modules are quasi-injective if and only if A is a finite direct product of full matrix rings over division rings and (right and left) invariant linearly compact rings whose prime ideals are maximal.

A module M is said to be *skew-injective* if each endomorphism of any its submodule is extendable to an endomorphism of M. Every quasi-injective module is skew-injective. Each semi-Artinian skew-injective module is quasi-injective (a module is said to be *semi-Artinian* if every of its nonzero factor modules is an essential extension of a semisimple module). An integrally closed commutative Noetherian domain A is either a field or a skew-injective A-module that is not quasi-injective. In particular, the ring of integers $\mathbb{Z}$ is a skew-injective $\mathbb{Z}$-module that is not quasi-injective. An abelian group M is skew-injective if and only if either M is quasi-injective or the torsion part $T(M)$ of M is a direct sum of quasi-cyclic groups and the factor group $M/T(M)$ is isomorphic to a subgroup of the additive group $\mathbb{Q}$. If A is a ring which is integral over its center, then all cyclic right (resp. left) A-modules are skew-injective if and only if A is a finite direct product of hereditary Noetherian invariant domains and full matrix rings over division rings.

A module M is said to be *π-injective* if each idempotent endomorphism of any of its submodules can be extended to an endomorphism of M. Every skew-injective module is π-injective. Each π-injective abelian group is skew-injective. If D is the quaternion division algebra over the field of real numbers, then the polynomial ring $D[x]$ is a π-injective module that is not skew-injective.

A module M is uniform if and only if M is an indecomposable π-injective module. All π-injective right modules over a right Noetherian ring are direct sums of uniform modules. If all 2-generated right A-modules are π-injective, then A is a classically semisimple ring. If all cyclic right A-modules are π-injective, then A is a finite direct product of full matrix rings over division rings and right uniform ring.

For information on the above classes of modules see Anderson and

Fuller (1992), (Cartan and Eilenberg 1956), (Faith 1973), (Faith 1982), (Sharpe and Vámos 1972), (Tachikawa 1973), (Tuganbaev 1998), and (Wisbauer 1991).

C.22 Flat Modules

by Askar A. Tuganbaev in Moscow, Russia

All rings are assumed to be associative and with nonzero identity element. Expressions such as a "Noetherian ring" mean that the corresponding right and left conditions hold. A right module M over a ring A is said to be **flat** if for each left A-module N and any submodule N' of N, the natural group homomorphism $M \otimes_A N' \to M \otimes_A N$ is a monomorphism. A right A-module M is flat if and only if for each finitely generated left ideal B of A, the natural group homomorphism $M \otimes_A B \to MB$ is an isomorphism. Each free module is flat (a module M_A is said to be **free** if M is isomorphic to a direct sum of copies of the module A_A). All direct sums and direct summands of flat modules are flat. Consequently, all projective modules are flat (a module is said to be **projective** if it is isomorphic to a direct summand of a free module). An Abelian group M is a flat module over the ring of integers $\mathbb{Z}$ if and only if M is torsion-free. The additive group of rational numbers $\mathbb{Q}$ is a countable flat $\mathbb{Z}$-module that is not projective. A right A-module M is flat if and only if the left A-module $\operatorname{Hom}(M_{\mathbb{Z}}, \mathbb{Q}/\mathbb{Z})_{\mathbb{Z}}$ is injective (an A-module X is said to be **injective** (see Section C.21) if for every module N and any submodule N' of N, each homomorphism $N' \to X$ can be extended to a homomorphism $N \to X$). For a module M, a projective module P is called a **projective cover** of M if there is an epimorphism $f\colon P \to M$ such that $\operatorname{Ker}(f)$ is a small submodule of P (a submodule X of a module P is said to be **small** in P if $X + Y \neq P$ for any proper submodule Y of P). If a module M has a projective cover, then M is flat if and only if M is projective.

A right A-module M is said to be **Hattori torsion-free** if for any two elements $m \in M$ and $a \in A$ with $ma = 0$, there are elements $m_1, \ldots, m_n \in M$ and $a_1, \ldots, a_n \in A$ such that $m = \sum_{i=1}^{n} m_i a_i$ and $a_i a = 0$ for all i. All Hattori torsion-free modules are torsion-free (a module M_A is said to be **torsion-free** if $ma \neq 0$ for any nonzero element $m \in A$ and each element $a \in A$ that is not a right or left divisor of zero). A right A-module M is Hattori torsion-free if and only if for any principal left ideal Aa of A, the natural group epimorphism $M \otimes Aa \to Ma$ is an isomorphism. A module M_A is flat if and only if M is Hattori torsion-free and $MB \cap MC = M(B \cap C)$ for any two left ideals B and C of A if and only if M is Hattori torsion-free and $MB \cap MC = M(B \cap C)$ for any two finitely generated left ideals B and C of A. A module M is said to be **distributive** (see Section C.36) if the lattice of all its submodules is distributive. A ring is said to be **right** (resp. **left**) **invariant** if all its right (resp. left) ideals are ideals. If M is a module over an invariant distributive ring, then M is flat if and only if M is Hattori torsion-free. A module is called a **Bezout module** if all its finitely generated submodules are cyclic. If M is a right module over a left Bezout ring, then M is flat if and only if M is Hattori torsion-free. If M is a right module over a left Bezout ring without divisors of zero, then M is flat if and only if M is torsion-free.

A module M is flat if and only if for each finitely generated submodule X of M, there is a flat submodule of M containing X (therefore, all submodules of M are flat if and only if all finitely generated submodules of M are flat).

C.22.1 Theorem *For a ring A, the following conditions are equivalent:*

(1) *All submodules of right A-modules are flat;*
(2) *All submodules of left A-modules are flat;*
(3) *For every finitely generated right ideal B of A and each finitely generated left ideal C of A, the natural group homomorphism $B \otimes_A C \rightarrow BC$ is an isomorphism;*
(4) *All finitely generated right ideals of A are flat;*
(5) *All finitely generated left ideals of A are flat.*

A ring is said to be **reduced** if it does not have nonzero nilpotent elements. For a commutative ring A, all A-modules are flat if and only if all 2-generated ideals are flat if and only if A is a distributive reduced ring.

C.22.2 Theorem *For an invariant reduced ring A, the following conditions are equivalent:*

(1) *All A-modules are flat;*
(2) *All 2-generated ideals are flat;*
(3) *A is a distributive ring;*
(4) *For every factor ring R of A, all endomorphisms of finitely generated right ideals of A can be extended to endomorphisms of R_R.*

For a ring A, its principal right ideal xA is flat if and only if for any element $y \in A$ with $xy = 0$, there is an element $a \in A$ such that $xa = 0$ and $(1 - a)y = 0$. For a ring A, all principal right ideals of A are flat if and only if all principal left ideals of A are flat if and only if for any two elements $x, y \in A$ with $xy = 0$, there are two elements $a, b \in A$ such that $a + b = 1$, $xa = 0$, and $by = 0$. If A is a reduced right or left Bezout ring, then all submodules of flat A-modules are flat. If A is a reduced right distributive ring that is a finitely generated module over its center, then all submodules of flat A-modules are flat.

A module M_A is flat if and only if for all elements $x_1, \ldots, x_n \in M$ and $y_1, \ldots, y_n \in A$ such that $\sum_{i=1}^{n} x_i y_i = 0$, there are elements $\bar{x}_1, \ldots, \bar{x}_s \in M$ and $a_{ik} \in A$ such that $1 \leq i \leq n$, $1 \leq k \leq s$, $x_i = \sum_{k=1}^{s} \bar{x}_k a_{ik}$, and $\sum_{i=1}^{n} a_{ik} y_i = 0$ for all i and k. If n is a positive integer and M is a right A-module generated by n elements $x_1, \ldots, x_n \in M$, then M is flat if and only if for every n-generated left ideal Y of A, the canonical group epimorphism $M \otimes Y \rightarrow MY$ is an isomorphism if and only if for all elements $y_1, \ldots, y_n \in A$ with $\sum_{i=1}^{n} x_i y_i = 0$, there are elements $a_{ij} \in A$ such that $1 \leq i, j \leq n$, $x_i = \sum_{j=1}^{n} x_j a_{ij}$ and $\sum_{i=1}^{n} a_{ij} y_i = 0$ for all i and j.

C.22.3 Theorem *Let Q be a submodule of a flat right A-module P. Then the following conditions are equivalent:*

(1) *The factor module P/Q is flat;*
(2) *For any finitely generated left ideal B of A, the natural group epimorphism $PB/QB \rightarrow (P/Q)B$ is an isomorphism;*
(3) *For any left ideal B of A, the natural group epimorphism $PB/QB \rightarrow (P/Q)B$ is an isomorphism;*
(4) *$Q \cap PB = QB$ for any finitely generated left ideal B of A;*
(5) *$Q \cap PB = QB$ for any left ideal B of A.*

Therefore, if Q is a right ideal of a ring A, then the cyclic right A-module A/Q is flat if and only if $Q \cap B = QB$ for any finitely generated left ideal B of A. If Q is a submodule of a free right A-module P, then the factor module P/Q is flat if and only if for any element $q \in Q$, there is a homomorphism $h \colon P \rightarrow Q$ such that $h(q) = q$ if and only if for any finite subset $\{q_1, \ldots, q_n\}$ of Q, there is a homomorphism $h \colon P \rightarrow Q$ such that $h(q_i) = q_i$ for all q_i.

If A and B are two rings and N is a (A, B)-bimodule that is a flat B-module, then for every flat right A-module M, the right B-module $M \otimes_A N$ is flat. For a commutative ring A, all tensor products of flat

A-modules are flat A-modules. If A is a unitary subring of a ring B and M is a finitely generated right A-module such that the right B-module $M \otimes_A B$ is projective, then the A-module M is projective. A module is said to be *semisimple* if it is a direct sum of simple modules. If A is a unitary subring of a semisimple ring B, then all finitely generated flat right A-modules are projective. A ring A with Jacobson radical $J(A)$ is said to be *semilocal* (cf. Section C.7) if the factor ring $A/J(A)$ is semisimple. If A is a semilocal ring and each idempotent of the semisimple ring $A/J(A)$ can be lifted to an idempotent of A, then all finitely generated flat A-modules are projective. For a ring A, all flat right A-modules are projective if and only if A is semilocal and for any sequence $a_1, a_2 \ldots$ of elements in $J(A)$, there is a subscript k such that $a_k a_{k-1} \cdots a_1 = 0$.

A finitely presented flat module is projective (a module M is said to be *finitely presented* if there are a finitely generated projective module P and a finitely generated submodule Q of M such that $M \cong P/Q$). A ring A is said to be (*von Neumann*) *regular* or *absolutely flat* if for any element $a \in A$, there is an element $b \in A$ with $a = aba$. A ring A is regular if and only if all right A-modules and all left A-modules are flat if and only if all finitely presented right A-modules are flat if and only if for any element $a \in A$, the cyclic finitely presented right A-module A/aA is flat. A ring A is said to be *left coherent* if A satisfies the following two equivalent conditions: (1) all finitely generated left ideals of A are finitely presented: (2) the intersection of any two finitely generated left ideals of A is finitely generated and for any element $a \in A$, its left annihilator $\ell(a)$ is a finitely generated left ideal of A. For a ring A, all direct products of flat right A-modules are flat if and only if all direct products of copies of the module A_A are flat if and only if the ring A is left coherent. For a ring A, all submodules of direct products of copies of the right A-module A are flat if and only if all finitely generated left ideals of the ring A are projective.

A right A-module M is said to be *finitely injective* (resp. *countably injective*) if for every finitely generated (resp. countably generated) right ideal B of A, each homomorphism $B_A \to A_A$ can be extended to a homomorphism $A_A \to M$. For a ring A, all injective right A-modules and all injective left A-modules are flat if and only if A is a coherent finitely injective ring. For a ring A, all flat right A-modules are injective if and only if all flat left A-modules are injective if and only if A is an Artinian injective ring.

Let A be a ring, φ be a ring automorphism of A, and let $A_\ell[[x, \varphi]]$

be the **left skew power series ring** (Section C.9) consisting of formal series $\sum_{i=0}^{\infty} a_i x^i$ ($a_i \in A$), where $x^i a = \varphi^i(a) x^i$ for any $a \in A$ and each positive integer i. If all idempotents of A are central and $\varphi(e) = e$ for any idempotent $e \in A$, then all submodules of flat $A_\ell[[x, \varphi]]$-modules are flat if and only if all 2-generated right ideals of $A_\ell[[x, \varphi]]$ are flat if and only if A is a regular countably injective ring.

For more information on the above classes of modules and rings, see Anderson and Fuller (1992), Cartan and Eilenberg (1956), Faith (1973), Faith (1976), Goodearl (1976), Lambek (1976), Stenström (1975), Tuganbaev (1998), and Wisbauer (1991).

C.23 Free Ideal Rings and Free Algebras

by Paul M. Cohn in London, UK

*In what follows, all rings are associative, with a unit-element 1, which is inherited by subrings, preserved by homomorphisms, and acts unitally on modules. If $1 = 0$, the ring is **trivial**, i.e., it reduces to $\{0\}$; generally our rings will be non-trivial: $1 \neq 0$.*

Polynomial rings $K[x_1, \ldots, x_r]$ in n variables over a field K are well known and have been much studied. For $r = 1$, the ring $K[x_1]$ has a division algorithm and so is a principal ideal domain, but no such properties persist for $r > 1$. Consider now the corresponding construction in non-commuting variables: this is the ring $K\langle x_1, \ldots, x_r \rangle$ of all expressions

$$ f = \sum_{i_1 \ldots i_n} a_{i_1 \ldots i_n} x_{i_1} \ldots x_{i_n}, \quad a_{i_1 \ldots i_n} \in K. $$

For $r = 1$, $K\langle x_1 \rangle = K[x_1]$, but for $r > 1$ we get a non-commutative ring, the **free associative algebra** in $x_1, \ldots, x_r$ over K; here, the number of free generators r is an invariant of the ring, called its **rank**. In a free algebra of rank 2, $K\langle x, y \rangle$, the elements xy^n, $n = 0, 1, 2, \ldots$, form a free set, and so they generate a subalgebra which is free of countable rank—cf. Section G.4.

Any free algebra has the property that every right ideal is free of uniquely determined rank, thus it is a **free right ideal ring** or **right fir** and similarly on the left. This makes it a left and right fir, briefly a **fir** (**free ideal ring**).

In the presence of the Ore condition (in particular, for a commutative ring), a fir reduces to a principal ideal domain. The fir property is usually proved by a generalisation of the division algorithm, the **weak algorithm**, which is defined below.

C.23.1 Definition Let R be a ring. A **degree function** is a function $d: R \to \mathbb{N}_0 \cup \{-\infty\}$ satisfying

D.1 $d(a) \in \mathbb{N}$ for $a \neq 0$ in R, $d(0) = -\infty$;

D.2 $d(a + b) \leq \max(d(a), d(b))$;

D.3 $d(ab) = d(a) + d(b)$;

for all $a, b \in R$. A family (a_i) of elements is then called **right d-dependent** if some a_i is zero or there exist elements $b_i \in R$, almost all zero, such that

$$d(\sum a_i b_i) < \max_i (d(a_i) + d(b_i)).$$

Note that any right linearly dependent family over R is right d-dependent and every element of degree 0 is a unit.

Secondly, an element a of R is said to be **right d-dependent** on a family (a_i) if $a = 0$ or if there exist $b_i \in R$, almost all 0, such that

$$d(a - \sum_i a_i b_i) < d(a), \quad d(a_i) + d(b_i) \leq d(a) \text{ for all } i.$$

A family $a_1, \ldots, a_m \in R$, where $d(a_1) \leq \cdots \leq d(a_m)$, is called **strongly right d-dependent** if some a_i is right d-dependent on $a_1, \ldots, a_{i-1}$. It is clear that every strongly right d-dependent family is right d-dependent; if, conversely, every right d-dependent family is strongly right d-dependent, the ring R is said to satisfy the **weak algorithm** with respect to d. It can be shown that if an element a is right d-dependent on a family (a_i), then a is already right d-dependent on the elements of the family of degree at most $d(a)$. As a consequence of the weak algorithm for a ring R, every invertible matrix over R is a product of diagonal and elementary matrices (i.e., matrices differing from the unit matrix only by one off-diagonal entry). The $n \times n$ matrices that are products of (invertible) diagonal and elementary matrices form a group $\mathrm{GE}(n; R)$, in general a subgroup of the general linear group $\mathrm{GL}(n; R)$, but for a ring with weak algorithm, $\mathrm{GE}(n; R) = \mathrm{GL}(n; R)$.

The free algebra has a degree-function in which each x_i has degree 1, and it turns out that it satisfies the weak algorithm relative to this degree-function (see Cohn 1985, 2.4)). This holds for any number of variables (even infinite); in the commutative case, i.e., for $r = 1$, *the weak algorithm reduces to the usual division algorithm*. However, unlike the latter, the weak algorithm is left-right symmetric, even for general rings. Any ring satisfying the weak algorithm relative to some filtration can be shown to be a fir; thus any free algebra (over a field) is a fir. Moreover, over a fir R every submodule of a free module is free and it can be shown to have unique rank. Thus every fir is projective-free (see Cohn 1985, Ch. 2)); moreover, every fir has global dimension 1.

Let R be any ring. An R-module M of homological dimension 1 has a presentation $M = F/G$, where F and G are free. Provided that the rank of any free R-module is unique (i.e., R has an invariant basis number), one can define the **characteristic** $\chi(M)$ of M as $\mathrm{rk}\,F - \mathrm{rk}\,G$. When R is a principal ideal domain, then $\mathrm{rk}\,G \leq \mathrm{rk}\,F$, and so $\chi(M) \geq 0$ in this case. By contrast, for a fir R, where $\chi(M)$ can be defined as before, it can assume any positive or negative integer value, or zero.

A free algebra $R = K\langle X \rangle$ may be regarded as a graded ring, by assigning to each free generator the degree 1. Then the homogeneous component G_n of degree n in a free algebra of rank r has dimension r^n over the ground field. As with every graded ring, we have a Hilbert series $H(R) = \sum \dim_K(G_n)t^n$, which is therefore $H(R) = \sum r^n t^n = (1-rt)^{-1}$.

If A is any ring and $R = K\langle X \rangle$ a free algebra, it is well-known that any mapping $f: X \to A$ extends to a homomorphism $f^*: R \to A$; hence we also obtain a homomorphism $\hat{f}: \mathrm{GL}(n; R) \to \mathrm{GL}(n; A)$. Here $\mathrm{GL}(n; R)$ is the same as $\mathrm{GE}(n; R)$, by what has been said earlier. Moreover, one can show that for $n = 2$, the homomorphism $\hat{f}$ exists even if f^* is merely a linear mapping preserving the unit-element (see Cohn 1966).

Generalising the notion of fir, one defines a **semifir** as a ring in which every finitely generated right ideal is free of unique rank; it comes to the same to assume this for finitely generated left ideals. Clearly, every fir is a semifir, but a right fir need not be a left fir (see Cohn 1985, 3.4)). In the commutative case (more generally, in the presence of the Ore condition), a semifir just reduces to a Bezout domain, i.e., an integral domain in which every finitely generated right (or left) ideal is principal.

If for some integer $n \geq 1$, every right ideal on at most n generators is free of unique rank, we have an *n-fir*. This notion is again left-right symmetric and clearly the class of n-firs becomes smaller with increasing n (in fact, for each $n \geq 0$, there are n-firs which are not $(n+1)$-firs, (see Cohn 1995, 5.7) Another useful way of characterizing n-firs is by the trivialisability of relations. A relation $xy = x_1 y_1 + \cdots + x_r y_r = 0$ in a ring R is called **trivial** if for each $i = 1, \ldots, r$, $x_i = 0$ or $y_i = 0$. If for this relation there exists an invertible $r \times r$ matrix $\mathbf{P}$ such that the relation $\mathbf{x}\mathbf{P}\,\mathbf{P}^{-1}\mathbf{y} = 0$ is trivial, we say that the relation is **trivialisable**.

Now an n-fir can be characterised as a non-trivial ring in which every relation of the above form, where $r \leq n$, can be trivialised. Thus in particular, a 0-fir is a ring $\neq \{0\}$ and a 1-fir is an integral domain (not necessarily commutative).

For each semifir R, there exists a skew field K in which R can be embedded; more precisely, K can be chosen to be a universal skew field of fractions for R, in the sense that it has all other skew fields containing R as specialisations (see Cohn 1995). Thus a free algebra has a skew field of fractions; of course its elements cannot all be written in the form ab^{-1} where $a, b \in R$, unless we are in the Ore case. The skew field K is formed by inverting all full matrices over R and taking their entries, where a matrix $\mathbf{A}$ is called **full** if it is square, say $n \times n$, and in any factorisation $\mathbf{A} = \mathbf{PQ}$, $\mathbf{Q}$ has at least n rows. For every $n \geq 1$, examples can be found of n-firs that are not $(n+1)$-firs. For example, $\mathbb{Z}[x]$, like any integral domain, is a 1-fir, but it is not a 2-fir. For a non-commutative example, take the ring on 12 generators, arranged as a 2×3 matrix $\mathbf{A}$ and a 3×2 matrix $\mathbf{B}$, with defining relations (in matrix form) $\mathbf{AB} = \mathbf{I}$, $\mathbf{BA} = \mathbf{I}$. This also provides an example of an integral domain not embeddable in a skew field. However, for every 2-fir R, the multiplicative monoid of non-zero elements can be embedded in a group; more generally, R can be embedded in a ring S in which every non-zero element of R has an inverse (see Cohn 1985, Theorem 7.11.22)).

For free algebras there is a very simple method of embedding them in skew fields, as follows: Let K be a field (possibly skew, it doesn't matter) with an endomorphism β which is *not* an automorphism (i.e., not onto). Then the skew polynomial ring $R = K[x; \beta]$ consisting of all polynomials with the commutation rule $ax = xa^\beta$ is a right Ore domain (because it is right principal), but not left Ore; e.g., if $c \in K \setminus K^\beta$, then $Rx \cap Rxc = 0$. As a consequence, one can show that it contains a free algebra of rank 2 over a field (the subfield of the centre of K which is fixed under β). Hence it contains a free algebra of countable rank (see above). But $K[x, \beta]$, being right Ore, has a skew field of right fractions ab^{-1} (see above), so a free algebra has been embedded in a skew field (see Cohn 1985 or Cohn 1995). Of course, this skew field is very far from being the universal skew field of fractions for the free algebra.

Another important example of a fir is the coproduct of skew fields: let F, K, L be skew fields such that K and L contain subfields isomorphic to F. Then we can form the coproduct $K \amalg_F L$, i.e., the free product of K and L amalgamating (the copies isomorphic to) F. This coproduct is a fir and hence has a universal skew field of fractions, which is called the **field coproduct** of K and L over F (see Cohn 1995, 5.3).

The elements of firs have a factorisation which can be described as follows. By an **atom**, we understand a non-unit which cannot be written as a product of two non-units. Two elements a, b of a ring R are called **similar** if $R/aR \cong R/bR$; when a, b are regular, this is equivalent to the condition $R/Ra \cong R/Rb$. In a fir, every element not zero or a unit can be written as a product of atoms and two such factorisations of a given element have the same number of atomic factors, which can be paired off into similar pairs; of course the order of the factors will not usually be the same. More generally, such factorisations can be defined for any full matrix and with appropriate definitions, we again obtain a relation between the different factorisations of a given matrix. All these results extend to the class of **(fully) atomic** semifirs, i.e., those in which non-zero elements (resp. full matrices) can be written as products of atoms (unfactorable matrices) (Cohn 1985, Ch. 3).

In any 2-fir, the set of all principal right ideals containing a given non-zero element is a lattice with respect to sum and intersection, called the **factor lattice**. For a commutative PID, this factor lattice is distributive; this property extends to certain quite restricted classes of noncommutative rings, but surprisingly, also to free algebras, thus: any free algebra $K\langle X \rangle$ has a distributive factor lattice (Cohn 1985, Ch. 4).

In the study of automorphisms of a polynomial ring $K[x, y]$, one singles out linear transformations $x \mapsto ax + by$, $y \mapsto cx + dy$ $(ad \neq bc)$ and elementary automorphisms $x \mapsto x + f(y)$, $y \mapsto y$. Any automorphism which can be expressed as a product of linear and elementary automorphisms is called **tame** (cf. Section C.10) and the Jung-van der Kulk theorem asserts that every automorphism of $K[x, y]$ is tame. Further, the Czerniakiewicz-Makar-Limanov theorem asserts that every automorphism of the free algebra $K\langle x, y \rangle$ is tame and the natural mapping from the automorphism group of $K\langle x, y \rangle$ to that of $K[x, y]$ is an isomorphism (Cohn 1985, Ch. 6).

C.24 Simple Rings

by Kostia Beidar in Tainan, Taiwan

It is well-known that in a skew field D, or, more generally, in the ring $M_n(D)$ of $n \times n$ matrices over D, the trivial ideal (0) is both prime and maximal. Rings, in which (0) is both prime and maximal, are called **simple**. In other words, simple rings are exactly the rings with nontrivial multiplications having no nonzero proper ideals. We mention several remarkable examples of simple rings:

(i) Sasiada's example of a simple Jacobson radical ring (see Herstein 1969);

(ii) an example of a simple Jacobson radical domain all of whose left (right) ideals are linearly ordered by inclusion (Dubrovin 1980);

(iii) Zalesskii's and Neroslavkii's example of a simple Noetherian ring with unity, with zero divisors and without nontrivial idempotents; their example of a simple Noetherian hereditary ring which is not a matrix ring over a domain (see Chatters and Hajarnavis 1980), cf. sections C.20 and C.26;

(iv) a recent example of a simple nil ring (Agata Smoktunowicz, preprint 1999).

We now describe an important construction which yields simple rings. First we recall that an additive endomorphism $d\colon R \to R$ of a ring R is called a **derivation** of R if $d(rs) = d(r)s + rd(s)$ for all $r, s \in R$. A derivation $d\colon R \to R$ is said to be **inner** if there exists an element $a \in R$ such that $d(r) = ar - ra$ for all $r \in R$. Given a derivation d of a ring R, we denote by $R[x; d]$ the ring whose additive group is equal to that of $R[x]$ and the multiplication is induced by $xr = rx + d(r)$ for all $r \in R$. The ring $R[x; d]$ is called the **differential polynomial ring** over R (cf. Section D.5).

C.24.1 Theorem (Amitsur) *Let R be a simple ring of characteristic 0 with derivation $d\colon R \to R$. Suppose d is not inner. Then $R[x; d]$ is a simple ring.*

Given a field F of characteristic 0 and a positive number n, we denote by $A_n(F)$ the algebra with $2n$ generators $x_1, \dots, x_n, y_1, \dots, y_n$ and relations $[x_i, y_j] = \delta_{ij}$, the Kronecker delta, and $[x_i, x_j] = 0 = [y_i, y_j]$ for all $1 \le i, j \le n$. The algebra $A_n(F)$ is called the nth **Weyl algebra** over F. It is known that each $A_n(F)$ is a simple Noetherian domain.

Given a ring R and $x, y \in R$, we set $[x, y] = xy - yx$ (the **Lie product**) and $x \circ y = xy + yx$ (the **Jordan product**). The nonassociative rings $R^{(-)} = (R; +, [,])$ and $R^{(+)} = (R; +, \circ)$ are a Lie ring and a Jordan ring, respectively. An additive subgroup S of R is called a Lie (Jordan) subring of R provided that $[x, y] \in S$ (respectively, $x \circ y \in S$) for all $x, y \in S$. For example, if R is a ring with involution $*$ (i.e., an antiautomorphism of order ≤ 2, see Section C.39), then $K(R) = \{x \in R \mid x^* = -x\}$, the set of **skew elements** of R, is a Lie subring of R while $S = \{x \in R \mid x^* = x\}$, the set of **symmetric elements** of R, is a Jordan one. Further, $[R, R]$, the subgroup of R generated by all elements of the form $[x, y]$, is a Lie subring of R.

C.24.2 Theorem (Herstein) *Let R be a simple ring with $2R \neq 0$. Then $R^{(+)}$ is a simple Jordan ring. If I is an ideal of $R^{(-)}$, then either I is contained in the center of R or $I \supseteq [R, R]$. Further, if C is the center of the Lie ring $[R, R]$, then the Lie ring $[R, R]/C$ is simple. Finally, if R is not commutative, then the subring of R generated by $[R, R]$ is equal to R.*

The Lie structure of simple (and prime) rings of any characteristic was studied by Lanski and S. Montgomery. Results in this vein on simple rings with involution are also known (Herstein 1969). The study of the Jordan and Lie structure of simple rings leads naturally to the consideration of **Jordan and Lie isomorphisms**, that is, bijective additive maps preserving Jordan (respectively Lie) products.

C.24.3 Theorem ((Herstein 1969)) *Any Jordan isomorphism of simple rings of characteristic not 2 is either an isomorphism or anti-isomorphism of rings.*

Obviously every (anti)isomorphism of rings is also a Jordan isomorphism. Transposing matrices is an example of a nontrivial ring anti-isomorphism. The resolution of the question on isomorphism of Jordan subrings of symmetric elements of simple rings with involution is much more difficult and is based on Zelmanov's powerful technique developed for the classification of prime Jordan algebras. Modulo cases of low dimension, every such Jordan isomorphism can be uniquely extended to an isomorphism of rings. Results in this direction are due to Lagutina, Martindale and McCrimmon (cf. Section C.53).

The description of isomorphisms between various important Lie rings arising from simple associative rings is even more challenging problem and has been completed just recently. Results in this direction are based on newly developed branch of Ring Theory - the theory of functional identities. Let $\alpha \colon R \to T$ be a Lie isomorphism. In order to describe it, one has to determine $\alpha(xy)$. Since $xy = \frac{1}{2}\{[x, y] + x \circ y\}$, it is enough to find $\alpha(x \circ y)$. Applying α to the identity $[\alpha^{-1}(t)^2, \alpha^{-1}(t)] = 0$, $t \in S$, and setting $B(t, z) = \alpha(\alpha^{-1}(t) \circ \alpha^{-1}(z))$, one gets

$$[B(t, t), t] = 0 \quad \text{for all } t \in S, \tag{1}$$

the functional identity investigated by Brešar. Loosely speaking, a functional identity is an identity involving a function and variables. Solution of such identities is the subject of the theory of functional identities. In the case of simple rings, the main result of the theory states that either every functional identity of degree n has only the standard (obvious) solutions or the ring in question is finite dimensional over its center (see (Beidar and Chebotar 2000) and references over there). For example, if S is a simple ring with 1 and with center C, then (1) has standard solutions means that there exist elements $\lambda_1, \lambda_2 \in C$, maps $\mu_1, \mu_2 \colon S \to C$, and $\nu \colon S \times S \to C$ such that

$$B(t, z) = \lambda_1 tz + \lambda_2 zt + \mu_1(z)t + \mu_2(t)z + \nu(t, z) \quad \text{for all } t, z \in S.$$

Obviously, every function of this form is a solution of (1). We now present two results on Lie isomorphisms (Beidar et al. 1996):

C.24.4 Theorem (Brešar) *If R and T are simple rings with 1, $\mathrm{char}(R) \neq 2$ and R is not 4-dimensional over its center, then every Lie isomorphism of R and T is either of the form $\alpha + \tau$, where α is an isomorphism of rings and τ is an additive map into the center of T vanishing on $[R, R]$, or $-\alpha + \tau$, where α is an anti-isomorphism of rings and τ is as above.*

C.24.5 Theorem (Beidar, A. V. Mikhalev, and Martindale) *If R and T are simple rings with 1 and with involution $*$ acting identically on their centers, then every Lie isomorphism of $K(R)$ onto $K(T)$ can be uniquely extended to an isomorphism of R and T, provided that $\mathrm{char}(R) \neq 2, 3$, and the dimension of R over its center is greater then 64.*

Some theorems known for skew fields were generalized to simple rings. Among them is the celebrated Cartan-Brauer-Hua theorem on invariant subrings of skew fields (Faith 1999).

C.24.6 Theorem (Amitsur) *Let R be a simple algebra over a field F with nonzero center C which is not the ring of 2×2 matrices over a field of characteristic 2. Suppose that R has an idempotent $e \neq 0, 1$. Then:*

(1) *for any C-subspace S invariant under all inner automorphisms induced by elements of the form $1 + u$, $u \in R$ with $u^2 = 0$, either $S \subseteq C$, or $S \supseteq [R, R]$;*

(2) *for any F-subalgebra S of R invariant under all inner automorphisms induced by elements of the form $1 + u$, $u \in R$ with $u^2 = 0$,
either $S = R$, or $S \subseteq C$.*

Bokut proved that every algebra over a field is embeddable into a
simple algebra. In order to obtain a definite information about the
structure of a simple ring one has to impose certain conditions on the
elements or/and left (right) ideals of the ring or/and modules over the
ring. The Litoff-Ánh theorem states that R is a simple ring with minimal
one-sided ideal if and only if for every finite subset S of R there exists
a subring A of R containing S such that $(RA) \cap (AR) \subseteq A$ and A is
isomorphic to a matrix ring over a division ring (Kertész 1987). Faith
and Szász independently characterized simple rings with minimal one-
sided ideal as prime rings with minimal condition on principal left (or
right) ideals (Kertész 1987). The classical Wedderburn-Artin theorem
describes simple left (right) Artinian rings as matrix rings over skew
fields.

One of the major direction in the study of simple left Noetherian
rings is finding sufficient conditions forcing the ring in question to be a
matrix ring over a domain or to be Morita-equivalent to a domain (see
Chatters and Hajarnavis 1980). The first example in (iii) shows that a
simple Noetherian ring with zero divisors does not necessarily contain
a nonzero idempotent. Stafford proved that the example is not Morita-
equivalent to a domain (Chatters and Hajarnavis 1980). The second
example in (iii) shows that even a simple hereditary Noetherian ring
may not be of the form $M_n(D)$, where D is a domain.

C.25 Principal Ideal Domains
by Paul M. Cohn in London, UK

By a **principal ideal domain** (**PID**), one understands a ring which
is an integral domain, not necessarily commutative, in which every left
ideal and every right ideal is **principal**, i.e., generated by a single element. Well known commutative examples are the ring $\mathbb{Z}$ of integers and
the polynomial ring in one variable over a field. In addition, the ring
of integers in an algebraic number field is frequently a PID. To show
that a ring of algebraic numbers is a PID, one usually verifies that it is
Euclidean (see below). We have

C.25.1 Theorem *The ring of integers in $\mathbb{Q}(\sqrt{d})$ is Euclidean if and*

only if $d \in \{2, 3, 5, 6, 7, 11, 13, 17, 19, 21, 29, 33, 37, 41, 57, 73, -1,$
$-2, -3, -7, -11\}$ *(Stewart and Tall 1987, p. 95).*

By contrast, the ring of integers in $\mathbb{Q}(\sqrt{-19})$ is a PID which is not Euclidean (see Cohn 1966). In general, the ring I of integers in an algebraic number field need not be a PID, but for any non-principal ideal $J \trianglelefteq I$, one can find a finite integral extension I' in which J becomes principal. However, there may not be a finite integral extension which is itself a PID.

More generally, one has the skew polynomial rings over a skew field: let K be a skew field with an automorphism α and an α-**derivation** δ, i.e., an endomorphism of the additive group of K satisfying in addition the relation

$$(ab)^\delta = ab^\delta + a^\delta b^\alpha$$

for all $a, b \in K$. The ring of all polynomials $\sum x^i a_i$ in x over K with the commutation rule $ax = xa^\alpha + a^\delta$ for all $a \in K$, is a PID, denoted by $K[x; \alpha, \delta]$ and called the **skew polynomial ring**. It is an integral domain because the degree is multiplicative. To prove that left and right ideals are principal, one uses the division algorithm. This states that there is a function $d(f)$ on $K[x; \alpha, \delta]$ taking positive integer values such that

DA for any f, g, where $g \neq 0$, there exist $q, r \in K[x; \alpha, \delta]$ such that

$$f = gq + r \quad \text{and} \quad d(r) < d(g).$$

More generally, any integral domain with a function defined on its elements taking positive integer values and satisfying DA is called a **Euclidean ring**, and it can be shown that in a Euclidean ring every right ideal is principal; we say it is a principal **right** ideal domain (**PRID**). In a skew polynomial ring as defined above, the ring itself as well as its opposite is Euclidean; however, when α is merely an endomorphism, the opposite ring may not be Euclidean and so the ring may be merely a PRID.

Another source of PIDs is the coproduct of skew fields. Let K, L be any skew fields with a common sub(skew)field F. Then one can form the coproduct $K \amalg_F L$, i.e., the free product amalgamating F; cf. Section B.10. This ring is always a fir (free ideal ring, see Cohn 1985 or Section C.23); when it is an Ore domain, it is actually a PID; this happens if and only if K, L are at most 2-dimensional over F. More precisely, when $[K : F], [L : F] \leq 2$, where K, L are considered as right F-spaces, then the coproduct is a PRID (see Cohn 1995, 5.10).

Any commutative PID is a **unique factorisation domain (UFD)**; this means that every element other than zero or a unit can be written as a product of **atoms** (= **irreducible elements**, i.e., non-units that cannot be factorised into non-units), and this factorisation is unique up to the order of the factors and multiplication by unit factors. For the proof one makes use of the **Bezout formula**: given elements a, b in a commutative PID R, their greatest common divisor (GCD) can be written as $d = au + bv$, for some $u, v \in R$—cf. Section C.54. This result remains true even in the non-commutative case, as long as we pay attention to the order of the products. Thus d is called a **highest common left factor (HCLF)** of a and b if d is a left factor of a and b and any common left factor of a and b is a left factor of d. To find d one observes that $aR + bR$ is principal, say $= dR$.

General PIDs have a similar property. Let R be a PID; two elements $a, b \in R$ are called **similar** if $R/aR \cong R/bR$ or, equivalently, $R/Ra \cong R/Rb$. In a commutative PID, a and b are similar if and only if they are associated. Now a general PID has the following unique factorisation property: any element, not zero or a unit, can be written as a product of atoms: $a = c_1 \ldots c_r$, and if $a = d_1 \ldots d_s$ is another such factorisation, then $r = s$ and for some permutation $i \mapsto i'$ of $\{1, 2, \ldots, r\}$, a_i is similar to $b_{i'}$ (see Cohn 1985, 3.3)).

Let R be any PRID. Then R is a right Ore domain (because it is right Noetherian, see Cohn 1985), and so it has a skew field of fractions whose elements can be written as ab^{-1}, where $a, b \in R$ and of course $b \neq 0$. If $a \neq 0$, we can assume a and b to be right coprime, i.e., to have no common right factor apart from units, and they are then determined by ab^{-1} up to a common right unit factor.

For any PID, one has the following normal form for matrices: Given an $m \times n$ matrix A over a PID R, there exist invertible matrices P, Q of orders m, n, respectively, such that PAQ is diagonal, $= \operatorname{diag}(d_1, d_2, \ldots, d_r, 0, \ldots)$, say, where d_i is a **total divisor** of d_{i+1}, i.e., there exists an element c_i such that $Rc_i = c_iR$ and $d_iR \supseteq c_iR \supseteq d_{i+1}R$. The d_i are unique up to similarity; they are called the **invariant factors** of A. When R is commutative, this condition is just ordinary divisibility. For a Euclidean ring, the matrices P, Q can be taken to be products of elementary matrices, i.e., matrices with only one off-diagonal entry different from zero and units along the main diagonal, but this may no longer hold for a general PID (see Cohn 1966). This normal form is used to obtain the rational canonical form for a square matrix over a field, and the Jordan normal form for a square matrix over an algebraically closed field.

The matrix reduction can also be used to describe finitely generated modules. Let R be a PID; an R-module M is called a **torsion module** if every element of M has a non-zero annihilator in R. In every R-module M, the set of elements with non-zero annihilator in R form a submodule tM, called the **torsion submodule** of M. If $tM = 0$, M is said to be **torsion-free**; a finitely generated R-module M is torsion-free if and only if it is free. Thus any finitely generated R-module M can be written in the form $M = tM \oplus F$, where F is free. Every submodule G of a free R-module F is again free, and if F has finite rank m, then G is of rank at most m. It follows that every finitely generated R-module is finitely presented; moreover, it can be expressed as a direct sum of a finite number of cyclic modules: $M \cong P_1 \oplus \cdots \oplus P_r \oplus F$, where the P_i are cyclic, P_i is a homomorphic image of P_{i+1}, and F is free. Here the isomorphism types of the P_i and the rank of F are uniquely determined by M.

Homologically, PIDs are particularly simple. By what has been said, every projective module is free (Section C.20), and it can be shown that its rank is unique; thus every PID is **projective-free** (see Cohn 1985)); it also follows that every PID has global dimension 1 (Section H.4).

C.26 Noetherian Rings
by Lance W. Small in San Diego, CA, USA

Commutative Noetherian rings have already been discussed in the articles of Gilmer and Lam, among others. Here we shall concentrate on the nocommutative situation. Noncommutative Noetherian rings are of importance in the representation theory of finite dimensional Lie algebras, quantum groups and various generalizations of these algebras. A ring R with unit element is **(right) Noetherian** if R satisfies one of the three familiar equivalent conditions:

(a) Any ascending chain of right ideals eventually terminates; that if $I_1 \subset I_2 \subset \cdots \subset I_t \subset \ldots$ is our chain of right ideals, then there is an n such that $I_n = I_{n+1} = \ldots$.
(b) Any right ideal has a finite generating set.
(c) Any nonempty set of right ideals has a maximal element.

Polynomials over right Noetherian rings remain Noetherian (**Hilbert Basis Theorem**) as do factor rings and matrix rings with entries in a right Noetherian ring. However, a right Noetherian ring need not be left Noetherian: The easiest example may be the ring

$$R = \begin{pmatrix} \mathbb{Z} & \mathbb{Q} \\ 0 & \mathbb{Q} \end{pmatrix}, \qquad 2 \times 2 \text{ matrices}$$

with entries in the rationals, where the $(1,1)$ entry is integral and the $(2,1)$ entry is 0. A global reference for this article is the encyclopedic book of McConnell and Robson (1987).

There are two models for the study of noncommutative Noetherian rings:

(1) Commutative Noetherian rings and the study of prime ideals and localization;
(2) Semisimple Artinian rings and the Wedderburn theory.

It turns out that both models are relevant and the most important result in the theory of Noetherian rings, *Goldie's Theorem*, is the bridge between the two.

A ring, R, is **prime** if the product of nonzero ideals is nonzero (see Section C.28). This is the appropriate generalization of integral domain to the noncommutative setting. The relevant example is the ring of 2×2 matrices over the integers. Here there are many zero divisors, but the product of nonzero ideals is nonzero. A ring is **semiprime** if it has no nonzero nilpotent ideals. A ring R has a (right) **classical quotient ring** (cf. Section C.15) Q if

(a) $R \subset Q$.
(b) If $q \in Q$, then $q = ac^{-1}$ where $a \in R$ and c is a regular element in R.
(c) Every regular element in R is invertible in Q.

An element $c \in R$ is **regular** if $dc = 0$ $(cb = 0)$ implies $d = 0$ $(b = 0)$. **Goldie's Theorem** asserts—among other things—that a semiprime right Noetherian ring R has a classical quotient ring Q that is a semisimple Artinian ring and that if R is prime, Q is a simple Artinian ring. Thus, via localization, semiprime Noetherian rings are embedded in semisimple Artinian rings, and results true for such rings can often be established for Noetherian rings.

For example, Jacobson showed that nil weakly-closed subsets, like Lie sets, of Artinian rings generate nilpotent associative subrings. Jacobson's result, applying Goldie's Theorem, holds for Noetherian rings.

The attentive reader will have noticed that in the statement of Jacobson's result no mention was made of "semisimple" or "semiprime". There is a well-trodden path via factoring out a nilpotent ideal to pass to a semiprime ring. For Artinian rings, this nilpotent ideal is the Jacobson radical. For Noetherian rings, **Levitzki's Theorem** says that the maximal nilpotent ideal contains all nil one-sided ideals.

One of Goldie's motivations for his investigations was a question of Jacobson. Motivated by analogy with the Krull intersection theorem, Jacobson asked whether the intersection of the powers of the Jacobson radical (Jacobson calls it the "radical"). Herstein, in 1965, dispatched the original question with an astonishingly simple example.

Set $R = \{ \begin{pmatrix} s & q_1 \\ 0 & q_2 \end{pmatrix} \mid$ with s a rational number with odd denominator, and q_1 and q_2 arbitrary rationals$\}$. Write $R = \begin{pmatrix} S & \mathbb{Q} \\ 0 & \mathbb{Q} \end{pmatrix}$. Now R is right Noetherian and $J(S)$ is the set of all rationals with even numerator and odd denominator. $J(R) = \begin{pmatrix} J(S) & \mathbb{Q} \\ 0 & 0 \end{pmatrix}$ and $\bigcap_{n=1}^{\infty} J(R)^n = \begin{pmatrix} 0 & \mathbb{Q} \\ 0 & 0 \end{pmatrix}$. Herstein's example is not left Noetherian. The Jacobson question remains open for right and left Noetherian rings.

Recent work in noncommutative Noetherian ring theory has been directed toward the study of particular classes of rings. The techniques range from Gel'fand-Kirillov dimension to algebraic geometry.

C.27 Artinian Rings

by Victor T. Markov in Moscow, Russia

Definition and examples

In what follows, all the rings are assumed to be associative but not necessarily with unity. When it is explicitly stated that some ring R possesses a unity, then all R-modules are assumed to be unitary.

A module M is called **Artinian** if it satisfies the **descending chain condition (DCC)** for submodules: any descending chain of submodules $M \supseteq M_1 \supset M_2 \supset \ldots \supset M_i \supset \ldots$ is eventually constant, i.e., there exists an integer n such that $M_i = M_{i+1}$ for all $i \geq n$. A ring R is called **left (right) Artinian**, if the module $_RR$ (resp., R_R) is Artinian. The descending chain condition for submodules of M is equivalent to the **minimal condition**: every non-empty set of submodules of M contains a minimal submodule (with respect to inclusion).

The books (Artin et al. 1944, Kertész 1987) are devoted to the theory of Artinian rings; the reader can find the information on them in any manual of ring theory (e.g., Faith 1973, 1976, Herstein 1968, Jacobson 1964, Lambek 1976).

The **ascending chain condition (ACC)** is defined similarly, the modules and rings satisfying ACC are called **Noetherian** (cf. Section C.26). An example of an Artinian but not Noetherian module is provided by the quasi-cyclic Abelian group $\mathbb{Z}_{p^\infty}$ considered as a module over the ring $\mathbb{Z}$ of integers. Any ring that belongs to one of the following classes is left and right Artinian (as well as left and right Noetherian): finite rings, skew fields, finite dimensional algebras with unity over commutative Artinian rings. Every proper quotient ring of a principal left and right ideal domain (cf Section C.25) is a left and right Artinian ring.

An example due to Small (cf. Herstein 1968, 1.4) shows that a left Artinian ring with unity is not necessarily right Artinian.

Every finitely generated Artinian module is Noetherian. In particular, a left Artinian ring with unity is left Noetherian. Moreover, any left Artinian ring whose additive group contains no quasi-cyclic subgroups (in particular, having right or left unity) is also Noetherian (Kertész 1987, 10.10).

The following theorem shows some ring constructions that preserve the Artinian property.

C.27.1 Theorem *Let R be a left Artinian ring.*

(1) *The full matrix ring $\mathrm{Mat}_n R$ is left Artinian for every $n \geq 1$;*
(2) *If $e = e^2 \in R$ then eRe is left Artinian.*

Theorem C.27.1 implies that the Artinian property is invariant under Morita equivalence of rings. It is also evident that any quotient ring of a left Artinian ring is left Artinian and any finite direct sum of left Artinian rings is left Artinian.

The following theorem provides other sufficient conditions on the ring to be Artinian.

C.27.2 Theorem *Let R be a ring with unity. Then R is left Artinian in the following two cases: (a) any left R-module is embeddable into a direct sum of finitely generated modules; (b) there exists a cardinal $\mathfrak{c}$ such that any R-module is a direct sum of $\mathfrak{c}$-generated modules.*

C.27.3 Theorem (Connel) *A group ring RG (cf. Section E.1) is left Artinian if and only if R is left Artinian and G is finite.*

Idempotents, radicals, semisimple Artinian rings

The Jacobson radical (cf. Section C.18) $J(R)$ of a left Artinian ring R is nilpotent. So all nil radicals of R coincide with $J(R)$. The structure of semisimple Artinian rings is determined by the following

C.27.4 Theorem (Wedderburn-Artin) *The following properties of a ring R are equivalent:*
(1) *R is left Artinian and $J(R) = 0$;*
(2) *R is isomorphic to a finite direct sum of full matrix rings over skew fields (we consider the zero ring as a sum of empty set of summands).*

Theorem C.27.4 implies that a left Artinian semisimple ring is right Artinian and has a unity.

C.27.5 Theorem (Herstein 1968, 1.4.3) *Is R is a left Artinian ring and its additive group is torsionfree then R contains a left unity.*

C.27.6 Theorem (Kertész 1987, 10.27) *A left Artinian ring is embeddable into a left Artinian ring with unity if and only if its additive group contains no quasi-cyclic subgroups.*

The connection between arbitrary left Artinian ring R and its semisimple quotient ring $R/J(R)$ is known as "lifting idempotents" property:

C.27.7 Theorem *If R is a left Artinian rind and $\{\bar{e}_1, \ldots, \bar{e}_k\}$ is a set of mutually orthogonal idempotents in the ring $\bar{R} = R/J(R)$ (i.e. $\bar{e}_i\bar{e}_j = \bar{0}$ for $i \neq j$) then there exist mutually orthogonal idempotents $e_1, \ldots, e_k \in R$ such that $\bar{e}_i = e_i + J(R)$ for $i = 1, \ldots, k$. Moreover, for any idempotents $e, f \in R$, $Re \cong Rf$ if and only if $\bar{R}\bar{e} \cong \bar{R}\bar{f}$.*

It follows from the proof of this theorem that any left (right) ideal of a left Artinian ring is either nilpotent or contains a non-zero idempotent element.

Subrings of Artinian rings

In this subsection, all rings are assumed to be rings with unity, and subrings of a ring have the same unity.

C.27.8 Theorem (Bjork) *Let R be a subring of a ring S such that S is finitely generated as a left R-module.*

(1) *If the associated Lie ring $R^{(-)}$ is (Lie) nilpotent and S is left Artinian then R is left and right Artinian;*
(2) *If R is left Noetherian and S is left Artinian then R is left Artinian.*

C.27.9 Theorem (Eisenbud) *Let R be a subring of a ring S such that S is generated as a left R-module by a finite set of elements that centralize R. If S is left Artinian (Noetherian) then R is left Artinian (Noetherian).*

An element r of a ring R is called **regular**, if it is neither left nor right zero divisor. A ring $Q \supseteq R$ is called a **left ring of quotients (or ring of fractions)** for the ring R (cf. Section C.15), and R a **left order** in Q, if (a) every regular element of R is invertible in Q and (b) any element $x \in Q$ has a representation $x = a^{-1}b$, where $a, b \in R$ and a is regular (some authors use different definitions of orders, cf. Section C.34).

C.27.10 Theorem (Goldie 1960) *A ring R is a left order in a semisimple Artinian ring if and only if R is nonzero semiprime (cf. Section C.28) ring and: (a) R contains no infinite direct sums of nonzero left ideals and (b) R satisfies the ACC for left annihilators.*

A ring satisfying conditions (a) and (b) of Theorem C.27.10 is said to be **left Goldie**. Let $\beta(R)$ be the prime radical of the ring R. For any ideal I of a ring R, denote by $C_R(I)$ the set of all elements $a \in R$ such that $a + I$ is regular in R/I. A ring R is called **left T-Goldie ring**, if R/T_k is a left Goldie ring for every integer $k > 0$, where $T_k = \beta(R) \cap r_R(\beta(R)^k)$. Now we can state Small's

C.27.11 Theorem *The following conditions on the ring R are equivalent:*
(1) *R is a left order in a left Artinian ring;*
(2) *R is a non-nilpotent left T-Goldie ring and $C_R(\beta(R)) = C_R((0))$.*

Generalizations of Artinian rings

In this subsection, all rings are rings with unity again.

Many classes of rings can be considered as generalizations of the class of Artinian rings, each of them making a good subject for a separate essay. We present a rather random selection of such classes.

1. *Semi-Artinian rings.* A module M over a ring R is called **semi-Artinian**, if every nonzero quotient module of M contains a simple

module. A ring R is called *left semi-Artinian*, if $_RR$ is a semi-Artinian R-module. A ring R is left semi-Artinian if and only if every left R-module is semi-Artinian. Every submodule of a semi-Artinian module is itself semi-Artinian. A subset A of a ring is called *left (resp. right) T-nilpotent* if for every infinite sequence $\{a_i\} \subseteq A$ there exists $n > 0$ such that $a_1 a_2 \ldots a_n = 0$ (resp. $a_n a_{n-1} \ldots a_1 = 0$). A ring R is left semi-Artinian if and only if $J(R)$ is right T-nilpotent and $R/J(R)$ is left semi-Artinian.

2. Perfect rings. The notion of left (semi)perfect ring was introduced by Bass (1960). A ring R is called *left (semi)perfect*, if every left (finitely generated) R-module has a projective cover (cf. Section C.20).

C.27.12 Theorem (Bass) *The following conditions on the ring R are equivalent:*

(1) *R is left perfect;*
(2) *R satisfies the DCC for principal right ideals;*
(3) *R is right semi-Artinian and the orthogonal sets of idempotents in R have bounded number of elements;*
(4) *$R/J(R)$ is Artinian and $J(R)$ is left T-nilpotent;*
(5) *Every flat left R-module is projective.*

C.27.13 Theorem (Bass) *The following conditions on the ring R are equivalent:*

(1) *R is left semiperfect;*
(2) *$R/J(R)$ is Artinian and the idempotents can be lifted modulo $J(R)$.*

3. Rings with Krull dimension. The **Krull dimension** KDim M of a module M is defined inductively as follows (cf. Section C.6). Let KDim $M = 0$ iff the module M is Artinian; let KDim $M = \alpha$ (an ordinal), if (1) KDim $M \not< \alpha$ and (2) for any infinite descending chain $M = M_0 \supset M_1 \supset \ldots$, KDim$(M_i/M_{i+1}) < \alpha$ for all except may be finite number of i (see Gordon and Robson 1973). The ordinal KDim$_R R$ is called *left Krull dimension* of a ring R. Here we quote only two properties of ring having left Krull dimension that are most closely connected to that of Artinian rings: (1) if a semiprime ring has left Krull dimension then it is left Goldie; (2) the prime radical of a ring with left Krull dimension is nilpotent.

C.28 Maximal, Prime, and Semiprime

by Gary F. Birkenmeier in Lafayette, LA, USA

In this article we discuss the fundamental concepts of the title. Some of the pioneers of these concepts were R. Dedekind, W. Krull, H. Fitting, E. Noether, and N. McCoy (Eisenbud 1995, Faith 1999, McCoy 1973). R denotes a ring not necessarily commutative nor necessarily having unity. If $X \subseteq R$, then $< X >$ denotes the ideal of R generated by X.

Players and preliminaries

Let $P \lhd R$. Then P is called a **completely prime** (**completely semiprime**) ideal ideal ideal if $a, b \in R$ such that $ab \in P$, then either $a \in P$ or $b \in P$ ($a^2 \in P \Rightarrow a \in P$). P is called a **prime** (**semiprime**) ideal if $A, B \unlhd R$ such that $AB \subseteq R$, then $A \subseteq P$ or $B \subseteq P$ ($A^2 \subseteq P \Rightarrow A \subseteq P$); equivalently, if $a, b \in R$ such that $aRb \subseteq P$, then $a \in P$ or $b \in P$ ($aRa \subseteq P \Rightarrow a \in P$). P is called a **maximal** ideal if $I \unlhd R$ such that $P \subseteq I$ then either $I = P$ or $I = R$. Observe that if R is a commutative ring, then P is prime (semiprime) iff P is completely prime (completely semiprime). For $R = R^2$, every maximal ideal is a prime ideal. A ring R is called a **domain** or a **completely prime ring**, a **reduced ring** or a **completely semiprime ring**, a **prime ring**, a **semiprime ring**, or a **simple ring** if the zero ideal is a completely prime ideal, a completely semiprime ideal, a prime ideal, a semiprime ideal, or a maximal ideal, respectively. (Some authors include $R^2 \neq 0$ in their definition of a simple ring). Some examples are: $< x >$ in $\mathbb{Z}[x]$ is a completely prime ideal which is not maximal; if p is a prime in $\mathbb{Z}$, then $< p >$ is a maximal ideal of $\mathbb{Z}$; if F is a field and p is an irreducible polynomial in $F[x]$, then $< p >$ is a maximal ideal in $F[x]$; if V is a vector space over F then the ring of linear operators $\mathrm{End}_F(V)$ (see Section C.16) is a simple ring if $\dim(V) < \infty$ and a prime ring (not simple) if $\dim(V) = \infty$; if R is a subdirectly irreducible ring whose heart is not square zero, then R is a prime ring. A subset M (N) of R is said to be an **m-system** (**n-system**) if for any $a, b \in M$ ($a \in N$) there exists $x \in R$ such that $axb \in M$ ($axa \in N$) (McCoy 1973).

C.28.1 Proposition *Let $P \lhd R$. T.F.A.E.:*

(1) P *is a prime (completely prime) ideal;*
(2) $R \setminus P$ *is an m-system (multiplicatively closed set);*

(3) R/P is a semiprime (reduced) ring and each of its nonzero ideals is ideal essential in R.

C.28.2 Proposition *Let $S \lhd R$. T.F.A.E.:*

(1) *S is a semiprime (completely semiprime) ideal;*
(2) *$R \setminus S$ is an n-system ($x \in R \setminus S \implies x^2 \in R \setminus S$);*
(3) *$S = \bigcap \{P$ is a prime ideal of $R \mid S \subseteq P\}$ ($S = \bigcap \{P$ is a completely prime ideal of $R \mid S \subseteq P\}$;*
(4) *if $A, B \unlhd R$ such that $AB \subseteq S$, then $A \cap B \subseteq S$.*

C.28.3 Proposition *Let $P \lhd R$. Then*

(1) *P is maximal iff $P + \langle a \rangle = R$, for any $a \notin P$.*
(2) *If P is a prime ideal and $a \notin P$, then $P + \langle a \rangle$ is right and left essential in R.*

C.28.4 Proposition *Let $P \lhd R$. Then P is a prime (semiprime) ideal iff P is maximal among ideals which have empty intersection with some nonempty m-system (n-system).*

Since the set of regular elements (i.e., elements which are not zero-divisors) and $\{x^n \mid x$ is not nilpotent and n is a positive integer $\}$ are multiplicatively closed, Proposition C.28.4 and Zorn's Lemma yield prime ideals having empty intersection with these sets.

Certain radical properties (Divinsky 1965) are associated with the concepts of prime and maximal. The **prime radical** (also known as the **Baer lower radical**) of R, $\mathbf{P}(R)$, is the intersection of all prime ideals of R. Observe that $\mathbf{P}(R) \subseteq \mathcal{N}(R)$, where $\mathcal{N}(R)$ is the set of nilpotent elements of R, in fact $\mathbf{P}(R)$ is the set of strongly nilpotent elements of R (Faith 1999). The **generalized nil radical** of R, $\mathbf{N_g}(R)$, is the intersection of all completely prime ideals of R. The **Brown-McCoy radical** of R, $\mathbf{G}(R)$, is the intersection of all maximal ideals M of R such that R/M is a simple ring with unity. These radicals provide tools for investigating the structure of a ring as indicated in the next result; see also Section C.18 and Section C.19.

C.28.5 Proposition *If $\rho(R) = \mathbf{P}(R)$, $\mathbf{N_g}(R)$, or $\mathbf{G}(R)$, then $R/\rho(R)$ is a subdirect product of prime rings, domains, or simple rings with unity, respectively.*

Tools

In this section we mention several *tool* concepts which are based on the notions of prime, semiprime, and maximal. First we consider *localization at a prime ideal*. In the case of commutative rings, one may intuitively think of this as adjoining to a ring the multiplicative inverses of the elements in the set complement of a prime ideal (Eisenbud 1995). This has been generalized in a variety of ways to the class of noncommutative rings (McConnell and Robson 1987, Rowen 1988a, Stenström 1975). Next various *topologies* can be defined using the sets of prime or maximal ideals. These topologies are useful in localization and in applying Sheaf Theory to Ring Theory (Lambek 1976, Rowen 1988a, Stenström 1975). Since prime, semiprime, and maximal are "Morita invariants", these properties can be used in investigating equivalent module categories (Lam 1999b, Rowen 1988a). Various dimensions are defined on rings. One of the prototypes is the (classical) **Krull dimension** (i.e., the maximal length of a chain of prime ideals in a ring) Eisenbud (1995), Rowen (1988a). The last tool concept we consider is Radical Theory. An important class of radicals are those called the *special radicals* which are each associated with a special class of prime rings. This class includes **P**, **N$_g$**, **G**, as well as most of the well known radicals (Divinsky 1965).

Results

In this section we present several results which illustrate the motivation for, the interconnections between, and the utility of the concepts of prime, semiprime, and maximal. However, many prominent results (e.g., Goldie's Theorem) have been omitted since they will be covered in great detail in subsequent sections. Our first result traces its origins to Dedekind and characterizes the rings called **Dedekind domains** (Eisenbud 1995)—cf. also Section C.5.

C.28.6 Theorem *Let R be a commutative domain with unity. Every nonzero ideal of R is uniquely a product of prime ideals iff R is Noetherian, integrally closed, and has Krull dimension ≤ 1.*

Our next result is a version of Hilbert's Nullstellensatz which is of crucial importance in algebraic geometry.

C.28.7 Theorem *Let F be an algebraically closed field. Then every prime ideal of $F[x_1, x_2, \ldots, x_n]$ is the intersection of maximal ideals.*

Rings with classical Krull dimension zero are exactly those which satisfy the **pm condition** (i.e., every prime ideal is maximal). Surprisingly, this condition has been related to various generalizations of von Neumann regularity (see (Birkenmeier 1995) for a survey and (Birkenmeier et al. 1997b) for recent results). The next theorem by Storrer is the prototype for these results (see (Birkenmeier et al. 1997b)).

C.28.8 Theorem *Let R be a commutative ring with unity. T.F.A.E.:*
(1) R *is π-regular;*
(2) $R/\mathbf{P}(R)$ *is von Neumann regular;*
(3) R *satisfies the pm condition.*

Every completely prime (completely semiprime) ideal is a prime (semiprime) ideal. In general the converse does not hold. However the converse does hold for several important classes of rings such as right (left) duo rings (i.e., every right (left) ideal is an ideal) and the enveloping algebras of finite dimensional solvable Lie algebras over fields of characteristic zero. The class of rings in which every prime ideal is completely prime is a radical class. A related class of rings is the class of **2-primal** rings (i.e., R is 2-primal if $\mathbf{P}(R) = \mathcal{N}(R)$, equivalently, $\mathbf{P}(R) = \mathbf{N_g}(R)$) (Birkenmeier et al. 1993). Our next result is due to Shin (see Birkenmeier et al. 1993).

C.28.9 Theorem
(1) *Every prime ideal is completely prime iff R/I is a 2-primal ring, for all $I \trianglelefteq R$.*
(2) *R is 2-primal iff every minimal prime ideal of R is completely prime.*

Since prime rings are *building blocks* (see Proposition C.28.5), their characterization in any class of rings is vital to the structure theory of that class. These characterizations have been provided for PI-rings, group rings, and many others (Lam 1991, Rowen 1988b). The following result follows from the classical Wedderburn-Artin theorem.

C.28.10 Theorem
R is a prime ring with DCC on right ideals iff R is a complete matrix ring over a division ring.

A natural question is: how do our players behave under various extensions?

C.28.11 Theorem

(1) *R is a prime (semiprime, domain, reduced) ring iff $R[X]$ satisfies the same condition, respectively, where X is a set of commuting indeterminates.*

(2) *R is a prime (semiprime, simple) ring iff $\mathrm{Mat}_{n \times n} R$ satisfies the same condition, respectively.*

(3) *If $\{R_i\}$ is a family of semiprime (reduced) rings, then any subdirect product of the R_i is semiprime (reduced).*

There are various *lying over* and *going up* results comparing the prime ideals in a ring with those of a subring (see (Rowen 1988a) for a overview of these results).

Our final result provides structural information about a large class of rings which includes all right PP rings with no infinite sets of orthogonal idempotents (hence all right hereditary right Noetherian or right hereditary semiperfect rings), right nonsingular right CS semiperfect rings, and all piecewise domains. Recall that a ring R with unity is **quasi-Baer** if the right annihilator of an ideal is generated by an idempotent.

C.28.12 Theorem ((Birkenmeier et al. 2000)) *Let R be a quasi-Baer ring with unity such that $\{bR \mid b = b^2 \text{ and } bRb = Rb\}$ satisfies the ACC and DCC. Then $R = A \oplus B$ (ring direct sum) where A is a finite direct sum of prime rings and B is isomorphic to a generalized triangular matrix ring whose diagonal rings are all prime rings.*

C.29 Regular Rings

by Alexander V. Mikhalev and Askar A. Tuganbaev in
Moscow, Russia

All rings are assumed to be associative and with nonzero identity element.

A ring A is said to be (**von Neumann**) **regular** or **absolutely flat** if for any element $a \in A$, there is an element $b \in A$ with $a = aba$. Examples of regular rings are classically semisimple rings and rings of linear transformations of vector spaces. Each factor ring of any direct product of regular rings is regular. All full matrix rings over regular rings are regular. For any nonzero idempotent e of a regular ring A, the ring eAe is regular. Therefore, the endomorphism ring of any finitely generated projective module over a regular ring is regular. The center of a regular ring is a regular ring. A module is said to be **semisimple** if it is a direct sum of simple modules. For any semisimple module M, the endomorphism ring $\text{End}(M)$ is a regular ring. For a ring A and a group G, the group ring $A[G]$ is regular if and only if A is regular, G is locally finite, and the order of each element of G is invertible in A.

A module is said to be **regular** if every its finitely generated submodule is a direct summand. A right A-module M is said to be **flat** if for each left A-module N and any submodule N' of N, the natural group homomorphism $M \otimes_A N' \to M \otimes_A N$ is a monomorphism.

C.29.1 Theorem *For a ring A, the following conditions are equivalent:*

(1) *A is regular;*
(2) *Every principal right ideal of A is generated by an idempotent;*
(3) *A_A is a regular module;*
(4) *All A-modules are flat;*
(5) *A is a semiprime ring, the union of any ascending chain of semiprime ideals is a semiprime ideals, and each prime factor ring of A is regular.*

Every countably generated regular module is a direct sum of cyclic modules. If M is a right module over a regular ring A, then $NB = N \cap MB$ for every submodule N of M and each left ideal B of A. A module is called a **Bezout module** if each its finitely generated module is cyclic. A regular ring is a Bezout ring.

A right A-module is said to be **nonsingular** if the right annihilator of any nonzero element of the module is not an essential right ideal of A. For any injective nonsingular module M, the ring $\text{End}(M)$ is regular. For any right nonsingular ring A with right injective hull Q_A, there is a natural ring monomorphism from A into the regular right self-injective ring $\text{End}(Q_A)$ which is the maximal right ring of quotients of Q.

A ring A is said to be **unit-regular** if for every element $a \in A$, there is an invertible element $b \in A$ with $a = aba$. Each unit-regular ring is regular and the endomorphism ring of a vector space M is unit-regular if and only if M is finite-dimensional. Therefore, every semisimple ring is unit-regular. A ring A is called a **ring of stable range 1** if for any four elements f, v, x, and y of A with $fx + vy = 1$, there is an element $h \in A$ such that $f + vh$ is an invertible element of A.

C.29.2 Theorem *For a module M, the following conditions are equivalent:*

(1) *The ring $\text{End}(M)$ is unit-regular;*

(2) End(M) *is a regular ring of stable range 1;*
(3) *Every endomorphism of M is a product of a projection and an automorphism of M;*
(4) *For every endomorphism f of M, there are two projections e_1 and e_2 and an automorphism u of M with $f = e_1 u = u e_2$;*
(5) *For every endomorphism f of M, the modules $f(M)$ and* Ker(f) *are direct summands of M and* Ker(f) $\cong M/f(M)$.

If A is a ring and P is a maximal element in the nonempty set of all proper ideals of A generated by central idempotents, then the factor ring A/P is called a **Pierce stalk** of A. A ring A is regular (resp. unit-regular) if and only if all Pierce stalks of A are regular (resp. unit-regular).

A ring A with Jacobson radical $J(A)$ is said to be **semiregular** if the factor ring $A/J(A)$ is regular and any idempotent of $A/J(A)$ is lifted to an idempotent of A. For any injective or algebraically compact module M, the ring End(M) is semiregular. Every projective right module over a semiregular ring A is isomorphic to $\bigoplus_{i \in I} e_i A$, where all the e_i are idempotents of A. A module M is said to be **amply f-supplemented** if for any two of its submodules U and V such that $U + V = M$ and U is finitely generated, there is a submodule V' of V such that $U + V' = M$ and $U \cap V'$ is a small submodule in V'. A ring A is semiregular if and only if A is right amply f-supplemented if and only if A is left amply f-supplemented.

A ring A is called a a **right V-ring** if all simple right A-modules are injective. A commutative ring A is regular if and only if A is a V-ring. There is a field F with derivation δ such that the differential polynomial ring $F[x, \delta]$ is a simple nonregular right V-domain. Let M be an infinite dimensional right vector space over a field F, S be the ideal of End(M_F) consisting of linear transformations of finite rank, and let A be the subring in End(M_F) generated by S and all scalar transformations. Then A is a regular right V-ring that is not a left V-ring.

A ring A is said to be **strongly regular** or **abelian regular** if A is a regular ring satisfying one of the following conditions that are equivalent for regular rings: (1) each idempotent of A is central; (2) A is a reduced ring (a ring is said to be *reduced* if it does not have nonzero nilpotent elements); (3) A is right distributive (a module is said to be *distributive* if the lattice of all its submodules is distributive); (4) each maximal right ideal of A is an ideal of A; (5) each right or left ideal of A is an ideal of A.

C.29.3 Theorem *For a ring A, the following conditions are equivalent:*

(1) *A is strongly regular;*
(2) *Each element of A is a product of a central idempotent and an invertible element;*
(3) *For every element $a \in A$, there is an element $b \in A$ with $a = ba^2$;*
(4) *For every element $a \in A$, there is an element $b \in A$ with $a = a^2b$.*

Each factor ring of any direct product of strongly regular (resp. unit-regular) rings is strongly regular (resp. unit-regular).

If M is an A-module and the factor ring $A/J(A)$ is strongly regular, then M is a distributive module if and only if M is a Bezout module. For any strongly regular ring A with automorphism α, the skew polynomial ring $A[x, \alpha]$ is a distributive reduced Bezout ring. If F is a field and A is the ring formed by all eventually constant sequences of elements of F, then A is a commutative regular ring such that the commutative ring $A[[x]]$ is not a distributive or Bezout ring.

A ring is said to be ***biregular*** if each of its ideals generated by one element is generated by a central idempotents. A ring A is biregular (resp. strongly regular) if and only if all Pierce stalks of A are simple rings (resp. skew fields). Each strongly regular ring is a biregular unit-regular ring.

A ring A is said to be ***right weakly regular*** if $B^2 = B$ for every right ideal B of A. Each right V-ring is right weakly regular. Any regular or simple ring is (right and left) weakly regular. If F is a field and A is the Weyl F-algebra with two generators x and y and one relation $xy - yx = 1$, then A is a biregular weakly regular domain that is not regular. The center of a right weakly regular ring is a regular ring. A ring A is regular if and only if A is right weakly regular and all prime factor rings of A are regular. If A is a right weakly regular ring and A/B is a factor ring of A, then every injective right A/B-module is an injective A-module. For a ring A and a group G, the group ring $A[G]$ is right weakly regular (resp. right V-ring) if and only if A is right weakly regular (resp. right V-ring), G is locally finite, and the order of each element of G is invertible in A.

A ring A is said to be ***π-regular*** if for each element $a \in A$, there is an element $b \in A$ such that $a^n = a^nba^n$ for some positive integer n. A ring A is said to be ***strongly π-regular*** if it satisfies the following equivalent conditions: (1) for every $x \in A$, there is a positive integer m with $x^m \in x^{m+1}A$; (2) for every $x \in A$, there is a positive integer n with $x^n \in Ax^{n+1}$. Every strongly π-regular ring is π-regular. If F is a division ring and M is a right vector F-space with infinite basis $\{e_i\}_{i=1}^{\infty}$, then $\text{End}(M_F)$ is a regular (and π-regular) ring that is not strongly π-regular. The factor ring of the ring of integers with respect to the ideal generated by the integer 4 is a strongly π-regular ring that is not regular. If A is a ring and $\{A_i\}_{i \in I}$ is a set of strongly π-regular unitary subrings of A, then the ring $\bigcap_{i \in I} A_i$ is strongly π-regular.

C.29.4 Theorem *For a ring A, the following conditions are equivalent:*

(1) *A is strongly π-regular;*
(2) *The factor ring A/N with respect to the prime radical N of A is strongly π-regular;*
(3) *Every prime factor ring of A is strongly π-regular;*
(4) *Each injective endomorphism of every cyclic right A-module is an automorphism;*
(5) *For every element $a \in A$, there is an element $x \in A$ such that $ax = xa$ and $a^n = a^{n+1}x = xa^{n+1}$.*

For a module M, the ring $\text{End}(M)$ is strongly π-regular if and only if for every endomorphism f of M, there is a positive integer n such that $M = \text{Ker}(f^n) \oplus f^n(M)$; in this case, every injective or surjective endomorphism of M is an automorphism. For a ring A, all matrix rings A_n are strongly π-regular if and only if all injective endomorphisms of every finitely generated right A-module are automorphisms.

Let F be a field, x be an indeterminate, $F(x)$ be the field of rational functions, and let $A = \begin{pmatrix} F & 0 \\ F(x) & F(x) \end{pmatrix}$. Then all matrix rings $\text{Mat}_n A$ are strongly π-regular (therefore, all injective endomorphisms of finitely generated right or left A-modules are automorphisms), all surjective endomorphisms of finitely generated left A-modules are automorphisms, and there are surjective endomorphisms of cyclic right A-modules are automorphisms.

For more information on the above classes of rings, see (Berberian 1972), (Goodearl 1991), (Jacobson 1964), (Kaplansky 1968), (von Neumann 1960), (Skornyakov 1964), (Tuganbaev 1998), and (Wisbauer 1991).

C.30 Quasi-Frobenius and Self-injective Rings

by Alexander V. Mikhalev and Askar A. Tuganbaev in Moscow, Russia

All rings are assumed to be associative and with nonzero identity element.

Expressions such as an "Artinian ring" mean that the corresponding right and left conditions hold. If M is a right (resp. left) module over a ring A and X is a subset of M, then we denote by $r(X)$ (resp. $\ell(X)$) the right (resp. left) annihilator of X in A. A ring A is called a **dual ring** if $B = r(\ell(B))$ and $C = \ell(r(C))$ for any right ideal B of A and each left ideal C of A. A duality between the left ideals and the right ideals of A follows from the above definition. A module with the minimum (resp. maximum) condition on submodules is called an **Artinian** (resp. **Noetherian**) **module**. A dual Artinian ring A is called a **quasi-Frobenius ring** or a **QF-ring**. All finite direct products of quasi-Frobenius rings are quasi-Frobenius. If F is the residue field $\mathbb{Z}_2$ and G is the direct product of two cyclic multiplicative groups $\langle x \rangle$ and $\langle y \rangle$ of order 2, then the group ring $F[G]$ is a commutative quasi-Frobenius ring and the factor ring $F[G]/(x + y)F[G]$ is not quasi-Frobenius.

There is a commutative dual ring that is not a quasi-Frobenius ring. Let B be the ring of all rational numbers with odd denominators, $\mathbb{Q}$ be the field of rational numbers, and let M be the R-module $\mathbb{Q}/R$. Then the group direct sum $B \oplus M$ with product given by $(b_1, m_1)(b_2, m_2) = (b_1 b_2, b_1 m_2 + m_1 b_2)$ is the required ring. A ring A with Jacobson radical $J(A)$ is a quasi-Frobenius ring if and only if A is a dual ring and $(J(A))^\alpha = 0$ for some ordinal α, where $(J(A))^\alpha = (J(A))^{\alpha-1} J(A)$ if α is not a limit ordinal and $(J(A))^\alpha = \bigcap_{\beta < \alpha} (J(A))^\beta$ if α is a limit ordinal.

A finite-dimensional algebra A over a field F is called a **Frobenius algebra** if the right A-modules A_A and $\mathrm{Hom}(_F A, _F F)$ are isomorphic to each other. All Frobenius algebras are quasi-Frobenius rings. For any finite group G and each field F, the group algebra $F[G]$ is a Frobenius algebra. If F is a field and A is the 3-dimensional F-algebra consisting of all upper triangular matrices $\left(\begin{smallmatrix} x & y \\ 0 & z \end{smallmatrix}\right)$ with $x, y, z \in F$, then A is not a quasi-Frobenius ring.

C.30.1 Definition A right A-module M is said to be **reflexive** if the natural A-module homomorphism $M \mapsto \mathrm{Hom}(\mathrm{Hom}(M, A_A), _A A)$ is an isomorphism.

C.30.2 Definition A ring A is said to be a *right Ikeda-Nakayama ring* if $\ell(B \cap C) = \ell(B) + \ell(C)$ for any two right ideals B and C of A.

C.30.3 Theorem *For a ring A, the following conditions are equivalent:*
(1) *A is quasi-Frobenius;*
(2) *A is Artinian and for every simple right A-module S and each simple left A-module T, the left A-module $\mathrm{Hom}(S_A, A_A)$ and the right A-module $\mathrm{Hom}(_AT, _AA)$ are simple;*
(3) *For every finitely generated right A-module S with composition length s and each finitely generated left A-module T with composition length t, the composition length of the left A-module $\mathrm{Hom}(S_A, A_A)$ is equal to s and the composition length of the right A-module $\mathrm{Hom}(_AT, _AA)$ is equal to t;*
(4) *A is left Noetherian and each cyclic right A-module is reflexive;*
(5) *A is a left perfect (right and left) Ikeda-Nakayama ring;*
(6) *A is an Ikeda-Nakayama ring with the maximum condition on right annihilators and all elements of A with zero right annihilators are invertible in A.*

Let F be a field, $F(x)$ be the field of rational functions over F in variable x, α be a ring endomorphism of the polynomial ring $F[x]$ which fixes F and maps x onto x^2, β be a natural extension of α to the field $F(x)$, and let A be the right $F(x)$-space on basis $\{1, t\}$ with multiplication given by $t^2 = 0$ and $gt = t\alpha(g)$ for all $g \in F(x)$. Then A is an Artinian right Ikeda-Nakayama ring that is not quasi-Frobenius.

A ring A is called a *right Kasch* ring if every simple right A-module is isomorphic to a minimal right ideal of A. A right self-injective right Kasch ring is called a *right pseudo-Frobenius* ring or a *right PF-ring*. A ring with the minimum condition on principal right ideals ring is called a *left perfect* ring.

C.30.4 Problem For a left perfect (right and left) Kasch ring A, find conditions to be quasi-Frobenius.

There is a commutative pseudo-Frobenius ring that is not a quasi-Frobenius ring. Let B be the ring of all 2-adic numbers, $\mathbb{Q}$ be the field of fractions of B, and let M be the R-module $\mathbb{Q}/R$. Then the group direct sum $B \oplus M$ with product given by $(b_1, m_1)(b_2, m_2) = (b_1b_2, b_1m_2 + m_1b_2)$ is the required ring.

If M and N are two right modules over a ring A, then M is said to be *injective with respect to* N (or *N-injective*) if for any submodule

N' of N, each homomorphism $N' \to M$ can be extended to a homomorphism $N \to M$. A right A-module M is said to be **injective** if M is injective with respect to every right A-module. A ring A is said to be **right self-injective** if the module A_A is injective. A group ring $A[G]$ is quasi-Frobenius (resp. right self-injective) if and only if A is a quasi-Frobenius (resp. right self-injective) ring and G is a finite group. Any infinite direct product of division rings is a self-injective ring that is not quasi-Frobenius.

C.30.5 Theorem *For a ring A, the following are equivalent:*

(1) *A is quasi-Frobenius;*
(2) *A is a self-injective Artinian ring;*
(3) *A is a right self-injective ring with the maximum condition on right annihilators*
(4) *A is a right self-injective ring with the maximum condition on left annihilators;*
(5) *The endomorphism ring of every free right A-module is right self-injective.*

All countable right self-injective rings are quasi-Frobenius. For a field F, each right self-injective F-algebra of countable dimension is quasi-Frobenius. If A is a left perfect right self-injective ring and $(J(A))^2$ is the right annihilator of a finite subset of A, then A is quasi-Frobenius. Therefore, a semilocal right self-injective ring with square-zero radical is quasi-Frobenius (a ring A is said to be **semilocal** if the factor ring $A/J(A)$ is Artinian) (cf. Section C.7).

For a module M, a submodule M' of M is said to be **essential** (in M) if $M' \cap X \neq 0$ for every nonzero submodule X of M; in this case, we say that M is an **essential extension** of M'. A ring is a quasi-Frobenius ring if and only if A is a right or left self-injective with the minimum condition on essential right ideals. A direct sum of simple modules is called a **semisimple module**. For a module M, its largest semisimple submodule is called the **socle** of M; it is denoted by $\mathrm{Soc}(M)$. For a module M, its submodule $\mathrm{Soc}_2(M)$ is defined by $\mathrm{Soc}(M/\mathrm{Soc}(M)) = \mathrm{Soc}_2(M)/\mathrm{Soc}(M)$. A ring A is quasi-Frobenius if and only if A is a self-injective left perfect ring if and only if A is a right self-injective left perfect ring and the ideal $\mathrm{Soc}_2(A_A)$ is a countably generated left ideal.

For a module N, a module M is said to be **projective with respect to N** (or **N-projective**) if for every epimorphism $h: N \to \overline{N}$ and each homomorphism $\overline{f}: M \to \overline{N}$, there is a homomorphism $f: M \to N$ with

$\overline{f} = hf$. A module M is said to be π-**injective** if each idempotent endomorphism of any its submodule can be extended to an endomorphism of M. A module M is said to be π-**projective** if for any two its submodules X and Y with $X + Y = M$, there is an endomorphism f of M such that $f(M) \subseteq X$ and $(1 - f)(M) \subseteq Y$.

C.30.6 Theorem *For a ring A, the following are equivalent:*
(1) *A is quasi-Frobenius;*
(2) *All injective right A-modules are projective;*
(3) *All projective right A-modules are injective;*
(4) *All flat right A-modules are injective;*
(5) *All injective right A-modules are π-projective;*
(6) *All projective right A-modules are π-injective.*

A module is said to be **quasi-injective** (resp. **quasi-projective**) if it is injective (resp. projective) with respect to itself. A module M is said to be **skew-injective** if each endomorphism of any its submodule can be extended to an endomorphism of M. A module M is said to be **skew-projective** if each endomorphism of any its factor module is lifted to an endomorphism of M.

C.30.7 Theorem *For a ring A, the following are equivalent:*
(1) *All factor rings of A are quasi-Frobenius rings;*
(2) *All factor rings of A are dual rings;*
(3) *All quasi-injective right A-modules are quasi-projective;*
(4) *All quasi-projective right A-modules are quasi-injective;*
(5) *All skew-injective right A-modules are skew-projective;*
(6) *All skew-projective right A-modules are skew-injective;*
(7) *All quasi-injective right A-modules are π-projective;*
(8) *All quasi-projective right A-modules are π-injective;*
(9) *A is an Artinian ring whose right or left ideals are principal.*

If F is the residue field $\mathbb{Z}_2$ and G is the direct product of two cyclic multiplicative groups $\langle x \rangle$ and $\langle y \rangle$ of order 2, then the group ring $F[G]$ is a commutative quasi-Frobenius ring and its ideal $(1+x)F[G]+(1+y)F[G]$ is not principal.

A ring A is called a **right QF-3-ring** if there is a faithful right A-module M such that every faithful right A-module has a direct summand that is isomorphic to M. All quasi-Frobenius rings are (right and left) QF-3-rings. There is a division ring with a division subring P such that $\dim({}_PD) = 2$ and $\dim D_P = \infty$. Let $D^* = \mathrm{Hom}({}_PD, {}_PP)$, A be the ring $\begin{pmatrix} P & 0 \\ D & D \end{pmatrix}$, B be the ring $\begin{pmatrix} P & 0 \\ D* & D \end{pmatrix}$, and let $_AU_B$ be the (A, B)-bimodule $\begin{pmatrix} P & D \\ D & D \end{pmatrix}/\begin{pmatrix} 0 & 0 \\ D & 0 \end{pmatrix}$. Then $\text{End}(_AU \oplus {}_AA)$ is a semiprimary QF-3-ring that is not quasi-Frobenius (a ring is said to be *semiprimary* if it is a semilocal ring whose Jacobson radical is nilpotent).

For more information on the above classes of rings, see Anderson and Fuller (1992), Curtis and Reiner (1962), Jans (1964), Kasch (1982), Faith (1982), Tachikawa (1973), Tuganbaev (1998), and Wisbauer (1991).

C.31 Gröbner Bases: The Commutative Case

by Bruno Buchberger in Linz, Austria

Many problems in algebraic geometry (polynomial ideal theory, commutative algebra) can be reduced to the problem of constructing Gröbner bases and, hence, can be solved algorithmically by the Gröbner bases method. Over the years, the Gröbner bases method has also found applications in many areas of mathematics other than algebraic geometry. Roughly, all problems that can be expressed, by some encoding, in terms of multivariate polynomials or relations on such polynomials are candidates for the application of the Gröbner bases method. Currently, the most comprehensive survey on applications is (Buchberger and Winkler 1998). The first, and still most comprehensive, monograph on Gröbner bases theory is (Becker and Weispfenning 1993). The most recent textbook is (Kreuzer and Robbiano 2000). This book also lists all other current monographs and textbooks on the subject. Most of these books contain extensive bibliographies on the research literature on Gröbner bases. The concept of Gröbner bases, the main theorem on Gröbner bases together with an algorithm for constructing Gröbner bases, and first applications were introduced in (Buchberger 1970). The early paper (Gröbner 1950), although not yet containing the essential ingredients of Gröbner bases theory, pointed into the right direction and motivated me, in 1976, to assign the name of my former PhD thesis advisor Professor W. Gröbner (1899–1980) to the theory.

In this section, we explain the key ideas of Gröbner bases theory for multivariate polynomials over a commutative field. The huge literature on the subject is mainly concerned with generalizations of the method to other algebraic domains, e.g., polynomial rings over various rings or to the noncommutative case, with improvements and tuning of the algorithm for constructing Gröbner bases, and with a permanently growing variety of applications. The main techniques of how Gröbner bases can be applied are described in Section C.32.

Admissible orderings of power products and remainder algorithms

Let $\mathbb{F}[x_1, \ldots, x_n]$ be the ring of polynomials over indeterminates $x_1, \ldots, x_n$ and a commutative coefficient field $\mathbb{F}$. We use f, g, h as variables ranging over $\mathbb{F}[x_1, \ldots, x_n]$; s, t, u as variables ranging over the set of power products in the indeterminates $x_1, \ldots, x_n$; and F, G as variables ranging over subsets of $\mathbb{F}[x_1, \ldots, x_n]$. By $I(F)$ we denote *the ideal generated by F*. Let $\prec$ be a total ordering on the power products that satisfies the following two properties: $t \neq 1 \implies 1 \prec t$, and $s \prec t \implies s.u \prec t.u$. We call such orderings **admissible**. Two important examples of such orderings are the lexicographic orderings and the total degree lexicographic orderings that, first, order power products by degree and, within a degree, by a lexicographic ordering. By $\mathrm{lp}(f)$ we denote the **leading power product** of f (i.e., the maximal power product of f w.r.t. $\prec$). A polynomial is called **monic** iff the coefficient at its leading power product is 1.

Furthermore, $f \to_F g$ (*f reduces to g modulo F*) iff g results from f by a division step using a polynomial in F as divisor. Instead of a formal definition of $\to_F$, we give an example: Let $\prec$ be the lexical ordering defined by $x \prec y$, let $f_1 := xy^2 - x$, $f_2 := x^2 y - x + 1$, $F := \{f_1, f_2\}$, and $f := x^2 y^3 - 5x^2 y + 1$. The leading power products of f_1, f_2, and f are xy^2, $x^2 y$, and $x^2 y^3$, respectively. Now, $g := f - xy f_1 = -4x^2 y + 1$ results from f by a division step using f_1 as a divisor, i.e., $f \to_F g$. Note that, using f_2 as a divisor, also $f \to_F f - y^2 f_2 = xy^2 - y^2 - 5x^2 y + 1$. Let $\to_F^*$ be the reflexive, transitive closure and $\leftrightarrow_F^*$ the reflexive, symmetric, transitive closure of $\to_F$. Furthermore, f *is reduced modulo F* iff f cannot be reduced using divisors in F. A function R is called a **remainder function** iff, for all F and f, $\mathrm{R}(F, f)$ is a polynomial in reduced form modulo F and $f \to_F^* \mathrm{R}(F, f)$.

Note that there exist algorithmic remainder functions: Given F and f, just iterate individual division steps using divisors from F until a reduced form is reached.

The notion of Gröbner bases

For a given F, it is possible that $f \to_F^* g$, $f \to_F^* h$, both g and h are in reduced form modulo F but $g \neq h$. Now, F is called a *Gröbner basis*, iff this situation is *not* possible:

C.31.1 Definition F is a **Gröbner basis** iff, for all f, g, h, if $f \to_F^* g$, $f \to_F^* h$, and both g and h are reduced modulo F, then $g = h$.

The above set F, w.r.t. the lexical ordering defined by $x \prec y$, is *not* a Gröbner basis: The polynomial $f := x f_1 - y f_2 = xy - y - x^2$ is, of course, in $\mathrm{I}(F)$ but it cannot be reduced modulo F because none of its power products is a multiple of the leading power product of a polynomial in F. In contrast, the set $G := \{y^2 - x, x^2 - x + 1\}$, w.r.t. to same ordering, *is* a Gröbner basis. This cannot be seen by ad-hoc inspection but can be proved using the main theorem below.

The notion of S-polynomial and the main theorem on Gröbner bases

The main problem of Gröbner bases theory consists in establishing an *algorithm* GB having the property that, for arbitrary finite F, GB(F) is a *finite* Gröbner basis and $\mathrm{I}(F) = \mathrm{I}(\mathrm{GB}(F))$. All applications of the Gröbner bases method, essentially, proceed by reducing problems in various areas of mathematics to this problem of Gröbner bases construction.

All current algorithms for constructing Gröbner bases hinge on the following main theorem of Gröbner bases theory that characterizes Gröbner bases by the result of the reduction of the so called *S-polynomials*: The S-polynomial of two (monic) polynomials f and g is defined to be $\mathrm{sp}(f, g) := uf - vg$, where $u := t/\mathrm{lp}(f), v := t/\mathrm{lp}(g)$, and t is the least common multiple of $\mathrm{lp}\, f$ and $\mathrm{lp}\, g$.

C.31.2 Theorem *For any (algorithmic) remainder function* R, *F is a Gröbner bases iff, for all f, g in F,* $\mathrm{R}(F, \mathrm{sp}(f, g)) = 0$.

Note that, in case F is finite, the condition on the right-hand side of the main theorem is *algorithmic*: S-polynomials and their remainders have to be computed only for *finitely* many pairs f and g! The proof of the main theorem involves Noetherian induction on $\prec$ and a careful analysis of the reduction process. The power of the Gröbner bases method hinges on the main theorem and its proof. By the main theorem, the above set F is *not* a Gröbner bases because $(f_1, f_2) = xy - y - x^2$ cannot be reduced any further modulo F. In contrast, the set G above *is* a Gröbner basis because $\mathrm{sp}(y^2 - x, x^2 - x + 1) = xy^2 - y^2 - x^3$ can be reduced to 0 modulo G.

The transformation of the main theorem to an algorithm for constructing Gröbner bases

Now the main theorem can be transformed easily into an algorithm for the Gröbner bases construction problem:

C.31.3 Algorithm
$$\mathrm{GB}(F) := \mathrm{GB0}(F, \{\{f, g\} \mid f, g \in F\}),$$

$$\mathrm{GB0}(F, P) :=$$

$$\begin{cases} \mathrm{GB0}(F, P - \{\{f, g\}\}), & \text{if } h = 0, \\ \mathrm{GB0}(F \cup \{h\}, P - \{\{f, g\}\} \cup \{\{h, f\} \mid f \in F\}), & \text{if } h \neq 0, \end{cases}$$

where $\{f, g\} \in P$ and $h :=$ the monic version of $\mathrm{R}(F, \mathrm{sp}(f, g))$.

One can use Dixon's lemma to prove that this algorithm always terminates. The correctness of the algorithm is clear by the main theorem. Improved versions of the main theorem that characterize Gröbner bases by weaker conditions (involving fewer S-polynomials) lead to computationally better versions of the algorithm. If we carry out the above algorithm for the example F, we obtain the reduced Gröbner basis $\{-1 + 2x - x^2 + x^4, \, 1 - x + x^2 + x^3 + y\}$. Note that this basis has the indeterminates separated, which makes it easy to determine all solutions of the system.

C.32 Gröbner Bases: Applications
by Bruno Buchberger in Linz, Austria

The importance of Gröbner bases theory is due to the fact that many problems of polynomial ideal theory (commutative algebra) and quite some other areas of mathematics (e.g., invariant theory, coding theory, automated geometrical theorem proving, systems theory, integer optimization, statistics, symbolic summation theory) can be reduced, by structurally simple algorithms, to the problem of constructing Gröbner bases. (Buchberger and Winkler 1998) contains tutorials on most of the application areas. (Lin and Xu 2001) is a recent source on applications in systems theory which are mainly based on the possibility of computing linear syzygies by Gröbner bases, see below.

In this section, we present the three techniques of Gröbner bases theory, which represent the core for most of the applications of Gröbner bases. We use the notation introduced in Section C.31. In addition, we write $f \equiv_F g$ for f congruent g modulo $\mathrm{I}(F)$.

The problem of computation in the residue class rings modulo polynomial ideals

This is the problem of finding an algorithm for the isomorphic representation of the residue class rings of polynomial ideals given by a finite set F of generators. Gröbner bases theory gives a complete answer to this problem starting from the following theorem.

C.32.1 Theorem *For the congruence relation modulo a Gröbner basis G any algorithmic remainder function R provides an **algorithmic canonical simplifier**, i.e., for all f, $R(G, f) \equiv_G f$, and, for all f, g, $f \equiv_G g$ implies $R(G, f) = R(G, g)$.*

As a consequence, for example, for arbitrary F, $f \in I(F)$ can be decided algorithmically by testing whether or not $R(GB(F), f) = 0$. As another consequence, by applying the main theorem on canonical simplifiers for congruence relations, we obtain the following theorem.

C.32.2 Theorem *The residue class ring modulo a Gröbner basis G is isomorphic to the algorithmic structure whose carrier is the set of all polynomials that are reduced modulo G, and whose ring operations are just the ring operations in the polynomial ring with subsequent reduction by R, an algorithmic remainder function. For example, addition is defined by $R(G, f + g)$.*

It is then clear how to obtain an algorithmic representation of the residue class ring modulo the ideal generated by an arbitrary finite F: Just apply the theorem to $GB(F)$. Based on the above theorems, it is also easy to see that, for arbitrary finite F, the power products that are reduced modulo $GB(F)$, form a linearly independent vector space basis for the residue class ring modulo $I(F)$. Furthermore, it is then also easy to obtain the "structure constants" (entries in the multiplication table) for the residue class ring modulo $I(F)$ considered as an associative algebra. Also, the number of power products of any given degree d that are reduced modulo $GB(F)$ determines the value of the Hilbert function for $I(F)$ and d. This observation allows to obtain an explicit formula for the Hilbert function of polynomial ideals. As a further consequence, one also obtains an algorithmic criterion for the solvability and the number of solutions of a system of equations determined by a finite F: the system is solvable iff $GB(F)$ does not contain a non-zero constant; in case the system has only finitely many solutions, the number of solutions (counted with multiplicities) coincides with the number of power products that are reduced modulo GB(F); the system has only finitely many solutions iff for all $i = 1, \ldots, n$, GB(F) contains a polynomial whose leading power product has the form $x_i^{k_i}$.

The problem of computing elimination ideals

This is the problem of establishing an algorithm for computing, for any finite F and $i = 1, \ldots, n$, a finite basis for the ideal $I(F) \cap \mathbb{F}[x_1, \ldots, x_i]$ ("the i-th elimination ideal of $I(F)$)"). This problem can be solved completely using the following theorem.

C.32.3 Theorem *If G is a (finite) Gröbner basis w.r.t. the lexicographic ordering determined by $x_1 \prec \ldots \prec x_n$ in ascending ordering and $1 \leq i \leq n$, then $G \cap \mathbb{F}[x_1, \ldots, x_i]$ is a basis for $I(G) \cap \mathbb{F}[x_1, \ldots, x_i]$.*

As a consequence, for arbitrary (finite) F, GB(F) $\cap \mathbb{F}[x_1, \ldots, x_i]$ is a basis for $I(F) \cap \mathbb{F}[x_1, \ldots, x_i]$.

A huge number of problems can be solved by reducing the problem to the above elimination property of Gröbner bases. The most prominent problem in this category is the problem of solving systems of equations described by sets F: First compute GB(F). Then start by finding all solutions α_1 of the univariate system GB(F) $\cap \mathbb{F}[x_1]$. Now, by the elimination property, it is guaranteed that any solution $(\alpha_1, \ldots, \alpha_i)$ of GB(F) $\cap \mathbb{F}[x_1, \ldots, x_i]$ can be extended to all solutions $(\alpha_1, \ldots, \alpha_{i+1})$ of GB(F) $\cap \mathbb{F}[x_1, \ldots, x_{i+1}]$ by substituting $(\alpha_1, \ldots, \alpha_i)$ into GB(F) $\cap \mathbb{F}[x_1, \ldots, x_i]$ and solving the resulting system, which is univariate in x_{i+1}.

Another important problem in this category is the problem of finding a finite basis for the ideal $\{g \in \mathbb{F}[y_1, \ldots, y_m] \mid g(f_1, \ldots, f_m) = 0\}$ (the ideal of **nonlinear syzygies** for $f_1, \ldots, f_m$ in $\mathbb{F}[x_1, \ldots, x_n]$). One knows that this ideal is identical to the ideal generated by $\{y_1 - f_1, \ldots, y_m - f_m\}$ in $\mathbb{F}[y_1, \ldots, y_m, x_1, \ldots, x_n]$ intersected with $\mathbb{F}[y_1, \ldots, y_m]$.

The problem of computing linear syzygies

This is the problem of devising an algorithm for finding (all) polynomial vectors $(z_1, \ldots, z_n)$ that satisfy a linear equation $g_1.z_1 + \cdots + g_m.z_m = h$ with given $g_1, \ldots, g_m, h \in \mathbb{F}[x_1, \ldots, x_n]$. (Such solution vectors are called **linear syzygies**.)

Gröbner bases theory provides a complete answer also for this problem. First, let us consider the case in which the set $G := \{g_1, \ldots, g_m\}$

is a Gröbner basis. Then, by the above test for ideal membership, using any algorithmic remainder function R, $g_1.z_1 + \cdots + g_m.z_m = h$ has a solution iff $R(G, h) = 0$. Collecting the monomial factors of the polynomials $g_1, \ldots, g_m$ that occur in the reduction of h modulo G using R, one obtains a representation of the form $g_1.z_1 + \cdots + g_m.z_m = h$, i.e., one particular solution of the linear equation. In order to obtain all solutions one has to add the solutions of the homogeneous equation $g_1.z_1 + \cdots + g_m.z_m = 0$.

By Hilbert's theorem one knows that the module of all the infinitely many solutions of $g_1.z_1 + \cdots + g_m.z_m = 0$ has a finite basis. Now, such a finite basis can be found by using the theorem below. In this theorem, we need the auxiliary function $SV(G, f, g)$ that is defined for Gröbner bases G and arbitrary f, g. Instead of a general definition, we explain SV in an example: Let $G := \{g_1, g_2\}$ with $g_1 := y^3 - y$ and $g_2 := xy - y$. G is a Gröbner basis w.r.t the lexicographic ordering determined by $y \prec x$. For computing $SV(G, g_1, g_2)$, we first compute $\text{sp}(g_1, g_2) = xg_1 - y^2 g_2 = -xy + y^3$. Now we reduce this polynomial to zero by adding, successively, $-1.g_1$ and $1.g_2$. Now, $SV(G, g_1, g_2) := (x - 1, -y^2 + 1)$. Note that $SV(G, g_1, g_2)$ is a syzygy for $g_1.z_1 + g_2.z_2 = 0$. SV is an algorithmic function.

C.32.4 Theorem *Let* R *be an algorithmic remainder function. If* $G = \{g_1, \ldots, g_m\}$ *is a Gröbner basis then the polynomial vectors* $SV(G, g_i, g_j)$ $(1 \leq i < j \leq m)$ *form a finite basis of the module* $\{(z_1, \ldots, z_m) \mid g_1.z_1 + \cdots + g_m.z_m = 0\}$.

The general case of linear equations $f_1.z_1 + \cdots + f_m.z_m = h$ whose coefficient set $F := \{f_1, \ldots, f_m\}$ is not a Gröbner basis can be reduced to the Gröbner basis case by, computing $G := GB(F)$, solving the linear equation with the coefficients taken from G, and then translating the solutions back.

C.33 Gröbner Bases: The Non-commutative Case

by Leonid Bokut' in Novosibirsk, Russia

Let $K\langle X \rangle$ be a free associative algebra with 1, generated by an alphabet X over a field K. A linear basis of the K-vector space $K\langle X \rangle$ consists of all monomials $u = x_{i_1} \ldots x_{i_n}$ ($n \geq 0$, $n = |u| =$ the length of u). Let X^* be the free monoid over X. Let us fix some linear order $\leq$ on X^*, such that $(X^*, \leq)$ is an ordered monoid with minimum condition. One of the most common examples is the **deg-lex order**, which compares two words first by their lengths and then lexicographically according to some order on X (in most cases, X is a finite set). Any polynomial $f \in X$ has a unique representation of the form

$$f = \alpha \overline{f} + \sum \alpha_i u_i, \quad \alpha, \alpha_i \in K; \ \overline{f}, u_i \in X^*; \text{ all } u_i < \overline{f}.$$

The word $\overline{f}$ is called the **leading word** of f, and f is **monic** if $\alpha = 1$.

C.33.1 Example Let $f = x_3 x_2 x_2 + 5 x_3 x_2 + x_1$, and $g = x_2 x_1 x_2 + x_1$ with $x_1 < x_2 < x_3$. Then $\overline{f} = x_3 x_2 x_2$ and $\overline{g} = x_2 x_1 x_2$, both being monic polynomials.

C.33.2 Definition Let f, g be two monic polynomials and w a word in X^*. The **composition of f, g relative to an ambiguity w** is given by

(1) If $w = \overline{f} a = b \overline{g}$ and $|\overline{f}| + |\overline{g}| > |\overline{w}|$, then $(f, g)_w := fa - bg$;
(2) If $w = \overline{f} = a \overline{g} b$ then $(f, g)_w := f - agb$.

Observe that $(f, g)_w$ may not exist nor be unique; a and/or b might be empty. In case (2), the transformation $f \to (f, g)_w$ is called the **elimination of the leading word (ELW)** of g in f.

C.33.3 Example Let f and g be as in Example C.33.1, $w = x_3 x_2 x_2 x_1 x_2$. Then $(f, g)_w = 5 x_2 x_2 x_1 x_2 + x_1 x_1 x_2 - x_3 x_2 x_1$, and both $\overline{f}$ and $\overline{g}$ are cancelled.

C.33.4 Example Let f be as above and $h = x_2 x_2 + x_3$. Then $w = x_3 x_2 x_2$ and $(f, g)_w = -x_3 x_3 + 5 x_2 x_3 + x_1$.

The main properties of any composition $(f, g)_w$ are:

$$(f, g)_w \in \mathrm{Id}(f, g) \quad \text{and} \quad \overline{(f, g)_w} < w.$$

Let $S \subset K\langle X \rangle$ be a set of monic polynomials. A polynomial f is called **trivial** (mod S), if f can be transformed to zero using ELW in S.

C.33.5 Definition Let $S \subset K\langle X \rangle$ be a set of monic polynomials.
S is called a **Gröbner-Shirshov (GS-)basis (non-commutative Gröbner basis, standard basis)** if
(1) the elements of S do not have any compositions of type (2) (in this case, S is called **reduced**).

(2) Any composition of type (1) is trivial (mod S).

The latter property is equivalent to the fact that for all $f, g \in S$ there are $\alpha_i \in K$, $a_i, b_i \in X^*$, $s_i \in S$ with $a_i \overline{s_i} b_i < w$ such that $(f, g)_w = \sum \alpha_i a_i s_i b_i$.

C.33.6 Composition lemma *Let $S \subset K\langle X \rangle$ be a Gröbner-Shirshov basis, $\mathrm{Id}(S)$ be the ideal generated by S. If $f \in \mathrm{Id}(S)$, then $\overline{f} = a\overline{s}b$ for some $s \in S$, $a, b \in X^*$.*

C.33.7 Corollary *S is a Gröbner-Shirshov basis iff the set of all S-reduced words u (which means $u \neq a\overline{s}b$ for $s \in S$) yields a linear basis of the factor algebra $K\langle X \rangle / \mathrm{Id}(S) = \langle X \mid S \rangle$.*

The Composition lemma is essentially equivalent to the following diamond lemma. A reduced set S in $K\langle X \rangle$ of monic polynomials has the **confluence (diamond) condition** if for any $f, g \in S$, $w \in X^*$, $w = \overline{f}a = b\overline{g}$, $|\overline{f}| + |\overline{g}| > |\overline{w}|$, the diagram $w \to f_1 = (-f + \overline{f})a$, $w \to g_1 = b(-g + \overline{g})$ can be completed (see Section G.7 by sequences of the ELW's of S to $f_1 \to \cdots \to \varphi, g_1 \to \cdots \to \varphi$.

C.33.8 Diamond lemma *Let $S \subset K\langle X \rangle$ be a reduced set of monic polynomials, and let $R = K\langle X \rangle / \mathrm{Id}(S)$. The following conditions are equivalent:*
(1) S has the confluence condition;
(2) the set of S-reduced words form a linear basis of R.

It is easy to see that S has the confluence condition iff S is a GS-basis. So, the Diamond lemma is equivalent to the corollary of the Composition lemma. One may call any (both) of these lemmas as the **Composition-Diamond (CD-)lemma**.

Essentially the same lemma is valid for Lie algebras. Let $\mathrm{Lie}(X)$ be the subspace of $K\langle X \rangle$ generated by X under the Lie bracket $[x, y] = xy - yx$. Then $\mathrm{Lie}(X)$ is the **free Lie algebra**, generated by X over a field K. A word $u \in X^*$ is called a **Lyndon-Shirshov associative word** or **LSA-word** if $u = u_1 u_2 > u_2 u_1$ for all $u_1, u_2 \neq 1$ in the lexicographic order.

C.33.9 Example The word $x_2 x_1 x_2 x_1 x_1$ is an LSA-word, as we have $x_2 x_1 x_2 x_1 x_1 > x_2 x_1 x_1 x_2 x_1 > \cdots$. Also, $x_3 x_2 x_1 x_2 x_1 x_1$ is an LSA-word.

Let us bracket u by induction: $[x] = x$, for $x \in X$, and $[u] = [[u_1], [u_2]]$, where u_2 is an LSA-word of maximal possible length such that $u_1 \neq 1$. Then u_1 is an LSA-word as well. A word $[u]$ is called a **Lyndon-Shirshov word** or **LS-word**.

C.33.10 Example Let $u = x_2 x_1 x_2 x_1 x_1$ and $v = x_3 x_2 x_1 x_2 x_1 x_1$. Then

$$[u] = [[x_2 x_1], [x_2 x_1 x_1]] = [[x_2 x_1], [[x_2 x_1], x_1]],$$
$$[v] = [x_3, [x_2 x_1 x_2 x_1 x_1]] = [x_3, [[x_2 x_1], [[x_2 x_1], x_1]]].$$

The set of all LS-words forms a linear basis of $\mathrm{Lie}(X)$. An important property of $[u]$ is that $\overline{[u]} = u$ (as a polynomial in $K\langle X \rangle$ under the deg-lex order of X^*). Then any polynomial $f \in \mathrm{Lie}(X)$ has the unique representation of the form $f = \alpha[\overline{f}] + \sum \alpha_i [u_i]$, where $\alpha, \alpha_i \in K$ and $\overline{f}, u_i$ are LSA-words with $u_i < \overline{f}$.

Let $u = avb$, where u, v are LSA-words. Then $[u] = [a[vc]d]$, where $b = cd$, c may by empty. As any other word, c can be represented in a unique way in the form $c = c_1 \ldots c_k$, where all the c_i are LSA-words with $c_1 \preccurlyeq \cdots \preccurlyeq c_k$. Here $u \prec v$ if $u = vw$, $w \neq 1$, or $u < v$ lexicographically. Let $[u]_{v,a} = [a[\ldots [[v][c_1] \ldots [c_k]]d]$. Then $\overline{[u]}_{v,a} = u$ as well.

C.33.11 Example Let $u = x_3 x_2 x_1 x_2 x_1 x_1$ and $v = x_3 x_2$. Then $[u]_v = [[[x_3 x_2], x_1], [[x_2 x_1], x_1]]$.

Let us define the **Lie compositions** of two monic Lie polynomials f, g relative to some word:

C.33.12 Definition Let f, g be two monic Lie polynomials and w be a word in X^*. The **composition of f, g relative to an ambiguity w** is given by

(1) if $w = \overline{f}a = b\overline{g}$ and $|\overline{f}| + |\overline{g}| > |\overline{w}|$, then $[f, g]_w = [fa]_{\overline{f},1} - [bg]_{\overline{g},b}$;

(2) if $w = \overline{f} = a\overline{g}b$ then $[f, g]_{w,a} = f - [agb]_{\overline{g},a}$.

Note that w in (1) is automatically an LSA-word. In (2), the transformation $f \to [f, g]_{\overline{g},a}$ is called the **Lie ELW** of g in f.

C.33.13 Example Let $f = [x_3[x_2 x_1]]$, $g = [[x_2 x_1][[x_2 x_1]x_1]]$, and $w = x_3 x_2 x_1 x_2 x_1 x_1$. Then $[f, g]_w = [f[[x_2 x_1]x_1]] - [x_3 g]$.

The main properties of any composition $[f, g]_w$ are

$$[f, g]_w \in \mathrm{Lie}\,\mathrm{Id}(f, g) \quad \text{and} \quad \overline{[f, g]}_w < w.$$

C.33.14 Definition Let $S \subset \mathrm{Lie}(X)$ be a set of monic Lie polynomials. S is called a ***Gröbner-Shirshov (GS-)basis***, if

(1) the elements of S do not have any compositions of type (2);

(2) any composition of type (1) results in zero by the Lie ELW's of S.

The latter is equivalent to the conditions

$$[f, g]_w = \sum \alpha_i [a_i s_i b_i]_{\overline{s_i}, a_i},$$

where $\alpha_i \in K$, $a_i \overline{s_i} b_i$ are LSA-words such that $a_i \overline{s_i} b_i < w$.

C.33.15 CD-lemma for Lie algebras *Let $S \subset \mathrm{Lie}(X)$ be a GS-basis and $\mathrm{Lie\,Id}(S)$ be the Lie ideal generated by S. If $f \in \mathrm{Lie\,Id}(S)$, then $\overline{f} = a\overline{s}b$, for some $s \in S$, $a, b \in X^*$.*

C.33.16 Corollary (Diamond lemma for Lie algebras) *S is a GS-basis in $\mathrm{Lie}(X)$ iff the set of all S-reduced LS-words $[u]$ (which means $u \neq a\overline{s}b$ for all $s \in S$, $a, b \in X^*$) consists of a linear basis of the factor algebra $\mathrm{Lie}(X)/\mathrm{Lie\,Id}(S) = \mathrm{Lie}\langle X \mid S \rangle$.*

C.33.17 Remark For Lie algebras, the theory of GS-bases and the CD-lemma were discovered by Shirshov (1962), for the commutative case it was done by Buchberger (1965) and Hironaka (1964)—cf. Section C.31. For associative algebras, the CD-lemma was formulated and proved by Bokut' (1976) and Bergman (1978)

Some applications

Universal enveloping algebras

Let $L = \mathrm{Lie}\langle \{a_i\} \mid [a_i a_j] - \sum \alpha_{ij}^k a_k = 0, i > j \rangle$ be a Lie algebra presented by a linear basis and its multiplication table. Let $U(L) = \mathrm{Ass}\langle \{a_i\} \mid a_i a_j - a_j a_i - \sum \alpha_{ij}^k a_k = 0, i > j \rangle$ be the ***universal enveloping algebra*** of L. It is not difficult to see that the set of defining relations of $U(L)$ is a GS-basis in $K\langle \{a_i\} \rangle$. By the CD-lemma, we get

C.33.18 PBW-Theorem *A linear basis of $U(L)$ consists of all words $a_{i_1} \ldots a_{i_n}$, $i_1 \leqslant \cdots \leqslant i_n$.*

More generally, let $L = \mathrm{Lie}\langle X \mid S \rangle$ be a Lie algebra presented by generators X and defining relations S (a set of monic Lie polynomials). Then $U(L) = \langle X|S \rangle$.

C.33.19 Theorem *S is a GS-basis in* $\mathrm{Lie}(X)$ *iff S is a GS-basis in* $K\langle X\rangle$.

Now, the PBW-theorem follows from the CD-lemma, because the multiplication table of any Lie algebra is a GS-basis by the inverse property in the CD-lemma.

Clifford algebras

Let $f(x_1, \ldots, x_n) = \sum a_{ij} x_i x_j$ be a quadratic form over a field K, char $K \neq 2$. Let $\mathrm{Cl}(f) = \langle e_1, \ldots, e_n \mid e_i e_j + e_j e_i - 2a_{ij} = 0 \rangle$ be the *Clifford algebra* of f (see Section C.38. It is easy to see that the set of defining relations of $\mathrm{Cl}(f)$ is a GS-basis in $K\langle \{e_1, \ldots, e_n\} \rangle$. From the CD-lemma, we have

C.33.20 Theorem *The words* $e_{i_1} \ldots e_{i_k}$, $i_1 < \cdots < i_k$, *form a linear basis of* $\mathrm{Cl}(f)$.

For $f = 0$, the Clifford algebra $\mathrm{Cl}(f) = \mathrm{Cl}_n$ is the *Grassman (Exterior) algebra* in n generators.

C.34 Orders and Maximal Orders

by T. Y. Lam in Berkeley, CA, USA

Definitions and examples

Loosely speaking, the theory of orders and maximal orders enables us to study certain rings and their representations in an integral, noncommutative setting, in generalization of the arithmetic study of number rings.

Let R be a commutative domain, with quotient field K, and let A be a finite-dimensional K-algebra. By a *full R-lattice* in A, we mean a finitely generated R-submodule $\Lambda \subseteq A$ such that $K\Lambda = A$. If such a Λ happens to be a subring of A, we say that Λ is an *R-order*. If, moreover, Λ is not properly contained in any R-order of A, we say it is a *maximal order* (with R understood). For instance, if $M \subseteq A$ is any full R-lattice, then $O_\ell(M) = \{a \in A \mid aM \subseteq M\}$ is an R-order, called the *left order* of M (and the *right order* $O_r(M)$ is defined similarly). If $a_1, \ldots, a_n \in K$ are integral over R, then $R[a_1, \ldots, a_n]$ is an R-order in K.

Here are some more examples. Let τ be the golden ratio $(1 + \sqrt{5})/2$. For any positive integer n, $\Lambda_n = \mathbb{Z} + \mathbb{Z} \cdot n\sqrt{5}$ is a $\mathbb{Z}$-order in $\mathbb{Q}(\sqrt{5})$, and $\mathbb{Z} + \mathbb{Z} \cdot \tau$ is the unique maximal $\mathbb{Z}$-order. More generally, for any number field A, the full ring of algebraic integers Λ is the unique maximal $\mathbb{Z}$-order. For some *noncommutative* examples, take $A = \mathbb{M}_2(\mathbb{Q})$. For any integer $n > 0$, $\{(a_{ij}) \in \mathbb{M}_2(\mathbb{Z}) \mid n \text{ divides } a_{21}\}$ is a $\mathbb{Z}$-order in $\mathbb{M}_2(\mathbb{Q})$, and it is contained in the maximal $\mathbb{Z}$-order $\mathbb{M}_2(\mathbb{Z})$. For any finite group G, the group ring $\Lambda = \mathbb{Z}G$ is a $\mathbb{Z}$-order in $\mathbb{Q}G$. In the case where G is abelian, Maschke's Theorem gives $\mathbb{Q}G = A_1 \times \cdots \times A_r$, where the A_i's are suitable number fields. Here, Λ is contained in the unique maximal order $\Lambda_1 \times \cdots \times \Lambda_r$, where Λ_i is the full ring of algebraic integers in A_i. In the $\mathbb{Q}$-algebra of quaternions $\mathbb{Q} + \mathbb{Q}i + \mathbb{Q}j + \mathbb{Q}k$, the $\mathbb{Z}$-span of $1, i, j, k$ is a $\mathbb{Z}$-order, and so is the $\mathbb{Z}$-span of $1, 2i, 2j$ and k, while the $\mathbb{Z}$-span of i, j, k and $(1 + i + j + k)/2$ (the "Hurwitz ring of integral quaternions") is a maximal $\mathbb{Z}$-order.

The term *order* used in this article is not to be confused with the same term used in the sense of an *ordering* (as, for instance, in Section C.44 and Section D.6). This confusion is unfortunate, but is also rather difficult to avoid. The word *order* as defined in the second paragraph above is a direct translation of the German word *Ordnung*, which was commonly used in the 19th century number theory literature to refer to rings of algebraic integers (see, e.g., Dedekind 1877). The extension of the notion of "Ordnung" to a *noncommutative* setting seemed to have stemmed from the work of Emmy Noether.

Orders in separable algebras

For R, K, A as above, we say A is a **separable K-algebra** if A is semisimple and the center of each simple component of A is a separable field extension of K. For such (and only such) A, $A \otimes_K E$ remains a semisimple E-algebra for any field $E \supseteq K$. Assuming henceforth that R is noetherian and integrally closed, *any R-order in A is contained in a maximal order*, and each maximal order arises as a direct product of maximal R-orders in the simple components of A.

Maximal orders behave well with respect to localizations and completions. First, an R-order Λ is maximal iff, for any maximal ideal $\mathfrak{p} \subset R$, $\Lambda_{\mathfrak{p}}$ is a maximal $R_{\mathfrak{p}}$-order in A. And, if R is already local, with maximal ideal $\mathfrak{m}$ and $\mathfrak{m}$-adic completion $\hat{R}$, then an R-order $\Lambda \subseteq A$ is maximal iff its completion $\hat{\Lambda} = \Lambda \otimes_R \hat{R}$ is a maximal $\hat{R}$-order in $\hat{A}$.

Maximal orders over Dedekind rings

From now on, we assume R is a Dedekind ring (and A is a separable K-algebra). The consideration of maximal R-orders in A is substantially reduced to the case where R is a complete DVR (discrete valuation ring). We first consider the key case where A is a division K-algebra D. Here, there is a *unique* maximal order, namely the integral closure Λ of R in D. This Λ is a "noncommutative DVR" in that it has an element π (called a **uniformizer**) such that the 1-sided and 2-sided nonzero ideals of Λ are all of the form $\pi^i \Lambda$. If π_0 is a uniformizer of R, then $\pi_0 \Lambda = \pi^e \Lambda$, where e is the **ramification index** of D over K. With the **inertial degree** f defined to be $\dim_{R/\pi_0 R} \Lambda/\pi\Lambda$, we have $ef = [D : K]$, exactly as in the case of a field extension of the complete field K. If, moreover, $R/\pi_0 R$ is a finite field and the center of D is exactly K, then in fact $e = f$, and D is a "cyclic algebra" completely determined by its index e together with another integer $r \in (0, e)$ relatively prime to e. (The image of r/e in $\mathbb{Q}/\mathbb{Z}$ is the **Hasse invariant** of $[D]$ in the Brauer group of K, and all elements in $\mathbb{Q}/\mathbb{Z}$ are realized this way). More generally, if $A = \mathbb{M}_n(D)$, all maximal orders in A are conjugate to $\mathbb{M}_n(\Lambda)$. In view of Wedderburn's Theorem, this completes the determination of maximal orders over a complete DVR.

Continuing with the case R is a DVR (but not necessarily complete), a maximal order Λ in a central simple K-algebra A is always a principal ideal ring (all 1-sided ideals are principal), and $\Lambda/\mathrm{Rad}(\Lambda)$ is a simple artinian ring. Finally, in the general case when R is a Dedekind ring, any maximal order Λ in a separable K-algebra A is a hereditary ring, in the sense that any left or right ideal of Λ is projective as a Λ-module. In the case when A is simple, that is, $A = \mathbb{M}_n(D)$ where D is a division K-algebra, any maximal order in A is Morita equivalent to one in D.

The ideal structure for a maximal R-order Λ in a separable K-algebra A bears a good resemblance to that in the Dedekind ring R itself. By the word "ideal", we mean here a 2-sided ideal in Λ that is a full lattice. With this understanding, prime ideals in Λ are maximal, the product of prime ideals is commutative, and any ideal is uniquely a product of prime ideals. For any ideal I, we also have $I \cdot I^{-1} = I^{-1} \cdot I = \Lambda$, and $(I^{-1})^{-1} = I$, where I^{-1} means $\{x \in A \mid I\,x\,I \subseteq I\}$, just as for nonzero ideals in Dedekind rings. In the case when A is simple, there is a natural one-one correspondence (defined by contraction) between the prime ideals of Λ and the (nonzero) prime ideals of R.

There is also a theory of one-sided ideals in A, but it is more technical.

A full R-lattice in A is called a **normal ideal** if its left (or equivalently, right) order in A is a maximal order. Certain normal ideals can be multiplied, and with respect to such a "proper" multiplication, all normal ideals form a groupoid, called the **Brandt groupoid** of A.

Lattices over orders

By a left Λ-*lattice* , we mean a left Λ-module that is finitely generated and torsion-free as an R-module. For instance, for a finite group G, the study of RG-lattices is basically the theory of integral representations of groups.

Certainly, every Λ-lattice breaks up into a direct sum of indecomposable ones. However, the isomorphism types of the intervening indecomposable lattices are not uniquely determined, as is already witnessed by the case $\Lambda = R$. We express this by saying that, for lattices, the Krull-Schmidt Theorem (generally) fails. Nevertheless, there are some important cases in which the KS Theorem holds; e.g., it does when

(1) R is a *complete* DVR, or
(2) R is a DVR and A is a split K-algebra (direct product of matrix rings over K).

Independently of whether the KS Theorem holds, it is of interest to determine the number $n(\Lambda)$ of indecomposable Λ-lattices (up to isomorphism). To demonstrate the subtlety of this problem, we just mention a sample result of A. Jones. For a group ring $\Lambda = \mathbb{Z}G$, $n(\Lambda)$ is finite iff for any prime p dividing $|G|$, a p-Sylow group of G is cyclic of order p or p^2.

An important tool used in classifying lattices is the notion of *genera*: two Λ-lattices M, N are said to be **in the same genus** (written $M \vee N$) if $M_{\mathfrak{p}} \cong N_{\mathfrak{p}}$ as $\Lambda_{\mathfrak{p}}$-lattices for every maximal ideal $\mathfrak{p} \subset R$. The latter condition can also be checked over the completions $\hat{R}_{\mathfrak{p}}$. This remark implies for instance that, if $M \oplus L \cong N \oplus L$ for some lattice L, or if $M^r \cong N^r$ for some integer r, then $M \vee N$. In case Λ is a *maximal* order, $M \vee N$ amounts to $KM \cong KN$ as A-modules. Another interesting problem is to compute the number of lattices in the genus of a given lattice M.

The remarks in the paragraph above linked the study of genera to questions on *cancellation* of lattices. Over arbitrary orders, however, we do not expect cancellation to hold. In fact, examples of Swan showed that cancellation can fail already for projective lattices over maximal orders.

Assume now that K is a global field (that is, either a number field or a function field in one variable over a finite field). A well-known result in number theory says that the Dedekind ring R has a finite class number, i.e., there are only finitely many R-isomorphism types of ideals. This famous result admits a strong generalization to Λ-lattices. According to the **Jordan-Zassenhaus Theorem**, if Λ is any R-order in a semisimple K-algebra A, then for any positive integer n, there are only finitely many isomorphism types of left Λ-lattices of R-rank $\leq n$. For instance, in the special case $n = \text{rank}_R \Lambda$, there are only finitely many isomorphism types of full left Λ-lattices in A. (These are known as "left Λ-ideals".) For detailed expositions (and a survey) on the general theory of lattices, see (Swan 1970, Reiner 1970, 1975).

Things we have no space for

There are many other directions in the research on orders that we have not touched upon. Let us close by mentioning just a few of the most fruitful directions.

(1) *Unit groups of orders* (for instance $\text{GL}(n; \mathbb{Z})$ in $\mathbb{M}_n(\mathbb{Z})$, or units in quaternionic $\mathbb{Z}$-orders) are basic examples of arithmetic groups. These groups are finitely presented — in generalization of Dirichlet's Unit Theorem for rings of algebraic integers. Units in group rings already constitute a big topic for research, and require techniques specific to the group-theoretic setting.

(2) *K-theory of orders* is the study of Grothendieck groups and White-head groups of orders, and their higher K-theoretic analogues. The computations of class groups and Picard groups of group rings, Milnor's K_2-groups of rings of algebraic integers, etc. are special cases. These are all finitely generated abelian groups. (For general background on the K-theory groups, see Section H.7 and Section H.8.)

(3) *A plethora of orders*: Besides maximal orders, many other kinds of orders have been investigated; e.g., clean orders, hereditary orders, Gorenstein orders, and Bass orders, etc. The main job is to "classify" each such kind of orders, to the extent possible.

(4) *Analytic methods*: For any $\mathbb{Z}$-order Λ, L. Solomon has defined the zeta function: $\zeta_\Lambda(s) = \sum_M [\Lambda : M]^{-s}$, where M ranges over the (full) left ideals in Λ. When Λ is a full ring of algebraic integers, this is Dedekind's zeta function, which in turn generalizes Riemann's zeta function. It is hoped that the study of ζ_Λ will be useful for noncommutative orders in the same way Dedekind's zeta function is

useful for classical algebraic number theory.

C.35 Representations of Orders

by Vladimir Kirichenko in Kiev, Ukraine

Let Λ be a **$\mathbb{Z}$-ring**, i.e., a ring with identity whose additive group is a free abelian group of a finite rank n with $\mathbb{Z}$-basis $e_1, \ldots, e_n$. Every $\mathbb{Z}$-ring Λ is naturally embedded in a $\mathbb{Q}$-algebra $\tilde{\Lambda} = \Lambda \otimes_{\mathbb{Z}} \mathbb{Q}$ with $\mathbb{Q}$-basis $e_1 \otimes 1, \ldots, e_n \otimes 1$, where $\mathbb{Q}$ is the field of rationals. Hence, a $\mathbb{Z}$-ring is also called a **$\mathbb{Z}$-order**, or simply an **order** (cf. Section C.34). A regular module Λ_Λ determines a regular integral representation of an order Λ. Modules which corresponds to integral representations are called either **representation modules**, **torsionless modules**, **Cohen-Macaulay modules** (Section C.8), or **Λ-lattices**.

As is well known (Faddeev 1964, see, for example), the determination of the integral representations of an arbitrary $\mathbb{Z}$-ring Λ actually falls into two stages: the determination of the p-adic representations of the ring Λ and the study of representations belonging to the same genus, that is, of representations that are p-adically equivalent for all prime p. The latter problem is nontrivial, even for the rings that are simplest from the point of view of integral representations. If Λ is the ring of integers of an algebraic number field, the problem is equivalent to calculating the ideal class group of the field. In the general case, rather little is known about the structure of genera. It can be shown there that if the $\mathbb{Z}$-ring Λ is semisimple (in the sense of Jacobson), then the number of integral representations of Λ belonging to the same genus does not exceed a certain constant which depends only on Λ (the late sixties: Roiter, Jakobinsky, Drozd, Platonov).

An order Λ is said to be a **ring of finite representation type** (*FRT*) if it has a finite number of indecomposable integral representations. One of the fundamental problems in the theory of integral representations is to find criteria for a given $\mathbb{Z}$-ring to be *FRT*. This problem has been solved in the following cases:

(A) group $\mathbb{Z}$-rings (the early sixties: Berman-Gudivok, Heller-Reiner, Higman, Jones). See (Roggenkamp 1990);

(B) commutative rings (the late sixties: Drozd-Roiter, Jakobinsky);

(C) primary orders (Drozd-Kirichenko-Hijikata-Nishida). See (Hijikata and Nishida 1997);

(D) semimaximal rings or tiled orders (see below) (Zavadskij-

Kirichenko).

The criterions in the cases (A) and (B) are very simple: *Case (A).* Let G be a finite group. Then group $\mathbb{Z}$-ring $\mathbb{Z}_G$ is *FRT* if and only if for every prime number $p \geq 2$, any Sylow p-subgroup G_p of G is cyclic of order p or p^2. *Case (B).* Let Λ be a commutative *FRT* $\mathbb{Z}$-ring. Then the $\mathbb{Q}$-algebra $\tilde{\Lambda} = \Lambda \underset{\mathbb{Z}}{\otimes} \mathbb{Q}$ is semisimple, and $\Lambda \subset \bar{\Lambda} \subset \tilde{\Lambda}$, where $\bar{\Lambda}$ is the unique maximal overorder of Λ, $I = \bar{\Lambda}/\Lambda$,, and $I' = rad(\bar{\Lambda}/\Lambda)$.

C.35.1 Theorem *The order Λ is FRT if and only if the following two conditions hold:*

(DR1) *I has two generators as a Λ-module;*
(DR2) *the Λ-module I' is cyclic.*

In the case (D), a criterion will be formulated in the partial case for the reduced tiled 0-1-orders. Let R be a discrete valuation ring (d.v.r.) with a prime element π and $\mathcal{E} = (\alpha_{ij})$ be a matrix of order n, $\alpha_{ij} \in \mathbb{Z}$. Moreover, assume that $\alpha_{ij} + \alpha_{jk} \geq \alpha_{ik}$ and $\alpha_{ii} = 0$ for all i, j, k. A ring $\Lambda = \sum\limits_{i,j=1}^{n} e_{ij} \pi^{\alpha_{ij}} R$ is called a **tiled order**. If Λ is reduced, then $\alpha_{ij} + \alpha_{ji} > 0$ for all i, j. Every tiled order is Morita-equivalent to a reduced order. A tiled order $\Lambda = \sum\limits_{i,j=1}^{n} e_{ij} \pi^{\alpha_{ij}} R$ is called a $0-1$-**order** if $0 \leq \alpha_{ij} \leq 1$. For any reduced $0-1$-order, there exists a finite partially ordered set (poset) $\mathcal{N}(\Lambda) = \{1, \ldots, n\}$ and $i \leq j$ if and only if $\alpha_{ij} = 0$. Conversely, after the same rule, we connect with every finite poset the reduced $0-1$-order. Let $(l_1, \ldots, l_s)$ be a cardinal sum of s chains containing $l_1, \ldots, l_s$ elements, respectively. Denote $K_1 = (1, 1, 1, 1)$, $K_2 = (2, 2, 2)$, $K_3 = (1, 3, 3)$, $K_4 = (1, 2, 5)$, $K_5 = (N, 4)$, where $N = \{a < b > c < d\}$. The posets $K_1, \ldots, K_5$ are called **critical**. The reduced 0-1-order Λ is *FRT* if and only if $\mathcal{N}(\Lambda)$ does not contain any of the critical subposets $K_1, \ldots, K_5$. It is natural to distinguish and study those classes of rings, where the structure of their integral representations is comparatively simple. Representations of hereditary rings are evidently the simplest ones, since in this (and only in this) case, any exact sequence of representation modules splits. Let Λ a commutative Noetherian ring of Krull dimension 1 without nilpotent elements, whose integral closure is a finitely generated Λ-module. Bass proved that if every ideal of Λ has two generators, then every torsionless Λ-module splits into a direct sum of ideals of the ring Λ. For integral domains, this condition is also necessary. Practically, the same class of orders was considered by Borevich and Faddeev under the name of orders with cyclic index. Bass's results and the methods have turned out to be very useful in the theory of integral representations. Drozd-Kirichenko-Roiter introduced *Bass orders* in 1967. For Bass orders, every representation module splits into a direct sum of ideals. In a certain sense, Bass orders form the smallest class of orders containing hereditary orders and orders which satisfy Bass' condition. Namely, it can be proved that in the local case every Bass order is Morita-equivalent to a direct product of a hereditary order and an order in which every ideal has two generators. The further development of this theory and refinement of the results about Bass orders, see (see Hijikata and Nishida 1992, Iyama 1998).

Let Λ be an R-ring where R is a complete commutative d.v.r. There is a Krull-Schmidt Theorem for representation Λ-modules and Λ is semiperfect. Tame-wild Dichotomy for Cohen-Macaulay modules, Auslander-Reiten quivers connections with representations of posets are considered for such orders (see Van Oystaeyen and Saorin 2000, Roggenkamp and Ştefănescu 1996, Simson 1992).

C.36 Distributive Modules and Rings

by Askar A. Tuganbaev in Moscow, Russia

All rings are assumed to be associative and with nonzero identity element. Expressions such as a "Noetherian ring" mean that the corresponding right and left conditions hold.

A module M is said to be **distributive** if the lattice $\mathrm{Lat}(M)$ of its submodules is distributive. A module M is **uniserial** if the lattice $\mathrm{Lat}(M)$ is a chain. All uniserial modules are distributive. A quasi-cyclic Abelian group is a uniserial module over the ring of integers $\mathbb{Z}$. The additive group $\mathbb{Q}$ of rational numbers is a distributive $\mathbb{Z}$-module which is not uniserial. If all simple subfactors of a module M are isomorphic to each other, then M is distributive if and only if M is uniserial.

A module is called a **Bezout** module if all its finitely generated submodules are cyclic. A ring is said to be **regular** (resp. **strongly regular**) if each of its principal right or left ideal is generated by an idempotent (resp. by a central idempotent). The class of distributive rings includes all commutative Dedekind domains, all commutative Bezout rings, and all strongly regular rings. Therefore, the ring of integers, the polynomial ring $F[x]$ over a field F, all factor rings of direct products of skew fields, and all commutative regular rings are examples of distributive rings. The subring $\mathbb{Z} + \mathbb{Z}\sqrt{-5}$ of the field of complex numbers is a commutative distributive domain that is not a Bezout ring. For any field F and each positive integer $n \geq 2$, the ring of all $n \times n$ matrices over F is a Bezout ring that is not right or left distributive.

A cyclic module M with Jacobson radical $J(M)$ is said to be **local** (cf. Section C.6) if the factor module $M/J(M)$ is simple. A module M over a local ring is distributive if and only if M is uniserial if and only if M is a Bezout module. A ring A is said to be **right quasi-invariant** (**right invariant**, respectively) if all maximal right ideals (resp. all right ideals) of A are ideals. Each Bezout module over a right quasi-invariant ring is distributive and every right distributive ring is right quasi-invariant. Therefore, a right Bezout ring is right distributive if and only if it is right quasi-invariant. If A is a ring such that the factor ring $A/J(A)$ is strongly regular, then an A-module is distributive if and only if it is a Bezout module. A ring A is said to be **semilocal** (cf. Section C.7) if the factor ring $A/J(A)$ is Artinian. Each distributive module over a semilocal ring is a Bezout module that does not contain an infinite direct sum of nonzero modules.

A module M is said to be **distributively generated** (resp. **semidistributive**) if M is a sum (resp. a direct sum) of distributive modules. All full matrix rings over right distributively generated (resp. right semidistributive) rings are right distributively generated (resp. right semidistributive). A right module M over a ring A is said to be **projective** if M is a direct summand of a direct sum of copies of the modules A_A. The endomorphism ring of a finitely generated projective right module over a right distributively generated ring is right distributively generated. If either A is a right distributively generated ring or A is the endomorphism ring of a finite direct sum of distributive modules, then the intersection of all prime ideals of A contains all right or left nil-ideals of A. If A is a ring and all right A-modules are semidistributive, then A is an Artinian ring and every right A-module is a direct sum of cyclic Noetherian Artinian distributive Bezout modules.

If F is a field and A is the 5-dimensional algebra over F generated by all 3×3 matrices of the form $\begin{pmatrix} f_{11} & f_{12} & f_{13} \\ 0 & f_{22} & 0 \\ 0 & 0 & f_{33} \end{pmatrix}$ with $f_{ij} \in F$, then each right A-module is semidistributive and the principal right ideal $e_{11}A$ of A is a distributive local Noetherian Artinian Bezout A-module that is not uniserial.

Assume that A is a ring, B is a right ideal of A, $T \equiv A \setminus B$, and there is a ring homomorphism $f : A \to A_B$ such that for every element $t \in T$, the element $f(t)$ is invertible in A_B, $A_B = \{f(a)f(t)^{-1}\}$, and $\mathrm{Ker}(f) = \{a \in A \mid at = 0 \text{ for some } t \in T\}$. In this case, A is called the **right localization of A with respect to B**. A module M over a commutative ring A is distributive if and only if for every maximal ideal B of A, the module of fractions M_B over the localization A_B with respect to B is uniserial. If A is a ring and either all nilpotent elements of A are central or A is a ring with maximum condition on right annihilators then A is right distributive if and only if for every maximal right ideal B of A, there exists the right localization A_B with respect to B and the ring A_B is uniserial.

A module M is said to be **injective** if for every module N and any submodule N' of N, each homomorphism $N' \to M$ is extended to a homomorphism $N \to M$. A module M over a ring A is distributive if and only if for any injective module E_A, the left $\mathrm{End}(E)$-module $\mathrm{Hom}(M, E)$ is distributive if and only if the lattice $\mathrm{Lat}(M)$ is isomorphic to the lattice of all open subsets of a topological T_0-space if and only if each factor module of M does not contain a direct sum of two isomorphic simple modules if and only if for any two elements $x, y \in M$, there is an element $a \in A$ with $xaA + y(1 - a)A \subseteq xA \cap yA$.

A right module M is said to be **endodistributive** if M is distributive left $\mathrm{End}(M)$-module.

C.36.1 Theorem *For a ring A, T.F.A.E.:*

(1) *A is right distributive;*
(2) *All injective right A-modules are endodistributive;*
(3) *The injective hull of the direct sum of all representatives of the classes of isomorphic simple right A-modules is endodistributive;*
(4) *The injective hulls of all simple right A-modules are endodistributive.*

A right A-module M is said to be *flat* if for each left A-module N and any submodule N' of N, the natural group homomorphism $M \otimes_A N' \to M \otimes_A N$ is a monomorphism. For a commutative ring A, all ideals of A are flat if and only if A is a distributive semiprime ring. For an invariant semiprime ring A, all right ideals of A are flat if and only if A is a distributive ring.

For a ring A, the polynomial ring $A[x]$ is right distributive if and only if the Laurent polynomial ring $A[x, x^{-1}]$ is right distributive if and only if A is a commutative regular ring. If F is a field and A be the ring formed by all eventually constant sequences of elements of F, then A is a commutative regular ring such that the commutative ring $A[[x]]$ is not distributive.

Let A be a ring, and let φ be an injective endomorphism of A. We denote by $A_\ell[[x, \varphi]]$ and $A_r[[x, \varphi]]$ the **left skew power series ring** and the **ight skew power series ring** (cf. Section C.9) consisting of formal series $\sum_{i=0}^\infty a_i x^i$ and $\sum_{i=0}^\infty x^i a_i$ $(a_i \in A)$, respectively; we note that for any $a \in A$ and each positive integer i, we have $x^i a = \varphi^i(a) x^i$ in the first case and $ax^i = x^i \varphi^i(a)$ in the second case. The ring $A_\ell[[x, \varphi]]$ is right distributive if and only if A is an strongly regular, φ is an automorphism, $\varphi(e) = e$ for each idempotent $e \in A$, and for any countably generated right ideal B of A, each homomorphism $B_A \to A_A$ can be extended to an endomorphism of A_A.

Let k be a field, F be the field of rational functions over k in countable number of variables $\{t_i\}_{i=0}^\infty$, and let $t \equiv t_0$. There exists an injective endomorphism α of F such that $\alpha(t_i) = t_{i+1}$ for all i and t is transcendental over $\alpha(F)$. Let A be the ring which is formed by all the fractions $f/(1+yg)$ such that f and g are arbitrary polynomials from the polynomial ring $F[y]$ in one variable y. By $\varphi(y) = t$, we extend α to an injective endomorphism φ of A. Then A is a commutative uniserial Noetherian domain such that A is not a field and $\varphi(a) \in U(A)$ for all nonzero $a \in A$. We denote by R and S the right skew power series ring $A_r[[x, \varphi]]$ and its ideal $xRJ(A)R$, respectively.

C.36.2 Theorem *For a ring R, the following assertions hold:*

(1) *R is a right uniserial right Noetherian domain (in particular, R is a right distributive right hereditary right Bezout domain);*

(2) *R is neither a left distributive ring nor a left Bezout ring;*

(3) *R/S is a right uniserial right Noetherian ring that is neither right Artinian nor semiprime.*

Let A be a ring, and let φ be an automorphism of A. We denote by $A_\ell((x, \varphi))$ the **left skew Laurent series ring** consisting of the series $f \equiv \sum_{i=m}^\infty a_i x^i$, where $m = m(f)$ is an integer and $x^i a = \varphi^i(a) x^i$. The ring $A_\ell((x, \varphi))$ is a semilocal right distributive ring if and only if $A_\ell((x, \varphi))$ is a finite direct product of right Artinian right uniserial rings if and only if A is a finite direct product of right Artinian right uniserial rings $A_1, \ldots, A_n$ and $\varphi(A_i) = A_i$ for all i.

A module M over a ring A is called a **multiplication module** if for any submodule N of M, there is an ideal B of A such that $N = MB$.

C.36.3 Theorem *Let A be a right invariant ring. For a right A-module M, the following conditions are equivalent:*

(1) *M is distributive;*
(2) *All finitely generated submodules of M are multiplication modules;*
(3) *All 2-generated submodules of M are multiplication modules;*
(4) *For any two finitely generated submodules X and Y of M, there is an element $a \in A$ such that $Xa \subseteq Y$ and $Y(1-a) \subseteq X$.*

A ring is said to be **hereditary** if all right ideals of A are projective. A right hereditary ring is right distributive if and only if it is right invariant. A right Noetherian ring A is right distributive if and only if each right ideal of A is a product of prime (two-sided) ideals of A. A ring A is Noetherian and right distributive if and only if A is a finite direct product of Artinian right uniserial rings and invariant hereditary domains.

For more information on the above classes of modules and rings, see Behrens (1972), Cohn (1985), Facchini (1998), Pierce (1982), Tuganbaev (1998), Tuganbaev (1999), and Xue (1992).

C.37 PI-Algebras

> *by Alexei J. Belov in Perth, Australia, and Louis H.*
>
> *Rowen in Ramat Gan, Israel*

R will always denote an algebra. Although many results hold for algebras over an arbitrary commutative base ring, or at least over a Noetherian ring, it is convenient in this survey to consider only algebras over a given field K. An **identity** of an algebra R is an element f of the free associative algebra $K\langle x_1, x_2, \dots \rangle$ (the polynomial algebra in non-commuting variables, see Section C.23) all of whose substitutions in R are 0. In this case, we also say that R **satisfies** the identity f. For example, any commutative algebra R satisfies the identity $x_1 x_2 - x_2 x_1$.

C.37.1 Definition R is called **algebra with polynomial identity** or, more shortly, a **PI-algebra** if R satisfies some identity, for which at least one of its coefficients is 1.

It is easy to see that any algebra R spanned by a finite number m of elements satisfies the **standard identity**

$$s_k(x_1, \dots, x_k) := \sum_{\pi \in S_k} \operatorname{sgn}(\pi)\, x_{\pi(1)} \dots x_{\pi(k)}$$

for any $k > m$. For $m = 1$ and $k = 2$, this is $x_1 x_2 - x_2 x_1$ as above. More generally, one can define the *Capelli polynomial*

$$c_k(x_1, \ldots, x_k, y_0, \ldots, y_k) = \sum_{\pi \in S_k} \mathrm{sgn}(\pi)\, y_0 x_{\pi(1)} y_1 \cdots x_{\pi(k)} y_k,$$

which is also an identity of such R for any $k > m$. In particular, c_{n^2+1} (and thus s_{n^2+1}) is an identity of the $n \times n$ matrix algebra $\mathrm{Mat}_n(K)$ (K arbitrary). In fact, Amitsur and Levitzki (1950) proved that s_{2n} is an identity of $\mathrm{Mat}_n(K)$, and has the lowest degree (as a polynomial) of all identities of $\mathrm{Mat}_n(K)$.

The PI-theory was initiated by Dehn (1922) in order to find algebraic formulations for intersection theorems in Desarguian planes, and has roots in 19th century invariant theory. Not much was done until the late 1940's and 1950's, when algebraists were searching for a workable generalization of commutative algebra. During that time, the groundwork for the structure theory was laid by Levitzki, Jacobson, Kaplansky, and Amitsur. Nevertheless, the crucial connection to commutativity came only with the discovery of central polynomials for $R = \mathrm{Mat}_n(K)$, independently (with different constructions) by Formanek and Razmyslov (1972). These are non-identities of R which take scalar values only. For example $(x_1 x_2 - x_2 x_1)^2$ is easily seen to be central for $n = 2$, but the constructions for arbitrary n are much more intricate. Razmyslov's idea can be explained by means of the isomorphism $R \otimes R^{\mathrm{op}} \simeq \mathrm{End}_K R$.

The structure theory of PI-algebras

C.37.2 Theorem *Let R be a unitary PI-algebra with center C.*

(1) (Kaplansky's Theorem) *All primitive images are simple and of finite dimension over their centers.*

(2) (Posner-Formanek-Razmyslov-Rowen) *If R is semiprime, then R has a central polynomial, and any nonzero ideal intersects C nontrivially. In particular, if R is prime, then localizing at $C \setminus \{0\}$ produces a simple PI-algebra. In general, any semiprime PI-algebra R can be embedded into a matrix algebra over a commutative algebra, and R is highly amenable to techniques of commutative localization.*

(3) (Levitzki-Amitsur) *The upper and lower nilradicals of R coincide.*

(4) (Artin-Procesi) *R is Azumaya of constant rank n iff R satisfies each identity of $\mathrm{Mat}_n(K)$ and s_{2n-2} is not an identity in any homomorphic image of R.*

(5) (Regev) *The tensor product of PI-algebras is again PI.*

These results show that PI's are closely related to finite dimensional representations of algebras; an irreducible representation of dimension n is just a simple homomorphic image satisfying s_{2n}. Isaacs-Passman classified those group algebras which satisfy a polynomial identity, i.e., groups of bounded representation degree.

Relatively free PI-algebras

Any set S of identities defines the variety $\mathscr{V}(S)$ of algebras satisfying the identities in S. Each variety contains a free object in the class, which is called a **relatively free**, or **generic algebra**. When S is the set of identities of $\mathrm{Mat}_n(K)$, then the generic algebra is a (noncommutative) domain, which localized at its center yields a division algebra denoted by $\mathrm{UD}(n, K)$. This division algebra has dimension n^2 over its center and has many important properties, cf. Section C.13.

In this context, it is convenient to think of identities as universal formulas in the language of algebras, since the relatively free algebra is the free object respecting that formula, in that language. Enriching the language yields theories of rational identities (if one permits inverses), or generalized identities (if one permits the constant symbols to include R). Amitsur proved that every division algebra satisfying a nontrivial rational identity is finite dimensional over its center, thereby concluding Dehn's project (since every intersection theorem in a Desarguian plane gives rise to a rational identity of the coordinate division algebra). Belov proved that every affine relatively free ring can be embedded in a ring which is a Noetherian module over its center, and that every relatively free algebra has a rational Hilbert series.

Affine PI-algebras

The previous paragraph hints at a geometric theory for PI-algebras. An algebra over K is **affine** if it is generated by a finite number of elements. In the 1940's Jacobson developed a structure theory of algebraic algebras, which enabled Kaplansky (1948) to prove that any affine algebraic PI-algebra is finite dimensional. Shirshov (1957) reworked this combinatorially, showing that for any affine algebra $R = \mathrm{K}\langle x_1, \ldots, x_t \rangle$ satisfying a PI of degree n, there exist $k \in \mathbb{N}$ and monomials $m_1, \ldots, m_k$ in the x_i, such that the power products $m_1^{j_1} \ldots m_k^{j_k}$ span R. This is a powerful result, which implies at once that R has Gel'fand-Kirillov dimension $\leq k$. Also, if R is algebraic, it immediately yields Kaplansky's theorem.

In case the affine PI-algebra R is also prime with center C, then the transcendence degree of C over K is some finite number d; also, R has Gel'fand-Kirillov dimension d, and all saturated chains of prime ideals have length d. In fact, embedding R into $\mathrm{Mat}_n(K)$ (for a field K) and adjoining the coefficients of the characteristic polynomials of the m_j in Shirshov's theorem produces a ring $\mathrm{T}(R)$, known as the **trace ring**, or **characteristic closure** of R. $\mathrm{T}(R)$ has a common nonzero ideal with R, but at the same time is finite as a module over its center, which is commutative affine and has a common ideal with C. This passage to commutative theory is the main technique of the proof.

The Jacobson radical of any affine PI-algebra is nilpotent, by theorems of Amitsur, Procesi, Razmyslov, Kemer, and Braun (who obtained the final result), and various formulations of the Hilbert Nullstellensatz have been proved, leading to a highly successful theory of algebraic geometry for affine PI-algebras based on work of M. Artin- W. Schelter, and which is still being developed.

Codimension theory

A **T-ideal** of the free associative algebra $\mathrm{K}\langle x \rangle$ is an ideal I which is the set of identities of a suitable PI-algebra A; for example, the identities of $\mathrm{Mat}_2(K)$ matrices is a T-ideal. In other words, I is closed under homomorphic images of $\mathrm{K}\langle x \rangle$. Thus the algebra $\mathrm{K}\langle x \rangle/I$ is a generic PI-algebra, so the T-ideals are in one-to-one correspondence with the varieties. Regev's proof of the tensor product theorem opened up a beautiful theory. Let V_n denote the subspace of the free associative algebra $\mathrm{K}\langle x \rangle$ consisting of the multilinear polynomials in $x_1, \ldots, x_n$. Given the T-ideal I of a PI-algebra A, define the **nth codimension** $\mathrm{codim}(A) = \dim_K V_n/(V_n \cap I)$. Regev (1972) proved $\mathrm{codim}_n(A)$ is exponentially bounded for all PI-algebras A, and his tensor product theorem follows at once. This raised the question of determining the precise asymptotic behavior of the codimensions as $n \to \infty$. Regev conjectured that $\mathrm{codim}_n(A)$ asymptotically approaches $cn^{\frac{1}{2}s}t^n$ for $s, t \in \mathbb{N}$ and $c \in \mathbb{Q}(\sqrt{\pi}, \sqrt{2})$; Giambruno and Zaicev (1999) verified that $\lim_{n \to \infty} \sqrt[n]{\mathrm{codim}_n(A)}$ always is an integer, whose precise value closely reflects the algebraic structure of A.

Chain conditions on T-ideals

Specht (1950) asked whether each T-ideal is finitely generated, i.e., is the consequence of a finite number of identities. For example, let E

denote the infinite dimensional exterior (Grassman) algebra. The T-ideal of the variety of E is generated by the single identity $[x_1, [x_2, x_3]]$. Culminating a great effort by many PI-theorists, Kemer (1988) answered Specht's question affirmatively in characteristic 0.

Kemer introduced powerful new methods, including passing to identities of 2-graded algebras by means of tensoring by E with the odd-even grade. Kemer's Theorem implies that R satisfies the ACC on T-ideals. Kemer also classified the T-prime T-ideals of $K\langle x \rangle$, proving any T-prime algebra satisfies the same identities as $\mathrm{Mat}_n(K)$, $\mathrm{Mat}_n(E)$, or another more technical class of algebras arising from the grade of $\mathrm{Mat}_n(E)$. Kemer also proved that any affine PI-algebra over an infinite field satisfies the same identities as a finite dimensional algebra (this is not true for finite fields); Belov proved that every affine relatively free ring can be embedded in a ring which is a Noetherian module over its center. Let A be a relatively free algebra over R and $a_1, \ldots, a_s$ are free generators; let $S(n)$ be the dimension of the vector space, generated by all words of length $\leq n$ over $a_1, \ldots, a_s$. Then the **Hilbert series** $\sum_{n=1}^{\infty} S(n)t^n$ is a rational function (a result of Belov). This is an answer to an open question, posed by C. Procesi in 1975. Note that there are representable algebras with a transcendental Hilbert series (Belov, Borisenko, and Latyshev 1997).

The picture is cloudier in characteristic p: Kemer proved finite generation of T-ideals over an infinite field with a finite number of generators, but Belov (based on work of Grishin and Shigolev), found a counterexample in general. Later, Shigolev and Grishin found additional examples.

Trace identities

Kemer's proof does not give the best explicit generation of T-ideals, and much effort was given to computing the T-ideals of generic matrices. However, Razmyslov and Procesi (1974) independently proved a different result which is, in retrospect, even more satisfying: If we consider **trace identities**, which are identities which also may use the trace function, then in characteristic 0 all trace identities are consequences of a single trace identity. (Donkin and Zubkov later extended this to characteristic p.) Razmyslov showed this can be taken to be the Hamilton-Cayley polynomial

$$x^n - \mathrm{tr}(x)x^{n-1} + \cdots \pm \det(x),$$

viewed as a trace identity when we rewrite the coefficients in terms of traces by means of Newton's formulae; as applications he gave a new derivation of the Amitsur-Levitzki theorem, and the following result,

improving the Nagata-Higman theorem: Any $\mathbb{Q}$-algebra without 1 satisfying the identity x^n is nilpotent and satisfies $x_1 \ldots x_{n^2}$. Trace identities
play a significant role in the theory of invariants.

Important areas which we have not been able to cover here are Noetherian PI-algebras, PI-algebras with involution, and actions of groups on
PI-algebras.

For further information, the reader might consult (Formanek 1991)
(for a fast and comprehensive introduction); (Jacobson 1975) (for an introduction including Amitsur's important application to division rings);
(Kemer 1991) (for a description of his results); (Rowen 1980) (for a treatment of results known through 1980), as well as (Beidar, Martindale, and
Mikhalev 1996), (Drensky 2000), (Procesi 1973), and (Razmyslov 1994).

C.38 Clifford Algebras

by Jaques Helmstetter in Grenoble, France

Definition and immediate properties

Let M be a module over a ring K (commutative, associative, with unit
element 1), and $q \colon M \to K$ a quadratic form on M. Here all rings and
algebras are associative, they contain unit elements acting as the identity on all concerned modules, and all algebra morphisms respect the
unit elements. The **Clifford algebra** $\mathrm{Cl}(M, q)$ (with unit 1_q) is characterized (up to isomorphism) by the existence of a K-linear mapping
$j \colon M \to \mathrm{Cl}(M, q)$ satisfying the equality
$$(1) \qquad\qquad j(a)^2 = q(a)1_q \quad \text{for all } a \in M,$$
and such that the elements $j(a)$ generate $\mathrm{Cl}(M, q)$ as an algebra with
unit, without more relations between them than those that are consequences of (1). The exact meaning of this last condition is the following
"universal property".

(2) *If $f \colon M \to A$ is a K-linear mapping from M into a K-algebra A
(with unit 1_A) such that $f(a)^2 = q(a)1_A$ for all $a \in M$, there exists
a unique algebra morphism $f' \colon \mathrm{Cl}(M, q) \to A$ such that $f = f' \circ j$.*

The precise construction of $\mathrm{Cl}(M, q)$ by means of a suitable quotient of the "tensor algebra" proves its existence, and also the fact that
it admits a parity grading, that is a grading over the group $\mathbb{Z}_2$; it is
the direct sum of the subalgebra $\mathrm{Cl}_0(M, q)$ of even elements and the
submodule $\mathrm{Cl}_1(M, q)$ of odd elements; an element x is said to be **homogeneous** if it is even or odd, and its degree ∂x is 0 or 1 according to its

parity.

From (2) we derive the existence of an involutive mapping $\rho\colon \mathrm{Cl}(M,q) \to \mathrm{Cl}(M,q)$ leaving invariant 1_q and all $j(a)$, and such that $\rho(xy) = \rho(y)\rho(x)$ for all x and y in $\mathrm{Cl}(M,q)$; this "involution" ρ (cf. Section C.39) is called *reversion*.

When q completely vanishes, $\mathrm{Cl}(M,q)$ is the *exterior algebra* $\Lambda(M)$.

Clifford algebras are automatically involved whenever relations similar to (1) appear in some algebra; this happens for instance when Dirac equations are introduced. But Clifford's motivations in 1878 were very far from such concerns; beside the fields $\mathbb{R}$ and $\mathbb{C}$ of real and complex numbers, he knew the division ring $\mathbb{H}$ of real quaternions, and he wished to construct more general algebras (not necessarily division rings). Indeed, $\mathbb{C}$ and $\mathbb{H}$ can be obtained either as Clifford algebras $\mathrm{Cl}(M,q)$ with q negative definite and M of dimension 1 and 2 over $\mathbb{R}$, or as even subalgebras $\mathrm{Cl}_0(M,q)$ with q (positive or negative) definite and M of dimension 2 and 3. Subsequent developments in this direction are presented beneath after (5).

For other people, Clifford algebras afford a geometrical calculus involving both exterior and inner multiplications; this is suggested in (4a,b). When Lipschitz discovered again these algebras in 1880 (independently of Clifford), he used them to represent orthogonal transformations (as explained in (7)). For many problems concerning orthogonal groups, Clifford algebras are still the best tool; they afford an elementary construction of spinorial groups and spin representations (see beneath).

Scalar products

A *scalar product* for q is any bilinear form $\beta\colon M \times M \to K$ such that $q(a) = \beta(a,a)$ for every $a \in M$. Scalar products always exist when M is a projective module; besides when the mapping $a \mapsto 2a$ is bijective from M onto M, there exists a unique symmetric scalar product. When a scalar product β exists, the Clifford multiplication can be carried onto the module $\Lambda(M)$ which thus becomes an algebra $\Lambda(M,\beta)$ equipped with two multiplications (the Clifford multiplication noted xy and the exterior one $x \wedge y$) related by the following equalities (for all a, b_1, $\ldots$, $b_k \in M$):

(4a) $\qquad a(b_1 \wedge \ldots \wedge b_k) = a \wedge b_1 \wedge \ldots \wedge b_k + \sum_{i=1}^{k}(-1)^{i-1} b_1 \wedge \ldots \beta(a,b_i) \ldots \wedge b_k;$

(4b) $\qquad (b_1 \wedge \ldots \wedge b_k)a = b_1 \wedge \ldots \wedge b_k \wedge a + \sum_{i=1}^{k}(-1)^{k-i} b_1 \wedge$

... $\beta(b_i, a) \ldots \wedge b_k$.

The equality $ab = a \wedge b + \beta(a, b)$ obviously implies $a^2 = q(a)$.

The existence of $\Lambda(M, \beta)$ implies that the algebra morphism $K \to$ $\mathrm{Cl}(M, q)$ and the mapping $j \colon M \to \mathrm{Cl}(M, q)$ are injective (whereas there are not always injective when no scalar products exist); therefore we may identify 1_q with 1, and $j(a)$ with a.

Clifford algebras of quadratic spaces

Many authors say that (M, q) is a **quadratic space** if M is a finitely generated projective module, and q a nondegenerate quadratic form that determines an isomorphism $M \to M^*$.

(5) *The Clifford algebra of a quadratic space (M, q) is a graded Azumaya algebra.*

From now on, we assume that K is a field and that M has dimension $m > 0$ over K. Then (5) means that $\mathrm{Cl}(M, q)$ is a **graded central simple algebra** ; in other words, all homogeneous $x \in \mathrm{Cl}(M, q)$ such that $xy = (-1)^{\partial x \partial y} yx$ for all homogeneous y, belong to K (identified with $K1_q$), and $\mathrm{Cl}(M, q)$ contains no two-sided ideal other than 0 or $\mathrm{Cl}(M, q)$ that is the direct sum of its intersections with $\mathrm{Cl}_0(M, q)$ and $\mathrm{Cl}_1(M, q)$.

Let us suppose that K is the field $\mathbb{R}$; let r (resp. s) be the maximal dimension of the subspaces of M on which q is positive definite (resp. negative definite); thus $r + s = m$ and $r - s$ is the signature of q; then $\mathrm{Cl}(M, q)$ and $\mathrm{Cl}_0(M, q)$ are respectively isomorphic to matrix algebras $\mathrm{Mat}_{n_1} K_1$ and $\mathrm{Mat}_{n_0} K_0$ over rings K_1 and K_0 belonging to the set $\{\mathbb{R}^2, \mathbb{R}, \mathbb{C}, \mathbb{H}, \mathbb{H}^2\}$; of course, when K_1 and K_0 are known, n_1 and n_0 are also known, since the dimensions of $\mathrm{Cl}(M, q)$ and $\mathrm{Cl}_0(M, q)$ are respectively 2^m and 2^{m-1}.

(6) *These rings K_1 and K_0 only depend on the value of $(r - s)$ modulo 8, according to this table:*

$(r - s) \bmod 8$	0	1	2	3	4	5	6	7
K_1	$\mathbb{R}$	$\mathbb{R}^2$	$\mathbb{R}$	$\mathbb{C}$	$\mathbb{H}$	$\mathbb{H}^2$	$\mathbb{H}$	$\mathbb{C}$
K_0	$\mathbb{R}^2$	$\mathbb{R}$	$\mathbb{C}$	$\mathbb{H}$	$\mathbb{H}^2$	$\mathbb{H}$	$\mathbb{C}$	$\mathbb{R}$

Lipschitz groups and orthogonal groups

Let (M, q) still be a quadratic space of dimension m over a field K. Each homogeneous and invertible element x of $\mathrm{Cl}(M, q)$ determines a **twisted inner automorphism** J_x of $\mathrm{Cl}(M, q)$:

$$J_x(y) = (-1)^{\partial x \partial y} x y x^{-1}.$$

The **Lipschitz group** $\mathrm{GLip}(M, q)$ (also called the **Clifford group**, despite of Lipschitz's paternity) is the group of all invertible homogeneous x such that $J_x(M) = M$. This equality implies that the restriction of J_x to M is an *orthogonal transformation* g_x of (M, q):

$$q(g_x(a)) = (xax^{-1})^2 = xa^2x^{-1} = q(a).$$

(7) *The mapping $x \mapsto g_x$ is a surjective group morphism from* $\mathrm{GLip}(M, q)$ *onto the group* $\mathrm{GO}(M, q)$ *of all orthogonal transformations of (M, q); its kernel is the group K_* of nonzero elements of K.*

The reversion ρ (see the first subsection) plays an important role here, since $x\rho(x)$ belongs to K for every $x \in \mathrm{GLip}(M, q)$. When K is the field $\mathbb{R}$, the subgroup of all $x \in \mathrm{GLip}(M, q)$ such that $x\rho(x) = \pm 1$ is called the **spinorial group**; when $r \geq 2$ or $s \geq 2$ (consequently whenever $m \geq 3$), the spinorial group is a 2-sheet covering group over $\mathrm{GO}(M, q)$, because 1 and -1 are both in its neutral connected component; the number of its connected components is 2 if q is definite, but 4 if $rs \neq 0$.

Spin representations

When (M, q) is a quadratic space over a field, the **spin representations** of $\mathrm{Cl}(M, q)$ are its minimal faithful representations; often they are irreducible. Important informations about them can be derived from (5) and (6) above. Here are the two most classical examples of spin representations.

First suppose that M is the direct sum $V^* \oplus V$ of a vector space V and the dual one, and that $q(\ell, v) = \ell(v)$, so that (M, q) is a hyperbolic (or neutral) quadratic space. There is an isomorphism Φ from $\mathrm{Cl}(M, q)$ onto the algebra $\mathrm{End}(\Lambda(V))$ of all endomorphisms of the vector space $\Lambda(V)$; the elements of V operate in $\Lambda(V)$ by exterior multiplication, whereas the elements of V^* operate by inner multiplication. According to the parity of x in $\mathrm{Cl}(M, q)$, $\Phi(x)$ leaves the even and odd components $\Lambda_0(V)$ and $\Lambda_1(V)$ invariant, or permutes them.

Secondly suppose that M is $(V^* \oplus V) \oplus K$ and that $q(\ell, v; \lambda) = \ell(v) + \lambda^2$; thus (M, q) is a quadratic space if K has not characteristic 2. There is an isomorphism Ψ from $\mathrm{Cl}(M, q)$ onto $(\mathrm{End}(\Lambda(V)))^2$ (whence two so-called **half-spin representations**); according to the parity of x in $\mathrm{Cl}(M, q)$, $\Psi(x)$ looks like (f, f) or $(f, -f)$ for some $f \in \mathrm{End}(\Lambda(V))$. The elements $(0, v, 0)$ or $(\ell, 0, 0)$ still operate by exterior or inner multiplication on $\Lambda(V) \times 0$, whereas $(0, 0, 1)$ operates by $w \mapsto (-1)^{\partial w} w$; put the opposite operation on $0 \times \Lambda(V)$ (the other half-spin space) since odd elements operate like $(f, -f)$.

For more information, see (Bokut' and Collins 1980), (Chevalley 1954) (a classical, but sometimes outdated book on Clifford algebras over fields), (Bourbaki 1959), (Porteous 1995) (an essential book for Clifford algebras over $\mathbb{R}$ or $\mathbb{C}$, and periodicity properties), and (Micali and Revoy 1979) as well as (Knus 1991).

C.39 Rings with Involution

by Henry E. Heatherly in Lafayette, LA, USA

An **involution** * on a ring is an additive endomorphism on R such that $(r^*)^* = r$ and $(rs)^* = s^* r^*$, for each $r, s \in R$. Early motivation for studying rings with involution (also called ***-rings**) came from rings of operators (see Section C.40). If $\mathcal{B}(H)$ is the set of all bounded linear operators on a (real or complex) Hilbert space H, then each $\Phi \in \mathcal{B}(H)$ has an adjoint $\mathrm{adj}\,\Phi$ in $\mathcal{B}(H)$, and $\Phi \mapsto \mathrm{adj}\,\Phi$ is an involution on the ring $\mathcal{B}(H)$. For the ring $\mathrm{Mat}_n R$ of $n \times n$ matrices over ring R, the transpose operation yields an involution on $\mathrm{Mat}_n R$. The point-of-view engendered by these and similar examples is exemplified in the monographs by Kaplansky (1968) and Berberian (1972), and even influenced Herstein (1976), which gave the state-of-the-art that time.

Throughout, R will always denote a *-ring. Several important subsets of R are suggested for study from the rings of operators and matrix ring origins. Let $S(R) := \{ x \in R \mid x = x^* \}$ and $\mathcal{K}(R) := \{ x \in R \mid x^* = -x \}$, the sets of **symmetric** and **skew-symmetric** elements, respectively, and let $K_0(R) = \{ x - x^* \mid x \in R \}$, the set of **skew trace** elements. If $x \in R$ implies $xx^* \neq 0$, whenever $x \neq 0$, then R is said to be **positive definite**. The following theorem illustrates how ring-of-operator results can be transformed into ones for *-rings.

C.39.1 Positive Definiteness Theorem (Herstein 1976, pp. 73, 78)
Let R be a prime ring. If either $S(R)$ or $K_0(R)$ contains no non-zero

nilpotents, then either R is positive definite or R is an order in $\mathrm{Mat}_2\, F$, where F is a field.

Let R have a unity. An idempotent $e \in R$ is a **projection** if $e \in S(R)$. If the right annihilator of each non-empty subset of R is a principal right ideal generated by a projection, then R is called a **Baer *-ring**. Motivated by the study of rings of operators, Kaplansky introduced this concept in 1955. He showed that AW*-algebras are exactly the Baer *-rings that are C^*-algebras (Berberian 1972, Kaplansky 1968). If the right annihilator of every singleton set in R is a principal right ideal generated by a projection, then R is called a **Rickart *-ring**. The set of all projections of a Rickart *-ring form a lattice (Berberian 1972, Ber, p. 14). The following are equivalent: (i) R is a Baer *-ring; (ii) R is a Rickart *-ring whose projections form a complete lattice; (iii) R is a Rickart *-ring in which every orthogonal set of projections has a supremum (Berberian 1972, p. 20).

Simple rings with involution have been intensely investigated. (See Herstein 1976, Chapters 4, 5 and the papers referenced there by S. A. Amitsur, J. Herstein, W. Martindale, and Lance Small.) This work is intimately connected with Lie and Jordan homomorphisms and commutativity theory (Herstein 1976, Rowen 1988a). In regard to the latter theory, if R is a simple ring with unity and if $S(R)$ is commutative, then R is at most 4-dimensional over its center. This can be considered as a launching point for the study of *-rings that satisfy a polynomial identity. Major results on this topic were achieved by S. A. Amitsur in the 1960's. For the details on this and further references, see (Herstein 1976, Chapter 5).

A substantial theory for prime and semiprime *-rings was developed by Herstein, Lanski, and Montgomery, an explication of which can be found in (Herstein 1976, Chapter 23). Symmetric and skew symmetric elements also play an important role in commutativity theorems for *-rings (Herstein 1976, Chapter 3). Semiprime rings with involution that contain a minimal left (right) ideal and Artinian *-rings have been thoroughly investigated; see (Rowen 1988a, pp. 301–303) and (Herstein 1976).

Let $Z(R)$ be the center of R. If $Z(R) \subseteq S(R)$, then * is called an **involution of the first kind**; otherwise it is called an **involution of the second kind**. The ring $\mathrm{Mat}_n(A)$, where A is a commutative ring with unity, has the transpose operation as an involution of the first kind. For any ring T, on the ring $T \oplus T^{\mathrm{op}}$ define the **exchange involution**:

$(a, b)^* = (b, a)$. If T is non-trivial, then the exchange involution is one of the second kind. The breakdown by kind of involution occurs naturally in the theory, e.g., (Herstein 1976, Section 4.1), (Rowen 1988a, p. 305). The exchange involution plays an indispensable part in certain structure theorems, e.g., (Herstein 1976, Sections 2.1, 2.2). Some other important special types of involutions are the **canonical symplectic involution** (Rowen 1988a, p .298); **involutions of symplectic type**, (Rowen 1988a, p .304); and those compatible in certain ways with norms, e.g., in Banach algebras, where the identity $\|xx^*\| = \|x\|^2$ yields C^*-algebras.

Involutions often arise in the study of skew fields, e.g., conjugation on the complex field or on the real quaternions. One of the most important results along these lines is a unitary version of the **Brauer-Cartan-Hua theorem**.

C.39.2 Theorem (Herstein 1976, Theorem 6.1.1) *A subskewfield A of a division ring R is either commutative or all of R when $uAu^{-1} \subseteq A$, for all $u \in R$ such that $uu^* = 1 = u^*u$, i.e., when u is **unitary**.*

For more on skewfields with involution, see (Herstein 1976, Chapters 3,6), (Draxl 1983, Dra, Chapter 16), (Knus, Merkurjev, Rost, and Tignol 1998).

The study of central simple algebras with involution is one of the most venerable parts of the theory, ranking with rings of operators as a motivating source. (Here *algebra* means linear associative algebra with unity.) Work on the subject in the 1930's by A. A. Albert led to the first systematic development of the theory of finite dimensional central simple algebras with involution in his 1939 monograph. One of his main motivations came from the theory of Riemann matrices and the multiplication algebras of Riemann surfaces (see Knus, Merkurjev, Rost, and Tignol 1998, pp. 67–68). Central simple algebras with involution are intimately connected with both bilinear and quadratic forms (Knus, Merkurjev, Rost, and Tignol 1998, Chapter I). Let E be a finite dimensional central simple algebra over a field F and let M be a finitely generated E-module. Then the algebra $\mathrm{End}_{E(M)}(M)$ can be used to define involutions on E (Knus, Merkurjev, Rost, and Tignol 1998, 4.2 Theorem). Analogous to the classical theory of invariants of quadratic forms, a rich theory of invariants for central simple algebras with involutions has been developed (Knus, Merkurjev, Rost, and Tignol 1998, Chapter II). This includes connections with Clifford algebras. Let E also have an involution and assume E has characteristic not two. Using $a \cdot b = \frac{1}{2}(ab + ba)$ yields a Jordan algebra on E with $S(E)$ a Jordan subalgebra. The relations between these Jordan algebras and the central simple algebra E are elaborate, (Knus, Merkurjev, Rost, and Tignol 1998, Chapter IX). An important relation between cubic Jordan algebras and cubic associative algebras was given by Jacques Tits, (Knus, Merkurjev, Rost, and Tignol 1998, Section 39).

In the late 1980's, the point-of-view of considering *-rings as objects in the category of Ring*, rings with involution, instead of in the category of rings, Ring, began to be developed and has since been under intense investigation. Richard Wiegandt played a seminal role in advocating this viewpoint. For $R \in$ Ring*, a ring ideal of R need not be an ideal in terms of the category Ring*. A similar situation occurs for many other concepts. Maximal, minimal, and prime ideals in Ring*, and simple prime, and semiprime rings in Ring*, have been thoroughly investigated in (Birkenmeier, Groenewald, and Heatherly 1997a) and (Birkenmeier and Groenewald 1997). Exemplary of this is the following

C.39.3 Theorem (Birkenmeier, Groenewald, and Heatherly 1997a)
Let R be simple in Ring. Then either (i) R is simple in Ring; or (ii) R contains a maximal ring ideal I such that $R = I \oplus I^*$, where I and $I^* = \{ x^* \mid x \in I \}$ are simple in Ring, $R^2 \neq 0$, and the only proper, non-zero ring ideals of R are I and I^*.*

More on the approach to *-rings via the category Ring* can be found in (Rowen 1988a, pp. 299-307).

C.40 Rings of Operators

by Henry E. Heatherly in Lafayette, LA, USA

The study of rings of operators began with von Neumann's work in 1929 and was further shaped by a series of papers by von Neumann and Murray in the 1930's. The original point-of-view was that a ring of operators is a unital self-adjoint subalgebra of bounded linear operators on a Hilbert space which is closed in the weak operator topology. The concept of *ring of operators* has broadened and become more abstract over time, including C*-algebras and AW*-algebras. The move to put the subject in an abstract algebraic setting was given major impetus in the 1950's by I. Kaplansky. This in turn led to several fruitful purely algebraic generalizations that have taken on lives of their own, e.g., Baer *-rings, Baer rings, and quasi-Baer rings.

Here, R will always be a ring and A will be a linear associative algebra over $\mathbb{R}$ or $\mathbb{C}$. Let S be a nonempty subset of R and define $S' = \{r \in R \mid rs = sr \ \forall s \in S\}$, the **commutant** of S in R, and let $S'' = (S')'$. Observe that S' is a subring of R and if R has unity, then $1 \in S'$. If R is an algebra, then so is S'. A ***-ring** R (i.e., a ring with involution, see Section C.39) which is an algebra as well is called a ***-algebra** if $(\lambda r)^* = \bar{\lambda} r^*$ for each $r \in R$ and each scalar λ.

Let $\mathcal{B}(H)$ be the set of all bounded linear operators on a complex Hilbert space H. With the usual operations and the supremum norm inherited from H, $\mathcal{B}(H)$ is a Banach algebra. Associated with each $\varphi \in \mathcal{B}(H)$ there is a unique $\operatorname{adj} \varphi \in \mathcal{B}(H)$, the **adjoint** of φ, such that $\langle \varphi(x) \mid y \rangle = \langle x \mid \operatorname{adj} \varphi(y) \rangle$ for each $x, y \in H$. The operation $^* : \varphi \to \operatorname{adj} \varphi$ is an involution on $\mathcal{B}(H)$.

C.40.1 Proposition *Let V be a unital *-subalgebra of $\mathcal{B}(H)$. Then the following are equivalent: (i) $V'' = V$; (ii) V is closed in the strong operator topology; (iii) V is closed in the weak operator topology.*

Von Neumann called such algebras *rings of operators*; following Dixmier they are now called **von Neumann algebras**. Von Neumann's work was part of his development of a mathematical foundational theory for the then new quantum mechanics, as well as reflecting his interest in the representation of continuous groups and in ergodic theory. (For an excellent synopsis of von Neumann's work on rings of operators, see Kadison 1958).

C.40.2 Theorem *For a complex Banach algebra G with involution * the following are equivalent: (C1) $\|x^*x\| = \|x^*\|\|x\|$ for each $x \in G$ and (C2) $\|x^*x\| = \|x\|^2$ for each $x \in G$, (Doran and Belfi 1986, Chapter III).*

It is immediate that (C2) implies: (C3) $\|x^*\| = \|x\|$ for each $x \in G$. An algebra G satisfying (C1) is called a **C^*-algebra**, a term introduced by I. E. Segal in 1947, where he used it for *-Banach subalgebras of $\mathcal{B}(H)$. In 1943, Gel'fand and Naĭmark began the study of unital complex *-Banach algebras which satisfy (C1),(C3), and (C4): for each $x \in G, 1 + x^*x$ is invertible in G. They called these algebras *normed *-rings*. The main thrust of their work was to embed such algebras in $\mathcal{B}(H)$. Gel'fand and Naĭmark also raised the question as to whether (C3) and (C4) are a consequence of the other axioms. Work in the early 1950's by Fukamiya, Kelley and Vaught, Kaplansky, and Schatz, resulted in a proof of (C4) for all unital C*-algebras. Glimm and Kadison proved condition (C3) holds in that same setting in 1960. Work continued on the matter, resulting in an elementary proof by Harris in 1972, and a very short proof by Gardner in 1984. In light of these developments the main result of Gel'fand and Naĭmark can be stated as follows:

C.40.3 Theorem *A unital C^*-algebra which satisfies (C1) is isometrically *-isomorphic to a *-Banach subalgebra of some $\mathcal{B}(H)$.*

Many C^*-algebras of interest do not have unity, for example see (Doran and Belfi 1986). The following result due to B. J. Vowden (1967) allows one to extend the Gel'fand-Naĭmark Theorem to this more general setting:

C.40.4 Theorem *A C^*-algebra without unity can be isometrically and *-isomorphically embedded in a C^*-algebra with unity.*

Gel'fand and Naĭmark obtained a result for commutative C^*-algebras that tie them intimately to another major motivating class of examples, function algebras. Let X be a locally compact Hausdorff space and let $\mathcal{C}(X)$ be the vector space of all continuous complex-valued functions on X. Under pointwise multiplication and with the involution $f^*(t) = \overline{f(t)}$, $\mathcal{C}(X)$ is a commutative *-algebra. Let $\mathcal{C}_0(X)$ be the set of all $f \in \mathcal{C}(X)$ that **vanish at infinity**, i.e., for each $\varepsilon > 0$, there is a compact set $K \subseteq X$ such that $|f(x)| < \varepsilon$ when $x \in X \setminus K$. If X is a locally compact (resp. compact) Hausdorff space, then with the supremum norm $\mathcal{C}_0(X)$ (resp. $\mathcal{C}(X)$) is a Banach *-algebra. The algebra $\mathcal{C}(X)$ has unity, but $\mathcal{C}_0(X)$ has unity if and only if X is compact, in which case $\mathcal{C}(X) = \mathcal{C}_0(X)$.

C.40.5 Theorem (Gel'fand-Naĭmark, cf. Doran and Belfi 1986, p. 22) *If G is a commutative C^*-algebra, then G is isometrically *-isomorphic to some $\mathcal{C}_0(X)$, where X is locally compact.*

Observe that when G has unity in this theorem, then G is isometrically *-isomorphic to some $\mathcal{C}(X)$, where X is compact.

A **W^*-algebra** is a C^*-algebra which is isometrically *-isomorphic to a von Neumann algebra. In the mid-1940's attempts began to abstractly characterize W^*-algebras without reference to an underlying Hilbert space of operators, and to give an algebraic characterization of W^*-algebras. Some important early results along these lines are due to C. Rickart in 1946. Let X be a nonempty subset of the *-algebra A and define $r(X) = \{a \in A \mid Xa = 0\}, l(X) = \{a \in A \mid aX = 0\}$, the right and left annihilators of X, respectively. An element $b \in A$ is said to be **symmetric** if $b^* = b$. A **projection** is a symmetric idempotent. The relation $e \preceq f$ if and only $ef = e$ yields a partial ordering on the set $\mathcal{P}(A)$ of all projections in A. The algebra A is a **Rickart *-algebra** if for each singleton set $X \subseteq A$ there exists $e \in \mathcal{P}(A)$ such that $r(X) = eA$.

C.40.6 Proposition (Berberian 1972, pp. 12, 14) *If A is a Rickart *-algebra, then A has unity and for each singleton set $X \subseteq A$, there exists $g \in \mathcal{P}(A)$ such that $l(X) = Ag$. Furthermore, $(\mathcal{P}(A), \preceq)$ is a lattice.*

A Rickart *-algebra which is also a C*-algebra is called a Rickart C*-algebra. Rickart observed that every von Neumann algebra is a Rickart C*-algebra; moreover the lattice of projections in a von Neumann algebra is complete (see Takesaki 1979, p. 290). There exist Rickart C*-algebras whose lattices of projections are not complete (Berberian 1972, p. 15).

Motivated in part by Rickart's work, in the early 1950's I. Kaplansky made the next major step in abstractly characterizing von Neumann algebras (see Kaplansky 1968). He observed that if V is a W*-algebra, then for each nonempty subset X of V, there is some $e \in \mathcal{P}(V)$ such that $r(X) = eV$. This in turn led him to the following general considerations. Let A be a *-algebra. If for each $\varnothing \neq X \subseteq A$ there exists $e \in \mathcal{P}(A)$ such that $r(X) = eA$, then A is called a **Baer *-algebra**. The terminology *Baer* came from the use by R. Baer of a closely related condition in the setting of algebras of linear transformations. It is routine to show that every Baer *-algebra is a Rickart *-algebra and for each $\varnothing \neq X \subseteq A$ there exists $f \in \mathcal{P}(A)$ such that $l(X) = Af$.

C.40.7 Theorem (Berberian 1972, p. 20) *The following are equivalent for a *-algebra A:*
(1) *A is a Baer *-algebra;*
(2) *A is a Rickart *-algebra and $(\mathcal{P}(A), \preceq)$ is a complete lattice;*
(3) *A is a Rickart *-algebra in which every orthogonal family of projections has a supremum.*

A C*-algebra which is also a Baer *-algebra is called an **AW*-algebra** (see Section C.39). Much of the theory of W*-algebras carries over to AW*-algebras. However, in 1951, J. Dixmier gave an example of a commutative AW*-algebra which is not a W*-algebra. Other examples of AW*-algebras which are not W*-algebras have been given since. Additional conditions are thus needed for an AW*-algebra to be a W*-algebra.

Technical complexity precludes their inclusion here; however see (Pederson 1979, p. 66) for details. In 1956 R. V. Kadison and S. Sakai independently obtained characterizations for a C*-algebra to be a W*-algebra.

C.40.8 Theorem (Pederson 1979, p. 67) *If the dual of a Banach space is a C*-algebra, then that algebra is also a W*-algebra.*

Let A be a unital real *-Banach algebra. Then A is a real C*-algebra if it satisfies (C2) and (C4). Unlike the complex case, (C4) does not follow from (C2), as can be seen by using $\mathbb{C}$ as a real Banach algebra with the identity function as involution and the absolute value as norm, and noting that $1 + i^*i = 0$. Some important examples of real C*-algebras are: (i) $\mathcal{B}_{\mathbb{R}}(H)$, the algebra of bounded linear operators on a real Hilbert space H; (ii) any norm-closed, self-adjoint subalgebra of $\mathcal{B}_{\mathbb{R}}(H)$; (iii) $\mathcal{C}(X)$, the continuous real-valued functions on a compact Hausdorff space X.

C.40.9 Theorem (Goodearl 1982, p. 108) *If A is a real C*-algebra then there exists a real Hilbert space H such that A is *-isomorphic to a norm-closed subalgebra of $\mathcal{B}_{\mathbb{R}}(H)$.*

A real unital Banach *-algebra B can be embedded isometrically and real *-algebra isomorphically in a complex unital *-Banach algebra in a natural way, called the **complexification** of B (Goodearl 1982, Chapter 9). Unfortunately, when B is a real C*-algebra the complexification process does not transfer condition (C2) to B. Nevertheless, the complexification process is useful in obtaining structural results for real C*-algebras.

Extensive bibliographic information on *rings of operators* can be found among (Dixmier 1977, 1981, Pederson 1979, Takesaki 1979, Doran and Belfi 1986), which altogether list on the order of 3000 papers. Also, see (Doran 1994).

C.41 The Gel'fand-Kirillov Dimension

by Vesselin Drensky in Sofia, Bulgaria

Let R be an associative algebra generated by a finite set of elements $\{r_1, \ldots, r_m\}$ over a field $\Bbbk$ and let V^n be the vector subspace of R spanned by all products $r_{i_1} \ldots r_{i_n}$ of length n of the generators $\{r_1, \ldots, r_m\}$ of R (we assume that $V^0 = \Bbbk$). The **growth function** of R with respect to the generating space $V = V^1$ is the function of a nonnegative integer argument defined by

$$g_V(n) = \dim_{\Bbbk}(V^0 + V^1 + \cdots + V^n), \quad n = 0, 1, 2, \ldots.$$

Let Φ be the class of all functions $f: \mathbb{N}_0 \to \mathbb{R}$ which are eventually monotone increasing and positive valued (i.e., for each $f \in \Phi$ there exists an $n_0 \in \mathbb{N}_0$ such that $f(n_2) \geq f(n_1) \geq f(n_0) > 0$ for all $n_2 \geq n_1 \geq n_0$). One defines a partial ordering in Φ assuming that $f \preceq g$ for $f, g \in \Phi$ if and only if there exist positive integers a and p such that for all sufficiently large n the inequality $f(n) \leq ag(pn)$ holds. The equivalence relation in Φ defined by $f \sim g$ for $f, g \in \Phi$ if and only if $f \preceq g$ and $g \preceq f$ removes the dependence of the growth function of the algebra R on the generating subspace V. The equivalence class $\mathcal{G}(f) = \{g \in \Phi \mid f \sim g\}$ is called the **growth** of f. In particular, $\mathcal{G}(R) = \mathcal{G}(g_V(n))$ is called the **growth** of the algebra R. The algebra has **polynomial growth** if $\mathcal{G}(R) = \mathcal{G}(n^k)$ for some nonnegative integer k; it has **exponential growth** if $\mathcal{G}(R) = \mathcal{G}(e^n)$. Finally, R has **sub-exponential** or **intermediate growth** if $\mathcal{G}(R) \npreceq \mathcal{G}(n^k)$ for any positive k and $\mathcal{G}(R) \prec \mathcal{G}(e^n)$.

C.41.1 Definition The **Gel'fand-Kirillov dimension** (or **GKdim**) of the finitely generated algebra R with respect to the finite dimensional generating space V is defined by

$$\mathrm{GKdim}(R) = \limsup_{n\to\infty}(\log_n g_V(n)) = \limsup_{n\to\infty} \frac{\log g_V(n)}{\log n}.$$

If the algebra R is not finitely generated, then

$$\mathrm{GKdim}(R) = \limsup\{\mathrm{GKdim}(S) \mid S \text{ finitely generated subalgebra of } R\}.$$

The GKdim does not depend on the choice of the set of generators of the algebra R and, when R is finitely generated,

$$\mathrm{GKdim}(R) = \inf\{\alpha \in \mathbb{R} \mid \mathcal{G}(R) \prec \mathcal{G}(n^\alpha)\}.$$

C.41.2 Examples
(1) The value of the growth function $g(n)$ of the polynomial algebra $\Bbbk[x_1, \ldots, x_m]$ with respect to the natural set of generators $\{x_1, \ldots, x_m\}$ is equal to the number of monomials of degree $\leq n$ in m commuting variables, $g(n) = (n+m)(n+m-1)\ldots(n+1)/m!$. Hence, $\mathcal{G}(\Bbbk[x_1, \ldots, x_m]) = \mathcal{G}(n^m)$ and $\mathrm{GKdim}(\Bbbk[x_1, \ldots, x_m]) = m$.

(2) The growth of the free associative algebra $\Bbbk\langle x_1, \ldots, x_m \rangle$, $m \geq 2$, with respect to the set of generators $\{x_1, \ldots, x_m\}$ is equal to the number of monomials of degree $\leq n$ in m noncommuting variables,

$$g(n) = 1 + m + m^2 + \cdots + m^n = \frac{m^{n+1} - 1}{m - 1},$$

$\mathcal{G}(\Bbbk\langle x_1, \ldots, x_m \rangle) = \mathcal{G}(m^n) = \mathcal{G}(e^n)$, i.e., the growth of R is exponential and $\mathrm{GKdim}(\Bbbk\langle x_1, \ldots, x_m \rangle) = \infty$.

(3) The simplest example of an algebra with intermediate growth was discovered by Martha Smith: Let G be the Lie algebra with basis $\{x, y_1, y_2, y_3, \ldots\}$ and multiplication given by $[y_i, x] = y_{i+1}$, $[y_i, y_j] = 0$, for all $i, j \geq 1$. Then G is generated by x, y_1. The value $g(n)$ of the growth function of the universal enveloping algebra $U(G)$ of G is equal to the coefficient of t^n in the expansion of the infinite product

$$\frac{1}{(1 - t)^2} \prod_{k \geq 1} \frac{1}{1 - t^k},$$

which is of intermediate growth. Later this result was generalized:

C.41.3 Theorem (Lichtman) *The universal enveloping algebra (Section C.49) of any finitely generated infinite dimensional Lie algebra which has a solvable ideal of finite codimension is of intermediate growth.*

C.41.4 Theorem

(1) *For finitely generated associative algebras, $\mathrm{GKdim}(R) = 0$ if and only if the algebra R is finite dimensional; for infinitely dimensional algebras, $\mathrm{GKdim}(R) \geq 1$. If the algebra R is not finitely generated, then $\mathrm{GKdim}(R) = 0$ if and only if every finitely generated subalgebra of R is finite dimensional.*

(2) (**Bergman Gap Theorem**) *There is no associative algebra R with $1 < \mathrm{GKdim}(R) < 2$.*

(3) (Borho and Kraft) *For any real $\alpha \geq 2$ there exists an algebra R such that $\mathrm{GKdim}(R) = \alpha$.*

The interest in the GKdim was inspired by Gel'fand and Kirillov (1966) in relation with their conjecture that the universal enveloping algebra of a finite dimensional algebraic Lie algebra has a division algebra of fractions which is isomorphic to the quotient division algebra D_n of a suitable Weyl algebra. Although, by a result of Makar-Limanov, D_n contains a free subalgebra and $\mathrm{GKdim}(D_n) = \infty$, using a variant of GKdim, Gel'fand and Kirillov showed that $D_m \cong D_n$ implies

$m = n$. (Alev, Ooms and Van den Bergh (1996) found a series of counterexamples to the Gel'fand-Kirillov conjecture.) Another source is the
corresponding notion of growth of groups introduced by Milnor (1968)
(which coincides with the growth in our sense of the group algebra) and
his result that the growth of a Riemannian manifold is related to its
curvature.

C.41.5 Theorem
(1) *For commutative algebras, the GKdim is equal to the transcendence
degree of R (i.e., to the maximal number of algebraically independent
elements).*
(2) *If S is a finitely generated subalgebra of the center of R and R is a
finitely generated S-module, then* $\mathrm{GKdim}(R) = \mathrm{GKdim}(S)$.

GKdim behaves sufficiently well under standard ring theoretic operations such as forming factor rings, skew polynomial rings, tensor products, and localizations.

C.41.6 Proposition
(1) If I is an ideal of R and S is a subalgebra of R, then $\mathrm{GKdim}(R/I) \leq
\mathrm{GKdim}(R)$ and $\mathrm{GKdim}(S) \leq \mathrm{GKdim}(R)$.
(2) For any two algebras R_1 and R_2,

$$\max\{\mathrm{GKdim}(R_1), \mathrm{GKdim}(R_2)\} \leq \mathrm{GKdim}(R_1 \otimes_k R_2)$$
$$\leq \mathrm{GKdim}(R_1) + \mathrm{GKdim}(R_2)$$

(3) (Warfield) The inequalities for the tensor products cannot be improved: For any $\alpha \geq \beta \geq 2$, there exist algebras R_1 and R_2 with
$\mathrm{GKdim}(R_1) = \alpha$, $\mathrm{GKdim}(R_2) = \beta$ such that $\mathrm{GKdim}(R_1 \otimes_k R_2) =
\alpha + 2$.

GKdim is an important tool in noncommutative ring theory, especially
in the study of certain classes of algebras sufficiently close to commutative algebras. (In particular, it shares many formal properties with
the Krull dimension in the sense of Gabriel and Rentscher.) Information about the growth of the algebra frequently has structural consequences. For example, a partial case of a theorem of Jategaonkar gives:
If $\mathrm{GKdim}(R) < \infty$ and R has no zero divisors, then it satisfies the Ore
condition.

C.41.7 Theorem (Small-Stafford-Warfield) *If the GKdim of the
finitely generated algebra R is equal to 1, then R satisfies a polynomial
identity.*

C.41.8 Theorem (Gromov) *A finitely generated group G is of polynomially bounded growth (e.g., the GKdim of its group algebra $\Bbbk G$ is finite) if and only if G has a nilpotent normal subgroup of finite index.*

It is difficult to judge how good is the behavior of GKdim for PI-algebras. For example, the algebras of Borho and Kraft with arbitrary non-integer GKdim all satisfy polynomial identities. On the other hand:

C.41.9 Theorem

(1) (Berele) *The GKdim of a finitely generated PI-algebra is finite.*
(2) (Markov) *For relatively free (or generic) PI-algebras the GKdim is integer.*
(3) *If R is a Noetherian PI-algebra, then $\mathrm{GKdim}(R)$ is an integer (Markov) and $\mathrm{GKdim}(R \otimes_\Bbbk S) = \mathrm{GKdim}(R) + \mathrm{GKdim}(S)$ for any $\Bbbk$-algebra S (Lorenz and Small).*

One has introduced, in a natural way, the notion of GKdim for modules and for other classes of (nonassocative) algebras. Considering Lie algebras, $\mathrm{GKdim}(G) = 0$ if and only if G is finite dimensional; otherwise $\mathrm{GKdim}(G) \geq 1$. Examples of Petrogradsky show that there exist Lie algebras with $\mathrm{GKdim}(G) = \alpha$ for any real $\alpha \geq 1$.

In order to measure the growth of algebras with superpolynomial growth, Borho and Kraft studied the notion of superdimension defined by $\limsup_{n \to \infty} \log \log g_V(n)/\log n$ where $g_V(n)$ is the growth function of the finitely generated algebra R with respect to the finite dimensional generating subspace V (but does not depend on the choice of V). Recently, Petrogradsky has developed the theory of higher level GKdim of associative and Lie algebras. It takes into account sub-exponential growths of algebras greater than any function $\exp(n^\beta)$, $\beta < 1$. The dimension of first level is equal to the dimension as a vector space over $\Bbbk$, this of second level coincides with the usual GKdim and that of level 3 (up to normalization) is the superdimension of Borho and Kraft. The dimensions of higher level are completely new and work especially successfully for solvable Lie algebras and their universal enveloping algebras.

For further reading on GKdim see the book by Krause and Lenagan (2000), the seminal paper by Borho and Kraft (1976), and the surveys by Ufnarovskij (1995), Belov et al. (1997), and Drensky (1998).

C.42 Coalgebras

by Vyacheslav A. Artamonov in Moscow, Russia

The usual definition of a monoid A can be presented in the following equivalent form. Define the map $m\colon A \times A \to A$ by setting $m(x,y) = xy$. Then the associativity of multiplication means that $m(x,m(y,z)) = m(m(x,y),z)$ for all $x,y,z \in A$, or equivalently that the following diagram is commutative

$$
\begin{array}{ccc}
A \times A \times A & \xrightarrow{\ 1 \times m\ } & A \times A \\
{\scriptstyle m \times 1}\big\downarrow & & \big\downarrow{\scriptstyle m} \\
A \times A & \xrightarrow{\ \ m\ \ } & A
\end{array}
\tag{1}
$$

Similarly, the identity element can be identified with a map $u\colon I \to A$ such that $u(I) = 1 \in A$ and the diagram

$$
\begin{array}{ccccc}
A & \xrightarrow{\ \simeq\ } & I \times A & \xrightarrow{\ u \times 1\ } & A \times A \\
{\scriptstyle \simeq}\big\downarrow & & & & \big\downarrow{\scriptstyle m} \\
A \times I & \xrightarrow{\ 1 \times u\ } & A \times A & \xrightarrow{\ \ m\ \ } & A
\end{array}
\tag{2}
$$

is commutative with identical resulting morphisms. If A is a group then there are maps $S\colon A \to A, \Delta\colon A \to A \times A, \varepsilon\colon A \to I$ such that $S(x) = x^{-1}$, $\Delta(x) = x \times x$, $\varepsilon(x) = I$. These maps satisfy the following equalities

$$m(1 \times S)\Delta(x) = m(S \times 1)\Delta(x) = u\varepsilon(x), \quad m(u\varepsilon \times 1)\Delta = m(1 \times u\varepsilon)\Delta.$$

Similarly, if we take the category $R-\mathrm{mod}$ of R-modules with tensor products $\otimes$, where R is a commutative associative base ring with a unit element, then an associative R-algebra A with a unit element 1 can be actually defined in the same way using the language of diagrams. In fact, multiplication in A can be viewed as a morphism $\mu\colon A \otimes A \to A$ in $R-\mathrm{mod}$ for which the diagram (1) is commutative where $\times$ and m are replaced by $\otimes$ and μ, respectively. The unit element can be identified with the morphism $u\colon R \to A$ in $R-\mathrm{mod}$ for which the diagram (2) is commutative, where $\times$, m, I, and m are replaced by $\otimes$, μ, R, respectively. It means that monoids in the category $R-\mathrm{mod}$ are precisely the associative algebras. These observations lead to the following

C.42.1 Definition Let C be a category whose objects form a monoid with respect to a product $\otimes$. The unit object of C with respect to $\otimes$ will denoted by I. An *algebra* in the category C is the pair m, u of morphisms $m\colon A \otimes A \to A$, $u\colon I \to A$ satisfying (1), (2) where $\times$ is replaced by $\otimes$.

In the categorical approach, it makes sense to consider dual objects. For example, an algebraic manifold V in algebraic geometry is a dual object to the coordinate algebra of functions on V, a compact abelian group is dual to the group of its characters, a monoid (group) and the R-module of R-valued functions on it, etc. Usually, instead of considering dual objects, we look at the dual category C^* having the same objects as C but all morphism have opposite directions. Thus we can quote the formal

C.42.2 Definition A *coalgebra* C over a commutative associative base ring R with a unit element is a R-module equipped with R-module morphisms of *comultiplication* $\Delta\colon C \to C \otimes C$ and of *counit* $\varepsilon\colon C \to R$, where $\otimes = \otimes_R$. It is assumed that comultiplication is coassociative that is

$$(\Delta \otimes 1_C)\Delta = (1_C \otimes \Delta)\Delta\colon C \to C \otimes C \otimes C,$$

where 1_C is the identity map on C. Moreover, it is required that

$$
\begin{aligned}
(\varepsilon \otimes 1_C)\Delta(a) &= 1 \otimes a \in k \otimes_k C \simeq C, \\
(1_C \otimes \varepsilon)\Delta(a) &= a \otimes 1 \in C \otimes_k k \simeq C
\end{aligned}
\tag{3}
$$

hold for every element $a \in C$.

The following *sigma-notation* has been suggested by Sweedler (1969). If $a \in C$, then $\Delta(x) = \sum_x x_{(1)} \otimes x_{(2)} \in C \otimes C$. Now (3) can, for $x \in C$, be written in the form $x = \sum_x \varepsilon(x_{(1)})x_{(2)} = \sum_x x_{(1)}\varepsilon(x_{(2)})$, for any element $x \in C$.

For example, let M_n be the n^2-dimensional vector space with the base $X_{ij}, 1 \le i, j \le n$. Then M_n is a coalgebra with comultiplication and counit

$$\Delta(X_{ij}) = \sum_{t=1}^{n} X_{it} \otimes X_{tj}, \quad \varepsilon(X_{ij}) = \delta_{ij}.$$

A coalgebra C is *cocommutative* if $\tau\Delta = \Delta$, where $\tau(x \otimes y) = y \otimes x$ for all $x, y \in C$.

Subcoalgebras and coalgebra morphisms are defined as expected. If C is a coalgebra then $C^* = \mathrm{Hom}(C, R)$ is an associative **dual** R-algebra with respect to the **convolution** $f * g$ as multiplication, where

$$(f * g)(x) = \sum_x f(x_{(1)}) g(x_{(2)})$$

for each element $x \in C$. The unit element of C^* is the counit morphism ε. The algebra C^* is commutative if and only if C is cocommutative. If $\varphi \colon C \to C'$ is a coalgebra morphism, then $\varphi^* \colon (C')^* \to C^*$ is an algebra morphism.

For example, $M_n^* \simeq \mathrm{Mat}_{n \times n} R$—the algebra of $(n \times n)$-matrices. In fact, an element $\varphi \in M_n^*$ is associated with the matrix whose (i, j)-entry is $\varphi(X_{ij})$.

If G is a finite monoid then the R-module $R[G]$ of all maps $G \to R$ is a coalgebra. Indeed $R[G] \otimes R[G] \simeq R[G \times G]$ and therefore we can define comultiplication Δ and counit u by setting

$$\Delta \colon R[G] \to R[G] \otimes R[G],$$
$$(\Delta f)(x \times y) = f(xy) \in R, \quad \varepsilon(f) = f(1). \tag{4}$$

Similarly, if G is an algebraic group defined over a field K then the algebra of regular functions $K[G]$ on G has the same structure of a coalgebra defined by (4).

An element $g \in C$ is a **group-like** element if $\Delta(g) = g \otimes g$, $\varepsilon(g) = 1$. The set of all group-like elements in C is denoted by $G(C)$. An element $p \in C$ is a **primitive element** if $\Delta(p) = p \otimes 1 + 1 \otimes p$. It follows that $\varepsilon(p) = 0$. The set of all primitive elements in C is denoted by $P(C)$.

A coalgebra A is a **bialgebra** if A is an associative R-algebra and the following equivalent conditions are satisfied:

(1) the maps Δ, ε are algebra homomorphisms;
(2) the R-morphisms of multiplication $m \colon A \otimes A \to A$ and the natural embedding $u \colon R \to A, u(\alpha) = \alpha 1 \in A$ are coalgebra morphisms.

The set $G(A)$ of all group-like elements in a bialgebra A is a multiplicative monoid in A. The set $P(A)$ of all primitive elements in A is a (restricted) Lie algebra with respect to Lie multiplication $[x, y] = xy - yx$ (and $x^{[p]} = x^p$). For example, a monoid algebra RS of a monoid S is a bialgebra with respect to comultiplication and counit $\Delta(s) = s \otimes s$, $\varepsilon(s) = 1$, for any $s \in S$. In particular, $S = G(RS)$.

A morphism of bialgebras is a morphism of algebras which is simultaneously a morphism of coalgebras. Thus a bialgebra is a coalgebra in the category of associative algebras and an algebra in the category of coalgebras.

Let R be a Noetherian ring and A a bialgebra which is a flat R-module. Then the **dual bialgebra** A^o is the R-module of all k-module morphisms $A \to R$ having a kernel I such that A/I is a finitely generated R-module. Then A^o is bialgebra. In fact, A^o is a subalgebra of A^* defined in terms of A as a coalgebra, see above. Comultiplication and a counit in A^o are introduced as in (4). This definition is correct since $\Delta(A^0) \subseteq A^0 \otimes A^0$. It is easy to observe that $G(A^*)$ is the set of all algebra homomorphisms $A \to R$ and there is an isomorphism of (restricted) Lie algebras $P(H^o) \simeq \ker \varepsilon / (\ker \varepsilon)^2$. If R is a field and A has finite dimension over R then $A^o = A^*$ and the bialgebras A^{**} and A are isomorphic.

Categories of modules over a bialgebra A are tensor categories: If M, N are left A-modules then $M \otimes N$ is again a left A-module. In fact, if $a \in A$, $m \in M$, $n \in N$, then we put $a(m \otimes n) = \Delta(a)(m \otimes n)$.

C.43 Hopf Algebras

by Vyacheslav A. Artamonov in Moscow, Russia

A bialgebra H (see Section C.42) over a commutative ring R with a comultiplication Δ and a counit ε is a **Hopf algebra** if there is a R-module endomorphism S in H, called **antipode**, such that for all $h \in H$,

$$\sum_h S(h_{(1)})h_{(2)} = \sum_h h_{(1)}S(h_{(2)}) = \varepsilon(h).$$

An antipode S satisfies the identities

$$S(ab) = S(b)S(a), \quad \Delta S = S\tau\Delta, \quad \varepsilon S = \varepsilon,$$

where $\tau: H \otimes H \to H \otimes H$, $\tau(h_1 \otimes h_2) = h_2 \otimes h_1$.

According to Definition C.42.1, Hopf algebras are groups in the category of associative algebras or a functor from a dual category to the category of groups to the category of associative algebras.

An antipode S in a finite dimensional Hopf algebra over a field has a finite order, which is equal either to 1 or to any even number.

If G is a group, then a group algebra RG is a Hopf algebra in which $\Delta(g) = g \otimes g$, $S(g) = g^{-1}$, $\varepsilon(g) = 1$, for any $g \in G$. If G is an algebraic (analytic) group and $R[G]$ is the algebra of regular (analytic) functions of G, then $R[G]$ is a Hopf algebra with respect to

$$\Delta(f)(g_1, g_2) = f(g_1 g_2), \quad \varepsilon(f)(g) = f(1), \quad S(f)(g) = f(g^{-1}),$$

for all $g_1, g_2, g \in G$. Note that $R[G] \otimes R[G] \simeq R[G \times G]$. Another series of examples of Hopf algebras are universal (restricted) enveloping algebras $U(L)$ for (restricted) Lie algebras L, where

$$\Delta(x) = x \otimes 1 + 1 \otimes x, \quad \varepsilon(x) = 0, \quad S(x) = -x,$$

for all $x \in L$. In this case, the Lie algebra $P(H^0)$ is the tangent Lie algebra at 1 of the group G. Other examples of Hopf algebra can be found in (Artamonov 1991, Montgomery 1993).

The set $G(H)$ of all group-like elements in a Hopf algebra is always a multiplicative group.

Suppose that $H = K[G]$ is the algebra of regular functions on an algebraic group G defined over a field K. Each element $g \in G$ determines a map $\tilde{g}: K[G] \to K$, namely $\tilde{g}(f) = f(g)$. It easy to see that $\tilde{g}$ is an algebra homomorphism. Indeed, $\tilde{g}(fh) = (fh)(g) = f(g)h(g) = [\tilde{g}(f)][\tilde{g}(h)]$. Thus G is embedded into the set of all algebra homomorphisms $K[G] \to K$. From algebraic geometry, we know that the set of all algebra homomorphisms $K[G] \to K$ coincides with the group of all K-points of G. From our point of view, this group coincides with the group $G(K[G])$ of all group-like elements.

The Lie algebra $P(K[G]^o) \simeq \ker \varepsilon / (\ker \varepsilon)^2$ is precisely the Lie algebra defined on the tangent space of G at the unit element (Serre 1992a, Part 2, Chapter 3, §8). It means that the theory of algebraic and Lie groups can be viewed as a part of the theory of Hopf algebras.

An element h of a Hopf algebra H is a **left (right) integral** if $xh = \varepsilon(x)h$, $(hx = h\varepsilon(x)$, respectively) for any $x \in H$. If R is a field, then the space $\int_l(H)$, $(\int_r(H)$, respectively) of all left (right) integrals in H has dimension at most 1. For example, if H is a group algebra KG of a finite group G, then $\int_l(H) = \int_r(H) = \sum_{g \in G} g \in KG$. If H has finite dimension then $\dim \int_l(H) = \dim \int_r(H) = 1$. The following result generalizes Maschke's theorem: a finite dimensional Hopf algebra H over a field is semisimple if and only if $\varepsilon(\int_r(H)) = R$ (or, $\varepsilon(\int_l(H)) = R$). If H is a finite dimensional Hopf algebra over a field with a Hopf subalgebra B, then H is a free left (right) B-module. In particular, $\dim B$ divides $\dim H$.

A Hopf algebra H over a field K is **pointed** if any simple subcoalgebra in H has dimension 1. Observe that a Hopf algebra is pointed if and only if each maximal ideal in the dual algebra H^* has codimension 1. If the field R is algebraically closed, then any finite dimensional cocommutative Hopf algebra is pointed. A pointed Hopf algebra H has a chain of subcoalgebras $C_0 \subset C_1 \subset \cdots \subset C_n \subset$, where C_0 is the group algebra $RG(H)$. Moreover, if $n \geq 1$ and $c \in C_n$, then c is a sum of elements $c_{g,h}$, where $g, h \in G(H)$, and

$$\Delta c_{g,h} = c_{g,h} \otimes g + h \otimes c_{g,h} + w_{g,h}, \qquad w_{g,h} \in C_{n-1} \otimes C_{n-1},$$

(Mishchenko 1990, p. 68). Every group algebra and universal (restricted) enveloping algebra is an example of a pointed Hopf algebra.

If we study an associative algebra A, then we can consider some derived structures on A, namely the automorphism group $\mathrm{Aut}\, A$ and the Lie algebra $\mathrm{Der}\, A$ of derivations of A. Then $\mathrm{Aut}\, A$ acts on A and on $\mathrm{Der}\, A$. Also, $\mathrm{Der}\, A$ acts on A and all these three actions are compatible. This situation is unified in the following notions. A Hopf algebra H **measures** an associative algebra A with a unit element (or H has a **weak action** on A) if there is given a map of R-modules $H \otimes A \to A$, $h \otimes a \mapsto h \cdot a$, such that

$$h(ab) = \sum_h (h_{(1)} \cdot a)(h_{(2)} \cdot b), \quad h \cdot 1 = \varepsilon(h),$$

for all $a, b \in A$, $h \in H$. A **2-cocycle** is a morphism $\sigma : H \otimes H \to A$ of R-modules such that $\sigma(h \otimes 1) = \sigma(1 \otimes h) = \varepsilon(h)$, for all $h \in H$, and

$$\sum_{h,h',h''} \left[h_{(1)} \cdot \sigma \left(h'_{(1)} \otimes h''_{(1)} \right) \right] \sigma \left(h_{(2)} \otimes h'_{(2)} h''_{(2)} \right) =$$

$$\sum_{h,h'} \sigma(h_{(1)} \otimes h'_{(1)}) \sigma \left(h_{(2)} h'_{(2)} \otimes h'' \right).$$

By (Mishchenko 1990, Lemma 7.1.2) it means that the R-module $A \otimes H$ with respect to the multiplication

$$(a \otimes h)(a' \otimes h') = \sum_{h,h'} a(h_{(1)} \cdot a')\sigma(h_{(2)} \otimes h'_{(1)}) \otimes h_{(3)} h'_{(2)}$$

is an associative algebra. This algebra is called the **crossed product** and is denoted by $A\natural_\sigma H$. Crossed products are used for developing Hopf-Galois extensions (Mishchenko 1990, chapter 8). A 2-cocycle is trivial if $\sigma(h \otimes g) = \varepsilon(h)\varepsilon(g)$. In this case, the crossed product is called a **skew product** and it is denoted by $A\natural H$. Any cocommutative Hopf algebra H over an algebraically closed field R of characteristic zero has the form $H \simeq U(P(H))\natural RG(H)$, where $U(P(H))$ is the universal envelope of $P(H)$ (Mishchenko 1990, §5.6).

If the 2-cocycle is trivial then the associative algebra A is a left H-module, which is called **left H-module** algebra. In this case, we say that H **acts** on the algebra A. In particular, H is a left H-module algebra with respect to the **left adjoint** action $\operatorname{ad} h(g) = \sum_h h_{(1)} g S(h_{(2)})$ Moreover, H is a left and right H^0-module algebra with respect to the actions

$$f \leftharpoonup h = \sum f(h_{(2)}) h_{(1)}, \quad h \rightharpoonup f = \sum_h f(h_{(1)}) h_{(2)}, \qquad f \in H^0, h \in H.$$

A left **coaction** of a Hopf algebra H on an associative algebra A is an R-algebra homomorphism $\rho \colon A \to H \otimes A$ such that $(1_H \otimes \rho)\rho = (\Delta \otimes 1_A)\rho$ and $(\varepsilon \otimes 1_A)\rho(a) = a$ for all $a \in A$. Then A is called a **left H-comodule** algebra. If H has finite dimension over a field R then A is a left H-comodule algebra if and only if A is a left H^0-module algebra (Montgomery 1993, § 1.6). For example, if an algebraic (analytic) group G acts on an algebraic (analytic) manifold M, then the algebra $R[M]$ of regular (analytic) functions on M is a left $R[G]$-comodule algebra.

An element a of a left H-module algebra A is an **invariant** if $h \cdot a = \varepsilon(h)a$ for every $h \in H$. The set A^H of all invariants in A is a subalgebra of A. If H is a finite dimensional cocommutative Hopf algebra over a field and A is a commutative H-module algebra, then the extension A/A^H is integral (Mishchenko 1990, p. 43-44). Dually, a **coinvariant** in a left H-comodule algebra A is an element $a \in A$ such that $\rho(a) = 1 \otimes a$. Each $\mathbb{Z}$-graded algebra $A = \oplus_{n \in \mathbb{Z}} A_n$ is a left comodule algebra over the group ring of an infinite cyclic group $A = R[X^{\pm 1}]$ (Artamonov 1991). In this case, coinvariants are precisely homogeneous elements of A.

Hopf algebras H and H-(co)module algebras are an important tool in the theory of algebraic and quantum groups (Abe 1980, Montgomery 1993, Manin 1988).

C.44 Ordered Rings

by Laszlo Fuchs in New Orleans, LA, USA

It is a familiar fact that the real numbers carry an order relation in which the sums and products of positive numbers are again positive. The complex numbers do not admit such an order because, in any totally ordered field, squares are always positive, and so both $1 = 1^2$ and $-1 = i^2$ must be positive, which is impossible. Thus the question is raised: when does a field (or more generally, a skewfield or a ring) admit a (total) order? For ordered fields, see Section D.6.

C.44.1 Definition A ring R is called **partially ordered** if a partial order $\leq$ is defined in R that 1) makes the additive group R^+ into a partially ordered group (i.e., the sum of positive elements is again positive), and 2) products of positive elements are always positive. The set $P = \{r \in R \mid r \geq 0\}$ of positive elements is called the **positive cone** of $(R, \leq)$. The ring $(R, \leq)$ is called **fully (totally) ordered** or **lattice-ordered** according as $\leq$ defines a linear or a lattice order.

The positive cone P of a partially ordered ring $(R, \leq)$ can be characterized as a semiring (i.e., $P + P \subseteq P$, $P \cdot P \subseteq P$) with the additional property that $a, -a \in P$ if and only if $a = 0$; in other words, $P \cap -P = \{0\}$. As in the group case, the positive cone defines the ordering by the familiar rule that, for $a, b \in R$, $a \leq b$ holds exactly if $b - a \in P$. Note that $(R, \leq)$ is fully ordered if and only if $P \cup -P = R$.

Examples of ordered rings are abundant. 1) The field $\mathbb{R}$ of real numbers is the most familiar example. 2) The ring of all real-valued continuous functions on a closed interval of the real line (or, more generally, on a topological space) is a lattice-ordered ring under the pointwise ordering of the functions. 3) If $(R, \leq)$ is an ordered domain, then the lexicographic order turns the polynomial ring $R[X]$ into an ordered domain; $R[X]$ will be fully ordered if and only if $(R, \leq)$ was fully ordered. 4) The same holds for the formal power series ring $R[[X]]$. 5) Calling a matrix positive if all of its entries are positive, the matrix ring $\mathrm{Mat}_{n \times n}(R)$ becomes lattice-ordered whenever $(R, \leq)$ is fully (or just lattice) ordered.

A partially ordered ring $(R, \leq)$ is called **archimedean** if its additive group is archimedean, i.e., $\{0\}$ is the only subgroup that is bounded in R.

C.44.2 Theorem (Pickert-Hion) *A fully ordered archimedean ring (that is not a zero ring) is order-isomorphic to a unique subring of the real number field* $(\mathbb{R}, \leq)$. *In particular, it must be commutative.*

When can a field or ring be fully ordered? The answer depends on the behavior of the element -1.

C.44.3 Definition A ring R is called **formally real** if -1 is not a sum of elements of the form $r_1^2 \cdots r_n^2$ where $r_i \in R$.

Note that if R is commutative, then $r_1^2 \cdots r_n^2 = (r_1 \cdots r_n)^2$, and so 'formally real' simply means that -1 is not a sum of squares.

The most important result tells us when a skewfield admits a full order; this is the precise explanation why the complex number field $\mathbb{C}$ cannot be fully ordered.

C.44.4 Theorem (Artin-Schreier) *A skewfield can be fully ordered if and only if it is formally real.*

More on the Artin-Schreier theory can be found in Section D.6. For rings without divisors of zero the situation is only slightly more complicated:

C.44.5 Theorem (Johnson-Podderyugin) *A ring without divisors of zero admits a full order exactly if no sum of even products vanishes (a product is even if it contains every factor $a_i \neq 0$ an even number of times).*

Various results are concerned with the extendibility of a partial order to a full order.

C.44.6 Theorem (Serre-Fuchs) *A partial order of a skewfield F can be extended to a full order if and only if no sum of elements of the form $r_1^2 \cdots r_n^2$ or of the form $pr_1^2 \cdots r_n^2$ vanishes; here $0 \neq a_i \in F$ and $0 < p \in F$. (If F is commutative, then it suffices to take $n = 1$.)*

Extension of a full order to a larger ring is sometimes possible. This is the case when the larger ring is the ring of fractions of an Ore domain.

C.44.7 Theorem (Albert) *Let R be left (or right) Ore ring without divisors of zero. Then every full order of R extends uniquely to the skewfield of left (resp. right) quotients.*

Of great importance is the subclass of lattice-ordered rings. In these rings R, the multiplication satisfies $c(a \vee b) \geq ca \vee cb$ and $c(a \wedge b) \leq ca \wedge cb$ for all $a, b, c \in R$ provided $c > 0$. Both inequalities become equalities in function rings.

C.44.8 Definition Let R^* denote the ring that is a direct product $R^* = \prod_{i \in I} R_i$ of fully ordered rings R_i, lattice-ordered under the pointwise ordering. By a *function ring* is meant a ring R that is both a subring and a sublattice of such an R^*.

The most important example for function rings is example 2) above. These particular rings have been studied extensively. Function rings R enjoy additional properties. e.g., every square is ≥ 0, and, for all $a, b \in R$, $a \wedge b = 0$ implies $ab = 0$.

Two relevant results on function rings read as follows.

C.44.9 Theorem (Birkhoff-Pierce) *A lattice-ordered ring $(R, \leq)$ is a function ring if and only if $a \wedge b = 0$ and $c > 0$ imply $ca \wedge b = 0$ and $ac \wedge b = 0$ for all $a, b, c \in R$.*

C.44.10 Theorem (Birkhoff-Pierce) *Archimedean function rings are commutative. They admit Dedekind completions, unique up to isomorphism.*

Finally, let us point out that the lattice-ordered rings as well as the function rings form equational classes. More on all of that can be found in (Fuchs 1963), for instance.

C.45 Topological Rings and Modules

by S. T. Glavatsky and A. V. Mikhalev in Moscow, Russia

A ring R is called **topological** if the set R is endowed with a topology such that R is a topological abelian group (i.e. the operation of addition $a + b$ is a continuous mapping from $R \times R$ endowed with the product topology to R, operation of taking the additive inverse $-a$ is a continuous mapping from R to R), and the operation of multiplication $a \cdot b$ is also a continuous mapping from $R \times R$ to R. Denote by (R, τ) a topological ring R endowed with the topology τ (this topology also is called a **ring topology** on R). A topological ring (TR) is called **Hausdorff** (**complete**) if the topological space R is Hausdorff (complete). If a continuous homomorphism $f : R_1 \to R_2$ is an open mapping, then it is called a **topological homomorphism**; if in addition it is a bijection, then it is called a **topological isomorphism**.

All the following results, if not mentioned otherwise, are valid for associative rings. Examples of TR are:

- Banach and Hilbert algebras;
- The fields $\mathbb{R}$ and $\mathbb{C}$ endowed with natural topologies;
- The ring $C([0; 1], R)$ of real-valued continuous functions, endowed with the topology induced by the pseudonorm $\xi(f) = \max\{|f(x)| \mid x \in$

[0; 1]} or with the topology of pointwise convergence (see Arnautov et al. 1996);

- For any two-sided ideal I of a ring R, the system $\{I^k \mid k = 1, 2, \ldots\}$ of ideals of the ring R as a basis of neighborhoods of zero defines a ring topology on R. This topology is called *I-adic topology*.

It is easy to see that any ring R being endowed with the discrete as well as anti-discrete topology (only $\varnothing$ and R are open sets) becomes topological. The problem *"Does any infinite associative ring admit a non-discrete Hausdorff ring topology"* is still open in a general case (the answer is positive for countable and for commutative rings, see (Arnautov et al. 1996)). Close to it is the problem of extending a topology on a ring to its over-rings. Several results are achieved here.

C.45.1 Theorem (Arnautov et al. 1996) *Let (R, τ_0) be a Hausdorff TR with the unitary element, (X, τ_1) be a completely regular topological space. Then on the ring $R[X]$ of polynomials with the set of variables X, there exists a ring topology τ such that $\tau\big|_R = \tau_0$ and $\tau\big|_X = \tau_1$, and X is a closed subset in $(R[X], \tau)$. Moreover, several properties of the topology τ_0 could be kept for τ_1.*

A left (right) module $_RM$ (M_R) over a TR R is called a **topological R-module** if it is endowed with such a topology that $(M, +)$ is a topological abelian group and multiplication $r \cdot m$ is a continuous mapping from $R \times M$ (endowed with the product topology) to M. Denote by (M, τ) the topological module M (endowed with the topology τ—this topology is also called a **module topology**) on M. A topological module (TM) is called **Hausdorff (complete)** if the topological space M is Hausdorff (complete). For TR's and TM's, there are widely used constructions of the completion, of the direct product with Tikhonoff topology (also, the more general variant of the m-topology could be used), of the inverse limit, of several subdirect products (local products, fan products, direct sums, etc.) as well as semidirect products (see Arnautov et al. 1996). Let (R, τ_0) be a Hausdorff TR with the unitary element and (X, τ_1) be a completely regular topological space. A topological (R, τ_0)-module (M, τ) is called a *free topological module* which is generated by (X, τ_1) over the topological ring (R, τ_0), if:

a) (X, τ_1) is a subspace of the topological space (M, τ);

b) M does not contain any proper R-submodules, which contain the set X;

c) for any topological (R, τ_0)-module (M', τ') every continuous mapping $\xi \colon (X, \tau_1) \to (M', \tau')$ has an extension $\hat{\xi} \colon (M, \tau) \to (M', \tau')$, which is a continuous homomorphism of modules.

In the similar way, a free TR is defined.

C.45.2 Theorem (see Arnautov et al. 1996) *For any Hausdorff TR* (R, τ_0) *with the unitary element and for arbitrary completely regular space* (X, τ_1), *there exists a free topological module* $(M_X, \hat{\tau})$ *over the ring* (R, τ_0), *which is generated by the topological space* (X, τ_1), *moreover,* X *is a closed subset in* $(M_X, \hat{\tau})$.

The analogous result is valid for a free TR. Varieties of TR's and of topological algebras with different approaches to the definition of the varieties are under investigation (see Taylor 1977)).

A general theory of topological radicals defined in specific classes of TR's has been developed (Arnautov 1998, Arnautov et al. 1992, Beidar et al. 1989). In particular, many situations were considered when the Jacobson radical of a TR R is closed in R, and topological analogues of discrete structure theorems were proved in these situations. Also, different topological analogues of the Jacobson radical were considered (Arnautov 1998), in particular, for the classes of TR's, bounded from the left (bounded from the right, bounded), a topologically quasi-regular radical was entered into consideration as a topological radical of Jacobson. In (Arnautov et al. 1992), it was proved that in many natural situations this radical is special (i.e., it is an upper radical defined by a special class of prime TR's). In the class of all TR's, analogues of the Jacobson radical could be used: the lower radical generated by the class of all quasi-regular TR's, or the lower radical generated by the class of all topologically quasi-regular TR's (see (Arnautov 1998) and Section C.18).

Most of the advanced results in the structure theory were achieved for compact, locally compact and for linearly compact rings (modules). In (Pontrjagin 1973), (Warner 1989), (Warner 1993), (Wieslaw 1988), a complete description of locally compact topological skew fields as finite extensions of the reals, or of p-adic numbers, or else of some Laurent series fields of one variable over a finite field $\mathbb{F}_q$ was obtained. In a series of papers (Bourbaki 1972c, Kaplansky 1948, 1951), it was proved that a compact commutative ring is the direct product of compact local rings and a compact Jacobson-radical ring; that if the square of the Jacobson radical of a compact local ring is open, then this ring is local noetherian whose topology is its natural (radical) topology. Here also, with the use of advanced techniques, it was proved that the Jacobson radical of locally compact rings is not only closed, but it is the intersection of the closed regular maximal right ideals. This yields a description of bounded from the right locally compact semisimple rings.

The ring of linear operators on a finite-dimensional vector space over a non-discrete, locally compact skew field K, endowed with its unique Hausdorff topology as a finite-dimensional algebra over the center of K, is a locally compact simple ring. So, it was a natural problem to determine conditions insuring that non-discrete locally compact primitive or simple rings have this description. In (Kaplansky 1951, Skornjakov 1966), such a result was proved for torsion free locally compact primitive rings, for simple locally compact rings having a minimal left ideal, and for simple locally compact rings without proper open left ideals.

C.45.3 Theorem (Kaplansky 1951) *A topological ring R is locally compact and **strongly regular** (for each $r \in R$ there exists $x \in R$ such that $r^2 x = r$) if and only if R is a topological direct sum of ideals R_1, R_2 and R_3, described as follows: R_1 is the direct sum of finitely many non-discrete locally compact skew fields; R_2 is topologically isomorphic to the local product of a family $\{S_i \mid i \in I\}$ of discrete strongly regular rings with identity relative to finite subfields $\{K_i \mid i \in I\}$, where K_i contains the identity of S_i for each $i \in I$; and R_3 is a discrete strongly regular ring.*

This theorem was generalized in (Skornjakov 1966) to any locally compact **biregular** (for each $x \in R$ there is a central idempotent $e \in R$ such that $RxR = ReR$) ring R.

A TM $_R M$ over a TR R is called

- *linearly topologized* (LT) if M has a basis of neighborhoods of zero consisting of submodules;

- *linearly compact* (LC) if it is LT and every linear filter on M has an adherent point;

- a *strictly linear compact* (SLC) if it is LC and any continuous homomorphism $\varphi\colon M \to M'$ on a LT R-module M' is open.

A TR R is called LC (SLC) if it is LC (SLC) as left R-module. A direct product with the Tikhonoff topology of any family of LC (SLC) modules is LC (SLC) module. The same is true for LC (SLC) rings.

C.45.4 Theorem (Warner 1993) *A TR R is semisimple and LC if and only if R is topologically isomorphic to the direct product of a family of*

TR with the Tikhonoff topology, each one being the ring of all linear operators on a discrete vector space over a discrete skew field, endowed with the topology of pointwise convergence.

The Wedderburn–Artin theorem for semisimple artinian rings is the discrete case of this theorem (see Warner 1993). The Jacobson radical of a SLC ring is transfinitely nilpotent.

Several results on the structure of LC rings were achieved if some conditions on Jacobson radical were considered, in particular, when an LC ring admits a Wedderburn decomposition (see Warner 1993). Also, SLC rings form a natural domain for extending theorems about discrete artinian rings.

Duality theories for certain classes of TM, especially locally compact or LT modules, were developed as for generalization of the classical Pontrjagin–van Kampen duality theory and for investigation of Morita dualities for TM. In (Müller 1971) was shown that if A and B be TR's with identity, then the duality induced by a bimodule $_AU_B$ (see Morita 1958) having certain properties, may be extended to a duality for the categories of all Hausdorff LT left A- and right B-modules (here, the **dual module** M^* is the module of all continuous homomorphisms from M to $_AU_B$, endowed with the topology of pointwise convergence). Also if the TR A has a Morita duality induced by $_AU_B$, then A and $_AU_B$ are LC A-modules for the discrete topology and reflexive A-modules are precisely those that are LC for the discrete topology.

The use of categorial concepts in the investigation of TM also has proved useful in duality theory and in the structure theory of TR's and other areas of topological algebra.

C.46 Semirings and Semifields

by Udo Hebisch in Freiberg and Hanns J. Weinert in

Clausthal, Germany

A semiring may be roughly described as a ring whose additive structure need not be a group. Obvious examples are the semirings N_0 and N of natural numbers, including the zero 0 or not, and the semifields H_0 and H of non-negative or positive rational numbers, respectively. In fact, the first abstract concept of a semiring was introduced in 1934 by H. S. Vandiver in order to investigate questions occurring in elementary arithmetic. Semirings have become important in Theoretical Computer

Science.

C.46.1 Definition a) A universal algebra $S = (S, +, \cdot)$ is called a *semiring* if $(S, +)$ and $(S, \cdot)$ are arbitrary semigroups such that $a(b + c) = ab + ac$ and $(b + c)a = ba + ca$ hold for all $a, b, c \in S$. If $(S, +)$ has a neutral element o, this is called the *zero* of S, and we speak about an *absorbing zero* if $oa = o = ao$ holds for all $a \in S$. The terms (left) *identity* and *zero-divisor* are used as for rings, and a semiring S with a zero o is called *zero-sum-free* if $a + b = o$ implies $a = b = o$.

b) For a semiring S we introduce S^* by $S^* = S \setminus \{o\}$ if S has a zero o and by $S^* = S$ otherwise. If $|S| \geq 2$ holds, S is called *multiplicatively (left) cancellative* if each $a \in S^*$ is left cancellable in $(S, \cdot)$, and a *semifield* if $(S, \cdot)$ or $(S \setminus \{o\}, \cdot)$ is a group.

C.46.2 Example a) Let $\mathfrak{m}$ be a transfinite cardinal number and $\mathfrak{K}$ the set of all cardinals less than or equal to $\mathfrak{m}$. Then, with the usual operations for cardinals, $(\mathfrak{K}, +, \cdot)$ is a semiring.

b) Each distributive lattice $(L, \vee, \wedge) = (L, +, \cdot)$ is a semiring where both operations are idempotent. In particular, all subsets of a given set X form a semiring $(\mathscr{P}(X), \cup, \cap)$. For $|X| = 1$ one obtains a copy of the *Boolean semiring* $(\mathbb{B}, +, \cdot)$ (in fact a *semifield*) consisting of an absorbing zero $o = \varnothing$ and an identity $e = X$ satisfying $e + e = e$. It is important for applications.

c) The set $\mathbb{P}_0$ of all non-negative real numbers yields a semifield $(\mathbb{P}_0, +, \cdot)$ with the usual operations, and an additively idempotent semifield $(\mathbb{P}_0, \oplus, \cdot)$ if one replaces the usual addition by $a \oplus b = \min(a, b)$. Then, for each $c \geq 1$, the set $S_c = \{ s \in \mathbb{P}_0 \mid s \geq c \}$ is a subsemiring with the zero c. All elements, including the zero c, are multiplicatively cancellable, and $(S_1, \oplus, \cdot)$ is a semiring with the same element $c = 1$ as zero and identity.

C.46.3 Definition An *ideal* A of a semiring S is a subsemigroup of $(S, +)$ satisfying $sa \in A$ and $as \in A$ for all $a \in A$ and $s \in S$. Then the *k-closure* $\overline{A} = \{ s \in S \mid s + a \in A \text{ for some } a \in A \}$ of A is also an ideal of S, and a *k-ideal* A of S is defined by $\overline{A} = A$.

Considering a semiring as a universal algebra $(S, +, \cdot)$, each homomorphism $\varphi \colon S \to T$ corresponds to a congruence κ of S, and the homomorphic image $\varphi(S)$ is isomorphic to the semiring $(S/\kappa, +, \cdot)$ of congruence classes $[s]_\kappa$ (see Section G.1). However, there is no correspondence between congruences and ideals as for rings, despite the fact that each ideal A of an additively commutative semiring S determines a congruence κ_A of S by $s\kappa_A s' \Leftrightarrow s + a_1 = s' + a_2$ for some $a_i \in A$, and the notation S/A for S/κ_A. Firstly, a lot of different ideals may define the same congruence, since $\kappa_A = \kappa_{\overline{A}}$ and thus $S/A = S/\overline{A}$ hold for all ideals A with the same k-closure $\overline{A}$. Secondly, a semiring S has in general congruences κ which cannot be described by an ideal of S. For example, all non-zero k-ideals of the semiring $\mathbb{N}_0$ are the sets $m\mathbb{N}_0$ for $m \geq 1$, and each of them is the k-closure of infinitely many other ideals. Moreover, $\mathbb{N}_0/m\mathbb{N}_0$ is isomorphic to the ring $\mathbb{Z}/(m)$, and there are infinitely many other homomorphic images of $\mathbb{N}_0$, e.g., the Boolean semifield.

C.46.4 Theorem *A semiring $(S, +, \cdot)$ with zero o and $|S| \geq 2$ is multiplicatively left cancellative iff either $(S, \cdot)$ is left cancellative or o is absorbing and $(S^*, \cdot)$ a left cancellative semigroup. Such a semiring has no zero-divisors, but the converse implication (well known for rings) is false.*

There are 6 non-isomorphic semifields with a zero o on a two-element set $S = \{o, c\}$. Firstly, the field of order 2 and the Boolean semifield, both multiplicatively cancellative with o as absorbing zero and $c = e$ as identity. Secondly, four rather pathological semifields, which do not satisfy any of the properties just mentioned. This is the reason for the assumption $|S^*| \geq 2$ in the following four theorems, also due to Weinert.

C.46.5 Theorem *Let S be a semiring satisfying $|S^*| \geq 2$. Then S is a semifield iff S has a left identity e_ℓ such that for each $a \in S^*$ there is some $y \in S$ satisfying $ya = e_\ell$. Such a semifield S is multiplicatively cancellative and has either no zero or an absorbing one; in both cases, $(S^*, \cdot)$ is a group and the identity of $(S^*, \cdot)$ is the identity of $(S, +, \cdot)$.*

C.46.6 Theorem *Let S be a semifield with a zero and $|S^*| \geq 2$. Then S is either a field or zero-sum-free and hence $(S^*, +, \cdot)$ a subsemifield of $(S, +, \cdot)$.*

C.46.7 Theorem *Let S be an additively commutative proper semifield (i.e., not a field) such that $|S^*| \geq 2$. Then the group $(S^*, \cdot)$ is torsion free and thus S of infinite order. Moreover, all finite semifields are the Galois fields and the semifields mentioned after Theorem C.46.4, or additively not commutative.*

C.46.8 Theorem *Each additively commutative and idempotent semifield* $(S, +, \cdot)$ *without a zero is a lattice ordered group* $(S, \cdot, \leq)$*, and conversely.*

C.46.9 Remark Contrasting the situation with fields, a proper semifield S may have non-trivial epimorphisms $\varphi \colon S \to T$, e.g., $\varphi \colon S \to \mathbb{B}$ if S has a zero and more than two elements. If S has no zero, each epimorphism of S is determined by a substructure N of S as its kernel. The latter are all invariant subgroups N of $(S, \cdot)$ which satisfy $s + t \in N \implies s + t \cdot n \in N$ for all $s, t \in S$ and $n \in N$ (cf. Hutchins and Weinert (1990)).

Semirings and semifields occur as positive cones of partially ordered rings (and semirings)(see Section C.44). Here we restrict ourselves to the following example.

C.46.10 Example Let $K = \mathbb{Q} + \mathbb{Q}\sqrt{k}$ be a quadratic field extension of $\mathbb{Q}$. Then K contains, in addition to $\mathbb{H}_0$, three proper subsemifields with a zero: $S_1 = K \cap \mathbb{P}_0$, its image S_2 under the non-trivial automorphism of K (both are the positive cones of the total orders of K), and their intersection $S_3 = S_1 \cap S_2$. A corresponding result holds for each algebraic number field.

Finally, we deal with additively commutative semirings and give some examples of semirings which are used in Theoretical Computer Science.

C.46.11 Example The set $\mathcal{L}_X$ of all formal languages over an alphabet X yields a semiring $(\mathcal{L}_X, +, \cdot) = (\mathcal{L}_X, \cup, \cdot)$ with the union $L_1 \cup L_2$ of languages as addition and their concatenation $L_1 L_2 = \{\, w_1 w_2 \mid w_i \in L_i \,\}$ as multiplication. Using also arbitrary unions $\bigcup L_i$, one obtains an example of a (complete) Σ-semiring $(\mathcal{L}_X, +, \Sigma, \cdot) = (\mathcal{L}_X, \cup, \bigcup, \cdot)$. In general, in a Σ-*semiring* $(S, +, \Sigma, \cdot)$ also infinite sums $\Sigma_{i \in I} a_i$ exist for certain summable families $(a_i)_{i \in I}$. *These infinite sums can be defined in an abstract way* (i.e., not necessarily by a topological concept of convergence), *subjected to some algebraic rules.* The latter can be adapted to different applications and in most cases implies that $(S, +, \cdot)$ is zero-sum-free and thus not a ring.

C.46.12 Example For a semiring S with an absorbing zero o and an identity $e \neq o$, let $S[[X]]$ be the semiring of formal powers series for a finite or infinite set $X = \{\, x_1, x_2, \dots \,\}$ of commuting or non-commuting indeterminates x_i over S (see Section C.9). The so-called formal infinite sums, often used to obtain the elements of $S[[X]]$, are in fact infinite sums as in Example C.46.11 if one considers $S[[X]]$ in a natural way as a Σ-semiring. *In particular, the semiring $\mathbb{B}[[X]]$ with non-commuting indeterminates (for $|X| \geq 2$) is isomorphic to the semiring $(\mathcal{L}_X, \cup, \bigcup, \cdot)$ of formal languages over X. This important fact allows to use algebraic calculations in the theory of formal languages and automata.* We also note that the semirings $S[[X]]$ are examples for *generalized semialgebras over a semiring S*, which are semirings or Σ-semirings over a free S-semimodule.

C.46.13 Example Let G be a finite directed graph, valuated by a **path algebra**, i.e., a semiring S with an absorbing zero and an identity $e \neq o$. For many applications, which clearly determine the choice of S, one needs information about paths in G which have an extreme value. This problem can be solved by several algorithms. However, to prove their correctness, one deals in most cases with infinite sums, hence one considers, tacitly or not, S as a Σ-semiring (cf. Example C.46.11).

C.46.14 Example For any semiring S and any nonempty set X, the set S^X of all mappings from X into S becomes a semiring, if addition and multiplication are defined pointwise. In particular, if S is a bounded distributive lattice (see Example C.46.2 b)), S^X is called the **semiring of S-valued fuzzy subsets of X**. This concept has found hundreds of applications since it was established by Zadeh in 1965 for the special case that $S = [0, 1]$.

For more details and further applications of semirings see the following books (Hebisch and Weinert 1996), (Hebisch and Weinert 1998), (Gunarwardena 1998), (Golan 1999b), (Golan 1999a), and (Glazek 2001).

C.47 Near-Rings and Near-Fields

by Günter F. Pilz in Linz, Austria

The set of all real functions, together with pointwise addition and composition, does not form a ring. Observe that $(f + g) \circ h = f \circ h + g \circ h$ holds by definition, while $f \circ (g + h) = f \circ g + f \circ h$ holds for all $g, h \in \mathbb{R}^{\mathbb{R}}$ if and only if f is *linear*, i.e., if $f(x + y) = f(x) + f(y)$ for all $x, y \in \mathbb{R}$. Structures of the type like $\mathbb{R}^{\mathbb{R}}$ are called **near-rings**:

C.47.1 Definition A set N, together with two operations $+$ and $\circ$, is a **near-ring** provided that $(N, +)$ is a group (not necessarily abelian), $(N, \circ)$ is a semigroup, and $(n + n') \circ n'' = n \circ n'' + n' \circ n''$ holds for all $n, n', n'' \in N$.

Of course, every ring is a near-ring; hence near-rings are generalized rings. Two ring axioms are missing: the commutativity of addition and (much more important) the other distributive law.

C.47.2 Examples We have met the near-ring $\mathbb{R}^{\mathbb{R}}$ already above. This and many other occasions show that near-rings form the "nonlinear counterpart to ring theory". More generally, take a group $(G, +)$; then $(G^G, +, \circ)$ is a near-ring, and every near-ring can be embedded in some $(G^G, +, \circ)$. More examples of near-rings are given by $(R[x], +, \circ)$, where R is a commutative ring with identity, by the collection of continuous real functions, by the differentiable ones, or by the polynomial functions. Other important examples (see below) are **centralizer near-rings** $(M_S(G), +, \circ) = \{ f \colon G \to G \mid f \circ s = s \circ f \text{ for all } s \in S \}$, where $(G, +)$ is a group and $S \leq \text{End}(G)$.

Actually, the first near-rings considered were **near-fields**, near-rings $(N, +, \circ)$ in which $(N^*, \circ)$ forms a group. In 1905, L. E. Dickson constructed the first *proper* near-fields by *distorting* the multiplication in a field. These types of near-fields are now called **Dickson near-fields**. Two years later, Veblen and Wedderburn applied near-fields to coordinatize geometric planes. In a monumental paper, H. Zassenhaus showed in 1936 that all finite near-fields are Dickson ones, with the exception of 8 (well-known) cases. 51 years later, Zassenhaus showed that there do exist non-Dickson infinite near-fields of every prime characteristic.

For near-rings, we now have a sophisticated structure theory: a "working" radical theory, primitivity, density theorems, and the like. Quite often, the results look similar to the ring case, but the proofs are completely different. For instance, matrix rings ("the stuff rings are made of") have to be replaced by centralizer near-rings $(M_S(G), +, \circ)$, where S is a fixed-point free automorphism group.

An interesting class of near-rings arises as follows. The sum of two endomorphisms (automorphisms, inner automorphisms) of a non-abelian group $(G, +)$ is usually not of this type any more. If one takes the additive closures, one gets near-rings, denoted by $E(G)$, $A(G)$, and $I(G)$, respectively. For instance, $I(G)$ is "primitive" iff G is simple. The suitable Density Theorem, combined with a result of Thomson on the nilpotency of a finite group with a fixed-point free automorphism group, gives the interesting result that precisely the finite, simple, non-abelian groups $(G, +)$ have the property that every map from G to G is the sum of inner automorphisms and a constant—a purely group-theoretic result.

Fixed-point free automorphisms turn up in another context as well; this one has nice applications to real-world problems.

C.47.3 Definition A near-ring N is called **planar** if all equations $x \circ a = x \circ b + c$ $(a, b, c \in N$, not $x \circ a = x \circ b$ for all $x)$, have exactly one solution $x \in N$.

This condition is motivated by geometry. It implies that two "non-parallel lines $y = x \circ a + c_1$ and $y = x \circ b + c_2$ have exactly one point of intersection". One example is $(\mathbb{C}, +, *)$, where $a * b := a|b|$. We need finite examples, however. There are various methods to construct planar near-rings (see Clay 1992, Pilz 1983, or Pilz 1996). Most of them use fixed-point free automorphism groups; we just present the easiest way.

C.47.4 Construction method for finite planar near-rings. Take a finite field $\mathbb{F}_q$ and choose a generator g for its multiplicative group. Choose a factorization $q - 1 = st$. Define a new multiplication $*_t$ in $\mathbb{F}_q$ as $g^a *_t g^b := g^{a+b-[b]_t}$, where $[b]_t$ denotes the remainder of b on division by t; if one factor is 0, the product should be 0. Then $(\mathbb{F}_q, +, *_t)$ is a planar near-ring.

C.47.5 Example We choose $\mathbb{F}_7$, and $t = 2$ as divisor of $7 - 1$, so $g = 3$ serves well as a generator. From this, we get the multiplication table

$*_2$	0	1	2	3	4	5	6
0	0	0	0	0	0	0	0
1	0	1	2	1	4	4	2
2	0	2	4	2	1	1	4
3	0	3	6	3	5	5	6
4	0	4	1	4	2	2	1
5	0	5	3	5	6	6	3
6	0	6	5	6	3	3	5

C.47.6 Example We want to test combinations of 7 ingredients for fertilizers. Testing all possible combinations of ingredients requires a huge amount of space and money. So we conduct an incomplete test; but this one should be *fair* to the ingredients (each ingredient should be applied the same number of times) and fair to the experimental fields (each test-field should get the same number of ingredients). We take the above near-ring of order 7 and form the sets $B_i = a *_2 N^* + b$, $a \neq 0$, and abbreviate $a *_2 b$ by ab:

$$1N^* + 0 = \{1, 2, 4\} =: B_1, \qquad 3N^* + 0 = \{3, 5, 6\} =: B_8,$$
$$1N^* + 1 = \{2, 3, 5\} =: B_2, \qquad 3N^* + 1 = \{4, 6, 0\} =: B_9,$$

$$\dots\dots\dots\dots\dots\dots\dots\dots \qquad \dots\dots\dots\dots\dots\dots\dots\dots$$

$$1N^* + 6 = \{0, 1, 3\} =: B_7, \qquad 3N^* + 6 = \{2, 4, 5\} =: B_{14}.$$

We see: these **blocks** form a so-called **balanced incomplete block design (BIB-design)** with $v = 7$ points (namely $0, 1, 2, 3, 4, 5, 6$) and $b = 14$ blocks; each point lies in exactly $r = 6$ blocks; each block contains precisely $k = 3$ elements, and every pair of different points appears in $\lambda = 2$ blocks. (v, b, r, k, λ) are called the **parameters** of the design. In order to solve our fertilizer problem, we divide the whole experimental area into 14 experimental fields, which we number by $1, 2, \dots, 14$. We then apply precisely the fertilizers F_i to a field k if $i \in B_k$, $i = 1, 2, \dots 6$.

Field Fertilizer	1	2	3	4	5	6	7	8	9	10	11	12	13	14
F_0				x		x	x		x	x		x		
F_1	x				x		x			x	x		x	
F_2	x	x				x					x	x		x
F_3		x	x				x	x				x	x	
F_4	x		x	x					x				x	x
F_5		x		x	x			x		x				x
F_6			x		x	x		x	x		x			
Yields:	2.8	6.1	1.2	8.5	8.0	3.9	7.7	5.5	4.0	10.7	3.2	4.9	3.8	5.5

The last row indicates the yields on the experimental fields after performing the experiment. Then we get: every field contains exactly 3 fertilizers, and every fertilizer is applied to 6 fields. Finally, every pair of different fertilizers is applied precisely twice in *direct competition*. Hence planar near-rings even have applications in agriculture!

The statistical analysis uses the incidence matrix $\mathbf{A}$ of the design: compute $(\mathbf{AA}^T)^{-1}\mathbf{Ay}^T =: (\beta_0, \dots, \beta_{v-1})^T$. Then Linear Algebra tells us that $\beta_0, \dots, \beta_{v-1}$ are the best estimates for the effects of the ingredients $I_0, \dots, I_{v-1}$. In our case, we get the best estimates for the effects of $F_0, \dots, F_6$ as $(3.6, 2.7, 0.3, 1.0, 0.1, 4.8, 0.1)$. If we take F_0, F_1, F_3, F_5, we can expect a yield of $3.6 + 2.7 + 1.0 + 4.8 = 12.1$, which is considerably better than the yields y_i which we got in our experiment. A more detailed statistical analysis gives confidence intervals for the effects of F_i, as well as information concerning theirsynergy effects. More on the analysis of experiments can be found in (Box, Hunter, and Hunter 1978).

Even more can be done with incidence matrices coming from BIB-Designs. The rows in the matrix $\mathbf{A}$ are $0-1$-sequences, so they can be taken as a binary code $C_{\mathbf{A}}$. This is a nonlinear, constant weight r and constant distance $d = 2(r - \lambda)$ code of length b with v codewords. These codes have the largest number of codewords among all codes with equal weight r, length b, and fixed minimal distance d. So they also solve discrete sphere packing problems!

For another natural show-up of near-rings, see Section C.55.

C.48 Conformal Algebras

by Efim Zelmanov in New Haven, CT, USA

The concept of a conformal algebra is closely related to the better known concept of a vertex operator algebra. The latter received wide recognition during the last twenty years due to deep connections with physics and the Moonshine representation of the Monster group.

Let A be an arbitrary (not necessarily associative) algebra over a ground field F of zero characteristic. By a **formal distribution** $a(z) = \sum_{i \in \mathbb{Z}} a(i) a^{-i-1} \in A[[z^{-1}, z]]$ we mean a power series over A which is infinite in both directions. Such are "fields" from two dimensional conformal field theory and such are Fourier decompositions of distributions.

Consider a countable family bilinear binary operations

$$a(z) \circ_n b(z) = \operatorname{Res}_w a(w) b(z) (w - z)^n, \quad n \geq 0,$$

where the residue is the coefficient of $a(w)b(z)(w - z)^n$ at w^{-1}.

C.48.1 Definition Two formal distributions $a(z), b(z)$ are said to be **mutually local** if there exists an integer $N = N(a, b) \geq 0$, such that

$$a(w)b(z)(w - z)^N = b(w)a(z)(w - z)^N = 0.$$

Consider the formal delta function $\delta(w - z) = \sum_{i \in \mathbb{Z}} w^i z^{-1-i}$. It is easy to see that $a(w)b(z)(w - z)^N = 0$ if and only if $a(w)b(z) = \sum_{j=0}^{N-1}(a \circ_j b)(z) \partial_w^j \delta(w - z)/j!$.

C.48.2 Examples

(1) Let g be an arbitrary algebra and let $A = g[t^{-1}, t]$ be the algebra of **Laurent polynomials** over g. For an arbitrary element $a \in g$, let $\tilde{a} = \sum_{i \in \mathbb{Z}} (at^i) z^{-i-1} \in A[[z^{-1}, z]]$. Any two formal distributions $\tilde{a}, \tilde{b}$ are mutually local with $N(\tilde{a}, \tilde{b}) = 1$.

(2) Let $\mathrm{Vir} = \mathrm{Der} F[t^{-1}, t]$ be the (centerless) **Virasoro algebra**. The formal distribution $L = \sum_{i \in \mathbb{Z}} t^i \frac{d}{dt} z^{-i-1} \in \mathrm{Vir}[[z^{-1}, z]]$ is mutually local with itself, $N(L, L) = 2$.

(3) Let $W = <t^{-1}, t, \frac{d}{dt}>$ be the (associative) **algebra of differential operators** on $F[t^{-1}, t]$. Let $J_k = \sum_{i \in \mathbb{Z}} t^i (\frac{d}{dt})^k z^{-i-1}, k \geq 0$. Any two formal distributions J_k, J_ℓ are mutually local, and $N(J_k, J_\ell) = \max(k, \ell) + 1$.

(4) If formal distributions $a(z), b(z)$ are mutually local then $\frac{da(z)}{dz}, b(z)$ are mutually local as well and $N(\frac{da(z)}{dz}, b(z)) \leq N(a, b) + 1$. Moreover, if the coefficient algebra A is an associative or a Lie algebra, then the following assertion is true.

C.48.3 Dong Lemma *Let* $a, b, c \in A[[z^{-1}, z]]$ *be formal distributions which are pairwise mutually local. Then for an arbitrary* $n \geq 0$*, the formal distributions* $a \circ_n b, c$ *are mutually local as well.*

It follows from the Dong Lemma that if $C \subseteq A[[z^{-1}, z]]$ is a vector space of formal distributions such that every two elements of C are mutually local, then the closure of C with respect to the action of $\frac{d}{dz}$ and to all operations $\circ_n, n \geq 0$, also has this property, which motivates the following definition.

C.48.4 Definition A vector space $C \subseteq A[[z^{-1}, z]]$ is called a **conformal algebra of formal distributions** if $\frac{d}{dz} C \subseteq C$, or $C \circ_n C \subseteq C$, for $n \geq 0$, and every two elements from C are mutually local.

The formal distributions of examples (1), (2), (3) generate conformal algebras which will be denoted as $\mathrm{Cur}(g)$, Vir, and W, respectively.

Now we are ready to introduce an abstract definition of a conformal algebra.

C.48.5 Definition Let C be a module over the polynomial algebra $F[\partial]$ with countable many binary bilinear operations $C \circ_n C \to C, n \geq 0$. We say that $(C, \partial, \circ_n)$ is an (abstract) **conformal algebra** if for all $a, b \in C$ and all $n \geq 0$ we have

(1) $\partial(a \circ_n b) = \partial a \circ_n b + a \circ_n \partial b$;

(2) $\partial a \circ_n b = -n a \circ_{n-1} b$; [for $n = 0$, the condition turns into $\partial a \circ_0 b = 0$]

(3) (*locality*) there exists an integer $N = N(a, b) \geq 0$ such that $a \circ_n b = 0$ for all $n \geq N$.

An arbitrary conformal algebra of formal distributions is an abstract conformal algebra. It is remarkable that the reverse is also true.

Let C be an abstract conformal algebra and let $C(i)$, $i \in \mathbb{Z}$, be a countable family of isomorphic copies of the vector space C. Consider the direct sum $A = \bigoplus_{i \in \mathbb{Z}} C(i)$. We define a multiplication on A via $a(i)b(j) = \sum_{k=0}^{N(a,b)} \binom{i}{k} (a \circ_k b)(i + j - k)$, where $a, b \in C$; $i, j \in \mathbb{Z}$. Let I be the linear span of all elements of the type $(\partial a)(i) + ia(i - 1)$, $a \in C$, $i \in \mathbb{Z}$. It is easy to see that $IA \subseteq I$ and $AI = (0)$. Hence I is an ideal in A.

Let $\mathrm{Coeff}(C) = A/I$. We will keep denoting $a(i) + I \in \mathrm{Coeff}(C)$ as $a(i)$ (an abuse of notation). The mapping

$$u \colon C \to \mathrm{Coeff}(C)[[z^{-1}, z]], \quad a \mapsto \sum_{i \in \mathbb{Z}} a(i) z^{-i-1},$$

is a homomorphism of conformal algebras. Moreover, $a(-1) = 0$ implies $a = 0$. Hence u is an embedding.

C.48.6 Definition Let $\mathscr{V}$ be a variety of algebras. A conformal algebra C is called a $\mathscr{V}$-*conformal algebra* if $\mathrm{Coeff}(C) \in \mathscr{V}$.

The inclusion $\mathrm{Coeff}(C) \in \mathscr{V}$ is equivalent to a certain family of identities on C involving circle products and the action of $F[\partial]$.

C.48.7 Example A conformal algebra C is associative if and only if for arbitrary non-negative integers n, m it satisfies the identity

$$(a \circ_n b) \circ_m c = \sum_{s=0}^{n} (-1)^s \binom{n}{s} a \circ_{n-s} (b \circ_{m+s} c).$$

C.48.8 Example A conformal algebra C is Lie conformal if and only if it satisfies all identities

$$(a \circ_n b) \circ_m c = \sum_{s=0}^{n} (-1)^s \binom{n}{s} (a \circ_{n-s} (b \circ_{m+s} c) - b \circ_{m+s} (a \circ_{n-s} c)),$$

which correspond to the Jacobi identity on $\text{Coeff}(C)$, and

$$a \circ_n b = -\sum_{s \geq 0} (-1)^{n+s} \frac{1}{s!} \partial^s (b \circ_{n+s} a),$$

which corresponds to $\text{Coeff}(C)$ being anticommutative.

An arbitrary associative conformal algebra $C = (C, \partial, \circ_n)$ has a structure of a Lie conformal algebra $C^{(-)} = (C, \partial, *_n)$, where

$$a *_n b = a \circ_n b - \sum_{s \geq 0} (-1)^{n+s} \frac{1}{s!} \partial^s (b \circ_{n+s} a).$$

However, not every Lie conformal algebra is embeddable into $C^{(-)}$, where C is an associative conformal algebra.

C.48.9 Definition A conformal algebra C is of *finite type* if C is a finitely generated module over $F[\partial]$.

A. D'Andrea and V. Kac developed a structure theory of Lie conformal algebras of finite type.

C.48.10 Theorem *A simple Lie conformal algebra of finite type is isomorphic to a current conformal algebra $Cur(g)$, where g is a simple finite dimensional Lie algebra or to Vir.*

Every conformal algebra C of finite type has the unique maximal solvable ideal $\text{Rad}(C)$. The algebra C is semisimple if $\text{Rad}(C) = (0)$. A semisimple Lie conformal algebra of finite type is not necessarily a direct sum of simple ones.

C.48.11 Theorem *An indecomposable semisimple Lie conformal algebra of finite type is either simple or isomorphic to a semidirect sum $Cur(g) \rtimes Vir$.*

Relations with vertex algebras. Let V be a $F[\partial]$-module equipped with a countable family of binary bilinear operations $\circ_n$, $n \in \mathbb{Z}$. For an arbitrary element $a \in V$ let $L_n(a)$ denote the operator of the n-th left multiplication by a, $L_n(a) : b \to a \circ_n b$, $L_n(a) \in \text{End}_F V$. Consider the formal distribution $Y(a) = \sum_{n \in \mathbb{Z}} L_n(a) z^{-n-1} \in (\text{End}_F V)[[z^{-1}, z]]$.

C.48.12 Definition A $F[\partial]$-module V above is called a *vertex algebra* if it contains a distinguished element **1** such that

(1) $\partial \mathbf{1} = 0$;
(2) $\mathbf{1} \circ_n a = \delta_{n,-1} a$;
(3) $a \circ_{-1} \mathbf{1} = a, a \circ_n \mathbf{1} = 0$ for $n \geq 0$;
(4) $\partial(a \circ_n b) = a \circ_n (\partial b) - na \circ_{n-1} b$;
(5) for arbitrary elements $a, b \in V$, the formal distributions $Y(a), Y(b)$ are mutually local.

It follows from the axioms that for a vertex algebra V, the system $C = (V, \partial, \circ_n, n \geq 0)$ is a Lie conformal algebra. In particular, $Y(a)(w)Y(b)(z)(w - z)^N = 0$ implies $a \circ_n b = 0$ for $n \geq N$. Moreover, the map $Y : V \to (\text{End}_F V)[[z^{-1}, z]]$ is an embedding.

Consider the coefficient Lie algebra $L = \text{Coeff}(C)$. Let $L_+ = \text{Span}(a(n), a \in C, n \geq 0)$, $L_- = \text{Span}(a(n), a \in C, n < 0)$, $L = L_- + L_+$. The space V becomes an L-module via $a(n)v = a \circ_n v$; $a \in C, v \in V$. This module is generated by $\mathbf{1}$ and $L_+ \mathbf{1} = (0)$.

Now let $C = (C, \partial, \circ_n, n \geq 0)$ be an arbitrary Lie conformal algebra, $L = \text{Coeff}(C) = L_- + L_+$. An L-module V is called a **highest weight module** over L if it is generated over L by a single element v such that $L_+ v = (0)$.

Every highest weight module over L has a natural structure of a vertex algebra. Thus one could say that vertex algebras correspond to highest weight modules over Lie conformal algebras.

Multivariable generalization. For an arbitrary Hopf algebra H, one can define a notion of a conformal algebra that corresponds to H (more generally, an H-pseudoalgebra). The conformal algebras considered above correspond to the Hopf algebra $H = F[\partial]$, $\Delta(\partial) = \partial \otimes 1 + 1 \otimes \partial$.

C.49 Lie Algebras

by Mikhael Zaicev in Moscow, Russia

C.49.1 Definition A non-associative algebra L over a field F is called a **Lie algebra** if it satisfies the following two identities:

$$x^2 = 0,$$
$$x(yz) + y(zx) + z(xy) = 0.$$

The first of this identities is called an **anticommutativity** and it is equivalent to the condition $xy = -yx$ if $\text{char} F \neq 2$. The second identity is the **Jacobian identity**. Lie algebras appear naturally; for example, if A is an arbitrary associative algebra over F then one can define a new product $[x, y] = xy - yx$, which is called the **Lie brackets** of x and y. Under this new operation, A becomes a Lie algebra. Derivations give us another natural class of Lie algebras. Recall that $\delta: R \to R$, $\delta \in \operatorname{End} R$, is called a **derivation** if $\delta(ab) = \delta(a)b + a\delta(b)$. For any (non-associative) algebra R, the derivations form a subspace $\operatorname{Der} R$ of the vector space $\operatorname{End} R$ of all linear transformations on R. As above, $\operatorname{Der} R$ with the product $[\delta, \mu] = \delta\mu - \mu\delta$ is a Lie algebra. Lie algebras also arise naturally in geometry. For any (real or complex) Lie group G, the family of all left-invariant vector fields on G form a Lie algebra $\operatorname{Lie}(G)$ corresponding to G.

A vector space M is said to be a **module** over a Lie algebra L if one has an action of L on M such that $(ab)x = a(bx) - b(ax)$ for any $a, b \in L, x \in M$ and each map $t_a: x \mapsto ax$ is a linear transformation on M. Clearly, the correspondence $\rho: L \to \operatorname{End} M$, $\rho(a) = t_a$, is a homomorphism from L to the Lie algebra $\operatorname{End} M$ with the multiplication $[\varphi, \psi] = \varphi\psi - \psi\varphi$. Any such a homomorphism is called a **representation** of a Lie algebra L. So, there is a one-to-one correspondence between modules and representations of a Lie algebra L. Among all representations of L, there is a standard one, which is called the **adjoint representation** and denoted by ad. For this representation, $M = L$ and the action is left multiplication. In other words, $\operatorname{ad}: L \to \operatorname{End} L, (\operatorname{ad} x)(y) = xy$.

For subspaces $A, B \leq L$, we define a **mutual commutator** $[A, B]$ as the linear span of all possible products ab where $a \in A, b \in B$. If both A and B are ideals of L then $[A, B]$ is also an ideal. In particular, a descending chain $L^{(0)} = L \supset L^{(1)} \supset L^{(2)} \supset \ldots$ where $L^{(k+1)} = [L^{(k)}, L^{(k)}]$ consists of ideals of L. This chain is called a **derived series** of L and $L^{(n)}$ is called a nth **commutator** or nth **derived subalgebra**. Another descending chain form ideals $L^1 = L, L^2 = [L, L], \ldots, L^{k+1} = [L^k, L], \ldots$. This chain is called a **lower central series**. From anticommutativity and the Jacobian identity, we get the inclusions $[L^k, L^m] \subset L^{k+m}$ and $L^{(k)} \subset L^{2^k}$.

A Lie algebra L is called **nilpotent of class** m if $L^{m+1} = 0$ and $L^m \neq 0$. In particular, if $m = 1$ then L is said to be an **abelian Lie algebra**. Similarly, if $L^{(m)} = 0, L^{(m-1)} \neq 0$, then L is called **solvable of length** m. A **metabelian** Lie algebra is a solvable algebra of length 2. If L is a nilpotent Lie algebra, i.e., $L^n = 0$, then L is also solvable and $(\operatorname{ad} x)^{n-1} = 0$ as a linear transformation on L for any $x \in L$. For a finite-dimensional Lie algebra these two conditions are equivalent.

C.49.2 Theorem (Engel) *If* $\dim L < \infty$ *and if for any* $x \in L$ *there* *exists a positive integer* $n = n(x)$ *such that* $(\operatorname{ad} x)^n = 0$, *then* L *is nilpotent.*

Another structural result about finite-dimensional Lie algebras is closely connected with representation theory (cf. Section E.3). A representation $\rho: L \to \operatorname{End} V$ is called **irreducible** if V has no nontrivial subspaces invariant under all transformations $\rho(x)$, $x \in L$. A representation ρ is **faithful** if $\operatorname{Ker} \rho = 0$. The dimension of ρ is defined as $\dim V$.

C.49.3 Theorem (Lie) *Let* L *be a finite-dimensional Lie algebra over a field of characteristic zero. If* $\rho: L \to \operatorname{End} V$ *is an irreducible finite-dimensional representation of* L, *then* $\rho(L^2) = 0$.

If, in addition, the field F in the previous Theorem is algebraically closed, then $\dim V = 1$. As a corollary of Lie's Theorem, we obtain a result on the structure of finite-dimensional solvable Lie algebras.

C.49.4 Corollary *Let* L *be a finite-dimensional solvable Lie algebra over a field of characteristic zero. Then its commutator* L^2 *is nilpotent.*

Obviously, if L is an arbitrary Lie algebra with nilpotent commutator L^2, then L itself is solvable. For another criterion of solvability, we need the following notion. Let $\dim L < \infty$ and $\operatorname{char} F = 0$. The bilinear form $B(x, y) = \operatorname{tr}(\operatorname{ad} x \, \operatorname{ad} y)$ is called the **Killing form of** L.

C.49.5 Theorem *A Lie algebra* L *is solvable if and only if its Killing form is the zero map.*

Any finite-dimensional Lie algebra L has a greatest solvable ideal (the **solvable radical**) $\operatorname{Rad} L$ and a greatest nilpotent ideal (the **nilpotent radical**). If $\operatorname{Rad} L = 0$ then L is called **semisimple**. If L has no non-trivial ideals then L is called **simple**. Any finite-dimensional Lie algebra L over a field of characteristic zero can be decomposed (as a vector space) into the direct sum $L = G \oplus R$, where $R = \operatorname{Rad} L$ and G is a semisimple subalgebra which is called the **Levi subalgebra** of L. Besides, $[L, R]$ is a nilpotent subalgebra. Any semisimple Lie algebra L is a direct sum $L = L_1 \oplus \cdots \oplus L_k$ of ideals $L_1, \ldots, L_k$ which are simple algebras. Over an algebraically closed field of characteristic zero, all finite-dimensional simple Lie algebras are classified. There is also a classification of the finite-dimensional simple Lie algebras over $\mathbb{R}$.

A linear map $\rho: L \to A$ from a Lie algebra L to an associative algebra A is called a **homomorphism** if $\rho(ab) = \rho(a)\rho(b) - \rho(b)\rho(a)$.

For any Lie algebra L, there exists a so-called **universal enveloping algebra**, that is an associative unitary algebra $U(L)$ with a canonical injective homomorphism $\varepsilon \colon L \to U(L)$ such that for any homomorphism $\varphi \colon L \to A$ from L to some associative algebra A there exists an unique associative homomorphism $\psi \colon U(L) \to A$ with $\psi\varepsilon = \varphi$. The correspondence between the bases of L and $U(L)$ gives

C.49.6 Theorem (Poincare-Birkhoff-Witt) *Let E be a totally ordered basis for L. Then 1 and all possible products $e_1 \ldots e_n$ with $e_1, \ldots, e_n \in E$, $e_1 \geq \cdots \geq e_n$, form a basis of $U(L)$.*

Any L-module M is also a module over the associative algebra $U(L)$, and vice versa.

For more information, see (Jacobson 1962, Serre 1992a, Bourbaki 1972b, Amayo and Stewart 1974, Goto and Grosshans 1978).

C.50 Simple Lie Algebras

by Andrej A. Zolotykh in Moscow, Russia

Let V be a finite dimensional Euclidean vector space over a field K under a non-degenerate symmetric bilinear form $(\ ,\ )$. For any nonzero element $\alpha \in V$, we consider the hyperplane that is orthogonal to α, and we let R_α be the corresponding reflection, i.e., $R_\alpha(\beta) = \beta - \langle \beta \mid \alpha \rangle \, \alpha$, where $\langle \beta \mid \alpha \rangle = 2\,(\beta, \alpha)/(\alpha, \alpha)$.

C.50.1 Definition A non-empty finite set Δ of nonzero elements of V is called a **root system** if the following conditions are fulfilled:
(1) $R_\alpha(\Delta) = \Delta$ for all $\alpha \in \Delta$;
(2) $\langle \alpha \mid \beta \rangle \in \mathbb{Z}$ for all $\alpha, \beta \in \Delta$;
(3) $2\alpha \notin \Delta$ for all $\alpha \in \Delta$.

Let V be a vector space of dimension at least 2 and $e_1, \ldots, e_n$ be an orthonormal basis of V. Then the set $\Delta_V := \{\, e_i - e_j \mid i \neq j \,\}$ is a root system.

A set $\Pi \subseteq \Delta$ is called a **system of simple roots** of a root system Δ if this set is linearly independent and any root of Δ can be presented as a linear combination of elements of Π with either all non-negative or all non-positive integer coefficients. A system of simple roots Π divides the root system Δ into the set of positive roots Δ_+ and the set of negative roots Δ_-, where $\Pi \subseteq \Delta_+$. We say that two sets M_1 and M_2 of elements of the vector space V are isomorphic if there are a metric preserving linear transformation φ of V and a nonzero element $a \in K$ such that $a\varphi(M_1) = M_2$. In this terminology, any two systems of simple roots of the same root system are isomorphic. Two root systems are isomorphic if and only if they have isomorphic systems of simple roots. The number of elements of a system of simple roots is called the **rank** of the corresponding root system.

A root system Δ is called **irreducible** if Δ is not a union of two mutually disjoint orthogonal subsystems. Any root system can be presented as a union of a finite number of irreducible root systems. The example Δ_V above is irreducible.

Any irreducible root system Δ consists of roots of not more that two different lengths. It means that either all the roots have the same length or there are roots of exactly two different lengths. In the second case, the root system is a unification of two subsets, namely the subset of the roots of the smaller length (short roots) and the subset of roots of the larger length (long roots). Let $\alpha, \beta \in \Pi$ be simple roots of Δ, $\alpha \neq \beta$, and let $|\alpha| \geq |\beta|$. Then one of the following conditions is fulfilled:

1) $\langle \alpha \mid \beta \rangle = \langle \beta \mid \alpha \rangle = 0$;

2) $\langle \alpha \mid \beta \rangle = \langle \beta \mid \alpha \rangle = -1$, $|\alpha| = |\beta|$;

3) $\langle \alpha \mid \beta \rangle = -2$, $\langle \beta \mid \alpha \rangle = -1$, $|\alpha|^2 = 2|\beta|^2$;

4) $\langle \alpha \mid \beta \rangle = -3$, $\langle \beta \mid \alpha \rangle = -1$, $|\alpha|^2 = 3|\beta|^2$.

We can draw a picture of a system of simple roots, presenting simple roots by circles and connecting circles of simple roots α, β by $\langle \alpha \mid \beta \rangle \langle \beta \mid \alpha \rangle$ lines. It means that for any two nonorthogonal simple roots their circles are connected by 1, 2 or 3 lines. If there are roots of two different lengths then we fill the circles of short roots. It is easy to see that two root systems are isomorphic if and only if their systems of simple roots have the same pictures.

The irreducible root systems are classified. There are four infinite series A_n, B_n, C_n, D_n of irreducible root systems of rank n ($n \geq 1$ for A_n, $n \geq 3$ for B_n, $n \geq 2$ for C_n, $n \geq 4$ for D_n), and five exceptional root systems E_6, E_7, E_8, F_4, G_2. All these root systems are drawn in the following picture.

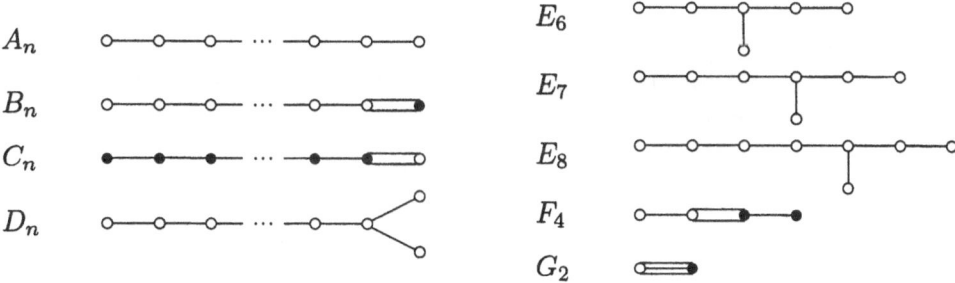

An example of an irreducible root system in a two dimensional vector space is drawn in the picture below. The 4 long roots are $\sqrt{2}$ times longer than the 4 short ones. The type of this root system is C_2. The roots α and β form a set of simple roots.

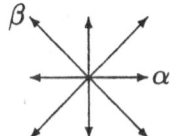

Another example of an irreducible root system is the following. The type of the example Δ_V above is A_{n-1}, and the set $\Pi = \{e_i - e_{i+1} \mid i = 1, \ldots, n-1\}$ is a set of simple roots. All these roots are of the same length.

A subalgebra H of an arbitrary finite dimensional Lie algebra is called a **Cartan subalgebra** if H is nilpotent and the condition $[x, H] \subseteq H$ is valid only for elements $x \in H$. All Cartan subalgebras of a finite dimensional Lie algebra have the same dimension. This dimension is called the **rank** of a Lie algebra.

Let V be a finite dimensional vector space of the dimension n over a field K. The Lie algebra $gl(V)$ of all linear transformations of V is called the general linear Lie algebra. For a given basis $E = \{e_1, \ldots, e_n\}$ of V, one can consider a set of linear transformations $H(E) = \{A \in gl(V) \mid Ae_i \in Ke_i \text{ for all } i\}$. Subalgebras $H(E) \subseteq gl(V)$ are examples of Cartan subalgebras.

Let L be a simple finite dimensional Lie algebra over an algebraically closed field of zero characteristic (for example, over the field of complex numbers $\mathbb{C}$). In this case, any Cartan subalgebra H of the Lie algebra L is commutative, i.e., $[H, H] = 0$. For any element α of the dual vector space H^* we denote by L_α the set of all $a \in L$ such that $[h, a] = \alpha(h)\, a$ for all $h \in H$. The Lie algebra L can be decomposed into a direct sum

$L = \bigoplus_{\alpha \in H^*} L_\alpha$. It is clear that $L_0 = H$, and that only a finite number of L_α are nonzero. A nonzero element $\alpha \in H^*$ is called a **root** of a simple Lie algebra L if $L_\alpha \neq 0$. The dimensions of the root spaces L_α are equal to 1 for all roots α of a Lie algebra L. We denote the set of all roots of a simple Lie algebra L by $\Delta(L)$.

The **Killing form** $(x, y) = \mathrm{Tr}\,(\mathrm{ad}x\, \mathrm{ad}y)$ is non-degenerate on a simple Lie algebra L as well as on a Cartan subalgebra H of L. It gives us a possibility to define a non-degenerate bilinear form on a dual space H^*.

C.50.2 Theorem *For any simple finite dimensional Lie algebra L over an algebraically closed field of characteristic 0, the set $\Delta(L)$ of roots of L is an irreducible root system. Two simple finite dimensional Lie algebras are isomorphic if and only if their root systems are isomorphic. For any irreducible root system there is a simple finite dimensional Lie algebra corresponding to this root system.*

This theorem and the classification of irreducible root systems show that a finite dimensional Lie algebra has one of the types A_n $(n \geq 1)$, B_n $(n \geq 3)$, C_n $(n \geq 2)$, D_n $(n \geq 4)$, E_6, E_7, E_8, F_4, G_2.

The subalgebra $sl(V)$ of $gl(V)$ of all transformations with zero trace is called the **special linear Lie algebra**. If $n \geq 2$ then $sl(V)$ is a simple Lie algebra of the type A_{n-1}.

Consider an arbitrary bilinear form $(\,,\,)$ on V. We say that a linear transformation $a \in gl(V)$ **preserves** the bilinear form if $(au, v) + (u, av) = 0$ for all $u, v \in V$. In the case of a non-degenerate symmetric bilinear form, the Lie subalgebra of all linear transformations preserving the form is called the **orthogonal Lie algebra** $so(V)$. In the case of a non-degenerate skew-symmetric bilinear form, the analogous Lie algebra is called the **symplectic Lie algebra** $sp(V)$. If n is odd and $n \geq 7$ then $so(V)$ is a simple Lie algebra of the type $B_{(n-1)/2}$. The Lie algebra $so(V)$ is a simple Lie algebra of the type $D_{n/2}$ for all even integers $n \geq 8$. In the case of a skew-symmetric form, the dimension n of V is even, and $sp(V)$ is a simple Lie algebra of the type $C_{n/2}$ for all $n \geq 4$.

There are descriptions of exceptional Lie algebras, too. For more information, see (Bourbaki 1971, 1998, 1968, 1975, Goto and Grosshans 1978, Jacobson 1979, Serre 1987)

C.51 Varieties of Lie Algebras

by Mikhael Zaicev in Moscow, Russia

Let $F(X)$ be the absolutely free non-associative algebra over a field F generated by countable a set X and let I be the ideal in $F(X)$ generated by all elements $ab - ba$ and $(ab)c + (bc)a + (ca)b$, with $a, b, c \in F(X)$. Then $\mathrm{Lie}(X) = F(X)/I$ is the **free Lie algebra** generated by X and any (countably generated) Lie algebra is its homomorphic image. Now let $0 \neq f = f(x_1, \ldots, x_n) \in \mathrm{Lie}(X)$. We say that the Lie algebra L **satisfies the identity** $f \equiv 0$ if $f(a_1, \ldots, a_n) = 0$ for any $a_1, \ldots, a_n \in L$. For a given set J of identities, we define a the variety $\mathscr{V} = \mathscr{V}(J)$ as the class of all Lie algebras satisfying all identities of J. For a variety $\mathscr{V}$, let $\mathrm{Id}(\mathscr{V})$ be the set of all identities of $\mathscr{V}$. We identify the identity $f \equiv 0$ with the element $f \in \mathrm{Lie}(X)$. Then $\mathrm{Id}(\mathscr{V})$ is an ideal of $\mathrm{Lie}(X)$, which is invariant under all endomorphisms of $\mathrm{Lie}(X)$. Such an ideal is said to be a **verbal ideal**. The factor algebra $\mathrm{Lie}(X, \mathscr{V}) = \mathrm{Lie}(X)/\mathrm{Id}(\mathscr{V})$ belongs to $\mathscr{V}$ and any (countably generated) algebra from $\mathscr{V}$ is its homomorphic image. All these notions can be extended to the case of an arbitrary set of generators X. The Lie algebra $\mathrm{Lie}(X, \mathscr{V})$ is called a **relatively free algebra** of $\mathscr{V}$, freely generated by X.

If $\mathscr{U}$ and $\mathscr{V}$ are two varieties and $I = \mathrm{Id}(\mathscr{U})$, $J = \mathrm{Id}(\mathscr{V})$ are their verbal ideals, then $I + J$, $I \cap J$, and the commutator $[I, J]$ are also verbal ideals of $\mathrm{Lie}(X)$. The corresponding varieties $\mathscr{U} \cap \mathscr{V}$, $\mathscr{U} \cup V$, and $[\mathscr{U}, \mathscr{V}]$ are called the intersection, union, and commutator of $\mathscr{U}$ and $\mathscr{V}$, respectively. Another important construction is the product of two varieties. Given varieties $\mathscr{U}, \mathscr{V}$, their product $\mathscr{U}\mathscr{V}$ is the class of all Lie algebras L such that $H \in \mathscr{U}$ and $L/H \in \mathscr{V}$ for some ideal H in L. If the base field F is infinite, then the family of all varieties over F with this multiplication forms a free monoid. Over a finite field, this multiplication might be non-associative.

An important property of the free Lie algebra $\mathrm{Lie}(X)$ is that every subalgebra is also a free Lie algebra. If $\mathscr{V}$ is an abelian variety, i.e., if it is defined by the identical relation $x_1 x_2 \equiv 0$, then $\mathrm{Lie}(X, \mathscr{V})$ is a vector space with basis X. Clearly, any subalgebra in $\mathrm{Lie}(X, \mathscr{V})$ is also relatively free with respect to $\mathscr{V}$, in this case. We say that $\mathscr{V}$ is a **Schreier variety** if any subalgebra in $\mathrm{Lie}(X, \mathscr{V})$ is also a relatively free algebra with respect to $\mathscr{V}$. It is known that only an abelian variety and the variety of all Lie F-algebras are Schreier varieties of Lie algebras.

An identity $g \equiv 0$ **follows** from $f \equiv 0$ if g belongs to the verbal ideal generated by f in $\mathrm{Lie}(X)$. One of the most important questions of the theory of varieties is the **Specht problem** or the "problem of a finite basis": *Is a given verbal ideal finitely generated?* If char $F = 0$, this problem is still open; for fields of positive characteristic, it has a negative solution. If L is finite-dimensional, then the answer to this question for the verbal ideal $\mathrm{Id}(L)$ of all identities of L depends on the base field. Namely, if char $F = 0$ then $\mathrm{Id}(L)$ has a finite basis of identities. If char $F = p > 0$ and F is infinite then there exists a Lie algebra L with $\dim L = 2p + 3$ such that $\mathrm{Id}(L)$ is infinitely generated as a verbal ideal. Finally, if $|F| < \infty$ and $\dim L < \infty$, then $\mathrm{Id}(L)$ can be defined by a finite number of identical relations. For example, if char $F = 0$ and $L = sl_2(F)$ is the Lie algebra of 2×2 traceless matrices then all its identities follows from two relations: $\sum_{\sigma \in S_4} (-1)^\sigma x_{\sigma(1)} \ldots x_{\sigma(4)} x_5 \equiv 0$; $(\mathrm{ad}\, x)^3 (yz) \equiv (\mathrm{ad}\, x)^3 (y) z + y((\mathrm{ad}\, x)^3 (z))$. Here $a \ldots bc$ means the right-normed product $a(\ldots(bc))$ of Lie elements.

Another central problem of the theory of varieties was the so-called **Engel's problem**: is it true that a Lie algebra L with an identity $(\mathrm{ad}\, x)^n \equiv 0$ is (locally) nilpotent?

C.51.1 Theorem (Zelmanov) *Let L be a Lie algebra over a field F with an identity $(\mathrm{ad}\, x)^n \equiv 0$. If L is finitely generated then L is nilpotent. If char $F = 0$ then L is nilpotent.*

Over fields of positive characteristic, there are examples of non-nilpotent (even non-solvable) Lie algebras with an identity $(\mathrm{ad}\, x)^n \equiv 0$.

Identities of Lie algebras play an important role in studying of their structure, we will illustrate it with a few examples. Let L and H be two finite-dimensional simple Lie algebras over an algebraically closed field F with char $F = 0$. Then L and H are isomorphic if and only if they have the same identities. Let L be a finitely generated Lie algebra over an infinite field F with an identity of the type $zy^n x \equiv \sum_{j \geq 1} \alpha_j y^j z y^{n-j} x$, where $y^k x = yx$ for $k = 1$ and $y^k x = y(y^{k-1}x)$ for $k \geq 2$. Then L is Noetherian and satisfies the **Hopf condition**, i.e., any surjective endomorphism of L is an automorphism. If, in addition, char $F > 0$ or char $F = 0$, but L also satisfies an identity $(x_1 y_1) \ldots (x_m y_m) \equiv 0$ then L is residually finite and **representable**, i.e., can be embedded into a matrix algebra $M_t(K)$ over some extension K of the base field F. Finally, let L be a **special** Lie algebra, that is, L can be embedded as a Lie algebra into some associative algebra satisfying a polynomial identity. If L is finitely generated and F is infinite, then L has a greatest solvable ideal R. Moreover, L/R is representable and is a subdirect product of finite-dimensional simple Lie algebras of restricted dimension. If, in addition, char $F = 0$ then $[L, R]$ is a nilpotent ideal.

In case of characteristic 0, any identity is equivalent to some set of multilinear identities and any variety $\mathscr{V}$ can be uniquely defined by the family of subspaces $\mathrm{Id}(\mathscr{V}) \cap P_n$, $n = 1, 2, \ldots$, where P_n denote a subspace of multilinear Lie polynomials in $\mathrm{Lie}(X)$ depending on $x_1, \ldots, x_n$. The symmetric group S_n acts on P_n by a natural way permuting variables. Under this action, $\mathrm{Id}(\mathscr{V}) \cap P_n$ is an S_n-submodule of P_n for any variety $\mathscr{V}$. The structure of $P_n(\mathscr{V}) = P_n/\mathrm{Id}(\mathscr{V}) \cap P_n$ plays an important role in studying $\mathscr{V}$. Its S_n-character $\chi_n(\mathscr{V}) = \chi(P_n(\mathscr{V}))$ is called nth **cocharacter** of $\mathscr{V}$ and the dimension $c_n(\mathscr{V}) = \dim P_n(\mathscr{V})$ is called nth **codimension** of $\mathscr{V}$. If $\chi_n(\mathscr{V}) = \sum_{\lambda \vdash n} m_\lambda \chi_\lambda$ is a decomposition of the nth cocharacter into a sum of irreducible S_n-characters χ_λ with multiplicities m_λ, then $l_n(\mathscr{V}) = \sum_{\lambda \vdash n} m_\lambda$ is another numerical characteristic of $\mathscr{V}$ which is called the **colength**. For associative varieties, it is well-known that $c_n(\mathscr{V})$ is an exponentially bounded function of n and $l_n(\mathscr{V})$ is polynomially bounded. It is also known that $\lim_{n \to \infty} \sqrt[n]{c_n(\mathscr{V})}$ always exists and is an integer. For a Lie variety $\mathscr{V}$, both $c_n(\mathscr{V})$ and $l_n(\mathscr{V})$ may have an overexponential growth. For example, if $\mathscr{V} = \mathbf{AN}_2$ is the variety of all abelian-by-nilpotent (of step two) Lie algebras then $c_n(\mathscr{V}) \simeq \sqrt{n!}$. For an arbitrary variety $\mathscr{V}$, one can restrict $c_n(\mathscr{V})$ only by $\frac{n!}{\ln^{(k)}(n)}$ for some $k \geq 1$, where $\ln^{(k)}(x) = \underbrace{\ln \ldots \ln(x)}_{k}$.

Even in case of exponentially bounded codimension growth, the existence of $\lim_{n \to \infty} \sqrt[n]{c_n(\mathscr{V})}$ is still an open question, but there exists a variety such that $3 < \liminf_{n \to \infty} \sqrt[n]{c_n(\mathscr{V})}$, $\limsup_{n \to \infty} \sqrt[n]{c_n(\mathscr{V})} < 4$. For any special variety $\mathscr{V}$, its codimension and colength are exponentially and polynomially bounded, respectively. In particular, this holds for any variety $\mathscr{V}$ generated by a finite-dimensional Lie algebra L. In this case, $\lim_{n \to \infty} \sqrt[n]{c_n(\mathscr{V})}$ exists and is an integer d such that $d \leq \dim L$. Moreover, $d = \dim L$ if and only if L is a central simple Lie algebra.

For more, see (Bahturin 1987, Razmyslov 1989, Bahturin et al. 1992, Mishchenko 1990, Zaicev 1993).

C.52 Lie Superalgebras

by Alexander A. Mikhalev in Moscow, Russia

Lie superalgebras appeared in homotopy theory in connection with the Whitehead product on homotopy groups, in particle physics for describing the boson-fermion systems, where a fundamental role is played by the supersymmetry mixing even and odd coordinates, and in the theory of deformations of graded algebraic systems. In fact, supersymmetry is described by Lie supergroups or, infinitesimally, by Lie superalgebras. Since the 1950s, the theory of Lie superalgebras was developed systematically (Killing-Cartan theory, representation theory, classification of simple Lie superalgebras, Lie superalgebras of vector fields, varieties of Lie superalgebras, etc.). At the same time, in many situations in physics we have to consider not only $\mathbb{Z}_2$- or $\mathbb{Z}$-gradings of algebras but also G-gradings, where G is an abelian group equipped with a skew symmetric bilinear form given by a 2-cocycle. In the 1970s, we got some generalized Lie superalgebras under different names: color Lie superalgebras, ε-Lie algebras, generalized Lie algebras, and so on. In fact, these algebras (generalized Lie algebras) were introduced by R. Ree in 1960.

Let K be a commutative associative ring with identity, K^* the group of invertible elements of K, and G a commutative semigroup. A mapping $\varepsilon \colon G \times G \to K^*$ is called a **commutation factor** (a **bicharacter**) on G if $\varepsilon(g,h)\varepsilon(h,g) = 1$, $\varepsilon(g,g) = \pm 1$, $\varepsilon(g, h + f) = \varepsilon(g,h)\varepsilon(g,f)$, $\varepsilon(g + h, f) = \varepsilon(g,f)\varepsilon(h,f)$, for all $f, g, h \in G$.

Also, let $G_+ = \{ g \in G \mid \varepsilon(g,g) = 1 \}$ and $G_- = \{ g \in G \mid \varepsilon(g,g) = -1 \}$. For instance, if $G = \mathbb{Z}_2 \oplus \mathbb{Z}_2$, $K = \mathbb{C}$, $f = (f_1, f_2)$, $g = (g_1, g_2) \in G$, then the following form is a commutation factor on G: $\varepsilon(f,g) = (-1)^{(f_1+f_2)(g_1+g_2)}$; $G_+ = \{ (0,0), (1,1) \}$, $G_- = \{ (0,1), (1,0) \}$. Suppose that $n = 2^k$, $k \geq 1$. Let ε_n be a primitive root of 1, $\varepsilon_n = e^{2\pi i/n}$, $G = \langle f \rangle_n \oplus \langle g \rangle_n$ the direct sum of cyclic groups of order n generated by a and b, respectively, $\varepsilon(f,f) = 1$, $\varepsilon(f,g) = \varepsilon_n$, $\varepsilon(g,g) = -1$. Then ε is a commutation factor on G.

A G-graded K-algebra $R = \bigoplus_{g \in G} R_g$ with the multiplication $[\,,\,]$ is a **color Lie superalgebra** if $[a,b] = -\varepsilon(d(a), d(b))[b,a]$ (ε-**anti-commutativity**), $\varepsilon(d(c), d(a))[a, [b,c]] + \varepsilon(d(a), d(b))[b, [c,a]] + \varepsilon(d(b), d(c))[c, [a,b]] = 0$ (ε-**Jacobi identity**), and $[v, [v,v]] = 0$ with $d(v) \in G_-$, for G-homogeneous elements $a, b, c, v \in R$, where $d(u) = g$ if $u \in R_g$. A color Lie superalgebra L with the trivial multiplication ($[a,b] = 0$ for all $a, b \in L$) is called **abelian** (ε-**commutative**). If $G = \mathbb{Z}_2$, $\varepsilon(g,h) = (-1)^{gh}$, then a color Lie superalgebra is an ordinary Lie superalgebra. If $\varepsilon \equiv 1$, then we have a G-**graded Lie algebra**.

Let $K = \mathbb{C}$, β a bilinear form on a G-graded vector space V, $\beta(x,y) = -\varepsilon(d(x), d(y))\beta(y,x)$, for G-homogeneous elements x, y of V. Let $H(V; \beta) = \bigoplus_{g \in G} H(V; \beta)_g$, $H(V; \beta)_0 = V_0 \oplus \mathbb{C} \cdot z$, and $H(V; \beta)_g = V_g$, for $g \in G$, $g \neq 0$, $[x,y] = \beta(x,y)z$, $[H(V; \beta), z] = [z, H(V; \beta)] = 0$, for all homogeneous elements $x, y \in V$. Then $H(V; \beta)$ is a color Lie superalgebra (so called ε-**Heisenberg algebra**).

Let p be a prime number, K a $\mathbb{Z}_p$-algebra. A color Lie superalgebra R over K is a **color Lie p-superalgebra** if on homogeneous components R_g, $g \in G_+$, we have a mapping $a \to a^{[p]}$, $d(a^{[p]}) = p\,d(a)$, such that for all $\alpha \in K$, $a, b, c \in R$, $d(a) = d(b) \in G_+$, the following conditions are satisfied: $(\alpha a)^{[p]} = \alpha^p a^{[p]}$; $(\operatorname{ad}(a^{[p]}))(c) = [a^{[p]}, c] = (\operatorname{ad} a)^p(c)$; $(a + b)^{[p]} = a^{[p]} + b^{[p]} + \sum s_i(a, b)$, where $j s_j(a, b)$ is the coefficient on t^{j-1} in the polynomial $(\operatorname{ad}(ta + b))^{p-1}(a)$ and $(\operatorname{ad} a)(z) = [a, z]$.

Let A be a G-graded associative K-algebra, ε a commutation factor on G. We consider a new multiplication $[,]$ on A given by $[a, b] = ab - \varepsilon(d(a), d(b))\, ba$ for homogeneous elements a, b of A. Then A, with this new multiplication, becomes a color Lie superalgebra and is denoted by $[A]$. If p is a prime number and K is a $\mathbb{Z}_p$-algebra, then A with the above defined new operation $[,]$ and the mapping $[p]: A_g \to A_{pg}$ for $g \in G_+$ given by $a^{[p]} = a^p$ for homogeneous elements of A with $d(a) \in G_+$, is a color Lie p-superalgebra (notation: $[A]^p$).

Let K be a field, $V = \bigoplus V_g$ a G-graded vector space over K. An element x of V is said to be **homogeneous** if $x \in V_g$ for some $g \in G$; in this case we set $d(x) = g$. By $\operatorname{END}_K(V)_g$, $g \in G$, we denote the set of all K-linear operators $f \in \operatorname{End}_K(V)$ such that the image of each homogeneous element $x \in V$ is a homogeneous element and $d(f(x)) = g + d(x)$. Let $\operatorname{END}_K(V) = \bigoplus_{g \in G} \operatorname{END}_K(V)_g$. It is clear that $\operatorname{END}_K(V)$ is a G-graded associative K-algebra, with the usual multiplications of operators (a K-subalgebra of $\operatorname{End}_K(V)$). The color Lie superalgebra $\operatorname{gl}(V) = [\operatorname{END}_K(V)_g]$ is called the **general linear color Lie superalgebra**. Note that, in the linear algebra of graded vector spaces, instead of the usual trace function, it is necessary to use the function grtr given by $\operatorname{grtr}(A) = \sum_{g \in G} \varepsilon(g, g)\operatorname{tr}(A_{gg})$, where A_{gg} is the block on the diagonal of the matrix A corresponding to the subspace V_g, tr is the usual trace function. Let $K = \mathbb{C}$, $G = \mathbb{Z}_2$, $\varepsilon(g, h) = (-1)^{gh}$. The Lie superalgebra $\operatorname{sl}(V) = \{A \in \operatorname{gl}(V) \mid \operatorname{grtr}(A) = 0\}$ is called a **special linear Lie superalgebra** if it is the commutator algebra of $\operatorname{gl}(V)$.

Let L be a color Lie superalgebra, U a G-graded associative K-algebra with 1, $\delta: L \to [U]$ a homomorphism of color Lie superalgebras. We met the enveloping algebra $U(L)$ already in Section C.49. Let K be a field, $p = \operatorname{char} K > 2$, L a color Lie p-superalgebra, U a G-graded associative K-algebra with 1, $\delta: L \to [U]^p$ a homomorphism of color Lie p-superalgebras. We say that U (with δ) is the **restricted universal enveloping algebra** of L (notation: $u(L)$) if for any homomorphism

$\sigma \colon L \to [R]^p$ of color Lie p-superalgebras, where R is a G-graded associative K-algebra with 1 (with the same ε and G), there exists a unique homomorphism $\theta \colon U \to R$ of G-graded associative K-algebras with 1 such that $\sigma = \theta\delta$. Suppose now that K has no 2-torsion, L is a color Lie superalgebra which is a free K-module with a G-homogeneous basis $X = X_+ \cup X_-$, $X_\pm = \{\, x \in X \mid d(x) \in G_\pm \,\}$, $A(X)$ is the free G-graded associative algebra, and I is the two-sided ideal in $A(X)$, generated by the elements of the form $ab - \varepsilon(d(a), d(b))\, ba - [a, b]$ for all homogeneous $a, b \in L$. Consider the canonical mapping $\delta \colon L \to A(X)/I = U(L)$. It is clear that $U(L)$ is the universal enveloping algebra of L. Let $\le$ be a total ordering of X.

C.52.1 Theorem *Let K have no 2-torsion. The universal enveloping algebra $U(L)$ constructed above is a free K-module with a G-homogeneous basis consisting of 1 and all monomials of the form $\delta(x_1) \cdots \delta(x_n)$ where $n \in \mathbb{N}$, $x_i \in X$, $x_i \le x_{i+1}$, $x_i \ne x_{i+1}$ with $x_i \in X_-$ for all $i = 1, \ldots, n - 1$. In particular, δ is an embedding.*

Let K be a field, char $K = p > 2$, L a color Lie p-superalgebra with a G-homogeneous linear basis X, J the two-sided ideal of $A(X)$ generated by elements $ab - \varepsilon(a, b)\, ba - [a, b]$ (where a, b are homogeneous, $a, b \in L$), and $a^p - a^{[p]}$ ($a \in L_g$, $g \in G_+$). Let $\delta \colon L \to A(X)/J$ be the canonical mapping. It is clear that $A(X)/J = u(L)$ is the restricted universal enveloping algebra for L.

C.52.2 Theorem *The monomials of the form $\delta(x_1)^{\lambda_1} \cdots \delta(x_n)^{\lambda_n}$, where $x_i \in X$, $x_1 < \cdots < x_n$, $0 \le \lambda_i \le p - 1$ for $x_i \in X_+$, $\lambda_i = 0, 1$ for $x_i \in X_-$ give us a linear basis of $u(L)$.*

C.52.3 Theorem *Let K be a field, L a color Lie superalgebra, B a subalgebra of L. Then B is differential separable, i.e., if J is the left ideal of $U(L)$ generated by B, then $L \cap J = B$.*

In the case where char $K = p > 2$, let M be a color Lie p-superalgebra, and D a subalgebra of M. Then D is differential separable relative to $u(M)$.

A Lie superalgebra of dimension greater than 1 is simple if it does not contain proper nontrivial ideals. For instance, let V be a finite dimensional $\mathbb{Z}_2$-graded vector space over $\mathbb{C}$, $\dim(V_0) \ge 1$, $\dim(V_1) \ge 1$, $\dim(V_0) \ne \dim(V_1)$, $G = \mathbb{Z}_2$, and $\varepsilon(g, h) = (-1)^{gh}$. Then the special linear Lie superalgebra $\mathrm{sl}(V)$ is a simple Lie superalgebra. The classification of simple finite dimensional ordinary Lie superalgebras over

an algebraically closed field of characteristic zero has been given by
V. G. Kac. One can define solvable and nilpotent Lie superalgebras
by analogy with Lie algebras definitions. A finite dimensional ordinary
Lie superalgebra $L = L_0 \oplus L_1$ is solvable if and only if this is true for L_0.
A Lie superalgebra L is called **semisimple** if it has no nonzero solvable
ideals (this is equivalent to having no nonzero abelian ideals). In the
case of finite dimensional Lie algebras, it is true that any semisimple Lie
algebra is a direct sum of its simple ideals. This is not true already in
the case of Lie algebras with positive characteristic, not mentioning Lie
superalgebras, where it is wrong already for finite dimensional ordinary
Lie superalgebras over a field of characteristic zero.

Finally, we define free Lie superalgebras. Given an abelian semigroup G, a field K, a commutation factor ε, and a non-empty G-graded
set $X = \bigcup_{g \in G} X_g$, we define the **free color Lie superalgebra** $L(X)$
with free generating set X as a Lie superalgebra generated by X whose
grading is compatible with that of X and such that any graded map-
ping $\varphi \colon X \longrightarrow M$ of X into a color Lie K-superalgebra M (with the
same G and ε) extends to a homomorphism $\bar{\varphi} \colon L(X) \longrightarrow M$ such that
$\bar{\varphi}\,|_X = \varphi$. Let K be a field, char $K \neq 2$, $A(X)$ the free G-graded associative algebra, and $[X]$ the subalgebra of the color Lie superalge-
bra $[A(X)]$ generated by X. The algebra $[X]$ is the free color Lie superalgebra. If char $K = p > 2$, $[X]^p$ is the subalgebra of the color Lie
p-superalgebra $[A(X)]^p$ generated by X, then $[X]^p$ is the free color Lie
p-superalgebra. Every Lie superalgebra is isomorphic to a quotient algebra of a suitable free Lie superalgebra. Subalgebras of free color Lie
superalgebras and free color Lie p-superalgebras are free. One can find
results on the structure and combinatorial properties of free Lie superalgebras in the monographs (Bahturin et al. 1992, Mikhalev and Zolotykh
1995).

Further bibliography: (Bahturin, Mikhalev, and Zaicev 2000, Berezin
1987, Kac 1977b,a, Leĭtes 1984, Marcinek 1991, Scheunert 1979).

C.53 Other Nonassociative Algebras

by Luiz A. Peresi in São Paulo, Brazil

Besides Lie algebras, there are several classes of nonassociative algebras.
Some of them have applications in other mathematical fields, physics,
and genetics. For instance, *Bernstein algebras* are motivated by the
problem of classifying populations that achieve equilibrium at the second

generation. In this section, we consider those nonassociative algebras which are most closely related to associative algebras, namely *alternative* and *Jordan* algebras.

All algebras in this section are nonassociative, i.e., not necessarily associative. If A is an algebra, its **center** $Z(A)$ is the set of elements of A that associate and commute with all elements of A.

An algebra is called **power-associative** if any of its elements generates an associative subalgebra. A power-associative algebra is a **nilalgebra** when its elements are nilpotent. Any power-associative algebra A contains a unique maximal nilideal $\mathrm{Nil}(A)$, called the **nilradical** of A.

We define the following powers of an algebra A inductively: $A^1 = A$, $\ldots$, $A^n = A^{n-1}A + A^{n-2}A^2 + \cdots + AA^{n-1}$; $A^{(0)} = A$, $\ldots$, $A^{(n)} = \left(A^{(n-1)}\right)^2$. We say that A is **nilpotent (solvable)** if $A^k = 0$ ($A^{(k)} = 0$) for some k. In other words, A is nilpotent if there exists a positive integer k such that any product of k elements of A (no matter how they are associated) is zero.

Albert conjectured that any finite-dimensional commutative power-associative nilalgebra is nilpotent. Suttles gave the following counter-example in 1972: let A be the commutative algebra with basis $\{e_1, \ldots, e_5\}$ and nonzero products given by $e_1 e_2 = e_2 e_4 = -e_1 e_5 = e_3$, $e_1 e_3 = e_4$, $e_2 e_3 = e_5$; A is a power-associative nilalgebra of index 4 that is solvable but not nilpotent (since $A^3 = A^2$ and $A^2 A^2 = 0$).

The following is an open problem, known as **Albert's Problem**: *Is any finite-dimensional commutative power-associative nilalgebra solvable?*

Alternative algebras

An algebra A over F (with an identity) is a **composition algebra** if there is a nondegenerate quadratic form $n: A \to F$ such that $n(xy) = n(x)n(y)$ ($\forall x, y \in A$). A composition algebra has dimension 1, 2, 4, or 8, and is isomorphic to one of the following algebras: F, $\mathrm{char}(F) \neq 2$; $F \oplus Fv_1$, $v_1^2 = v_1 + \mu$ ($\mu \in F$, $4\mu + 1 \neq 0$); a four-dimensional associative algebra that generalizes quaternions over $\mathbb{R}$; a *Cayley-Dickson algebra* (i.e., an eight-dimensional algebra that generalizes octonions over $\mathbb{R}$).

An algebra A is **alternative** if $(xx)y = x(xy)$ and $(yx)x = y(xx)$ ($\forall x, y \in A$). Cayley-Dickson algebras are alternative.

Let A be an alternative algebra. Artin proved that any two elements of A generate an associative subalgebra. In particular, A is power-associative. Kleinfeld showed that if A is simple and not associative then

$Z(A)$ is a field and A is a Cayley-Dickson algebra over $Z(A)$. When A is finite-dimensional, $\mathrm{Nil}(A)$ is nilpotent, and $A/\mathrm{Nil}(A)$ is isomorphic to a finite direct sum of simple algebras which are either matrix algebras over associative division algebras or Cayley-Dickson algebras.

The alternative algebra A is called *primitive* if it contains a maximal *modular* right ideal M (i.e., there is an $e \in A$ such that $x - ex \in M$ for all $x \in A$) and M does not contain nonzero two-sided ideals. A Cayley-Dickson algebra is primitive. Any primitive alternative algebra is associative primitive or a Cayley-Dickson algebra. An ideal I of A is *primitive* when A/I is primitive.

For the alternative algebra A, consider $A^{\sharp} = F \oplus A$ with multiplication $(\alpha + a)(\beta + b) = (\alpha\beta) + (\alpha b + \beta a + ab)$. The *radical* $\mathrm{Rad}(A)$ of A is the largest *quasi-regular* right ideal I of A (i.e., $\forall x \in I$, $1 - x$ has an inverse in $A^{\sharp}$), and it is also the intersection of all maximal modular right ideals of A.

When A is alternative and Artinian, $\mathrm{Rad}(A)$ is the largest nilpotent ideal and $A/\mathrm{Rad}(A)$ is a finite direct sum of matrix algebras over associative division algebras and Cayley-Dickson algebras.

C.53.1 Theorem (Kleinfeld) *Let A be an alternative algebra. Then $\mathrm{Rad}(A)$ is the intersection of all primitive ideals of A, and $A/\mathrm{Rad}(A)$ is isomorphic to a subdirect product of associative primitive and Cayley-Dickson algebras.*

Jordan algebras

Jordan algebras were introduced by Jordan, von Neumann and Wigner (in a paper published in Annals of Mathematics in 1934) with the aim that this new algebraic system would give a better interpretation of quantum mechanics. This aim was not achieved, but Jordan algebras became an interesting topic.

An algebra J is a *Jordan algebra* if $xy = yx$ and $(x^2 y)x = x^2(yx)$, for all $x, y \in J$. Any Jordan algebra is power-associative.

Let A be an associative or an alternative algebra. If we replace the multiplication xy in A by the *Jordan product* $x \circ y = \frac{1}{2}(xy + yx)$ (cf. Section C.24), we obtain a Jordan algebra denoted by $A^{(+)}$.

Let V be a vector space over F, $f : V \times V \to F$ a symmetric bilinear form, and $J(V, f) = F \oplus V$. With the multiplication $(\alpha + x)(\beta + y) = (\alpha\beta + f(x, y)) + (\beta x + \alpha y)$, the vector space $J(V, f)$ is a Jordan algebra.

Let A be an algebra with involution $^{-} : a \in A \to \bar{a} \in A$. Let

$H(A, {}^-) = \{\, a \in A \mid \bar{a} = a \,\}$. If A is associative or alternative, then $H(A, {}^-)$ is a subalgebra of $A^{(+)}$, hence a Jordan algebra. For the matrix algebra $\mathrm{Mat}_{n \times n}(A)$, we consider the involution given by $\overline{(a_{ij})}^t = (\overline{a_{ji}})$ and the algebra $H(\mathrm{Mat}_{n \times n}(A), {}^{-t})$ that is not necessarily a Jordan algebra. In particular, if C is a Cayley-Dickson algebra and $^-$ denotes its involution, then $H(C, {}^-)$, $H(\mathrm{Mat}_{2 \times 2}(C), {}^{-t})$ and $H(\mathrm{Mat}_{3 \times 3}(C), {}^{-t})$ are Jordan algebras.

The Jordan algebra J is **special** if J is a subalgebra of $B^{(+)}$ for some associative algebra B. Otherwise, J is **exceptional**. The Jordan algebras $A^{(+)}$, $J(V, f)$, $H(C, {}^-)$, and $H(\mathrm{Mat}_{2 \times 2}(C), {}^{-t})$ are special. Also in 1934, Albert proved that $H(\mathrm{Mat}_{3 \times 3}(C), {}^{-t})$ is exceptional. In 1956, Shirshov proved that any Jordan algebra with two generators is special, and since $H(\mathrm{Mat}_{3 \times 3}(C), {}^{-t})$ has three generators, this result cannot be improved.

C.53.2 Theorem (Albert) *Let J be a finite-dimensional Jordan algebra. Then $\mathrm{Nil}(J)$ is nilpotent, and $J/\mathrm{Nil}(J)$ is isomorphic to a finite direct sum of simple algebras. Over an algebraically closed field F, each one of these simple algebras is isomorphic to one of the following algebras: F; $J(V, f)$, where f is nondegenerate; $H(\mathrm{Mat}_{n \times n}(A), {}^{-t})$, $n \geq 3$, where A is a composition algebra that is associative for $n > 3$.*

An **Albert algebra** is a Jordan algebra J over F such that (for some extension K of F) $J \otimes_F K \cong H(\mathrm{Mat}_{3 \times 3}(C), {}^{-t})$. An Albert algebra is exceptional, simple, and has dimension 27 over its center.

C.53.3 Theorem (Zelmanov) *A simple Jordan algebra is isomorphic to one of the following algebras: $J(V, f)$, where f is nondegenerate; $A^{(+)}$, where A is a simple associative algebra; $H(A, {}^-)$, where A is a simple associative algebra with involution $^-$; an Albert algebra.*

As in the case of alternative algebras, we can define the quasi-regular radical $\mathrm{Rad}(J)$ of a Jordan algebra J. A subspace K of J is a **quadratic ideal** if $2(ka)k - k^2 a \in K$ for all $k \in K$ and $a \in J$.

C.53.4 Theorem *Let J be a Jordan algebra. Suppose that J satisfies the minimal condition for quadratic ideals. Then we have:*

(1) (Slinko-Zelmanov) *The radical $\mathrm{Rad}(J)$ is nilpotent and has finite dimension.*

(2) (Jacobson-Osborn) *The quotient algebra $J/\mathrm{Rad}(J)$ is isomorphic to a finite direct sum of simple Jordan algebras of one of the following*

forms: a division Jordan algebra; $H(A,^-)$, where A is an associa- tive Artinian algebra with involution $^-$, and A is $^-$-simple (i.e., A does not contain proper ideals I such that $\overline{I} \subset I$); $J(V, f)$ for f nondegenerate; an Albert algebra.

For more information, see (Jacobson 1968, Kuz'min and Shestakov 1994, Zhevlakov, Slinko, Shestakov, and Shirshov 1982, Schafer 1995, Lyubich 1992).

C.54 Computational Ring Theory

by Franz Winkler in Linz, Austria

Whenever R is a ring, then by R^* we denote $R \setminus \{0\}$.

C.54.1 Definition A *Euclidean domain* D is a commutative inte- gral domain together with a *degree function* deg: $D^* \to \mathbb{N}_0$, such that

(1) $\deg(a \cdot b) \geq \deg(a)$ for all $a, b \in D^*$,
(2) (division property) for all $a, b \in D$, $b \neq 0$, there exists a *quotient* q and a *remainder* r in D such that $a = q \cdot b + r$ and ($r = 0$ or $\deg(r) < \deg(b)$).

Any Euclidean domain is a unique factorization domain.

C.54.2 Theorem *Any two non-zero elements a, b of a Euclidean do- main D have a greatest common divisor g which can be written as a linear combination $g = s \cdot a + t \cdot b$ for some $s, t \in D$.*

The elements s, t in the previous theorem are called the *Bézout co- factors* of a and b.

A Euclidean domain in which quotient and remainder are computable by algorithms QUOT and REM admits an algorithm for computing the greatest common divisor g of any two elements a, b. This algorithm has originally been stated by Euclid for the domain of the integers. In fact, it can be easily extended to compute not only the gcd but also the Bézout cofactors.

In a Euclidean domain we can solve the *Chinese remainder prob- lem (CRP):*
given: $r_1, \ldots, r_n \in D$ (remainders)
$\qquad m_1, \ldots, m_n \in D^*$ (moduli), pairwise relatively prime
find: $r \in D$, such that $r \equiv r_i \bmod m_i$ for $1 \leq i \leq n$.

Algorithm 1 Extended Euclidean algorithm

Require: a, b are elements of the Euclidean domain D;
Ensure: g is the greatest common divisor of a, b and $g = s \cdot a + t \cdot b$;
1: $(r_0, r_1, s_0, s_1, t_0, t_1) := (a, b, 1, 0, 0, 1)$;
2: $i := 1$;
3: **while** $r_i \neq 0$ **do**
4:     $q_i := \text{QUOT}(r_{i-1}, r_i)$;
5:     $(r_{i+1}, s_{i+1}, t_{i+1}) := (r_{i-1}, s_{i-1}, t_{i-1}) - q_i \cdot (r_i, s_i, t_i)$;
6:     $i := i + 1$;
7: **end while**
8: $(g, s, t) := (r_{i-1}, s_{i-1}, t_{i-1})$;

There are basically two solution methods for the CRP. The first one is usually associated with the name of J. L. Lagrange. The second one is associated with I. Newton and is a recursive solution.

In the Lagrangian solution one first determines u_{kj} such that $1 = u_{kj} \cdot m_k + u_{jk} \cdot m_j$, for $1 \leq j, k \leq n, j \neq k$. This can obviously be achieved by the extended Euclidean algorithm. Next one considers the elements $l_k := \prod_{j=1, j \neq k}^{n} u_{jk} m_j$, for $1 \leq k \leq n$. Clearly $l_k \equiv 0 \bmod m_j$ for all $j \neq k$. On the other hand $l_k = \prod_{j=1, j \neq k}^{n}(1 - u_{kj} m_k) \equiv 1 \bmod m_k$. So $r = \sum_{k=1}^{n} r_k \cdot l_k$ solves CRP.

The disadvantage of the Lagrangian approach is that it yields a static algorithm, i.e., it is virtually impossible to increase the size of the problem by one more pair r_{n+1}, m_{n+1} without having to recompute everything from the start.

The Newton approach is recursive in the sense that one first solves the problem with 2 remainders and 2 moduli, yielding the solution $r_{1,2}$, and then has to solve the problem with remainders $r_{1,2}, r_3, \ldots, r_n$ and moduli $m_1 \cdot m_2, m_3, \ldots, m_n$. So for given remainders r_1, r_2 and moduli m_1, m_2 we want to find an $r \in D$ such that $r \equiv r_i \bmod m_i$ for $i = 1, 2$. The solution of this CRP of size 2 is computed by the Chinese remainder algorithm CRA.

A special case of the CRP in the polynomial ring $K[x]$, K a field, is the *interpolation problem*. All the moduli m_i are linear polynomials of the form $x - \beta_i$.

given: $\alpha_1, \ldots, \alpha_n \in K$,

    $\beta_1, \ldots, \beta_n \in K$, such that $\beta_i \neq \beta_j$ for $i \neq j$,
find: $u(x) \in K[x]$, such that $u(\beta_i) = \alpha_i$ for $1 \leq i \leq n$.

Algorithm 2 Chinese remainder algorithm(Newtonian solution of the CRP)

Require: r_1, r_2, m_1, m_2 are elements of the Euclidean domain D, m_1, m_2 are non-zero and relatively prime;

Ensure: $r \equiv r_1 \bmod m_1$ and $r \equiv r_2 \bmod m_2$;

1: $c := m_1^{-1} \bmod m_2$;
2: $r_1' := r_1 \bmod m_1$;
3: $\sigma := (r_2 - r_1')c \bmod m_2$;
4: $r := r_1' + \sigma m_1$;

Since $p(x) \bmod (x - \beta) = p(\beta)$ for $\beta \in K$, the interpolation problem is a special case of the CRP. The inverse of $p(x)$ in $K[x]_{/\langle x - \beta \rangle}$ is $p(\beta)^{-1}$. So CRA yields a solution algorithm for the interpolation problem, namely the Newton interpolation algorithm. Similarly, the Lagrangian solution to the CRP leads to a Lagrangian solution of the interpolation problem.

The Chinese remainder problem can, in fact, be described in greater generality. Let R be a commutative ring with unity. The abstract Chinese remainder problem is the following:

given $r_1, \ldots, r_n \in R$ (remainders),
$I_1, \ldots, I_n$ ideals in R (moduli), such that $I_i + I_j = R$ for all $i \neq j$;

find $r \in R$, such that $r \equiv r_i \bmod I_i$, for $1 \leq i \leq n$.

The abstract Chinese remainder problem can be treated basically in the same way as the CRP over Euclidean domains. Again there is a Lagrangian and a Newtonian approach and one can show that the problem always has a solution and if r is a solution then the set of all solutions is given by $r + I_1 \cap \cdots \cap I_n$. i.e., the map $\varphi : r \mapsto (r + I_1, \ldots, r + I_n)$ is a homomorphism from R onto $\prod_{j=1}^n R_{/I_j}$ with kernel $I_1 \cap \cdots \cap I_n$. However, in the absence of the Euclidean algorithm it is not possible to compute a solution of the abstract CRP. See (Lauer, 1983).

Instead of solving a problem in several homomorphic images and then combining these modular solutions to a solution of the original problem by the Chinese remainder algorithm, one can also attempt to solve a problem in one homomorphic image, say modulo an ideal I, and then *lift* this solution to a solution modulo I^t, for high enough t. The basis for such an approach is the Lifting Theorem.

C.54.3 Theorem (*Lifting Theorem*) *Let I be the ideal generated by $p_1, \ldots, p_l$ in the commutative ring with unity R, $f_1, \ldots, f_n \in R[x_1, \ldots, x_r]$, $r \geq 1$, and $a_1, \ldots, a_r \in R$ such that $f_i(a_1, \ldots, a_r) \equiv 0$ (mod I) for $i = 1, \ldots, n$. Let U be the Jacobian matrix of $f_1, \ldots, f_n$ evaluated at $(a_1, \ldots, a_r)$, i.e., $U = (u_{ij})$, where $u_{ij} = \partial f_i / \partial x_j (a_1, \ldots, a_r)$. Assume that U is right-invertible modulo I, i.e., there is an $r \times n$ matrix $W = (w_{jl})$ such that $U \cdot W \equiv E_n$ (mod I) (E_n is the $n \times n$ identity matrix). Then for every $t \in \mathbb{N}$ there exist $a_1^{(t)}, \ldots, a_r^{(t)} \in R$ such that $f_i(a_1^{(t)}, \ldots, a_r^{(t)}) \equiv 0$ (mod I^t) for $1 \leq i \leq n$, and $a_j^{(t)} \equiv a_j$ (mod I) for $1 \leq j \leq r$.*

If the ideal I is generated by the prime element p, then the Lifting Theorem guarantees a p-adic approximation of the solution. An important special case of the Lifting Theorem is the Hensel Lemma on p-adic approximation of factors of a polynomial.

C.54.4 Theorem (*Hensel Lemma*) *Let p be a prime number and $a(x), a_1(x), \ldots, a_r(x) \in \mathbb{Z}[x]$. Let $(a_1 \bmod p), \ldots, (a_r \bmod p)$ be pairwise relatively prime in $\mathbb{Z}_p[x]$ and $a(x) \equiv a_1(x) \cdot \ldots \cdot a_r(x) \bmod p$. Then for every natural number k there are polynomials $a_1^{(k)}(x), \ldots, a_r^{(k)}(x) \in \mathbb{Z}[x]$ such that $a(x) \equiv a_1^{(k)}(x) \cdot \ldots \cdot a_r^{(k)}(x) \bmod p^k$ and $a_i^{(k)}(x) \equiv a_i(x) \bmod p$ for $1 \leq i \leq r$.*

The lifting procedure, and also the Hensel Lemma, can be made quadratic, i.e., lifting from equality modulo I^t to equality modulo I^{2t}.

Further details on computational ring theory can be found in a variety of books on computer algebra, e.g., (Gathen von zur and Gerhard 1999) or (Winkler 1996).

C.55 Applications of Rings

by Günter F. Pilz in Linz, Austria

Many results in ring theory enable us to present a ring R as a direct sum of simpler rings R_i. If the transitions between R and $\bigoplus R_i$ is reasonably "smooth", *a direct sum decomposition means the ability to do parallel computations*. Examples are the use of Chinese remainder theorem, the decomposition of the group algebra $\mathbb{C}[G]$, and the Fast Fourier and the Fast Hadamard Transformations. One can see that all these cases are closely related. Real-life applications include compact disc players, which use these fast transforms to decode music signals very efficiently.

But we now turn to another area which seems to be less well-known: rings and dynamical systems. More on this can be found in (Lidl and Pilz 1998).

C.55.1 Example Let a point of mass $m = 1$ be the end of a pendulum of length r (cf. Figure 1). Here, g is gravity and x the radiant measure of α. If b denotes air resistance, Newton's second law tells us that $x = x(t)$) is governed by the differential equation

$$r\ddot{x}(t) + br\dot{x}(t) + g\sin x(t) = 0,$$

For small x, this can be replaced by the linear equation

$$r\ddot{x}(t) + br\dot{x}(t) + gx(t) = 0.$$

For $\mathbf{x}(t) := \left(\begin{smallmatrix} x(t) \\ \dot{x}(t) \end{smallmatrix} \right)$, we then get $\dot{\mathbf{x}}(t) = \left(\begin{smallmatrix} 0 & 1 \\ -g/r & -b \end{smallmatrix} \right) \mathbf{x}(t) =: \mathbf{F}\mathbf{x}(t)$.

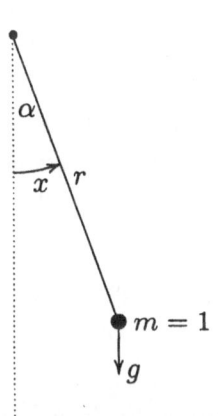
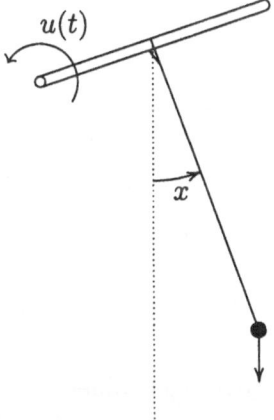

Figure 1: Free pendulum   Figure 2: Manipulated pendulum

Suppose now that the pendulum is hanging on a bar which can be turned at an angular speed of $u(t)$ and has the friction coefficient k (between the bar and the rope, cf. Figure 2). For small x, the linear equation has the form

$$r\ddot{x}(t) + br\dot{x}(t) + gx(t) = k(u - \dot{x}).$$

With $\bar{\mathbf{F}} := \begin{pmatrix} 0 & 1 \\ -g/r & -b - k/r \end{pmatrix}$ and $\mathbf{G} := \begin{pmatrix} 0 \\ k/r \end{pmatrix}$, the equation becomes

$$\dot{\mathbf{x}}(t) = \bar{\mathbf{F}}\mathbf{x}(t) + \mathbf{G}u(t).$$

This is a typical example of a *linear continuous system*. We shall not introduce a formal definition of a system as a 7-tuple here; see e.g., (Kalman, Falb, and Arbib 1969) for precise definitions. The main terms are *states* $\mathbf{x}(t)$ depending on a *time* t via a difference equation (discrete systems) or a differential equation (continuous systems), involving *input functions* $\mathbf{u}(t)$. If the system is in state $\mathbf{x}$ at time t_0, and if the input $\mathbf{u}$ arrives, the state will change to $\Phi_{t_0,t}(\mathbf{x}, \mathbf{u})$ at time t. In the linear continuous finite dimensional case, we have $\dot{\mathbf{x}}(t) = \mathbf{F}\mathbf{x}(t) + \mathbf{G}\mathbf{u}(t)$, with suitable matrices $\mathbf{F}$ and $\mathbf{G}$. Often we also have an *output function* $\mathbf{y}(t)$, which, in the linear case, is given by $\mathbf{y}(t) = \mathbf{H}\mathbf{x}(t)$. Here, $\mathbf{x}, \mathbf{y}, \mathbf{u}$ come from properly chosen vector spaces. In Example C.55.1, we might take $\mathbf{y}(t) = \mathbf{x}(t)$, i.e., $\mathbf{H} = \mathbf{I}$.

C.55.2 Definition Consider the system described by $\dot{\mathbf{x}}(t) = \mathbf{F}\mathbf{x}(t) + \mathbf{G}\mathbf{u}(t)$, with appropriate time set T. This system is *stable* if for all $t_0 \in T$ and states $\mathbf{x}$ the function $t \mapsto \Phi_{t_0,t}(\mathbf{x}, \mathbf{0})$ is bounded for $t \geq t_0$. The system is *asymptotically stable* if moreover $\Phi_{t,t_0}(\mathbf{x}, \mathbf{0}) \to \mathbf{0}$ as $t \to \infty$.

C.55.3 Theorem *Consider such a system, and let $\lambda_1, \lambda_2, \ldots$ be the eigenvalues of $\mathbf{F}$. Then the system is asymptotically stable iff each $\Re(\lambda_i) < 0$; it is stable iff each $\Re(\lambda_i) \leq 0$ and, in the case $\Re(\lambda_i) = 0$, the algebraic and geometric multiplicities of λ_i coincide.*

For a proof, see e.g., Szidarovszky and Bahill (1992). Observe that these are purely algebraic characterizations. In Example C.55.1, we get the real eigenvalues $-\frac{b}{2} \pm \sqrt{(\frac{b}{2})^2 - \frac{g}{r}}$. Since $\sqrt{(\frac{b}{2})^2 - \frac{g}{r}} < \frac{b}{2}$, both eigenvalues have negative real parts. Hence the system is asymptotically stable, as everybody knows: after a while, the pendulum will stop.

One of the main topics in systems theory is control theory: given an unstable system $\mathscr{S}$, can we modify it so that it will become stable? In the linear case, we have seen that stability only depends on $\mathbf{F}$. So we might try a "closed-loop configuration" as in Figure 3. $\mathscr{C}$ is another system, called the *controller*, *compensator*, *regulator*, or the *feedback*. It takes $\mathbf{x}$, transforms it properly, and subtracts the result from the input $\mathbf{u}$. In the linear case, we might represent $\mathscr{C}$ by a matrix $\mathbf{C}$, so that the new equation is $\dot{\mathbf{x}}(t) = (\mathbf{F} - \mathbf{G}\mathbf{C})\mathbf{x}(t) + \mathbf{G}\mathbf{u}(t)$. So, given $\mathbf{F}$ and $\mathbf{G}$, we look for a matrix $\mathbf{C}$ such that the real parts of all eigenvalues of $\mathbf{F} - \mathbf{G}\mathbf{C}$ are (strictly) negative. In fact, *the more negative, the better*, since "very negative" real parts will quickly bring the system to the desired stable

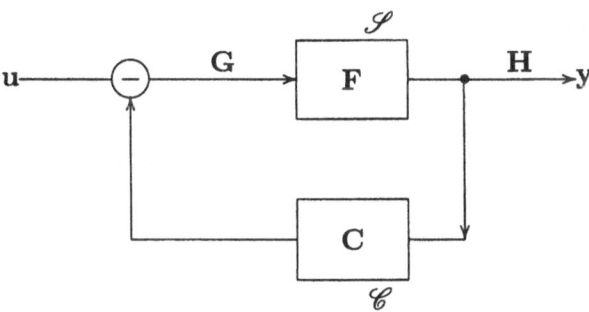

Figure 3:

state. Or we can start with a stable system and improve its behavior by switching from $\mathbf{F}$ to $\mathbf{F} - \mathbf{GC}$. An unskillful choice of $\mathscr{C}$, however, can turn a stable system into an unstable one. Efficient methods are available to actually construct $\mathbf{C}$. So, given two matrices $\mathbf{F}$ and $\mathbf{G}$, we want to find out whether there is a matrix $\mathbf{C}$ such that all eigenvalues of $\mathbf{F} - \mathbf{GC}$ have negative real parts. This can be formulated if we work over $\mathbb{C}$. Often, however, one works with fields other than $\mathbb{C}$, or even with rings. In applications, one very soon needs $\mathbf{F}$ and $\mathbf{G}$ to be matrices over rings like $\mathbb{R}[x]$, $\mathbb{C}[x]$, or $\mathbb{Z}_n$. Then "negative real parts" have no meaning any more. We now give some conditions which assure the existence of efficient controllers $\mathscr{C}$.

C.55.4 Definition Let R be a commutative ring with identity, $\mathbf{F}$ an $n \times n$ matrix and $\mathbf{G}$ an $n \times m$ matrix over R. The pair $(\mathbf{F}, \mathbf{G})$ has the *pole assignment property* (PA) if for all $\lambda_1, \ldots, \lambda_n \in R$ there is some $m \times n$ matrix $\mathbf{C}$ with characteristic polynomial $(x - \lambda_1) \ldots (x - \lambda_n)$.

C.55.5 Theorem *(PA) implies controllability.*

Controllability and (PA) can be shown to be equivalent for $R = \mathbb{R}$ and $R = \mathbb{R}[x]$. They are not equivalent for $\mathbb{R}[x, y]$ and $\mathbb{Z}[x]$. More generally:

C.55.6 Theorem *If R is a principal ideal domain then controllability and PA are equivalent.*

Because it is often impossible to measure all the relevant components of the states, it is often desirable to compute the output sequence (function) directly from the input sequence (function) without referring to the states. For instance, transmitting to the cockpit all relevant information on closed parts inside the jet engines in an airplane would result in such heavy radio noise that no communication between the aircraft and the ground stations would be possible any more. For linear discrete systems, this can be done in the following way:

C.55.7 Theorem *For a linear constant discrete system* $(\mathbf{F}, \mathbf{G}, \mathbf{H})$ *with* $T \subseteq \mathbb{N}_0$ *and initial state* $\mathbf{0}$*, we write the sequence* $\mathbf{u}$ *as* $\mathbf{u} = \sum u_t x^t$ *and the output sequence as* $\mathbf{y} = \sum y_t x^t$*. Then* $\mathbf{y} = \mathbf{H}(x\mathbf{I} - \mathbf{F})^{-1}\mathbf{G}\mathbf{u} = T(\mathbf{u})$*.*

T is called the **transfer function** of the system $(\mathbf{F}, \mathbf{G}, \mathbf{H})$.

C.55.8 Theorem *If two systems have transfer functions* T_1, T_2*, respectively, then their parallel connection has the transfer function* $T_1 + T_2$*, while their series connection has the transfer functions* $T_1 T_2$*. Linear time-invariant systems themselves form a ring w.r.t. parallel and series connection.*

For a proof, see e.g., Szidarovszky and Bahill (1992). Observe that this ring is closed w.r.t. feedbacks. So it seems that ring theory can contribute a lot to the theory of linear systems.

What about generalizations? The matrices $\mathbf{F}, \mathbf{G}, \mathbf{H}$ might be time-dependent, the changes of states might be given by partial differential equations, and/or the system might be non-linear. Sometimes, one can approximate a non-linear system by a linear one; we did that in Example C.55.1, when we replaced $\sin x$ by x. Studying these more general (but more realistic) systems leads to numerous open questions, and makes up a vast area of active research. A class of "tame" nonlinear systems is the one of **affine-input systems** (**AI-systems**), where the corresponding equations are

$$\dot{\mathbf{x}}(t) = \mathbf{f}(\mathbf{x}(t)) + \mathbf{g}(\mathbf{x}(t))\mathbf{u}(t),$$
$$\mathbf{y}(t) = \mathbf{h}(\mathbf{x}(t)),$$

with sufficiently smooth functions $\mathbf{f}, \mathbf{g}, \mathbf{h}$ in several variables. Many realistic models of industrial plants can be described in this way.

C.55.9 Theorem *AI-systems form a near-ring (Section C.47) w.r.t. parallel and series connections. This near-ring is closed w.r.t. AI-feedbacks.*

Chapter D

FIELDS

D.1 Field Extensions

by Shreeram S. Abhyankar in Lafayette, IN, USA

After the initial impetus given to it by Dedekind and Weber (1882), the theory of field extensions was first systemized by Steinitz (1910). Chevalley (1951) is, in effect, an updated version of the Dedekind-Weber paper. Ultimately, traveling through the edifice of algebraic geometry, it acquired the form presented in Zariski and Samuel (1975). Indeed, it is not easy to decipher where the theory of field extensions ends and algebraic geometry begins. Its ubiquitous use in algebraic geometry can be seen in Abhyankar's resolution of singularities book (Abhyankar 1998) and in the more heuristic introduction to it in his engineering book (Abhyankar 1990). The intimate relationship of field extensions to group theory is expounded in his recent survey article (Abhyankar 2001).

Polynomials in one or more indeterminates $X_1, X_2, \ldots$, over a field K form a domain $K[X_1, X_2, \ldots]$ with rational functions ($=$ quotients of polynomials) in them as the quotient field $K(X_1, X_2, \ldots)$. Quite generally if $x_1, x_2, \ldots$ are elements in an overring B of a ring A, then the smallest subring of B which contains A and the elements $x_1, x_2, \ldots$ is called the **ring generated** by $x_1, x_2, \ldots$ over A and is denoted by $A[x_1, x_2, \ldots]$, and if A and B are fields then the smallest field which contains A and the elements $x_1, x_2, \ldots$ is called the **field generated** by $x_1, x_2, \ldots$ over A and is denoted by $A(x_1, x_2, \ldots)$. If A is a subring of a ring B, then B is called an overring or a **ring extension** of A; letting A and B be fields we get the notion of an overfield or a **field extension**. A ring extension B of a field A is automatically a vector space over A and so we may talk about the vector space dimension $[B : A]$.

An ideal I in a ring A is **prime** if A/I is a domain, and I is **maximal** means A/I is a field. The set of all prime ideals in A is called the

spectrum of A and denoted by spec(A). The ***dimension*** of A is the maximum length d of strictly increasing chains of prime ideals $P_0 \subset P_1 \subset \cdots \subset P_d$ in A.

To get an interesting example of this, let $A = K[X_1, \ldots, X_n]$ where K is a field. This corresponds to the affine n-space over K. A maximal ideal in A has n polynomial generators of the form $g_1(X_1)$, $g_2(X_1, X_2)$, $\ldots, g_n(X_1, \ldots, X_n)$. If K is algebraically closed, i.e., if every univariate polynomial over K has a root in K as in the case of complex numbers, then it is possible to arrange g_i to be of the form $X_i - a_i$ with $a_i \in K$, giving rise to the point $(a_1, \ldots, a_n)$ in the affine space. The ***variety*** $V(I)$ associated to an ideal I is the set of all common solutions of $f_i(X_1, \ldots, X_n) = 0$ for $1 \le i \le r$ where $f_1, \ldots, f_r$ is a basis of I. If K is not algebraically closed then $V(I)$ may be thought of as the set of all maximal ideals in A containing I. The dimension of A/I is called the ***dimension*** of the variety $V(I)$. Varieties defined by prime ideals are ***irreducible***. The correspondence between spec(A) and irreducible varieties is inclusion reversing, i.e., $P \subset Q$ in spec(A) gives $V(Q) \subset V(P)$; here $V(Q)$ is called a ***subvariety*** of $V(P)$. The dimension of an irreducible variety varies between 0 and n; points are zero dimensional, curves are one dimensional, surfaces are two dimensional, $\ldots$, the affine space is n-dimensional, and a ***hypersurface***, i.e., a variety defined by a single polynomial equation $f = 0$, is $n - 1$ dimensional.

Let $V = V(P)$ where P is a prime ideal. The ring $K[V] = A/P$ is called the ***affine coordinate ring*** of V, and its quotient field $K(V)$ is called the ***function field*** of V. Note that $K[V] = K[x_1, \ldots, x_n]$ and $K(V) = K(x_1, \ldots, x_n)$ where x_i is the residue class of X_i modulo P. The dimension of V turns out to be equal to the ***transcendence degree*** of $K(V)$ over K, i.e., the maximum number of elements $y_1, \ldots, y_d$ in $K(V)$ which are ***algebraically independent*** over K which means that they do not satisfy any nontrivial polynomial relation over K. Such elements constitute a ***transcendence basis*** of $K(V)$ over K, and the ***field degree*** $[K(V) : K(y_1, \ldots, y_d)]$ is finite. In case K is algebraically closed, the maximum of this field degree taken over all transcendence bases which are K-linear combinations of $x_1, \ldots, x_n$ turns out to be equal to the ***degree*** of V, i.e., the maximum number of points in which an $n - d$ dimensional linear space (not having a line common with V) meets V. For $W = V(Q)$ with $P \subset Q \in$ spec(A), we define the ***local ring*** $R(W, V)$ of W on V to be the ***localization*** $K[V]_{Q/P}$ of $K[V]$ at the prime ideal Q/P, i.e., $R(W, V) = \{ x/y \mid x, y \in K[V] \text{ with } y \notin Q/P \}$ (cf. Section C.6). Note that the ring $R(W, V)$ is ***Noetherian***, i.e., every ideal in it is finitely generated. Also, $R(V, W)$ has the unique maximal ideal $M(R(W, V)) = \{\, x/y \mid x, y \in K[V] \text{ with } x \in Q/P \text{ and } y \notin Q/P \,\}$. Thus $R(W, V)$ is indeed a *local ring*, which means a Noetherian ring R with a unique maximal ideal $M(R)$. Let $\mathfrak{V}(K[V])$ denote the set of localizations of $K[V]$ at various prime ideals in $K[V]$. We call $\mathfrak{V}(K[V])$ the **affine model** associated with V. Note that $W \mapsto R(W, V)$ gives an inclusion preserving bijection of the set of all irreducible subvarieties of V onto $\mathfrak{V}(K[V])$. To take care of **points at infinity** we introduce the **projective model** $\mathfrak{W}(z_0, \ldots, z_m; B) = \bigcup_{0 \leq i \leq m \text{ with } z_i \neq 0} \mathfrak{V}(B[z_0/z_i, \ldots, z_m/z_i])$ where $z_0, \ldots, z_m$ are elements of a field containing the ring B. Let $\overline{V}$ be the **projective closure** of V, i.e., V together with its points at infinity. When K is algebraically closed, a point in projective n-space is represented by proportional $(n + 1)$ tuples $(a_0, \ldots, a_n)$ with $a_i \in K$. Let $\overline{P}$ be the homogeneous prime ideal in $K[X_0, \ldots, X_n]$ obtained by homogenizing P, i.e., generated by all homogeneous polynomials $\overline{f}(X_0, \ldots, X_n)$ such that $\overline{f}(1, X_1, \ldots, X_n) \in P$. Now $\overline{V}$ consists of all points in projective n-space which are common solutions of all $\overline{f}(X_0, \ldots, X_n) = 0$ with $\overline{f} \in \overline{P}$. Points of $\overline{V}$ which are not points of V are called the **points at infinity** of V. The **dimension** of $\overline{V}$ is defined to be the same as the dimension d of V, and its **function field** $K(\overline{V})$ is defined to be the same as $K(V)$. By taking $x_0 = 1$, the above bijection $W \mapsto R(W, V)$ extends to an inclusion preserving bijection $\overline{W} \mapsto R(\overline{W}, \overline{V})$ of the set of all irreducible subvarieties $\overline{W}$ of $\overline{V}$ onto $\mathfrak{W} = \mathfrak{W}(x_0, \ldots, x_n; K)$, with an obvious extension of the notations; moreover, $K(\mathfrak{W}) = K(\overline{V})$ is called the **function field** of $\mathfrak{W}$, and $\mathfrak{W}$ is called a **projective model** of $K(\mathfrak{W})$ over K. In this bijection, an e-dimensional subvariety of $\overline{V}$ corresponds to a local ring of dimension $d - e$, so in particular, points of $\overline{V}$ correspond to maximal dimensional members of $\mathfrak{W}$; moreover, we call d to be the **dimension** of $\mathfrak{W}$. Recall that a local ring R is **regular** if its maximal ideal is generated by as many elements as its dimension. The time honored Jacobian Criterion for **simple points** shows that $\overline{W}$ is singular for $\overline{V}$, i.e., all its points are **singular** for $\overline{V}$, if and only if $R(\overline{W}, \overline{V})$ is not regular. The projective model $\mathfrak{W}$ is said to be **nonsingular** if all its members are regular.

The function field $K(\mathfrak{W}) = K(\overline{V})$ is clearly a finitely generated field extension of K. Conversely, a finitely generated field extension of K is the function field of several projective varieties and hence projective models over K. The projective variety $\overline{V}$ is said to be **birationally equivalent** to a projective variety $\overline{V}'$ if there is an algebraically defined correspondence between them which is one to one almost everywhere;

this is evidently so if and only if their function fields are isomorphic. Thus, properties of finitely generated field extensions are nothing but properties of projective varieties which are invariant under birational equivalence. Thus birational algebraic geometry is simply another name for the study of (finitely generated) field extensions. Now if this birational correspondence between $\overline{V}$ and $\overline{V}'$ is truly one to one then they are said to be *biregularly equivalent*; in our set up this comes down to saying that the corresponding projective models are actually equal. Examples of birationally equivalent varieties are a conic and a line, or the nodal cubic and a line, or more generally any rational curve and a line. Likewise a quadric or a nonsingular cubic surface are rational surfaces and hence they are birationally equivalent to a plane. Similarly a hyperquadric in n-space is birationally equivalent to the $(n-1)$-space.

Many interesting numerical objects related to field extensions, such as dimensions of spaces of differential forms, can be only studied on nonsingular models. Thus, it is of paramount importance to investigate whether every finitely generated field extension has a nonsingular projective model. This is the problem of *Resolution of Singularities*, which is still an open problem for dimension bigger than three in positive characteristic. To verify the birational invariance of various numerical objects associated with nonsingular projective models of the same function field, it is important to solve the *domination problem*. Given two nonsingular projective models $\mathfrak{W}$ and $\mathfrak{W}'$ of a common function field, the domination problem asks if there exists an iterated monoidal transform $\mathfrak{W}^*$ of $\mathfrak{W}$ which dominates $\mathfrak{W}'$.

To introduce the above concepts, we say that a local ring R^* *dominates* a local ring R' if $R' \subset R^*$ and $M(R') \subset M(R^*)$, and we say that a projective model $\mathfrak{W}^*$ *dominates* a projective model $\mathfrak{W}'$ if every $R^* \in \mathfrak{W}^*$ dominates a unique $R' \in \mathfrak{W}'$. The *blow-up* $\mathfrak{W}(R, I)$ of a local domain R by a nonzero ideal I in R is defined by putting $\mathfrak{W}(R, I) = \mathfrak{W}(z_0, \ldots, z_m; R)$ where $(z_0, \ldots, z_m)$ is a basis of I, and noting that this is independent of the basis. By an *irreducible ideal* on a projective model $\mathfrak{W}^\sharp$ which corresponds to an irreducible projective variety $\overline{V}^\sharp$, we mean a family $I^\sharp = I^\sharp(R)_{R \in \mathfrak{W}^\sharp}$, where $I^\sharp(R)$ is an ideal in R, such that for some irreducible subvariety $\overline{W}^\sharp$ of $\overline{V}^\sharp$ different from $\overline{V}^\sharp$ we have: $I^\sharp(R) \neq R \Leftrightarrow R \subset R(\overline{W}^\sharp, \overline{V}^\sharp)$ and $I^\sharp(R) = R \cap M(R(\overline{W}^\sharp, \overline{V}^\sharp))$. By a *nonsingular ideal* on a nonsingular projective model $\mathfrak{W}^\dagger$ we mean an irreducible ideal $I^\dagger$ on $\mathfrak{W}^\dagger$ such that $R/I^\dagger(R)$ is a regular local ring for all $R \in \mathfrak{W}^\dagger$ with $I^\dagger(R) \neq R$, and we put $\mathfrak{W}(\mathfrak{W}^\dagger, I^\dagger) = \bigcup_{R \in \mathfrak{W}^\dagger} \mathfrak{W}(R, I^\dagger(R))$;

we call $\mathfrak{W}(\mathfrak{W}^\dagger, I^\dagger)$ a **monoidal transform** of $\mathfrak{W}^\dagger$ and we note that $\mathfrak{W}(\mathfrak{W}^\dagger, I^\dagger)$ is a nonsingular projective model which is birationally equivalent to $\mathfrak{W}^\dagger$. Finally, by an **iterated monoidal transform** of $\mathfrak{W}^\dagger$ we mean a nonsingular projective model $\mathfrak{W}^\ddagger$ for which there exists a finite sequence $\mathfrak{W}^\dagger = \mathfrak{W}_0^\dagger, \mathfrak{W}_1^\dagger, \ldots, \mathfrak{W}_r^\dagger = \mathfrak{W}^\ddagger$ such that $\mathfrak{W}_i^\dagger$ is a monoidal transform of $\mathfrak{W}_{i-1}^\dagger$ for $1 \leq i \leq r$.

To illustrate the idea of monoidal transformations locally, let R be the localization of $A = K[X_1, \ldots, X_n]$ at the maximal ideal generated by $X_1, \ldots, X_n$, and let I be the ideal in R generated by $X_1, \ldots, X_m$ for some $m \leq n$. Assuming K to be algebraically closed, a typical n-dimensional member S of $\mathfrak{W}(R, I)$ is obtained by letting $Y_1 = X_1$, $Y_i = (X_i - \lambda_i X_1)/X_1$ for $2 \leq i \leq m$ with $\lambda_i \in K$, $Y_j = X_j$ for $m < j \leq n$, and taking S to be the localization of $R[Y_2, \ldots, Y_m]$ at the maximal ideal generated by $(Y_1, \ldots, Y_n)$. Clearly S is an n dimensional regular local ring which dominates the regular local ring R.

The details of the above material can be found in (Abhyankar 1998, 1990, 2001).

D.2 Finite Fields

by Harald Niederreiter in Singapore

A **finite field** is, not unexpectedly, a field with finitely many elements. The cardinality of a finite field is called its **order**. The standard example of a finite field is the **finite prime field** $\mathbb{Z}_p$ with a prime p. Any finite field $\mathbb{F}$ has a uniquely determined prime characteristic p and the order of $\mathbb{F}$ is p^n, where n is the dimension of $\mathbb{F}$ as a vector space over its finite prime field $\mathbb{Z}_p$.

The question whether for every prime power $q = p^n$ there exists a finite field of order q has an affirmative answer. The classical argument following Gauss and Dedekind proceeds as follows. If $I_p(n)$ is the number of monic irreducible polynomials over $\mathbb{Z}_p$ of degree n, then with μ denoting the Moebius function we have

$$I_p(n) = \frac{1}{n} \sum_{d|n} \mu(d) p^{n/d} \geq \frac{1}{n} \left(p^n - \sum_{d|n, d>1} p^{n/d} \right) > 0.$$

Pick an $f \in \mathbb{Z}_p[x]$ counted by $I_p(n)$, then the residue class ring $\mathbb{Z}_p[x]/(f)$ is a finite field of order q.

D.2.1 Theorem (Moore) *Any two finite fields of the same order are isomorphic as fields.*

In this sense, it is meaningful to speak of the finite field $\mathbb{F}_q$ of prime-power order q. Some authors use the term ***Galois field*** GF(q), in recognition of the pioneering work of Galois in this area. The finite field $\mathbb{F}_q$ of characteristic p may be viewed as the splitting field of the polynomial $x^q - x$ over $\mathbb{Z}_p$. In fact, $\mathbb{F}_q$ consists exactly of all roots of this polynomial.

D.2.2 Theorem *Every subfield of $\mathbb{F}_q$, $q = p^n$, has order p^d with a positive divisor d of n. Conversely, if d is a positive divisor of n, then there is a unique subfield of $\mathbb{F}_q$ of order p^d.*

The roots of irreducible polynomials over a finite field are simple and enjoy a special relationship among themselves.

D.2.3 Theorem (Galois) *The roots of an irreducible polynomial over $\mathbb{F}_q$ of degree m are given by m distinct elements α^{q^j}, $j = 0, 1, \ldots, m-1$, of $\mathbb{F}_{q^m}$.*

It follows that any finite extension $\mathbb{F}_{q^m}/\mathbb{F}_q$ is a Galois extension with a cyclic Galois group generated by the ***Frobenius automorphism*** $\sigma(\beta) = \beta^q$ for $\beta \in \mathbb{F}_{q^m}$.

A famous classical theorem describes the multiplicative structure of $\mathbb{F}_q$.

D.2.4 Theorem (Gauss, Dedekind) *The multiplicative group $\mathbb{F}_q^*$ of nonzero elements of $\mathbb{F}_q$ is cyclic.*

D.2.5 Definition A generator of the cyclic group $\mathbb{F}_q^*$ is called a ***primitive element*** of $\mathbb{F}_q$.

Primitive elements of a finite extension field $\mathbb{F}_{q^m}$ of $\mathbb{F}_q$ can be viewed as roots of a ***primitive polynomial*** f over $\mathbb{F}_q$, i.e., of a monic f of degree m with $f(0) \neq 0$ and the property that $q^m - 1$ is the least positive integer e such that $f(x)$ divides $x^e - 1$. The number of primitive polynomials over $\mathbb{F}_q$ of degree m is $\varphi(q^m - 1)/m$, where φ is Euler's totient function.

D.2.6 Theorem (Hensel) *For any finite extension $\mathbb{F}_{q^m}/\mathbb{F}_q$ there exists a normal basis of $\mathbb{F}_{q^m}$ over $\mathbb{F}_q$, i.e., a basis of the form $\{\alpha, \alpha^q, \alpha^{q^2}, \ldots, \alpha^{q^{m-1}}\}$ with a suitable $\alpha \in \mathbb{F}_{q^m}$.*

A normal basis B of $\mathbb{F}_{q^m}$ over $\mathbb{F}_q$ is very convenient for fast arithmetic. For instance, raising $\beta \in \mathbb{F}_{q^m}$ to the qth power just amounts to a cyclic shift of the coordinate vector of β with respect to B. A refinement of the above theorem, due to Lenstra and Schoof, says that there always exists a normal basis of $\mathbb{F}_{q^m}$ over $\mathbb{F}_q$ consisting of primitive elements of $\mathbb{F}_{q^m}$. The following recent theorem provides even more information.

D.2.7 Theorem (Cohen-Hachenberger) *For any finite extension $\mathbb{F}_{q^m}/\mathbb{F}_q$ with $m \geq 2$ there exists a normal basis of $\mathbb{F}_{q^m}$ over $\mathbb{F}_q$ generated by a primitive element α of $\mathbb{F}_{q^m}$ with a prescribed nonzero trace.*

Another refinement of the notion of a normal basis is given in the following definition.

D.2.8 Definition A normal basis of a finite extension $\mathbb{F}_{q^m}/\mathbb{F}_q$ is called a **completely normal basis** of $\mathbb{F}_{q^m}$ over $\mathbb{F}_q$ if it is a normal basis of $\mathbb{F}_{q^m}$ over every intermediate field of the extension.

D.2.9 Theorem (Blessenohl-Johnsen) *For any finite extension $\mathbb{F}_{q^m}/\mathbb{F}_q$ there exists a completely normal basis of $\mathbb{F}_{q^m}$ over $\mathbb{F}_q$.*

The number of (ordered) normal bases of $\mathbb{F}_{q^m}$ over $\mathbb{F}_q$ is given by $\Phi_q(x^m - 1)$, where Φ_q is Euler's totient function for the polynomial ring $\mathbb{F}_q[x]$. In detail, if $f \in \mathbb{F}_q[x]$ with $\deg(f) = m \geq 1$, then $\Phi_q(f)$ is the number of $g \in \mathbb{F}_q[x]$ with $\deg(g) < \deg(f)$ and $\gcd(g, f) = 1$. We have the explicit formula

$$\Phi_q(f) = q^m \prod_{j=1}^{r} \left(1 - q^{-m_j}\right),$$

where $m_1, \dots, m_r$ are the degrees of the distinct monic irreducible factors of f in $\mathbb{F}_q[x]$. The number of normal bases of $\mathbb{F}_{q^m}$ over $\mathbb{F}_q$ is also equal to the number of nonsingular $m \times m$ circulant matrices over $\mathbb{F}_q$.

Finite fields are fundamental algebraic structures for discrete mathematics and they allow many applications to combinatorial design theory, finite geometries, algebraic coding theory, cryptography, pseudorandom number generation, and so on. An introductory text to the theory as well as to applications is given by (Lidl and Niederreiter 1994); (Lidl and Niederreiter 1997) is more complete, and (Shparlinski 1999) emphasizes computational aspects.

D.3 Galois Theory

by Joseph J. Rotman in Urbana, IL, USA

One of the most famous outstanding problems at the beginning of the nineteenth century was that of finding the roots of polynomials. The quadratic formula had been known since Babylonian times, and the cubic and quartic formulas had been known since the 1500s. Newton's method could find roots to any desired degree of accuracy, but mathematicians hoped to find a formula generalizing the known classical formulas; that is, they wanted the roots of a polynomial to be obtained from its coefficients using only field operations and extraction of roots. We call a polynomial admitting such a formula *solvable by radicals*. In 1799, Gauss proved that the cyclotomic polynomials $x^n - 1$ are solvable by radicals (and he used this to construct regular 17-gons by straightedge and compass). However, by the late 1700s, especially with the work of Lagrange, some began to doubt that quintics are solvable by radicals. Indeed, in 1799, Ruffini announced a proof of this, but his proof had gaps and was not accepted by his contemporaries. In 1815, Cauchy established the calculus of permutations, and in 1824, Abel used this calculus to prove that the general quintic is not solvable by radicals. Galois realized that if a polynomial is solvable by radicals, then the formula for its roots may depend on the polynomial. For example, if $f(x)$ and $g(x)$ are quintics which are solvable by radicals, then the formulas may depend on $f(x)$ and $g(x)$ (in contrast to the quadratic formula, which applies uniformly to all quadratic polynomials). Galois invented groups in order to give a necessary and sufficient condition when a polynomial (of any degree) is solvable by radicals. To each $f(x) \in \mathbb{Q}[x]$ of degree n with no repeated roots, he defined a group (nowadays called its *Galois group*, as a certain subgroup of the symmetric group $S_X \simeq S_n$, where X is the set of all the roots of $f(x)$. The Galois group consists of permutations defined using a variation of Lagrange's proof of the Theorem of the Primitive Element, which expresses each root as a rational function of a certain complex number. We state the main results in modern language. We refer the reader to (Tignol 1988) for an excellent account of the history of this problem up to and including Galois.

D.3.1 Theorem (Galois) *A polynomial $f(x) \in K[x]$, where K is a field of characteristic 0, is solvable by radicals if and only if its Galois group is a solvable group.*

Galois proved that S_4 is a solvable group. Since every subgroup of a solvable group is solvable, every polynomial of degree at most 4 is solvable by radicals. On the other hand, Galois proved that S_5 is not solvable, and so a quintic having Galois group S_5 is not solvable by radicals.

Galois's work was submitted to the French Academy in 1830, but it was not published until the 1840s. There are several reasons for this. First, Galois was killed in 1832; second, for political reasons, Cauchy, the most knowledgeable person in the French Academy, left the country for a decade. But, perhaps the most important reason is that Galois's paper is full of new abstract ideas (which ultimately developed into modern algebra) that were very difficult to understand in those times.

Gradually, the work of Galois began to be known. In 1850, Puiseux studied the monodromy groups of certain functions of two complex variables and, in 1851, Hermite proved that the monodromy group is isomorphic to the Galois group of a polynomial with coefficients in the function field $\mathbb{C}(x)$ (Miller et al. 1916 (Dover reprint, 1961). Group theory developed from the 1840s, as did the study of fields and rings. About 100 years after its discovery, E. Artin redefined the Galois group, as we know it today. He recognized that a permutation of the roots of a polynomial with coefficients in a field K can be viewed as an automorphism of its splitting field E, and it is fruitful to define the Galois group of a field extension (generalizing the Galois group of a polynomial).

D.3.2 Definition The *Galois group* of a field extension E/K, denoted by $\mathrm{Gal}(E/K)$, is the subgroup of $\mathrm{Aut}(K)$ consisting of all those automorphisms which fix K pointwise.

In particular, if $f(x) \in K[x]$ has splitting field E/K, then the Galois group of $f(x)$ over K is $\mathrm{Gal}(E/K)$. Using fixed fields, one characterizes those field extensions E/K which are splitting fields of some polynomial in $K[x]$.

D.3.3 Definition If E is a field and if $G \leq \mathrm{Aut}\, E$, then the *fixed field* is

$$E^G = \{\, a \in E \mid \sigma(a) = a \text{ for all } \sigma \in G \,\}.$$

The following technical definition was introduced to compute the order of the Galois group.

D.3.4 Definition Let E/K be a field extension, and let $G = \mathrm{Gal}(E/K)$ be its Galois group. One calls E/K a *separable extension* if every

$a \in E$ is a root of an irreducible polynomial in $K[x]$ having no repeated roots.

If K has characteristic 0, then every extension E/K is separable, and if K is a finite field, then every finite extension E/K is separable. The easiest example of a nonseparable extension is obtained by adjoining a pth root of x to the function field $K(x)$, where $K = \mathbb{F}_p$.

D.3.5 Theorem *A finite field extension E/K is the splitting field of some $f(x) \in K[x]$ if and only if $K = E^{\mathrm{Gal}(E/K)}$. One calls E/K a* **Galois extension** *in this case.*

It can be proved that if E/K is a Galois extension, then $|\mathrm{Gal}(E/K)| = [E : K]$, the dimension of E as a vector space over K.

The fundamental theorem of Galois theory classifies all the intermediate fields F, where $K \leq F \leq E$ and E/K is a finite Galois extension.

D.3.6 Theorem (Fundamental Theorem) *Let E/K be a Galois extension with Galois group $G = \mathrm{Gal}(E/K)$. Then there is an order-reversing bijection between the lattice of all subgroups of G and the lattice of all the intermediate fields given by $H \mapsto E^H$. Moreover, E^H/K is a Galois extension if and only if $H \trianglelefteq \mathrm{Gal}(E/K)$.*

There have been many variations of the fundamental theorem. For example, the splitting field E of a polynomial is of finite degree over the base field K. But there are interesting field extensions that are infinite-dimensional. For example, if $\overline{K}$ is the algebraic closure of K, one may ask about $\mathrm{Gal}(\overline{K}/K)$. It turns out that the best way to view this group is as a **profinite group**, that is, as an inverse limit of finite groups (so the Galois group is, in fact, a compact topological group). There is a generalized fundamental theorem in which the *closed* subgroups of $\mathrm{Gal}(\overline{K}/K)$ classify the intermediate fields. The structure of the infinite Galois group $\mathrm{Gal}(\overline{\mathbb{Q}}, \mathbb{Q})$ is of great interest to algebraic number theorists. For example, it is intimately involved in Galois representations.

There are versions of Galois theory in other areas of algebra. For example, there is a Galois theory classifying division algebras due to Jacobson (1964), and a Galois theory classifying commutative rings due to Chase et al. (1965).

There are variations of the Galois correspondence in other branches of mathematics. In terms of diagrams, one can describe an automorphism σ of a field E that fixes a subfield K as a map making the following diagram

commute,

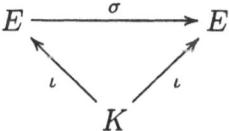

where $\iota \colon K \to E$ is the inclusion. If $p \colon \widetilde{X} \to X$ is the projection of a covering space (for example, if G is a Lie group and $p \colon G \to G/H$ is the natural map, where H is a closed normal subgroup), then a **covering map** is a continuous map $\sigma \colon \widetilde{X} \to \widetilde{X}$ making the following diagram commute:

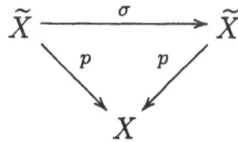

Denote the group of all covering maps by $\mathrm{Cov}(\widetilde{X}/X)$. In the special case that $\widetilde{X}$ is simply connected, then $\mathrm{Cov}(\widetilde{X}/X) \cong \pi_1(X)$, the fundamental group. Moreover, every subgroup G of the fundamental group acts on $\widetilde{X}$, and there is a bijection between such subgroups G and intermediate covering spaces $\widetilde{X}/G$ (which is the orbit space) (Spanier 1966).

There is also a Galois theory for differential equations, due to Ritt and Kolchin, see Section D.4

Given a finite group G, is $G \simeq \mathrm{Gal}(E/K)$ for some field extension E/K? In this generality, the answer is easily seen to be affirmative. However, if one restricts K to be $\mathbb{Q}$, this problem is still an open one, and it is called the **Inverse Galois Problem**. Let us call a finite group G **realizable** if $G \simeq \mathrm{Gal}(E/\mathbb{Q})$ for some finite Galois extension of $\mathbb{Q}$. In 1892, Hilbert proved, for all n, that the symmetric groups S_n and the alternating groups A_n are realizable (it is not known whether any subgroup of a realizable group is realizable). In 1937, Scholz and Reichardt showed that every finite p-group is realizable, where p is an odd prime, and in 1954, Šafarevič proved that every finite solvable group is realizable (although there are some obscure passages in this proof, there now clear and complete versions (Neukirch et al. 2000)). The link with the monodromy group mentioned earlier (Puiseux and Hermite) led Shih, Fried, Belyi, Matzat, Thompson, and Walter to find group theoretic criteria for realizability. This approach, usually called *rigidity,* coupled with the wealth of information provided by the Classification of Finite Simple Groups has led to many more results. For example, of the 26 sporadic simple groups, all but the Mathieu group M_{23} are known to be realizable. Many (not necessarily simple) linear groups are known to be realizable, if not over $\mathbb{Q}$, then over certain *cyclotomic* extensions of $\mathbb{Q}$. For example, all the groups $GL(n; \mathbb{F}_q)$ are realizable in this restricted sense, as are all the finite simple groups of Lie type. For a more detailed account, we refer the reader to (Malle and Matzat 1999)

D.4 Differential Galois Theory
by Andy R. Magid in Norman, OK, USA

Introduction

The goal of **Differential Galois Theory** is a Fundamental Theorem which sets up a bijective correspondence between the intermediate differential subfields of an extension of differential fields and certain subgroups of the group of differential automorphisms of the field extension (the **differential Galois group**). In the most basic case (a *Picard-Vessiot* extension), the extension is obtained by adjoining a full set of solutions of a linear homogeneous differential equation to the base differential field. In this case the differential Galois group is a linear algebraic group and the subgroups in the correspondence are the Zariski closed subgroups. The general case (a *strongly normal* extension) is defined so that the differential Galois group is a not necessarily linear algebraic group, and the fundamental correspondence is again with Zariski closed subgroups.

Differential field extensions

A **differential field** F is a field equipped with a **derivation** D_F; that is, an additive map $D_F \colon F \to F$ satisfying the product rule $D_F(ab) = D_F(a)b + aD_F(b)$ for all $a, b \in F$. The **constants** of F is the kernel C_F of D_F. It is a subfield of F. A **differential extension** E of F is a differential field E which contains F and such that D_E restricted to F is D_F. If E is a differential extension of F then C_E contains C_F. If they coincide, then we say that $E \supset F$ is a **no new constants** extension.

It is conventional in discussing differential extension fields to omit the subscript on the derivations. Thus if $a_0, \ldots, a_n$ are elements of a differential field F, we can regard the expression $L = a_n D^n + \cdots + a_0 D^0$ as a (linear, homogeneous, order n) **differential operator** on any differential extension E of F, so for $y \in E$, $L(y) = a_n D^n(y) + \cdots + a_0 y$. The set of solutions $V = L^{-1}(0) = \{ y \in E \mid L(y) = 0 \}$ in E of the corresponding differential equation $L = 0$ is a vector space over the field C_E.

A differential field extension $E \supset F$ is *finitely generated* if there is some finite subset S of F such that the smallest differential subfield of E containing F and S is E itself. A differential field extension $U \supset F$ is *universal* if all finitely generated differential field extensions of differential subfields of U finitely generated over F embed in U.

Although there are extensions of differential Galois theory to wider contexts, we will now fix a base differential field F of *characteristic zero*, denote its field of constants C and assume that C is *algebraically closed*.

Picard-Vessiot extensions

Let L be a linear homogeneous differential operator over F of order n.

D.4.1 Definition An extension $E \supset F$ is a *Picard-Vessiot extension of F for L* if

(1) E is generated as a differential field over F by $L^{-1}(0)$.
(2) $E \supset F$ is a no new constants extension.
(3) $L^{-1}(0)$ is a vector space of dimension n over $C_E = C$.

A Picard-Vessiot extension E for L over F always exists, and is unique up to differential automorphism. It follows from the fact that E is differentially generated over F by $L^{-1}(0)$ that differential automorphisms of E over F are determined their restriction to $L^{-1}(0)$. And since $L^{-1}(0)$ is an n dimensional vector space over C, these restrictions are given by elements of $GL_n(C)$. Let $G(E/F)$ denote the differential Galois group. Thus there is an injective morphism $G(E/F) \to GL_n(C)$. It turns out that the image is Zariski closed. This provides a structure of algebraic group on $G(E/F)$; it can be shown to depend only on E and not on the L giving rise to it. With this structure, the following Fundamental Theorem holds (Kaplansky 1976),(Magid 1994):

D.4.2 Theorem (Fundamental Theorem for Picard-Vessiot Extensions) *Let $E \subset F$ be a Picard-Vessiot extension. Then*

$$K \mapsto G(E/K) \quad and \quad H \mapsto E^H = \{\, a \in E \mid \sigma(a) = a \; \forall \sigma \in H \,\}$$

give inverse bijections between the set of intermediate differential fields $E \supset K \supset F$ and the set of Zariski closed subgroups of $G(E/K)$. An intermediate field K is itself a Picard-Vessiot extension of F if and only *if* $G(E/K)$ *is a normal subgroup of* $G(E/F)$, *and, if so, then the quotient* $G(E/F)/G(E/K)$ *is isomorphic to* $G(K/F)$.

For an example, let $F = \mathbb{C}(t)$ be the field of rational functions in one complex variable with derivation $D_F = \frac{d}{dt}$. Consider the differential equation $Y'' + \frac{1}{t}Y' = 0$. The field $E = \mathbb{C}(\log(t), t)$ generated over F by $\log(t)$ is a Picard-Vessiot extension of F for the equation. Any differential automorphism σ of E over F has to send $\log(t)$, which satisfies the equation $Y' = \frac{1}{t}$, to another solution of this equation, so that $(\sigma(\log(t)) - \log(t))' = 0$, which means that $\sigma(\log(t)) - \log(t)$ is a constant, say $s \in \mathbb{C}$, and then $\sigma(\log(t)) = \log(t) + s$. The map $\sigma \mapsto \sigma(\log(t)) - \log(t)$ is a group homomorphism from $G(E/F)$ to the additive group $\mathbb{C}$. It is in fact a group isomorphism. Since $\mathbb{C}$ has no proper algebraic subgroups, the Fundamental Theorem implies that there are no intermediate differential fields between F and E.

Strongly normal extensions

Universal differential field extensions F can be shown to exist and we fix one, U. Let E be a subfield of U containing F. A differential injection $\sigma \colon E \to U$ inducing the identity on F (*an isomorphism of E over F*) is called **strong** if σ is the identity on C_E and if the differential composita $E \cdot C_U$ and $\sigma(E) \cdot C_U$ coincide. When E is differentially finitely generated over F any strong isomorphism of E over F extends to a differential automorphism of $E \cdot C_U$ over $F \cdot C_U$.

D.4.3 Definition A differential field extension $E \supset F$ contained in the universal differential field $U \supset F$ is called **strongly normal** if

(1) Every isomorphism of E over F is strong; and

(2) E is finitely generated as a differential field over F.

Suppose $E \supset F$ is a strongly normal extension. Then every isomorphism of E over F extends to an automorphism of $E \cdot C_U$ over $F \cdot C_U$. It turns out that $G(E \cdot C_U / F \cdot C_U)$ is the C_U points of an algebraic group defined over C whose C points are $G(E/F)$. Then there is the following Fundamental Theorem (Borel 1999):

D.4.4 Theorem (Fundamental Theorem for Strongly Normal Extensions) *Let $E \subset F$ be a strongly normal extension. Then*

$$K \mapsto G(E/K) \quad and \quad H \mapsto E^H = \{\, a \in E \mid \sigma(a) = a \forall \sigma \in H \,\}$$

give inverse bijections between the set of intermediate differential fields
$E \supset K \supset F$ *and the set of Zariski closed subgroups of* $G(E/K)$. *An*
intermediate field K *is itself a strongly normal extension of* F *if and*
only if $G(E/K)$ *is a normal subgroup of* $G(E/F)$, *and, if so, then the*
quotient $G(E/F)/G(E/K)$ *is isomorphic to* $G(K/F)$.

D.5 Differential Dimension Polynomials

by A. V. Mikhalev and E. V. Pankratiev in Moscow, Russia

Consider a differential field $\mathcal{F}$, i.e., a field in the usual sense (which is supposed to have characteristic 0) together with a finite set of mutually commuting derivation operators $\Delta = \{\delta_1, \ldots, \delta_m\}$; when $m = 1$ the differential field is **ordinary**, and when $M > 1$ it is **partial**. A **differential polynomial** over $\mathcal{F}$ in a finite number n of differential indeterminates is a polynomial, with coefficients in $\mathcal{F}$, in the derivatives

$$\delta_1^{i_1} \ldots \delta_m^{i_m} Y_j \quad (0 \leq i_1 < \infty, 0 \leq i_m < \infty, 1 \leq j \leq n). \tag{1}$$

The set $\mathcal{R} = \mathcal{F}\{Y_1, \ldots, Y_n\}$ of all differential polynomials over $\mathcal{F}$ in $Y_1, \ldots, Y_n$ is a differential ring, which is called the **ring of differential polynomials**.

An ideal $\mathcal{A}$ of $\mathcal{R}$ is **differential** if $\delta_i \mathcal{A} \subset \mathcal{A}$ $(1 \leq i \leq m)$ and is **perfect** if the condition $a^2 \in \mathcal{A}$ implies that $a \in \mathcal{A}$. The smallest perfect differential ideal containing the set Σ is denoted by $\{\Sigma\}$. A fundamental fact is that Σ always has a finite subset Φ such that $\{\Phi\} = \{\Sigma\}$; this is the **Ritt-Raudenbush theorem**. As a consequence, every perfect differential ideal $\mathcal{A}$ is the intersection of finitely many prime differential ideals none of which contains another.

A prime differential ideal $\mathfrak{p}$ of $\mathcal{F}\{Y_1, \ldots, Y_n\}$ has a **generic zero**: this is a tuple $(\eta_1, \ldots, \eta_n)$ in F^n such that every element of $\mathcal{F}\{Y_1, \ldots, Y_n\}$ that vanishes at $(\eta_1, \ldots, \eta_n)$ is an element of $\mathfrak{p}$. The extension $\mathcal{G} = \mathcal{F}\langle \eta_1, \ldots, \eta_n \rangle$ has a finite differential transcendence degree that is denoted by $d(\mathfrak{p})$. This number $d(\mathfrak{p})$ is called the **differential dimension** of $\mathfrak{p}$, and is the "correct" definition of what in the classical literature is called the number of arbitrary functions of m variables on which the solution of the system $P = 0$ $(P \in \mathfrak{p})$ depends; it measures the size of the set of zeros of $\mathfrak{p}$. A finer measure of the size is provided by the **differential dimension polynomial** $\omega_{\mathfrak{p}}$ of $\mathfrak{p}$. Its definition was given by Kolchin (1964). It should be noted that earlier, in 1952, Einstein (1953) introduced a notion of the "strength" of systems of partial differential equations and gave examples of its calculations. In essence, Einstein's strength coincides with the Kolchin dimension polynomial.

For any natural number s, the ring

$$\mathcal{R}_s = \mathcal{F}\left[(\delta_1^{i_1}\dots\delta_m^{i_m}Y_j)_{i_1+\cdots+i_m\leq s,\ 1\leq j\leq n}\right] \tag{2}$$

is a polynomial algebra over $\mathcal{F}$ in $\binom{s+m}{m}n$ indeterminates. The intersection $\mathfrak{p}\cap\mathcal{R}_s$ is a prime ideal of $\mathcal{R}_s$ and, therefore, has a dimension in the usual sense. There exists a unique polynomial ω_s in one variable over $\mathbb{Q}$ such that $\dim(\mathfrak{p}\cap\mathcal{R}_s) = \omega_{\mathfrak{p}}(s)$ for all sufficiently large natural numbers s. This polynomial has degree $\leq m$, and, therefore, can be written in the form $\omega_{\mathfrak{p}}(s) = \sum_{0\leq i\leq m}\alpha_i(\mathfrak{p})\binom{s+i}{i}$. The connection with the differential dimension is given by the equation $\alpha_m(\mathfrak{p}) = d(\mathfrak{p})$.

It may happen that $\alpha_m(\mathfrak{p}) = 0$. The largest natural number $\tau = \tau(\mathfrak{p})$ such that $\alpha_\tau(\mathfrak{p}) \neq 0$ is called the **differential type** of $\mathfrak{p}$ and $\alpha(\mathfrak{p}) \neq 0$ is called the **typical differential dimension** of $\mathfrak{p}$. This terminology is justified by the fact that if we regard $\mathcal{F}$ and $\mathcal{G}$ as differential fields with respect to τ new derivation operators $\delta'_{i'} = \sum_{1\leq i\leq m}c_{ii'}\delta_i$ $(1 \leq i' \leq \tau)$, where the $c_{ii'}$ are constants in $\mathcal{F}$ subject to a certain inequation, then $\mathcal{G}$ becomes a finitely generated extension of $\mathcal{F}$ of differential transcendence degree $\alpha_\tau(\mathfrak{p})$.

The differential dimension polynomial $\omega_{\mathfrak{p}}$ is a birational invariant, but not a differential birational invariant; this means that if η and ζ are generic zeros of $\mathfrak{p}$ and $\mathfrak{q}$, respectively, then the condition $\mathcal{F}(\eta) = \mathcal{F}(\zeta)$ implies $\omega_{\mathfrak{p}} = \omega_{\mathfrak{q}}$ but the weaker condition $\mathcal{F}\langle\eta\rangle = \mathcal{F}\langle\zeta\rangle$ (i.e., the differential field extension generated by η coincides with this generated by ζ) does not. Nevertheless, $\omega_{\mathfrak{p}}$ carries certain differential birational invariants with it. An obvious example is the differential dimension $\alpha_m(\mathfrak{p})$; two others are $\tau(\mathfrak{p})$ and $\alpha_{\tau(\mathfrak{p})}(\mathfrak{p})$.

The differential dimension polynomials can be ordered by defining $\omega_{\mathfrak{p}} \geq \omega_{\mathfrak{q}}$ if and only if $\omega_{\mathfrak{p}}(s) \geq \omega_{\mathfrak{q}}(s)$ for all sufficiently large integers s, i.e., lexicographically with respect to $(\alpha_m(\mathfrak{p}),\dots,\alpha_0(\mathfrak{p}))$. Sit (1975) proved that among all differential dimension polynomials corresponding to a given differential field extension $\mathcal{G}/\mathcal{F}$ there exists minimal with respect to this ordering. Thus, this minimal polynomial is a differential birational invariant of the extension.

Very few results concerning the minimal differential dimension polynomial are known. We only mention the following

D.5.1 Theorem ((Mikhalev and Pankratiev 1980)) *Let the cardinality* $|\Delta| = m$ *be fixed. The polynomial* $\omega(s) = \binom{s+m}{m} - \binom{s+m-r}{m}$ *is minimal*

in the set of all differential dimension polynomials for differential field extensions of differential type $m - 1$ and typical differential dimension r.

There is a close connection between the differential dimension polynomials and the Hilbert polynomials of graded polynomial modules. Namely, for any differential field extension $\mathcal{G}/\mathcal{F}$, the module of Kähler differentials is a filtered left module over the ring of linear differential operators $\mathcal{F}[\Delta]$. The filtration of this module depends on the choice of the generators η such that $\mathcal{G} = \mathcal{F}\langle\eta\rangle$ or, equivalently, of the prime differential ideal $\mathfrak{p}$ whose general zero is η. The associated module is a graded polynomial module, and the Hilbert polynomial of this module coincides with the differential dimension polynomial $\omega_{\mathfrak{p}}$ as is proved by Johnson (1969).

The calculation of the differential dimension polynomial $\omega_{\mathfrak{p}}$ can be divided into two stages. At the first one, either the so-called characteristic set of $\mathfrak{p}$ or the Gröbner basis of the appropriate module of differentials should be found. Unfortunately, no universal algorithms for computation of the characteristic sets are known. For the computation of the Gröbner bases, the completion algorithm is usually used, however, in the case of modules of Kähler differentials, the coefficient field is, as a rule, non-constructive. Thus, only linear differential systems over constructive differential fields can be treated. At the second stage, a combinatorial problem should be solved, which is the same as in the case of calculation of the Hilbert polynomials for a polynomial ideal with a known Gröbner basis. Algorithms for solution of this combinatorial problem are investigated in detail (see, for instance Kondratieva, Levin, Mikhalev, and Pankratiev 1999, Chap. 2).

There is a number of problems connected with differential dimension polynomials. For example, there are grounds for conjecturing that a so-called *Jacobi number* provides an upper estimate for the coefficient α_{m-1} in the case, where $\alpha_m = 0$. The Jacobi bound is proved in some partial cases but the problem remains open in the general case.

Similar to the differential dimension polynomial, the notion of difference dimension polynomials and differential-difference dimension polynomials is introduced (Kondratieva, Levin, Mikhalev, and Pankratiev 1999). The similar problems, e.g., the Jacobi bound, are under investigation.

D.5.2 Examples

1. (***Pure transcendental differential extensions***) Let $\mathcal{F}$ be a dif-

ferential field with a set of derivation operators $\Delta = \{\delta_1, \ldots, \delta_m\}$ and $\mathcal{G} = \mathcal{F}\langle y_1, \ldots, y_n \rangle$ be its pure transcendental differential extension. Then, the differential dimension polynomial is $\omega(t) = n\binom{t+m}{m}$.

2. (***Wave equation***) Let $\mathcal{F}$ be a differential field with the set of derivation operators $\Delta = \{\delta_1, \ldots, \delta_4\}$ and $\mathcal{G} = \mathcal{F}\langle \varphi \rangle$, where φ satisfies the wave equation $\delta_1^2 \varphi + \delta_2^2 \varphi + \delta_3^2 \varphi - \delta_4^2 \varphi = 0$, be its differential extension. Then, the differential dimension polynomial is $\omega(t) = \binom{t+4}{4} - \binom{t+2}{4} = 2\binom{t+3}{3} - \binom{t+2}{2}$. One can prove that this dimension polynomial is minimal.

3. (***Irreducible ordinary differential polynomials***) Let $\mathcal{F}$ be an ordinary differential field and $A = y'^2 + y$ be a differential polynomial. Then, the radical differential ideal $\{A\}$ is the intersection of two prime differential ideals. The differential dimension polynomials for the corresponding extensions are constants and are equal to 0 and 1, respectively.

4. (***Maxwell's equations***) Consider the Δ-equations (the Maxwell equations for empty space)

$$\varphi_{ij} = -\varphi_{ji}, \quad 1 \le i \le j \le 4,$$

$$U_i = \sum_{j=1}^{4} \delta_j \varphi_{ij} = 0, \quad 1 \le i \le 4,$$

$$V_{ikl} = \delta_l \varphi_{ik} + \delta_i \varphi_{kl} + \delta_k \varphi_{li} = 0, \quad i \ne k \ne l \ne i.$$

They define a differential field extension and the corresponding differential dimension polynomial is $\omega(t) = 4\binom{t+3}{3} + 2\binom{t+2}{2}$.

5. (***Maxwell's equations through the potential***) The electromagnetic field may be defined by the Δ-equations for its potential: $\mathcal{G} = \mathcal{F}\langle \psi \rangle$, where $\psi_i, i = 1, 2, 3, 4$ satisfy the following system of Δ-equations:

$$U_i = \sum_{j=1}^{4} (\delta_j^2 \psi_i - \delta_i \delta_j \psi_j) = 0,$$

$$B = \sum_{j=1}^{4} \delta_j \psi_j = 0.$$

They define a differential field extension and the corresponding differential dimension polynomial is $\omega(t) = 6\binom{t+3}{3} - \binom{t+2}{2} - (t+1)$.

D.6 Ordered Fields

by T. Y. Lam in Berkeley, CA, USA

Definitions, examples, and early history

A field (or a division ring) F is said to be **ordered** if there is given a subset $P \subseteq F$ (called a **positive cone**) closed under addition and multiplication such that $P \cup (-P) = F$ and $P \cap (-P) = \{0\}$ (cf. Section C.44). Writing $a \geq b$ to mean $a - b \in P$, we get a total ordering of the elements of F. An ordered division ring must have characteristic 0; its positive cone $P = \{a \mid a \geq 0\}$ cannot contain -1, but must contain $\sum F^2$ (all sums of squares in F). Prototype examples are $\mathbb{Q}$ and $\mathbb{R}$ with their usual orderings, while fields such as $\mathbb{Q}(i)$ and $\mathbb{C}$ cannot be ordered. The ordering on $\mathbb{R}$ is *archimedean* in the sense that any element in the field is bounded by an element of $\mathbb{Q}$. Subfields of $\mathbb{R}$ with their inherited orderings turn out to be, in fact, *all* archimedean ordered division rings. A typical nonarchimedean ordered field is $\mathbb{R}(x)$, with $P \setminus \{0\}$ consisting of $f(x)/g(x)$ where $f, g \in \mathbb{R}[x]$ have positive leading coefficients. In this ordering, the variable t is "infinitely large" with respect to the subfield $\mathbb{R}$. By the same token, we can extend this ordering to one on $\mathbb{R}(x, y)$ in which y is infinitely large with respect to $\mathbb{R}(x)$.

Ordered division rings were first considered by Hilbert in his work on the foundations of geometry. Roughly speaking, an ordering corresponds to a "betweenness" notion for the points on a line in a geometry coordinatized by the division ring in question. Hilbert constructed the first example of a noncommutative (and hence nonarchimedean) ordered division ring, from which he obtained a geometry that satisfies Desargues' Theorem but not Pappus' Theorem. Historically, Hilbert's example also provided the first division ring that is infinite dimensional over its center.

Artin-Schreier theory

Artin and Schreier (1927) defined a field to be **formally real** if $-1 \notin \sum F^2$, and showed that such fields are precisely the ones that can be ordered. Artin showed further that, if $\mathrm{char}(F) \neq 2$, then $a \in \sum F^2$ iff $a \geq 0$ w.r.t. all orderings of F. Using this result as one of his tools,

Artin gave an affirmative answer to Hilbert's 17th Problem: for $F = \mathbb{R}(x_1, \ldots, x_n)$, $f \in \sum F^2$ iff $f(a_1, \ldots, a_n) \geq 0$ for all $(a_1, \ldots, a_n) \in \mathbb{R}^n$ where it is defined (and the same holds if $\mathbb{R}$ is replaced by $\mathbb{Q}$).

Central to the Artin-Schreier theory is the notion of a ***real-closed field***: a formally real field F is real-closed if any proper algebraic extension of F is non-real. Such a field F captures the main algebraic features of the real field $\mathbb{R}$. There are several algebraic characterizations for such fields; for instance,

(1) F is formally real, $|F^*/F^{*2}| = 2$, and every odd-degree polynomial has a root in F.

(2) F has finite codimension > 1 in its algebraic closure.

(3) $F(\sqrt{-1}) \supsetneq F$ is an algebraically closed field.

Real-closed fields are ubiquitous as every ordered field (F, P) has a unique "real-closure", that is, a real-closed algebraic extension F_P whose unique ordering extends P.

Real valuations and real places

A (Krull) valuation v on a field F is said to be ***compatible*** with an ordering P if its valuation ring A is convex w.r.t. P. (Alternatively, we can also say that A is compatible with P.) In this case, P "pushes down" to an ordering on the residue class field of A. The principal example is the following. If E is any subfield of F, the convex hull A of E w.r.t. P is the smallest valuation ring $\supseteq E$ compatible with P, and its maximal ideal $\mathfrak{m}$ is the set of elements infinitely small w.r.t. E. Here, P pushes down to an ordering of $A/\mathfrak{m}$ with respect to which $A/\mathfrak{m}$ is archimedean over $\bar{E}$ (the image of E). The value group Γ of the valuation is the ordered group of "E-archimedean classes" on F. Taking $E = \mathbb{Q}$, we get in particular the ***Hahn valuation*** of the ordered field (F, P), together with a natural ***Hahn embedding*** of (F, P) into the ordered Laurent series field $\mathbb{R}((\Gamma))$.

Given any valuation v on F with residue field k and value group Γ, the work of Baer and Krull provides a description of the set of all orderings on F compatible with v, in terms of the orderings on k and the characters on the group Γ/Γ^2.

Theorems on real valuations are often related to those on places into real-closed fields. The systematic study of such real places began with S. Lang, who proved that, for any formally real function field F over a real-closed constant field k, there exists a k-place $\lambda \colon F \to k \cup \{\infty\}$, and λ can be chosen to be finite on a finite number of prescribed points in F.

An important consequence of this is **Lang's Homomorphism Theorem**: *If A is a commutative finitely generated k-domain with quotient field F, then there exists a k-algebra homomorphism from A to k.* These results constitute the beginning of a theory of real function fields.

Tarski's transfer principle and Robinson's nonstandard reals

The algebraic theory of ordered fields has played a very substantial role in the creation of the model theory of fields in logic. In 1948, Tarski showed that the theory of real-closed ordered fields admits quantifier elimination in the language of ordered rings (every "formula" is equivalent to a quantifier-free formula). From this, it follows that the theory of real-closed fields is "model-complete", that is, if $E \subseteq F$ are real-closed fields, then an elementary sentence about ordered fields with parameters from E holds in $(F, \leq)$ iff it holds in $(E, \leq)$. A more popular form of these results is **Tarski's Transfer Principle**, to the effect that every elementary sentence about ordered fields which holds in $(\mathbb{R}, \leq)$ also holds in every real-closed field $(F, \leq)$. (This would quickly imply, for instance, an affirmative answer to Hilbert's 17th Problem.) These famous results of Tarski set an important paradigm for much of the subsequent model-theoretic research in logic.

Another possibly surprising development is A. Robinson's invention of nonstandard analysis in the 1960s. By using an ultrapower of the real number system $\mathbb{R}$ (w.r.t. a non-principal ultrafilter (Section G.1) on $\mathbb{N}$), Robinson constructed a "nonstandard" number domain $^*\mathbb{R} \supsetneq \mathbb{R}$, which is real-closed, nonarchimedean, and such that every real function $y = f(x)$ extends canonically to a function on $^*\mathbb{R}$. In the nonstandard reals $^*\mathbb{R}$, infinitesimals exist, and can be used to efficiently develop the differential and integral calculus on $\mathbb{R}$, much in the way Leibniz originally did in the 17th century, but now with fully justified proofs!

Orderings and quadratic forms

For a field F of characteristic $\neq 2$, the **Witt ring** $W(F)$ consists of *Witt classes* of nonsingular quadratic forms over F. (Two such forms belong to the same Witt class if their anisotropic parts are isometric.) The $+$ and $\times$ operations on $W(F)$ are given by the $\perp$-**sum** and **tensor product** of quadratic forms. (For instance, for the unary forms ax^2 and by^2 $(a, b \in F^*)$, the $\perp$-sum is $ax^2 + by^2$, and the tensor product is

abz^2.) The nicest nontrivial example of a Witt ring is that for a real-closed field K: in this case, the *signature map* (assigning to any quadratic form its signature) gives a ring isomorphism $W(K) \cong \mathbb{Z}$, according to Sylvester's Law of Inertia. More generally, if (F, P) is any ordered field, taking signatures defines a ring surjection $W(F) \to \mathbb{Z}$. This is essentially the functorial map $W(F) \to W(F_P)$, if we identify $W(F_P)$ with $\mathbb{Z}$.

For any formally real field F, the set of orderings X_F carries the *Harrison topology*, with a subbasis given by $\{ P \mid a \in P \}$ where a ranges over F. This makes X_F into a Boolean (compact, Hausdorff, totally disconnected) space. By taking the "total signature", we get a ring homomorphism σ from $W(F)$ to $\mathcal{C}(X_F, \mathbb{Z})$, the ring of continuous functions from X_F to $\mathbb{Z}$ (the latter with the discrete topology). According to *Pfister's Local-Global Principle*, the kernel of σ is precisely $W(F)_t$, the ideal of torsion elements in $W(F)$. Pfister also showed that $W(F)_t$ is a 2-primary group. In case F is nonreal, all of this remains formally true as $W(F)$ itself is 2-primary in that case. In fact, if $-1 = a_1^2 + \cdots + a_s^2$ in F with s chosen minimal, then the "level" s of the nonreal field F has the form 2^n for some n (by another theorem of Pfister), and one has $2^{n+1}W(F) = 0$.

The quotient ring $W(F)/W(F)_t \cong \mathrm{im}(\sigma) \subseteq \mathcal{C}(X_F, \mathbb{Z})$ is called the *reduced Witt ring* of F. More generally, if T is any *preordering*, that is, an intersection of orderings, one can construct a reduced Witt ring $W_T(F)$, whose elements have the form $\langle a_1, \ldots, a_n \rangle$, where two such forms are identified if they have the same dimension n, and the same signature with respect to all orderings containing T. We get back the classical reduced Witt ring $W(F)/W(F)_t$ by letting $T = \sum F^2$ in case F is formally real.

The consideration of quadratic forms has also led to the study of many special kinds of formally real fields, such as euclidean fields, SAP fields (fields with a Strong Approximation Property), pythagorean and superpythagorean fields (and some of their hereditary versions), etc. For more detailed information on these (and on other material surveyed in this article), see the references (Lam 1980, 1983, 1984, Prestel 1984).

Generalizations

The theory of ordered fields sketched above has been generalized in many directions. E. Becker initiated a theory of orderings of higher level: in this theory, a positive cone P satisfies only the weaker condition that F^*/P^* is a cyclic group of even order (instead of order 2). A large part of the Artin-Schreier theory and Krull-Baer theory generalizes to this setting, and the theory can be used to study the structure of sums of even powers in a field. Other generalizations include the theory of preorderings on fields, formally p-adic fields, ordered and lattice-ordered rings (see Section C.44), *-ordered division rings, and the theory of real spectra of commutative rings with its various applications to Nullstellensätze and real algebraic geometry.

D.7 Applications of Fields

by Franz Binder in Linz, Austria and Rudi Lidl in Hobart, Australia

There is a wide range of applications of fields, but arguably the most interesting applications within branches of mathematics and outside mathematics are applications involving finite fields. Some aspects of codes are dealt with in sections A.17, I.2, and I.3.

For a more comprehensive treatment of these and other topics see (Lidl and Pilz 1998, Lidl and Niederreiter 1997). In this section, we sketch some more applications: discrete logarithms and cryptology, Hadamard matrices, and Latin squares. In addition, we give some not so well-known examples of applications of infinite fields in the areas of computational geometry and polynomial arithmetic.

Cryptography

Methods to ensure the secrecy, authenticity, and integrity are essential for modern techniques of communications. The concept of public-key ciphers (or cryptosystems) is due to Diffie and Hellman. In 1976 they also proposed a key exchange system for establishing a common key between A and B. Let α be a primitive element of $\mathbb{F}_q$. Users A and B choose random integers a and b, respectively, where $2 \leq a, b \leq q - 2$. Then A sends α^a to B, while B transmits α^b to A. Both take α^{ab} as their common key. This system can be broken if one can solve the discrete logarithm problem. We describe this problem first in a more general setting.

The *discrete logarithm problem* in a group G to the base $g \in G$ is the problem, given $y \in G$, of finding an integer x such that $g^x = y$, provided that such an integer exists. (When the group G is written additively then an integer x has to be found such that $xg = y$.)

There are several specific discrete logarithm problems depending on the nature of the group G. If $G = \mathbb{F}_q^*$, the multiplicative group of a finite field of order q, then the discrete logarithm problem for a well chosen finite field is hard. The first public-key cryptosystem for the exchange of keys, by Diffie and Hellman, was based on the difficulty of solving this particular discrete logarithm problem.

Other discrete logarithm problems and variations on the Diffie-Hellman key exchange system can be obtained from $G = \mathbb{F}_p^*$, p a prime; or $G = E$ where E is an elliptic curve over $\mathbb{F}_q$; or G is the Jacobian group of a hyperelliptic curve, see (Koblitz 1998) for details. Laywine and Mullen (1998) describe a system that is based on row-Latin squares and the difficulty of computing the discrete logarithm in the group G of all $n \times n$ row-Latin squares.

The running time for most (subexponential time) discrete logarithm algorithms for $G = \mathbb{F}_q$ is of the form $\exp(O(\sqrt{\ln q \ln \ln q}))$. It seems to require about the same amount of time as factorization of an integer of approximately the same size as q. No subexponential time algorithm is known in the case of non-supersingular elliptic curves.

If a cryptosystem is based on the discrete logarithm problem in a group G, attacks on the cryptosystem can be defeated by choosing the order of the group such that it is divisible by a very large prime. (In practice, at present, a prime of 40 or more decimal digits suffices.)

In the case where $G = \mathbb{F}_q^*$ and $q - 1$ has only small prime factors, the ***Silver-Pohlig-Hellman algorithm*** allows us to compute discrete logarithms, see (Lidl and Pilz 1998) for details and examples. Let α be a primitive element of $\mathbb{F}_q$, $q - 1 = \prod_{i=1}^k p_i^{e_i}$ where $p_1 < p_2 < \cdots < p_k$ are the distinct prime factors. Given $a \in \mathbb{F}_q^*$, the algorithm determines the value r for which $a = \alpha^r$.

First consider $r \equiv \sum_{j=0}^{e_i-1} r_j p_i^j \pmod{p_i^{e_i}}$, $0 \leq r_j \leq p_i - 1$. In order to determine r_j, $j = 0, 1, \ldots, e_i - 1$, for each prime factor p_i, we form

$$a^{(q-1)/p_i} = \alpha^{(q-1)r/p_i} = c_i^r = c_i^{r_0}$$

where $c_i = \alpha^{(q-1)/p_i}$ is a primitive p_i-th root of unity. There are only p_i possible values for $a^{(q-1)/p_i}$ corresponding to $r_0 = 0, 1, \ldots, p_i - 1$. Evaluating these is feasible if p_i is small. The next digit r_1 in the representation of $r \bmod p_i^{e_i}$ is obtained from $a\alpha^{-r_0} = \alpha^{t_1}$, where $t_i = \sum_{j=1}^{e_i-1} r_j p_i^j$. Then $\alpha^{t_1(q-1)/p_i^2} = c_i^{t_1/p_i} = c_i^{r_1}$ uniquely determines r_1, etc. Finally the value of $r \bmod p_i^{e_i}$, $i = 1, 2, \ldots, k$, will be combined by the Chinese Remainder Theorem to obtain $r \bmod q - 1$. It can be shown that this algorithm has a running time of order at most $p_k^{1/2}(\log q)^2$, where p_k is the largest prime factor of $q - 1$.

Hadamard matrices

An $n \times n$ Matrix H with entries in $\{1, -1\}$ is a **Hadamard matrix** of order n if $H^T H = HH^T = nI$. Thus, any two distinct rows (or columns) are orthogonal. These matrices have applications in coding theory, information science, physics, and pattern recognition, mainly because of Hadamard transforms. We refer to (van Lint and Wilson 1992).

A Hadamard matrix H of order n has $|\det H| = n^{n/2}$. It can be shown that if H exists then n is equal to 1, 2, or n is a multiple of 4. It is, however, an unresolved problem whether or not Hadamard matrices of orders $n = 4k$ exist for all $k \in \mathbb{N}$.

The **Payley construction** is a basic method of constructing a Hadamard matrix. First we define the Legendre symbol χ by

$$\chi(a) = \begin{cases} 1 & \text{if } a \neq 0 \text{ is a square,} \\ -1 & \text{if } a \text{ is not a square,} \\ 0 & \text{if } a = 0, \end{cases}$$

for any $a \in \mathbb{F}_q$. Let $\mathbb{F}_q = \{a_o = 0, a_1, \ldots, a_{q-1}\}$, $q = p^e = 4t - 1$. Then the following $(q + 1) \times (q + 1)$ matrix is a Hadamard matrix:

$$\begin{pmatrix} 1 & 1 & 1 & \cdots & 1 & 1 \\ 1 & -1 & \chi(a_1) & \cdots & \chi(a_{q-2}) & \chi(a_{q-1}) \\ 1 & \chi(a_{q-1}) & -1 & \cdots & \chi(a_{q-3}) & \chi(a_{q-2}) \\ \multicolumn{6}{c}{\cdots\cdots\cdots\cdots\cdots\cdots\cdots\cdots\cdots\cdots\cdots\cdots} \\ 1 & \chi(a_1) & \chi(a_2) & \cdots & \chi(a_{q-1}) & -1 \end{pmatrix}$$

Latin Squares

As a third area of applications of $\mathbb{F}_q$ we refer to Latin squares. A standard book covering Latin squares is (Dénes and Keedwell 1991), a more recent book including a wide range of applications in geometry, combinatorics, statistics and coding is (Laywine and Mullen 1998).

An $n \times n$ matrix $L = (a_{ij})$ over $A = \{a_1, \ldots, a_n\}$ is called a **Latin square** of order n if each row and column contains each element of A exactly once. Two Latin squares (a_{ij}) and (b_{ij}) over A of order n are called **orthogonal** if all n^2 ordered pairs, (a_{ij}, b_{ij}) of entries are distinct.

The famous Euler's officers problem goes back to Leonard Euler, who posed the following question: is it possible to arrange 36 officers of six regiments with six ranks into a 6×6 square in a parade, such that each row and each column of the square has exactly one officer of each regiment and each rank? In other words, do there exist two orthogonal Latin squares of order 6? Euler conjectured: there are no two orthogonal Latin squares of order n, where $n \equiv 2 \pmod 4$. In 1899, G. Tarry proved that Euler's officers problem cannot be done. But Euler's conjecture was totally disproved in 1959 for $n \neq 6$.

We conclude with a construction of mutually orthogonal Latin squares (MOLS). Let $0, 1, a_2, \ldots, a_{q-1}$ be the elements of $\mathbb{F}_q$. Then it can be verified that the squares L_i, $1 \leq i \leq q-1$, form a set of $q-1$ MOLS, where

$$
L_i = \begin{pmatrix}
0 & 1 & \cdots & a_{q-1} \\
a_i & a_i + 1 & \cdots & a_i + a_{q-1} \\
a_i a_2 & a_i a_2 + 1 & \cdots & a_i a_2 + a_{q-1} \\
\cdots\cdots\cdots\cdots\cdots\cdots\cdots\cdots\cdots\cdots \\
a_i a_{q-1} & a_i a_{q-1} + 1 & \cdots & a_i a_{q-1} + a_{q-1}
\end{pmatrix}
$$

Simplification of parametrizations

Let $x(t), y(t) \in \mathbb{Q}(t)$. Then $(x(t), y(t))$ is a (rational) parametrization of a real plane curve. Suppose that these coordinate functions have a **common component**, i.e., that there are $r(t), u(t), v(t) \in \mathbb{Q}(t)$ such that $x(t) = u(r(t))$ and $y(t) = v(r(t))$. Then $(u(t), v(t))$ is another parametrization of the same curve (rational functions are surjective), and if $r(t)$ is chosen maximal, $u(t)$ and $v(t)$ have no non-trivial common component. This guarantees, that the parametrization is simple, i.e., only finitely many points of the curve are obtained more than once.

The situation is clarified by a classical result:

D.7.1 Theorem (Lüroth) *Let K be a field, t transcendental over K, and $K < L < K(t)$. Then there is a $w \in K(t)$ such that $L = K(w)$.*

Note that $K(f(t)) \leq K(g(t))$ iff there is a rational function $h(t)$ such that $f(t) = h(g(t))$, with equality iff h is invertible with respect to functional composition. Thus, any generator of $\mathbb{Q}(x(t), y(t))$ is a maximal common component. Moreover, there is a version of Bezout's theorem here: the maximal common component can also be expressed as a rational function in $x(t)$ and $y(t)$, i.e., there is a rational function φ, such that $r(t) =$

$\varphi(x(t), y(t))$. As an example, consider $x(t) = t^3 - t^2 + 3t - 4$ and $y(t) = t^5 + t^4 - t^3 + 5t^2 + 2t - 3$, then t is a maximal common component and

$$t = \frac{191 + 76x + 12x^2 + x^3 + 17y + 3xy}{286 + 101x + 11x^2 + y + xy}.$$

Binder (1996) contains methods to compute this, as well as a fast algorithm, similar to Euclid's algorithm, to compute maximal common components in the polynomial case. The latter has been generalized by Hong and Schicho (1998) to the important case of trigonometric parametrizations, which often occur in mathematics, physics, and engineering, including Fourier analysis, linear differential equations, representations of groups, electrical circuit analysis, fracture mechanics, and robotics.

Generalized functional decomposition

A functional decomposition of a polynomial (or rational function) like $f(t) = h(g(t))$ has various advantages, including equation solving, Gröbner bases computation, or splitting a mechanical system into simpler parts. For example, to solve $h(g(t)) = 0$, we first solve $h(x) = 0$, obtaining solutions ξ_i, and then solve all equations $g(t) = \xi_i$. Unfortunately, such decompositions are possible only in rather special cases. However, many benefits of functional decomposition can be preserved to the more general case of **_norm-decompositions_**, which mean that at least some multiple of $f(t)$ can be functionally decomposed in a nontrivial way. Because norm-decompositions correspond exactly to the subfields of $K(\alpha)$, where $f(\alpha) = 0$, the problem can be reduced to determining subfields. The available algorithms have exponential worst-case running time, but are efficient enough to norm-decompose polynomials of degree 40 within seconds (cf. Klüners 1999).

Chapter E

REPRESENTATION THEORY

E.1 Group Rings

by Donald S. Passman in Madison, WI, USA

Introduction

If R is a ring and G is a multiplicative group, then we let $R[G]$ denote
the **group ring** of G over R. Thus $R[G]$ is a free R-module with basis
$\{x \mid x \in G\}$ and with multiplication defined distributively using the
group multiplication in G. When $R = K$ is a field, then $K[G]$ is the
group algebra of G over K, and our main concern here is with group
algebras of infinite groups. Nevertheless, we are frequently forced to deal
with more general objects such as twisted group rings, skew group rings,
crossed products and group-graded rings. In the following, we discuss a
selection of fairly well-developed topics which relate the group structure
of G to the ring-theoretic structure of $K[G]$. Basic references for this
material include the books by Karpilovsky (1987, 1989), Passi (1979),
Passman (1971, 1977, 1989), and Sehgal (1978, 1993), and the papers
by Passman (1984, 1998), Roseblade (1978), and Zalesskiĭ (1995).

Linear identities

A **linear identity** is an equation of the form $\alpha_1 x \beta_1 + \alpha_2 x \beta_2 + \cdots +
\alpha_n x \beta_n = 0$, with $\alpha_1, \alpha_2, \ldots, \alpha_n, \beta_1, \beta_2, \ldots, \beta_n \in K[G]$, that holds for all
$x \in G$ or perhaps for just a *large subset* of G. A basic result asserts
that if some term α_j is not zero, then the β_i are linearly dependent,
in a precise manner, over $K[\Delta]$. Here $\Delta = \Delta(G)$ is the **f. c. (finite
conjugate) center** of G, namely the subgroup consisting of all elements
of G having only finitely many conjugates. As a consequence of this and
a certain amount of group theory, we can characterize when $K[G]$ is

prime, semiprime, one-sided Artinian, or satisfies a polynomial identity. In addition, these techniques, known as the Δ-*method*, allow us to study annihilator ideals, to describe $\mathcal{N}K[G]$, the sum of all nilpotent ideals of the group algebra, and to decide when $\mathcal{N}K[G]$ is nilpotent.

Semiprimitivity

Much more difficult is the problem of determining when $K[G]$ is semiprimitive, and more generally, describing the Jacobson radical $\mathcal{J}K[G]$. If char $K = 0$, then $\mathcal{J}K[G] = 0$ if K is not algebraic over the rationals and, in all cases, $K[G]$ has no nontrivial nil ideals. It appears that group algebras in characteristic 0 are always semiprimitive. On the other hand, if char $K = p > 0$, then the situation is much different. To start with, if G is a finitely generated solvable or linear group, then $\mathcal{J}K[G] = \mathcal{N}K[G]$, and one suspects that this equality may be true for all finitely generated groups. If this conjecture is correct, then $\mathcal{J}K[G] = \mathcal{J}K[\Lambda] \cdot K[G]$ for all G, where $\Lambda = \Lambda(G)$ is a characteristic locally finite subgroup of G. This would then solve the problem since the structure of the radical for locally finite groups is now well-understood, using work based on the Classification of the Finite Simple Groups.

Primitive and simple rings

We have very little information on when a group algebra is primitive. Unlike the previously mentioned properties, primitivity seems to depend on the size of the field and not just on its characteristic. There are numerous interesting examples and results, but no reasonable conjecture. For the second topic, note that if $N \lhd G$, then there is a natural epimorphism $K[G] \to K[G/N]$. When $N = G$, the kernel of this map is the *augmentation ideal* $\omega K[G]$, and we say that $K[G]$ is *almost simple* if there are no other nontrivial ideals. We know that certain rather exotic groups have almost simple group algebras. Furthermore, in recent years, the lattices of ideals of complex group algebras of infinite locally finite simple groups have been studied in detail. It seems that for *most* such groups, the group algebra is almost simple.

Noetherian rings

Let G be a *polycyclic-by-finite group*, so that G has a finite subnormal series with factors which are either cyclic or finite. If R is a (right) Noetherian ring, then any crossed product $R*G$ is also Noetherian. In particular, this applies to the group algebra $K[G]$. The prime spectrum of these group algebras is now quite well understood. To start with, let us assume that G is **orbitally sound** (a rather technical condition). Then every prime ideal of $K[G]$ is essentially generated by its intersection with $K[\Delta]$. As a consequence, the prime and primitive lengths of $K[G]$ are computable from G, and $K[G]$ is a catenary ring. Furthermore, since an arbitrary polycyclic-by-finite group G has a characteristic orbitally sound subgroup, $\text{nio}(G)$, of finite index, many properties can be lifted from $K[\text{nio}(G)]$ to $K[G]$. In particular, the primes can be described, their heights can be computed, and $K[G]$ is catenary.

Zero divisors

If $x \in G$ has finite order $n > 1$, then $(1 - x)(1 + x + \cdots + x^{n-1}) = 0$ shows that $K[G]$ has nontrivial zero divisors. On the other hand, if G is torsion free, then there are no obvious zero divisors. Based on this, and with very little supporting evidence, it was conjectured that the group algebra of a torsion-free group is always a domain. The best results on this most intractable problem use homological methods. For example, if D is a division ring and G is a polycyclic-by-finite group, then the **Grothendieck group** $K_0(D*G)$ is generated by the images under induction of $K_0(D*F)$, as F runs through the finite subgroups of G. In particular, if all such $D*F$ are domains, then so is $D*G$. It follows that if G is a torsion-free group having a finite subnormal series with locally polycyclic-by-finite factors, then $K[G]$ is a domain.

Trace and augmentation

The **trace** of a group ring element is defined to be its identity coefficient. If char $K = 0$, then the trace of an idempotent e is a rational number between 0 and 1, while the trace of a nilpotent element η is 0. In particular, if α is a unit of finite order in the integral group ring $Z[G]$, then either $\alpha = \pm 1$ or $\text{tr}\,\alpha = 0$. When char $K = p > 0$, we have $\text{tr}\,e \in \mathbb{F}_p$, and $\text{tr}\,\eta = 0$ if G is a p'-group. Powers of the augmentation ideal have also been extensively studied. For example, the **dimension subgroups** are defined by $D_n(R[G]) = \{x \in G \mid x - 1 \in (\omega R[G])^n\}$. If $R = K$ is a field, then the relationship between $D_n(K[G])$ and the nth term of the lower central series of G is well understood and just depends on the characteristic of K. On the other hand, there is no known formula for $D_n(Z[G])$.

Isomorphism and units

To what extent does the group ring $R[G]$ determine the finite group G? Since there exist nonisomorphic metabelian groups G and H with $K[G] \cong K[H]$ for all fields K, it is appropriate to consider this question in the context of integral group rings. Here, if $Z[G] \cong Z[H]$ and if G is metabelian or nilpotent, then $G \cong H$. Indeed, when G is nilpotent, then any two **group bases** for $Z[G]$ of augmentation 1 are conjugate in the rational group ring $Q[G]$. On the other hand, we now know that there exist two nonisomorphic even ordered groups G and H of derived length 4 with $Z[G] \cong Z[H]$. Thus only the isomorphism question for finite groups of odd order is still open. Since, G is a subgroup of the group of units $U(Z[G])$, the latter group has been studied in detail. In particular, for most groups G, there is a natural set of units that is known to generate a subgroup of $U(Z[G])$ of finite index.

E.2 Character Theory

by I. Martin Isaacs in Madison, WI, USA

Definitions and basics

Let G be a finite group and recall that a (complex) representation of G is a homomorphism $\mathcal{X}$ from G into the group $GL(n; \mathbb{C})$ of invertible $n \times n$ matrices over $\mathbb{C}$. The **character** of G **afforded** by $\mathcal{X}$ is the function $\chi: G \to \mathbb{C}$ defined by $\chi(g) = \mathrm{tr}(\mathcal{X}(g))$. The size of the representing matrices can be recovered from the character since $\chi(1) = n$, and this positive integer is referred to as the **degree** of the character. The characters of degree 1 are exactly the homomorphisms from G into the multiplicative group of $\mathbb{C}$. It is easy to show that the number of these **linear** characters is exactly equal to $|G : G'|$.

Note that a character χ is a **class function** on G, which means that it is constant on each conjugacy class of G. We can thus write $\mathrm{Char}(G) \subseteq \mathrm{cf}(G)$, where $\mathrm{Char}(G)$ is the set of all characters of G and $\mathrm{cf}(G)$ is the vector space of all complex-valued class functions on G. (The dimension of this space is clearly equal to the number $k(G)$ of conjugacy classes of G.)

It is easy to see that the set $\mathrm{Char}(G)$ is closed under addition. A character $\chi \in \mathrm{Char}(G)$ is **irreducible** if it cannot be written as a sum of two characters, and we write $\mathrm{Irr}(G)$ to denote the set of irreducible characters of G. Since character degrees are positive integers, it follows that the characters of G are exactly the functions that can be written as sums of irreducible characters.

In fact, $\text{Irr}(G)$ is a basis for the space $\text{cf}(G)$, and hence $|\text{Irr}(G)|$ is equal to $k(G)$ and each character χ is uniquely a sum of irreducible characters. Those members of $\text{Irr}(G)$ that actually occur in the decomposition of χ are the irreducible **constituents** of χ. We mention also that the degrees of the irreducible characters divide $|G|$ and the sum of their squares is equal to $|G|$.

The **character table** of G is the $k(G) \times k(G)$ matrix with rows indexed by $\text{Irr}(G)$ and columns indexed by the classes of G, and where the (i, j)-entry is the value of the ith character at the jth class (see Section E.5 and Section E.7, for instance). The character table of G clearly determines all characters of G, and although it provides a great deal of information about the group itself, it does not determine the group uniquely up to isomorphism. (The two nonabelian groups of order 8, for example, have identical character tables.)

General references for character theory are (Berkovich and Zhmud' 1998, 1999, Huppert 1998, Isaacs 1994a).

Normal subgroups

If $\chi \in \text{Char}(G)$ is afforded by a representation $\mathcal{X}$, it is easy to see that the set $\ker(\chi) = \{ x \in G \mid \chi(x) = \chi(1) \}$ is the kernel of the homomorphism $\mathcal{X}$, and hence $\ker(\chi)$ is a normal subgroup of G. It is also not hard to see that the set $\mathbf{Z}(\chi) = \{ x \in G \mid |\chi(x)| = \chi(1) \}$ is always a normal subgroup. Since characters can detect normal subgroups, they can be used to prove nonsimplicity theorems such as those of W. Burnside and G. Frobenius.

Burnside showed that if $\chi \in \text{Irr}(G)$ has degree coprime to the size of the conjugacy class of an element x in G, then either $\chi(x) = 0$, or else $x \in \mathbf{Z}(\chi)$. From this, it is not hard to show that a nonabelian simple group cannot have a nonidentity conjugacy class of prime-power size. An easy corollary is that a group whose order involves just two primes must be solvable. This is Burnside's famous "$p^a q^b$-theorem", which has now been proved by a (difficult) character-free argument. But no character-free proof has been found for the underlying fact about prime-power class sizes in simple groups.

Frobenius considered transitive permutation groups G in which the identity is the only element that fixes as many as two points. He showed that the subset consisting of the identity and all elements of G that fix no points is a normal subgroup of G. No character-free proof for this result is known.

Every normal subgroup of G is an intersection of kernels of irreducible characters, and thus the character table completely determines the lattice of normal subgroups and the order of each normal subgroup. It follows that the character table determines whether or not the group is simple, solvable or nilpotent. (In fact, it is known that just the column of degrees in the character table is enough to determine nilpotence.) The character table does not determine the derived length of a nilpotent group, however.

Restriction and induction

Suppose $\chi \in \operatorname{Char}(G)$ and that $H \subseteq G$ is a subgroup. Then the restriction χ_H of χ to H is clearly a character of H. Similarly, given a character $\theta \in \operatorname{Char}(H)$, there is a simple construction of a character $\theta^G \in \operatorname{Char}(G)$ induced by θ. The restriction and induction maps between $\operatorname{Char}(G)$ and $\operatorname{Char}(H)$ are related by the elementary but useful *Frobenius reciprocity* formula, which asserts that if $\chi \in \operatorname{Irr}(G)$ and $\theta \in \operatorname{Irr}(H)$, then the multiplicity of θ as a constituent of χ_H is equal to the multiplicity of χ as a constituent of θ^G.

Induction provides a potential tool for constructing the characters of G. For example, it might happen that every irreducible character $\chi \in \operatorname{Irr}(G)$ has the form $\chi = \lambda^G$, for some linear character λ of some subgroup of G. If G has this property it is said to be a *monomial* group or *M-group*. It is easy to show that all supersolvable groups are M-groups and that M-groups are always solvable. In general, subgroups of M-groups need not be M-groups, and a number of problems about such subgroups remain open. (It is unknown, for example, if Hall subgroups of M-groups must be M-groups.)

Remarkably, every character is an integer linear combination of characters induced from linear characters of subgroups. This is the induction theorem of R. Brauer, which he used to show that every character of a group G is afforded by a representation $\mathcal{X}$ such that all entries in the matrices $\mathcal{X}(g)$ lie in the cyclotomic field $\mathbb{Q}_m$, where m is the exponent of G. Closely related to this is Brauer's powerful and useful characterization of characters: a class function of G is a difference of characters if and only if its restriction to every nilpotent subgroup N is a difference of characters of N.

Correspondences and solvable groups

Suppose that $A \subseteq \text{Aut}(G)$, where $|A|$ and $|G|$ are coprime. The Glauberman-Isaacs correspondence is a natural bijection from the set $\text{Irr}_A(G)$ of A-fixed irreducible characters of G onto $\text{Irr}(C)$, where $C \subseteq G$ is the subgroup consisting of the A-fixed elements of G. In the case where A is solvable, the correspondence was constructed by G. Glauberman and in the remaining case, where (by the odd-order theorem) G is solvable of odd order, the author used the character structure theory of solvable groups to construct the correspondence. Although the two construction methods are unrelated, it was shown by T. Wolf that in cases where both constructions apply, they both yield exactly the same map. One application of the Glauberman-Isaacs correspondence is to show that the actions of A on $\text{Irr}(G)$ and on the set of classes of G are permutation isomorphic.

Some references for solvable group character theory are (Isaacs 1973a, 1994b, 1995, Manz and Wolf 1993). To state just one result from this theory, let $H \subseteq G$ be a subgroup and let $\theta \in \text{Char}(H)$. An obvious necessary condition for θ to be the restriction to H of a character of G is that θ is constant on the intersection of H with each class of G. The author showed that if G is solvable and H is a Hall subgroup of G, then this condition is also sufficient. But if G is not solvable, this result can fail, even when H is a Sylow subgroup.

Degree sets

Write $\text{cd}(G) = \{ \chi(1) \mid \chi \in \text{Irr}(G) \}$. A surprising amount of information about G can be recovered from the set $\text{cd}(G)$ of character degrees, even without knowing how many irreducible characters there are of each degree. It is known, for example, that G has an abelian normal Sylow p-subgroup if and only if the prime p divides no member of $\text{cd}(G)$. If p divides every member of $\text{cd}(G)$ exceeding 1, then G has a normal p-complement. (The first of these results is the Ito-Michler theorem, which depends on the simple group classification; the second is due to J. Thompson and is more elementary.) Finally, we mention that if G is solvable, then its derived length is bounded above by some linear function of $|\text{cd}(G)|$. Examples suggest that perhaps a logarithmic bound holds, but that is not known even for p-groups.

E.3 Ordinary Representations I

by Albert Fässler in Biel, Switzerland

Basic definitions and facts

Representation theory tries to replace "abstract" groups by "concrete" ones. Especially important "concrete" classes are matrix groups $\mathrm{GL}(n; K)$ or automorphism groups $\mathrm{Aut}(V)$ of vector spaces. Clearly, there is not much difference between these two versions. More formally, a **representation** of a group G (by matrices) is a homomorphism $\vartheta: G \to \mathrm{GL}(n; K)$. If ϑ is injective, it is called **faithful**; n is the **degree** of ϑ.

Similarly, a representation of an R-algebra A (by matrices) is a homomorphism from A into $\mathrm{Mat}_{n \times n}(R)$.

Surprisingly enough, one learns much more on the representation of a group G by making it more complicated: one extends G to the group algebra $R[G]$ (and one extends ϑ in the obvious way). The big advantage is that $R[G]$ has much more structure than G itself.

For the sake of simplicity, we will assume that R is a field, mostly an algebraically closed one. If $\mathrm{char}\, R$ does not divide $|G|$, we study **ordinary representations**, otherwise **modular** ones. So, for ordinary representations, $R = \mathbb{C}$ is the canonical choice. If ϑ_1 and ϑ_2 are two (matrix) representations of G or of $K[G]$, their sum $\vartheta_1 \oplus \vartheta_2$ is defined to be the representation $\alpha \mapsto \begin{pmatrix} \vartheta_1(\alpha) & 0 \\ 0 & \vartheta_2(\alpha) \end{pmatrix}$ for $\alpha \in G$ or $\in K[G]$.

Every representation ϑ of $K[G]$ in $\mathrm{Mat}_{n \times n}(K)$ makes K^n into a $K[G]$-module via $\left(\sum k_g g \right)\mathbf{x} := \sum k_g \vartheta(g)\mathbf{x}$, and conversely, for each $K[G]$-module structure on K^n, we get a representation $\vartheta: \sum k_g g$ is mapped to the matrix corresponding to the map $\mathbf{x} \mapsto \left(\sum k_g g \right)\mathbf{x}$. Hence, representations of $K[G]$ correspond to $K[G]$-modules, and representations which cannot be written as the sum of two non-trivial ones (**irreducible representations**) correspond to irreducible $K[G]$-modules. Isomorphic $K[G]$-modules correspond to **equivalent representations**.

Symmetry adapted basis

Let ϑ be a finite-dimensional representation of a compact Lie group G (examples: all finite groups and the matrix groups $\mathrm{U}(n)$, $\mathrm{SU}(n)$, $\mathrm{O}(n)$, $\mathrm{SO}(n)$) over the complex field $\mathbb{C}$ (see Section B.6). Maschke's theorem guarantees that ϑ is completely reducible, i.e.,

$$\vartheta = c_1\vartheta_1 \oplus c_2\vartheta_2 \oplus \cdots \oplus c_N\vartheta_N, \tag{1}$$

where the ϑ_j are irreducible, inequivalent representations of G with multiplicities $c_j \in \mathbb{N}$. Accordingly, the N-dimensional complex representation space V of ϑ decomposes into a direct sum of so-called *isotypic components*, the $c_j \cdot n_j$-dimensional subspaces V_j:

$$V = V_1 \oplus V_2 \oplus \cdots \oplus V_N \tag{2}$$

The V_j in turn decompose into c_j irreducible subspaces $V_j^1, V_j^2, \ldots, V_j^{c_j}$ of dimension n_j:

$$V_j = V_j^1 \oplus V_j^2 \oplus \cdots \oplus V_j^{c_j} \tag{3}$$

The goal is to completely reduce the given matrix representation $\vartheta: s \mapsto D(s)$ associated with a given basis. More precisely: We wish to generate a set of *symmetry adapted basis vectors* for all of the $\sum_{j=1}^{N} c_j$ irreducible invariant subspaces $V_j^1, V_j^2, \ldots, V_j^N$. The representing matrices $D^{ad}(s)$, relative to the symmetry adapted basis, have the following block diagonal structure:

$$D^{ad}(s) = \bigoplus_{j=1}^{N} \bigoplus_{k=1}^{c_j} D_j(s) \tag{4}$$

Thus if we arrange the $c_j n_j$ basis vectors of V_j in rows by the irreducible subspaces V_j^k

$$V_j \begin{cases} V_j^1: & \mathbf{b}_1^1 & \mathbf{b}_2^1 & \mathbf{b}_3^1 & \cdots & \mathbf{b}_{n_j}^1 \\ V_j^2: & \mathbf{b}_1^2 & \mathbf{b}_2^2 & \mathbf{b}_3^2 & \cdots & \mathbf{b}_{n_j}^2 \\ \vdots & \vdots & \vdots & \vdots & \vdots & \vdots \\ V_j^{c_j}: & \mathbf{b}_1^{c_j} & \mathbf{b}_2^{c_j} & \mathbf{b}_3^{c_j} & \cdots & \mathbf{b}_{n_j}^{c_j} \end{cases} , \tag{5}$$

then under ϑ each of the c_j rows transform in exactly the same way. In (5), the label j for the different basis vectors has been omitted for simplicity.

Operators with symmetry

Linear operators with symmetry can be decomposed using the following *Generalization of Schur's Lemma*:

E.3.1 Theorem *Assume that a linear operator L is symmetric w.r.t. a representation $\vartheta: s \to D(s)$ of a compact or finite group G, that is,*

$$LD(s) = D(s)L \qquad \forall s \in G \tag{6}$$

*Then relative to a favorably labeled symmetry adapted basis, the opera-
tor L has the block diagonal form:*

$$L = \bigoplus_{i=1}^{N} \bigoplus_{k=1}^{n_i} L_i \qquad (7)$$

where $\dim L_j = c_j$. *In comparing ϑ with L, multiplicities and dimensions
are interchanged. For every isotypic component, each of the n_j subspaces
of dimensions c_j, invariant under L, is spanned by the vectors in the
columns of (5), hence labeling favorably means labeling along columns.*

Proof With respect to a symmetry adapted basis, the matrices $D(s)$
have the structure (4). If both sides of (6) are blockwise multiplied in the
sense of this structure, then by observing the orthogonality relations (4
in Section E.4) of the irreducible, inequivalent parts and Schur's lemma,
we obtain

$$L = \bigoplus_{i=1}^{N} (L_i \otimes I_{n_i})$$

with $c_i \times c_i$-matrices L_i and identity matrices I_{n_i}. Relabeling along
columns yields equation (7).

Fields of applications

A frequent situation is the need to get information about the symmetry
group of some "object", which can be a molecule, a geometric object,
a differential equation (cf. Section B.7), etc. Symmetry concepts within
group theory (and in particular representation theory) are used in almost every field: arts, geometry, solid state physics, elementary particle
physics, chemistry, classical mechanics, quantum mechanics, relativity
theory, bifurcations and symmetry breaking, crystallography, ordinary
and partial differential equations, signal processing, and so on. See e.g.,
Section B.7, (Fässler and Stiefel 1992), and (Lidl and Pilz 1998).

E.4 Ordinary Representations II
by Albert Fässler in Biel, Switzerland

The algorithm

In a first approach, assume that the group G is finite. Also, the irreducible pairwise inequivalent n_j-dimensional matrix representations

$$\vartheta_j : s \mapsto D^{(j)}(s) = (d_{\sigma\mu}^{(j)}(s)) \qquad \text{where} \quad j = 1, 2, \ldots, N$$

are available in the form of a table and likewise their characters $\chi_j(s)$. Then the matrix representation $\vartheta \colon s \mapsto D(s)$ can be completely reduced by use of the

E.4.1 Algorithm (Generating a symmetry adapted basis)

I. Compute the multiplicities $c_j = \frac{1}{|G|} \sum_{s \in G} \overline{\chi_j(s)} \chi(s)$, where χ is the character of ϑ and χ_j the character of ϑ_j. A multiplicity c_j is 0, if ϑ_j does not appear in ϑ.

II. Compute the matrix $P_{11}^{(j)} = \sum_{s \in G} d_{11}^{(j)}(s^{-1}) D(s)$. The columns of $P_{11}^{(j)}$ span a c_j-dimensional subspace of V. Take an arbitrary basis $\mathbf{b}_1^1, \mathbf{b}_1^2, \ldots, \mathbf{b}_1^{c_j}$ from this subspace.

III. Compute the matrices $P_{1k}^{(j)} = \frac{n_j}{|G|} \sum_{s \in G} d_{1k}^{(j)}(s^{-1}) D(s)$, $k = 2, 3, \ldots, n_j$ of rank c_j.

IV. Generate a symmetry adapted basis for the irreducible subspaces $V_j^1, V_j^2, \ldots, V_j^{c_j}$ of V_j in the following way:
$$\mathbf{b}_k^i = P_{1k}^{(j)} \mathbf{b}_1^i, \; k = 2, 3, \ldots, n_j, \; i = 1, 2, \ldots, c_j.$$

This procedure must be carried out for each ϑ_j that occurs in ϑ. With respect to this basis, the representing matrices have the desired form ((4) in Section E.3).

E.4.2 Remarks

(1) The decomposition of V_j into irreducible subspaces V_j^1, V_j^2, …, $V_j^{c_j}$ is not unique, due to the arbitrary choice of the c_j linearly independent vectors in IV.

(2) The linear operators $P_{11}^{(j)}, P_{12}^{(j)}, \ldots, P_{1n}^{(j)}$ with $j = 1, 2, \ldots, N$ generate the basis. They are called *generators*.

(3) The equations in II and III show that only the first rows of the tabulated matrices of ϑ_ρ are used each time.

(4) For unitary representations, II and III can be simplified by using

$$d_{1k}^{(j)}(s^{-1}) = \overline{d_{k1}^{(j)}(s)} \qquad k = 1, 2, \ldots, n_j$$

where the bar denotes complex conjugation.

A simpler tool for decomposing V into its isotypic components V_j is described by the

E.4.3 Theorem *The projection that maps V onto the $c_j n_j$-dimensional subspace V_j is given by*

$$\mathsf{P}^{(j)} = \frac{n_j}{|G|} \sum_{s \in G} \chi_j(s^{-1}) \mathsf{D}(s) \tag{1}$$

For unitary representations, we can simplify $\chi_j(s^{-1})$ to $\overline{\chi_j(s)}$.

Proof Let us denote the basis-independent notions of linear operators by sans serif type style $\mathsf{T}, \mathsf{D}, \mathsf{P}$. We consider the linear operators (where i, k, ℓ are fixed):

$$\mathsf{P}^{(i)}_{k\ell} = \frac{n_i}{|G|} \sum_{s \in G} d^{(i)}_{k\ell}(s^{-1}) \mathsf{D}(s) \qquad k, \ell \in \{1, 2, \ldots, n_j\} \tag{2}$$

Its matrices relative to our symmetry adapted basis are in block diagonal structure as in ((4) in Section E.3):

$$\mathsf{P}^{(i)\mathrm{sp}}_{k\ell} = \frac{n_i}{|G|} \sum_{s \in G} d^{(i)}_{k\ell}(s^{-1}) \mathsf{D}^{\mathrm{ad}}(s) \qquad k, \ell \in \{1, 2, \ldots, n_j\} \tag{3}$$

By the *orthogonality relations for the irreducible representations* ϑ_j, the small $n_j \times n_j$-blocks of (3) become

$$\frac{n_i}{|G|} \sum_{s \in G} d^{(i)}_{k\ell}(s^{-1}) \, d^{(j)}_{pq}(s) = \delta_{ij} \delta_{qk} \delta_{p\ell} \qquad \forall p, q, k, \ell \tag{4}$$

Therefore, only the $c_i n_i \times c_i n_i$ block is not vanishing. Each of its c_i small blocks contains exactly one nonzero entry (namely 1) at row number $\ell = p$ and column number $q = k$. Hence the operator $\mathsf{P}_{k\ell}$ with rank c_i has the properties

$$\mathsf{P}^{(i)}_{k\ell} : V \to V_i \quad \text{with} \qquad \left. \begin{aligned} \mathsf{P}^{(i)}_{k\ell} \mathbf{b}^\alpha_\mathbf{k} &= \mathbf{b}^\alpha_\ell \\[2mm] \mathsf{P}^{(i)}_{k\ell} \mathbf{b}^\alpha_{\mathbf{k'}} &= \mathbf{0} \qquad \text{if } k \neq k' \end{aligned} \right\} \alpha = 1, 2, \ldots, c_j \tag{5}$$

The equation in step I of Algorithm E.4.1 is a consequence of the orthogonality of the characters χ_j, which completes the proof of Algorithm E.4.1. Furthermore: The sum

$$\mathsf{P}^{(i)\mathrm{ad}} = \sum_{k=1}^{n_i} \mathsf{P}^{(j)\mathrm{ad}}_{kk} \tag{6}$$

is the matrix with the ith block being a $c_i n_i \times c_i n_i$ identity matrix and all other entries zero. Therefore it describes a projector onto the isotypic component V_i. Interchanging summation yields

$$P^{(j)} = \frac{n_j}{|G|} \sum_{k=1}^{n_j} \sum_{s \in G} d_{kk}^{(j)}(s^{-1}) D(s) = \frac{n_j}{|G|} \sum_{s \in G} \chi_j(s^{-1}) D(s).$$

This completes the proof of Theorem E.4.3.

Generalization for continuous representations of compact Lie Groups: When replacing

$$\frac{1}{|G|} \sum_{s \in G} \rightarrow \underset{s \in G}{\mathcal{M}},$$

where $\mathcal{M}$ is the Haar measure, Algorithm E.4.1 and Theorem E.4.3 remain correct. $\mathcal{M}$ is linear, positive and normalized for real-valued functions f of G and, most of all, it is invariant under translations (See for instance Miller (1972)):

$$\underset{s \in G}{\mathcal{M}} f(s) = \underset{s \in G}{\mathcal{M}} f(as) = \underset{s \in G}{\mathcal{M}} f(sa) \qquad \forall a \in G$$

E.4.4 Remark Explicit tables for symmetry adapted bases of permutation representations of finite point groups in dimension 2 and 3 are given and cited in Fässler and Stiefel (1992).

Examples

Let $\mathcal{P}_n$ be the $\frac{1}{2}(n+1)(n+2)$- dimensional vector space of the homogeneous polynomials $p_n(x, y, z)$ of degree $n \in \{0, 1, 2, \ldots\}$ in cartesian coordinates x, y, z of R^3, spanned by the monomials

$$x^\alpha y^\beta z^\gamma \quad \text{with} \quad \alpha + \beta + \gamma = n.$$

We consider the following representation ϑ_n of the orthogonal group $O(3)$ with the general matrix S:

$$p_n(x, y, z) \mapsto p_n(x', y', z') \quad \text{with} \quad S^{-1}(xyz)^T = (x'y'z')^T.$$

Let $\mathcal{H}_\ell$ be the $(2\ell + 1)$- dimensional vector space of the homogeneous polynomials in x, y, z of degree ℓ, the so called spherical functions. It is well known that, with $x^2 + y^2 + z^2 = r^2$,

$$\mathcal{P}_n = \mathcal{H}_n \oplus r^2 \mathcal{H}_{n-2} \oplus r^4 \mathcal{H}_{n-4} \oplus \cdots \oplus \begin{cases} r^n \mathcal{H}_0 & \text{if } n \text{ even} \\ r^{n-1} \mathcal{H}_1 & \text{if } n \text{ odd} \end{cases}$$

The subspaces $r^{n-\ell}\mathcal{H}_\ell$ are invariant and irreducible subspaces of ϑ_n (rotational invariance of the Laplacian), and transformed under the irreducible representations $\vartheta^{(\ell)}$. The restriction $\vartheta_G^{(\ell)}$ of $\vartheta^{(\ell)}$ to a subgroup G of $O(3)$ may split the $2\ell+1$-dimensional vector space into smaller irreducible and invariant subspaces of $r^{n-\ell}\mathcal{H}_\ell$.

The Full Tetrahedral Group T_d

The full tetrahedral group T_d is isomorphic to the symmetric group S_4 with $4! = 24$ elements. Let us consider a regular tetrahedron with the following vertices in (x, y, z)-coordinates: no.1:$(1, -1, 1)$, no.2:$(-1, -1, -1)$, no.3:$(1, 1, -1)$, no.4:$(-1, 1, 1)$. The following isomorphism between the generating matrices $\mathbf{P}, \mathbf{Q}$ (describing isometric mappings, that leave the tetrahedron invariant) of T_d and its corresponding permutations of S_4 of the numbers of vertices is given by

$$\mathbf{P} = \begin{pmatrix} 0 & 1 & 0 \\ 0 & 0 & 1 \\ 1 & 0 & 0 \end{pmatrix} \leftrightarrow (143)(2) \quad \text{and} \quad \mathbf{Q} = \begin{pmatrix} -1 & 0 & 0 \\ 0 & 0 & 1 \\ 0 & -1 & 0 \end{pmatrix} \leftrightarrow (2143).$$

The first step of the algorithm E.4.1, together with the character table of S_4, gives the following multiplicities for $\vartheta_{S_4}^{(\ell)}$ with $\ell = 0, 1, \ldots, 9$:

irred. rep. & ℓ	0	1	2	3	4	5	6	7	8	9
ϑ_1	1	0	0	1	1	0	1	1	1	1
ϑ_1'	0	0	0	0	0	0	1	0	0	1
ϑ_3	0	1	1	1	1	2	2	2	2	3
$\vartheta_3' = \vartheta_3 \otimes \vartheta_1'$	0	0	0	1	1	1	1	2	2	2
ϑ_2	0	0	1	0	1	1	1	1	2	1

The next steps of the algorithm generate the irreducible invariant subspaces for $\vartheta_{S_4}^{(\ell)}$. For the example

$$\vartheta_{S_4}^{(5)} = 2\vartheta_3 \oplus \vartheta_3' \oplus \vartheta_2,$$

we obtain the following symmetry adapted basis:

- V_3^1: $2x^5 + 5xy^4 + 5xz^4 - 10x^3y^2 - 10x^3z^2$ and cyclically in x, y, z.

- V_3^2: $x^5 + 15xy^2z^2 - 5x^3y^2 - 5x^3z^2$ and cyclically in x, y, z.

- V_3': $2x^3z^2 - 2x^3y^2 + xy^4 - xz^4$ and cyclically in x, y, z.

- V_2: $xy^3z - x^3yz$, $xyz^3 - x^3yz$.

The Icosahedral Group I

The icosahedral group I consists of the 60 proper rotations carrying a regular icosahedron onto itself. It is isomorphic to the alternating group A_5, as can be proven by 5 inscribing cubes.

I and A_5 have the following corresponding classes beside the identity:

$\pm 72°$ rotations with axis about opposite vertices $\leftrightarrow$ (12345), containing twelve elements.
$\pm 144°$ rotations with axis about opposite vertices $\leftrightarrow$ (21345), containing twelve elements.
$\pm 120°$ rotations about the midpoints of faces $\leftrightarrow$ (123), containing twenty elements.
$180°$ rotations about the midpoints of edges $\leftrightarrow$ (12)(34), containing fifteen elements.

The two matrices

$$\frac{1}{\sqrt{5}}\begin{pmatrix} 1 & 2\cos\varphi & -2\sin\varphi \\ 2 & -\cos\varphi & \sin\varphi \\ 0 & -\sqrt{5}\sin\varphi & -\sqrt{5}\cos\varphi \end{pmatrix} \quad \text{and} \quad \begin{pmatrix} 1 & 0 & 0 \\ 0 & \cos\varphi & \sin\varphi \\ 0 & -\sin\varphi & \cos\varphi \end{pmatrix},$$

with $\varphi = 72°$, describing a $120°$-rotation and a $72°$-rotation generate I. The algorithm, with the character table of A_5 (Section E.7), and its notations gives the following results for $\ell = 0, 1, 2, 3$:

$$\vartheta_{T_d}^{(0)} = \vartheta_5 \qquad \vartheta_{T_d}^{(1)} = \vartheta_{3,1,1}^+ \qquad \vartheta_{T_d}^{(2)} = \vartheta_{2,2,1} \qquad \vartheta_{T_d}^{(3)} = \vartheta_{3,1,1}^- \oplus \vartheta_{4,1},$$

For the non-trivial case $\vartheta_{T_d}^{(3)}$, it yields the following invariant and irreducible subspaces:

- $3 - \dim$: $2x^3 - 3xy^2 - 3xz^2$, $y^3 + 3xy^2 - 3xz^2 - 3yz^2$, $6xyz + z^3 - 3y^2z$.

- $4 - \dim$: $4x^2y - y^3 - yz^2$, $4x^2z - z^3 - y^2z$, $-y^3 + 3yz^2 + 2xy^2 - 2xz^2$, $-z^3 + 3y^2z + 4xyz$.

E.5 Modular Representations: General theory

by Peter Fleischmann in Kent, UK

Basic facts

Let G be a finite group and $\mathbb{F}$ a field of characteristic p, dividing the group order $|G|$. A group homomorphism ρ from G into $\mathrm{GL}(n; \mathbb{F})$ is called a **modular representation** (compare with Section E.2). Every representation of G can be extended to a module of the group algebra $\mathbb{F}G$, and conversely, every $\mathbb{F}G$-module defines a representation of G in a natural way.

E.5.1 Example The representation $\rho\colon G = \langle g \rangle \simeq \mathbb{Z}_p \to \mathrm{GL}(2; \mathbb{F})$ of the cyclic group of order p, defined by $g \mapsto \left(\begin{smallmatrix} 1 & 1 \\ 0 & 1 \end{smallmatrix}\right)$, is modular and has a proper one-dimensional sub-representation without being isomorphic to a direct sum of one-dimensional representations. Hence it is indecomposable but not irreducible, which shows that the group algebra $\mathbb{F}G$ is *not semisimple* and modular representations are in general *not completely reducible*, in contrast to ordinary representations. But their composition factors and their indecomposable summands are unique up to isomorphism and multiplicity (Jordan-Hölder theorem, Krull-Schmidt theorem).

From now on, we assume that $\mathbb{F}$ is algebraically closed. Let $S_1, S_2, \ldots, S_{\ell_{p'}}$ be representatives of the isomorphism classes of irreducible $\mathbb{F}G$-modules; then $\ell_{p'}$ is the number of p'-conjugacy classes of G, i.e., conjugacy classes of elements with order coprime to p. The projective covers P_i of the S_i form a cross section of iso-types of indecomposable direct summands of the regular module ${}_{\mathbb{F}G}\mathbb{F}G$ and are called the **principal indecomposable modules**. Each P_i has a unique maximal and a unique minimal submodule such that the quotient of the first is isomorphic to the latter and both are isomorphic to S_i. Let c_{ij} denote the multiplicity of S_j as a composition factor of P_i, then the **Cartan matrix** $\mathbf{C} := (c_{ij})$ is symmetric, with its determinant being a power of p.

Brauer characters

Let $A \subseteq \mathbb{C}$ be the ring of algebraic integers, $R := A_{\mathfrak{p}}$ the localization at a maximal ideal $\mathfrak{p} \trianglelefteq A$ containing p and $\mathbb{F} := R/\mathfrak{p}$ the algebraically closed residue class field. Let $G_{p'}$ denote the set of p'-elements in G and V a finite dimensional (irreducible) $\mathbb{F}G$-module with corresponding representation ρ. Then for each $g \in G_{p'}$, the value of the trace $\mathrm{tr}(\rho(g))$ is a sum of roots of unity in $\mathbb{F}$ of order coprime to p and thus can be *lifted* to a unique value $\varphi(g) \in \mathbb{C}$. The resulting class function $\varphi \colon G_{p'} \to \mathbb{C}$ is called the (irreducible) **modular character** or **Brauer character** of V. The irreducible Brauer characters of G are linearly independent and form a basis of the space $\mathcal{Cl}_{p'}$ of complex valued class functions which are non-zero only on p'-classes. The Brauer character of an $\mathbb{F}G$-module does not determine its isomorphism type, but it determines its composition factors up to isomorphism and multiplicity.

Brauer characters are an important tool to relate modular representations and complex characters. For each finite dimensional $\mathbb{C}G$-module M with representation ρ, one can find a basis of M in such a way that the corresponding matrices describing all $\rho(g)$ have entries in R. Applying the canonical map $R \to \mathbb{F}$ to these entries yields an $\mathbb{F}G$-module M_p, the **mod-p-reduction** of M, whose Brauer character $\chi_{p'}$ coincides with the restriction of the character of ρ to p'-classes, and thus can be written as linear combination of the irreducible Brauer characters φ_j, $j = 1, \ldots, \ell_{p'}$. Let χ_i, $i = 1, \ldots, \ell := k(G)$ be the the irreducible complex characters (see Section E.2). Then there are unique coefficients d_{ij} such that

$$(\chi_i)_{p'} = \sum_{i,j} d_{ij} \varphi_j.$$

The **p-decomposition matrix** $\mathbf{D} := (d_{ij}) \in \mathbb{N}^{\ell \times \ell_{p'}}$ has maximal rank and satisfies $\mathbf{D}^{\mathrm{tr}} \circ \mathbf{D} = \mathbf{C}$ (**Brauer's reciprocity theorem**). For a modern account on the theory of Brauer characters, see (Navarro 1998); for information on Brauer characters of sporadic simple groups see (Jansen et al. 1995).

E.5.2 Example The alternating group $G := A_5$ has the five conjugacy classes $C_1 = \{\mathrm{id}\}$, $C_2 := (12)(34)^G$, $C_3 := (123)^G$, $C_{5_1} := (12345)^G$, $C_{5_2} := (13245)^G$, and the ordinary character table

	C_1	C_2	C_3	C_{5_1}	C_{5_1}
χ_1	1	1	1	1	1
χ_2	3	-1	0	z_1	z_2
χ_3	3	-1	0	z_2	z_1
χ_4	5	1	-1	0	0
χ_5	4	0	1	-1	-1

with $z_1 := \frac{1-\sqrt{5}}{2}$ and $z_2 := \frac{1+\sqrt{5}}{2}$.

For $p = 2$ there are the four p'-classes $C_1, C_3, C_{5_1}, C_{5_2}$, and hence there are four irreducible 2-Brauer-characters, which turn out to be: $\varphi_1 = (\chi_1)_{2'}$, $\varphi_2 = (\chi_2)_{2'} - (\chi_1)_{2'}$, $\varphi_3 = (\chi_3)_{2'} - (\chi_1)_{2'} = (\chi_4)_{2'} - (\varphi_2)_{2'} - (\chi_1)_{2'}$, $\varphi_4 = (\chi_5)_{2'}$. Hence we get the following 2-decomposition-matrix and Cartan-matrix, respectively.

	φ_1	φ_2	φ_3	φ_4
χ_1	1	0	0	0
χ_2	1	1	0	0
χ_3	1	0	1	0
χ_4	1	1	1	0
χ_5	0	0	0	1

$$\begin{pmatrix} 4 & 2 & 2 & 0 \\ 2 & 2 & 1 & 0 \\ 2 & 1 & 2 & 0 \\ 0 & 0 & 0 & 1 \end{pmatrix}$$

E.6 Modular Representations: Brauer's Theorems

by Peter Fleischmann in Kent, UK

The group algebra can be decomposed into a direct sum $\mathbb{F}G = \bigoplus_{i=1}^{s} B_i$ of **blocks**, i.e., two-sided ideals B_i, which themselves are not properly decomposable into a direct sum of smaller two-sided ideals. Let $1 = \sum_{i=1}^{s} e_i$ be the corresponding decomposition of 1 into central-primitive idempotents $e_i = e_i^2 \in B_i := \mathbb{F}Ge_i$. Then $B_i B_j = B_i$ iff $i = j$ and zero otherwise. For each block B_i of G, there is a unique algebra homomorphism λ_{B_i} from the center $Z(\mathbb{F}G)$ to $\mathbb{F}$ with $\lambda_{B_i}(e_j) = \delta_{ij}$. An $\mathbb{F}G$-module M is said to **belong to** B_i ($M \in B_i$) if $B_i M = M$. Similarly, the complex character χ is said to belong to B_i if its mod p-reductions M_p belong to B_i. This partitions the sets $\mathrm{IBr}(G)$ of Brauer Characters and $\mathrm{Irr}_{\mathbb{C}}(G)$ of ordinary irreducible characters into pairwise disjoint **blocks of irreducible (Brauer) characters** $\mathrm{IBr}_i := \mathrm{IBr}(G) \cap B_i$ and $(\mathrm{Irr}_{\mathbb{C}})_i := \mathrm{Irr}_{\mathbb{C}}(G) \cap B_i$ of sizes $\ell(B_i)$ and $k(B_i)$, respectively. It also decomposes the matrices **D** and **C** into corresponding block diagonal forms, with the matrix blocks being the decomposition matrices and Cartan matrices defined accordingly for the B_i. For any natural number $m \in \mathbb{N}$, let m_p denote the maximal p-power dividing m and set $p^a := |G|_p$. Then the exponent $d(B_i)$ with $p^{a-d(B_i)} := \min\{\chi(1)_p \mid \chi \in B_i\}$ is called the **defect** of B_i. Notice that $d(B_i) \geq 0$, since $\chi(1)$ divides $|G|$ (see Section E.2).

Example E.5.2 shows that A_5 has two 2-blocks: $B_1 \leftrightarrow \{\chi_1, \chi_2, \chi_3, \chi_4\} \leftrightarrow \{\varphi_1, \varphi_2, \varphi_3\}$ of defect 2 and $B_2 \leftrightarrow \{\chi_5\} \leftrightarrow \{\varphi_4\}$ of defect 0. The fact that blocks of defect zero contain exactly one irreducible ordinary character and one irreducible Brauer-character is the main content of

E.6.1 Theorem (R. Brauer, 1956) *The following are equivalent:*

(1) $d(B_i) = 0$;
(2) $k(B_i) = 1$;
(3) $p^a = (\chi(1))_p$ *for some* $\chi \in (\mathrm{Irr}_\mathbb{C})_i$;
(4) $k(B_i) = \ell(B_i)$.

If one (and hence all) of (1)–(4) are satisfied, then $(\mathrm{Irr}_\mathbb{C})_i = \{\chi\}$ *for some irreducible character* χ, *which is called of "defect zero" and has an irreducible mod-p-reduction.*

Each block idempotent $e_i \in Z(\mathbb{F}G)$ is an $\mathbb{F}$-linear combination $\sum \lambda_K K^+$, where $K^+ := \sum_{x \in K} x$ with K ranging through the conjugacy classes of G (in fact p'-classes, due to a theorem of Osima). One can show that for each e_i there exists a unique conjugacy class $\delta(B_i)$ of p-subgroups of G such that the following holds: if $\lambda_K \neq 0$, then each Sylow p-group of a centralizer of $g \in K$ lies in some $P \in \delta(B_i)$ with equality up to conjugacy for some K_0 with $\lambda_0 \neq 0$. Any $P \in \delta(B_i)$ is called a **defect group** of B_i. It turns out that $p^{d(B_i)} = |P|$ and that every defect group is an intersection of two Sylow p-subgroups of G. In particular, every normal p-subgroup of G is contained in any defect group.

Now let $H \leq G$ be a subgroup, b a block of $\mathbb{F}H \subseteq \mathbb{F}G$, and $\lambda_b^G : Z(\mathbb{F}G) \to \mathbb{F}$ the linear map defined by $K^+ \mapsto \lambda_b(\sum_{x \in K \cap H} x)$. If λ_b^G happens to be an algebra homomorphism, there is a unique block $B = b^G$ of G such that $\lambda_b^G = \lambda_B$. In this case, one says that the **induced block** b^G **is defined** (e.g., if $PC_G(P) \leq H \leq N_G(P)$ with p-subgroup P.)

E.6.2 First Main Theorem *Let P be a p-subgroup of G. Then the map $b \mapsto b^G$ defines a bijection between those blocks of $N_G(P)$ and of G which have a defect group P.*

Let $\gamma \in \mathrm{Irr}_\mathbb{C}(N_G(P)/P)$ be the complex character in a block of defect zero of $N_G(P)/P$, if such a block exists. Then the pair (P, γ) is called a p-**weight** of G. A prominent conjecture of J. Alperin states that the number of G-conjugacy classes of p-weights coincides with the number of p'-conjugacy classes in G. This conjecture, as well as some generalizations due to E. Dade and G. Robinson, have great impact on the current lines of research in modular representation theory. They have been proved for important classes of finite groups, e.g., p-solvable groups (Robinson 2000).

Let $x \in G$ be a p-element, i.e., an element whose order is a power of p, and let $H := C_G(x)$ be its centralizer. Then for each block b of H, the induced block $B := b^G$ is defined, and b is often called a *root* of B. Let $\chi \in \mathrm{Irr}_{\mathbb{C}}(G)$ and $\varphi \in \mathrm{IBr}(G)$; then the **generalized decomposition numbers** are defined as $d^x_{\chi,\varphi} := \sum_{\psi \in \mathrm{Irr}_{\mathbb{C}}(H)} \frac{\langle \chi_H, \psi \rangle_H \psi(x)}{\psi(1)} d_{\psi,\varphi}$, where $\langle \chi_H, \psi \rangle_H$ denotes the inner product of H-characters (see Section E.2) and $d_{\psi,\varphi}$ is a p-decomposition number of H.

E.6.3 Second Main Theorem *For each p'-element $y \in H$, one has*

$$\chi(xy) = \sum_{\varphi \in \mathrm{IBr}(H)} d^x_{\chi,\varphi} \varphi(y),$$

with $d^x_{\chi,\varphi} = 0$ if $\varphi \in b$ and $\chi \notin b^G$.

If $g \in G$ is an arbitrary element, there is a unique decomposition $g = g_p g_{p'} = g_{p'} g_p$ into a product of commuting p-and p'-elements. Taking $x := g_p$ and $y := g_{p'}$, Brauer's Second Main Theorem describes the character value $\chi(g)$ in terms of this decomposition, the blocks of G, and their roots in H.

Let B_0 be the unique p-block of G containing the trivial representation (and the trivial character 1_G). Then the defect groups of B_0 are precisely the Sylow p-groups of G. There is also an explicit formula in terms of G for the block idempotent $e_0 \in B_0$, due to Külshammer (1991). Moreover, there is a close connection between B_0 and its roots:

E.6.4 Third Main Theorem *Let $H \leq G$ be a subgroup with principal block b_0. If b is a block of H such that b^G is defined, then $b^G = B_0$ if and only $b = b_0$.*

Important recent results (Külshammer 1991, Puig 1999) clarify the structure of **nilpotent blocks**, which generalize blocks of p-nilpotent groups. Current modular representation theory is mainly concerned with a series of important conjectures such as the ones of Alperin, Dade, and Robinson. Restricting to abelian defect groups, some of them can be formulated as follows: Let $B = b^G$ be a block of $\mathbb{F}G$ with abelian defect group P and *Brauer correspondent* b in $\mathbb{F}N_G(P)$:

E.6.5 Conjecture (Alperin) *The numbers of irreducible modules of B and b coincide.*

E.6.6 Conjecture (Alperin-McKay) *The dimension of the centers of B and of b coincide.*

E.6.7 Conjecture (Broué) *If B and b are the principal blocks, then their module categories have equivalent* derived categories.

For further information on these conjectures and on derived categories see (Broué 1995, König and A.Zimmermann 1998).

E.7 Characters of the Symmetric and Alternating Groups

by Albert Fässler in Biel, Switzerland

More information on the subject and proofs in this section can be found in (Fulton and Harris 1991).

The number of irreducible representations of the symmetric group S_d with d elements equals the number of its conjugacy classes, which ist the number of ordered partitions of d:

$$d = \lambda_1 + \lambda_2 + \cdots + \lambda_m \qquad \text{with} \quad \lambda_1 \geq \lambda_2 \geq \ldots \geq \lambda_m > 0$$

There are bijections between the ordered partitions $\lambda = (\lambda_1, \lambda_2, \ldots, \lambda_m)$, the irreducible representations of S_d and the Young diagrams. Here are two examples $(3, 1, 1)$ and $(3, 2)$ for S_5 (please ignore the numbers for a moment):

5	2	1
2		
1		

4	3	1
2	1	

The characters of the irreducible representations ϑ_λ are denoted by χ_λ. They are constant on each conjugacy class. A conjugacy class C_α ist indexed by a vector

$$\alpha = (\alpha_1, \alpha_2, \ldots, \alpha_d) \qquad \text{with} \quad \sum_{j=1}^{d} j\,\alpha_j = d \quad \text{and} \quad \alpha_j \in \{0, 1, 2, \ldots, d\}$$

and consists of those permutations that have α_j cycles of length j.

We consider the power sums $P_j(x)$, the discriminant $\Delta(x)$ of the in-dependent variables $x_1, x_2, \ldots, x_m$, defined by

$$P_j(x) = x_1^j + x_2^j + \cdots + x_m^j, \qquad \Delta(x) = \prod_{k<\ell}(x_k - x_\ell)$$

and the m-tuple $(\mu_1, \mu_2, \ldots, \mu_m)$ of strictly positive integers with

$$\mu_1 = \lambda_1 + m - 1, \qquad \mu_2 = \lambda_2 + m - 2, \quad \ldots \quad , \mu_m = \lambda_m$$

Now we can formulate the important ***Frobenius Formula***:

$$\chi_\lambda(C_\alpha) = [\Delta(x) \cdot \prod_{j=1}^d P_j(x)^{\alpha_j}]_{(\mu_1, \mu_2, \ldots, \mu_m)}$$

where $[f]_{(\mu_1, \mu_2, \ldots, \mu_m)}$ denotes the coefficient of $x_1^{\mu_1} x_2^{\mu_2} \ldots x_m^{\mu_m}$ in the poly-nomial f.

As a special case, the dimension of ϑ_λ equals the character of the identity with $\alpha = (d, 0, \ldots, 0)$:

$$\dim(\vartheta_\lambda) = [\Delta(x) \cdot (x_1 + x_2 + \ldots x_m)^d]_{(\mu_1, \mu_2, \ldots, \mu_m)}.$$

Calculations, using that $\Delta(x)$ is the Vandermonde determinant, deliver the desired coefficient of $x_1^{\mu_1} x_2^{\mu_2} \ldots x_m^{\mu_m}$ in two forms:

$$\dim(\vartheta_\lambda) = \frac{d!}{\mu_1! \cdot \mu_2! \cdot \ldots \cdot \mu_m!} \prod_{k<\ell}(\mu_k - \mu_\ell) = \frac{d!}{\prod \text{hook lengths}}$$

The latter is the practical ***hook length formula***: The hook length of a box in a Young diagram is $1 +$ number of squares directly below $+$ number of squares directly to the right. For the two examples of S_5 given above, the hook lenghts are noted in the boxes, thus yielding $\dim \vartheta_{3,1,1} = 6$ and $\dim \vartheta_{3,2} = 5$.

Here is the complete **character table of S_5**:

repres & class	id	(12)	(12)(34)	(123)	(12)(345)	(1234)	(12345)
χ_5	1	1	1	1	1	1	1
$\chi_{1,1,1,1,1}$	1	-1	1	1	-1	-1	1
$\chi_{4,1}$	4	2	0	1	-1	0	-1
$\chi_{2,1,1,1}$	4	-2	0	1	1	0	-1
$\chi_{3,1,1}$	6	0	-2	0	0	0	1
$\chi_{2,2,1}$	5	1	1	-1	1	-1	0
$\chi_{3,2}$	5	-1	1	-1	-1	1	0
cardinality	1	10	15	20	20	30	24

Each of the three pairs with the same dimensions have conjugate partitions, which means that reflecting such a Young diagram at the diagonal gives the other diagram. We have $\chi_{2,1,1,1} = \chi_{1,1,1,1,1} \cdot \chi_{4,1}$ and $\chi_{3,2} = \chi_{1,1,1,1,1} \cdot \chi_{2,2,1}$, for instance.

The restriction of an irreducible representation of S_d to the alternating group A_d may become reducible; two distinct irreducible representations may become isomorphic. The odd conjugacy classes disappear, but some of the even conjugacy classes may split into two classes (since conjugation by a transposition becomes an outer automorphism). More precisely: A class splits, if and only if it consists of disjoint cycles of odd and distinct lengths, otherwise the conjugacy class remains intact for A_d.

For $d = 5$, each of the conjugate pairs have identical restrictions and remain irreducible, because $(\chi, \chi) = 1$. But the self-conjugate representation $\vartheta_{3,1,1}$ decomposes into two irreducible parts of dimension 3. They can be realized by the group of motions of an icosahedron. Their character can be completed with the traces $1+2\cos(2\pi/5) = (1\pm\sqrt{5})/2$ of rotations. And the only class that splits is (12345).

Here is the complete **character table of A_5:**

rep & class	id	(12)(34)	(123)	(12345)	(21345)
χ_5	1	1	1	1	1
$\chi_{4,1}$	4	0	1	-1	-1
$\chi^+_{3,1,1}$	3	0	-1	$(1+\sqrt{5})/2$	$(1-\sqrt{5})/2$
$\chi^-_{3,1,1}$	3	0	-1	$(1-\sqrt{5})/2$	$(1+\sqrt{5})/2$
$\chi_{2,2,1}$	5	1	-1	0	0
potency	1	15	20	12	12

In the general case for S_d and A_d, we have to deal with restricted and induced representations.

E.8 Representation Theory of Lie Groups

by Joachim Hilgert in Clausthal, Germany

Finite dimensional representations

A ***representation*** of a Lie group G on a finite dimensional (real or complex) vector space V is a continuous group homomorphism $\pi\colon G \to \mathrm{GL}(n; V)$. Then π is automatically smooth (even real analytic), and its derivative at the unit element e of G is a Lie algebra representation $d\pi\colon \mathfrak{g} \to \mathrm{End}(V)$ of the associated Lie algebra $\mathfrak{g}$ satisfying $d\pi(X) = \frac{d}{dt}|_{t=0}\pi(\exp tX)$. Conversely, if G is simply connected, any

Lie algebra representation $\mathfrak{g} \to \mathrm{End}(V)$ is the derivative of a uniquely determined group representation $\pi \colon G \to \mathrm{GL}(n; V)$. Thus Lie group representations can be studied via Lie algebra representations. Usually one considers $d\pi$ via a linear extension as a representation of the complexification $\mathfrak{g}_\mathbb{C}$ of the Lie algebra $\mathfrak{g}$, or even as a representation of the universal enveloping algebra $U(\mathfrak{g}_\mathbb{C})$ of $\mathfrak{g}_\mathbb{C}$ (cf. Section C.49) via the universal property of $U(\mathfrak{g}_\mathbb{C})$.

If G is compact, one can average over G using the uniquely determined G-invariant probability measure (i.e., normalized Haar measure) on G and thus obtain a G-invariant inner product on V. If V is a complex vector space, then $\pi(G)$ consists of unitary operators so that π is a *unitary representation*. Then the orthogonal complement $W^\perp$ of a G-invariant subspace W of V is again G-invariant and induction shows that V is *completely reducible*, i.e., the direct sum of *irreducible* G-invariant subspaces.

There is an elegant theory due to E. Cartan and H. Weyl which describes and classifies the (automatically finite dimensional) irreducible unitary representations of (connected) compact Lie groups.

It is based on the observation that the restriction of π to a *maximal torus*, i.e., a maximal (connected) abelian subgroup, completely determines π. These restrictions are best described on the Lie algebra level. They lead to a classification of the finite dimensional representations of connected simply connected compact Lie groups. To classify the representations of arbitrary connected compact Lie groups one has to check the restrictions to the center of the simply connected covering groups.

For general Lie groups, such a satisfying theory does not exist. In fact, it is easy to construct examples of representations which are not completely reducible once unitarity does not hold.

Infinite dimensional representations; examples

Representations of Lie groups on infinite dimensional vector spaces naturally occur in diverse situations, often in geometric or number theoretic contexts. Typically they are obtained from spaces of functions or sections of vector bundles. So, for instance, if a Lie group G acts smoothly on a manifold M, then G acts linearly on the space $C^\infty(M)$ of smooth functions on M via $(\pi(g)f)(x) = f(g^{-1} \cdot x)$. If M carries a G-invariant measure μ then the same formula defines a unitary representation of G on $L^2(M, \mu)$. If $M = G$ (and μ a left Haar measure) we speak about the (left) *regular representation*. The regular representation is not irreducible, but it may happen that $L^2(G)$ does not contain any G-invariant closed irreducible subspaces. If it does, these irreducible subrepresentations are called the ***discrete series representation.***

If H is a subgroup of G and ρ is a representation of H on a vector space V then the ***induced representation*** $\pi = \mathrm{Ind}_H^G(\rho)$ of G acts on the space of V-valued functions on G satisfying the identity $f(gh) = \rho(h)^{-1}f(g)$ for $g \in G$ and $h \in H$. The action is given by $(\pi(g)f)(x) = f(g^{-1}x)$. If one considers Banach space or unitary representations, one looks at completed versions of this *algebraic* induced representation. Induced representations play a crucial role in the theory. In the case of semisimple G, particular examples where H is a so called *parabolic* subgroup of G are the ***principal series representations.*** They are the main ingredients in the decomposition of the regular representations into irreducibles.

Lie algebra representations

In order to be able to build a reasonable representation theory one has to make some assumptions on the representation space V and the representation π. Usually the minimum one wants is that V is a complete metrizable vector space and π is ***strongly continuous,*** i.e., $G \to V, g \mapsto \pi(g)v$ is continuous for any $v \in V$.

The step from Lie group to Lie algebra representations which proved so useful in the finite dimensional case is more complicated in the infinite dimensional setting. In general, one cannot define $d\pi$ as a derivative and $d\pi(X)v = \frac{d}{dt}|_{t=0}\pi(\exp tX)v$ makes sense only for a linear subspace V^∞ of V called the space of ***smooth vectors.*** Under suitable hypotheses, V^∞ is dense in V and $d\pi$ is a Lie algebra representation of $\mathfrak{g}$ on V^∞. In general, representations of $\mathfrak{g}$ on V^∞ need not come from group representations, even if G is simply connected.

The semisimple case; Harish-Chandra modules

If V is a Banach space and $G \to V, g \mapsto \pi(g)v$ is real analytic, then v is called an analytic vector. For semisimple G (with finite center) pick a maximal compact subgroup K and call $v \in V$ ***K-finite*** if $\pi(K)v$ spans a finite dimensional space. For a wide class of representations (called the ***admissible*** ones), the space V_K of K-finite vectors is dense in V, consists of analytic vectors and is invariant under $d\pi(\mathfrak{g})$ as well as $\pi(K)$. The formula $d\pi(X)v = \frac{d}{dt}|_{t=0}\pi(\exp tX)v$ for $X \in \mathfrak{k}$, the Lie algebra of K, yields a compatibility condition between the $\mathfrak{g}$- and the K-action on V_K.

Spaces with compatible K- and $\mathfrak{g}$-actions in which all vectors are K-finite are called **Harish-Chandra modules**. One can show that the category of Harish-Chandra modules is equivalent to the category of admissible representations. Thus the basic questions of representation theory like classification of irreducibles and decomposition into irreducibles can be dealt with in terms of Harish-Chandra modules. This also makes plausible that in the study of admissible representations the **K-spectrum**, i.e., the decomposition of $\pi|_K$ into irreducible subrepresentations, plays a fundamental role.

The orbit method

Unitary representations are the most important class of infinite dimensional representations. One reason for this is the fact that they show up as symmetry operations in quantum mechanics. It is still an open problem to classify all the irreducible unitary representations (up to the natural isomorphism of such representations). This is true even for the class of simple Lie groups. Kirillov's **orbit method**, also known under the name **geometric quantization**, is a very general method for the construction and classification of unitary representations which works best for simply connected nilpotent Lie groups (where it achieves the desired classification) but it provides general insights and specific results also in many other contexts. The basic idea is to construct unitary representations of G from linear functionals on $\mathfrak{g}$ via induction. It turns out that functionals which are conjugate under the **coadjoint action**, i.e., the dual of the adjoint action, lead to isomorphic representations. For simply connected nilpotent groups one obtains in this way a bijection between the set of coadjoint orbits and isomorphism classes of unitary representations.

Unitarizability

Langlands obtained a classification of irreducible admissible representations of semisimple Lie groups via parabolic subgroups and induced representations. The *Langlands parameters*, however, give no clue whether the representation is **unitarizable**, i.e., whether the representation space can be given an inner product such that the representation is unitary. There are important classes of representations and groups for which the unitarizability problem has been solved, but the general problem is still open.

Applications

Very often, the representation theory of Lie groups is used for *harmonic analysis* type questions: Given a natural group action on a space, describe the building blocks of the representations on function or section spaces associated with this action. Such questions occur also in number theory (cf.Langlands program), geometry and invariant theory. Important applications of finite and infinite dimensional Lie group representations can be found in quantum mechanics (e.g., classification of elementary particles). Analytic properties of representations can be used in ergodic theory and its applications to metric number theory. Even though representations of Lie groups are still studied for their own sake, one observes a recent shift of emphasis towards applications of representation theory to other fields.

More on this topic can be found in (Bröcker and tom Dieck 1985, Corwin and Greenleaf 1990, Howie 1992, Knapp 1986, Wallach 1988,1992).

E.9 Groups in 2D/3D vision problems

by Kenichi Kanatani in Okayama, Japan and Günter F. Pilz in Linz, Austria

Image understanding is a task which shows up in a vast variety of applied mathematics: automatic traffic control via a camera (how can the camera distinguish between cars and other objects?), Computer Tomography, recognition of aircrafts from their silhouettes, and so on.

All of these problems are 2D/3D vision problems: how to extract knowledge about 3D objects from 2D images. Similarly, 2D/2D problems (*pattern recognition*) turn out in automatic text readers, and the like. For the details, see (Kanatani 1990).

Suppose we have a map from a 3D-scene to a 2D-(camera) image. *Image parameters* are numerical data extracted from the 3D-scene (like the grey level of a point). *Object parameters* are *properties* of the 3D-scene, like the speed of a certain object. *Recovery equations* relate the (known) image parameters $c_1, \ldots, c_m$ to the (unknown) object parameters $\alpha_1, \ldots, \alpha_n$ via

$$c_1 = F_1(\alpha_1, \ldots, \alpha_n), \ldots, c_m = F_m(\alpha_1, \ldots, \alpha_n)$$

These recovery equations are usually obtained from laws of geometry, of optics, etc.; they may be implicit and hard to solve. Often free parameter turn up which have to be understood. We will see now via an example how group theory can help.

E.9.1 Example Consider a plane P with equation $z = px + qy + r$ which travels through the 3D-space in a combination of translations and rotations (Figure 1). A camera is fixed high up on the z-axis (to avoid perspectives). Here p, q, r, a, b, c, w_1, w_2, and w_3 are suitable object

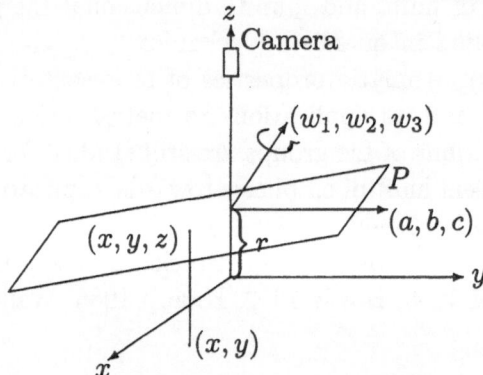

Figure 1: Translation and rotation of a plane

parameters which completely describe the journey of P. The camera measures the velocity field $\dot{x} = u(x, y)$, $\dot{y} = v(x, y)$. Physics gives us

$$\begin{pmatrix} u(x, y) \\ v(x, y) \end{pmatrix} = \begin{pmatrix} a \\ b \end{pmatrix} + \begin{pmatrix} A & B \\ C & D \end{pmatrix} \begin{pmatrix} x \\ y \end{pmatrix}$$

with $A := pw_2$, $B := qw_2 - w_3$ $C := -pw_1 + w_3$, $D := -qw_1$. So a, b, A, B, C, D can be computed and taken as image parameters, and $A = pw_2$ etc. (together wirh $a = a$, $b = b$) constitute the recovery equations. Apart from $a = a$ and $b = b$, we thus have 4 equations for the 5 unknowns p, q, w_1, w_2, w_3; we have to expect a free parameter. Observe that r and c cannot be recovered at all.

Not every combination of image parameters is geometrically *meaningful*; a has no meaning, while (a, b) has one. Some thinking leads to the definition of *meaning*. A simple image parameter is meaningful if it does not change its value after a coordinate rotation. So the grey level of the darkest point is a meaningful parameter, while the x-coordinate of a point is not.

If $\mathbf{R}_\theta$ is the rotation of the xy-plane around the origin by angle θ, a pair (c_1, c_2) of image parameters is meaningful if they transform to (c'_1, c'_2) under a rotation like points, i.e., $\begin{pmatrix} c'_1 \\ c'_2 \end{pmatrix} = \mathbf{R}_\theta \begin{pmatrix} c_1 \\ c_2 \end{pmatrix} = \begin{pmatrix} \cos\theta & -\sin\theta \\ \sin\theta & \cos\theta \end{pmatrix} \begin{pmatrix} c_1 \\ c_2 \end{pmatrix}$.

Finally, (c_1, c_2, c_3, c_4) is defined to be meaningful if $\mathbf{M} := \left(\begin{smallmatrix} c_1 & c_2 \\ c_3 & c_4 \end{smallmatrix}\right)$ transforms to $\mathbf{M}' = \left(\begin{smallmatrix} c_1' & c_2' \\ c_3' & c_4' \end{smallmatrix}\right)$ as $\mathbf{M}' = \mathbf{R}_\theta \mathbf{M} \mathbf{R}_\theta^{-1}$ for all θ.

If $(c_1, \ldots, c_k)$ are image parameters, we can also take $\begin{pmatrix} C_1 \\ \vdots \\ C_k \end{pmatrix} = \mathbf{P} \begin{pmatrix} c_1 \\ \vdots \\ c_k \end{pmatrix}$, provided that $\mathbf{P}$ is regular.

Suppose we find some $\mathbf{P}$ such that the coordinate rotation induces a linear mapping in a block diagonal form $\mathrm{diag}\,(Q_1, \ldots, Q_r)$, for all θ. Then

$$\begin{pmatrix} C_1' \\ \vdots \\ C_k' \end{pmatrix} = \mathrm{diag}\left(Q_1 \begin{pmatrix} C_1 \\ \vdots \\ C_{r_1} \end{pmatrix}, Q_2 \begin{pmatrix} C_{r_1+1} \\ \vdots \\ C_{r_2} \end{pmatrix}, \ldots \right)$$

so $C_1, \ldots, C_{r_1}$ transforms independently of $C_{r_1+1}, \ldots, C_{r_2}$, and so on. Hence $C_1, \ldots, C_{r_1}$ and $C_{r_1+1}, \ldots, C_{r_2}$ and ... indicate **single** and **independent properties**.

Now we are right in representation theory. We have a representation h of the group $\mathrm{SO}\,(2)$ of rotations of $\mathbb{R}^2$ by the subgroup $\{\,\mathbf{R}_\theta \mid 0 \leq \theta < 2\pi\,\}$ of all regular 2×2 matrices, and we want to decompose h into irreducible representations $h_1, \ldots, h_k$ to get the irreducible components $Q_1, \ldots, Q_k$ above. But this is one of the main purposes of representation theory!

The following statements are sometimes called *Weyl's Thesis* (Hermann Weyl, 1888-1955):

1. *Reducing a representation amounts to decomposing a complex property into independent ones.*

2. *Irreducible representations correspond to single properties.*

Example (revisited). In our example, (a, b) and (A, B, C, D) can be seen to be meaningful. With $\mathbf{P} := \left(\begin{smallmatrix} 1 & i \\ 1 & -i \end{smallmatrix}\right) \in \mathrm{GL}(2; \mathbb{C})$, we have $\mathbf{P}\mathbf{R}_\theta\mathbf{P}^{-1} = \left(\begin{smallmatrix} e^{i\theta} & 0 \\ 0 & e^{-i\theta} \end{smallmatrix}\right)$ for all θ. Hence we transform (a, b) into $a + ib$ and $a - ib$. Similarly, from (A, B, C, D) we get the single properties $A + D$, $B - C$, and $(A - D) \pm i\,(B + C)$; the recovery equations can now be solved, and the free parameter just turns out to be a scaling parameter. For the details, see (see Kanatani 1990) or (Lidl and Pilz 1998).

The resulting flows can also be characterized now: if $A = B = C = D = 0$, the flow is just a translation $\left(\begin{smallmatrix} u \\ v \end{smallmatrix}\right) = \left(\begin{smallmatrix} a \\ b \end{smallmatrix}\right)$; if only $B - C \neq 0$, we get a rotational flow; if all parameters except $A + D$ are zero, we have a divergent flow; and if only $(A - D) \pm i\,(B + C)$ is non-zero, we have a shear flow.

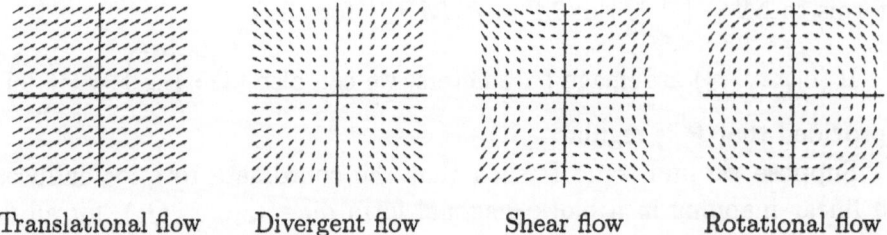

| Translational flow | Divergent flow | Shear flow | Rotational flow |

In summary, *reducing representations means decoupling overlapping influences!*

E.10 Representation Theory and Statistics
by Albert Fässler in Biel, Switzerland

Data sometimes come in the form of rankings or preferences, for example in marketing polls or in certain types of elections. Here we analyze a representative problem, which of course can be generalized. The American Psychological Association elects its president by asking its members to rank $n = 5$ candidates, denoted by A, B, C, D, E. Many members cast incomplete ballots, voting for their favorite $q = 1, 2$ or 3 out of 5 candidates. Here is a summary of some selected data of the 1980 election, with the case $q = 5$ describing the fully ranked data:

$q = 5$	$q = 1$	$q = 2$	$q = 3$
$f(12345) = 30$	$f(00001) = 1022$	$f(00012) = 196$	$f(00123) = 15$
$f(12354) = 28$	$f(00010) = 1145$	$f(00021) = 143$	$f(00132) = 19$
$f(12435) = 27$	$f(00100) = 1198$	$f(00102) = 93$	$f(01023) = 55$
	$f(01000) = 881$		
...	$f(10000) = 895$	...	...
$f(54321) = 29$		$f(21000) = 72$	$f(32100) = 57$
$\sum f(x) = 5738$	$\sum f(x) = 5141$	$\sum f(x) = 2462$	$\sum f(x) = 2108$
$\dim f = 120$	$\dim f = 5$	$\dim f = 20$	$\dim f = 60$
$f(x) \in [11, 172]$	$f(x) \in [881, 1198]$	$f(x) \in [48, 547]$	$f(x) \in [8, 79]$

Interpretations: $f(54321) = 29$ means that 29 members voted for candidate E in first, candidate D in second position, ... Zeros indicate unranked data: $f(00102) = 93$ means that 93 members ranked candidate C in first and candidate E in second position. In every column, the data vector is called f, even though they are all different. The symbol x stands for rankings.

A first order analysis of the fully ranked data ($q = 5$) showed that candidate C received the highest number of first position votes, while candidate A was "lowest" in average rank, due to few "hate votes". Now, should A or C be the president?

In order to make an adequate decision, further analysis is needed. The following group theoretical investigation will show that there is more structure in the data as one might think, allowing for better statistical interpretation.

Let $\lambda = (\lambda_1, \lambda_2, \ldots \lambda_r)$ be a partition of n with $\lambda_1 \geq \lambda_2 \geq \cdots \geq \lambda_r$.

A **ranking of shape** λ is specified as follows: Choose your favorite λ_1 candidates out of all n candidates without internal ranking, then choose your next λ_2 favorite candidates, again without internal ranking, and so on.

For example, data based on people's favorite and least favorite item are indexed by $(1, n{-}2, 1)$, this is equivalent to $(n{-}2, 1, 1)$. The Partition $(1, 1, 1, 1, 1)$ means complete rankings, $(3, 1, 1)$ means a choice of ordered subsets 2 out of 5. Each permutation $\pi \in S_n$ transforms the set R^λ of the $n!/(\lambda_1! \lambda_2! \cdots \lambda_r!)$ rankings of shape λ.

Here are the transforms for a ranking of shape $(3, 2)$ of the election problem:

$$\pi \left(\frac{\underline{A} \ \underline{B} \ \underline{E}}{\underline{C} \ \underline{D}} \right) = \frac{\pi(A) \ \pi(B) \ \pi(E)}{\pi(C) \ \pi(D)} \tag{1}$$

In the argument of π, candidates A, B, E are ranked first and C, D are ranked second. Underlining denotes unordered rows.

Let us consider the representations $\vartheta^\lambda \colon \pi \to D^\lambda(\pi)$ of S_n:

$$D^\lambda(\pi)\varphi(x) = \varphi(\pi^{-1}(x)), \quad \text{where } \varphi \text{ is a real function on } R^\lambda. \tag{2}$$

The table below shows the **spectral analysis** of the four data vectors: The corresponding representations ϑ^λ have irreducible parts ϑ_μ, labeled by the ordered partitions μ of 5. Their multiplicities and dimensions can be calculated by use of Young's rule and the hooklength formula (see Section E.7). SS means the square sums of the different isotypic components f_μ of the data vectors f, where the f_μ's can be obtained with the projection in Theorem E.4.3.

$q = 5$	$\vartheta^{(1,1,1,1,1)} =$	ϑ_5	$\oplus 4\vartheta_{4,1}$	$\oplus 5\vartheta_{3,2}$	$\oplus 6\vartheta_{3,1,1}$	$\oplus 5\vartheta_{2,2,1}$	$\oplus 4\vartheta_{2,1,1,1}$	$\oplus \vartheta_{1,1,1,1,1}$
dim f_μ	120 $=$	1^2	$+$ 4^2	$+$ 5^2	$+$ 6^2	$+$ 5^2	$+$ 4^2	$+$ 1^2
SS/120	3155 $=$	2286	$+$ 298	$+$ 459 $+$	78	$+$ 27	$+$ 7	$+$ 0

$q = 1$	$\vartheta^{(4,1)} =$	ϑ_5	$\oplus \vartheta_{4,1}$
dim f_μ	5 $=$	1	$+$ 4
SS/5	1073579 $= 1057195 + 16384$		

$q = 2$	$\vartheta^{(3,1,1)} =$	ϑ_5	$\oplus 2\vartheta_{4,1}$	$\oplus \vartheta_{3,2}$	$\oplus \vartheta_{3,1,1}$
dim f_μ	20 $=$	1	$+ 2 \cdot 4 +$	5 $+$	6
SS/20	27855 $=$	15154	$+$ 4268	$+7781+$	652

$q = 3$	$\vartheta^{(2,1,1,1)} =$	ϑ_5	$\oplus 3\vartheta_{4,1}$	$\oplus 3\vartheta_{3,2}$	$\oplus 3\vartheta_{3,1,1}$	$\oplus 2\vartheta_{2,2,1}$	$\oplus \vartheta_{2,1,1,1}$
dim f_μ	60 $=$	1	$+ 3 \cdot 4$	$+ 3 \cdot 5 +$	$3 \cdot 6 +$	$2 \cdot 5 +$	4
SS/60	1630 $=$	1234	$+$ 123	$+$ 243 $+$	20	$+$ 8	$+$ 2

The first line describes the regular representation of S_5. It contains every type of irreducible parts.

As is seen from the table, in all four data vectors the components f_μ that are clearly dominant correspond to the first three isotypic components belonging to $\mu = (5); (4,1); (3,2)$.

The f_5 components are not of interest, they correspond to the 1-dimensional subspaces spanned by the constant functions, and the $f_{4,1}$ describe the first order analysis mentioned earlier. But there are additional *unordered second order effects* that are *relevant*, described by $f_{3,2}$. Let us introduce the functions $\pi \mapsto \delta_{\{\pi(X),\pi(Y)\}\{i,j\}}$ on S_5. They are 1, if candidates X and Y are ranked in positions $\{i,j\}$ in either order. They are 0 otherwise.

For the case of $q = 5$, each of the pairs $\{i,j\}$ and $\{X,Y\}$ can be chosen in 10 ways. Thus there are a total of 100 such functions in the isotypic component of $5 \cdot \vartheta_{3,2}$ of dimension 5^2. For the case of $q = 3$, we have $10 \cdot 3$ such functions, and for $q = 2$, there are 10 functions, see the table below. If the data components lie in the vicinity of easily interpretable functions, we have a simple interpretation. The following table shows a characteristic part of the 100, the 30 and the 10 inner products with the data components $f_{3,2}$:

candi-dates	full ranks q=5										ranks q=3			q=2
	1,2	1,3	1,4	1,5	2,3	2,4	2,5	3,4	3,5	4,5	1,2	1,3	2,3	1,2
A,B	−137	−20	18	140	111	22	4	6	−97	−46	−50	6	12	−107
A,C	476	−88	−179	−209	−147	−169	−160	107	128	241	150	−3	−41	385
...	...	...	...	...	...	...	...	...	...	...	...	...	...	...
D,E	296	−24	−142	−130	−5	−163	−128	38	−9	267	97	26	0	197

Most of the omitted lines start or end with negative values. If they do not, they start and end with much smaller numbers than do the (A, C)- and (D, E)-lines.

Interpretations of the second order, unordered effects are given below:

- $q = 5$: The two groups (A, C) and (D, E) are forerunners. There is an enormous support for ranking (A, C) in the best positions $(1, 2)$, with a fair size of "hate votes", shown in the worst positions $(4, 5)$, while (D, E) show large values at both ends. The remaining groups (not shown here), had the opposite effect: Only few people liked or hated the two. Thus the rows begin and end with negative values.

- $q = 2$ or $q = 3$: The group (A, C) is way ahead of the second best group (D, E).

The four first order effects can be analyzed in a similar way; the interpretations for the four different cases vary. More information can be found in (Diaconis 1989).

Chapter F

LATTICES

F.1 Congruences and Constructions

by George Grätzer in Winnipeg, Canada and E. Tamás
Schmidt in Budapest, Hungary

Let Con **L** denote, up to isomorphism, the class of congruence lattices of lattices and let **DA** denote the class of all distributive algebraic lattices. For every lattice L, it it clear that the congruence lattice Con L is algebraic. By a 1942 result of N. Funayama and T. Nakayama, Con L is also distributive, so Con **L** $\subseteq$ **DA**. Is the converse true: *Is every distributive algebraic lattice isomorphic to the congruence lattice of a suitable lattice?* This is one of the most famous open questions of lattice theory. We shall briefly review this topic here, together with its related results; for a more complete overview (up to 1998), see Appendix C in (Grätzer 1998); we shall only reference later papers here.

The finite case

The question is answered for the finite case by a result of R. P. Dilworth, first published by G. Grätzer and E. T. Schmidt in 1962. These original results, for a finite distributive lattice D with n join-irreducible elements, constructed a lattice L with Con $L \simeq D$ of $O(2^{2^n})$ elements. A series of papers (concluding with a 1995 paper of G. Grätzer, H. Lakser, and E. T. Schmidt) improved this number to $O(n^2)$ (in fact, the optimal lattice constructed happens to be planar). G. Grätzer, I. Rival, and N. Zaguia proved in 1995 that this number is "best possible" in the sense that size $O(n^2)$ cannot be replaced by size $O(n^\alpha)$, for any $\alpha < 2$.

A finite distributive lattice D can be represented as the congruence lattice of a lattice L with very special properties:

(1) A *finite sectionally complemented* lattice L (Grätzer and Schmidt 1962).
(2) A *finite semimodular* lattice L (Grätzer, Lakser, and Schmidt 1998).
(3) A *relatively complemented* lattice L (Grätzer, Lakser, and Wehrung 2000).
(4) A *modular* lattice L, in fact, a sublattice of all subspaces of a countably infinite dimensional vector space over the two element field (Schmidt 1974).

The general case

In the general case, it is more convenient to consider $\text{Con}_c L$, the distributive semilattice with zero of compact congruences of the lattice L and the corresponding class **DS** of distributive semilattice with zero. The original question can be rephrased: Is $\text{Con}_c \mathbf{L} = \mathbf{DS}$?

Let us call $S \in \mathbf{DS}$ *representable* if $S \in \text{Con}_c \mathbf{L}$. Each one of the following conditions implies that S is representable:

(1) S is a *lattice* (E. T. Schmidt 1968).
(2) S is *locally countable* (that is, for every $s \in S$, $(s]$ is countable) (A. P. Huhn and H. Dobbertin in the early eighties).
(3) $|S| \leq \aleph_1$ (A. P. Huhn in the early eighties).

It was hoped for a long time that the two successful approaches solving the case for a lattice S can be used to answer the general question. F. Wehrung in 1998 proved that neither method can answer the general question even for lattices of size $\aleph_2$.

Complete congruences

For complete lattices, we have complete congruences, and the complete lattice of complete congruences. These lattices were characterized by G. Grätzer in 1990:

F.1.1 Theorem *Every complete lattice K can be represented as the lattice of complete congruence relations of a complete lattice L.*

In a series of papers, much sharper results have been obtained, culminating in the 1995 statement by G. Grätzer and E. T. Schmidt:

F.1.2 Theorem *Every complete lattice L can be represented as the lattice of complete congruence relations of a complete distributive lattice K.*

Congruence-preserving extensions

A lattice K is a ***congruence-preserving extension*** of the lattice L, if K is an extension and every congruence of L has *exactly one* extension to K. Of course, then the congruence lattice of L is isomorphic to the congruence lattice of K; we could say that the congruence lattice of K is *naturally isomorphic* to the congruence lattice of L or that the algebraic reasons determining the congruence lattice of L are carried over to K.

Every finite lattice has a congruence-preserving extension to a finite

(1) *atomistic* lattice (M. Tischendorf 1992);
(2) *sectionally complemented* lattice (Grätzer and Schmidt 1999);
(3) *semimodular* lattice (Grätzer and Schmidt 2000).

Let A and B be $\vee$-semilattices with zero. We denote by $A \otimes B$ the ***tensor product*** of A and B, defined as the free $\vee$-semilattices with zero generated by the set $(A - \{0\}) \times (B - \{0\})$ and subject to the relations

$$l(a, b_0) \vee l(a, b_1) = l(a, b_0 \vee b_1), \quad \text{for } a \in A - \{0\}, \ b_0, b_1 \in B - \{0\};$$
$$l(a_0, b) \vee l(a_1, b) = l(a_0 \vee a_1, b), \quad \text{for } a_0, a_1 \in A - \{0\}, \ b \in B - \{0\}.$$

The classical 1981 isomorphism of G. Grätzer, H. Lakser, and R. W. Quackenbush on tensor products of finite lattices:

$$\operatorname{Con} A \otimes \operatorname{Con} B \simeq \operatorname{Con}(A \otimes B)$$

shows that it is easy to construct a congruence-preserving extension in the finite case: $A \otimes B$ congruence-preserving extension of B, for any simple A. This has been extended to wide classes of infinite lattices (substituting Con_c for Con) in Grätzer and Wehrung (2000b) and to arbitrary lattices with zero using *box products* (a variant of tensor products) in Grätzer and Wehrung (1999a). Among the many applications, you will find the statement that every lattice has a proper congruence-preserving extension (Grätzer and Wehrung 1999b), every lattice has a congruence-preserving extension into a regular lattice, that is, into a lattice in which any two congruences sharing a congruence class are equal (Grätzer and Schmidt 2001), and the Strong Independence Theorem for automorphism groups and congruence lattices of arbitrary lattices (Grätzer and Wehrung 2000a).

Recent developments

A great deal has happened in this field in the last few years. We did not discuss the results of F. Wehrung and J. Tůma on congruence amalgamations, the results of H. Lakser and the authors on simultaneous representations, and many others. We would like to conclude with a very recent result of F. Wehrung of tremendous importance and depth:

F.1.3 Theorem *Let L be a lattice with the property that the meet of any two compact congruences is compact again. Then L has a congruence-preserving extension to a relatively complemented lattice.*

F.2 Modular Lattices

by Ralph Freese in Honolulu, HI, USA

A lattice is **modular** if it satisfies the following implication:

$$x \geq z \quad \text{implies} \quad x \wedge (y \vee z) = z \vee (x \wedge y)$$

This is called the **modular law** and is sometimes known as Dedekind's modular law as Dedekind (1897) had the original work on the subject. This law is equivalent to the identity

$$x \wedge (y \vee (x \wedge z)) = (x \wedge y) \vee (x \wedge z)$$

and since modular lattices are defined by an identity (a universally quantified equation) they are closed under homomorphic images, sublattices and direct products. Modular lattice derive their importance from the fact that most of the lattices associated with classical algebra systems are modular: submodule lattices, ideal lattices, normal subgroup lattices, etc.

Early work in modular lattice theory was used to gain insight into classical algebra with at least moderate success. For example if N and M are normal subgroups of a group G with trivial intersection and whose join is G then G is isomorphic to direct product of G/N and G/M. Thus the direct decompositions of a group are determined by its lattice of normal subgroups, that is, its congruence lattice. Ore (1935, 1936) noted this and was able to extend the Krull-Schmidt theorem on the uniqueness of direct decompositions of finite groups to general algebraic systems:

F.2.1 Theorem *Suppose that A is an algebra with permuting congruence relations. If A has a one-element subalgebra and its congruence lattice has finite height then A can be uniquely written as the direct product of directly indecomposable algebras.*

Jónsson (1966) was able to weaken the assumption of permuting congruence relations to just assuming the congruence lattice is modular under the stronger assumption that A is finite. McKenzie, McNulty, and Taylor (1987) contains a nice discussion of these results. Extensions to general algebras without chain conditions and strengthening of the exchange of the factors of two decompositions are explored by Crawley and Dilworth (1973).

In Jónsson' investigation of torsion free abelian groups of finite rank, which he showed did not have unique direct decompositions, he made extensive use of modular lattice theory in defining quasi-isomorphism and proving that the decompositions are unique modulo quasi-isomorphism. This theory starts with the fact that on any modular lattice the relation $x \sim y$ if the interval $[x \wedge y, x \vee y]$ has finite length is a congruence relation on the lattice. If L is the lattice of subgroups of a torsion free abelian group (which is the congruence lattice) then using $L/\sim$ Jónsson is able to show direct decompositions are unique.

One of the important developments of universal algebra was the discovery that every variety of algebras with modular congruence lattices has a commutator operation generalizing ideal multiplication in ring theory and the commutator of normal subgroups in group theory. This leads to a powerful general theory for congruence modular varieties in which modular lattice theory plays a significant role (Freese and McKenzie 1987). Gross, Herrmann, and Moresi (1987) made extensive use of modular lattice theory in their classification of Hermitian forms.

Combinatorics There several important combinatorial results in modular lattices. We mention two. Dilworth (1954) settled a 20 year old conjecture by proving that in a finite modular lattice the number of join irreducible elements equals the number of meet irreducible elements. In fact the number of elements with exactly k upper covers equals the number with exactly k lower covers. Kung (1985, 1987) showed that there is a bijection f between the join and meet irreducibles which is a matching, that is, $a \leq f(a)$ (see also Reuter 1987).

Complemented modular lattices and geometry Under certain finiteness conditions, complemented modular lattices can be represented as the direct product of lattices of subspaces of a projective geometry. For example if the the lattice is compactly generated, that is, algebraic. Since every complemented modular lattice can be naturally embedded into such a lattice this result is quite general (see Crawley and Dilworth

1973). In a different direction, von Neumann (1960) coordinatizes continuous geometries.

Intractability of modular lattices   The strong results for complemented modular lattices do not extend to lattices in general. Hall and Dilworth (1944) gave examples of modular lattices not embeddable in complemented modular lattices. P. Whitman, who solved the word problem for free lattices, reportedly spend 20 years trying to find a solution to the word problem for free modular lattices. Work of Gelfand and Ponomarev (1970) and of Day, Herrmann, and Wille (1972) let to the hope that it might be possible to classify all 4-generated, subdirectly irreducible modular lattices and thereby solve the word problem for the 4-generated free modular lattice. (The free modular lattice on 3 generators is finite, as Dedekind showed, and so certainly has a solvable word problem.)

Jónsson had shown that if the five element modular, nondistributive lattice M_3 is embedded into a modular lattice L then the interval $[0, a]$ satisfies the stronger arguesian law (a lattice equation closely related to Desargues law), where a is an atom of M_3. Intrigued by this, I produced a modular lattice M with elements $a < b$ such that whenever M was embedded into a modular lattice L the interval $[a, b]$ in L is distributive (Freese 1979). Building on this construction, it was possible to settle several outstanding problems. In particular, the word problem for free modular lattices with at least 5 generators is unsolvable (Freese 1980). Herrmann (1983) extended this result to 4 generators.

F.2.2 Theorem *The word problem from free modular lattices with four or more generators is recursively unsolvable.*

Herrmann (1984) proved a very strong negative result about varieties of modular lattices. He showed that under fairly weak conditions a variety (that is, equational class) of modular lattices either fails to have a finite basis for its equations or is not generated by its finite dimensional members (see also Freese 1987a).

As mentioned above, modular lattices derive much of their importance from the fact that most classical algebraic systems have modular lattices associated with them. Given their intractability one might might ask if there are actually stronger properties true of the lattices associated with classical algebraic systems. Freese and Jónsson (1976) proved that if a variety of algebraic systems has modular congruence lattices these lattices are in fact, arguesian. In fact even stronger laws hold. The following result (Freese 1994) shows there is no nice equational description of the lattices associated with classical algebraic systems.

F.2.3 Theorem *If a variety of algebras has modular, nondistributive congruence lattices then the equations of these lattices do not have a finite basis.*

F.3 Distributive Lattices; Heyting and Post Algebras

by Viacheslav N. Saliĭ in Saratov (Russia)

F.3.1 Definition A lattice $(L, \wedge, \vee)$ is said to be **distributive** if it satisfies the identity $x \wedge (y \vee z) = (x \wedge y) \vee (x \wedge z)$.

This identity is equivalent to its dual. The first example is the lattice $\mathscr{P}(A), \cap, \cup)$ of all subsets of a set A. Any distributive lattice can be embedded in the lattice of all subsets of some set (Birkhoff, 1933; Stone, 1934; see (Birkhoff 1967)).

The smallest nondistributive lattices are five-element lattices N_5 (**pentagon**, the smallest nonmodular lattice—see Section F.2), and M_3 (**diamond**, which has three atoms). In fact, any nondistributive lattice contains N_5 or M_3 as a sublattice. Every distributive lattice is modular. A finitely generated sublattice of a distributive lattice is finite.

F.3.2 Examples Distributive lattices are:
(1) Any chain.
(2) The lattice of open subsets and the lattice of closed subsets of a topological space.
(3) The lattice $(\mathbb{N}, \text{g.c.d.}, \text{l.c.m.})$ of all positive integers ordered by divisibility. Every finite distributive lattice can be embedded in it.
(4) The lattice $\text{Sub}G$ of all subgroups of a group G is distributive if and only if every finite subset of G generates a cyclic subgroup (Ore, 1938; see Birkhoff 1967).
(5) The congruence lattice $\text{Con}L$ of an arbitrary lattice L. Every finite distributive lattice is isomorphic to the lattice $\text{Con}L$ of some finite lattice L (Dilworth, see Crawley and Dilworth 1973).

For a lattice L the following are equivalent: (i) L is distributive; (ii) L satisfies the **median law** $(x \wedge y) \vee (y \wedge z) \vee (z \wedge x) = (x \vee y) \wedge (y \vee z) \wedge (z \vee x)$;

(iii) L satisfies the **cancellation rule** $x \wedge y = x \wedge z$ & $x \vee y = x \vee z \rightarrow$ $y = z$.

The class D of all distributive lattices is a variety and hence is closed under sublattices, homomorphic images, and direct products. The variety D is the smallest nontrivial variety of lattices, it is generated by the two-element lattice $\mathbb{B} = \{0, 1\}$.

An element $a \neq 0$ of a distributive lattice L is **join-irreducible** when $a = x \vee y$ implies $x = a$ or $y = a$. So the join-irreducible elements in Example 3 are the powers of primes. If a distributive lattice L satisfies the descending chain condition, then every element of L has a unique irredundant representation as a join of a finite number of join-irreducible elements. A finite distributive lattice is determined up to isomorphism by the poset of its join-irreducible elements. Any maximal chain of a finite distributive lattice L has the length equal to the number of join-irreducible elements of L. If all chains of a distributive lattice L are finite, then L itself is finite.

If φ is a homomorphism of a distributive lattice L onto some lattice with 0, then the inverse image $\varphi^{-1}(0)$ is an ideal of L, called the **kernel** of φ. A lattice L is distributive if and only if every ideal of L is the kernel of a suitable homomorphism. Two different homomorphisms of a distributive lattice onto the same lattice with 0 can have coinciding kernels.

Ideals of a distributive lattice L form a complete distributive lattice $J(L)$, the **ideal lattice** of L, in which L is embeddable (see Section F.7).

A proper ideal J of a lattice L is called a **prime ideal** if $a, b \in L$ and $a \wedge b \in J$ imply $a \in J$ or $b \in J$. The prime ideals of a lattice L are the kernels of homomorphisms from L onto $\{0, 1\}$. Every prime ideal of a sublattice K of a distributive lattice L is the restriction to K of some prime ideal of L. A lattice L is distributive if and only if for all elements $a, b \in L$ with $a \not\geq b$ there is a prime ideal containing a but not b (Stone, 1936; Iséki, 1951; see Szász 1963). In a distributive lattice any prime ideal contains a minimal prime ideal. Every (proper) maximal ideal of a distributive lattice L is a prime ideal of L.

The category of all bounded (that is, with 0 and 1) distributive lattices and all lattice homomorphisms preserving 0 and 1 is dually equivalent to the category of all compact totally order-disconnected ordered topological spaces and all continuous order preserving maps. Prime ideals play a central role in establishing this **Priestley duality** (Priestley 1972). A special case is the Stone duality between Boolean algebras and Boolean topological spaces (see Section F.6).

An element of a distributive lattice can have at most one complement. A complemented distributive lattice is called a **Boolean lattice**. A distributive lattice L is a Boolean lattice if and only if every prime ideal of L is a maximal ideal. A Boolean lattice is treated usually as an algebra $(L, \wedge, \vee, ', 0, 1)$, where $a \mapsto a'$ is a unary operation which assigns to each element its unique complement. For more on Boolean Algebras, see Section F.6.

Distributive lattices with additional operations arise in logic. Boolean algebras are closely related to classical logic, Heyting algebras characterize algebraically intuitionistic logic.

A **relatively pseudocomplemented lattice** is a lattice L in which, for any given elements a and b, the set of all $x \in L$ such that $a \wedge x \leq b$ contains a greatest element $a \to b$, the **relative pseudocomplement** of a in b. Every relatively complemented lattice is distributive and has a greatest element 1, because $x \to x = 1$ for all $x \in L$.

F.3.3 Definition A **Heyting lattice** (or a **pseudoboolean lattice**) is a relatively complemented lattice which has a least element 0.

Every Boolean lattice is a Heyting lattice as well as every bounded chain (here $a \to b = 1$ if $a \leq b$ and $a \to b = b$ otherwise). Each element a of a Heyting lattice L has the **pseudocomplement** $a^* = a \to 0$.

A complete lattice L is a Heyting lattice if and only if it satisfies the **infinite distributive law** $x \wedge \bigvee_{i \in I} x_i = \bigvee_{i \in I} (x \wedge x_i)$. Complete Heyting lattices are also referred to as **frames**.

F.3.4 Examples The following are Heyting lattices:
(1) All finite distributive lattices;
(2) The lattice $O(T)$ of open subsets of a topological space T;
(3) The ideal lattice $J(L)$ of a distributive lattice L;
(4) The congruence lattice $\mathrm{Con}L$ of any lattice L.

A Heyting lattice is usually considered as an algebra $(L, \wedge, \vee, \to, {}^*, 0, 1)$. Heyting algebras form a variety which is finitely based.

Every Heyting algebra can be embedded in the Heyting algebra $O(X)$ of open subsets of some topological space X (McKinsey and Tarski, 1946). Every Heyting algebra is embeddable in a complete Heyting algebra in such a way that all meets and joins are preserved (Rasiowa,

1951). For more on Heyting algebras, see (Balbes and Dwinger 1974, Rasiowa and Sikorski 1970).

The following construction leads to Post algebras. In the category of all bounded distributive lattices and all lattice homomorphisms preserving 0 and 1, a lattice L is called a *free product* of lattices L_1 and L_2 if there exist embeddings $\varepsilon_i \colon L_i \to L$ such that whenever $\varphi_i \colon L_i \to K$ are homomorphisms into a distributive lattice K there exists a unique homomorphism $\varphi \colon L \to K$ with $\varphi_i = \varphi \circ \varepsilon_i$.

F.3.5 Definition The *Post algebra* of order m is defined as the free product of a Boolean algebra and m-element chain E_m.

Every Post algebra is a Heyting lattice (Rousseau, 1970). The list of operations usually considered in Post algebras of order m includes, besides Heyting operations, $m - 1$ further unary operations and m constants which are elements of E_m. Post algebras of each order m form a finitely based variety (Traczyk, 1964).

F.3.6 Examples The simplest Post algebra of order m is P_m, the free product of the two-element Boolean lattice $\{0, 1\}$ and the chain E_m. Let A be a non-empty set. The set P_m^A of all P_m-valued functions on A with pointwisely defined operations is a Post algebra.

Every Post algebra can be embedded in a suitable Post algebra of the form P_m^A and is isomorphic to a Post field of subsets of some Post topological space (Traczyk, 1964; Dwinger, 1966). Post algebras correspond to the m-valued logics of Post. More on Post algebras can be found in (Balbes and Dwinger 1974, Rasiowa 1974).

F.4 Complemented and Orthocomplemented Lattices

by Gudrun Kalmbach H. E. in Ulm, Germany

Complements

In this article, $L, M, \ldots$ are *bounded lattices*, which contain bounds $0, 1 \in L$ such that for all $x \in L$, $0 \le x \le 1$. A *complement* c of an element a in a bounded lattice L satisfies $a \vee c = 1$, $a \wedge c = 0$. L is *complemented* if every element of L has a complement. The lattice of equivalence relations on a set is complemented. In a bounded distributive lattice, the set of elements with a complement form a subalgebra.

A lattice L is ***relatively complemented*** if every closed interval in L is complemented. Examples are all complemented modular and orthomodular lattices. The best known example is $\mathbb{B} = \{\,0,1\,\}$, the Boolean algebra with two elements, used in logic and computer science. A ***Boolean lattice*** B is a complemented distributive lattice. When the unique complementation in B is considered as additional operation $'$ with algebraic rules, B is a ***Boolean algebra***. They are exactly the distributive ortholattices (see below and Section F.6).

Every lattice is a sublattice of a lattice with *unique complements*. Such sublattices can be nondistributive, which means that not every lattice with unique complements is a Boolean algebra. Every complete uniquely complemented lattice is isomorphic to a direct product of a complete atomic Boolean algebra and a complete atomless uniquely complemented lattice. This holds also for (lattice) unions of uniquely complemented lattices.

Complemented modular lattices M have a close connection to subspace lattices of projective coordinate spaces, which for finite dimension are projective geometries. Such a finite projective plane with 7 points and 7 lines is the ***Fano plane***:

Every complemented modular lattice M can be embedded in a complete, atomic modular lattice L; L is the subdirect union of projective planes and irreducible projective coordinate spaces. For finite dimensions (i.e., there is a bound for the maximal chain length in M), M is isomorphic to the direct product of a finite Boolean algebra and a finite number of projective geometries. Lattice identities concerning geometric theorems of Pappus, Desargues are investigated. Von Neumann's *continuous geometries* $\mathcal{G}$ without atoms are complemented modular lattices. Each $\mathcal{G}$ can be represented as the continuous sections of a bundle $\mathcal{A}$ whose stalks are irreducible continuous geometries. A continuous map from $\mathcal{A}$ to $\mathbb{R}$ is given by the dimension functions on the stalks.

Some results are: The MacNeille completion of a Boolean lattice is Boolean. Every relatively complemented lattice L with the ascending chain condition is isomorphic to a direct product of simple relatively complemented lattices. L has permutable congruences. The lattice of all equational theories containing a fixed theory is complemented if it is lower semicomplemented. This lattice is a finite Boolean algebra if one of the following conditions holds: (i) upper or lower sectionally complemented, (ii) lower semicomplemented and lower semimodular, (iii) atomistic and upper semimodular.

Orthocomplements and states

An **ortholattice** is a bounded lattice L with an **orthocomplementation** $': L \to L$, such that $a'' = a$, a' is a complement of a $\forall a \in L$, and $'$ reverses the order. An orthocomplementation on L is an example of a Galois connection (correspondence) on L. The Galois-closed subsets $B^\perp$ of a set A with a symmetric, irreflexive orthogonality relation $\perp$ form an ortholattice. There are several simple conditions on $\perp$ which imply orthomodularity. A finite example of a non-distributive, modular ortholattice is the **Chinese Lantern** MO2 (drawn as a Hasse-diagram with $0 < a, a', b, b' < 1$):

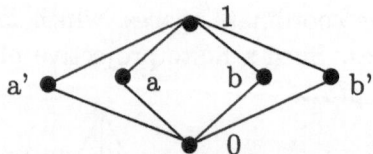

Recently automatic proofs by MACE or EQP are used to discover counterexamples among ortholattices for identities which hold in orthomodular lattices, or to prove them also for ortholattices. For constructing particular examples of finite ortho-structures, for their graphs or the (non-)existence of states (see below) on them, computer programs are in use. An example of a state on MO2 is: attach the real values 1 to 1 $\in$ MO2, 0 to 0 $\in$ MO2 and 1/2 to all elements in MO2\{ 0, 1 }. Another example: The order on the Boolean algebra $\mathbb{B} = \{ 0, 1 \}$ induces a componentwise defined order on the Boolean algebra $\mathbb{B}^3 = \{ (x, y, z) \mid x, y, z \in \mathbf{2} \}$. Choose as state the real values 0, 1 resp. for $0, 1 \in 2^3$, 1/3 for the triples with one coordinate 1, and 2/3 for the triples with two coordinates 1.

A **state** on an ortholattice L is a map $s: L \to [0, 1] \subset \mathbb{R}$ with $s(1) = 1$ and $s(a \vee b) = s(a) + s(b)$ for all $a \perp b$, $a, b \in L$. If in the last condition the set $\{ a, b \}$ is replaced by any countable subset A of L consisting of pairwise orthogonal elements, and supremas $\vee A$ of such subsets exist in L, then s is called a **probability**. On the orthomodular Hilbert lattice $\mathcal{C}$ of closed subspaces of a Hilbert space H, the set of states $\mathcal{S}$ is **full**, which means that the order $A \subseteq B$ for closed subspaces of H is determined by $s(A) \leq s(B)$ for all $s \in \mathcal{S}$. In Dvurečenskij (1993) many interesting applications of Gleason's theorem can be found. We mention just one version of this theorem:

F.4.1 Theorem *Probabilities m on on a separable Hilbert space H are in one-to-one correspondence with positive, selfadjoint, bounded operators T of the trace class with trace 1, such that for all $N \in \mathcal{C}$, $m(N) = \text{tr}(TP_N)$, where P_N is the projection of H onto N and $\text{tr}(TP_N)$ is the trace of TP_N.*

In the extensive research area on *quantum structures* and *logics*, the lattice structure of L is weakened. For such posets L only partially defined binary operations are requested.

Some results are: A **Jauch-Piron state** is a state s on an ortholattice L with the property $s(a) = 0 = s(b)$ imply $s(a \vee b) = 0$. L is **concrete** if it has a set representation. There exist concrete orthomodular posets with a set of two-valued Jauch-Piron states which are not Boolean. Every concrete (quantum) logic can be enlarged to one with a given group as automorphism group and a given state space (or set of dispersion-free states) and center. For σ-orthomodular posets P (P must not be a lattice) the following holds: If the center of P is complete then P is the direct product of a Boolean algebra and another lattice Q without Boolean factor. Q can further be decomposed into a **horizontal sum**, consisting of a union of Boolean algebras in which all 0's and all 1's are each identified. The free word problem for ortholattices is solvable. Ortholattices have the strong amalgamation property. The MacNeille completion of an ortho(modular) poset (lattice) is an ortholattice.

F.5 Orthomodular Lattices

by Gudrun Kalmbach H. E. in Ulm, Germany

Orthomodularity

Let L be an ortholattice. The **orthomodular law** is: $x \leq y$, $x, y \in L$, implies $x \vee (x' \wedge y) = y$. L is **orthomodular** (an **OML**) if this law holds in L. Equivalent conditions are: (i) $x \leq y, y \wedge x' = 0$ imply $x = y$, (ii) any $x \leq y$ generate a Boolean subalgebra of L, (iii) xCy iff yCx. Here, $x, y \in L$ are said to **commute** (denoted by xCy) if $x \wedge y = x \wedge (x' \vee y)$. This name is chosen because it applies to the following example: The lattice $\mathcal{C}$ of closed subspaces of a classical Hilbert space H is orthomodular, but modular only if $\dim H$ is finite. Commuting of its closed subspaces U in C corresponds to commuting of their associated projection operators $P_U \colon H \to U$.

Every interval $[a, b] = \{\, x \in L \mid a \le x \le b \,\} \subseteq L$ in an OML L is again an OML with the induced structure. The **center** $C(L)$ of L consists of those elements that commute with every element of L. $C(L)$ is a Boolean algebra. The central elements $c \in C(L)$ give rise to isomorphic direct decompositions $[0, c] \times [0, c']$ of L. Typical finite examples with nontrivial center are $L_B = B \times MO2$, the product of the Chinese Lantern MO2 (see Section F.4) with a finite Boolean algebra B or $L_A = MO2 \times MO2$. L_A is an example of an astroid. The complements of $y \in L$ are of the form $(x \wedge (x' \vee y')) \vee (x' \wedge y')$, $x \in L$. If L is uniquely complemented, it is a Boolean algebra, i.e., L is distributive. Additional operations and relations on OML's and their elementary arithmetic are investigated, for instance logical **implications** $a \to b = a' \vee (a \wedge b)$ or $a \to_1 b = (a' \wedge b) \vee (a' \wedge b') \vee (a \wedge (a' \vee b))$, the **covering property** (if a covers $a \wedge b$ then $a \vee b$ covers b) or the **exchange property**: $[a \wedge b, a]$ is isomorphic to $[b, a \vee b]$. The orthomodular lattice C from above has the properties of a **Hilbert lattice**: it is irreducible (not a direct product), complete (all suprema exist), atomic (every $U \in C$ is the supremum of the one-dimensional subspaces contained in it), and it has the exchange property.

Orthomodular lattices L have a **block-structure**, which means that L is the set-theoretic union of its maximal Boolean subalgebras (**blocks**). In *block-diagrams*, the finite Boolean algebras B_i of the block-structure of L are represented by hypergraphs G with its edges consisting of $|B_i|$ vertices for each B_i. L_A, $L_B = B \times MO2$ with $B = \mathbb{B}$ are represented by the first two hypergraphs in figure 1. The last hypergraph is a nonmodular example: Dilworth has pasted 3 copies B_1, B_2, B_3 of the 8-element Boolean algebra $\mathbb{B}^3$ together at subalgebras with $|B_i \cap B_{i+1}| = 4$ (indices mod 3).

The lattice-structure of L requires from G that in 4-circles C of 4 or less B_i, where for instance for some $m_i \in L$, $[0, m_i] \times [0, m'_i]$ is isomorphic to $B_i \cap B_{i+1}$, $|B_i \cap B_{i+1}| \ge 4$, $i = 1, 2, \ldots$, indices mod 4, with $m_{i-1} \le m'_i$, additional blocks in L exist which form "astroids" (Kalmbach 1983): Every four-circle has to have a central astroid. In a weaker form, Greechie's **Loop Lemma** says that in a Greechie-diagram for a suitable L, no circles consisting of 3 or 4 blocks are allowed. **p-ideals** I are ideals as in lattices which are closed under perspectivity: If $a, b \in L$ have a common complement then $a \in I$ implies $b \in I$. p-ideals (or dually **p-filters**) in an OML L correspond exactly to the 0-class (1-class) of congruence relations ψ on L. ψ is determined by each one of these

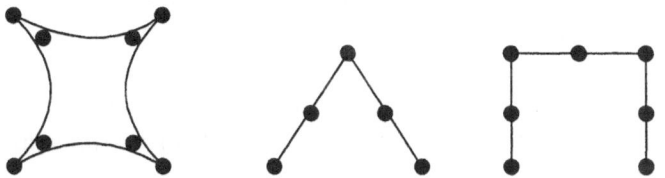

Figure 1: block-diagrams

classes. The congruences on L permute (cf. Section G.6). Some theorems in this theory are: Orthomodular lattices do not have the amalgamation property. The structure of free modular ortholattices (see Section F.2 and Section F.8) with $n \geq 3$ generators is described by duality theory, decomposing them into certain canonical intervals in the variety generated by MOn, $n \geq 2$, (modular ortholattice with n blocks $\mathbb{B}^2$) covering the variety of Boolean algebras. An explicite table exists for $n \leq 10$ generators. For every set L_i, $i \in I$, of OML and a Boolean algebra B there exists an OML L with center B and $\bigcup L_i \subseteq L$. Every OML is ortho-isomorphic to $CO(\mathcal{L})$, the OML of clopen sets in a compact Hausdorff closure space $\mathcal{L}$ with suitable closure (not set-theoretic) lattice-operations.

Orthomodular spaces

This research was initiated by the standard *quantum logic*, the orthomodular lattice $\mathcal{C}$ of closed subspaces U of a separable Hilbert space H over the complex numbers. In physics, the Hilbert space formalism of quantum mechanics attributes to a (physical) system P some H, and P is characterized by its (measured) state. To such physical quantities, observables or states correspond linear, positive, self-adjoint, bounded operators of trace 1 on H. Projection operators of H onto one-dimensional subspaces are examples of states. Since measuring in quantum mechanics deals with such operators which may not commute, the associated noncommutative measure theory is orthomodular, not classical Boolean. *Pure* states m cannot be written as a convex sum, *superposition*, $\sum_I w_i \cdot m_i$, $0 < w_i \in \mathbb{R}$, of two or more different states m_i, $\sum w_i = 1$, $|I| \geq 2$.

Examples: In the Stern-Gerlach experiment it is demonstrated that for a sample set of atoms, their arbitrary located 3D-spin state $s =$

(s_x, s_y, s_z) can turn in a magnetic field of a measuring apparatus, such that for instance 50% of the atoms have after the measurement a pure state *spin up* $s = (0, 0, s_z)$ and 50% *spin down* $s = (0, 0, -s_z)$ in direction of a vertical z-axis. If another experiment is made, where a laser beam is shot through them, they may alternatively get for instance all (100%) into a (scaled) mixed spin state (superposition) $s = (\frac{1}{2}s_x, \frac{1}{2}s_y, 0)$, often written as $s = \frac{1}{2}s_x + \frac{1}{2}s_y$.

The projection theorem for Hilbert spaces H has been investigated: Let K be a skew field with a normed (by some $0 \neq \varepsilon \in K$) antiautomorphism θ (with $\varepsilon\theta(\varepsilon) = 1, \theta(\theta(a)) = \varepsilon^{-1}a\varepsilon$), and φ a (θ) *Hermitian form* on a K-left vector space V, which is bilinear and satisfies $\varphi(y, x) = \varepsilon\theta(\varphi(x, y))$. V is an **orthomodular space** if it satisfies the **projection theorem**: for any subspace $U \subseteq V$, we have $V = U^{\perp\perp} + U^{\perp}$. This condition is equivalent to the orthomodular law in the lattice L of $\perp$-closed subspaces $(U = U^{\perp\perp})$ of L.

The real, complex and quaternionic fields for H are generalized to skew fields K, and K-left vector spaces V with Hermitian forms, for which orthomodular orthogonal-closed (or topologically closed) subspace lattices $\mathcal{L}(V)$ have similar properties. Assume that $\dim V \geq 4$ is finite and K carries some involutory anti-automorphism θ. There is a one-to-one correspondence between orthocomplementations on $\mathcal{L}(V)$ and pairs (θ, φ), where φ is a normed (θ) Hermitian form on V. Under additional conditions on $\mathcal{L}(V)$, this result can be extended to the case that $\dim V$ is infinite. There are several characterizations of classical infinite-dimensional Hilbert spaces among these orthomodular spaces, for instance in case an infinite, orthonormal sequence exists. Non-archimedian normed orthomodular spaces V exist over commutative fields with a non-archimedian valuation of rank 1. Bounded countable orthogonal families in more general such V produce non-orthomodular cases, where *topological closed* and *orthogonally closed* for subspaces differ.

F.6 Boolean Algebras

by Carlton J. Maxson in College Station, TX, USA and

Günter F. Pilz in Linz, Austria

Boolean algebras (complemented distributive lattices) are among the most thoroughly studied and most applicable algebraic structures. By definition, they are bounded (by 0 and 1, say), and complements are unique; we denote the complement of $b \in B$ by b'. If one considers

Boolean algebras as universal algebras of type $(2, 2, 0, 0, 1)$, they form a variety. The smallest non-trivial Boolean algebra is $\mathbb{B} = \{0, 1\}$ with the operations $a \vee b = \max\{a, b\}$, $a \wedge b = \min\{a, b\}$. By **Stone's Theorem**, every finite Boolean algebra is isomorphic to some power set Boolean algebra $(\mathscr{P}(M), \cup, \cap)$, which in turn is isomorphic to $\mathbb{B}^{|M|}$. Here, isomorphic means **Boolean isomorphic**: isomorphisms must preserve $0, 1$, and complements. So every finite Boolean algebra has as its order a power of 2, and two finite Boolean algebras of the same order are isomorphic. Every (infinite) Boolean algebra can at least be Boolean embedded into some $\mathscr{P}(M)$. A Boolean algebra $(B, \wedge, \vee)$ determines a **Boolean ring** $(B, \Delta, \wedge)$ (a ring with identity in which every element is idempotent), and conversely, every Boolean ring $(B, +, \cdot)$ induces a Boolean algebra $(B, \wedge, \vee)$ with $a \wedge b := ab$, $a \vee b := a + b + ab$, and $b' = 1 + b$.

The subgroup lattice of a finite abelian group is Boolean iff $G \cong \mathbb{Z}_n$, where n is square-free (see e.g., Fuchs 1960).

Boolean algebras were first used around the middle of the 19th century by George Boole to *formalize thinking*; generalizations to Heyting and Post algebras are treated in Section F.3. Beside these connections, the relations to set theory, to probability theory (σ-algebras) and the applications to switching circuits (basically due to C. E. Shannon) are well known. Most of the applications run via **Boolean polynomials**, which are defined to be the symbols $0, 1, x_1, x_2, \ldots, x_n$ and all words which can be constructed in finitely many steps from these symbols using $\wedge$, $\vee$ and $'$. Equality is formal, so, e.g., $x_1 \wedge x_2 \neq x_2 \wedge x_1$. For $\mathbb{B}$, polynomials and terms (Section G.1) coincide. The set P_n of Boolean polynomials in $x_1, \ldots, x_n$ is therefore an algebra of type $(2, 2, 0, 0, 1)$, but not a Boolean algebra. Each $p \in P_n$ induces a polynomial function $\bar{p} \colon \mathbb{B}^n \to \mathbb{B}$ in the obvious way. We call p, q equivalent ($p \sim q$) if $\bar{p} = \bar{q}$. Hence $x_1 \wedge x_2 \sim x_2 \wedge x_1$, and $P_n/\sim$ is a Boolean algebra, the free algebra over $\{x_1, x_2, \ldots, x_n\}$, and $P_n/\sim$ has 2^{2^n} elements. Infinite free Boolean algebras F are more complicated, e.g., they are atomless, but every chain in F is countable.

For every $p \in P_n$, we can find all $(b_1, \ldots, b_n) \in \mathbb{B}^n$ where $\bar{p}$ has value 1. These $(b_1, \ldots, b_n)$ form the *ON*-set, $ON(p)$, of p. If $x^1 := x$ and $x^0 := x'$, p is equivalent to the sum p_d of all $x_1^{b_1} x_2^{b_2} \ldots x_n^{b_n}$ where $(b_1, \ldots, b_n) \in ON(p)$, since this sum has value 1 precisely at all $(b_1, \ldots, b_n) \in ON(p)$. Then p_d is called the **disjunctive normal form** of p. For instance, if $p = x_1(x_1 + x_2') \in P_2$ then $\bar{p}(b_1, b_2) = 1$ iff $(b_1, b_2) \in ON(p) = \{(1, 0), (1, 1)\}$; so $p_d = x_1 x_2' + x_1 x_2$. This also shows that each function

$f \colon \mathbb{B}^n \to \mathbb{B}$ can be written as $f = \bar{p}$ for some $p \in P_N$ i.e., $\mathbb{B}$ is functionally (or polynomially) complete.

For the remainder of this section, we denote $p \wedge q$ as pq and $p \vee q$ as $p + q$. In the interpretation of switching circuits, $p \neq q$, but $p \sim q$ means that the circuits describing p and q are different (and one may be cheaper than the other one), but they function the same way. The functional completeness of $\mathbb{B}$ means that every desired switching behavior can actually be built. Thus the need arose to find the *simplest* of all Boolean polynomials q which are equivalent to a given $p \in P_n$. If *simplest* means sum of product expression with a minimal number of summands and then with a minimal number of appearing x_i, the Quine-McCluskey algorithm (see e.g., Dornhoff and Hohn 1978, Lidl and Pilz 1998) gives a method to find the simplest q to a given p_d in disjunctive normal form.

More general are **irredundant polynomials** which are sums of products in which every omission of a symbol yields an inequivalent polynomial. Using the definition above, a simplest polynomial is irredundant, but $p = x_1 x_2' + x_1' x_2 + x_2 x_3' + x_2' x_3 \sim x_1 x_2' + x_1' x_3 + x_2 x_3' \sim x_1 x_3' + x_1' x_2 + x_2' x_3$ shows that the converse is not true and that simplest forms are not unique. Nevertheless, finding irredundant polynomials is an important step in many simplification processes.

One of the main technique in all reductions is to use the rule

(1) $pqr + pq'r \sim pr$.

The bad thing in the Quine-McCluskey algorithm is that p must be in disjunctive normal form, so it might have up to 2^{2^n} summands. If p is already "pretty much reduced", better techniques are at hand:

(2) Absorption: If $p \to q$, apply $p + q \sim p$

(3) Consensus: If $p = p_1 x_i p_2$ and $q = q_1 x_i' q_2$ so that x_i and x_i' do not appear in p_1, p_2, q_1, q_2 then $r := p_1 p_2 q_1 q_2$ is called the **consensus** of p and q w.r.t. x_i. It is easy to show that $p + q \sim p + q + r$. This often allows one to apply (1) and (2) again.

(4) Ghazala-Reusch: Since for $p \in P_n$ there are $q, r \in P_{n-1}$ with $p \sim q x_n + r x_n'$, a recursive treatment of q and r and a skillful use of what is called the *Ghazala-table* finally yields all irredundant polynomials equivalent to p (see Reusch 1975).

Often in switching algebra, certain combinations $(b_1, \ldots, b_n)$ do not make practical sense, so the value of p at $(b_1, \ldots, b_n)$ is irrelevant (**don't-care** situations). So one can try if $\bar{p}(b_1, \ldots, b_n) = 0$ or $\bar{p}(b_1, \ldots, b_n) = 1$ gives shorter minimal forms. For many don't-care cases there are special techniques (see Hammer and Kogan 1992).

Tests performed by J. Gahleitner in his masters thesis showed that in the area of $n = 10$, the Consensus and the Ghazala-Reusch method needed similar computing times, but they were about 1000 times faster than the Quine-McCluskey algorithm. This is very essential in the case of many don't-care situations. Altogether, minimizing Boolean polynomials is known to be an NP-hard problem in the number of variables.

There are also some nice connections to topology. If B is a Boolean algebra then $B^* := \mathrm{Hom}(B, \mathbb{B})$ has the structure of a **Boolean space** (a compact Hausdorff space in which every open set is the union of clopen (closed and open) sets). Conversely, in a Boolean space X, the set of all clopen subsets forms a Boolean algebra. For a particular result, we have X is metrizable iff B is countable and X is discrete iff B is finite. For more on this topic, see (Halmos 1963).

$\mathbb{B}$ is not only polynomially complete, but also primal (see Section G.3), since polynomials and terms coincide. By Stone's theorem, the variety $\mathcal{B}$ of Boolean algebras is generated by $\mathbb{B}$.

F.6.1 Theorem (*Hu's theorem*) *A variety $\mathcal{V}$ is categorically equivalent to $\mathcal{B}$ iff $\mathcal{V}$ is generated by a finite primal algebra with at least 2 elements.*

See Section G.3 for a slightly different wording of this result. As an application, let G be a finite, simple, non-abelian group. Adjoin all unary operations. Then the variety generated by G is categorically equivalent to $\mathcal{B}$.

In every Boolean algebra $(B, \wedge, \vee, 0, 1, ')$, there is a natural order relation: $a \leq b : \Longleftrightarrow a \wedge b = a$. A subset F of B containing 1 is a **filter** if it is closed w.r.t. $\wedge$ and if $a \geq f$, $f \in F$, implies $a \in F$. Maximal filters are ultrafilters (see Section G.1), and $b \in B$ is contained in an ultrafilter iff $b > 0$ (using Zorn's Lemma). From this it is easy to derive the completeness theorem of propositional logic. Staying in this area, we have a couple of surprising results.

F.6.2 Theorem
(1) (Tarski) *The first order theory of Boolean algebras is decidable.*
(2) (Rubin) *The first order theory of Boolean algebras with a distinguished subalgebra is undecidable.*

Other important examples of Boolean algebras include **interval algebras** (formed by finite unions of clopen intervals) and tree algebras.

For a tree $(T, \leq)$ and $t \in T$, let $\langle t \rangle := \{ s \in T \mid s \geq t \}$. The **tree algebra** $B(T)$ is the Boolean subalgebra of $\mathscr{P}(T))$, generated by all $\langle t \rangle$. A Boolean algebra is isomorphic to an interval algebra iff it is generated by a chain. Interval algebras are not closed with respect to subalgebras, but every tree algebra can be embedded into an interval algebra. Tree algebras can be **rigid** (i.e., they have no non-identity automorphisms) and they are closed with respect to taking quotients and finite products. If A and B have isomorphic endomorphism semigroups then they are (Boolean) isomorphic (Schein). Interval algebras always have non-trivial endomorphisms, and if B is countable, then $|\mathrm{End}(B)| \geq 2^{\omega}$. However, there does exist a Boolean algebra C of order 2^{ω} with $\mathrm{End}(C) = \{\, \mathrm{id}\,\}$.

F.7 Complete Lattices

by Viacheslav N. Saliĭ in Saratov, Russia

Mathematical practice frequently leads to partially ordered sets (or posets for short), in which every non-empty subset X has a greatest lower bound $\inf X$ and a least upper bound $\sup X$. Such a poset is obviously a lattice and is called a **complete lattice** (Birkhoff, 1933). In complete lattices, the elements $\inf X$ and $\sup X$ are usually denoted by $\bigwedge X$ (**meet of** X) and $\bigvee X$ (**join of** X), respectively.

A first example of a complete lattice is $(\mathscr{P}(A), \subseteq)$, the collection of all subsets of a set A, partially ordered by set-inclusion. Here for any family $\{\, X_i \mid i \in I \,\} \subseteq \mathscr{P}(A)$ the meet $\bigwedge_{i \in I} X_i$ is the set-intersection $\bigcap_{i \in I} X_i$ and the join $\bigvee_{i \in I} X_i$ is the set-union $\bigcup_{i \in I} X_i$. Clearly, every finite lattice is complete.

Any complete lattice L has a least element and a greatest element, $\bigwedge L$ and $\bigvee L$, respectively. Many lattices, although not complete, can be made so by adjoining the missing least or greatest element (or both). In this way, the chain of reals $(\mathbb{R}, \leq)$ can be completed by adjoining the universal bounds $-\infty$ and $+\infty$. We have the following result.

F.7.1 Theorem (1) *A lattice L is complete iff $\bigwedge C$ and $\bigvee C$ exist for every chain C of L.*

(2) *If a poset P has a greatest element and every non-empty subset of P has a greatest lower bound, then P is a complete lattice.*

Part 2 of this theorem implies, for example, the completeness of such lattices as $\mathrm{Equ}(A)$, the lattice of all equivalence relations on an arbitrary set A (often referred to as the **partition lattice** of A), the **subalgebra** *lattice* Sub A and the **congruence lattice** Con A of an algebra A, and many others.

A sublattice L^* of a complete lattice L is called a **complete sublattice** of L if $\bigwedge_L X \in L^*$ and $\bigvee_L X \in L^*$ for every nonempty subset $X \subseteq L^*$. For example, each closed interval $[a, b]$ of a complete lattice is a complete sublattice. The closed subsets of a topological space T form a complete lattice which is a sublattice of the lattice $\mathscr{P}(T)$ but not a complete sublattice (it is not closed under arbitrary unions). The congruence lattice Con A of an algebra A is a complete sublattice of the partition lattice Equ A.

Any homomorphic image of a complete chain is complete, but, for example, if A is a countable set and $\mathcal{F}$ is the ideal of finite subsets in the complete Boolean lattice $\mathscr{P}(A)$, then the homomorphic image $\mathscr{P}(A)/\mathcal{F}$ is not a complete lattice.

A lattice homomorphism φ is called a **complete homomorphism** if it preserves joins and meets of arbitrary nonempty sets, i.e., $\varphi(\bigwedge X) = \bigwedge \varphi(X)$, $\varphi(\bigvee X) = \bigvee \varphi(X)$. Each isomorphism of complete lattices is a complete isomorphism. The inverse image of any element under any complete homomorphism of a complete lattice L is a closed interval of L.

The direct product of complete lattices is a complete lattice.

F.7.2 Theorem (1) *Every isotone mapping φ of a complete lattice to itself has a fixed point (Tarski, 1955).*

(2) *The set of fixed points of φ forms a complete lattice in itself.*

(3) *If in a lattice L every isotone mapping has a fixed point, then L is complete (Davis, 1955, see Crawley and Dilworth 1973).*

Any poset can be embedded in a complete lattice in such a way that all meets and joins are preserved. The best known method is the so called completion by cuts (Mac Neille, 1937). A **closed ideal** of a poset $(P, \leq)$ is a subset of P which contains all lower bounds to the set of all its upper bounds. For example, each principal ideal $J(a) = \{ x \in P \mid x \leq a \}$ is closed. Closed ideals of P form a complete lattice $CJ(P)$. **Completion by cuts** of a poset P is its embedding into the complete lattice $CJ(P)$ which associates to each element $a \in P$ the principal ideal $J(a)$. The term *completion by cuts* is also referred to the lattice $CJ(P)$ itself.

If L is a complete lattice, then $L \cong CJ(L)$, i.e., L is isomorphic to $CJ(L)$, its completion by cuts. Dedekind's construction of reals is nothing more than completion by cuts of the chain of rationals.

F.7.3 Theorem *The completion by cuts of any chain is again a chain.*

The class of Boolean algebras is also closed under completion by cuts (Glivenko, 1929), but there exist distributive lattices for which completion by cuts is even nonmodular.

The completion by cuts is a minimal completion in the sense that any order embedding of a poset P into a complete lattice L can be extended to an order embedding of $CJ(P)$ into L.

F.7.4 Definition An *ideal* of a lattice L is a sublattice J such that $a \in J$, $x \in L$, and $x \leq a$ imply $x \in J$.

The set $J(L)$ of all ideals of the lattice L is a complete lattice under the set-inclusion. Of course, $CJ(L) \subseteq J(L)$, but closed ideals of L do not even form a sublattice of the ideal lattice $J(L)$ in general. The mapping $a \mapsto J(L)$ is an embedding of L into $J(L)$. This embedding, and the lattice $J(L)$ itself, is called the *ideal completion* of the lattice L. The lattice $J(L)$ satisfies every lattice identity that L satisfies (Sachs, 1961, see Crawley and Dilworth 1973). Thus any distributive (resp. modular) lattice can be embedded in a complete distributive (resp. modular) lattice. But the ideal completion of a Boolean lattice is not always a Boolean lattice. The ideal completion, in contrast to the completion by cuts, does not preserve in general existing infinite joins (though all meets are preserved). Complete lattices are often related to closure operations.

F.7.5 Definition A *closure operation* on the set $\mathscr{P}(A)$ is a mapping $\mathscr{P}(A) \to \mathscr{P}(A)$, $X \mapsto \overline{X}$, which satisfies the conditions $X \subseteq \overline{X}$ (*extensionality*), $\overline{\overline{X}} = X$ (*idempotency*), $X \subseteq Y \Rightarrow \overline{X} \subseteq \overline{Y}$ (*isotoncity*). A subset $X \subseteq A$ is called a *closed subset* if $\overline{X} = X$.

The set of all closed subsets of A is a complete lattice in which $\bigwedge_{i \in I} X_i = \bigcap_{i \in I} X_i$, $\bigvee_{i \in I} X_i = \overline{\bigcup_{i \in I} X_i}$. Any complete lattice is isomorphic to the lattice of all closed subsets of some set with a suitable closure operation. Moreover, the closure operation can be chosen in such a way that each closed subset is the closure of some point (Saliĭ, 1984, see Saliĭ 1988).

A closure operation is an *algebraic closure operation* if $\overline{X} = \bigcup \{ \overline{F} \mid F \subseteq X$ and F is finite $\}$. A complete lattice is called an *algebraic lattice* if it is isomorphic to the lattice of all subsets of some set that are closed under a suitable algebraic closure operation.

F.7.6 Theorem *For a lattice L, the following are equivalent:*

(1) L *is algebraic;*
(2) $L \cong \operatorname{Sub} A$ *for some algebra A (Birkhoff and Frink, 1948, see Birkhoff 1967);*
(3) $L \cong \operatorname{Con} A$ *for some algebra A (Grätzer and Schmidt, 1963, see (Grätzer 1979)).*

There is another approach to algebraic lattices. A **compact element** of a complete lattice L is an element c for which $c \le \bigvee X$ implies the existence of a finite subset $F \subseteq X$ with $c \le \bigvee F$. If each element of a complete lattice L can be expressed as the join of a suitable set of compact elements, then L is said to be a **compactly generated lattice**. Algebraic lattices, and only these, are compactly generated lattices. The compact elements of $\operatorname{Sub} A$ are precisely the finitely generated subalgebras. The unit interval $[0, 1]$ of the real line, though a complete lattice, is not algebraic.

In a complete lattice L write $x \ll y$ (in words: "x is **well below** y" or "x is **way below** y") if for any upward directed subset $D \subseteq L$ the inequality $y \le \bigvee D$ implies $x \le d$ for some $d \in D$. A complete lattice L is called a **continuous lattice** if $x = \bigvee \{ u \in L \mid u \ll x \}$ for all $x \in L$. Complete chains, the direct products of complete chains, and algebraic lattices all are continuous lattices. One more basic example is the lattice of open subsets of a locally quasicompact space. There is a canonical bijection between continuous lattices and injective topological T_0-spaces (Scott, 1972, see Gierz, Hofmann, Keimel, Lawson, Mislove, and Scott 1980).

A continuous lattice in which every element has one and only one complement is distributive (see Saliĭ 1988)). It is not known if there exist complete nondistributive uniquely complemented lattices.

F.8 Free Lattices
by Ralph Freese in Honolulu, HI, USA

We survey some of the major developments in the theory of free lattices and certain related topics. The book by Freese, Ježek, and Nation (1995) is a complete reference on the subject.

Whitman's solutions

In his classic papers Whitman (1941, 1942) solved the word problem for free lattices, that is, he gave an algorithm for determining if two lattice terms were equal in all lattices. He showed that each element of the free lattice has a shortest term representing it, known as the **canonical form**, and gave an algorithm to put an arbitrary term into canonical form. This canonical form is closely connected with the arithmetic of the free lattice and Whitman exploited this connection to obtain important results about free lattices (cf. Section G.4).

The word problem (Section G.7) for free lattices, in fact for finitely presented lattices, was first solved by Skolem (1920, 1970). What was interesting about Skolem's solution is that it is polynomial time, unlike some of the later solutions (McKinsey 1943, Evans 1951b,a). The computational complexity of Whitman's algorithm is discussed in (Freese 1987b) and of (Freese et al. 1995, Chapter XI). Even though there is such a nice canonical form in free lattices and an easy algorithm for obtaining it, there is no term rewrite system for lattice theory (Freese, Ježek, and Nation 1993). This is also proved in (Freese et al. 1995, Chapter XII) along with some further results in this area.

The key ingredient of Whitman's solution is the following condition known as **Whitman's condition**:

$$\text{If } v = v_1 \wedge \cdots \wedge v_r \leq u_1 \vee \cdots \vee u_s = u,$$
$$\text{then either } v_i \leq u, \text{ for some } i, \text{ or } v \leq u_j, \text{ for some } j. \tag{1}$$

The aforementioned connection between the canonical form and the arithmetic of free lattices is summarized in the following theorem which shows that the canonical form corresponds to the best way to express an element a free lattice as a join or meet. For any lattice $\mathcal{L}$ and finite subsets A and B of L we say that A **join refines** B and we write $A \ll B$ if for each $a \in A$ there is a $b \in B$ with $a \leq b$. We let $\mathcal{FL}(X)$ denote the free lattice generated by a set X. For n a cardinal, $\mathcal{FL}(n)$ denotes $\mathcal{FL}(X)$ with $|X| = n$.

F.8.1 Theorem *Let* $w = w_1 \vee \cdots \vee w_n$ *canonically in* $\mathcal{FL}(X)$. *If also* $w = u_1 \vee \cdots \vee u_m$, *then*

$$\{ w_1, \ldots, w_n \} \ll \{ u_1, \ldots, u_m \}.$$

Thus $w = w_1 \vee \cdots \vee w_n$ *is the unique minimal join representation of* w.

Jónsson observed that there is a close connection between canonical form and a weak form of distributivity known as semidistributivity. A lattice is **semidistributive** if it satisfies the following condition and it dual.

$$a \vee b = a \vee c \quad \text{implies} \quad a \vee b = a \vee (b \wedge c). \tag{2}$$

Jónsson showed free lattices are semidistributive.

Major results

There are several deep results in the subject of free lattices. Whitman was able to use his solution to the word problem to prove that a subset of a free lattice generates a free sublattice if and only if it is both join and meet irredundant. A subset S is ***join irredundant*** if no element $a \in S$ is below the join of any subset of S not containing a. Of course ***meet irredundant*** is defined dually. Using this he was able to show $\mathcal{FL}(\aleph_0)$ can be embedded into $\mathcal{FL}(3)$.

F.8.2 Theorem (Whitman 1942) *A subset of a free lattice generates a free sublattice if and only if it is both join and meet irredundant. $\mathcal{FL}(\aleph_0)$ can be embedded into $\mathcal{FL}(3)$.*

Jónsson's conjecture   The properties (W), (SD$_\vee$) and (SD$_\wedge$) are of course inherited by sublattices of free lattices. Jónsson conjectured that these properties characterize finite sublattices of free lattices. For about 20 years work on free lattices was dominated by this problem. This problem was finally solved by Nation (1982).

F.8.3 Theorem *A finite lattice is isomorphic to a sublattice of a free lattice if and only if it is semidistributive and satisfies (W).*

The covering relation in free lattices   Whitman was able to show certain elements ***cover*** certain other elements and in an unpublished result Richard Dean was able to show there are elements which do not have any lower covers. Substantial progress in this area was made by McKenzie (1972) who exhibited a wide class of covers and showed how covers in free lattices were related to equational theories of lattices. He conjectured that free lattices with a finite generating set are ***weakly atomic***, that is, every proper interval contains a cover. This was shown to be the case by Day (1977).

F.8.4 Theorem *Finitely generated free lattices are weakly atomic.*

Despite this result, covers remained mysterious. Problems such as

- Is there an algorithm to decide if an element has a lower cover?
- Are there elements with no upper and no lower cover?
- How long can a chain of covers be?

These problems motivated Freese and Nation (1985) to make a thorough study of the subject. We were able to give an algorithm for the first question above and give a relatively simple formula for the lower covers of an element. With these results it was possible to solve the other problems above and many more. There are elements with no lower and no upper covers. In fact, for most $a < b$ in a free lattice, there is a maximal chain from a to b such that each element of the chain has no lower and no upper covers. A chain of covers in a free lattice can have length at most four and chains of length three and four are extremely limited, only occurring at the top or bottom of the free lattices. While there more chains of covers of length two, even they are limited and it is possible to precisely characterize them.

Tschantz's Theorem Nation and I felt that it would be relatively easy to show that every infinite interval of a free lattice would contain $\mathcal{FL}(\omega)$ as a sublattice. We were surprised when we were not even able to rule out the possibility of an infinite interval which is a chain. (See the introduction and Chapter IX of Freese et al. 1995 for more of the interesting history of this problem.) This problem was solved by Tschantz (1990) with the following deep theorem. Interestingly the theory of covers in free lattices can be used to greatly simplify the proof of this theorem (Freese, Ježek, and Nation 1995).

F.8.5 Theorem *Every infinite interval of a free lattice contains a sublattice isomorphic to $\mathcal{FL}(\omega)$.*

Applications The results and techniques of free lattices have surprisingly far reaching applications. In lattice theory to such areas as projective lattices, finitely presented lattices, congruence lattices of lattices, transferable lattices, and varieties (equational classes) of lattices. Applications to general algebra includes the study of the equational theory of the congruence lattices of algebras in a variety of algebras.

F.9 Varieties of Lattices
by George Grätzer in Winnipeg, Canada

This is a huge, active, and very technical field; see (Jipsen and Rose 1992) for book-length coverage up to 1992 and Appendix F by the same authors in (Grätzer 1998) for an updated account.

A class $\mathscr{K}$ of lattices is a **variety** iff $\mathscr{K}$ is defined by a set of lattice identities. Applying the universal algebraic results in Section G.5, we obtain (Birkhoff's theorem) that a class $\mathscr{K}$ of lattices is a variety iff $\mathscr{K}$ is closed under **H** (homomorphic images), **S** (sublattices), and **P** (direct products); or by A. Tarski's formulation: for a class $\mathscr{K}$ of lattices, $\mathbf{Var}(\mathscr{K}) = \mathbf{H\,S\,P}(\mathscr{K})$ is the smallest variety containing $\mathscr{K}$. For lattices, we get a much stronger 1967 result of B. Jónsson (which he proved not only for lattices but for any congruence distributive variety of algebras):

F.9.1 Theorem
(1) *For a class $\mathscr{K}$ of lattices ($\mathbf{P_S}$ and $\mathbf{P_U}$ are the subdirect product and ultra product operators, see Section G.1),*

$$\mathbf{Var}(\mathscr{K}) = \mathbf{P_S\,H\,S\,P_U}(\mathscr{K}).$$

(2) *For a finite class $\mathscr{K}$ of finite lattices (**Si** stands for the subdirectly irreducible members),*

$$\mathbf{Si\,Var}(\mathscr{K}) \subseteq \mathbf{H\,S}(\mathscr{K}).$$

(3) *If $\mathscr{K}_1$ and $\mathscr{K}_2$ are lattice varieties and $\mathscr{K}_0 \vee \mathscr{K}_1$ is the smallest lattice variety containing both, then*

$$\mathbf{Si}(\mathscr{K}_0 \vee \mathscr{K}_1) = \mathbf{Si}\,\mathscr{K}_0 \cup \mathbf{Si}\,\mathscr{K}_1.$$

Theorem F.9.1(i) is called **Jónsson's lemma** in the literature. Its importance lies in the fact that the construction $\mathbf{P_U}$ preserves many more properties than does $\mathbf{P}$. Theorem F.9.1(ii) and Theorem F.9.1(iii) are, thus, easy consequences.

All varieties of lattices form a lattice Λ; by Theorem F.9.1(iii), this lattice is distributive. The zero of this lattice is $\mathbf{T}$ (the trivial variety), consisting of all one-element lattices, and the unit element is $\mathbf{L}$, the class of all lattices. The lattice Λ is sketched in Figure 2. How big is the lattice Λ? Since there are only $\aleph_0$ identities, there are at most $2^{\aleph_0}$ varieties.

F.9.2 Theorem *There are $2^{\aleph_0}$ varieties of (modular) lattices.*

This result was proved by K. A. Baker, R. N. McKenzie (without modularity), and R. Wille 1967–1972.

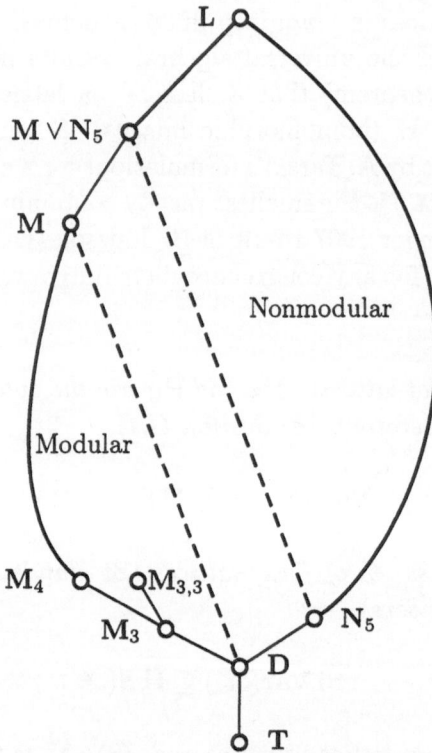

Figure 2: The lattice Λ

The bottom of Λ

Theorem F.9.1(ii) is the main tool for investigating the *bottom* of Λ. The variety of distributive lattices, **D**, is the only atom of Λ, and every nonzero member contains it. **D** is covered by exactly two varieties, $\mathbf{N_5}$ (generated by the five-element nonmodular lattice N_5) and $\mathbf{M_3}$ (generated by the five-element modular nondistributive lattice M_3), and every variety properly containing **D** contains $\mathbf{N_5}$ or $\mathbf{M_3}$. Let M_4 be like M_3 but with one more atom, and let $M_{3,3}$ be two copies of M_3 glued together over an edge; then it follows from Theorem F.9.1(ii) that $\mathbf{M_4} := \mathbf{Var}\, M_4$ and $\mathbf{M_{3,3}} := \mathbf{Var}\, M_{3,3}$ cover $\mathbf{M_3}$ in Λ. The following result is due to G. Grätzer and B. Jónsson (1966, 1968):

F.9.3 Theorem $\mathbf{M_3}$ *is covered by* $\mathbf{M_4}$ *and* $\mathbf{M_{3,3}}$. *If* $\mathscr{K}$ *is any variety of modular lattices and* $\mathbf{M_3} \subset \mathscr{K}$, *then* $\mathbf{M_4} \subseteq \mathscr{K}$ *or* $\mathbf{M_{3,3}} \subseteq \mathscr{K}$.

In 1970, R. N. McKenzie exhibited a list of 16 covers of $\mathbf{N}_5$. B. Jónsson and I. Rival in 1979 verified that the list is complete:

F.9.4 Theorem *In Λ, the variety $\mathbf{N}_5$ has 16 covers and any variety properly containing $\mathbf{N}_5$ contains one of these covers.*

There is a very large number of results describing the covering relations near the bottom of Λ, see the references in P. Jipsen and H. Rose for detail.

It follows immediately from Theorem F.9.1(ii) that a finite lattice generates a variety which is of finite height in Λ. It was a long standing conjecture that the converse may also be true. This was disproved in 1996 by J. B. Nation:

F.9.5 Theorem *There is a variety $\mathscr{K}$ of finite height in Λ that cannot be generated by a finite lattice.*

Finitely defining varieties

$\mathbf{L}$ is generated by its finite members (R. A. Dean 1956), in fact (A. Day 1977),

F.9.6 Theorem *The variety $\mathbf{L}$ is generated by the finite bounded homomorphic images of the free lattice on $\aleph_0$ generators.*

However, R. Freese in 1979 proved the following:

F.9.7 Theorem *The variety $\mathbf{M}$ of all modular lattices is not generated by its finite (dimensional) members.*

A variety is **finitely based** if it can be defined by a finite set of identities. The following result is due to R. N. McKenzie (1970):

F.9.8 Theorem *For any finite lattice K, the variety $\mathbf{Var}\, K$ is finitely based.*

(The analogous statement holds in any congruence distributive variety of algebras by K. A. Baker.) Also, $\mathbf{M} \vee \mathbf{N}_5$ is finitely based by a 1974 result of B. Jónsson.

Lattices generating non-finitely-based varieties were constructed by K. A. Baker, R. Freese, C. Herrmann, R. N. McKenzie, and R. Wille.

B. Jónsson showed in 1974 that the join of two finitely based lattice varieties need not be finitely based, and K. A. Baker did the same for two finitely based modular varieties.

The amalgamation property

A $\mathbb{V}$-*formation* (A, B_0, B_1) in a variety $\mathscr{K}$ is a pair of lattices B_0 and B_1 in $\mathscr{K}$ with a lattice $A \in \mathscr{K}$ that is a sublattice of both B_0 and B_1. The $\mathbb{V}$-formation (A, B_0, B_1) is *amalgamated by* $C \in \mathscr{K}$ if C contains B_0 and B_1 as sublattices. A variety $\mathscr{K}$ has the *Amalgamation Property* iff every $\mathbb{V}$-formation can be amalgamated.

L, **D**, and **T** have the Amalgamation Property. That these three are the only lattice varieties with the Amalgamation Property was a long standing conjecture, confirmed in two steps: for modular varieties by G. Grätzer, B. Jónsson, and H. Lakser in 1973, and for nonmodular varieties by A. Day and J. Ježek in 1984.

F.9.9 Theorem *A lattice variety $\mathscr{K}$ has the amalgamation property iff $\mathscr{K}$ is one of* **L**, **D**, *and* **T**.

Following G. Grätzer and H. Lakser, for a variety $\mathscr{K}$ of lattices, let **Amal**$(\mathscr{K})$ be the class of all those $A \in \mathscr{K}$ for which all $\mathbb{V}$-formations in $\mathscr{K}$ of the form (A, B_0, B_1) can be amalgamated in $\mathscr{K}$. Obviously, $\mathscr{K}$ has the Amalgamation Property iff **Amal**$(\mathscr{K}) = \mathscr{K}$. The following result of M. Yasuhara from 1974 is quite surprising.

F.9.10 Theorem *In a lattice variety $\mathscr{K}$, every lattice A in $\mathscr{K}$ can be embedded in a lattice B in* **Amal**$(\mathscr{K})$.

So, for instance, **Amal(M)** is huge. We have yet to discover a single nontrivial member of **Amal(M)**.

F.10 Applications of Lattices: Formal Concept Analysis

by Rudolf Wille in Darmstadt, Germany

Lattice theory and its applications

The rise of *lattice theory* in the 19th century had two main origins: Schröder's *algebra of logic* and Dirichlet's *algebraic number theory* (Mehrtens 1979). Eventually in the 1930's, lattice theory became established as a mathematical discipline in the spirit of *modern algebra*, mainly by the pioneering work of Birkhoff (1967). Besides the broad development of the internal theory since then, results on lattices have been applied in different parts of mathematics, but only little out-side mathematics. Only when *computer science* reached its eminent significance in our world, lattice theory found substantial applications in non-mathematical fields. Since more than thirty years the *semantics of programming languages* is such an application domain for which D. Scott did the initial work in creating *denotational semantics* based on continuous lattices and more general structures (Scott 1993, Abramsky, Gabbay, and Maibaum 1994). Here we concentrate, because of the lack of space, on a second field of rich lattice-theoretical applications, namely *formal concept analysis* which has been developed in the last twenty years mainly to support *data analysis* and *knowledge processing* (Ganter and Wille 1999, Stumme and Wille 2000).

Mathematical concept analysis

The mathematical foundation of *Formal Concept Analysis* (Wille 1982) is grounded on the primitive notion of a **formal context** understood as an incidence structure $\mathbb{K} := (G, M, I)$ for which G and M are sets while I is a binary relation between G and M; the elements of G and M are called **objects** (in German: *Gegenstände*) and **attributes** (in German: *Merkmale*), respectively, and gIm is read: the object g **has** the attribute m. A Galois connection between the power sets of G and M are given by the following **derivation operators** ($X \subseteq G$ and $Y \subseteq M$):

$$X \quad \mapsto \quad X' := \{ m \in M \mid gIm \text{ for all } g \in X \},$$
$$Y \quad \mapsto \quad Y' := \{ g \in G \mid gIm \text{ for all } m \in Y \}.$$

A **formal concept** of a formal context $\mathbb{K} := (G, M, I)$ is a pair (A, B) with $A \subseteq G$, $B \subseteq M$, $A = B'$, and $B = A'$; A and B are called the **extent** and the **intent** of the formal concept (A, B), respectively. The **subconcept-superconcept-relation** is formalized by

$$(A_1, B_1) \leq (A_2, B_2) :\Longleftrightarrow A_1 \subseteq A_2 \ (\Longleftrightarrow B_1 \supseteq B_2).$$

The set of all formal concepts of $\mathbb{K}$ together with the defined order relation is denoted by $\mathcal{B}(\mathbb{K})$.

A general method of constructing formal concepts uses the derivation operators to obtain, for $X \subseteq G$ and $Y \subseteq M$, the formal concepts (X'', X') and (Y', Y''). For an object $g \in G$, its **object concept** $\gamma g := (\{g\}'', \{g\}')$ is the smallest concept in $\mathcal{B}(\mathbb{K})$ whose extent contains g and, for an attribute $m \in M$, its **attribute concept**

$\mu m := (\{m\}', \{m\}'')$ is the greatest concept in $\underline{\mathcal{B}}(\mathbb{K})$ whose intent contains m.

F.10.1 Basic theorem on concept lattices *Let $\mathbb{K} := (G, M, I)$ be a formal context. Then $\underline{\mathcal{B}}(\mathbb{K})$ is a complete lattice, called the **concept lattice** of $\mathbb{K}$, for which infima and suprema can be described by*

$$\bigwedge_{t \in T}(A_t, B_t) = (\bigcap_{t \in T} A_t, (\bigcup_{t \in T} B_t)''),$$

$$\bigvee_{t \in T}(A_t, B_t) = ((\bigcup_{t \in T} A_t)'', \bigcap_{t \in T} B_t).$$

*In general, a complete lattice L is isomorphic to $\underline{\mathcal{B}}(\mathbb{K})$ if and only if there exist mappings $\tilde{\gamma} : G \longrightarrow L$ and $\tilde{\mu} : M \longrightarrow L$ such that $\tilde{\gamma}G$ is **supremum-dense** in L (i.e., $L = \{\bigvee X \mid X \subseteq \tilde{\gamma}G\}$), $\tilde{\mu}M$ is **infimum-dense** in L (i.e., $L = \{\bigwedge X \mid X \subseteq \tilde{\mu}M\}$), and $gIm \Longleftrightarrow \tilde{\gamma}g \leq \tilde{\mu}m$ for $g \in G$ and $m \in M$; in particular, $L \cong \underline{\mathcal{B}}(L, L, \leq)$, i.e., the concept lattices are, up to isomorphism, the complete lattices.*

The following theorem stands as an example of the many applications of formal concept analysis within mathematics:

F.10.2 Theorem (Wille 1991) *For a finite-dimensional vector space V, its dual V^*, and a scalar $r \neq 0$, let $\mathbb{K}(V) := (V, V^*, \perp_r)$ be the formal context with $v \perp_r \varphi :\Leftrightarrow \varphi(v) = r$. Then the derivation operators of $\mathbb{K}(V)$ yield inverse antiisomorhisms between the lattices consisting of the total space of V resp. V^* and of all affine subspaces not containing zero; in particular, if $V = \mathbb{R}^n = V^*$, $v \perp_r w :\Leftrightarrow v \cdot w = r$, and $r > 0$, the derivation operators yield the well-known inversion in the hypersphere of radius $\sqrt{r}$ and center 0 in the euclidean space $\mathbb{R}^n$.*

Applications of formal concept analysis

The driving force of most real world applications of formal concept analysis is due to the method of effectively representing concept lattices by *labelled line diagrams*, which particularly support exploring, analyzing, and processing of the information coded in the given formal context. The basic idea is to attach in a line diagram of a concept lattice the name of each object g to its represented object concept γg and the name of each attribute m to its represented attribute concept μm; then, by the basic theorem, extents, intents, and the underlying formal context can be recovered from the diagram and the relationships of the represented concepts become transparent.

How this can be performed shall be demonstrated by an example taken from the *evaluation of repertory tests* in the treatment of anorectic patients (Spangenberg 1990). The formal context, represented in Figure 3 by a cross-table, codes the information given by a patient during a test.

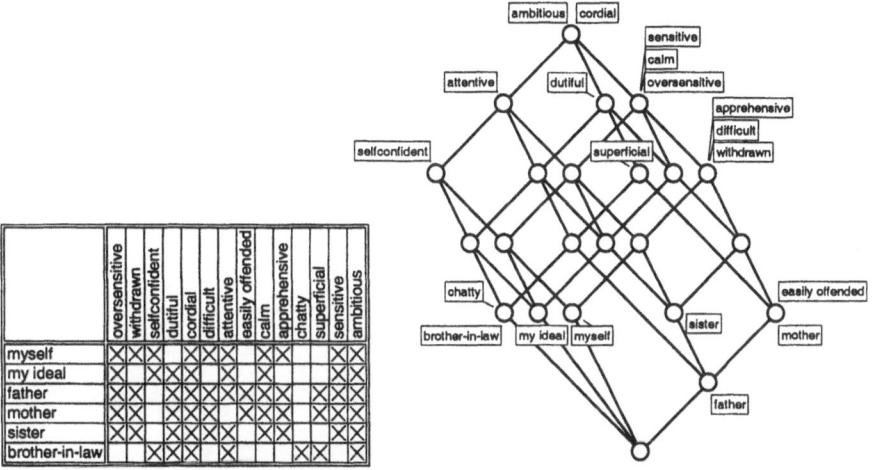

Figure 3: A context and its concept lattice used for the evaluation of a test concerning an anorectic patient

This information is discussed with the patient so that she becomes conscious of conflicts which might cause her anorexia. The diagram of the concept lattice unfolding the test information is used to support this discussion. For instance, the attributes *apprehensive, difficult,* and *withdrawn* at the supremum of the object concepts of *myself, sister, father,* and *mother* show how negative the patient views her family and, seeing this materialized in the diagram, the patient might gain more consciousness about that view and its reasons.

Formal concept analysis has been *academically* and *commercially* applied in a wide range of domains such as medicine, biology, psychology, musicology, archeology, linguistics, politics, sociology, economics, commerce, law, civil and industrial engineering, library and information science, computer science, and mathematics. In those applications, the main achievements of concept lattices are due to the support of *general tasks* like exploring, searching, recognizing, identifying, analyzing, investigating, deciding, improving, restructuring, and memorizing (Wille

1999), for which suitable software has been developed. The broad success in supporting general tasks is founded on the *mathematization of concepts* as basic units of thought constituted by their extension and intension. This mathematization opens a door for activating mathematical notions, results, and methods within human thinking in a wide range; it may therefore be understood as a first step in mathematizing the traditional philosophical logic, understood as the doctrine of the forms and functions of thinking based on concepts, judgments, and conclusions, which is leading to a Contextual Logic (Wille 2000) and thereby to an even broader field for applications of lattices.

Chapter G

UNIVERSAL ALGEBRA

G.1 Constructions in Universal Algebras

by Vyacheslav A. Artamonov in Moscow, Russia, and

Günter F. Pilz in Linz, Austria

An *n-**ary** operation* ($n \in \mathbb{N}_0$) on a set A is a map $\omega\colon A^n \to A$, where $A^0 := \{\varnothing\}$. The number n is called the **arity** of ω. A **universal algebra** is a pair $\mathcal{A} = (A, \Omega)$ consisting of a non-empty set A and a set Ω of operations on A. The set A and the members of Ω are called the **universe** and the **fundamental operations** of the algebra $\mathcal{A}$, respectively. In practice, one usually is not interested in a single, isolated algebra but in a class of algebras of the *same type*. Therefore it is more customary to consider the set Ω not as the set of operations on the given set A but rather as the set of operation symbols. Formally this is achieved by first introducing the notion of type. The **type** is a set Ω together with a partition $\Omega = \Omega_0 \cup \Omega_1 \cup \Omega_2 \cdots$. (Empty Ω_i are allowed.) Alternatively, the type is a set Ω together with a mapping $r\colon \Omega \to \mathbb{N}_0$. A universal algebra of type Ω, or simply an Ω-algebra is a pair $\mathcal{A} = (A, \Omega)$ where A is a non-empty set and to every $\omega \in \Omega_n$ it is assigned an n-ary operation on A, denoted by the same symbol ω. Cf. Section G.10 for a generalization of these concepts.

Usually, universal algebras are called just *algebras*, and the universe of algebras $\mathcal{A}, \mathcal{B}, \ldots$ is denoted by $A, B, \ldots$, respectively.

It is clear how *subalgebras, direct products* $\prod_{i \in I} \mathcal{A}_i$, *homomorphisms* (between algebras of the same type), *congruences*, and *factor algebras* (w.r.t. congruences) are defined. Every subset of the universe of the algebra $\mathcal{A}$ which is closed with respect to all operations $\omega \in \Omega$ is called a **subuniverse** of $\mathcal{A}$. The empty set is a subuniverse iff $\Omega_0 \neq \varnothing$. The non-empty subuniverses of $\mathcal{A}$ are precisely universes of subalgebras of $\mathcal{A}$.

It is less clear how a certain class of algebras should be defined in the terms above. For instance, a group $\mathcal{G}$ can be defined as an algebra of type (2), where $\Omega = (\circ)$ consists of a single binary operation, or as one of type $(2,0)$, where $\Omega = (\circ, \mathbf{1})$ and $\mathbf{1}\colon G^0 \to \{1\}$. Finally, a group can also be regarded as an algebra of type $(2,0,1)$, if we add the inversion $^{-1}\colon g \to g^{-1}$. In these cases, the subalgebras of $\mathcal{G}$ are in the subsemigroups, submonoids, and subgroups, respectively; hence the third version is usually the preferred one. An $\mathcal{R}$-module $\mathcal{M}$ can be considered as an algebra of type $(2,0,1)\cup(1)_{m\in M}$, where we add all unary operations $x \to mx$ $(m \in M)$ to the group structure of $(M,+)$. Rings $\mathcal{R}$ with identity can be considered as algebras of type $(2,0,1,2)$ or of type $(2,0,1,2,0)$ in the obvious way. In the latter case, the subalgebras are the subrings with the same identity as $\mathcal{R}$ (thus $\{0\}$, as well as any ideal different from R, are not subalgebras), and homomorphisms must preserve the identity.

In the case of Ω-*groups*, i.e., additive groups with (possibly) additional operations ω such that $\omega(0,\dots,0) = 0$, we have the **direct sum** $\bigoplus_{i\in I}\mathcal{A}_i$ of algebras $\mathcal{A}_i$ (of the same type), consisting of all tuples $(\dots, a_i, \dots)$ in which almost all $a_i = 0$ (i.e., with a finite number of exceptions).

By far not every algebra can be decomposed into a direct sum (or product) of *indecomposable* algebras, but in another context, a more general decomposition is always possible. A subalgebra $\mathcal{A}$ of $\prod_{i\in I}\mathcal{A}_i$ is called a **subdirect product** of the $\mathcal{A}_i$ if all projections $\mathcal{A} \to \mathcal{A}_i$ are surjective, i.e., if each $a_i \in A_i$ appears in some tuple $(\dots, a_i, \dots) \in A$ at least once. Direct sums of Ω-groups are examples of subdirect products. An algebra $\mathcal{A}$ is a subdirect product of algebras $\mathcal{A}_i$ iff there is a family $(\Theta_i)_{i\in I}$ of congruences of $\mathcal{A}$ whose intersection is the equality. Then, with proper indexing, $\mathcal{A}_i \simeq \mathcal{A}/\Theta_i$ for all i. We call $\mathcal{A}$ **subdirectly irreducible** if in each of its subdirect decomposition at least one $\mathcal{A}_i$ has to be isomorphic to $\mathcal{A}$. Equivalently, this is the case iff $\mathcal{A}$ has a least congruence $\neq$ identity.

G.1.1 Theorem (Birkhoff) *Every algebra is isomorphic to a subdirect product of subdirectly irreducible algebras (of the same type).*

This result is equivalent to the Axiom of Choice. Other constructions can be obtained as follows. The factor algebra $\prod_{i\in I}\mathcal{A}_i/\bigoplus_{i\in I}\mathcal{A}_i$ of Ω-groups is a so-called **filter product**, since two elements $[(\dots, a_i, \dots)]$ and $[(\dots, b_i, \dots)]$ in the factor are equal iff $\{i \mid a_i = b_i\}$ is an element of the cofinite filter on I. Especially important are ultraproducts: Take an ultrafilter $\mathscr{F}$ on the index set I and form the congruence $\sim_{\mathscr{F}}$ on $\prod_{i \in I} A_i$ via

$$(\dots, a_i, \dots) \sim_{\mathscr{F}} (\dots, b_i, \dots) : \iff \{ i \mid a_i = b_i \} \in \mathscr{F}.$$

Then $\prod_{i \in I} A_i / \sim_{\mathscr{F}} =: \prod_{\mathscr{F}} A_i$ is called the **ultraproduct** of the algebras A_i w.r.t. $\mathscr{F}$. If $\mathscr{F} = \{ J \subseteq I \mid i_0 \notin J \}$ for some fixed i_0, then $\mathscr{F}$ is called **fixed** or **principal**, and the corresponding ultraproduct is isomorphic to A_{i_0}, hence uninteresting. Non-fixed ultrafilters exist by Zorn's Lemma (for instance, by extending the cofinite filter on $\mathbb{N}$ to a maximal one), but nobody has ever seen one. Nevertheless, they are very attractive due to

G.1.2 Theorem (Łos) *A (first-order) formula φ holds in $\prod_{\mathscr{F}} A_i$ iff $\{ i \in I \mid \varphi$ holds in $A_i \} \in \mathscr{F}$.*

As a consequence, an ultraproduct of fields is again a field. Less trivial are the next results which heavily use ultraproducts; they should give a flavor of typical results.

G.1.3 Theorem

(1) *A non-fixed ultraproduct of the fields $\mathbb{F}_p$ $(p \in \mathbb{P})$ has characteristic 0.*
(2) *Ultraproducts of general linear groups are again general linear.*
(3) *Ultraproducts of primitive rings are primitive.*
(4) *(Robinson-Amitsur) If a prime ring $\mathcal{R}$ can be embedded in a product of skew fields then $\mathcal{R}$ can be embedded in a skew-field.*
(5) *(Sabbagh) There does not exist a set of first-order sentences which characterizes the multiplicative groups of fields.*

Non-fixed products are quite big: a countable non-fixed ultraproduct of countable algebras is uncountable. For this and many more results, see, e.g., (Eklof 1977).

There are some other useful constructions generalizing direct products. Let $\mathscr{C}$ be a category of algebras of the same type whose objects $\{ A_i \mid i \in I \}$ are indexed by a set I. A **direct limit** $\varinjlim \mathscr{C}$ is an algebra $A \in \mathscr{C}$ together with the family of homomorphisms $f_i \colon A_i \to A$ such that

1. if $g \colon A_i \to A_j$ is a morphism in $\mathscr{C}$ then $f_j g = f_i$;

2. (the universality property) if there exists a family of morphisms $f_i' \colon A_i \to A'$ for some algebra A' such that $f_j' g = f_i'$ for any morphism in $\mathscr{C}$ then there exists a unique homomorphism $f' \colon A \to A'$ such that $f' f_i = f_i'$ for any i.

Dually, one can introduce the notion of an *inverse limit* $\varprojlim \mathscr{C}$ by inverting all arrows in the definition of a direct limit. In the category of all Ω-algebras, there exist unique (up to isomorphisms) direct and inverse limits of any subcategory $\mathscr{C}$. Inverse limits of (non-void) algebras might be void, but this cannot happen if the algebras involved are finite (this is part of **König's Lemma**, see (Grätzer 1979). In the following lines, it is assumed that the described concepts exist.

If the category $\mathscr{C}$ contains only identical morphisms then the inverse limit $\varprojlim \mathscr{C}$ is precisely the direct product $\prod_{i \in I} \mathcal{A}_i$. If the set I is linearly ordered, then all morphisms $f_{ij} \colon \mathcal{A}_i \to \mathcal{A}_j$ in $\mathscr{C}$ are defined iff $i \leq j$ and all of them are embeddings then the direct limit $\varinjlim \mathscr{C}$ is the union of all algebras $\bigcup_{i \in I} \mathcal{A}_i$. If the category $\mathscr{C}$ contains only identical morphisms of algebras $\{\, \mathcal{A}_i \mid i \in I \,\}$ then the direct limit $\varinjlim \mathscr{C}$ is called a *free product* or a *coproduct* of algebras $\{\, \mathcal{A}_i \mid i \in I \,\}$, denoted by $\bigstar_{i \in I} \mathcal{A}_i$ or by $\coprod_I \mathcal{A}_i$. In the category of $\mathcal{R}$-modules, the free product of modules $\mathcal{A}_i$ coincides with direct sum $\bigoplus_{i \in I} \mathcal{A}_i$.

A category $\mathcal{C}$ of algebras is an *amalgam* if some of its objects are intersections $\mathcal{A}_{i_1} \cap \cdots \cap \mathcal{A}_{i_m}$ of objects $\mathcal{A}_{i_j} \in \mathscr{C}$ and non-identical morphisms in $\mathscr{C}$ are embeddings of intersections. An *amalgamated product* (cf. Section B.10) is a direct limit $\varinjlim \mathscr{C}$ of this category $\mathscr{C}$. It is interesting to determine whether each morphism $f_i \colon \mathcal{A}_i \to \varinjlim \mathscr{C}$ is an embedding. For example, this is the case in the class of all groups when $\mathscr{C}$ consists of three objects $\mathcal{A}_1, \mathcal{A}_2, \mathcal{A}_1 \cap \mathcal{A}_2$.

Finally, an important construction is the one of *terms*:

G.1.4 Definition Let X be a set (which is considered disjoint from Ω). A *term* of type Ω over X by the following inductive way:

(1) each element of $X \cup \Omega_0$ is a term;

(2) if $\omega \in \Omega_n$ and $t_1, \ldots, t_n$ are terms, then the *string* $\omega(t_1, \ldots, t_n)$ is a term.

In the class of commutative rings with identity, the terms coincide with the polynomials with integer coefficients. For Boolean algebras (with the usual collection of operations), terms and polynomials agree completely. On the other hand, for lattices (with two binary operations only) and $X = \{x\}$, not very much can be built from x by $\cap$ and $\cup$: only expressions like $(x \cup (x \cap (x \cap x)))$.

The set $T_\Omega(X)$ of all terms forms again an algebra in $\mathscr{K}(\tau)$, which turns out to be the free algebra in $\mathscr{K}(\tau)$ (see Section G.4 and Section G.3).

Much more can be found in (Cohn 1981, Grätzer 1979), for example.

G.2 Automorphisms and Endomorphisms in Universal Algebra

by Boris I. Plotkin in Jerusalem, Israel

The notions of automorphisms and endomorphisms are naturally defined in terms of universal algebra. Besides, automorphisms and endomorphisms are widely used in the applications of universal algebra to computer science and other fields.

Let us fix a **signature** (a family of symbols of operations) Ω. Algebras with the signature Ω are called Ω-**algebras**. A mapping of Ω-algebras $\varphi\colon H \to H'$ is called a **homomorphism** if it commutes with all operations from Ω. This means that if $\omega \in \Omega$ is an n-ary operation then

$$\varphi(\omega(a_1,\ldots,a_n)) = \omega(\varphi(a_1),\ldots,\varphi(a_n))$$

If φ is a bijection then φ is called an **isomorphism**. Similarly, the notions of homomorphism and isomorphism are defined for arbitrary algebraic systems where not only operations but also relations are considered. If $H = H'$ then homomorphisms and isomorphisms are called **endomorphisms** and **automorphisms**, respectively. See also Section G.1.

Universal algebra often deals with **axiomatized classes** of algebraic systems, i.e., with classes defined by first-order formulas. We often consider groups of automorphisms and semigroups of endomorphisms rather than individual automorphisms and endomorphisms.

G.2.1 Definition An arbitrary homomorphism $\rho\colon G \to \mathrm{Aut}(H)$ is called a **representation** of the group G in the group of automorphisms $\mathrm{Aut}(H)$. A representation is called **faithful** if the homomorphism ρ is injective.

A representation of a semigroup G in the semigroup of endomorphisms $\mathrm{End}(H)$ is defined similarly.

One of the principal problems is to find out which abstract groups admit faithful representations as a group of automorphisms of an algebra from the given axiomatized class. Let us mention the main theorems on this problem.

G.2.2 Theorem (Mostowski-Ehrenfeucht) *Let Θ be an axiomatized class of algebras, and let G a group which acts faithfully and order-preserving on some ordered set I. If Θ contains infinite systems then*

*there exists $H \in \Theta$, $I \subset H$, with a faithful representation $G \to \mathrm{Aut}(H)$
which extends the action of G on the set I.*

This theorem says that in Θ there are systems H with highly non-trivial groups of automorphisms. Automorphisms move elements from H to other elements of H with the same properties. Therefore, the axioms of the class Θ cannot individualize elements in algebraic systems in Θ.

In the following theorem, the class Θ is not arbitrary and is determined by some special set of axioms (see Plotkin 1966, Mal'cev 1973).

G.2.3 Theorem (Mal'cev) *Let G be a group and let every finitely generated subgroup of G have a faithful representation as a group of automorphisms of an algebra of Θ. Then the group G also has such a representation.*

Long ago, G. Birkhoff showed part (1) of

G.2.4 Theorem
(1) (Birkhoff) *Every group is the group of all automorphisms of some algebra.*
(2) (deGroot) *Every group is the group of all automorphisms of some ring.*
(3) (Sabidussi) *Every group is the group of all automorphisms of some graph.*

All these and similar facts can be found in (Plotkin 1966).

For a given category C consider the new category denoted by Rep-C. Its objects are representations of groups $\rho: G \to Aut(H)$, where H is an object in C and G is an arbitrary group. One can treat these objects as triples (H, G, ρ). Morphisms in Rep-C have the form

$$\mu = (\alpha, \beta): (H, G, \rho) \to (H', G', \rho'),$$

where $\alpha: H \to H'$ is a morphism in C, $\beta: G \to G'$ is a homomorphism of groups and α, β are **coordinated**. This means that for every $g \in G$ the following diagram is commutative:

$$
\begin{array}{ccc}
H & \xrightarrow{\alpha} & H' \\
\downarrow{\scriptstyle\rho(g)} & & \downarrow{\scriptstyle\rho'\beta(g)} \\
H & \xrightarrow{\alpha} & H'
\end{array}
$$

We consider here the definition of C for an arbitrary category; but for the sake of simplicity, one might think of C as the category of modules over a commutative ring with identity. If the category C is a variety of algebras Θ (see Section G.5) then we come to a new variety of algebras Rep-Θ. This is the variety of two-sorted algebras (H, G). The representation ρ is replaced by the action $\circ : H \times G \to H$ according to the rule $(a \circ g) = \rho(g)(a)$, where $a \in H$, $g \in G$. In the variety Rep-Θ, one can consider axiomatized classes of representations of groups.

G.2.5 Definition Two representations (H_1, G_1, ρ_1) and (H_2, G_2, ρ_2) in Rep-Θ are called *similar* if the corresponding faithful representations are isomorphic.

G.2.6 Definition A class of representations $\mathfrak{X}$ is called *saturated* if $(H, G, \rho) \in \mathfrak{X}$ implies that every similar representation (H_1, G_1, ρ_1) also belongs to $\mathfrak{X}$.

For every saturated class of representations $\mathfrak{X}$, denote by $\overrightarrow{\mathfrak{X}}$ a class of all groups G having a faithful representation $(H, G, \rho) \in \mathfrak{X}$.

G.2.7 Theorem (Plotkin)

(1) *If $\mathfrak{X}$ is a variety or a quasivariety of representations, then $\overrightarrow{\mathfrak{X}}$ is a quasivariety of groups.*
(2) *If $\mathfrak{X}$ is a universal (i.e., quantifier-free) axiomatizable class of representations, then $\overrightarrow{\mathfrak{X}}$ is a universal axiomatizable class of groups.*

The proof of this theorem uses well known criteria of quasivarieties and universal classes (for details, see Plotkin and Vovsi 1983).

Universal algebra can be used to build database or knowledge base models, as well as to define the notion of informational equivalence of models. The algorithm of recognition of such equivalence is based on a specific Galois theory. In this theory, the semigroup of endomorphisms of a free algebra is involved in the signature of the corresponding logic, while the group of automorphisms of a data algebra plays the crucial role in the Galois theory. For applications of universal algebra in computer science see, for instance, (Beidar and Brešar 2001, Birkhoff and Lipson 1970, Plotkin 2000) and Section G.10.

G.3 Polynomials and Polynomial Completeness
by Kalle Kaarli in Tartu, Estonia

We shall denote the set of all terms of type Ω over X by $T_\Omega(X)$ (see Section G.1). If $X = \{x_1, \ldots, x_n\}$ then we write $T_\Omega(n)$ instead of $T_\Omega(X)$

and call its elements n-**ary terms** of type Ω. Let $\mathcal{A}$ be an Ω-algebra, $\mathbf{a} = (a_1, \ldots, a_n) \in A^n$, and let t be a term of type Ω over $X = \{\, x_1, \ldots, x_n \,\}$. Then we may substitute every occurrence of x_i in t by a_i, $i = 1, \ldots, n$, and find $t(\mathbf{a})$, the value of t in $\mathbf{a}$. As a result, the term t defines a function $t^{\mathcal{A}} \colon A^n \to A$. Such functions are called **term operations** on algebra $\mathcal{A}$. In particular, if $t = x_i$ then $t^{\mathcal{A}}$ is the i'th **projection operation** which assigns to the n-tuple $\mathbf{a}$ its i'th component a_i, and if $t = \omega \in \Omega_n$ then $t^{\mathcal{A}}$ is the basic operation $\omega^{\mathcal{A}}$. If $\mathcal{A}$ is a ring then the commutator function $[x_1, x_2] = x_1 x_2 - x_2 x_1$ is a term function $t^{\mathcal{A}}$ determined by the term $t = x_1 x_2 - x_2 x_1$.

If $\mathcal{A}$ is an Ω-algebra then we may extend Ω by nullary operation symbols, each for every element of A, and interpret the symbol corresponding to $a \in A$ as the same element a. The resulting algebra will be denoted by $\mathcal{A}^+$.

G.3.1 Definition The terms (term operations) of the algebra $\mathcal{A}^+$ are called **polynomials** (**polynomial operations**) of the algebra $\mathcal{A}$.

Alternatively, polynomial operations are precisely the ones obtainable from term operations by fixing some of the variables. Classical polynomials are polynomials of commutative rings with identity. Every polynomial operation of a vector space $\mathcal{A}$ can be written in the form $\alpha_1 x_1 + \cdots + \alpha_n x_n + a$ where α_i are scalars and a is a fixed element of A. An n-ary operation f on an algebra $\mathcal{A}$ is called **local term operation** if for any finite subset X of A^n there exists a term operation t of $\mathcal{A}$ such that the restrictions of f and $t^{\mathcal{A}}$ to X coincide. Local term operations of $\mathcal{A}^+$ are **local polynomial operations** of $\mathcal{A}$.

G.3.2 Definition An algebra $\mathcal{A}$ is said to be *(locally) primal* if all finitary operations on A are (local) term operations. An algebra $\mathcal{A}$ is said to be *(locally) functionally complete* if the algebra $\mathcal{A}^+$ is (locally) primal.

In earlier literature, functionally complete algebras were referred to as *polynomially complete*. Now the latter term is mainly used in a more general sense. An algebra is called **polynomially complete** if it is rich in polynomial or term operations, in some sense. Thus functional completeness and primality are two examples of polynomial completeness. In the sequel, some other examples are introduced. For more examples and results, see (Kaarli and Pixley 2001).

Usually primal algebras are assumed to be finite. The most common primal algebra is the 2-element Boolean algebra $\mathbb{B}$. The next theorem (see also Section F.6) says that any (finite) primal algebra is categorically equivalent to $\mathbb{B}$.

G.3.3 Theorem (***Hu's theorem***) *A finite algebra $\mathcal{A}$ is primal iff there exists a categorical equivalence between the variety generated by $\mathcal{A}$ and the variety of Boolean algebras under which $\mathcal{A}$ corresponds to $\mathbb{B}$.*

A useful criterion for primality is given by the following theorem.

G.3.4 Theorem (Pixley) *A finite algebra $\mathcal{A}$ is primal iff it is simple, has no proper subalgebras, has only one automorphism, and generates a variety which is **arithmetical**, that is, congruence permutable and congruence distributive.*

The most powerful primality criterion is the test due to Rosenberg (1970). This is given in terms of 6 ***Rosenberg classes*** each of which consists of special subsets of powers A^k. The first three of them are binary: 1) bounded (partial) orders, 2) nontrivial equivalence relations, 3) graphs of permutations with all cycles of equal prime length p. The fourth consists of the sets $\{\,(x, y, u, v) \in A^4 \mid x+y = u+v\,\}$ where $(A; +)$ is an elementary abelian group. The definitions of the two remaining classes need more space than we can afford here.

G.3.5 Theorem (Rosenberg) *A finite algebra $\mathcal{A}$ is primal iff none of the subuniverses of any finite power A^k belongs to one of Rosenberg classes.*

All locally functionally complete algebras are simple. In case of congruence permutable varieties the converse is *almost true*. An operation on an abelian group $\mathcal{A}$ is called **affine** if it is the sum of a homomorphism $A^k \rightarrow A$ and a constant. We call an algebra **affine over an abelian group** $\mathcal{A}$ if its universe is A and all basic operations (but then also all polynomial operations) are affine over $\mathcal{A}$.

G.3.6 Theorem (McKenzie, Gumm) *A simple algebra $\mathcal{A}$ in a congruence permutable variety is either locally functionally complete or affine over an abelian group which is either elementary or torsion free divisible.*

In particular, all (skew) fields and all simple nonabelian groups are locally functionally complete. All Galois fields and finite simple nonabelian groups are functionally complete. Surprisingly, functional completeness of a finite algebra can be tested with just a single operation. For any set A we put $d^A(x, x, z) = z$ and $d^A(x, y, z) = x$ if $x \neq y$. This operation d^A is called **ternary discriminator** on A.

G.3.7 Theorem (Werner) *A finite algebra $\mathcal{A}$ is functionally complete iff d^A is a polynomial operation of $\mathcal{A}$.*

Sometimes the ternary discriminator can be realized even by a term operation. Such finite algebras are said to be **quasiprimal**. For example, a ternary discriminator on $\mathrm{GF}(p^n)$ is realized by the operation $f(x, y, z) = z + (x + (p-1)z)(x + (p-1)y)^{p^n-1}$, so all Galois fields are quasiprimal. They are not primal, in general, because they may have nontrivial subfields. The prime field $\mathrm{GF}(p)$ is primal if $1 \in \Omega_0$. The following theorem shows that quasiprimality is a polynomial completeness property.

G.3.8 Theorem (Werner) *A finite algebra $\mathcal{A}$ is quasiprimal iff its term operations are exactly the operations on A which permute with all isomorphisms between subalgebras of $\mathcal{A}$.*

Natural generalizations of (local) functional completeness to nonsimple algebras are given in the next definition.

G.3.9 Definition An algebra $\mathcal{A}$ is called *(locally) affine complete* if an operation on A is a (local) polynomial of $\mathcal{A}$ iff it preserves all congruences of $\mathcal{A}$. An algebra $\mathcal{A}$ is called **strictly locally affine complete** if a partial operation f on A with finite domain $D \subseteq A^k$ is the restriction of some polynomial operation of $\mathcal{A}$ iff f preserves all congruences of $\mathcal{A}$.

Note that every affine complete algebra and every strictly locally affine complete algebra is locally affine complete, while the former two notions are incomparable.

(Locally) affine complete members have been described in several important classes of algebras. Here are some examples. More results and an extensive bibliography is available in Kaarli and Pixley 2001. A vector space is affine complete iff its dimension is not 1 (Werner). Every Boolean algebra is affine complete (Grätzer). A distributive lattice is locally affine complete iff it has no nontrivial Boolean intervals (Grätzer, Dorninger, Eigenthaler).

Strictly locally affine complete algebras were characterized by Hagemann and Herrmann (1982). Here we present their result for an important special case.

G.3.10 Theorem (Hagemann, Herrmann) *An algebra A in a congruence permutable variety is strictly locally affine complete iff it is neutral.*

Hence, a group G is strictly locally affine complete iff every normal subgroup of G coincides with its commutator subgroup, and a ring R is strictly locally affine complete iff every two-sided ideal of R coincides with its square. In particular, no soluble group or nilpotent ring can be strictly locally affine complete. On the other hand, von Neumann regular rings are strictly locally affine complete.

It follows from Hagemann and Herrmann (1982) that all strictly locally affine complete algebras are arithmetical. We conclude with a remarkable result of Pixley, which asserts that a variety is arithmetical iff all of its members are strictly locally affine complete. Important examples of arithmetical varieties are the varieties generated by quasiprimal algebras, in particular the variety of Boolean algebras, and the variety of lattice ordered groups.

G.4 Free Algebras

by Lev N. Shevrin and Evgeny V. Sukhanov in

Ekaterinburg, Russia

The concept of a free algebra is one of the most useful in abstract algebra. Its essence is the **universal mapping property** for algebraic objects, which is disclosed by the definition below.

Let $\mathscr{K}$ be a class of algebras of the same type. A $\mathscr{K}$-*algebra* is any algebra of $\mathscr{K}$. Let F be a $\mathscr{K}$-algebra and X be a generating set of F. Then F is called a **free algebra** in $\mathscr{K}$ (or a **free $\mathscr{K}$-algebra**) over X if for every mapping $\varphi \colon X \to A$ there exists a homomorphism $\bar{\varphi} \colon F \to A$ which extends φ (i. e., $\bar{\varphi}(x) = \varphi(x)$ for $x \in X$).

Here X is called a **free** (more precisely, $\mathscr{K}$-**free**) **base** for F, and F is said to be **freely generated by** X. If F is a free $\mathscr{K}$-algebra over X, then X is a minimal generating set of F, i. e., any proper subset of X does not generate F. The cardinal number $|X|$ is called the **rank** of F. The question about uniqueness of the rank of a free algebra will be discussed in the last paragraphs of the article.

It is straightforward from the definitions given above that if a class $\mathscr{K}$ possesses a free algebra of rank r, then any $\mathscr{K}$-algebra with a generating set of cardinality $\leq r$ is a homomorphic image of such an algebra. Hence if $\mathscr{K}$ possesses free algebras of every rank, then any $\mathscr{K}$-algebra is a homomorphic image of an appropriate free $\mathscr{K}$-algebra. A nontrivial (i. e., containing a nonsingleton algebra) class $\mathscr{K}$ has this property if it is closed under subalgebras and direct products. So any nontrivial variety and, more generally, quasivariety has free algebras of every rank. There is a substantial connection between free algebras and varieties (see Section G.5). In particular, if F is a free $\mathscr{K}$-algebra, then F is at the same time a free **Var** $\mathscr{K}$-algebra, where **Var** $\mathscr{K}$ is the variety generated by F. Free algebras in the variety of all algebras of a given type are called *absolutely free*. Free algebras in other varieties are termed *relatively free*.

A free $\mathscr{K}$-algebra over X is commonly denoted by $F_{\mathscr{K}}(X)$. Since two free $\mathscr{K}$-algebras over the same set X are isomorphic, and there is an isomorphism between them which acts identically on X, the algebra $F_{\mathscr{K}}(X)$ is determined up to such an isomorphism. If a class $\mathscr{K}$ in some considerations is fixed, then one may omit the letter $\mathscr{K}$ and write simply $F(X)$. If $|X| = |Y|$, then $F_{\mathscr{K}}(X) \cong F_{\mathscr{K}}(Y)$. This permits to denote a free $\mathscr{K}$-algebra of rank r by $F_{\mathscr{K}}(r)$ as well as by $F(r)$ or F_r if it is not so important to indicate its free base explicitly. This notation determines a corresponding algebra uniquely up to isomorphism. If $\mathscr{K}$-algebras have 0-ary operations (constants), then it makes sense to consider a free $\mathscr{K}$-algebra of rank 0; an algebra $F_{\mathscr{K}}(0)$ consists of elements corresponding to all these constants.

Many applications of free algebras use only the universal mapping property, the fact that they exist in varieties and their relation to the identities holding in varieties. However, besides this, it is natural to desire to have some transparent description of the structure of free algebras under consideration; such a description can be useful in different questions. Unfortunately, in many cases it is rather hard or even impossible to find such a description. In particular, free algebras in certain varieties of semigroups, groups, rings and lattices can be very complicated. But for quite a number of large *basic* classes of algebras, free objects have a transparent structure. Below we present several such classes from this point of view.

All algebras of a given type   Let $\mathscr{K}$ be the class (which is a variety) of all algebras of type Ω. The set of all terms (Section G.1) of type Ω

over X is denoted by $T_\Omega(X)$. So $T_\Omega(X)$ is the smallest set with the following properties: it contains $X \cup \Omega_0$; if $f \in \Omega_n$ and $p_1, \ldots, p_n$ belong to it, then it contains $f(p_1, \ldots, p_n)$ as well (for Ω_n, see again Section G.1).

From the definition of $T_\Omega(X)$ it is clear that, if $T_\Omega(X) \neq \varnothing$, the set $T_\Omega(X)$ turns into an algebra of type Ω. This algebra is called the **term algebra of type Ω over X**. The term algebra is just the absolutely free algebra of type Ω, i. e., in the case under consideration, $F_{\mathscr{K}}(X) = T_\Omega(X)$.

Semigroups For the variety of all semigroups, $F(X)$ is the semigroup of all (finite) words over X (see Section A.3).

Groups For the variety of all groups, a canonical presentation for $F(X)$ is the group of all reduced words over X (see Section B.9).

Vector spaces Let $\mathscr{K}$ be the class of all vector spaces over a fixed field $\Bbbk$. Such spaces can be considered as algebras of a type consisting of one binary operation symbol $+$ (which corresponds to addition) and a family of unary operation symbols (which correspond to scalar multiplication by elements from $\Bbbk$). Under this interpretation, $\mathscr{K}$ turns into a variety. Every algebra of this variety is free, and its rank coincides with its dimension. In order to have the right to speak about rank 0, we must include in the type 0-ary symbol corresponding to the zero of a space.

For the more general situation of modules, see Section C.20.

Rings For the variety of all associative and commutative rings with unit, $F(X)$ is the ring $\mathbb{Z}[X]$ of polynomials over $\mathbb{Z}$ in indeterminates from X. If we omit the condition of commutativity, then, instead of $\mathbb{Z}[X]$, we obtain the ring $\mathbb{Z}\langle X \rangle$ of polynomials in non-commuting indeterminates from X. See also Section C.23.

Lattices For the variety of all lattices, the structure of its free algebras is more complicated. But they can be described constructively. It was done by Whitman (1941). His algorithm effectively determines when two terms of the type of lattices represent the same element of a free lattice. It leads to a set of reduced terms, one for each element of the free lattice (see details in Section F.8).

Boolean algebras　　For the variety of all Boolean algebras, there is an extremely transparent description of the free algebras of finite rank: F_r is isomorphic to the Boolean algebra of all subsets of a 2^r-element set. So $|F_r| = 2^{2^r}$; conversely, if B is a Boolean algebra having 2^{2^r} elements, then B is free of rank r. Another representation of F_r is the algebra of all Boolean (switching) functions of r variables (see Section F.6).

As it was noted above, for any class $\mathcal{K}$ having free algebras, an algebra $F_{\mathcal{K}}(r)$ is uniquely determined up to isomorphism by its rank r. Is the converse also true? In other words, does the condition $F_{\mathcal{K}}(r) \cong F_{\mathcal{K}}(s)$ imply $r = s$? In many cases it is so; in particular, this holds for all varieties presented in the preceding paragraphs. Most of these cases can be covered by the following sufficient condition (Fujiwara 1955, Jónsson, B. and Tarski, A. 1961): if a variety $\mathcal{K}$ contains a finite nontrivial algebra, then free algebras of different ranks in $\mathcal{K}$ are not isomorphic. However, it is rather surprising that in general this is not the case. If $F_{\mathcal{K}}(r) \cong F_{\mathcal{K}}(s)$ and $r \neq s$, then r and s must be finite (Fujiwara 1955). There are varieties in which this "irregular" situation takes place. Moreover, there exists a variety in which even all free algebras of finite ranks are isomorphic. To indicate such a variety, let us consider a type consisting of three operation symbols: a binary one $\circ$ and two unary ones φ_1, φ_2. The variety of algebras of this type, given by the identities

$$\varphi_1(x \circ y) = x, \;\; \varphi_2(x \circ y) = y, \;\; \varphi_1(x) \circ \varphi_2(x) = x,$$

just has the property under discussion (Jónsson, B. and Tarski, A. 1961).

The indicated example can be regarded as a partial case of the following construction (Świerczkowski 1960). Let m and n be natural numbers such that $m < n$. Let us consider a type consisting of m-ary operation symbols $\varphi_1, \ldots, \varphi_n$ and n-ary operation symbols $\psi_1, \ldots, \psi_m$. In the variety of this type given by the identities

$$\varphi_i(\psi_1(x_1, \ldots, x_n), \ldots, \psi_m(x_1, \ldots, x_n)) = x_i,$$
$$\psi_j(\varphi_1(x_1, \ldots, x_m), \ldots, \varphi_n(x_1, \ldots, x_m)) = x_j,$$

where $i = 1, \ldots, n$ and $j = 1, \ldots, m$, the algebras F_r and F_s with different r and s are isomorphic if and only if $r \equiv s \pmod{n-m}$, $r \geq m$, $s \geq m$.

The identities just featured are in a certain sense characterizing the situation when free algebras of different ranks are isomorphic. Namely, in a class $\mathcal{K}$ of type Ω, the algebras F_r and F_s, with different r and s, are isomorphic if and only if there exist terms $f_i(x_1, \ldots, x_r)$, $i = 1, \ldots, s$, and $g_i(x_1, \ldots, x_s)$, $j = 1, \ldots, r$, of type Ω, such that in $\mathscr{K}$ the identities

$$f_i(g_1(x_1, \ldots, x_s), \ldots, g_r(x_1, \ldots, x_s)) = x_i, \quad i = 1, \ldots, s,$$
$$g_j(f_1(x_1, \ldots, x_r), \ldots, f_s(x_1, \ldots, x_r)) = x_j, \quad j = 1, \ldots, r,$$

are valid (Świerczkowski 1960).

Some further general information on free algebras can be found, for instance, in (Grätzer 1979, Ch. 4), (Burris and Sankappanavar 1981, Ch. II, §§10, 11), and (McKenzie, McNulty, and Taylor 1987, §4.11).

G.5 Varieties and Quasi-varieties

> *by Lev N. Shevrin and Mikhail V. Volkov in Ekaterinburg, Russia*

Basic definitions

Let Ω be a type and $T_\Omega(X)$ the corresponding term algebra over a set X (see Section G.4). An *identity* of type Ω is merely a pair of elements $u, v \in T_\Omega(X)$, usually written as $u = v$. The best known identities are certainly the commutative and the associative laws: $xy = yx$ and $(xy)z = x(yz)$.

A *quasi-identity* of type Ω is an expression of the form

$$u_1 = v_1 \ \& \ u_2 = v_2 \ \& \ \ldots \ \& \ u_n = v_n \implies u = v \qquad (1)$$

where $u_1, v_1, u_2, v_2, \ldots, u_n, v_n, u, v \in T_\Omega(X)$. Natural examples of quasi-identities are the cancellation laws, $x_1 y = x_2 y \implies x_1 = x_2$ and $xy_1 = xy_2 \implies y_1 = y_2$, which were arguably the very first quasi-identities considered in algebra.

Let A be an Ω-algebra, $u, v \in T_\Omega(X)$. Then A is said to **satisfy the identity** $u = v$ if every substitution of elements of A for variables from X yields equal values to the terms u and v. More formally, this means that $\varphi(u) = \varphi(v)$ for all homomorphisms $\varphi \colon T_\Omega(X) \to A$. Similarly, A **satisfies the quasi-identity** (1) if for every homomorphism $\varphi \colon T_\Omega(X) \to A$ such that $\varphi(u_1) = \varphi(v_1)$, $\varphi(u_2) = \varphi(v_2)$, $\ldots$, $\varphi(u_n) = \varphi(v_n)$, one also gets $\varphi(u) = \varphi(v)$. A class $\mathscr{C}$ of Ω-algebras is a *[quasi-]variety* if there is a set Σ of [quasi-]identities such that $\mathscr{C}$ consists of all Ω-algebras that satisfy all [quasi-]identities in Σ.

Since the identity $u = v$ is equivalent to the quasi-identity $x = x \implies u = v$, each variety is a quasivariety; the converse is not true. It is

clear from the very definition that many natural classes of groups, rings,
semigroups, lattices studied in abstract algebra since its early age form
varieties. Certain quasivarieties which are not varieties also arise quite
naturally: we mention, for example, the quasivariety of all torsion-free
groups or the quasivariety of all semigroups embeddable into a group.
Several sections of this book (A.15, B.21, C.37, C.51, F.9) are devoted
to varieties and identities of "classical" algebras. Here we concentrate
on a few general features which underlie the study of concrete varieties
and quasivarieties.

Characterization theorems

The theory of varieties has started with their characterization in terms of
class operators discovered by Birkhoff (1935). A similar characterization
of quasivarieties has been found by Mal'cev (1966) and then simplified
by Grätzer and Lakser (1973). The characterizations employ the opera-
tors $\mathbf{H}$, $\mathbf{S}$, $\mathbf{P}$, and $\mathbf{P_U}$ for formation of homomorphic images, subalgebras,
direct products, and ultraproducts, respectively (see Section G.1 for the
definition of an ultraproduct). In what follows, we adopt the standard
convention that every class of algebras is meant to consist of algebras
of the same type. For such a class $\mathscr{C}$, we denote by $\mathbf{Var}\,\mathscr{C}$ [$\mathbf{QVar}\,\mathscr{C}$]
the [quasi-]variety *generated by* $\mathscr{C}$, that is, the least [quasi-]variety
containing $\mathscr{C}$.

The HSP theorem (Birkhoff 1935) *A class $\mathscr{K}$ of algebras is a variety
if and only if $\mathscr{K}$ is closed under the operators $\mathbf{H}$, $\mathbf{S}$, and $\mathbf{P}$. For any
class $\mathscr{C}$, $\mathbf{Var}\,\mathscr{C} = \mathbf{H\,S\,P}\,\mathscr{C}$.*

The SPP$_U$ theorem (Grätzer and Lakser 1973) *A class $\mathscr{K}$ of algebras
is a quasivariety if and only if $\mathscr{K}$ is closed under the operators $\mathbf{S}$, $\mathbf{P}$,
and $\mathbf{P_U}$. For any class $\mathscr{C}$, $\mathbf{QVar}\,\mathscr{C} = \mathbf{S\,P\,P_U}\,\mathscr{C}$.*

For a collection of characterization theorems for varieties and quasivarieties in terms of other class operators, cf. (Gorbunov 1998, Section 2.3).

Free algebras and the role of congruences

As the proof of the HSP theorem reveals, every *non-trivial*
[quasi-]variety $\mathscr{K}$ (i.e., one which includes a non-singleton algebra) possesses a free algebra $F_{\mathscr{K}}(X)$ over any non-empty set X (see Section G.4).
In particular, every X-generated algebra in $\mathscr{K}$ is a homomorphic image
of $F_{\mathscr{K}}(X)$. If $\mathscr{K}$ is a variety, the converse is also true: each homomorphic image of a $\mathscr{K}$-free algebra lies in $\mathscr{K}$. Thus, in view of the

well-known correspondence between homomorphisms and congruences (see Section G.1), X-generated algebras in the variety $\mathscr{K}$ correspond to congruences of the free algebra $F_{\mathscr{K}}(X)$. This explains the fundamental role congruences play in the theory of varieties; in fact, the most important classes of varieties are distinguished by certain properties of congruences of their algebras such as *congruence-permutability*, *congruence-modularity* (see Section G.6), *congruence-distributivity* etc. If $\mathscr{K}$ is a quasivariety, then it is natural to impose similar restrictions to the lattice $\mathrm{Con}_{\mathscr{K}} A$ of $\mathscr{K}$-congruences on an algebra A rather than to the whole congruence lattice $\mathrm{Con}\, A$ (recall that $\theta \in \mathrm{Con}\, A$ is called a $\mathscr{K}$-*congruence* if A/θ belongs to $\mathscr{K}$). In this way, *relatively congruence modular/distributive* quasivarieties come into play and become objects of study, which to some extent parallel the profound theories of congruence modular/distributive varieties.

Lattices of varieties and quasivarieties

The concepts of a variety and of a quasivariety serve to create a convenient framework to classify algebras; conversely, in order to classify varieties and quasivarieties, it is natural to endow the collection of all sub[quasi-]varieties of a given [quasi-]variety $\mathscr{K}$ with an algebraic structure, namely, with the structure of a complete lattice (with respect to the class-theoretical inclusion). We denote this lattice by $L_{\mathrm{v}}(\mathscr{K})$ [respectively $L_{\mathrm{q}}(\mathscr{K})$]. If $\mathscr{K}$ is a variety, then $L_{\mathrm{v}}(\mathscr{K})$ is dually isomorphic to the lattice of all fully invariant congruences on the free $\mathscr{K}$-algebra $F_{\mathscr{K}}(X)$ over a countably infinite set X (recall that a congruence θ on an algebra A is said to be *fully invariant* if $(a, b) \in \theta$ implies $(\varphi(a), \varphi(b)) \in \theta$ for all $a, b \in A$ and for all endomorphisms φ of A). Therefore, the subvariety lattice of a congruence modular/distributive variety is modular/distributive; in particular, groups and rings have modular variety lattices, while the lattice of lattice varieties is distributive. Similarly, the lattice $L_{\mathrm{q}}(\mathscr{K})$ for a quasivariety $\mathscr{K}$ is dually isomorphic to the lattice of certain fully invariant subsets consisting of $\mathscr{K}$-congruences on $F_{\mathscr{K}}(X)$ with a countably infinite X—cf. (Gorbunov and Tumanov 1982), (Hoehnke 1986), (Gorbunov 1995) for various results of this kind.

The [*quasi-*]*equational theory* of a [quasi-]variety $\mathscr{K}$ is the collection of all [quasi-]identities that hold true in $\mathscr{K}$. Equational and quasi-equational theories also form complete lattices under the set-theoretical inclusion. This gives yet another way (preferred by some authors) to look at lattices of [quasi-]varieties since the lattice $L_{\mathrm{v}}(\mathscr{K})$ [respectively $L_q(\mathcal{K})$] is easily seen to be dually isomorphic to the lattice of all [quasi-]equational theories containing that of $\mathcal{K}$.

Sub[quasi-]variety lattices are *co-algebraic* (see Section F.7) and **atomic**. The latter property means that every non-trivial [quasi-]variety contains a minimal non-trivial sub[quasi-]variety, alias an **atom**. Minimal non-trivial varieties are sometimes called **equationally complete** (because the equational theory of such a variety is a maximal proper subtheory in the collection of all identities of the given type). Several well known classes of algebras—such as semilattices, distributive lattices, Boolean algebras—turn out to form equationally complete varieties.

The subquasivariety lattice of an arbitrary quasivariety $\mathcal{K}$ satisfies the following implication of **complete upper semidistributivity**:

$$(\forall\, i \in I : \mathscr{A} = \mathscr{B} \vee \mathscr{C}_i) \implies \mathscr{A} = \mathscr{B} \vee \bigwedge_{i \in I} \mathscr{C}_i, \qquad (2)$$

see (Gorbunov 1976). For further general properties of subquasivariety lattices cf. (Gorbunov 1998, Chapter 5). An implication similar to (2) holds also in the lattice $L_v(\mathcal{K})$ for an arbitrary variety $\mathcal{K}$ — one only has to add the premise that the meet $\bigwedge_{i \in I} \mathscr{C}_i$ is equal to the trivial variety $\mathscr{T}$ (Lampe 1986):

$$\bigwedge_{i \in I} \mathscr{C}_i = \mathscr{T} \ \& \ (\forall i \in I : \mathscr{A} = \mathscr{B} \vee \mathscr{C}_i) \implies \mathscr{A} = \mathscr{B}.$$

The problem (proposed by Birkhoff and Mal'tsev) of characterizing all lattices which are representable as $L_v(\mathcal{K})$ or $L_q(\mathcal{K})$ for a suitable [quasi-]variety $\mathcal{K}$ still remains open.

The finite basis property

Classifying algebras with respect to various properties of the [quasi-]varieties they generate constitutes—along with studying sub-[quasi-]variety lattices—one of the major directions in the theory of [quasi-]varieties. An important and intriguing property here is the finite basability. A [quasi-]variety is said to be **finitely based** if it can be defined by a finite system of [quasi-]identities; we call an algebra A **finitely [q]-based** if it generates a finitely based [quasi-]variety. Every 2-element algebra is both finitely based (Lyndon 1951) and finitely q-based (Rautenberg 1981, Gorbunov 1983), and, in a sense, "almost all" finite algebras are so (Murskiĭ 1975) because "almost all" of them are quasi-primal, see Section G.3. Surprisingly enough, there exists a

3-element groupoid [semigroup] which is not finitely [q]-based (Murskiĭ 1964, Sapir 1980). A finite algebra A is finitely [q]-based if it generates a [relatively] congruence distributive [quasi-]variety (Baker 1977, Pigozzi 1988). In general (and even for groupoids), the question of whether a given finite algebra is finitely based or not is algorithmically undecidable (McKenzie 1996); the corresponding question for finite q-basability still seems to remain open.

Further references

The monographs (Burris and Sankappanavar 1981, McKenzie et al. 1987, Smirnov 1989) nicely represent various aspects of the general theory of varieties, while the lecture notes (Taylor 1988) and the surveys (Taylor 1979, Bakhturin and Ol'shanskiĭ 1988) provide an overview and may serve as a guide for those approaching the area. For a comprehensive survey of algorithmic problems in varieties of "classical" algebras, we recommend (Kharlampovich and Sapir 1995).

G.6 Congruence Modular Varieties

by A. G. Pinus in Novosibirsk, Russia and Yefim Katsov in Hanover, IN, USA

Mal'tsev conditions

An algebra A is said to be ***congruence permutable*** iff $\alpha \circ \beta = \beta \circ \alpha$ for every two congruences α and β of A. Groups, rings, and modules are examples of congruence permutable algebras. An algebra whose congruence lattice is distributive (modular) is called ***congruence distributive (modular)***. Lattices constitute principal examples of congruence distributive algebras, which also include Boolean and Heyting algebras. The class of congruence modular algebras includes congruence distributive as well as congruence permutable algebras, and many of the deepest results in universal algebra involve algebras of those subclasses of the class of congruence modular algebras. A variety is ***congruence permutable*** (or ***congruence distributive, modular***) iff each algebra in the variety has this property. In 1954, A. I. Maltsev discovered the very important connections—***Maltsev conditions***—between properties of congruence lattices of algebras of a variety and the existence of (what are now often called) ***Maltsev terms*** satisfying special identities; and he obtained the classic result—a characterization of congruence permutable varieties in terms of Maltsev conditions. Similarly, the congruence distributive varieties were characterized by Jónsson (1995). As an illustration of the Maltsev conditions, we give B. Day's (1969) characterization of congruence modular varieties: A variety $\mathscr{V}$ is congruence modular iff there exist a natural number $n \geq 1$ and 4-ary terms $p_0, \ldots, p_n$ such that for $i = 0, \ldots, n - 1$ the identities $p_0(x, y, z, u) = x$, $p_n(x, y, z, u) = u$, $p_i(x, y, y, x) = x$, $p_i(x, y, y, u) = p_{i+1}(x, y, y, u)$ (for even i), and $p_i(x, x, u, u) = p_{i+1}(x, x, u, u)$ (for odd i) hold in $\mathscr{V}$. Using different Maltsev conditions, P. Gumm (1981) also provides a characterization of congruence modular varieties that explicitly distinguishes the roles played by the congruence permutability and congruence distributivity properties.

Commutator theory

As is well known, in the structure theory of algebras of a variety $\mathscr{V}$, the class $\mathscr{V}_{\mathrm{SI}}$ of subdirectly irreducible algebras of $\mathscr{V}$ is of great importance. Therefore, the following observation (*Jónsson's lemma* (1969)) is crucial: For a congruence distributive variety $\mathbf{Var}(\mathscr{K})$, generated by the class $\mathscr{K}$ of algebras, $\mathbf{Var}(\mathscr{K})_{\mathrm{SI}} \subseteq \mathbf{HSP_P}(\mathscr{K})$, where $\mathbf{HSP_U}(\mathscr{K})$ is the class of all homomorphic images of subalgebras of ultraproducts of some members of $\mathscr{K}$. Having observed this, Fraser and Horn (1970) established the following important result

G.6.1 Theorem *For any algebras A_1, A_2 in a congruence distributive variety $\mathscr{V}$, and a congruence $\theta \in \mathrm{Con}(A_1 \times A_2)$, there exist congruences $\theta_1 \in \mathrm{Con}\, A_1$ and $\theta_2 \in \mathrm{Con}\, A_2$ such that $\theta = \theta_1 \times \theta_2$.*

Influenced by these facts, the notion of the *commutator* was introduced—a commutative, completely join-preserving binary operation preserved by homomorphisms and defined on the set of congruences of any algebra of a variety.

As one knows, in the theory of groups, the concepts of abelian group, the center of a group, the centralizer of a normal subgroup, and solvable and nilpotent groups can all be defined in terms of the commutator operation on group elements as well as, alternatively, in terms of the commutator operation on normal subgroups, i.e., the commutator. Analogous concepts, based on the multiplication of ideals, proved to be important in ring theory. An extension of these concepts to arbitrary universal algebras by means of the commutator operation, defined on the set of congruences of any algebra, constitutes a very exciting direction of research

in universal algebra—general commutator theory. The pioneering works
in this area are those by J. Smith (1976) for congruence permutable
varieties and by J. Hagemann and C. Herrmann (1979) for congruence
modular varieties. Although the commutator operation can be defined
in any variety, it is very well-behaved and proved to be the most useful in
congruence modular varieties. Therefore, limiting ourselves to introducing two definitions of the commutator—axiomatic (Herrmann-Gumm,
1983) and via term conditions (Taylor, 1982)—which are equivalent in
congruence modular varieties, we refer the reader to (Freese and McKenzie 1987) for thorough, interesting presentations of many topics in this
field.

G.6.2 Definition

(1) In a congruence modular variety $\mathcal{V}$, the **commutator of con-
    gruences** is defined as the greatest (with respect to the order in
    Con A) binary operation $f(x, y)$, defined on Con A ($A \in \mathcal{V}$), such
    that for any $\alpha, \beta, \gamma \in$ Con A: (i) $f(\alpha, \beta) \leq \alpha \wedge \beta$, (ii) $f(\alpha, \beta \vee
    \gamma) = f(\alpha, \beta) \vee f(\alpha, \gamma)$, (iii) $f(\alpha \vee \beta, \gamma) = f(\alpha, \gamma) \vee f(\beta, \gamma)$, and
    (iv) for any surjective $\mathcal{V}$-homomorphism $\varphi \colon B \to A$, $\varphi^{-1}(f(\alpha, \beta)) =
    f(\varphi^{-1}(\alpha), \varphi^{-1}(\beta)) \vee \ker \varphi$;
(2) For $\alpha, \beta, \gamma \in$ Con A, $A \in \mathcal{V}$, α *centralizes* β *modulo* γ (in
    notation, $C(\alpha, \beta; \gamma)$) if for all $n \geq 1$, and every $t \in \text{Clo}_{n+1} A$
    (the clone of $(n + 1)$-ary term operations of A), $\langle a, b \rangle \in \alpha$, and
    $\langle c_1, d_1 \rangle, \ldots, \langle c_n, d_n \rangle \in \beta$, we have $\langle t(a, c_1, \ldots, c_n), t(a, d_1, \ldots, d_n) \rangle \in
    \gamma \iff \langle t(b, c_1, \ldots, c_n), t(b, d_1, \ldots, d_n) \rangle \in \gamma$.
(3) The **commutator** $[\alpha, \beta]$ of the congruences $\alpha, \beta \in$ Con A is defined
    as the smallest congruence $\gamma \in$ Con A such that $C(\alpha, \beta; \gamma)$.

For groups, this notion coincides with the usual commutator for normal subgroups (using the identification between congruences and normal subgroups, of course), for commutative rings, it is the well-known
ideal product IJ for ideals, whereas for non-commutative rings, we get
$IJ + JI$.

An algebra A is called **neutral** if $[\alpha, \beta] = \alpha \wedge \beta$ for any $\alpha, \beta \in$ Con A;
and a variety is **neutral** iff all its algebras are such. J. Hagemann and
C. Herrmann (1979) proved that a variety is neutral iff it is congruence
distributive.

Abelian and nilpotent algebras

A congruence $\alpha \in \mathrm{Con}\, A$ is called **abelian** if $[\alpha, \alpha] = 0_A$; and A is said to be an **abelian algebra** if the universal congruence $1_A \in \mathrm{Con}\, A$ is abelian. An **abelian variety** is a variety whose algebras are abelian. The notion of abelian algebra obviously is the polar opposite of that of neutral algebra, and it is easy to see that abelian groups are just commutative groups as well as that only rings with zero multiplication are abelian algebras in ring theory. Many results in congruence modular varieties are based on the following deep theorem of C. Herrmann (1979): Every abelian algebra in a congruence modular variety is polynomially equivalent to a module. Next, R. Freese and R. McKenzie obtained the following important result: An abelian congruence modular variety $\mathscr{V}$ is polynomially equivalent to the variety $\mathrm{Mod}_{R(\mathscr{V})}$ of unitary modules over an appropriate associative ring $R(\mathscr{V})$ with 1. Also, one can easily show that the class $\mathscr{V}_{\mathrm{ab}}$ of abelian algebras of a congruence modular variety $\mathscr{V}$ forms a subvariety of $\mathscr{V}$.

A congruence $\alpha \in \mathrm{Con}\, A$ is called **central** if $[\alpha, 1_A] = 0_A$; and A is **nilpotent** iff there exists a series $0_A = \theta_0 < \theta_1 < \cdots < \theta_n = 1_A$ of congruences of A such that $\theta_i/\theta_{i-1} \in \mathrm{Con}(A/\theta_{i-1})$ is central for any $i < n$. Many familiar facts about nilpotent groups have been extended to the context of congruence modular varieties by R. Freese and R. McKenzie. Here are some of them: every congruence on a nilpotent algebra is defined by any of its congruence classes (**congruence regularity**); the congruence classes of every congruence on a nilpotent algebra have the same cardinality (**congruence uniformity**); the order of any subalgebra of a finite nilpotent algebra is a divisor of the order of the algebra; and a nilpotent algebra of a prime power order is finitely based.

Properties of congruence modular varieties

Among numerous interesting and important results regarding the structure and/or representations of congruence modular varieties, we mention the following results, which make essential use of commutator theory.

G.6.3 Theorem (R. Freese, R. McKenzie) *For a finite algebra A generating the congruence modular variety $\mathscr{V}(A)$, the following statements are equivalent:*

(1) *The algebras of $\mathscr{V}_{\mathrm{SI}}$ form (up to isomorphism, of course) a set (i.e., $\mathscr{V}(A)$ is **residually small**);*

(2) *The cardinalities of $\mathscr{V}_{SI}$-algebras are less than or equal to n, where $n = (m + l!)$, $l = m^{m+3}$, and $m = |A|$ ($\mathscr{V}(A)$ is n-residually small)*;

(3) *For any $B \in \mathscr{V}(A)$, and $\alpha, \beta \in \mathrm{Con}\, B$, we have $\beta \leq [\alpha, \alpha] \Rightarrow \beta[\beta, \alpha]$.*

G.6.4 Theorem (R. Freese, R. McKenzie) *$|B| \leq |A|$ for any simple algebra B of the congruence modular variety* **Var**(A) *generated by a finite algebra A. If the length of $\mathrm{Con}\, B$ is less than or equal to n, then $|B| \leq |A|^n$.*

A finitely generated variety containing (up to isomorphism, of course) only a finite number of directly indecomposable finite algebras is said to be **directly representable**.

G.6.5 Theorem (R. McKenzie) *A congruence modular variety* **Var**$(\mathscr{K})$, *generated by the finite set $\mathscr{K}$ of finite algebras, is directly representable iff a)* **Var**$(\mathscr{K})$ *is congruence permutable; b) any algebra in* **S**$(\mathscr{K})$ *is a direct product of abelian and simple algebras; and c) the variety generated by the abelian factors of the representations by direct products of algebras of* **S**$(\mathscr{K})$, *is directly representable. In particular, a finitely generated, semisimple, congruence permutable variety is directly representable. Also, a congruence distributive variety is directly representable iff it is a finitely generated, semisimple, congruence permutable variety.*

G.6.6 Theorem (S. Burris, R. McKenzie) *For a finitely generated congruence modular variety $\mathscr{V}$ of a finite signature, the elementary theory* $\mathrm{Th}(\mathscr{V})$ *is decidable iff $\mathscr{V}$ is a direct product of a discriminator variety with a decidable elementary theory and an abelian variety $\mathscr{V}_{ab}$ with a decidable elementary theory of the $R(\mathscr{V}_{ab})$-modules.*

Finally, using commutator theory, one can describe the spectral functions as well as obtain a series of results concerning counting of the non-isomorphic algebras of the same cardinality in congruence modular varieties. For those results and many other important facts about congruence modular varieties, we encourage the reader to consult (Pinus 1986). Also, we refer the reader to work by Gumm (1983) in which very attractive and effective relations between commutator theory and affine geometry are considered, to the survey by Jónsson (1995) for many well-presented results on congruence distributive varieties, and to the paper by Pedicchio (1995) to explore a categorical perspective on commutator theory.

Our basic references for notions and notations of Universal Algebra include (Burris and Sankappanavar 1981) and (McKenzie, McNulty, and Taylor 1987)

G.7 Word Problems and Rewriting Systems

by Leonid Bokut' in Novosibirsk, Russia

Let Ω be some set of functional symbols (a signature). Let $\mathscr{K}$ be a variety of Ω-algebras with free objects $\mathscr{K}[X]$. Let $R = \mathscr{K}[X \mid \Phi]$ be a finitely presented (f. p.) $\mathscr{K}$-algebra. Here, X is a finite set of generators (an alphabet), and Φ is a finite set of defining relations $u_i = v_i$ ($i \in I$), where u_i, v_i are terms in $\Omega \cup X$. This means that R is the factor-algebra of the free algebra $\mathscr{K}[X]$ by the congruence relation generated by Φ. The **word problem (WP)** for R is the question: *Does there exist an algorithm that, for any two terms u, v in $\Omega \cup X$, gives an answer whether the relation $u = v$ is true in R or not.* R has **solvable** (or **decidable**) word problem if this kind of algorithm exists.

A classical result going back to Gauss states: The WP is solvable for any f. p. abelian group G (this is the same as finitely generated (f. g.) abelian group). Let us remark that much later W. Szmielew (1949) proved the more general result that the elementary theory $\text{Th}(G)$ (the set of all closed first order formulas valid on G) is solvable. The elementary theory $\text{Th}(S)$ of any f. p. abelian semigroup is solvable as well (M. A. Taitzlin, 1966). The WP is solvable for any f. p. commutative algebra (again this is the same as a f. g. commutative algebra by the Hilbert Basis Theorem) over an *effective* field (e.g., finite or $\mathbb{Q}$). It was partially known to MacCauley and Gröbner and in all generality it was proved by B. Buchberger using his Gröbner bases algorithm (1965). Between 1910 and 1950, the solvability of the WP was proved for some important groups:

- The fundamental groups of closed oriented 2-dimensional manifolds (M. Dehn).
- The braid groups (E. Artin; essentially S_n and A_n).
- The one-relation groups (W. Magnus). Later analogous results were proved for any semigroup with one defining relation of the form $A = 1$ (S. I. Adian) and for any one-relation Lie algebra (A. I. Shirshov).
- The 1/8-groups (V. A. Tartakovskii). Let $G = \langle X \mid R \rangle$ be a finitely presented group, where R is a symmetric set (it means that for any $r \in R$, words r^{-1} and any cyclic permutation of r belong to R).

Then G is called a λ-*group*, where $\lambda \in \mathbb{R}_+$, if for any different words $r_1, r_2 \in R$, if $r_1 = bc_1$, $r_2 = bc_2$, then $|b| \le \lambda|r_1|$ (here $|u|$ means the length of a word u). Later this result was generalized by M. Greendlinger to 1/6-groups and by R. Lyndon to 1/5-groups. The last statement is the best possible.

- Nilpotent groups (A. I. Mal'cev).

In the context of universal algebras, an important result (as it became clear some time later) was proved by T. Evans (1951). He found the *normal form* algorithm (a rewriting system) for the solvability of the WP for any f. p. *quasigroup* (this is an algebra with binary operations $\Omega = (\cdot, /, \backslash)$ and identities $(xy)/y = y \backslash (yx) = (x/y)y = y(y \backslash x) = x$). Later, Knuth and Bendix (1965) generalized the Evans quasigroup algorithm to a general Knuth-Bendix universal algebra algorithm. Now the Knuth-Bendix algorithm is an important tool in many branches of computer science, especially in theoretical programming.

After the development of the notion of an algorithm and adopting Church's Thesis, i.e., that this notion is adequate to the intuitive notion of an algorithm, it became possible to prove that the WP can be unsolvable. The unsolvability of the WP for semigroups (the *Thue problem*) was proved by A. A. Markov and E. Post (1947). The same result for groups (*Dehn problem*) was proved by P. S. Novikov in 1955. A much simpler proof of the Novikov theorem was obtained by W. W. Boone (1959). Based on the Novikov theorem, A. A. Markov proved the unsolvability of the homeomorphism problem for n-dimensional manifolds for $n \ge 4$; S. I. Adian proved the unsolvability of the isomorphism problem to any f. p. group (*Tietze problem*). Later, A. A. Markov, S. I. Adian, and M. Rabin proved the unsolvability of any Markov property (Section A.4) of f. p. semigroups and f. p. groups.

The previous unsolvability results marked a heroic period in the history of WP's in algebra. At the end of this period, P. S. Novikov (1959) announced a solution of the famous Burnside problem (cf. Section B.13). It took 10 years to publish a proof by P. S. Novikov and S. I. Adian (1968). At the heart of the proof there was a method for solving the WP for a free Burnside group $B(m,n) = \langle x_1, \ldots, x_m \mid x^n \equiv 1 \rangle$ for $m \ge 2$, n odd and sufficiently large. Later, stimulated by this result and the Mal'cev nilpotent group theorem (above), V. N. Remeslennikov (1973) proved that the WP is unsolvable for solvable groups. A much stronger result was proved by O. G. Kharlampovich (1981): There exists a f. p. group which is 3-solvable and has an unsolvable WP.

In the 1960's, the Markov-Post and Novikov theorems have been made more precise by results that the Turing degrees of unsolvability of the word problem for f. p. semigroups and the word and conjugacy problems for f. p. groups can be arbitrary (two algorithmic problems have the same Turing degree of unsolvability if each of them is solvable relative to the other one).

An important direction in the theory of word problems in algebra was opened by G. Higman (1961). Let us call a variety $\mathcal{K}$ a **Higman variety**, if any *recursively presented* $\mathcal{K}$-algebra $\mathcal{K}[X \mid \Phi]$ (it means that X is a finite set, and Φ is a recursively enumerable set, i.e., the set of values of some recursive function) is embeddable into a f. p. $\mathcal{K}$-algebra. G. Higman proved this property for the variety of groups. It gave another proof of the Novikov theorem. The following varieties are known to be Higman varieties:

• semigroups (V. L. Murskii),
• associative rings and algebras (over a "good" field, e.g., finite or $\mathbb{Q}$) (V. Ya. Belyaev),
• inverse semigroups (V. Ya. Belyaev).

An interesting situation is with the variety of Lie algebras. For almost 20 years it was supposed to be proved that it is a Higman variety. But recently a gap in the proof had been found (see Kharlampovich and Sapir 1995, pp. 560–566). So it is still an open problem (posed by L. A. Bokut').

In the context of universal algebras, an important direction is to study residually finite algebras. A universal algebra A in a class $\mathcal{M}$ is called **residually finite**, if for all elements $a \neq b$ in A there is a homomorphism φ of A onto a finite algebra in $\mathcal{M}$ such that $\varphi(a) \neq \varphi(b)$. Any finitely presented residually finite algebra has a solvable WP. This was noticed by McKenzie (1943). Any f. g. commutative ring is residually finite. It gives another proof of the solvability of the WP for this rings (different to the one which used the Gröbner bases technique). Any f. g. commutative semigroup has this property, as well as a f. g. linear group (A. I. Mal'cev, 1940). See also the survey (Kharlampovich and Sapir 1995) on the word and another algorithmic problems for varieties and pseudovarieties of groups, semigroups, Lie and associative algebras.

Algorithms of the Knuth-Bendix type (for example, Evans algorithm for quasigroups, Buchberger and Shirshov algorithms for algebras (Section C.31 and Section C.33), Dehn's algorithm for groups—Section B.11) are connected with an idea of a rewriting system. Let us give the definition of a rewriting system for a group presentation $G = \langle X \mid \Phi \rangle$.

The defining relations S of a **Semi-Thue system** $\mathrm{ST}[X, S]$ have the form $u \rightarrow v$, where u, v are words in the alphabet X (we can transform aub into avb for all words a, b but not in the opposite way). Let $S = \mathrm{ST}[X \cup X^{-1}|\Psi]$ be a Semi-Thue system on the group alphabet over X, where $\Psi = \{u_\lambda \rightarrow v_\lambda \,|\, \lambda \in \Lambda\}$. Let us define $u \rightarrow^* v$, if $u \rightarrow \cdots \rightarrow v$ can be deduced in S.

G.7.1 Definition A word $u \in S$ is called **canonical** if no transformation from Ψ (except the identity) can be applied to u.

S is called a **rewriting system** for G, if the following conditions are valid:

(1) S is a subsystem of the system of all the relations of the group G, i.e., if $u \rightarrow^* v$ in S, then the relation $u = v$ is true in G;
(2) if $u = v$ is true in G then there exists a word w such that $u \rightarrow^* w$ and $v \rightarrow^* w$ can be deduced in S (the **Church-Rosser property**);
(3) for any word u there exists a canonical word $C(u)$ such that $u \rightarrow^* C(u)$ is deduced in to S (it is equivalent to the minimality condition in S).

Then it follows that $u = v$ is true in G iff $C(u) \equiv C(v)$ (*graphically equal* as associative words).

One of the first rewriting systems for a group was constructed by M. H. A. Newman (1942). Let $F = \mathrm{Gr}[X]$ be a free group over a set X,

$$S = \mathrm{ST}[X \cup X^{-1} \,|\, xx^{-1} \rightarrow 1, \ x^{-1}x \rightarrow 1 \ (x \in X)].$$

Then S satisfies the **diamond (confluence) condition**:

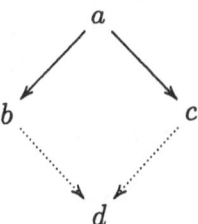

This means that if $a \rightarrow b$ and $a \rightarrow c$ can be deduced in S, then there exists a word d such that $b \rightarrow d$ and $c \rightarrow d$ can be also deduced in S. It follows that the same condition holds for $\rightarrow^*$ instead of $\rightarrow$.

One of the most complicated rewriting systems so far were constructed for universal groups (groups of quotients) of multiplicative semigroups of some rings (Bokut' and Collins 1980). It gave a positive solution of the Mal'cev's problem of the existence of an associative (semigroup) ring which is not embeddable into any skew-field, but its multiplicative semigroup is embeddable into a group.

For more informations, see also (Book and Otto 1993).

G.8 Relational Algebras

by Hajnal Andréka, Judit X. Madarász, and István Németi

in Budapest, Hungary

Boolean algebras (BA's for short) can be regarded as algebras of unary relations; i.e., the elements of a BA, say $\mathcal{B}$, are unary relations and the operations of $\mathcal{B}$ are the natural operations on unary relations. The purpose of relational algebra is to expand the natural algebras of unary relations (i.e., BA's) to natural algebras of relations of higher ranks, i.e., of relations in general. What will be the elements of the new algebras? The elements of BA's can be visualized as sets of *points*. Then, the elements of the new algebras will be *sets of sequences* (the reason for this is that the elements of relations are sequences independently of whether our relations are binary, ternary or n-ary).

The simplest case is when we concentrate on binary relations. For a set U, $\mathscr{P}(U)$ denotes its power set (the set of all subsets of U) while $\mathcal{P}(U)$ denotes the BA $(\mathscr{P}(U); \cup, \cap, -)$ with universe $\mathscr{P}(U)$. The *full relation algebra* over the set U is defined to be the algebra

$$\mathcal{R}e(U) = \left(\mathcal{P}(U \times U), \circ, {}^{-1}, \mathrm{Id}_U \right)$$

where "$\circ$" is the usual composition of two relations, R^{-1} is the usual converse (or inverse) of the relation R and $\mathrm{Id} = \mathrm{Id}_U$ is the identity relation on U. The class RRA of *representable relation algebras* is defined as

$$\mathrm{RRA} = \mathbf{S}\mathbf{P}\{ \mathcal{R}e(U) \mid U \text{ is a set} \}$$

where $\mathbf{S}$ and $\mathbf{P}$ are the operators on classes of algebras corresponding to taking isomorphic copies of subalgebras and direct products, respectively.

G.8.1 Theorem (Tarski) RRA *is a discriminator variety. The equational theory of* RRA *is recursively enumerable but not decidable.*

Before discussing RRA's further, let us look at algebras of relations of higher ranks (e.g., ternary, n-ary relations). The natural algebras are called cylindric algebras. In the following, n denotes a natural number. The **full cylindric algebra** of n-ary relations over a set U is defined as

$$\mathcal{R}el_n(U) = \big(\mathcal{P}(U^n), c_0, \ldots, c_{n-1}, \mathrm{Id}\big)$$

where Id is the n-ary identity relation $\mathrm{Id}_{n,U} = \{(a, \ldots, a) \mid a \in U\}$ and c_i is a unary operation for each $i < n$ defined by $c_i(R) = c_i^{(U)}(R) = \{(b_0, \ldots, b_{i-1}, a, b_{i+1}, \ldots, b_{n-1}) \mid (b_0, \ldots, b_{n-1}) \in R$ and $a \in U\}$, for any $i < n$ and $R \subseteq {}^nU$. We will omit the superscript U. Let $R \subseteq U^n$ be a relation. Then the relation $c_i(R)$ is called the smallest i-**cylinder** containing R. Choosing $n = 3$ and U the real numbers, we obtain the greatest element $U \times U \times U$ of our algebra as the usual Cartesian space, and i-cylinders appear as *cylinders* parallel to the i-th axis. Let $n = 2$ and $R \subseteq U \times U$. Then $c_0(R) = U \times \mathrm{Rg}(R)$ and $c_1(R) = \mathrm{Dom}(R) \times U$. This example shows that the operations c_i are natural ones (on relations). The class RCA_n of n-ary **representable cylindric algebras** is defined as

$$\mathrm{RCA}_n = \mathbf{S}\,\mathbf{P}\{\mathcal{R}el_n(U) \mid U \text{ is a set }\}.$$

G.8.2 Theorem (Tarski) RCA_n *is a discriminator variety. The equational theory of* RCA_n *is recursively enumerable, and if* $n > 2$ *then undecidable.*

To have all finitary relations over U in a single algebra, we need to extend cylindric algebras to α-ary relations with α an arbitrary ordinal. For this, we need to replace our single (α-ary) identity relation Id with $\alpha \times \alpha$ many identity relations $\mathrm{Id}_{ij} = \{q \in {}^\alpha U \mid q_i = q_j\}$, for $i, j \in \alpha$. Throughout, α is an arbitrary (possibly finite) ordinal. Now, we define the full algebra of α-ary relations as

$$\mathcal{R}el_\alpha(U) = \big(\mathcal{P}(U^\alpha), c_i, \mathrm{Id}_{ij} \mid i, j < \alpha\big),$$

where $c_i(R)$ and Id_{ij} are defined as above. Thus, besides the Boolean operations, $\mathcal{R}el_\alpha(U)$ has α many unary operations c_i (one for each $i < \alpha$) and $\alpha \times \alpha$ many constants Id_{ij}. Now, for $\alpha < \omega$ we have two versions for $\mathcal{R}el_\alpha(U)$ but they are polynomially equivalent. Indeed, if e.g., $\alpha = 3$ then $\mathrm{Id}_{1,2} = c_0(\mathrm{Id})$ while $\mathrm{Id} = \mathrm{Id}_{01} \cap \mathrm{Id}_{12}$, $\mathrm{RCA}_\alpha = \mathbf{S}\,\mathbf{P}\{\mathcal{R}el_\alpha(U) \mid U$ is a set $\}$.

G.8.3 Theorem (Tarski) RCA_α *is an arithmetical variety. The equational theory of* RCA_α *is recursively enumerable, but it is undecidable if* $\alpha > 2$.

So far, the greatest elements of our algebras were Cartesian spaces, i.e., of the form U^α (both in the cases of RRA's and RCA's). However, this restriction is not always convenient (cf. e.g., Andréka et al. (1998), van Benthem (1996), Monk (2000), Henkin et al. (1981)). Removing this restriction motivates the definition of cylindric-relativized set algebras. Let $V \subseteq U^\alpha$ be an arbitrary α-ary relation. Then the **algebra of subrelations** of V is defined as

$$\mathcal{Rel}(V) = \left(\mathcal{P}(V), c_i^V, \mathrm{Id}_{ij}^V \mid i, j < \alpha\right)$$

where $c_i^V(R) = V \cap c_i(R)$ and $Id_{ij}^V = V \cap \mathrm{Id}_{ij}$. The class of α-ary **cylindric-relativized set algebras** is defined as

$$\mathsf{Crs}_\alpha = \mathbf{S}\{\, \mathcal{Rel}(V) \mid V \subseteq U^\alpha \text{ for some set } U \,\}.$$

The finite algebra part of the next theorem is the result of a joint work with Hajnal Andréka and Ian Hodkinson.

G.8.4 Theorem (Németi) *Let $\alpha \neq 1$. Then Crs_α is an arithmetical variety. The equational theory of Crs_α is decidable. A finite Crs_n is isomorphic to one with finite greatest element.*

It is natural to ask whether any one of the distinguished kinds RRA, RCA_α, Crs_α of algebras of relations is axiomatizable by a finite set of equations. If $\alpha \geq \omega$, then having a finite set of axioms is impossible because there are infinitely many basic operations, but we still could hope for a finite scheme of equations like the scheme $c_i \, \mathrm{Id}_{ij} = 1$, for all $i, j < \alpha$.

G.8.5 Theorem (Monk, Monk, Németi, Jónsson, Andréka)
Assume $\alpha > 2$. None of the varieties $\mathsf{RRA}, \mathsf{RCA}_\alpha, \mathsf{Crs}_\alpha$ is axiomatizable by a finite scheme of equations. None of RRA or RCA_α is axiomatizable by a scheme Σ of universally quantified formulas such that Σ involves only finitely many variables.

The negative result above motivates the definition of the **finitely axiomatizable approximations** RA and CA_α of RRA and RCA_α. The axioms for RA are **(R1)** − **(R3)** below.

(R1) The Boolean axioms; and the operations $\circ$, $^{-1}$ are "$\cup$"-distributive, i.e., they commute with the Boolean join "$\cup$".

(R2) $\circ, ^{-1}, \mathrm{Id}$ form an involuted monoid , where an **involuted monoid** is a monoid with an extra unary operation $^{-1}$ satisfying the two equations $x^{-1-1} = x^{-1}$, $(x \circ y)^{-1} = y^{-1} \circ x^{-1}$.

(R3) $x^{-1} \circ -(x \circ y) \leq -y$.

The axioms for CA_α are **(E1)** − **(E5)** below.

(E1) The Boolean axioms.

(E2) The c_i's are commuting complemented $\cup$-distributive closure operations (e.g., $c_i c_j x = c_j c_i x$, $c_i - c_i x = -c_i x$ etc.).

(E3) $\mathrm{Id}_{ii} = 1$ and $\mathrm{Id}_{ij} = \mathrm{Id}_{ji}$ (i.e., notational trivialities)

(E4) $\mathrm{Id}_{ik} = c_j(\mathrm{Id}_{ij} \cap \mathrm{Id}_{jk})$ if $j \notin \{i, k\}$.

(E5) $x \leq \mathrm{Id}_{ij} \Rightarrow c_i(x) \cap \mathrm{Id}_{ij} = x$.

Clearly, RA $\supseteq$ RRA and $\mathsf{CA}_\alpha \supseteq \mathsf{RCA}_\alpha$. CA_1's are also called **monadic algebras**. Both approximations RA and CA_α were introduced by Tarski. In some sense, RA is close to RRA and CA_α is close to RCA_α. However, it is hard to make it precise what we mean by *close* here. It is possible to introduce natural properties such that

$$\mathsf{RA} \cap \text{``property''} \subseteq \mathsf{RRA} \quad \text{and} \quad \mathsf{CA}_\alpha \cap \text{``property''} \subseteq \mathsf{RCA}_\alpha.$$

However, one can replace RA by a bigger class RA^- and CA_α with CA_α^- such that all the above style representation theorems remain true. Using the well established connections between logic and algebraic logic, one can argue that the axioms for RA and CA_α are optimal in some sense. E.g., the CA_α axioms correspond to one of the usual axiomatizations of first order logic and most of the equations separating RCA_α from CA_α would look strange to the logician as a possible extra axiom (unless he is trying to axiomatize the finite-variable fragments L_n of first order logic).

CA_α's correspond to first order logic L_α with equality. If we *algebraize* the same logic, but without equality, we obtain **substitution-cylindrification algebras** which are obtained from CA_α by throwing away the constants Id_{ij} and replacing them with the term-functions $s_j^i(x) = c_i(\mathrm{Id}_{ij} \cap x)$. **Quasi-polyadic algebras** (of Halmos) are almost the same as these, cf. Henkin et al. (1985) for both kinds of algebras.

Connections with logic are in Tarski and Givant (1987), Andréka et al. (2001), Henkin et al. (1985), Németi (1991) except for new kinds of recent applications of Crs_α-theory to the finite-variable fragments, finite model theory, the bounded fragments, and the guarded fragment, cf. e.g., Hoogland and Marx (2001), Andréka et al. (1998), van Benthem (1996). Cf. also Craig (1974), Henkin et al. (1971, 1985).

The negative result Theorem G.8.5 gave rise to the Finitization Problem which asks whether we could define our algebras of relations in such a way that they would form a finitely axiomatizable variety. There has been extensive research work on this problem recently, cf. e.g., Németi and Sain (2000) for further references.

Algebras of relations have been extensively applied in computer science, AI, linguistics and other areas, cf. e.g., Bergman et al. (1990), Marx et al. (1996), van Benthem (1996).

This research was supported by the Hungarian Foundation for Basic Research, Grants T30314 and T35192.

G.9 Partial Algebras

by Peter Burmeister in Darmstadt, Germany

Introduction

Quite often (e.g., in Computer Science, but also for the multiplicative inverse in fields) the operations in an "algebra" are not everywhere defined. Moreover, quite often constructions for "*total algebras*" make use of **partial algebras** and some general construction principles like *universal solutions*. For such "*partial algebras*" , a highly developed theory has been worked out. This *theory of partial algebras* lies in between those of *total algebras* and *relational systems* (see Section G.8). From *total algebras* it inherits in particular the concepts of *terms*, *direct products* (with the structure defined componentwise whenever possible in *all* components), *closed subsets* (and in connection with them *(closed) subalgebras* as the partial algebras obtained by restricting the structure to closed subsets, and the concept of *generation*). From *relational systems* it inherits the wealth of possible concepts, since partial algebras can model relational systems (cf. Burmeister 1986, 13.4.2). Moreover, *many-sorted (partial) algebras* can easily be considered as partial algebras on the disjoint union of the carriers of the different sorts, and their homomorphisms then have just to be compatible with the canonical homomorphisms into the set of sorts with the specification of the many-sorted (partial) operations as fundamental partial operations (cf. Burmeister 1986). Here we can only introduce some of the basic concepts and applications of a language for partial algebras.

The first order language

G.9.1 Definition Let X be some set, called the set of **variables**, which is disjoint with the set $\Omega := \bigcup_{n \in \mathbb{N}_0} \Omega_n$ (Ω_n the set of all n-ary operation names), $T_\Omega(X)$ the corresponding set of terms, $\mathcal{A} = (A; ((\omega^\mathcal{A} : A^n \supseteq \mathrm{dom}\,\omega^\mathcal{A} \to A)_{\omega \in \Omega_n})_{n \in \mathbb{N}_0})$ a partial Ω-algebra and $v : X \to A$ any mapping (**valuation**).

(PI): The **partial interpretation** $\tilde{v}$ induced by v is the partial mapping with smallest domain out of $T_\Omega(X)$ into A such that **(i):** $X \subseteq \mathrm{dom}\,\tilde{v}$ and $\tilde{v}(x) := v(x)$ for all $x \in X$; and **(ii):** for $n \in \mathbb{N}_0$, $\omega \in \Omega_n$, $t_1, \ldots, t_n \in T_\Omega(X)$ one has: If $\tilde{v}(t_i) = a_i$ for $1 \leq i \leq n$, and if $\omega^\mathcal{A}(a_1, \ldots, a_n)$ is defined in $\mathcal{A}$ and has value a, then $\tilde{v}(\omega(t_1, \ldots, t_n))$ is defined (i.e., $\omega(t_1, \ldots, t_n) \in \mathrm{dom}\,\tilde{v}$) with value a.

(Eeq): For $t, t' \in T_\Omega(X)$ we say that the **existence equation** (briefly: **E-equation**) $t \overset{e}{=} t'$ is satisfied in $\mathcal{A}$ w.r.t. the valuation v — in symbols $\mathcal{A} \models t \overset{e}{=} t'[v]$ —, iff $(t, t') \in \ker \tilde{v} := \{(p, q) \in (\mathrm{dom}\,\tilde{v})^2 \mid \tilde{v}(p) = \tilde{v}(q)\}$. In particular, for so-called **term existence statements** $t \overset{e}{=} t$ the assertion $\mathcal{A} \models t \overset{e}{=} t[v]$ gets a special meaning as: $\tilde{v}(t)$ *exists*. An E-equation $t \overset{e}{=} t'$ *is valid* in a partial algebra $\mathcal{A}$ iff it is satisfied w.r.t. every valuation $v : X \to A$.

(Form): **First order formulas** and their satisfaction and validity are then defined as usual. In particular, one needs **ECE-equations** (short for **existentially conditioned existence equations**):
$$\iota := (t_1 \overset{e}{=} t_1 \wedge t_2 \overset{e}{=} t_2 \wedge \ldots \wedge t_k \overset{e}{=} t_k \Rightarrow t \overset{e}{=} t').$$
The satisfaction of ι w.r.t. a valuation v means that whenever v interprets t_1 through t_k, then v has to interpret t, and t' and $\tilde{v}(t) = \tilde{v}(t')$. They are special cases of **elementary implications** (which are called **QE-equations**—short for **quasi-existence equations**—for finite I):
$$\eta := (\bigwedge_{i \in I} t_i \overset{e}{=} t'_i \Rightarrow t \overset{e}{=} t')..$$
(TO): Let $t \in T_\Omega(X)$ and $\mathcal{A} \in \mathrm{PAlg}_\Omega$, i.e. a partial Ω-algebra. Then the **partial term operation** $t^\mathcal{A}$ induced by t has as domain $\mathrm{dom}\,t^\mathcal{A} := \{v \in A^X \mid t \in \tilde{v}\}$ and for $v \in \mathrm{dom}\,t^\mathcal{A}$ one has $t^\mathcal{A}(v) := \tilde{v}(t)$.

ECE-equations are the basic axioms for describing partial algebras, e.g. in computer science. All concepts of identities conceived so far for partial algebras are (conjunctions of) special cases of them.[1] E-equations always require that the corresponding terms induce total term

---

[1] E.g. for the the so-called **strong equality** or **Kleene equality** $\overset{K}{=}$ of partial term operations one has $\mathcal{A} \models t \overset{K}{=} t'[v]$ iff $t^\mathcal{A}(v) = t'^\mathcal{A}(v)$, iff $\mathcal{A} \models ((t \overset{e}{=} t \Rightarrow t \overset{e}{=} t') \wedge (t' \overset{e}{=} t' \Rightarrow t \overset{e}{=} t'))[v]$.

operations;[2] yet they provide a good foundation of the theory.

Properties of homomorphisms

The concepts of *preservation* and *reflection of formulas* and of *factorization structures* allow to control the properties of structure preserving mappings.

G.9.2 Definition Let $\mathcal{A}, \mathcal{B} \in \text{PAlg}_\Omega$, Φ a first order formula with variables in X, and $f : A \to B$ any mapping. f **preserves the formula** Φ, iff, for every valuation $v : X \to A$ one has that $\mathcal{A} \models \Phi[v]$ implies that $\mathcal{B} \models \Phi[f \circ v]$. And f **reflects** Φ, iff it preserves its negation $\neg\Phi$. f is a **homomorphism** from $\mathcal{A}$ to $\mathcal{B}$ (denoted by $f : \mathcal{A} \to \mathcal{B}$), iff f preserves all E-equations.[3] Let TAlg_Ω ($\subseteq \text{PAlg}_\Omega$) and Hom_Ω designate the classes of all total Ω-algebras and of all homomorphisms between algebras in PAlg_Ω, respectively.

G.9.3 Remark Special classes of homomorphisms defined by reflection of formulas are (for X a countably infinite set of variables):[4]

notation	class of all ...	reflected formulas
Mono_Ω	injective homomorphisms	$x \overset{e}{=} y \ (x, y \in X)$
Closed_Ω	closed homomorphisms	$t \overset{e}{=} t \ (t \in T_\Omega(X))$
$\text{Mono}_{\Omega,c}$	closed injective homomorphisms	$x \overset{e}{=} y, \ t \overset{e}{=} t \ (t \in T_\Omega(X))$
$\text{Mono}_{\Omega,f}$	full injective homomorphisms	$x \overset{e}{=} y, \ \omega(x_1, \ldots, x_n) \overset{e}{=} y,$
		$x, y, x_i \in X, \ \omega \in \Omega_n, n \in \mathbb{N}_0$

A homomorphism $f : \mathcal{A} \to \mathcal{B}$ is an **isomorphism**, iff it is closed and bijective. And it is an **epimorphism**, iff $f(A)$ generates $\mathcal{B}$. Let Epi_Ω designate the class of all epimorphisms between algebras in PAlg_Ω. In order to describe *epimorphisms* with possibly additional properties, one

---

[2] Considering for fields $\Omega_0 := \{0, 1\}$, $\Omega_1 := \{-, ^{-1}\}$, and $\Omega_2 := \{+, \cdot\}$, the commutativity axioms $x + y \overset{e}{=} y + x$ and $x \cdot y \overset{e}{=} y \cdot x$ as well as the coresponding associativity or distributivity (with other) laws imply that $+$ and $\cdot$ are total, $x + (-x) \overset{e}{=} 0$, $x + 0 \overset{e}{=} x$ and $x \cdot 1 \overset{e}{=} x$ also imply that 0 and 1 are always defined (being subterms of terms in E-equations). However the axioms $\neg(0 \overset{e}{=} 1)$ and $\neg(x \overset{e}{=} 0) \Rightarrow x \cdot x^{-1} \overset{e}{=} 1$ cannot be replaced by ECE-equations, since the direct product of fields is no longer a field (see Theorem G.9.4).

[3] This implies that homomorphisms are exactly those mappings which map the structure of the start object compatibly into the one of the target object.

[4] Recall, e.g. from (Grätzer 1979) or (Burmeister 1986), that a homomorphism is called *full*, if it fully induces the structure on its image set.

needs *epi-factors* in *factorization structures* as described in (Adámek, Herrlich, and Strecker 1990), where the *mono-factors* are defined by reflection of QE-equations. Factorization structures of special interest for partial algebras are among others:[5]

$$
\begin{array}{rcl}
(\ \text{all full and surjective homomorphisms} & , & \text{Mono}_\Omega\ ), \\
(\ \text{all TAlg}_\Omega\text{-extendable epimorphisms} & , & \text{Closed}_\Omega\ ), \\
(\ \text{Epi}_\Omega & , & \text{Mono}_{\Omega,c}\ ), \\
(\ \text{all surjective homomorphisms} & , & \text{Mono}_{\Omega,f}\ ).
\end{array}
$$

Birkhoff-type theorems

G.9.4 Theorem *Let $\mathscr{K} \subseteq \text{PAlg}_\Omega$ (to avoid trivialities, let us exclude here the partial algebra with empty carrier).*

$\mathscr{K}$ is definable by E-equations, iff it is closed w.r.t. the formation of homomorphic images (w.r.t. surjective homomorphisms) of (closed) subalgebras of direct products of $\mathscr{K}$-algebras (i.e., iff $\mathscr{K} = \mathbf{HS_cP}\mathscr{K}$).

*$\mathscr{K}$ is definable by ECE-equations, iff it is closed w.r.t. the formation of closed homomorphic images (w.r.t. closed and surjective homomorphisms) of subalgebras of **reduced products** (in the usual model theoretic sense as for relational systems) of $\mathscr{K}$-algebras (i.e., iff $\mathscr{K} = \mathbf{H_cS_cP_r}\mathscr{K}$).*

Universal solutions

G.9.5 Definition Consider $\mathscr{K} \subseteq \text{PAlg}_\Omega$ and $\mathcal{A}, \mathcal{B} \in \text{PAlg}_\Omega$. An epimorphism $e : \mathcal{A} \to \mathcal{B}$ is called

$\mathscr{K}$-extendable, if, for every $g : \mathcal{A} \to \mathcal{K}$ with $\mathcal{K} \in \mathscr{K}$ there exists an $h : \mathcal{B} \to \mathcal{K}$ such that $h \circ e = g$. The epimorphism e is called

$\mathscr{K}$-universal, if it is $\mathscr{K}$-extendable, and if, for every $\mathscr{K}$-extendable epimorphism $f : \mathcal{A} \to \mathcal{C}$, there exists some $h' : \mathcal{C} \to \mathcal{B}$ such that $h' \circ f = e$. A pair $(\mathcal{F}, r : \mathcal{A} \to \mathcal{F})$ is called a *$\mathscr{K}$-universal solution* of $\mathcal{A}$, if r is a $\mathscr{K}$-universal epimorphism.

G.9.6 Theorem *For every class $\mathscr{K} \subseteq \text{PAlg}_\Omega$ and for every $\mathcal{A} \in \text{PAlg}_\Omega$ a $\mathscr{K}$-universal solution exists and is unique up to isomorphism (isomorphic to a subalgebra of a direct power of $\mathscr{K}$-algebras). If $\mathscr{K}$ is*

---

[5]See below for the definition of TAlg$_\Omega$-extendable epimorphisms.—Because of the factorization structure (all TAlg$_\Omega$-extend. epis, Closed$_\Omega$) one can define partial interpretations within the category with PAlg$_\Omega$ as object class and Hom$_\Omega$ as morphism class; and one does not need partially defined homomorphisms in this connection.

*closed w. r. t. isomorphic copies of subalgebras of direct products (i. e. a **pseudo-primitive class**), and if $(\mathcal{F}, r : \mathcal{A} \to \mathcal{F})$ is a $\mathcal{K}$-universal solution of $\mathcal{A}$, then $\mathcal{F} \in \mathcal{K}$.*

G.9.7 Remark Let $\mathcal{K} \subseteq \mathrm{PAlg}_\Omega$ be a non-trivial pseudo-primitive class, then, for every set M, the $\mathcal{K}$-**free algebra**, $\mathcal{K}$-freely generated by M exists in $\mathcal{K}$ as $\mathcal{K}$-universal solution of the **discrete partial algebra** on M (empty structure); in particular, **data types** in the **initial algebra semantics** of Computer Science are such $\mathcal{K}$-free algebras with $\emptyset$ as set of free generators.[6] Moreover, all pseudo-primitive classes—e.g., classes defined by E-, ECE- or QE-equations—have **coproducts** in the sense of category theory and these are the $\mathcal{K}$-universal solutions of the coproducts of the corresponding families of partial algebras for PAlg_Ω (which are either disjoint unions or quotients of them w.r.t. the necessary identifications of the existing nullary constants). Also *presentations* of groups or other algebras are just universal solutions of some (usually easily describable) partial algebras.

G.10 Abstract Data Types

by Hans-Dieter Ehrich in Braunschweig, Germany

In computing, a **data type** is given by a domain of data values and the operations applicable to them. In mathematical terms, this is an algebra. Since data operations are often many-sorted, a data type is a many-sorted (or heterogeneous) algebra as introduced by Birkhoff and Lipson (1970).

An **abstract data type** is a data type where some details are considered as irrelevant, like internal memory representation. In mathematical terms, this is a class of algebras. Differences between isomorphic data types are always considered irrelevant, so an abstract data type is a class of algebras that is closed under isomorphism. It is called **monomorphic** if it consists of just one isomorphism class, otherwise it is called **polymorphic**.

A problem extensively studied in an algebraic setting is abstract data type **specification**. We concentrate on equational specification using

---

[6]E.g., the interval $[-l, k]$ of integers $(k, l \in \mathbb{N}_0)$ of a type with $\Omega_0 := \{0\}$ and $\Omega_1 := \{p, s\}$ is the free $\mathcal{K}$-algebra on $\emptyset$, when $\mathcal{K}$ is defined by the ECE-equations $p^l(0) \stackrel{e}{=} p^l(0)$, $s^k(0) \stackrel{e}{=} s^k(0)$, $p(x) \stackrel{e}{=} p(x) \Rightarrow s(p(x)) \stackrel{e}{=} x$ and $s(x) \stackrel{e}{=} s(x) \Rightarrow p(s(x)) \stackrel{e}{=} x$.

initiality, but briefly mention other approaches at the end. We largely follow the presentation in Loeckx et al. (1996, 2000).

A *signature* $\Sigma = (S, \Omega)$ is given by sets S of *sorts* and $\Omega \subseteq N \times S^* \times S$ of *operation symbols*; N is a set of operation names, and S^* denotes the set of finite sequences over S (including the empty sequence). A triple $(f, \langle s_1, \ldots, s_n \rangle, s_0) \in \Omega$ is often written as $f \colon s_1 \times \cdots \times s_n \to s_0$.

Sorts are names of data domains, for example bool for that of boolean values, nat for natural numbers and natlist for lists of natural numbers. Examples of operation names are $\vee \colon \text{bool} \times \text{bool} \to \text{bool}$ and $\leq \colon \text{nat} \times \text{nat} \to \text{bool}$. An example of a signature Σ_{bool} is given by the set $S_{\text{bool}} = \{\text{bool}\}$ of sorts and the set $\Omega_{\text{bool}} = \{ \mathit{false} \mid \to \text{bool}, \vee \colon \text{bool} \times \text{bool} \to \text{bool} \}$ of operation symbols.

A Σ-*algebra* assigns a set $A(s)$ to each sort $s \in S$, and a function $A(\omega) \colon A(s_1) \times \cdots \times A(s_n) \to A(s_0)$ to each operation symbol $\omega = (f \colon s_1 \times \cdots \times s_n \to s_0) \in \Omega$. For example, the usual Σ_{bool}-algebra *Bool* is given by $\mathit{Bool}(\text{bool}) = \{\mathit{true}, \mathit{false}\})$ and $\mathit{Bool}(\vee \colon \text{bool} \times \text{bool} \to \text{bool})$ is logical disjunction. Given a signature Σ, the class of all Σ-algebras is denoted by $\mathsf{Alg}(\Sigma)$.

A Σ-*algebra morphism* $h \colon A \to B$ is a family of maps $h = (h_s \colon A(s) \to B(s))_{s \in S}$ such that, for any operation symbol $\omega = (f \colon s_1 \times \cdots \times s_n \to s_0) \in \Omega$, we have $h_{s_0}(A(\omega)(a_1, \ldots, a_n)) = B(\omega)(h_{s_1}(a_1), \ldots, h_{s_n}(a_n))$, for all elements $a_1 \in A(s_1), \ldots, a_n \in A(s_n)$.

Good candidates for monomorphic abstract data types are isomorphism classes of initial algebras in certain subclasses of $\mathsf{Alg}(\Sigma)$.

G.10.1 Definition Let $\mathscr{C} \subseteq \mathsf{Alg}(\Sigma)$ be a class of Σ-algebras. $A \in \mathscr{C}$ is called *initial* in $\mathscr{C}$ iff there is exactly one Σ-algebra morphism $h \colon A \to B$ from A to any other algebra B in $\mathscr{C}$ (cf. H.2.1).

The definition of many-sorted terms is a straightforward generalization of the classical one, cf. Definition G.1.4. $T_{\Sigma(X)} = (T_{\Sigma(X),s})_{s \in S}$ denotes the S-indexed set family of terms over Σ and variables $X = (X_s)_{s \in S}$. For the *ground terms* over Σ (i.e., $X = \varnothing = \{ \varnothing_s \}_{s \in S}$), we write $T_\Sigma = \{ T_{\Sigma,s} \}_{s \in S}$. $T_{\Sigma(X)}$ can be made a Σ-algebra $T(\Sigma(X))$ by defining the operations as term constructors: $T(\Sigma(X))(\omega)(t_1, \ldots, t_n) = f(t_1, \ldots, t_n)$ for all operation symbols $\omega = (f \colon s_1 \times \cdots \times s_n \to s_0) \in \Omega$ and all terms $t_1 \in T_{\Sigma(X),s_1}, \ldots, t_n \in T_{\Sigma(X),s_n}$.

For $A \in \mathsf{Alg}(\Sigma)$, any variable assignment $\alpha \colon X \to A$ can be uniquely extended to a Σ-algebra morphism $A(\alpha) \colon T(\Sigma(X)) \to A$, defining term evaluation. $T(\Sigma(X))$ is a *free Σ-algebra* over X, cf. Section G.4. If $X = \varnothing$, then there is just one variable assignment $\varnothing \colon \varnothing \to A$ to every

Σ-algebra A. Its unique extension $A(\varnothing)\colon T(\Sigma) \to A$ is the one and only morphism from $T(\Sigma)$ to A. Thus, $T(\Sigma)$ is initial in $\mathsf{Alg}(\Sigma)$. We write $A(t)$ for $A(\varnothing)(t)$.

Given a class $\mathscr{C} \subseteq \mathsf{Alg}(\Sigma)$, the **congruence relation** of $\mathscr{C}$, denoted by $\equiv_\mathscr{C}$, is defined as $(\equiv_{\mathscr{C},s})_{s \in S}$ where $\equiv_{\mathscr{C},s} = \{\, (t,u) \mid t,u \in T_{\Sigma,s}$ and $A(t) = A(u)$ for each $A \in \mathscr{C} \,\}$. $T(\Sigma,\mathscr{C}) = T(\Sigma)/{\equiv_\mathscr{C}}$ is called the **quotient term algebra** of the class $\mathscr{C}$. $[t]_\mathscr{C}$ denotes the congruence class of t in $T(\Sigma,\mathscr{C})$.

G.10.2 Theorem *Let Σ be a signature and $\mathscr{C} \subseteq \mathsf{Alg}(\Sigma)$. There is a unique Σ-algebra morphism $h\colon T(\Sigma,\mathscr{C}) \to A$ to each algebra $A \in \mathscr{C}$, namely $h([t]_\mathscr{C}) = A(t)$ for each ground term $t \in T_\Sigma$.*

G.10.3 Corollary $T(\Sigma,\mathscr{C})$ *is initial in* $\mathscr{C}$ *iff* $T(\Sigma,\mathscr{C}) \in \mathscr{C}$.

The equational logic over a signature Σ is given by the set $\mathsf{EL}(\Sigma)$ of Σ-equations of the form $\forall X.t = u$ where X is a set of variables for Σ and $t,u \in T_{\Sigma(X),s}$ for some sort s of Σ. An equation $\varphi \in \mathsf{EL}(\Sigma)$ is **satisfied in a Σ-algebra** $A \in \mathsf{Alg}(\Sigma)$, denoted by $A \vDash_\Sigma \varphi$, iff $A(\alpha)(t) = A(\alpha)(u)$ for every assignment $\alpha\colon X \to A$.

If $\Phi \subseteq \mathsf{EL}(\Sigma)$, $A \vDash_\Sigma \Phi$ is defined to hold iff $A \vDash_\Sigma \varphi$ for every $\varphi \in \Phi$. In this case, we say that A is a **model** of Φ. $\mathsf{Mod}_\Sigma(\Phi)$ denotes the class of all models of Φ in $\mathsf{Alg}(\Sigma)$.

For any set $\Phi \subseteq \mathsf{EL}(\Sigma)$, $\mathsf{Mod}_\Sigma(\Phi)$ is an abstract data type. For practical purposes, however, abstract data types of this kind are corrupted by too much polymorphism: they contain the trivial algebras with one element per sort which are hardly ever acceptable as representatives of intended data types 'up to irrelevant details'. Thus, equational logic alone is not powerful enough for specifying useful abstract data types, especially monomorphic ones.

Adding initiality as a further specification concept helps. Let $T(\Sigma,\Phi) = T(\Sigma,\mathsf{Mod}_\Sigma(\Phi))$.

G.10.4 Theorem *Let Σ be a signature and $\Phi \subseteq \mathsf{EL}(\Sigma)$ a set of equations. Then $T(\Sigma,\Phi)$ is initial in $\mathsf{Mod}_\Sigma(\Phi)$.*

According to corollary G.10.3, all what has to be proved is that $T(\Sigma,\Phi) \in \mathsf{Mod}_\Sigma(\Phi)$, cf. Loeckx et al. (1996, theorem 7.2). Since $\mathsf{Mod}_\Sigma(\Phi)$ therfore has initial algebras, the following definition makes sense.

G.10.5 Definition An *initial abstract data type specification* in equational logic, or *initial specification* for short, consists of (i) the abstract syntax: an equational specification $D = (\Sigma, \Phi)$ where Σ is a signature and $\Phi \subseteq \mathsf{EL}(\Sigma)$ is a set of equations, and (ii) the semantics or meaning: the class $\mathcal{M}(D)$ of initial algebras in $\mathsf{Mod}_\Sigma(\Phi)$.

For any equational specification D, $\mathcal{M}(D)$ is a monomorphic abstract data type. $T(\Sigma, \Phi)$ is an obvious representative. A salient feature of initial semantics is that, in most cases of practical interest, there is a compatible operational semantics, namely term rewriting (cf. Section G.7).

A *reduction system* is a pair $(R, \rightarrow)$ where R is a set and $\rightarrow$ a binary relation on R. A *reduction sequence* of $(R, \rightarrow)$ is a possibly infinite sequence $r_1, \ldots, r_k, \ldots$ of elements of R such that $r_i \rightarrow r_{i+1}$ for each $i \geq 1$; for any $k \geq 1$ we write $r_1 \rightarrow^* r_k$. A *normal form* of r is an element $s \in R$ such that $r \rightarrow^* s$ and there is no $t \in R$ such that $s \rightarrow t$. An *equivalence sequence* is defined like a reduction sequence but with $r_i \rightarrow r_{i+1}$ *or* $r_{i+1} \rightarrow r_i$ for each $i \geq 1$; for each $k \geq 1$, we write $r_1 \simeq r_k$. It is easy to see that $\simeq$ is an equivalence relation. A reduction system is called *Noetherian* if it possesses no infinite reduction sequences; it is called *confluent* if for all $r, s, t \in R$ the following holds: if $r \rightarrow^* s$ and $r \rightarrow^* t$, then there exists an element $u \in R$ such that $s \rightarrow^* u$ and $t \rightarrow^* u$. If $(R, \rightarrow)$ is Noetherian and confluent, then each element of R has exactly one normal form, and for any two elements $r, s \in R$, $r \simeq s$ holds iff r and s have the same normal form.

G.10.6 Definition The *term rewriting system* for an initial specification $D = (\Sigma, \Phi)$ is a reduction system $(T_\Sigma, \rightarrow)$ where $\rightarrow$ is inductively defined by (i) $v\sigma \rightarrow w\sigma$ for each equation $\forall X. v = w \in \Phi$ and for each substitution $\sigma: X \rightarrow T_\Sigma$, and (ii) if $t \rightarrow u$, then $s[t/y] \rightarrow s[u/y]$ for all terms $s \in T_{\Sigma(\{y\})}$ containing at least one occurrence of the variable y.

G.10.7 Theorem *Let $(T_\Sigma, \rightarrow)$ be the term rewriting system of an equational specification $D = (\Sigma, \Phi)$. Let $v, w \in T_\Sigma$ be two ground terms of the same sort. Then $\Phi \models v = w$ iff $v \simeq w$.*

Noetherian and confluent term rewriting systems provide a useful operational semantics: in order to prove that $v \simeq w$, it is sufficient to prove that u and v have the same normal form. Both properties, however, are undecidable. There is a sufficient condition for being Noetherian (see, e.g., Loeckx et al. (1996, subsection 7.5.5)). Confluence may often be achieved by the *Knuth-Bendix completion algorithm* (see, e.g., Klop (1992)) which, where applicable, transforms a specification with a Noetherian but nonconfluent term rewriting system into another specification with the same initial semantics but with a Noetherian and confluent term rewriting system.

Computation by term rewriting does not precisely take place in $T(\Sigma, \Phi)$ where the carrier elements are congruence classes of terms, but in a **characteristic term algebra** $\mathcal{C}(\Sigma, \Phi)$ for $T(\Sigma, \Phi)$ where the carrier elements are the normal forms of $(T_\Sigma, \rightarrow)$. $\mathcal{C}(\Sigma, \Phi)$ is isomorphic to $T(\Sigma, \Phi)$.

Equational initial specification may be generalized to **conditional equations** (cf. Section G.9) of the form $\forall X.t_1 = u_1 \wedge \cdots \wedge t_n = u_n \Rightarrow t_{k+1} = u_{k+1}$. Most results go through, but the operational semantics of term rewriting is far more complex.

Among other approaches to abstract data type specification, we mention **loose specification** $D = (\Sigma, \Phi)$ using first-order logic, defining just the model class $\mathrm{Mod}_\Sigma(\Phi)$ as its semantics. While the degenerate models of equational logic can be avoided in first-order logic, the disadvantage is that there are non-generated models. This may be remedied by loose specifications with **free constructors** where a subset of sorts and operations may be specified with the intended meaning that the carriers of these sorts are (freely) generated. All these approaches enjoy a clean mathematical semantics but do not have an equivalent of the operational semantics of the initial approach. The latter is remedied (at the expense of mathematical semantics) by **constructive specifications**. These are particular cases of loose or initial specifications that have an *abstract programming* flavor, they allow rapid prototyping.

So far, we considered specification-in-the-small, i.e., how to create a specification by giving a signature and axioms. There is also a body of theory dealing with specification-in-the-large, i.e., how to derive a new specification from a given one by renaming, extending, forgetting or restricting sorts and operations; the algebraic essence is to handle relationships between algebras with different signatures. Modularization and parameterization concepts deal with possibly generic specification fragments and how to put them together. Further topics deal with behavioral abstraction, implementation, ordered sorts, exceptions, dynamic data types and objects. Loeckx et al. (1996) gives an in-depth treatment of these topics and references for further reading.

Chapter H

HOMOLOGICAL ALGEBRA

H.1 Foundations of Homological Algebra

by Peter Hilton in Binghampton, NY, USA

The topologist Witold Hurewicz observed in 1935 that, if X is an aspherical path-connected polyhedron, then the homotopy type of X is determined by the fundamental group $\pi_1 X$ of X. Thus, in particular, the homology groups of X are functions of $\pi = \pi_1 X$. The Swiss topologist Heinz Hopf realized that this must mean that there was a purely algebraic procedure for passing from the group π to the homology groups of X, which we may now think of as the homology groups of π. Hopf then proceeded to invent such an algebraic procedure, guided by his own experience working on the homology theory of topological spaces, and bearing in mind his seminal work on the influence of the fundamental group of a path-connected topological space on its second homology group. Hopf, working in the early 1940's, considered a *free resolution* of $\mathbb{Z}$, the additive group of integers regarded as a trivial π-module; that is, an exact sequence of π-modules and π-module homomorphisms

$$\cdots \xrightarrow{\partial} F_n \xrightarrow{\partial} F_{n-1} \xrightarrow{\partial} \cdots \xrightarrow{\partial} F_0 \xrightarrow{\varepsilon} \mathbb{Z} \tag{1}$$

where each F_n is free. One then tensors (1) with a π-module B, i.e., one takes the tensor product over π of (1) with B, to obtain a chain-complex of abelian groups

$$\cdots \xrightarrow{\partial \otimes 1} F_n \otimes_\pi B \xrightarrow{\partial \otimes 1} F_{n-1} \otimes_\pi B \xrightarrow{\partial \otimes 1} \cdots \xrightarrow{\partial \otimes 1} F_0 \otimes_\pi B \tag{2}$$

and calculates the homology groups of this chain-complex. Hopf proved that these homology groups are the homology groups of π with coefficients in the π-module B. This means that, reverting to the aspherical space X, they are the homology groups of X with local coefficients in B. Hopf's argument was this: First, the homology groups of (2) do not depend on the free resolution of $\mathbb{Z}$ chosen. Second, if $\tilde{X}$ is the universal cover of the aspherical space X and $C(\tilde{X})$ is its simplicial chain-complex, then $C(\tilde{X})$ may be given the structure of a π-complex and is then a free π-resolution of $\mathbb{Z}$ (with ε the augmentation). Third, it is then known that the homology groups of $C(\tilde{X}) \otimes_\pi B$ are the homology groups of X with local coefficients in B.

It should be noted that the Dutch topologist Hans Freudenthal, working under extremely difficult wartime conditions, appears to have followed the same path as Hopf quite independently; and that, in the United States, Samuel Eilenberg and Saunders MacLane were also working on very similar lines. We should also remark:

(a) One can, of course, carry out an analogous procedure to obtain the cohomology groups of π with coefficients in a π-module A. Indeed, this was done, independently, by Beno Eckmann in Switzerland and Eilenberg and MacLane in the United States. Moreover, Eckmann introduced a resolution based on a classical construction of algebraic topology and so obtained the cohomology ring of π.

(b) One may use a projective resolution of $\mathbb{Z}$ instead of insisting on the more restrictive free resolution—this was the point of view adopted by Henri Cartan and Samuel Eilenberg in their classical text (see References).

In that text, we find the homology of groups presented as a special case of the theory of derived functors. Moreover, we now view the category of modules over a unitary ring $\wedge$ as an important example of an abelian category with sufficient projectives and injectives. Thus an abelian category is an additive category in which one can carry out the usual operations of module theory; and the category has **enough projectives** if every object is the image of a projective object, and **enough injectives** if every object can be embedded in an injective object. One then considers an additive functor $F \colon \mathscr{A} \to \mathscr{B}$ from the abelian category $\mathscr{A}$ to the abelian category $\mathscr{B}$, and an object A of the category $\mathscr{A}$. If $\underline{P} \to A$ is a projective resolution of A, then $F\underline{P}$ is a chain-complex in $\mathscr{B}$ and its homology groups $H_n F\underline{P}$, $n = 0, 1, \ldots$, are the values of the derived functors of the functor F, evaluated on the object A. In our description of the homology groups of π with coefficients in the π-module B, we have $\mathscr{A}$ = category of (right) π-modules, $\mathscr{B}$ = category of abelian groups, $A = \mathbb{Z}$ as trivial π-module, and $FX = X \otimes_\pi B$, where $X \in \mathscr{A}$, and B is a fixed (left) π-module.

With these developments, homological algebra ceased to be closely tied to algebraic topology and became an autonomous discipline. Moreover, other areas of algebra acquired a homological aspect. Especially noteworthy are the results obtained in the cohomology theory of Lie algebras and the cohomology theory of associative algebras.

One of the great advantages of the use of the methods of homological algebra is that it yields very systematic statements and methods of proof. Let two examples suffice.

A celebrated theorem due to Maschke asserts that if G is a group of order m and K a field whose characteristic does not divide m, then the K-representations of G are completely reducible. This theorem turns out to be a special case of the result that $H^n(G : W) = 0$, $n \geq 1$, for any $K[G]$-module W. But this in turn merely depends on the observation that $m \colon W \to W$ is an automorphism and $mH^n(G : W) = 0$.

A second example concerns the famous lemmas due to Henry Whitehead in the theory of Lie algebras. The first lemma asserts that $H^1(\mathfrak{g}, A) = 0$ if $\mathfrak{g}$ is a finite-dimensional semi-simple Lie algebra and A is a finite-dimensional $\mathfrak{g}$-module; the second lemma asserts that $H^2(\mathfrak{g}, A) = 0$ if $\mathfrak{g}$ is a semi-simple Lie algebra and A is a finite-dimensional $\mathfrak{g}$-module.

For more information, consult (Brown 1982), (Cartan and Eilenberg 1956), and (Hilton and Stammbach 1997)

H.2 Universal Constructions

by Yefim Katsov in Hanover, IN, USA

Categorical algebra provides a common description of certain standard constructions—*universal constructions*, widely appearing and productively used throughout mathematics. Particularly in algebra, from given algebraic systems there can be constructed a new algebraic system, with some desirable property and morphisms (from the given systems into this new system), which is universally constructed in the sense that any other morphisms into a system with this property factor uniquely through the universal morphisms. As a rule, those universals are constructed via limiting procedures; and in many important cases, the idea of universality itself is expressed in terms of special universals—*universal elements*, which are also important for the structural theory of functors. Basic references for these topics include (Katsov 2000, Mac Lane 1998, Schubert 1972, Adámek, Herrlich, and Strecker 1990).

Colimits and limits

H.2.1   Let $|\mathcal{C}|$ denote the class of objects of the category $\mathcal{C}$. An *initial (terminal) object* in a category $\mathcal{C}$ is an object I (T) such that to every object $X \in |\mathcal{C}|$ there is exactly one morphism from I to X (from X to T). Initial and/or terminal objects might not exist; but if they do exist, all initial objects must be isomorphic as well as all terminal object being isomorphic. For example, in the category Sets of all sets and functions the empty set is the initial object, and any one-element set is a terminal object. Sometimes an object is both initial and terminal as, for instance, any one-element group in the category of groups, or the zero modules in module categories.

Let $\mathcal{A}^\mathcal{C}$ be the category of functors (or diagrams in $\mathcal{A}$, based on $\mathcal{C}$) from a *small category* $\mathcal{C}$ (*i.e.*, $|\mathcal{C}|$ is a set (not just a class)) to the category $\mathcal{A}$. If we assign to an object $A \in |\mathcal{A}|$ a *constant functor* $\Delta(A)\colon \mathcal{C} \to \mathcal{A}$ $(\Delta(A)(X) = A$, and $\Delta(A)(\alpha) = 1_A$ for any $X \in |\mathcal{C}|$, and any $\alpha \in \mathcal{C}(X,Y))$, then this assignment can be clearly extended to the full inclusion functor $\Delta\colon \mathcal{A} \to \mathcal{A}^\mathcal{C}$. For the sake of simplicity, $\Delta(A)$ is denoted just by A. Thus, a natural transformation $\varphi \in \mathcal{A}^\mathcal{C}(F, A)$ from F to $A = \Delta(A)$ is a collection $\{\varphi_X\}$ of morphisms $\varphi_X\colon FX \to A$ for each object $X \in |\mathcal{C}|$ such that $\varphi_X = \varphi_Y \circ F\alpha$ for any morphism $\alpha \in \mathcal{C}(X,Y)$.

H.2.2 Definition   Given a functor $F\colon \mathcal{C} \to \mathcal{A}$, consider the category $\mathcal{A}^F$ of *objects under* F with objects all natural transformations $\varphi \in \mathcal{A}^\mathcal{C}(F, A)$, and with morphisms f from $\varphi\colon F \to A$ to $\psi\colon F \to B$ being those morphisms $f\colon A \to B$ of $\mathcal{A}$ for which $\psi = f \circ \varphi$ in $\mathcal{A}^\mathcal{C}$. Then, a *colimit* (a *direct limit* or *inductive limit*) (Section G.1) of the functor (or diagram) F is defined to be an initial object in $\mathcal{A}^F$ and is denoted as $F \to \operatorname{Colim} F$, or just $\operatorname{Colim} F$. Like any initial objects, the $F \to \operatorname{Colim} F$ and $\operatorname{Colim} F$ are determined uniquely by the functor F, up to isomorphism in $\mathcal{A}$. Dualizing gives the category $\mathcal{A}_F$ of *objects over* F, and the *limit* (the "inverse limit" or "projective limit") $\operatorname{Lim} F$ of F is defined as a terminal object in $\mathcal{A}_F$ (cf. Section G.1). It is easy to see that the constructions $\operatorname{Colim} F$ and $\operatorname{Lim} F$ themselves can be obviously extended to functors $\operatorname{Colim} F, \operatorname{Lim} F\colon \mathcal{A}^\mathcal{C} \to \mathcal{A}$. Consider some special colimits and limits.

Coproducts and *products*. If $\mathcal{C}$ is a *discrete category* I—*i.e.*, every morphism is an identity, and thus the category can be identified with the set of its objects I—then a functor $F \in |\mathcal{A}^\mathcal{C}|$ is a collection $\{A_i \mid i \in I\}$ of objects in $\mathcal{A}$. One easily recognizes that (if they exist) $\operatorname{Colim} F$ and $\operatorname{Lim} F$ are the familiar constructions of the coproduct $\coprod_i A_i$

and product $\prod_i A_i$ in $\mathcal{A}$, respectively. We admit the ***empty category*** $\varnothing$ (it contains no objects) as discrete, and hence the colimit and limit of the diagram $\varnothing$ (if they exist) are initial and terminal objects in $\mathcal{A}$, respectively.

Coequalizers and ***equalizers***. If the category $\mathcal{C}$ is $\bullet \rightrightarrows \bullet$, then a functor $F \in |\mathcal{A}^{\mathcal{C}}|$ is a diagram $FA \rightrightarrows FB$ and it is clear that (if they exist) $\mathrm{Colim}\, F$ and $\mathrm{Lim}\, F$ are the familiar constructions of the coequalizer and equalizer of this pair of morphisms in $\mathcal{A}$, respectively.

Those special (co)limits are important inasmuch as—by taking (co)equalizers of ***universally constructed*** pairs of morphisms between appropriate (co)products—one can readily construct (co)limits of functors and prove:

H.2.3 Theorem *If in a category $\mathcal{A}$ there exist (co)products of sets of objects and (co)equalizers of pairs of morphisms, then in $\mathcal{A}$ there exist all (co)limits of functors from small categories. If $\mathcal{A}$ has a terminal (an initial) object, (co)equalizers of pairs of morphisms, and (co)products of pairs of objects, then $\mathcal{A}$ has all finite (co)limits.*

Universal elements and adjoint functors

Let $\mathsf{Sets}^{\mathcal{C}}$ be the category of functors from a small category $\mathcal{C}$ to Sets, and $F \in |\mathsf{Sets}^{\mathcal{C}}|$. There is the *category* $\int_{\mathcal{C}} F$ *of elements of* F (or the ***Grothendieck construction*** on F), having as objects all pairs (A, a) with $A \in |\mathcal{C}|$ and $a \in FA$, and as morphisms from one such pair (A, a) to another (B, b) those morphisms $\alpha \colon A \to B$ for which $F\alpha(a) = b$. F is said to be ***representable*** if $\int_{\mathcal{C}} F$ has an initial object (A, a), called a ***representation*** (or a ***universal element***) of F, and F is then said to be ***represented*** by the object A. ($\int_{\mathcal{C}} F$ can be viewed as a functorial analogue of the familiar Cayley table, and a representable functor as a one-generated free algebra.) Clearly, all Hom functors $\mathcal{C}(A, -)$ are representable. Furthermore, it is easy to show that (A, a) is a universal element of F iff $F \cong \mathcal{C}(A, -)$ in $\mathsf{Sets}^{\mathcal{C}}$ and $a \in FA$ is the isomorphic image of 1_A. From these observations, one may easily deduce the existence of natural isomorphisms $\mathsf{Sets}^{\mathcal{C}}(F, T) \cong \mathsf{Sets}^{\mathcal{C}}(\mathcal{C}(A, -), T) \cong TA$ for any $T \in |\mathsf{Sets}^{\mathcal{C}}|$ and F, represented by A (the ***Yoneda Lemma***), and extend the assignment $A \mapsto \mathcal{C}(A, -)$ to the *contravariant Yoneda full embedding* $Y^* \colon \mathcal{C} \to \mathsf{Sets}^{\mathcal{C}}$. It is clear that taking (co)products in Sets at each object $A \in |\mathcal{C}|$, one has (co)products in $\mathsf{Sets}^{\mathcal{C}}$. Therefore, ***free functors***

Sets$^{\mathcal{C}}$ are naturally defined as coproducts of representable functors, and **projective functors** are just retracts of free functors.

H.2.4 Remark One can obviously adjust all the constructions and results of **2.1.** to the *contravariant* setting, *i.e.*, for $F \in |\text{Sets}^{\mathcal{C}^{\text{op}}}|$, as well as to additive functors from a small additive category $\mathcal{C}$ (see Section H.1) to the category Ab of abelian groups.

H.2.5 Universal elements and representable functors are crucial to understanding one of the most useful concepts of categorical algebra permeating mathematics—the concept of an adjoint of a functor. Thus, given a functor $L\colon \mathcal{C} \to \mathcal{D}$ between categories $\mathcal{C}$ and $\mathcal{D}$ and given $X \in |\mathcal{D}|$, there is the composite functor $\mathcal{D}(-, X) \circ L = \mathcal{D}(L(-), X)\colon \mathcal{C}^{\text{op}} \to \text{Sets}$. Assume that for any $X \in |\mathcal{D}|$ this functor has a universal element, i.e., there are an object $A_X \in |\mathcal{C}|$ and a natural isomorphism $\alpha^X\colon \mathcal{C}(-, A_X) \xrightarrow{\sim} \mathcal{D}(L(-), X)$. Denoting $\alpha^X_{A_X}(1_{A_X})$ by $\varepsilon_X\colon L(A_X) \to X$, one observes that the pair $(\mathcal{D}(L(A_X), X), \varepsilon_X)$ is a universal element of $\mathcal{D}(L(-), X)$, and therefore for any $Y \in |\mathcal{C}|$ and $\gamma\colon L(Y) \to X$ there is a unique morphism $\overline{\gamma}\colon Y \to A_X$ such that $\varepsilon_X L(\overline{\gamma}) = \gamma$. Then, it is easy to see that the assignment $X \mapsto A_X$ is uniquely extended to a functor $R\colon \mathcal{D} \to \mathcal{C}$. These observations lead to the following important notion.

H.2.6 Proposition *A functor* $L\colon \mathcal{C} \to \mathcal{D}$ *is said to possess a* **right adjoint** *functor* $R\colon \mathcal{D} \to \mathcal{C}$ *if for any object* $X \in |\mathcal{D}|$ *the functor* $\mathcal{D}(L(-), X)$ *is representable, i.e., has a universal element* $(\mathcal{D}(L(A_X), X), \varepsilon_X)$. *The functor* $R\colon \mathcal{D} \to \mathcal{C}$ *is an extension of the assigning* $X \mapsto A_X$, *and therefore* $(\mathcal{D}(LRX, X), \varepsilon_X)$ *is a universal element of* $\mathcal{D}(L(-), X)$. *Then, for any object* $Y \in |\mathcal{C}|$ *the functor* $\mathcal{C}(Y, R(-))\colon \mathcal{D} \to \text{Sets}$ *is also representable, and* $(\mathcal{C}(Y, RLY), \eta_Y))$ *with uniquely defined* $\eta_Y\colon Y \to RLY$ *is a universal element of* $\mathcal{C}(Y, R(-))$; *the functor* L *is called a* **left adjoint** *of* R, *and this situation is denoted as* $L \dashv R$. *Finally, it is easy to show that* $L \dashv R$ *iff there is a natural isomorphism* $\mathcal{D}(L(-), -)) \xrightarrow{\sim} \mathcal{C}(-, R(-))$ *of bifunctors from* $\mathcal{C}^{\text{op}} \times \mathcal{D}$ *to* Sets.

H.2.7 Example Again, all considerations of H.2.5 and H.2.6 can obviously be adjusted to the additive setting, namely, to additive functors between additive categories $\mathcal{C}$ and $\mathcal{D}$. Thus, of numerous examples of adjoints, we mention the following classical one. Let $\mathcal{C}$ be the category

$_\Lambda\mathcal{M}$ of left modules over a ring Λ, and L_Λ a fixed right module, then there are two familiar additive functors

$$L_\Lambda \otimes - : {}_\Lambda\mathcal{M} \rightleftarrows \mathsf{Ab} : \mathsf{Ab}(L_\Lambda, -) \quad \text{and} \quad L_\Lambda \otimes - \dashv \mathsf{Ab}(L_\Lambda, -).$$

Among other useful properties of adjoint functors which can be deduced from the description of adjoints, the following is perhaps the most practical. If $L : \mathcal{C} \rightleftarrows \mathcal{D} : R$, $L \dashv R$, and for functors $F : \mathcal{A} \to \mathcal{C}$ and $G : \mathcal{B} \to \mathcal{D}$ the colimit and limit $C = \mathrm{Colim}\, F$ and $D = \mathrm{Lim}\, G$ exist, then $L(C) = \mathrm{Colim}\, LF$ and $R(D) = \mathrm{Lim}\, RG$; in other words, left adjoints and right adjoints **preserve colimits and limits**, respectively. However, left (right) adjoints in general do not preserve limits (colimits). For example, the functor $L_\Lambda \otimes -$ preserves limits ($=$ embeddings) iff the module L_Λ is *flat* (flat modules constitute an extremely important homological class of modules).

Finally, universal constructions are successfully used for obtaining structural algebraic results—namely, theorems that assert that an algebraic structure we are interested in can be obtained and/or described as a colimit of a special diagram consisting of well known and described structures. For instance, a module is flat iff it is a colimit of a **filtered diagram** of finitely generated free modules (Govorov and Lazard); also, a functorial version of this result in a non-additive setting can be found in Katsov (2000).

H.3 Projective and Injective Resolutions
by Yefim Katsov in Hanover, IN, USA

The theory of derived functors of an additive functor from the category $_R\mathcal{M}$ of left modules over a ring R to the category Ab of abelian groups constitutes the main motif and core of homological algebra. This theory—based on the concepts of complexes and resolutions of R-modules—can be also regarded as a far-reaching generalization of the general algebraic idea of extensions (constructions) of algebraic structures with given sub- and factor-structures. Since the key points of this section obviously can be presented in the realms of abelian categories (Section H.1), the reader should not be confused by free interchanges of categorical and module contexts in our discussion. For thorough expositions of these topics, we recommend the classics (Cartan and Eilenberg 1956) and (Mac Lane 1963), as well as (Osborne 2000), (Hilton and Stammbach 1997), and (Weibel 1994). The tenth chapters of the last two books also reflect the more recent theory of derived categories, which are of general mathematical interest and importance and a detailed exposition of which can be found in (Gel'fand and Manin 1996).

Chain complexes and homotopies

H.3.1 Definition A *chain complex* $\mathbf{C}$ is a sequence of R-modules C_n and homomorphisms d_n: $\cdots \to C_{n+1} \overset{d_{n+1}}{\to} C_n \overset{d_n}{\to} C_{n-1} \to \cdots$, $n \in \mathbb{Z}$, such that $d_n d_{n+1} = 0$ for each n. The submodule $Ker\ d_n$ of C_n is denoted by $Z_n(\mathbf{C})$; its elements are called *cycles*. The submodule $Im\ d_{n+1}$ of C_n is denoted $B_n(\mathbf{C})$; its elements are called *boundaries*. $B_n(\mathbf{C}) \subset Z_n(\mathbf{C})$ since $d_n d_{n+1} = 0$, and $H_n(\mathbf{C}) = Z_n(\mathbf{C})/B_n(\mathbf{C})$ is called the nth *homology module* of $\mathbf{C}$. A *chain map* $\varphi \colon \mathbf{C} \to \mathbf{D}$ between chain complexes $\mathbf{C}$ and $\mathbf{D}$ is a family $\{\varphi_n \mid C_n \to D_n\}$, $n \in \mathbb{Z}$, of homomorphisms such that, for every $n \in \mathbb{Z}$, $\varphi_{n-1} d_n = \tilde{d}_n \varphi_n$.

Thus, we obtain the category $\mathbf{CC}(_R\mathcal{M})$ of chain complexes and maps over $_R\mathcal{M}$. Clearly, $\varphi \colon \mathbf{C} \to \mathbf{D}$ maps the boundaries $B_n(\mathbf{C})$ and cycles $Z_n(\mathbf{C})$ into $B_n(\mathbf{D})$ and $Z_n(\mathbf{D})$, respectively, and hence induces the map $\varphi_* \colon H_n(\mathbf{C}) \to H_n(\mathbf{D})$ between the homology modules and additive functors $H_n \colon \mathbf{CC}(_R\mathcal{M}) \to _R\mathcal{M}$.

H.3.2 Examples

1. Let $\mathbf{C}$ be a complex with *zero differentiation*, i.e., $d_n = 0$ for all n. Then, clearly $H_n(\mathbf{C}) \approx C_n$ for all n.

2. A complex $\mathbf{C}$ is an *exact sequence* (or *acyclic sequence*) iff $\ker d_n = \operatorname{im} d_{n+1}$ for every $n \in \mathbb{Z}$. In particular, any short exact sequence $0 \to A \rightarrowtail B \twoheadrightarrow C \to 0$ can be obviously viewed as an acyclic complex. Clearly, a complex $\mathbf{C}$ is acyclic iff $H_n(\mathbf{C}) = 0$ for all n.

3. If $\mathbf{C}$ is a complex and F is a functor, then $F(\mathbf{C})$: $\cdots \to F(C_{n+1}) \overset{F(d_{n+1})}{\to} F(C_n) \overset{F(d_n)}{\to} F(C_{n-1}) \to \cdots$ is also a complex. However, $F(\mathbf{C})$ is usually not acyclic even $\mathbf{C}$ is such. For example, let $R = \mathbb{Z}_4$, and consider a projective resolution of $\mathbb{Z}_2 \in |_{\mathbb{Z}_4}\mathcal{M}|$ (see Definition H.3.10): $(\mathbf{P}, \varepsilon)$: $\cdots \to \mathbb{Z}_4 \overset{\times 2}{\to} \mathbb{Z}_4 \overset{\times 2}{\to} \mathbb{Z}_4 \overset{\times 2}{\to} \mathbb{Z}_4 \overset{\varepsilon}{\twoheadrightarrow} \mathbb{Z}_2 \to 0$. Tensoring with $\mathbb{Z}_2$, i.e., applying the functor $\mathbb{Z}_2 \otimes_{\mathbb{Z}_4} - \colon _{\mathbb{Z}_4}\mathcal{M} \to \mathsf{Ab}$, and dropping the last $\mathbb{Z}_2 \otimes_{\mathbb{Z}_4} \mathbb{Z}_2$, we obtain the chain complex $\mathbb{Z}_2 \otimes_{\mathbb{Z}_4} \mathbf{P}$: $\cdots \to \mathbb{Z}_2 \overset{0}{\to} \mathbb{Z}_2 \overset{0}{\to} \mathbb{Z}_2 \overset{0}{\to} \mathbb{Z}_2 \to 0$. Therefore, calculating its homology modules (groups) $H_n(\mathbb{Z}_2 \otimes_{\mathbb{Z}_4} \mathbf{P})$, we

have $H_n(\mathbb{Z}_2 \otimes_{\mathbb{Z}_4} \mathbf{P}) \approx \mathrm{Tor}_n^{\mathbb{Z}_4}(\mathbb{Z}_2, \mathbb{Z}_2) \approx \mathbb{Z}_2$ for all n (see H.3.12 and H.3.13).

4. Now let $R = \mathbb{Z}$, and consider a projective resolution of $\mathbb{Z}_p \in$ $|_{\mathbb{Z}}\mathcal{M} = \mathrm{Ab}| : (\mathbf{P}, \varepsilon) : \cdots \to 0 \to \mathbb{Z} \xrightarrow{\times p} \mathbb{Z} \twoheadrightarrow \mathbb{Z}_p \to 0$. Similarly to the previous example, but applying now the contravariant Hom functor $\mathrm{Ab}(-, A)$, we obtain $\mathrm{Ab}(\mathbf{P}, A) : \cdots \leftarrow 0 \leftarrow 0 \leftarrow \cdots \leftarrow$ $0 \leftarrow A \xleftarrow{\times p} A \leftarrow 0$. Thus, we have $H_0(\mathrm{Ab}(\mathbf{P}, A)) \approx \mathrm{Ext}_{\mathbb{Z}}^0(\mathbb{Z}_p, A) \approx$ $\{x \in A \mid px = 0\}$ and $H_1(\mathrm{Ab}(\mathbf{P}, A)) \approx \mathrm{Ext}_{\mathbb{Z}}^1(\mathbb{Z}_p, A) \approx A/pA$ (cf. H.3.14).

H.3.3 Remark Considering $\mathbb{Z}$ as a category consisting of the arrows $\cdots \to n + 1 \to n \to n - 1 \to \cdots$, all their composites, and the identity arrows for each integer-object, one may easily observe that a chain complex $\mathbf{C}$ of R-modules can be defined as a functor, from $\mathbb{Z}$ to the category $_R\mathcal{M}$, carrying each composite of non-identity arrows in $\mathbb{Z}$ to 0 in $_R\mathcal{M}$; and a chain map $\varphi : \mathbf{C} \to \mathbf{D}$ between chain complexes $\mathbf{C}$ and $\mathbf{D}$ is a natural transformation of the functors. Also, it is easy to see that $\mathrm{CC}(_R\mathcal{M})$ is a full abelian subcategory of the abelian category $_R\mathcal{M}^{\mathbb{Z}}$ of functors from $\mathbb{Z}$ to $_R\mathcal{M}$.

Accepting the functorial viewpoint, one may readily see that the sequence $\mathbf{A} \xrightarrow{\mu} \mathbf{B} \xrightarrow{\pi} \mathbf{C}$ of complexes is short exact iff $0 \to A_n \xrightarrow{\mu_n} B_n \xrightarrow{\pi_n} C_n \to 0$ is exact for all n. Then, by diagram chasing, one may establish the existence of the long exact homology sequences—one of the most important computational tools of homological algebra.

H.3.4 Theorem *Let* $\mathbf{A} \xrightarrow{\mu} \mathbf{B} \xrightarrow{\pi} \mathbf{C}$ *be a short exact sequence in* $\mathrm{CC}(_R\mathcal{M})$. *Then there exist natural (connecting) maps* $\partial_n : H_n(\mathbf{C}) \to$ $H_{n-1}(\mathbf{A})$ *such that*

$$\cdots \xrightarrow{\pi_*} H_{n+1}(\mathbf{C}) \xrightarrow{\partial_{n+1}} H_n(\mathbf{A}) \xrightarrow{\mu_*} H_n(\mathbf{B}) \xrightarrow{\pi_*} H_n(\mathbf{C}) \xrightarrow{\partial_n} H_{n-1}(\mathbf{A}) \xrightarrow{\mu_*} \cdots$$

is an exact sequence.

H.3.5 Definition Let $\varphi, \psi : \mathbf{C} \to \mathbf{D}$ be chain maps. A **chain homotopy** s from φ to ψ consists of maps $s_n : C_n \to D_{n+1}$ such that $d_{n+1}s_n + d_n s_{n-1} = \varphi_n - \psi_n$ for all n. φ and ψ are **chain homotopic**, $\varphi \simeq \psi$, if there exists a chain homotopy s from φ to ψ. $\mathbf{C}$ and $\mathbf{D}$ are **chain homotopy equivalent**, $\mathbf{C} \approx \mathbf{D}$, if there are chain maps $\varphi : \mathbf{C} \rightleftarrows \mathbf{D} : \psi$, called **chain homotopy equivalences**, such that $\psi\varphi \simeq 1_{\mathbf{C}}$ and $\varphi\psi \simeq 1_{\mathbf{D}}$.

One easily observes:

H.3.6 Proposition
(1) $\simeq$ is an equivalence relation on $\mathbf{CC}(_R\mathcal{M})$;
(2) $\varphi \simeq \psi \Rightarrow \varphi_* = \psi_* \colon H_n(\mathbf{C}) \to H_n(\mathbf{D})$ for all n. In other words, the chain homotopy classes of maps form a quotient category of the category $\mathbf{CC}(_R\mathcal{M})$, and the homology functors $H_n \colon \mathbf{CC}(_R\mathcal{M}) \to {_R\mathcal{M}}$ will factor through the quotient functor.

H.3.7 Remark All of the above obviously can be dualized for *cochain complexes* **C** of R-modules, just regarding them as contravariant functors from $\mathbb{Z}$ to $_R\mathcal{M}$. Also, it is clear that all of the concepts and conclusions of the preceding items carry over intactly for any abelian category $\mathcal{A}$ substituted for $_R\mathcal{M}$.

Resolutions and derived functors

H.3.8 Definition An object P in an abelian category $\mathcal{A}$ is *projective* if for every epi $\varepsilon \colon B \twoheadrightarrow C$ and a map $\gamma \colon P \to C$ there is a map $\beta \colon P \to B$ such that $\varepsilon\beta = \gamma$. Dualizing gives the notion of an *injective* I in $\mathcal{A}$. $\mathcal{A}$ *has enough projectives (injectives)* if to every object A of $\mathcal{A}$ there is an epi $P \twoheadrightarrow A$ (a mono $A \rightarrowtail I$) with P projective (with I injective, respectively).

H.3.9 Example (cf. C.20 and C.21) $_R\mathcal{M}$ has enough projectives and injectives, and any free module is projective. Moreover, $\mathbf{CC}(_R\mathcal{M})$ has enough projectives and injectives. In the category of abelian groups, any projective is free, and the injective groups are precisely the divisible ones. The abelian category of finite abelian groups has no projective objects.

H.3.10 Definition The category $\mathbf{CC}(\mathcal{A})$ of chain complexes over $\mathcal{A}$ contains the full subcategory $\mathbf{CC}(\mathcal{A})^+$ of *positive* chain complexes, i.e., the complexes with $C_n = 0$ for $n < 0$. A *left resolution* of an object $M \in |\mathcal{A}|$ is a positive complex **P** together with an epi $\varepsilon \colon P_0 \twoheadrightarrow M$ such that the augmented complex $(\mathbf{P}, \varepsilon) \colon \cdots \to P_2 \overset{d_2}{\to} P_1 \overset{d_1}{\to} P_0 \overset{\varepsilon}{\twoheadrightarrow} M \to 0$ over M is exact. It is a *projective resolution* if each P_n is projective. Consider the dual situation, namely: A *right resolution* of an object $M \in |\mathcal{A}|$ is a *positive cochain complex* **I** together with a mono $\mu \colon M \rightarrowtail I_0$ such that the augmented *cochain complex*

$(\mu, \mathbf{I})\colon 0 \to M \rightarrowtail I_0 \to I_1 \to I_2 \to \cdots$ is exact; it is an *injective resolution* if each I_n is injective.

By induction one obtains the following important observation:

H.3.11 Comparison Theorem *If γ is in the collection $\mathcal{A}(M, N)$ of all morphisms from M to N in $\mathcal{A}$, and if $(\mathbf{P}, \varepsilon)$ is a complex over M with each P_n projective and $(\mathbf{Q}, \varepsilon')$ is a resolution of N, then there is (unique up to chain homotopy equivalence) a chain map $\varphi\colon \mathbf{P} \to \mathbf{Q}$ in $\mathbf{CC}(\mathcal{A})(\mathbf{P}, \mathbf{Q})$ such that $\varepsilon'\varphi_0 = \varepsilon\gamma$.*

Thus, using the Comparison Theorem and the iteration procedure, one easily observes that if an abelian category $\mathcal{A}$—as for example $_R\mathcal{M}$—has enough projectives (injectives), then every object $M \in |\mathcal{A}|$ has a projective (injective) resolution (unique up to homotopy type). One of the main applications of projective and injective resolutions is defining derived functors.

H.3.12 Definition Let $F\colon \mathcal{A} \to \mathcal{B}$ be an additive functor between two abelian categories, with $\mathcal{A}$ having enough projectives. Pick a projective resolution $(\mathbf{P}, \varepsilon)$ of $M \in |\mathcal{A}|$. Then $F(\mathbf{P})$ is a chain complex in $\mathcal{B}$. If F is exact, *i.e.,* images under F of exact sequences are exact, then $(F(\mathbf{P}), F(\varepsilon))$ is a resolution of $F(M)$ in $\mathcal{B}$, and $H_0(F(\mathbf{P})) = F(M)$— which is not true for arbitrary F. Then, define $L_n F(M) = H_n(F(\mathbf{P}))$, and call it the nth **left derived functor** of F. Using the Comparison Theorem, it is easy to see that the functor $L_n F\colon \mathcal{A} \to \mathcal{B}$ is additive and is well defined (in the sense of being independent, up to isomorphism, of the choice of the projective resolution $(\mathbf{P}, \varepsilon)$). Intuitively, $L_n F$ measures the deviation of F from preserving exactness, i.e., from preserving kernels and cokernels. A functor preserving kernels (cokernels) is called *left (right) exact*. Left derived functors are most useful when F is right exact, as then $L_0 F = F$.

Let F be right exact, and $A \overset{\mu}{\rightarrowtail} B \overset{\pi}{\twoheadrightarrow} C$ be a short exact sequence in $\mathcal{A}$. Then, slightly varying Theorem H.3.4. one has the long exact sequence of derived functors: $\cdots \overset{\pi_*}{\to} L_{n+1}F(C) \overset{\partial_{n+1}}{\to} L_n F(A) \overset{\mu_*}{\to} L_n F(B) \overset{\pi_*}{\to} L_n F(C) \overset{\partial_n}{\to} L_{n-1}F(A) \overset{\mu_*}{\to} \cdots$ that is natural in F and in short exact sequences in $\mathcal{A}$. It can be shown that this condition together with $L_0 F = F$ and $L_n F(A) = 0$ for all n and projective A characterize the functors $L_n F$.

Dualizing, one comes to the concept of the nth **right derived functor** of F, namely: $R_n F(M) = H_n(F(\mathbf{I}))$, where $(\mu, \mathbf{I})$ is an injective resolution of $M \in |\mathcal{A}|$.

The two most important instances of derived functors are Tor and Ext, which are derived functors of the tensor product and Hom functors, respectively. Measuring, as derived functors, the deviation of the tensor product and Hom functors from preserving exactness, they actually measure how much modules depart from being projective, injective, or flat, which constitute the extremely important and well-studied in mathematics classes of modules (cf. C.20, C.21, C.22). They also play a crucial role in developing homological dimension theory and homological classification of rings and small additive categories.

H.3.13 Tor Let $A \in |\mathcal{M}_R|$ be a fixed right R-module. Then (see H 1–2), the functor $A \otimes_R - : {}_R\mathcal{M} \rightleftarrows \mathsf{Ab} : \mathsf{Ab}(A, -)$, and $A \otimes_R -$ is left adjoint to the Hom functor $\mathsf{Ab}(A, -)$. Hence, $A \otimes_R -$ is right exact, and we define the abelian groups $\mathrm{Tor}_n^R(A, M) = L_n(A \otimes_R -)(M)$. Thus, $\mathrm{Tor}_0^R(A, M) \cong A \otimes_R M$; and because these groups are computed by taking a projective resolution $(\mathbf{P}, \varepsilon)$ of M, and then the homology of $A \otimes_R \mathbf{P}$, for any projective module M we have $\mathrm{Tor}_n^R(A, M) = 0$ for $n > 0$.

H.3.14 Ext Given a fixed module $A \in |\mathcal{M}_R|$, there exists the left exact Hom functor $\mathcal{M}_R(A, -) : \mathcal{M}_R \to \mathsf{Ab}$, and we define the abelian groups $\mathrm{Ext}_R^n(A, M) = R_n(\mathcal{M}_R(A, -))(M)$. Again, $\mathrm{Ext}_R^0(A, M) = \mathcal{M}_R(A, M)$. Also, it can be shown that $\mathrm{Ext}_R^1(A, M)$ is isomorphic to the abelian group of Yoneda equivalence classes of short exact sequences $M \rightarrowtail E \twoheadrightarrow A$ (the extensions of A by M).

H.4 Homological Dimension

by Peter Hilton in Binghampton, NY, USA

Let $\wedge$ be a unitary ring and let A, B be (left) $\wedge$-modules. Then we may form the derived functors of the functors $\mathrm{Hom}_\wedge(A, -)$, $\mathrm{Hom}_\wedge(-, B)$ to obtain the groups $\mathrm{Ext}_\wedge^q(A, B)$. These Ext-groups allow us to introduce important ideas of dimension into homological algebra.

Thus we say that the **projective dimension** of the module A is less than or equal to m, written

$$\mathrm{proj\,dim}\, A \leq m,$$

if $\text{Ext}^n_\wedge(A, B) = 0$ for all $n > m$ and all modules B. It is easy to see that it suffices to assume that $\text{Ext}^{m+1}_\wedge(A, B) = 0$ for all modules B. It is important to remark that the property $\text{proj dim } A \leq m$ is characterized by the condition that, in every projective resolution of A,

$$\cdots \to P_n \to P_{n-1} \to \cdots \to P_1 \to P_0 \to A,$$

the image of P_m in P_{m-1} is projective ($P_{-1} = A$), so that we can find projective resolutions which terminate with the term P_m, in the sense that $P_n = 0, n > m$. Of course, we say that $\text{proj dim } A = m$ if $\text{proj dim } A \leq m$ but $\text{proj dim } A \not\leq m - 1$. Plainly, we have a dual notion of $\text{inj dim } B$. We bring these two notions together to define the **global dimension** of the ring $\wedge$ (gl dim $\wedge$). Thus we say that $\text{gl dim } \wedge \leq m$ if $\text{Ext}^q_\wedge(A, B) = 0$ for all $q > m$ and all $\wedge$-modules A, B. Once again, it suffices to assume that $\text{Ext}^{m+1}_\wedge(A, B) = 0$ for all $\wedge$-modules A, B. It is also clear that $\text{gl dim } \wedge \leq m$ precisely if all modules A satisfy $\text{proj dim } A \leq m$, or if all modules B satisfy $\text{inj dim } B \leq m$. Again, we say that $\text{gl dim } \wedge = m$ if $\text{gl dim } \wedge \leq m$ but $\text{gl dim } \wedge \not\leq m - 1$. We notice that a field $\wedge$ has global dimension 0 and a principal ideal domain $\wedge$ has global dimension 1. Let us consider the case in which $\wedge = \mathbb{Z}G$, the integral group ring of a group G. If we regard $\mathbb{Z}$ as a trivial G-module, we may discuss $\text{proj dim } \mathbb{Z}$ and we define the cohomological dimension of G, cd G, by the rule

$$\text{cd } G = \text{proj dim } \mathbb{Z}.$$

This means that, equivalently, $\text{cd } G \leq m$ if $H^q(G; B) = 0$ for all $q > m$ and all G-modules B, or if $H^{m+1}(G; B) = 0$ for all G-modules B. Thus if G is a free group, then $\text{cd } G = 1$.

H.4.1 Theorem *For the free product $G_1 * G_2$ of two groups G_1, G_2, we have*

$$\text{cd}(G_1 * G_2) = \max(\text{cd } G_1, \text{cd } G_2),$$

while, for the direct product $G_z \times G_2$, we have

$$\text{cd}(G_1 \times G_2) \leq \text{cd } G_1 + \text{cd } G_2.$$

If U is a subgroup of G of finite index and if $\text{cd } G < \infty$, then we may use the **corestriction** Cor: $H^n(U, B) \to H^n(G, B)$ to show that $\text{cd } U = \text{cd } G$. Now certainly a finite cyclic group C_k, $k \geq 2$, has infinite cohomological dimension, so that any projective resolution of $\mathbb{Z}$ as trivial C_k-module must be infinite. But this implies that, if G contains an element of finite order k, $k \geq 2$, then any projective resolution of $\mathbb{Z}$ as trivial G-module must be infinite. We conclude that, if G is not torsion free, then $\operatorname{cd} G = \infty$. Much work has been done in recent years on the homological study of torsion free groups. It is apparent that, if the group G admits an Eilenberg-MacLane space $K(G, 1)$ which is a *finite* CW-complex, then $\operatorname{cd} G < \infty$; we recall that $K(G, 1)$ is an aspherical topological space whose fundamental group is G, so that the homology (cohomology) groups of $K(G, 1)$ are, by definition, the homology (cohomology) groups of G. Of course, if G admits a *finite-dimensional* (but not necessarily finite) $K(G, 1)$, we may still conclude that $\operatorname{cd} G < \infty$. It is thus to be expected that recent research has seen a strong concentration on the study of groups G by means of properties of $K(G, 1)$ and of projective resolutions of $\mathbb{Z}$ as G-module.

Classical results of algebra have in some cases been reformulated, and reproved, in terms of homological concepts, including, of course, dimensional concepts. One of the most famous of these is Hilbert's Chain-of-Syzygies Theorem. Hilbert himself formulated the theorem in terms of presentations of polynomial ideals, but it is not difficult to recast it in homological terms.

Thus let K be a field and let P be the ring of polynomials over K in the indeterminates $x_1, x_2, \ldots, x_m$. We regard P as *internally graded* with the indeterminates x_i having grade 1. Then Hilbert's Theorem may be interpreted as follows:

H.4.2 Theorem $\operatorname{gl} \dim P = m$. *Moreover, projective (graded) P-modules are free.*

In fact, we can strengthen the second part of the statement of the theorem. For we prove by homological methods that if A is a (graded) P-module such that $\operatorname{Tor}_1^P(A, K) = 0$, then A is free. Here $\operatorname{Tor}_n^\wedge(A, B)$, $n = 0, 1, 2, \ldots$, are the derived functors of the tensor product $A \otimes_\wedge B$; of course, $\operatorname{Tor}_n^\wedge(A, B) = 0$, $n \geq 1$, if A is projective.

For more information, see (Alperin and Shalen 1982, Brown 1982, Brown and Geoghegan 1984, Cartan and Eilenberg 1956, Hilton and Stammbach 1997)

H.5 Homological Characterizations of Rings: The Commutative Case

by Sarah Glaz in Storrs, CT, USA

A large number of finiteness properties of commutative rings have homological characterizations. For example, it is well known that for a ring to be Noetherian—a condition most commonly described by the finite generation of the ideals of the ring, it is necessary and sufficient that arbitrary direct sums of injective modules be injective modules. One might speculate that this is the reason why homological algebra approaches in Noetherian settings yield such deep and beautiful results.

The same phenomena can be observed in another large class of rings, the class of coherent rings. Chase (1960) attempted to answer the homological question: for what rings arbitrary direct products of flat modules are flat modules. The answer is that this holds true precisely when the ring is coherent. Chase provides no less then seven equivalent characterizations of this homological condition. The most well known are the two equivalent finiteness conditions below:

A ring R is called a *coherent ring* (cf. Section C.22) if every finitely generated ideal of R is finitely presented.

Equivalently, R is a coherent ring if and only if for every element a of R and any two finitely generated ideals I and J of R, the ideals $I \cap J$ and $(0 : a) = \{ r \in R \mid ra = 0 \}$ are finitely generated.

Examples of coherent rings include all Noetherian rings, as well as many non-Noetherian rings, among them $k[x_1, x_2, \ldots]$ - the polynomial ring in infinitely many variables over a field k.

A number of other rings characterized by homological conditions were introduced or further developed in an attempt to answer the question: over which coherent rings are the polynomial rings in finitely many variables coherent rings. In contrast to the Noetherian situation, the coherent condition is not always inherited by the polynomial rings. We will follow here the thread of this investigation. A more detailed account of many of the results described below and other results involving coherent and related rings can be found in the books: (Vasconcelos 1976), (Glaz 1989), (Chapman and Glaz 2000), and in the article (Glaz 1992).

A ring R is called a *regular ring* (see Section C.29 and Section H.6) if every finitely generated ideal of R has finite projective dimension.

This definition coincides with the usual definition of regularity when the ring is Noetherian. Coherent rings of finite weak global dimensions are regular coherent rings, although, contrary to the Noetherian case, the converse is not necessarily true for quasi-local rings. An example of a quasi-local regular coherent ring of infinite weak global dimension is $k[[x_1, x_2, \ldots]]$ - the power series ring in infinitely many variables over a field k. The class of coherent rings of finite weak global dimension includes many of the classical rings defined by homological conditions, such as Von Neumann regular rings, semihereditary rings, and hereditary rings.

A ring R is called a ***von Neumann regular ring*** if for every element $a \in R$ there is an element $b \in R$ such that $a^2b = a$. Von Neumann regular rings have a simple homological description. R is a von Neumann regular ring if and only if the weak global dimension of R is zero; equivalently, every R module is flat. Von Neumann regular rings are coherent. The class of von Neumann regular rings includes the very applicable class of the Boolean rings. A von Neumann integral domain is a field.

A ring R is called a ***semihereditary ring*** if every finitely generated ideal of R is projective. Semihereditary rings are precisely those rings that are coherent and of weak global dimension one. A semihereditary integral domain is a Prüfer domain. The class of semihereditary rings includes valuation domains and Bezout domains.

A ring R is called ***hereditary ring*** (cf. Section C.36) if every ideal of R is projective, that is, R is a ring of global dimension one. Hereditary rings are coherent. A hereditary integral domain is a Dedekind domain.

Any coherent ring of finite weak global dimension described above, as well as any coherent ring of global dimension two, satisfies that the polynomial rings in finitely many variables over it is coherent. But there are examples of coherent integral domains R of weak global dimension two over which the polynomial ring in one variable is not a coherent ring. An example of such a ring R, constructed by Soublin and Alfonsi, is a localization at a prime ideal of the ring $S = \prod Q[[t, u]]$ - the direct product of countably many power series rings in two variables t and u over the rationals Q.

At this point we are compelled to ask to what extent the coherence or regular coherence of the base ring is reflected in coherent-like or regular-like conditions of the polynomial ring over it. The investigation in this direction is of much more recent vintage. The answers involve the definition and development of more rings characterized by homological conditions.

A ring R is called a ***PP ring (Principal Projective ring)*** if every principal ideal of R is projective.

This homological condition is actually a zero-divisor controlling condition on a ring R. In particular, a PP ring R is locally an integral domain, it satisfies that $MinR$—the space of minimal prime ideals of R, is compact in the Zariski topology, and $Q(R)$—the total ring of fractions of R, is a von Neumann regular ring.

Every regular coherent ring is a PP ring. A regular coherent ring is also locally a GCD (Greatest Common Divisor) domain, that is an integral domain in which every two elements possess a greatest common divisor. A class of rings that generalizes GCD domains is the class of G-GCD rings defined in (Glaz 2001). It has the following homological characterization:

A ring R is called *a **G-GCD ring (Generalized GCD ring)*** if R is a PP ring and the intersection of any two finitely generated flat ideals of R is a finitely generated flat ideal of R.

The class of G-GCD rings includes UFDs (Unique Factorization Domains), GCD domains and G-GCD domains – that is integral domains where finite intersections of principal ideals are invertible ideals. Regular coherent rings are G-GCD rings as well. But there are examples of coherent integral domains which are not G-GCD rings, such as $k[x^2, x^3, y, xy]$, where k is a field and x and y are variables over k. And there are examples of G-GCD rings which are not coherent, such as the polynomial ring in one variable over the Soublin-Alfonsi ring R mentioned above.

A ring R is called a ***finite conductor ring*** if for any two elements a and b of R the ideals $(a : b) = \{\, r \in R \mid rb \in aR \,\}$ are finitely generated.

Thus, the class of G-GCD rings and the class of coherent rings include the class of regular coherent rings and are included in the class of finite conductor rings. Moreover, a regular coherent ring and a G-GCD ring have in common the properties of being locally GCD domains and of possessing von Neumann regular total rings of fractions.

Glaz (2001) has shown that if R is a regular coherent ring, then $R[x]$ – the polynomial ring in one variable over R, is a G-GCD ring.

If R is a G-GCD ring and an integral domain then the polynomial rings in finitely many variables over R are G-GCD rings. It is not yet known if this result holds true for G-GCD rings which are neither regular coherent rings nor integral domains.

The diagram below provides an overview of the various containment relationships between the rings mentioned in this section. An arrow signifies containment of the class of rings at the start of the arrow into the class of rings to which the arrow points.

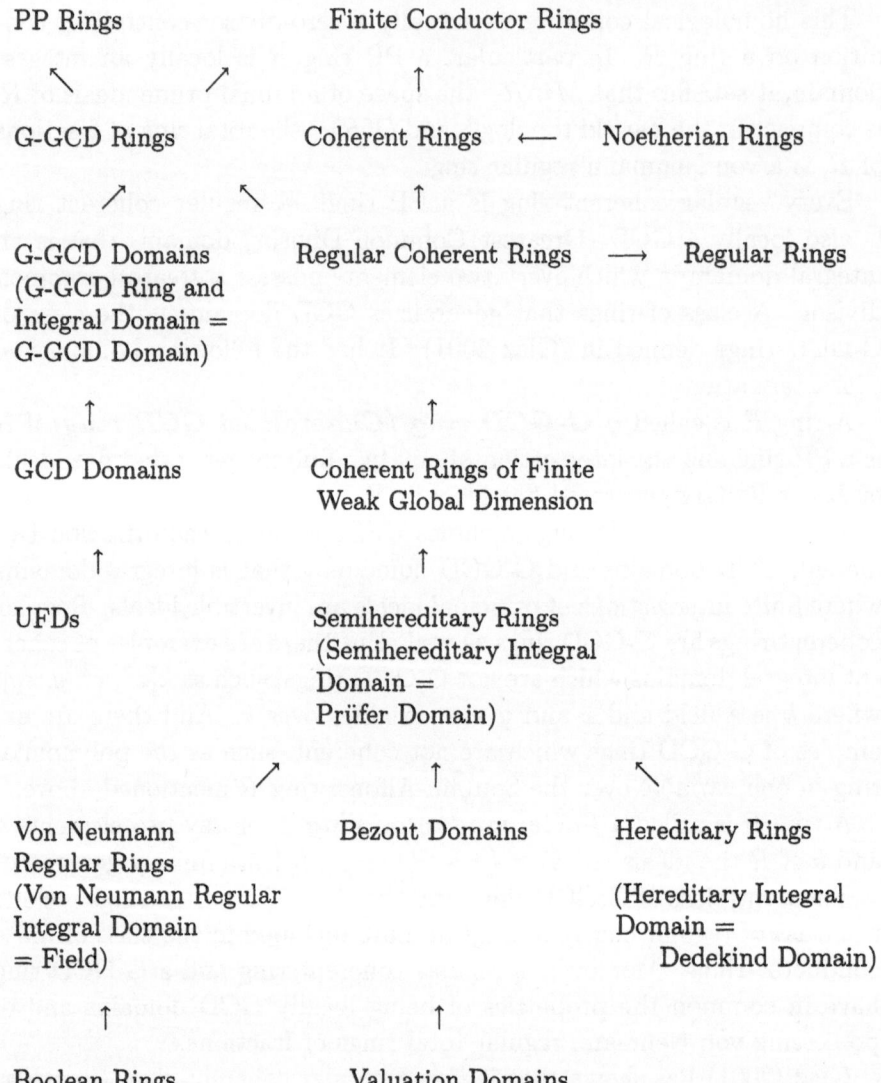

PP Rings   Finite Conductor Rings

G-GCD Rings   Coherent Rings   ⟵   Noetherian Rings

G-GCD Domains   Regular Coherent Rings   ⟶   Regular Rings
(G-GCD Ring and
Integral Domain =
G-GCD Domain)

GCD Domains   Coherent Rings of Finite
                            Weak Global Dimension

UFDs   Semihereditary Rings
                            (Semihereditary Integral
                            Domain =
                            Prüfer Domain)

Von Neumann   Bezout Domains   Hereditary Rings
Regular Rings
(Von Neumann Regular   (Hereditary Integral
Integral Domain   Domain =
= Field)   Dedekind Domain)

Boolean Rings   Valuation Domains

H.6 Homological Characterizations of Rings: The General Case

by Alexander V. Mikhalev and Askar A. Tuganbaev in Moscow, Russia

All rings are assumed to be associative and with nonzero identity element. Expressions such as a "Noetherian ring" mean that the corresponding right and left conditions hold. Homological classification of rings is the part of ring theory that consider characterizations of a ring by properties of the category of all its right (left) modules.

A ring is said to be *semisimple* or *classically semisimple* if it is isomorphic to a finite direct product of full matrix ring over division rings (Section C.16). A nonzero module M is said to be *simple* if all its nonzero submodules coincide with M. A module is said to be *semisimple* if it is a direct sum of simple modules. A ring A is semisimple if and only if A is right semisimple if and only if A is left semisimple if and only if all right A-modules and all left A-modules are semisimple.

A module M is said to be *injective* (Section C.21) if for every module N and any submodule N' of N, each homomorphism $N' \to M$ can be extended to a homomorphism $N \to M$. A ring A is said to *right self-injective* if the module A_A is injective. For a ring A, a right A-module M is said to be *free* if there is a set J such that M is isomorphic to a direct sum of J isomorphic copies of the module A_A. A right A-module M is said to be *projective* (Section C.20) if M satisfies the following equivalent conditions: (1) M is a direct summand of a free module; (2) for every A-module epimorphism $h : N \to \overline{N}$ and each homomorphism $\overline{f} : M \to \overline{N}$, there is a homomorphism $f : M \to N$ with $\overline{f} = hf$.

A ring A is semisimple if and only if all right A-modules and all left A-modules are injective and projective if and only if all simple right A-modules are projective if and only if all cyclic right A-modules are injective. A module M is said to be π-*injective* if each idempotent endomorphism of any its submodule can be extended to an endomorphism of M. A module M is said to be π-*projective* if for any two its submodules X and Y with $X + Y = M$, there is an endomorphism f of M such that $f(M) \subseteq X$ and $(1 - f)(M) \subseteq Y$. A ring A is semisimple if and only if all 2-generated right A-modules are π-injective if and only if all 2-generated right A-modules are π-projective.

A right A-module M is said to be *flat* if for each left A-module N and any submodule N' of N, the natural group homomorphism $M \otimes_A N' \to M \otimes_A N$ is a monomorphism. A module M is said to be *finitely presented* if there are a finitely generated projective module P and a finitely generated submodule Q of M such that $M \cong P/Q$. A ring A is said to be (*von Neumann*) *regular* or *absolutely flat* if for any element $a \in A$, there is an element $b \in A$ with $a = aba$. A ring is regular if and only if all right A-modules and all left A-modules are flat if and only if all finitely presented cyclic right A-modules are flat.

H.6.1 Theorem *For a ring A, the following conditions are equivalent:*

(1) *All submodules of right A-modules are flat;*
(2) *All submodules of left A-modules are flat;*
(3) *For every finitely generated right ideal B of A and each finitely generated left ideal C of A, the natural group homomorphism $B \otimes_A C \to BC$ is an isomorphism;*
(4) *All finitely generated right ideals of A are flat;*
(5) *All finitely generated left ideals of A are flat.*

If M is a right (resp. left) module over a ring A and X is a subset of M, then we denote by $r(X)$ (resp. $\ell(X)$) the right (resp. left) annihilator of X in A. A right A-module M is said to be **coherent** if M satisfies the following equivalent conditions: (1) each finitely generated submodule of M is finitely presented; (2) the intersection of any two finitely generated submodules of M is finitely generated and $r(m)$ is a finitely generated right ideal of A for any element $m \in M$. A ring A is left coherent if and only if all direct products of flat right A-modules are flat.

For a module P, a submodule X of P is said to be **small** (in M) if $X + Y \neq P$ for every proper submodule Y of P. If M is a module and there is an epimorphism $f : P \to M$ such that the module P is projective and $\mathrm{Ker}(f)$ is a small submodule of P, then P is called a **projective cover** of M. A finitely generated module M with Jacobson radical $J(M)$ is said to be **semilocal** if the factor module $M/J(M)$ is semisimple. A ring A is said to be **semiperfect** if A is semilocal and all idempotents of the factor ring $A/J(A)$ can be lifted to an idempotent of A. A ring A is semiperfect if and only if every finitely generated right A-module has a projective cover.

A subset B of a ring A is said to be **right t-nilpotent** if for any sequence $b_1, b_2 \ldots$ of elements in B, there is a subscript k such that $b_k b_{k-1} \cdots b_1 = 0$. A ring A is said to be **right perfect** if A is semilocal and the ideal $J(A)$ is right t-nilpotent.

H.6.2 Theorem *For a ring A, the following statements are equivalent:*

(1) A *is right perfect;*
(2) *All flat right A-modules are projective;*
(3) *Every right A-module has a projective cover;*
(4) A *is semilocal and every nonzero right A-module has a maximal submodule;*

(5) *A is semilocal and every nonzero left A-module has a minimal submodule.*

A module with the minimum condition on submodules is called an **Artinian module**. An injective module M is called an **injective hull** of its submodule N if $N \cap X \neq 0$ for every nonzero submodule X of M. A ring is right Artinian if and only if all nonzero injective right A-modules are direct sums of injective hulls of simple modules. A module is said to be **Noetherian** if all its submodules are finitely generated.

H.6.3 Theorem *For a ring A, the following statements are equivalent:*

(1) *A is right Noetherian;*

(2) *All finitely generated right A-modules are Noetherian;*

(3) *All direct sums of injective right A-modules are injective;*

(4) *All countable direct sums of injective right A-modules are injective;*

(5) *All countable direct sums of injective right A-modules are π-injective;*

(6) *For any injective right A-module M, the module $M^{(\aleph_0)}$ is injective;*

(7) *All injective right A-modules are direct sums of indecomposable modules;*

(8) *There is a cardinal $t(A)$ such that each injective right A-module is a direct sum of modules whose cardinalities do not exceed $t(A)$.*

A module is said to be **hereditary** (resp. **semihereditary**) if all its submodules (resp. all its finitely generated submodules) are projective.

H.6.4 Theorem *For a ring A, the following statements are equivalent:*

(1) *A is right hereditary;*

(2) *All projective right A-modules are hereditary;*

(3) *All submodules of projective right A-modules are π-projective;*

(4) *All factor modules of injective right A-modules are injective;*

(5) *All factor modules of injective right A-modules are π-injective.*

A ring A is right semihereditary if and only if all projective right A-modules are semihereditary.

A module is said to be **uniserial** if any two of its submodules are comparable with respect to inclusion. A module is said to be **serial** if it is a direct sum of uniserial modules. A ring A is serial if and only if all finitely presented right A-modules are serial if and only if all finitely presented left A-modules are serial. A ring A is serial and Artinian if and only if all right A-modules are serial if and only if all left A-modules are serial.

H.6.5 Theorem *For a ring A, the following conditions are equivalent:*

(1) *All finitely generated right A-modules are serial;*

(2) *All 2-generated right A-modules are serial;*

(3) *A is a right serial ring and for every isomorphism $f : M \to N$ between arbitrary submodules M and N of any primitive cyclic A-modules xA and yA, respectively, either f or f^{-1} can be extended to a homomorphism between xA and yA.*

A ring A is called a **quasi-Frobenius** ring or a **QF-ring** if A satisfies the following equivalent conditions: (1) A is a self-injective Artinian ring; (2) A is an Artinian ring, $B = r(\ell(B))$ for each right ideal B of A, and $C = \ell(r(C))$ for each left ideal C of A.

H.6.6 Theorem *For a ring A, the following conditions are equivalent:*

(1) *A is a QF-ring;*

(2) *All injective right A-modules are projective;*

(3) *All projective right A-modules are injective;*

(4) *All flat right A-modules are injective;*

(5) *All injective right A-modules are π-projective;*

(6) *All projective right A-modules are π-injective.*

A module M is said to be **quasi-injective** if for any submodule M' of M, each homomorphism $M' \to M$ can be extended to an endomorphism of M. A module M is said to be **quasi-projective** if for every epimorphism $h : M \to \overline{M}$ and each homomorphism $\overline{f} : M \to \overline{M}$, there is an endomorphism f of M with $\overline{f} = hf$. A module M is said to be **skew-injective** if each endomorphism of any its submodule can be extended to an endomorphism of M. A module M is said to be **skew-projective** if each endomorphism of any its factor module is lifted to an endomorphism of M.

H.6.7 Theorem *For a ring A, the following conditions are equivalent:*

(1) *A is a QF-ring;*

(2) *All quasi-injective right A-modules are quasi-projective;*

(3) *All quasi-projective right A-modules are quasi-injective;*

(4) *All skew-injective right A-modules are skew-projective;*

(5) *All skew-projective right A-modules are skew-injective;*

(6) *A is an Artinian ring whose right or left ideals are principal.*

A ring A is called a **right V-ring** if A satisfies the following equivalent conditions: (1) all simple right A-modules are injective; (2) each right ideal of A is the intersection of maximal right ideals of A. A commutative ring A is a V-ring if and only if A is regular. There is a field F with derivation δ such that the differential polynomial ring $F[x, \delta]$ is a simple nonregular right V-domain. Let M be an infinite dimensional right vector space over a field F, S be the ideal of $\mathrm{End}(M_F)$ consisting of linear transformations of finite rank, and let A be the subring in $\mathrm{End}(M_F)$ generated by S and all scalar transformations. Then A is a regular right V-ring that is not a left V-ring.

For more information on the above classes of rings, see (Anderson and Fuller 1992), (Cartan and Eilenberg 1956), (Kasch 1982), (Faith 1976), (Tuganbaev 1998), and (Wisbauer 1991).

H.7 Algebraic K-Theory I: K_0 and K_1

by Jonathan M. Rosenberg in College Park, MD, USA

Introduction

K-theory arose out of Grothendieck's study of *class groups* as part of his work on the generalized Riemann-Roch Theorem. Indeed, Grothendieck has claimed (Bak 1987) that he took the letter K from the German word *Klasse*. Many of the classical examples of K-groups go back to the 19th century, though of course the modern categorical approach is much more recent. For the early history of K-theory, we recommend Weibel (1999a). For general expositions of algebraic K-theory, some standard references are the classics Bass (1968) and Milnor (1971) for the pre-Quillen era, and for the more recent theory, Lluis-Puebla et al. (1992), Rosenberg (1994), Srinivas (1996), and Weibel (1999b). The journal literature on K-theory up through 1984 is also nicely indexed and summarized in Magurn (1985).

The K_0 functor

H.7.1 Definition If R is a ring, the isomorphism classes of finitely generated projective (left) R-modules, with the operation $\oplus$, form a monoid $\mathrm{Proj}\,R$. In general this monoid can be very complicated, so the best way to make sense of it is to introduce its **Grothendieck group** or **group completion**, denoted $K_0(R)$. This is the group of formal differences $[P] - [Q]$, where P and Q are finitely generated projective R-modules, subject to the relations that $[P \oplus Q] = [P] + [Q]$. Two projective by means of elementary row and column operations, in the sense of linear algebra.

The definition of K_1 is motivated by the notion of *Whitehead torsion* in topology (see Milnor (1966)), which takes its values in a quotient of $K_1(\mathbb{Z}[G])$, where G is the fundamental group of the spaces being studied.

When R is commutative, the determinant induces a homomorphism $K_1(R) \rightarrow R^\times = GL(1; R)$, which is split by the obvious inclusion $GL(1; R) \hookrightarrow GL(R)$. Thus, in this case, we have a splitting $K_1(R) \cong R^\times \times SK_1(R)$, where $SK_1(R)$ is the image in $K_1(R)$ of $SL(R) = \varinjlim SL(n; R)$.

H.7.10 Example If F is a field, or more generally a Euclidean ring, every square matrix over F with determinant 1 can be reduced to the identity by means of elementary row and column operations. Thus $SK_1(F)$ vanishes and $K_1(F) \cong F^\times$.

H.7.11 Remark However, there are PID's for which SK_1 is non-trivial, so the computation of $K_1(R)$ can be difficult even for relatively well-behaved rings.

The fundamental theorem and lower K

At first sight, the definitions of K_0 (Definition H.7.1) and K_1 (Definition H.7.9) appear to be quite different from one another. Nevertheless, the two functors are closely linked through an exact sequence (Theorem H.7.12) and through the *Fundamental Theorem* (Theorem H.7.13).

H.7.12 Theorem *Suppose R is a ring and I is a two-sided ideal in R. Let $i: I \hookrightarrow R$ be the inclusion (a map of rngs!) and let $q: R \rightarrow R/I$ be the quotient map. Then there is a long exact sequence*

$$K_1(R, I) \rightarrow K_1(R) \xrightarrow{q_*} K_1(R/I) \xrightarrow{\partial} K_0(I) \xrightarrow{i_*} K_0(R) \xrightarrow{q_*} K_0(R/I).$$

The sequence is functorial for pairs (R, I).

H.7.13 Theorem (Bass-Heller-Swan) *If R is a left regular ring, then $K_0(R[t]) \cong K_0(R)$, $K_1(R[t]) \cong K_1(R)$, and $K_1(R[t, t^{-1}]) \cong K_1(R) \oplus K_0(R)$. Even if R is not left regular, $K_1(R) \oplus K_0(R)$ is a split summand in $K_1(R[t, t^{-1}])$, and the other two summands are the cokernels of $K_1(R) \hookrightarrow K_1(R[t])$ and $K_1(R) \hookrightarrow K_1(R[t^{-1}])$.*

modules P and Q define the same class in $K_0(R)$ exactly when they are ***stably isomorphic***, that is, $P \oplus R^n \cong Q \oplus R^n$ for some n.

Note that the finitely generated *free* R-modules always define a canonical cyclic subgroup of $K_0(R)$. (This cyclic subgroup is isomorphic to $\mathbb{Z}$ if R is commutative, but can vanish for highly non-commutative rings.) The quotient of $K_0(R)$ by this distinguished subgroup is called the ***reduced K_0-group*** $\tilde{K}_0(R)$. When R is commutative, $K_0(R)$ splits as $\mathbb{Z} \oplus \tilde{K}_0(R)$.

H.7.2 Example If F is a field (or a division ring), a finitely generated projective F-module is just a finite-dimensional F-vector space, and is determined up to isomorphism by its dimension. So $\operatorname{Proj} F \cong \mathbb{N}_0$ and $K_0(F) \cong \mathbb{Z}$.

H.7.3 Example If R is a PID or a local ring, every projective R-module is free. So again in these cases, it is easy to see that $\operatorname{Proj} R \cong \mathbb{N}_0$ and $K_0(R) \cong \mathbb{Z}$. In particular, $K_0(\mathbb{Z}) \cong \mathbb{Z}$.

H.7.4 Remark Since every finitely generated projective left R-module is of the form $R^n p$ for some n and for some idempotent matrix $p \in \operatorname{Mat}_{n \times n}(R)$, $\operatorname{Proj} R$ can also be described as a group of *stable isomorphism classes* of idempotent matrices, where idempotent matrices lie in the same class if they are conjugate under $\operatorname{GL}(n; R)$. If one uses this point of view, the semigroup operation in $\operatorname{Proj} R$ corresponds to addition of orthogonal idempotents. Since finitely generated projective right R-modules are also described the same way in terms of idempotent matrices, using right modules instead of left modules would make no difference as far as $K_0(R)$ is concerned.

H.7.5 Definition It is easy to see that K_0 is in fact a *functor* Ring $\to$ AbGrp. We can extend it to the category Rng of rngs (i.e., rings which may not have an identity) as follows. If R is a rng, adjoin an identity to it to make the ring $\tilde{R}$ (which as an abelian group is just $R \oplus \mathbb{Z} \cdot 1$). We have a split exact sequence

$$0 \to R \to \tilde{R} \overset{\leftharpoonup}{\to} \mathbb{Z} \to 0$$

and define $K_0(R) = \operatorname{Ker}(K_0(\tilde{R}) \to K_0(\mathbb{Z})) = \tilde{K}_0(\tilde{R})$. An easy check shows that this is functorial and that it agrees with Definition H.7.1 if R already has a unit.

Two of the main motivations for studying K_0 come from vector bundles and from number theory. The following theorem, Theorem H.7.6, provides a useful dictionary for going back and forth between topology and algebra. And Theorem H.7.8 provides a link with the class group of a number field.

H.7.6 Theorem (Swan) *Let X be a compact Hausdorff space, and let $K = \mathbb{R}$ or $\mathbb{C}$. Then there is an equivalence of categories between the category of K-vector bundles over X and the category of finitely generated projective modules over $C(X)$, the ring of continuous K-valued functions on X. In particular, $K_0(C(X))$ coincides with the Grothendieck group $K^0(X)$ of vector bundles over X.*

H.7.7 Remark Similarly, if X is a smooth affine variety and R is its coordinate ring (so that $X = \operatorname{Spec} R$), then $K_0(R)$ coincides with the Grothendieck group of *algebraic* vector bundles over X.

H.7.8 Theorem *If R is a Dedekind ring, then $\widetilde{K}_0(R)$ is canonically isomorphic to the class group of R. In particular, if R is the ring of integers in a number field K, then $|\widetilde{K}_0(R)| = h_K$, the class number of K.*

If G is a cyclic group of prime order p, then the group ring $\mathbb{Z}[G]$ surjects onto the ring of integers in the cyclotomic number field $F = \mathbb{Q}[e^{2\pi i/p}]$. Using this, Theorem H.7.8, and the exact sequences of K-theory, one can show that $\widetilde{K}_0(\mathbb{Z}[G])$ is isomorphic to the class group of F, which is non-trivial if $p \geq 23$. The group $\widetilde{K}_0(\mathbb{Z}[G])$ appears in topology, where it is the receptacle for the *Wall finiteness obstruction* for spaces with fundamental group G.

The K_1 functor

H.7.9 Definition If R is a ring, $K_1(R)$ is defined to be the abelianization of the **stable general linear group** $GL(R) = \varinjlim GL(n; R)$, or $GL(R)/[GL(R), GL(R)]$. It is easy to see that $R \mapsto K_1(R)$ defines a functor Ring $\rightarrow$ AbGrp.

The commutator subgroup $[GL(R), GL(R)]$ turns out to have another description as $E(R)$, the group generated by **elementary matrices**, that is, matrices with 1's down the diagonal and with only one non-zero off-diagonal entry. Thus $K_1(R)$ measures the (stable) obstruction to being able to reduce an arbitrary invertible matrix over R to the identity

It is natural to want to extend the exact sequence of Theorem H.7.12 to the right, and Theorem H.7.13 provides a way to do this: one can define functors of rngs K_{-n} for all $n \in \mathbb{N}$, so that the exact sequence continues with:

$$\xrightarrow{\partial} K_0(I) \xrightarrow{i_*} K_0(R) \xrightarrow{q_*} K_0(R/I) \tag{1}$$

$$\xrightarrow{\partial} K_{-1}(I) \xrightarrow{i_*} K_{-1}(R) \xrightarrow{q_*} K_{-1}(R/I) \tag{2}$$

$$\xrightarrow{\partial} K_{-2}(I) \xrightarrow{i_*} K_{-2}(R) \xrightarrow{q_*} K_{-2}(R/I) \tag{3}$$

$$\xrightarrow{\partial} \cdots . \tag{4}$$

Here $K_{-n}(R)$ is a split summand in $K_{-n+1}(R[t, t^{-1}])$, just as in Theorem H.7.13. The functors K_{-n} vanish for left regular rings.

H.7.14 Example If $I = (m)$ is the rng of integers divisible by m, then $K_{-1}(I)$ is free abelian of rank $k - 1$, where k is the number of prime divisors of m, and $K_{-n}(I) = 0$ for $n \geq 2$. It is also known that $K_{-n}(\mathbb{Z}[G]) = 0$ for G a finite group and for $n \geq 2$.

H.8 Algebraic K-theory II: K_2, Milnor K-theory, and Symbols

by Jonathan M. Rosenberg in College Park, MD, USA

Milnor's K_2 functor

As we saw in Theorem H.7.12 and (1), given any ring R and (two-sided) ideal I, there is a long exact sequence of K-groups, starting with K_1, functorial in the pair (R, I). This behaves much like the long exact homology sequence of a pair of spaces. The problem of extending this sequence to the left was solved by Milnor (1971) with his K_2 functor, which turns out to have many applications in number theory, quadratic forms, and the structure theory of the classical groups.

H.8.1 Definition Let G be a group. A *central extension* of G is a short exact sequence of groups

$$1 \to A \to E \xrightarrow{q} G \to 1$$

with A central in G. The group A is called the *kernel* of the extension. Such a central extension is called *universal* if, for any other central extension $E' \xrightarrow{q'} G$, there is a unique homomorphism $E \to E'$ making the diagram

$$
\begin{array}{ccccccccc}
1 & \longrightarrow & A & \longrightarrow & E & \xrightarrow{q} & G & \longrightarrow & 1 \\
& & & & \downarrow & & \| & & \\
1 & \longrightarrow & A' & \longrightarrow & E' & \xrightarrow{q'} & G & \longrightarrow & 1
\end{array}
$$

commute. (In this case there is an induced map $A \to A'$.)

H.8.2 Proposition (Kervaire, Milnor) *A group G has a universal central extension if and only if it is perfect (i.e., $G = [G,G]$). In this case, the universal central extension is unique and its kernel is canonically isomorphic to $H_2(G, \mathbb{Z})$.*

H.8.3 Example The alternating group A_5 is the smallest perfect group. Its universal covering group is the binary icosahedral group, and the kernel is $H_2(A_5, \mathbb{Z}) \cong \mathbb{Z}_2$.

H.8.4 Definition For any ring R, the group of elementary matrices $E(R) = [GL(R), GL(R)]$ is perfect. Its universal covering group is called the **Steinberg group**, and the kernel of this covering, $H_2(E(R), \mathbb{Z})$, is called $K_2(R)$.

H.8.5 Example For $n \geq 3$, the Lie group $SL(n; \mathbb{R})$ has fundamental group $\mathbb{Z}_2$ (as a manifold), so it has a unique double cover $\widetilde{SL}(n; \mathbb{R})$ (as a Lie group). The inverse image of $SL(n; \mathbb{Z})$ in $\widetilde{SL}(n; \mathbb{R})$ turns out to be its universal central extension, so $K_2(\mathbb{Z}) \cong \mathbb{Z}_2$.

Matsumoto's theorem and symbols

The most important fact about K_2 is:

H.8.6 Theorem (Matsumoto's Theorem) *Let F be a field. Then $K_2(F)$ is canonically isomorphic to the free (multiplicative) abelian group on **Steinberg symbols** $\{a, b\}$, $a, b \in F \setminus \{0, 1\}$, subject to the relations $\{a, b\}\{c, b\} = \{ac, b\}$ (bilinearity), $\{a, b\} = \{b, a\}^{-1}$ (skew-symmetry), and $\{a, 1 - a\} = 1$. (These relations also imply that $\{a, -a\} = 1$.)*

H.8.7 Example From Theorem H.8.6 (or by an alternative direct argument), one can deduce that $K_2(F)$ is trivial when F is a finite field.

A calculation of Tate, based on Theorem H.8.6, shows that $K_2(\mathbb{Q})$ is the infinite direct sum of finite cyclic groups, one generated by $\{-1, -1\}$ (which has order 2), and the others isomorphic to $\mathbb{F}_p^\times$ (which of course has order $p - 1$), as p runs over the odd primes.

H.8.8 Remark Theorem H.8.6 shows that one can construct a homomorphism out of $K_2(F)$ any time one has a **symbol** satisfying the Matsumoto relations. Several examples arise in number theory, for example the classical **Hilbert symbol**, the **quaternion algebra symbol**:

$$K_2(F)/2K_2(F) \to {}_2\mathrm{Br}(F) \colon \{a, b\} \mapsto [\text{quaternion algebra gen. by } a, b]$$
$$\tag{1}$$

and the **norm residue symbol**:

$$K_2(F)/nK_2(F) \to H^2\left(\mathrm{Gal}(\bar{F}/F), \mu_n^{\otimes 2}\right).$$

H.8.9 Theorem (Suslin-Merkurjev) *For any field of characteristic $\neq 2$, the quaternion algebra symbol (1) is an isomorphism from $K_2(F)/2K_2(F)$ to the 2-torsion in the Brauer group of F.*

Higher Milnor K-Theory

H.8.10 Definition Motivated by Matsumoto's Theorem, Milnor (1969/1970) defined the [**Milnor**] **K-theory** ring, $K_*^M(R)$, of a commutative ring R to be the graded commutative ring with generators $a \in R^\times$ all of degree 1, subject to the relation that $a \cdot (1 - a) = 0$ if a, $1 - a \in R^\times$. (For convenience we have switched to additive notation for the groups $K_n^M(R)$.) By Theorem H.8.6, $K_2^M(F) = K_2(F)$ for F a field.

In an extension of the Suslin-Merkurjev Theorem (Theorem H.8.9), Voevodsky (1999) and his collaborators (see also Morel (1999)) have now proved the following conjectures of Milnor (1969/1970):

H.8.11 Theorem (Voevodsky et al.) *For F a field of characteristic $\neq 2$, the generalized norm residue symbol is an isomorphism*

$$K_*^M(F) \otimes_{\mathbb{Z}} \mathbb{Z}_2 \to H^*\left(\mathrm{Gal}(\bar{F}/F), \mathbb{Z}_2\right).$$

There is also a natural isomorphism

$$K_*^M(F) \otimes_{\mathbb{Z}} \mathbb{Z}_2 \to \mathrm{Gr}\, W(F),$$

where $\operatorname{Gr} W(F) = \bigoplus_{n=0}^{\infty} I^n(F)/I^{n+1}(F)$ *is the associated graded ring of the Witt ring of* F *with respect to the filtration by powers of the ideal* $I(F)$ *of classes of even-dimensional quadratic forms.*

H.9 Algebraic K-Theory III: Higher Algebraic K-Theory

by Jonathan M. Rosenberg in College Park, MD, USA

Introduction to the higher K-groups

Classical K-theory, due in large part to Grothendieck, Bass, and Milnor, and based on a direct study of projective modules and of the general linear linear group, succeeded in defining functors K_n, $n \leq 2$. Given any ring R and (two-sided) ideal I, there is a long exact sequence of these classical K-groups, functorial in the pair (R, I), behaving much like the long exact homology sequence of a pair of spaces. But making the link with homology theories, and extending this sequence to the left, required sophisticated machinery of category theory and homotopy theory. The definition and basic properties of the higher K-groups are the contribution of Quillen (1973)—see also Grayson (1976), Lluis-Puebla et al. (1992), Rosenberg (1994), Srinivas (1996), and Weibel (1999b).

H.9.1 Theorem (Quillen) *For any ring R and any two-sided ideal I in R, there are abelian groups $K_n(R)$ and $K_n(R, I)$, $n \in \mathbb{N}_0$ (functorial in R and in the pair (R, I)) with the following properties:*

(1) $K_0(R)$ *is as defined in Definition H.7.1, $K_1(R)$ is as defined in Definition H.7.9, and $K_0(R, I) = K_0(I)$ in the sense of Definition H.7.5.*

(2) $K_2(R)$ *is as defined in Definition H.8.4.*

(3) *There is a functorial long exact sequence*

$$\cdots \to K_{n+1}(R/I) \xrightarrow{\partial} K_n(R, I) \to K_n(R) \xrightarrow{q_*} K_n(R/I) \xrightarrow{\partial} K_{n-1}(I) \to \cdots$$

with q_* induced by the quotient map $R \to R/I$.

H.9.2 Remark The actual construction of the higher K-groups is quite complicated and we will not discuss it here. However, much of the power of K-theory comes from the fact that there are several equivalent *machines* for producing these groups out of the category of finitely generated projective R-modules. These machines, known as the Q-construction, $+$-construction, S-construction, and infinite loop machine, can also be applied to other categories with similar properties. Here are a few important examples:

1. the category of finitely generated (left) modules over a (left) Noetherian ring R. The resulting groups are denoted $G_n(R)$ or $K'_n(R)$.

2. the category of algebraic vector bundles over a smooth algebraic variety X. The resulting groups are denoted $K_n(X)$. When $X = \operatorname{Spec} R$, $K_n(X) \cong K_n(R)$.

3. more generally, the category of locally free sheaves over a scheme X. Again the resulting groups are denoted $K_n(X)$.

H.9.3 Theorem *If R is a commutative ring, then there is a functorial graded commutative graded ring structure on $\bigoplus_{n=0}^{\infty} K_n(R)$. There is a natural map of graded rings (usually far from an isomorphism) $K_*^M(R) \to \bigoplus K_*(R)$, with $K_*^M(R)$ as in* Definition H.8.10.

Basic properties of the higher K-groups

H.9.4 Theorem (Quillen) *$K_{2n}(F) = 0$ and $K_{2n-1}(F)$ is a finite cyclic group of order $q^n - 1$, for any finite field $F = \mathbb{F}_q$ and for all $n \in \mathbb{N}$.*

H.9.5 Theorem (Dirichlet, Borel, Quillen) *Suppose R is the ring of integers in a number field F. Then $K_n(R)$ is finitely generated for all n, and*

$$
\dim_{\mathbb{Q}} K_n(R) \otimes_{\mathbb{Z}} \mathbb{Q} = \begin{cases} 0 & \text{if } n \text{ is even,} \\ r_1 + r_2 - 1 & \text{if } n = 1, \\ r_1 + r_2 & \text{if } n \equiv 1 \mod 4 \text{ and } n > 1, \\ r_2 & \text{if } n \equiv 3 \mod 4. \end{cases}
$$

Here r_1 is the number of distinct embeddings $F \hookrightarrow \mathbb{R}$ and r_2 is the number of distinct conjugate pairs of non-real embeddings $F \hookrightarrow \mathbb{C}$, so that $r_1 + 2r_2 = [F : \mathbb{Q}]$.

H.9.6 Remark The 2-torsion in the groups $K_n(R)$ of Theorem H.9.5 has now been computed by Rognes and Weibel (2000), using Voevodsky's proof of the Milnor Conjecture, Theorem H.8.11.

H.9.7 Theorem (Quillen) *If R is a left regular ring, then the natural maps $K_n(R) \to G_n(R)$ (induced by the forgetful functor from projective modules to general modules) are isomorphisms. Furthermore, in this case, $K_n(R[t]) \cong K_n(R)$ and $K_n(R[t,t^{-1}]) \cong K_n(R) \oplus K_{n-1}(R)$. (Cf. Theorem H.7.13).*

H.9.8 Theorem (*Localization*) *If R is a Dedekind ring and $i: R \hookrightarrow F$ is the injection of R into its field of fractions, then there is an exact sequence*

$$\cdots \to K_{n+1}(F) \xrightarrow{\partial} \bigoplus_{\substack{\mathfrak{p} \lhd R \\ 0 \neq \mathfrak{p} \text{ maximal}}} K_n(R/\mathfrak{p}) \to K_n(R) \xrightarrow{i_*} K_n(F) \xrightarrow{\partial} \cdots,$$

functorial in R.

H.9.9 Remark The K-groups of varieties and schemes are closely related to the **Chow ring** as defined in algebraic geometry and to S. Bloch's higher Chow groups. When F is a field, Suslin constructed a surjective homomorphism $K_n(F) \to K_n^M(F)$ such that the composite $K_n^M(F) \to K_n(F) \to K_n^M(F)$ is multiplication by $(-1)^{n-1}(n-1)!$. Thus, modulo torsion, the K-theory of a field splits as the direct sum of Milnor K-theory and another *indecomposable* piece.

Chapter I

MISCELLANEOUS

I.1 Model Theory for Algebraists

by H. Jerome Keisler in Madison, WI, USA

Introduction

Model theory is a large and active area, and there is room here for only a few of its most basic results. For further reading and references, see the books of Chang and Keisler (1990) and of Hodges (1993).

The main focus is on first order formulas with quantifiers, rather than on equations or quantifier-free formulas (Boolean combinations of equations and inequalities). The central notions are those of a *sentence being true* in an algebraic structure, and of a *formula being satisfied* by a tuple of elements in an algebraic structure. We confine our attention to the case of a finite or countable vocabulary. The cardinality of a structure $\mathcal{B}$ is denoted by $|\mathcal{B}|$.

A set of first order sentences is called a ***theory***. A ***model*** of a theory T is an algebraic structure $\mathcal{A}$ in which every sentence in T is true. Two theories T and U are ***equivalent*** if they have exactly the same models.

Compactness theorem

I.1.1 Theorem (Compactness Theorem (Gödel, Mal'cev)) *If every finite subset of a set T of sentences has a model, then T has a model.*

Here are some typical applications.

I.1.2 Corollary
(1) *If a theory T has arbitrarily large finite models, then T has an infinite model.*

(2) *If T contains the field axioms and T has models of arbitrarily large prime characteristic, then T has a model of characteristic zero.*

(3) *If every model of T is a torsion group, then there is a finite bound on the orders of elements of models of T.*

(4) *(Henkin) If every finitely generated substructure of A can be extended to a model of a theory T, then A can be extended to a model of T.*

The compactness theorem leads to many preservation theorems, which give necessary and sufficient conditions for theory to be preserved under some algebraic operation.

A **universal formula** is a formula of the form $\forall \vec{x}\varphi$ where φ is quantifier-free, and a **universal theory** is a set of universal sentences. Existential formulas and $\forall\exists$ formulas are defined similarly.

I.1.3 Theorem (Los, Tarski) *For any theory T, there is a universal theory S whose models are exactly the substructures of models of T.*

I.1.4 Theorem (Los and Suszko, Chang) *A theory is preserved under unions of chains if and only if it is equivalent to an $\forall\exists$ theory.*

A **positive formula** is a first order formula built from atomic formulas using quantifiers are the logical connectives *and* and *or*.

I.1.5 Theorem (Lyndon) *A theory is preserved under surjective homomorphic images if and only if it is equivalent to a positive theory.*

Elementary equivalence and extensions

A and B are **elementarily equivalent**, in symbols $A \equiv B$, if they satisfy the same sentences. Given $X \subseteq A$, $(A, x)_{x \in X}$ is the structure A with a constant symbol for each $x \in X$. B is an **elementary extension** of A, in symbols $A \prec B$, if $A \subseteq B$ and $(A, x)_{x \in A} \equiv (B, x)_{x \in A}$. An **elementary embedding** $f \colon A \prec B$ is defined analogously.

Thus $f \colon A \cong B$ implies $f \colon A \prec B$, and $f \colon A \prec B$ implies $A \equiv B$.

I.1.6 Theorem (Löwenheim, Skolem, Tarski) *For every infinite $X \subseteq B$, there exists $A \prec B$ such that $X \subseteq A$ and $|A| = |X|$.*

I.1.7 Theorem (Elementary Chain (Tarski and Vaught)) *Suppose $A_0 \prec A_1 \prec \cdots \prec A_n \prec \cdots$. Then for each n, $A_n \prec \bigcup_m A_m$.*

I.1.8 Theorem (Ehrenfeucht and Mostowski) *Let $\mathcal{A}$ be infinite. For each infinite cardinal κ there exists $\mathcal{B} \equiv \mathcal{A}$ with $|\mathcal{B}| = \kappa$ such that for each n and countable $X \subseteq \mathcal{B}$, the relation $\{(\vec{a}, \vec{b}) \in \mathcal{B}^n \times \mathcal{B}^n \mid (\mathcal{B}, \vec{a}, x)_{x \in X} \cong (\mathcal{B}, \vec{b}, x)_{x \in X}\}$ has at most countably many equivalence classes.*

$\mathcal{A}$ is ***prime*** if $\mathcal{A}$ is elementarily embeddable in every model $\mathcal{B} \equiv \mathcal{A}$. $\mathcal{A}$ is ***universal*** if every $\mathcal{B}$ such that $\mathcal{B} \equiv \mathcal{A}$ and $|\mathcal{B}| \leq |\mathcal{A}|$ is elementarily embeddable in $\mathcal{A}$. $\mathcal{A}$ is ***saturated*** if $(\mathcal{A}, x)_{x \in X}$ is universal for each set $X \subseteq \mathcal{A}$ of cardinality less than $|\mathcal{A}|$.

I.1.9 Theorem (Vaught) *Suppose $\mathcal{A}$ is countable and universal. Then there is a prime model $\mathcal{B} \equiv \mathcal{A}$ and a countable saturated model $\mathcal{C} \equiv \mathcal{A}$.*

I.1.10 Theorem (Morley and Vaught) *Let $\mathcal{A}$ be infinite. For each uncountable inaccessible cardinal κ, there is a saturated model $\mathcal{C} \equiv \mathcal{A}$ of cardinality κ, and $\mathcal{C}$ is unique up to isomorphism.*

Model completions and quantifier elimination

This topic began with Tarski's Quantifier elimination theorem for the real ordered field, and Abraham Robinson's introduction of the model completion of a theory.

T ***admits quantifier elimination*** if every first order formula is T-equivalent to a quantifier-free formula with the same free variables.

Note that if T admits quantifier elimination, then $\mathcal{A} \prec \mathcal{B}$ whenever $\mathcal{A}, \mathcal{B}$ are models of T and $\mathcal{A} \subseteq \mathcal{B}$. Moreover, T has a set of $\forall\exists$ axioms.

I.1.11 Theorem (Shoenfield) *If T admits quantifier elimination, then the class of substructures of models of T has the amalgamation property.*

In the following let S be an $\forall\exists$ theory. T is a ***model completion*** of S if $T \supseteq S$, T admits quantifier elimination, and every model of S can be extended to a model of T. $\mathcal{A}$ is ***existentially closed*** in S if $\mathcal{A}$ is a model of S, and every existential formula with parameters from $\mathcal{A}$ which has a solution in some model $\mathcal{B} \supseteq \mathcal{A}$ of S has a solution in $\mathcal{A}$.

I.1.12 Theorem
(1) *(Robinson) Every model of S can be extended to an existentially closed model of S.*
(2) *(Eklof and Sabbagh) If T is a model completion of S then the class of models of T is equal to the class of all existentially closed models of S. (Hence the model completion of S is unique).*

I.1.13 Theorem (Tarski, Robinson) *In each of the following, the theory T in the right column is the model completion of the theory in the left column, and thus T admits quantifier elimination.*

integral domains,	*algebraically closed fields*
differential fields of char. 0	*differentially closed fields of char. 0*
ordered Abelian groups	*divisible ordered Abelian groups*
ordered integral domains	*real closed ordered fields*
Boolean algebras	*atomless Boolean algebras*
groups	*does not exist (Eklof and Sabbagh)*

Many other examples of this kind are also known. The first four T's above were the points of departure for two newer areas of model theory which are taken up in the next section—stable theories and o-minimal theories.

I.2 Linear Codes over Fields

by Günter F. Pilz in Linz, Austria

Let a message consist of words $\mathbf{w} = a_1 \ldots a_k$ over some alphabet A. The main goal of coding theory is to transform the words $\mathbf{w}_1, \mathbf{w}_2, \ldots$ of a message into (longer) words $\mathbf{w}'_1, \mathbf{w}'_2, \ldots$ which are sent through a transmission channel; there they might be distorted into $\mathbf{w}''_1, \mathbf{w}''_2, \ldots$ The idea is to make $\mathbf{w}'_1, \mathbf{w}'_2, \ldots$ *so distinct* that each $\mathbf{w}''_i$ will still be more *similar* to $\mathbf{w}'_i$ than to any other $\mathbf{w}'_j$. In this way, errors in the transmission of the $\mathbf{w}'_i$ (if there are not too many of them) can be detected and the original words be reconstructed.

Throughout this section, all $\mathbf{w}_i$ have the same length (k, say), and all $\mathbf{w}'_i$ and $\mathbf{w}''_i$ will have a constant length $n > k$. The number of digits in which two words $\mathbf{v}$ and $\mathbf{w}$ differ is called the **Hamming distance** $d(\mathbf{v}, \mathbf{w})$ of $\mathbf{v}$ and $\mathbf{w}$. This function d is a metric.

I.2.1 Definition Let A be a set, W a set of words over A of length k, and W' a set of words over A of length $n > k$. An injective function enc: $W \to W'$ is called an **encoding**, and $C := \text{enc}(W) \subset W'$ the corresponding **code**; more precisely, C is an (n, k)-code over A of **minimal distance** $d_{\min}(C)$, the minimum of $d(\mathbf{v}, \mathbf{w})$ for all $\mathbf{v}, \mathbf{w} \in W$, $\mathbf{v} \neq \mathbf{w}$.

In all interesting cases, the likelihood of 2 errors in an incoming word $\mathbf{w}_i''$ is much smaller than the one of 1 error; 3 errors are even less likely, and so on. Hence the **maximum likelihood decoding** decodes $\mathbf{w}_i''$ into the nearest word $\mathbf{w}_i' \in W'$.

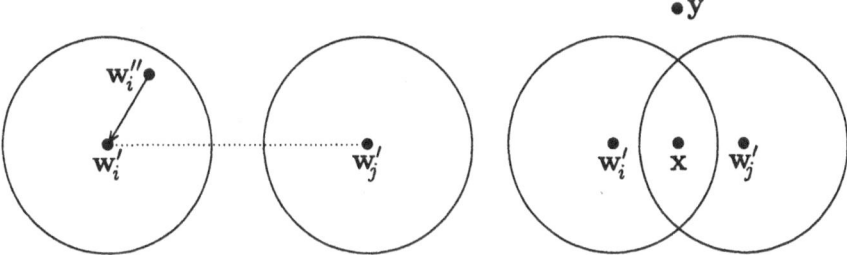

Figure 1: correct decoding   Figure 2: ambiguous decoding

In Figure 2, neither x nor y can be decoded uniquely. So we want to have balls (w.r.t. d) with radius r around all $\mathbf{w}' \in W'$ which do not intersect, and they should fill most of the set of all words over A of length n. So coding theory is related to sphere packing problems. Also, r should be maximal so the the balls do not intersect. In Figure 1, the dotted line has length $\geq d_{\min}(C)$, so we get

I.2.2 Proposition *A code C can detect up to $d_{\min}(C) - 1$ errors and correct up to $\left\lfloor \frac{d_{\min}(C)-1}{2} \right\rfloor$ errors.*

I.2.3 Main Goal of Coding Theory Given n, find an (n, k)-code C such that $n - k$ is small and $d_{\min}(C)$ is large.

These two goals contradict each other, so we look for good compromises. Other important goals of the theory are efficient decodings, and so on. Limitations to the main goal come soon. Each ball of radius r around a codeword has $\sum_{i=0}^{r} \binom{n}{i}(|A| - 1)^i$ elements; so we get

I.2.4 Theorem (Hamming bound)
$|A|^n \geq |C| \sum_{i=0}^{r} \binom{n}{i}(|A| - 1)^i$ *holds for each code C of length n, where* $r = \left\lfloor \frac{d_{\min}-1}{2} \right\rfloor$.

If we have equality in this bound, the code is **perfect**; then there are no words outside the balls. Many more (in-)equalities relating the code parameters are on the market. Shannon's *Main Theorem on Channel*

Coding says that for sufficiently large n there exist "arbitrarily good" codes; the proof is not constructive, however.

A good idea is to take for A the finite field $\mathbb{F}_q$, and to encode $\mathbf{w} = a_1 \ldots a_k$ into $\mathbf{w}' = a_1 \ldots a_k a_{k+1} \ldots a_n$, where $a_{k+1}, \ldots, a_n$ are computed linearly from $a_1, \ldots, a_k$. For instance, a_{k+1} might be $= a_1 + a_2 + a_3$, which can be written as $a_1 + a_2 + a_3 - a_{k+1} = 0$. In this way, the codewords in C are precisely the solutions of a homogeneous system of linear equations. If H is the coefficient matrix of this system, H is called the **check matrix**, the code C is **linear** and is the null space of C. So C is a subspace of $\mathbb{F}_q^n$, usually of dimension k. A very easy example is $\mathbb{F}_q = \mathbb{Z}_2$ with the encoding pc: $a_1 \ldots a_k \rightarrow a_1 \ldots a_k a_{k+1}$ with $a_{k+1} = a_1 + a_2 + \cdots + a_k$ (**parity-check code**). Another example is rep: $a_1 \ldots a_k \rightarrow a_1 \ldots a_k a_1 \ldots a_k$ (**repetition code**).

For a linear code, the **weight of a word** is $d(\mathbf{w}, \mathbf{o})$, i.e., the number of non-zero digits. Since, in a linear code, the difference between two codewords is again a codeword, $d_{\min}(C)$ is just the weight of the lightest codeword $\neq \mathbf{o}$. A linear code of length n, dimension k, and minimal distance d is called an $[n, k, d]$**-code**.

If $\mathbf{H}$ has k rows and if the columns are all non-zero vectors in $(\mathbb{F}_q)^k$, we get the $\left[\frac{q^k - 1}{q - 1}, n - k, 3\right]$ **Hamming code** $\mathbf{H}_n(q)$, which is 1-error correcting and easily seen to be perfect. Decoding is very easy: if the word $\mathbf{w}''$ arrives without error then $\mathbf{H}\mathbf{w}'' = \mathbf{o}$. If $\mathbf{w}''$ arrives with a single error at position i then $\mathbf{H}\mathbf{w}$ is the i-th row of $\mathbf{H}$; so it is a good idea to take the rows as the q-ary numbers $1, 2, \ldots, q^k - 1$.

More involved are the two famous **Golay-codes** G_2 (a [23,12,7]-code over $\mathbb{F}_2$) and G_3 (an [11,6,5]-code over $\mathbb{F}_3$) which are also linear and perfect. In 1973, Tietäväinen, Zinov'ev, and Leontiev showed

I.2.5 Theorem *If C is a non-trivial perfect code (linear or not) over $\mathbb{F}_q$ then C has the parameter of G_2, G_3, or of some $H_n(q)$.*

I.2.6 Example Let α be a generator of $\mathbb{F}_q$, $d, n \in \mathbb{N}$ with $2 \leq d \leq n \leq q$ and let M have the rows $(1, \alpha, \alpha^2, \ldots, \alpha^{q-2})$, $(\alpha, \alpha^2, \ldots, \alpha^{2(q-2)})$, $\ldots$, $(\alpha^{d-2}, \ldots, \alpha^{(d-2)(q-2)})$. Then the nullspace of M is an $[n, n - d + 1, d]$-code, the **Reed-Solomon code** $\mathrm{RS}_{n,d,q}$. These codes are used in compact discs to give us "error-free music".

I.2.7 Example The **Reed-Muller code** RM_m over $\mathbb{F}_2$ encodes $\mathbf{w} = a_1 a_2 \ldots a_k$ into

$$\left(f_{\mathbf{w}}(0, \ldots, 0), \, f_{\mathbf{w}}(0, \ldots, 0, 1), \, f_{\mathbf{w}}(0, \ldots, 1, 0), \, \ldots, \, f_{\mathbf{w}}(1, \ldots, 1)\right),$$

where $f_{\mathbf{w}}(x_1, \ldots, x_k) = a_1 x_1 + a_2 x_2 + \cdots + a_k x_k$.

RM_m is a $[2^m, m+1, 2^{m-1}]$-code; RM_5 was used in the Mariner missions to mars; without efficient codes, all photos transmitted from outer space would be completely useless. Decoding RM_m works very fast using Hadamard transforms.

For every linear code $C \leq V$,

$$C^\perp := \{\, v \in V \mid c \cdot v = 0 \text{ for all } c \in C \,\}$$

is a linear code as well, the **dual code** to C. The dual code to $H_{n,q}$ is the **simplex code** $S_{n,q}$, a $\left[\frac{q^k-1}{q-1}, k, q^{k-1}\right]$-code. Basically, $\mathrm{RS}^\perp_{n,d,k}$ is $\mathrm{RS}_{n,n-d+2,q}$, and if $q = 2$, $2d = n+2$, then $\mathrm{RS}_{n,d,q}$ is self-dual. Finally, $\mathrm{RM}^\perp_m$ is a "higher-order RM-code".

If a code C has A_i $(0 \leq i \leq n)$ words of weight i then $\sum A_i z^i =: A_C(z)$ is the **weight enumerator polynomial** of C. For the simplex code $S_{n,q}$ we get $1 + \left(q^k - 1\right) z^{q^{k-1}}$, for G_3 we have $1 + 132z^5 + 330z^8 + 110z^9 + 24z^{11}$, for example.

I.2.8 Theorem (MacWilliams) *Let C be a linear $[n, k, d]$-code over $\mathbb{F}_q$. Then $A_{C^\perp}(z) = q^{-k} \left(1 + (q-1) z\right)^n A_C \left(\frac{1-z}{1+(q-1)z}\right)$.*

Weight enumerators carry a lot of information on C and often allow to compute $d_{\min}\left(C^\perp\right)$.

A code C is **cyclic** if $(c_0, \ldots, c_{n-1}) \in C$ implies $(c_1, \ldots, c_{n-1}, c_0) \in C$. In the linear case, we might identify $(c_0, \ldots, c_{n-1})$ with $c_0 + c_1 x + \cdots + c_{n-1} x^{n-1}$. Then C is cyclic iff C is an ideal in the PID $\mathbb{F}_q[x]/(x^n - 1)$; this adds an enormous amount of structure to C, and C can be generated by a single polynomial (**generator polynomial**) $g \in C$. If α is a generator for F_q, the generator polynomial of $\mathrm{RS}_{n,d,q}$ is $\prod_{i=1}^{d-1} (x - \alpha^i)$, for instance. Hamming, Golay, and RM-codes are cyclic as well.

Special cyclic codes, given by Bose, Chaudhuri and Hocquenghem (1959/60), the **BCH-codes**, are of basic importance in coding theory; they can be decoded using the Berlekamp-Massey algorithm.

For more information, see e.g., the classic work (MacWilliams and Sloane 1977) or (van Lint 1992) or (Lidl and Pilz 1998).

I.3 Linear Codes over Rings

by V. T. Markov, A. V. Mikhalev, and A. A. Nechaev in
Moscow, Russia

Introduction

Let R be a finite commutative ring with identity e, and $_RM$ a finite
faithful module. Any submodule $\mathcal{K} < {_R}M^n$ is called a **linear n-code**
over $_RM$, its **Hamming distance** defined as $d(\mathcal{K}) = \min\{\,\|\alpha\| \mid \alpha \in
\mathcal{K} \setminus \mathbf{0}\,\}$, where $\|\alpha\|$ is the Hamming weight of α.

As for codes over fields, we get the concepts of the parity-check matrix
and the code $\mathcal{K}^o$ dual to the given code $\mathcal{K}$, defined in such a way that
in particular the equality $\mathcal{K}^{oo} = \mathcal{K}$ and the MacWilliams identity for
complete weight enumerators of codes $\mathcal{K}$ and $\mathcal{K}^o$ will be valid.

In the most general case, while studying linear codes over an arbitrary
finite module $_RM$, to obtain the results close enough to that of the theory
of linear codes over fields, the code dual to the given one must be defined
over the module $_RM^*$, which is the Morita dual to $_RM$. Now we pass
to the exact statements.

Quasi-Frobenius modules and the reciprocal modules

For any ideal $I \lhd R$ and any submodule $K < {_R}M$ we define their
annihilators:

$$\mathrm{Ann}_M(I) = \{\,\alpha \in M \mid I\alpha = 0\,\} < {_R}M;$$
$$\mathrm{Ann}_R(K) \ = \{\,r \in R \mid rK = 0\,\} \lhd R.$$

A module $_RM$ is called **quasi-Frobenius** (a **QF-module**), if
$\mathrm{Ann}_R(\mathrm{Ann}_M(I)) = I$ and $\mathrm{Ann}_M(\mathrm{Ann}_R(K)) = K$ for all $I \lhd R$
and $K < {_R}M$. For any finite commutative ring R there exists a unique
(up to isomorphism) QF-module $_RQ$ (cf. Section C.30). It might be described as the character group $Q = \mathrm{Hom}(R, \mathbb{Q}/\mathbb{Z})$ of the group $(R, +)$,
where the product $r\omega \in Q$ of an element $\omega \in Q$ by an element $r \in R$
is defined by the condition $r\omega(a) = \omega(ra)$ for any $a \in R$. We have
$(Q, +) \cong (R, +)$ and $|Q| = |R|$. A ring R is called **quasi-Frobenius**, if
$_RR$ is a QF-module.

We call the module $_RM^* = \mathrm{Hom}_R(M, Q)$ of all homomorphisms
$_RM \to {_R}Q$ **reciprocal** to $_RM$ (or **Morita-dual** to $_RM$). It may
be presented also as $\mathrm{Hom}(M, \mathbb{Q}/\mathbb{Z})$. It is important to note that

$(M^*, +) \cong (M, +)$, so in particular, M and M^* are equivalent alphabets.

Let us define the product of $\alpha \in M$ by $\varphi \in M^*$ as $\varphi\alpha = \varphi(\alpha) \in Q$. Then for a fixed $\alpha \in M$, the correspondence $\varphi \to \varphi\alpha$ induces a homomorphism $_R M^* \to Q$ belonging to $M^{**} = \mathrm{Hom}_R(M^*, Q)$. We identify this homomorphism with α, obtaining the equality $M^{**} = M$. Note that if $_R M = {}_R Q$ is a QF-module over R then the R-module $M^* = Q^* = \mathrm{Hom}_R(Q, Q)$ is isomorphic to R (see Faith 1976, Nechaev 1995). In particular, if R is a QF-ring and $M = R$, then there are natural identifications: $Q = R$ and $M^* = M = R$.

The dual code and the parity-check matrix

Consider the example. Suppose we define the dual code $\mathcal{L}^o$ for a linear code $\mathcal{L} < {}_R R^n$ over a ring R in the usual way as $\mathcal{L}^o = \{\beta \in R^n \mid \beta\mathcal{L} = 0\}$. Then $\mathcal{L}^{oo} \supseteq \mathcal{L}$, but the equality $\mathcal{L}^{oo} = \mathcal{L}$ is guaranteed if and only if R is a quasi-Frobenius ring (Nechaev 1995) (for example principal ideal rings and in particular $R = \mathbb{Z}_m$ are quasi-Frobenius). In such cases, the codes over R can be studied via codes over fields. However, if the ring R is not quasi-Frobenius, then for deriving deep enough results the dual code should be built not over R but over $_R Q$.

Let us define the product of the row $\mathbf{a} = (a_1, \ldots, a_n) \in R^n$ by the row $\alpha = (\alpha_1, \ldots, \alpha_n) \in Q^n$ as $\mathbf{a}\alpha = a_1\alpha_1 + \cdots + a_n\alpha_n \in Q$, and say that the code $\mathcal{L}^0 = \{\alpha \in Q^n \mid \mathcal{L}\alpha = 0\}$ over $_R Q$ is *dual* to the linear code $\mathcal{L} < R^n$, In turn, the code *dual* to a linear code $\mathcal{K} < {}_R Q^n$ is built over the ring R in the form $\mathcal{K}^0 = \{\mathbf{a} \in R^n \mid \mathbf{a}\mathcal{L} = 0\}$ over R. Then, in particular, we have $\mathcal{L}^{00} = \mathcal{L}, \mathcal{K}^{00} = \mathcal{K}, |\mathcal{L}| \, |\mathcal{L}^0| = |\mathcal{K}| \, |\mathcal{K}^o| = |R|^n = |Q|^n$ (cf. Nechaev 1995, 1996b).

Let now $\mathcal{K} <_R M^n$ be a linear code over a finite module $_R M$. To define the dual code we note that any element of the reciprocal module $(M^n)^* = \mathrm{Hom}_R(M^n, Q)$ might be considered as a row $\varphi = (\varphi_1, \ldots, \varphi_n) \in (M^*)^n = \mathrm{Hom}_R(M, Q)^n$, acting on elements $\alpha = (\alpha_1, \ldots, \alpha_n) \in M^n$ by the rule $\varphi(\alpha) = \varphi\alpha = \varphi_1\alpha_1 + \cdots + \varphi_n\alpha_n \in Q$. Then $(M^n)^* = (M^*)^n$. Now if $\mathcal{K} < {}_R M^n$ then the code $\mathcal{K}^o <_R (M^*)^n$ defined as $\mathcal{K}^o = \mathrm{Ann}_{(M^*)^n}(\mathcal{K}) = \{\varphi \in (M^*)^n) \mid \varphi\mathcal{K} = 0\}$ is called the *dual code* to the code $\mathcal{K}$. Our above conventions give the inclusions $\mathcal{K}^{oo} < (M^n)^{**} = M^n, \mathcal{K} \subseteq \mathcal{K}^{oo}$. If we consider $\mathcal{K}$ only as a subgroup of $(M^n, +)$, then $\mathcal{K}^o$ is the code dual to $\mathcal{K}$, as defined in Delsarte (1973). The presented construction allows additionally to study $\mathcal{K}^o$ as an R-module if $\mathcal{K}$ is a submodule of $_R M^n$.

I.3.1 Theorem *Delsarte (1973), Faith (1976) There is a group isomorphism* $\mathcal{K}^o \cong M^n/\mathcal{K}$, *and* $|\mathcal{K}^o| \cdot |\mathcal{K}| = |M|^n$, $\mathcal{K}^{oo} = \mathcal{K}$.

Let $\varphi_i = (\varphi_{i1}, \ldots, \varphi_{in}) \in (M^*)^n$, $1 \leq i \leq l$, be a generating system of the module $_R\mathcal{K}^o$. Let us call the matrix $\Phi = (\varphi_{ij})_{l \times n}$ over M^* a *parity-check matrix* of the code $\mathcal{K}$. It may be considered as a homomorphism $\Phi: {}_RM^n \to {}_RQ^{(l)}$ (here $_RQ^{(l)}$ is the module of all l-columns over Q) acting on $\alpha \in M^n$ by the rule $\Phi(\alpha) = (\varphi_1\alpha, \ldots, \varphi_l\alpha)^T$. Any column $\varphi_j^{\downarrow} = (\varphi_{1j}, \ldots, \varphi_{lj})^T$ of the matrix Φ is a homomorphism $\varphi_j^{\downarrow}: {}_RM \to {}_RQ^{(l)}$. Let us define the *guaranteed rank* $\kappa_M(\Phi)$ of the matrix Φ relatively to M as the maximal $k \in \mathbb{N}$ such that any system $\varphi_{j_1}^{\downarrow}, \ldots, \varphi_{j_k}^{\downarrow}$ of k columns of Φ is linearly independent over M, i.e., $\varphi_{j_1}^{\downarrow}(\alpha_1) + \cdots + \varphi_{j_k}^{\downarrow}(\alpha_k) \neq 0$ for any $(\alpha_1, \ldots, \alpha_k) \in M^k \setminus 0$. As for codes over fields we have

I.3.2 Theorem *(Nechaev 1996a)* $\mathcal{K} = \mathrm{Ker}\,\Phi$ *and* $d(\mathcal{K}) = \kappa_M(\Phi) + 1$.

If a code $\mathcal{K} < {}_RM^n$ has a parity-check matrix over the ring R, i.e., such a matrix $\Phi_{l \times n}$ over R that $\mathcal{K} = \{\alpha \in M^n \mid \Phi\alpha^T = 0\}$, then it is called *R-closed*. Proposition I.3.2 is true for such matrices. Note that all linear codes over the QF-module $_RQ$ are R-closed as well as all linear codes over the QF-ring R (cf. Nechaev 1995).

The MacWilliams identity

Let $M = \{\mu_1 = 0, \mu_2, \ldots, \mu_m\}$ and $M^* = \{\mu_1^* = 0, \mu_2^*, \ldots, \mu_m^*\}$. The *complete weight enumerators* of the codes $\mathcal{K} < {}_RM^n$ and $\mathcal{K}^o < {}_RM^{*n}$ are polynomials of m variables over $\mathbb{Z}$

$$W_\mathcal{K}(x_\mu : \mu \in M) = \sum_{\alpha \in \mathcal{K}} x_{\alpha_1} \cdots x_{\alpha_n},$$

$$W_{\mathcal{K}^o}(y_{\mu^*} : \mu^* \in M^*) = \sum_{\varphi \in \mathcal{K}^o} y_{\varphi_1} \cdots y_{\varphi_n}.$$

According to Nechaev (1995) there exists a *distinguishing character* $\chi: (Q, +) \to (\mathbb{C}^*, \cdot)$ of the module $_RQ$, i.e., such a character that $\chi(K) \neq 1$ for every nonzero submodule $K < {}_RQ$.

I.3.3 Theorem *(MacWilliams Identity*, Nechaev (1996a))
If, for $1 \leq s \leq m$,

$$\tilde{\mu}_s(y) = \sum_{t=1}^{m} \chi(\mu_t^*\mu_s) y_{\mu_t^*}$$

then

$$W_{\mathcal{K}^\circ}(\boldsymbol{y}) = \frac{1}{|\mathcal{K}|} W_{\mathcal{K}}(\tilde{\mu}_1(\boldsymbol{y}), \ldots, \tilde{\mu}_m(\boldsymbol{y})).$$

The (Hamming) *weight enumerator of a code* $\mathcal{K} < M^n$ *satisfies the equalities*

$$W_{\mathcal{K}}^H(x, y) = W_{\mathcal{K}}(x, y, \ldots, y), \quad W_{\mathcal{K}^\circ}^H(x, y) = \frac{1}{|\mathcal{K}|} W_{\mathcal{K}}^H(x + (m-1)y, x - y).$$

For a particular case of this result for linear codes over noncommutative QF-rings, see Wood (1999). A slightly different approach to the concept of duality was proposed in Zinov'ev and Èrikson (1996).

Reduction to local rings. Lifting of codes over fields to codes over modules.

A finite commutative ring R is called *local* if its nilradical $\mathfrak{N} = \mathfrak{N}(R)$ (the set of all nilpotent elements) is a maximal ideal of R. Then $\mathfrak{N}$ is the unique maximal ideal of R. Any ring R we consider has decomposition into a direct sum of local subrings: $R = R_1 \dotplus \ldots \dotplus R_t$. If $e = e_1 + \cdots + e_t$, where $e_s \in R_s$, then e_s is the identity of R_s and $R_s = e_s R$, $s \in \overline{1,t}$. The module M and the code $\mathcal{K} < {}_R M^n$ have corresponding decompositions: $M = M_1 \dotplus \ldots \dotplus M_t$, $\mathcal{K} = \mathcal{K}_1 \dotplus \ldots \dotplus \mathcal{K}_t$, where $M_s = e_s M$ is an R_s-module and $\mathcal{K}_s = e_s \mathcal{K}$ is a linear n-code over M_s. Then the equality $d(\mathcal{K}) = \min\{d(\mathcal{K}_1), \ldots, d(\mathcal{K}_t)\}$ is true [Nechaev (1996a), Nechaev et al. (1997)].

Let now R be a local ring with nil radical $\mathfrak{N}$. Then $\bar{R} = R/\mathfrak{N}$ is a field of elements $\bar{r} = r + \mathfrak{N}$, $r \in R$. The *socle* $\mathfrak{S}(M)$ of the module ${}_R M$ (of the code $\mathcal{K} < {}_R M^n$) is defined as the sum of all its irreducible submodules, and $\mathfrak{S}(M) = \mathrm{Ann}_M(\mathfrak{N})$ ($\mathfrak{S}(\mathcal{K}) = \mathrm{Ann}_{\mathcal{K}}(\mathfrak{N})$, respectively). We may consider $\mathfrak{S}(M)$ as a vector space over the field $\bar{R}$, where $\bar{r}\alpha = r\alpha$ for all $\bar{r} \in \bar{R}$ and $\alpha \in \mathfrak{S}(M)$.

I.3.4 Theorem (Nechaev (1996a), Nechaev et al. (1997)) *Let R be a local ring and $\mathcal{K} < {}_R M^n$. Then $\mathfrak{S}(\mathcal{K})$ is a linear n-code over the space ${}_{\bar{R}}\mathfrak{S}(M)$ and $d(\mathcal{K}) = d(\mathfrak{S}(\mathcal{K}))$.*

This allows us to build linear codes over ${}_R M$ which inherit the properties of linear codes over the residue field $\bar{R}$. We say that $\mathcal{K} \subseteq M^n$ is an $[n, k, d]$-*code* over M if $|\mathcal{K}| = |M|^k$, $d(\mathcal{K}) \doteq d$.

I.3.5 Theorem (Nechaev (1996a)) *Let R be a local ring. If there exists a linear $[n, k, d]$-code $\mathcal{L}$ over the field $\bar{R}$, then there exists a linear $[n, k, d]$-code $\mathcal{K}$ over $_RM$. If $(n, |\bar{R}|) = 1$ and $\mathcal{L}$ is a cyclic code, then $\mathcal{K}$ can be chosen to be cyclic as well.*

Linear codes over fields, spaces and modules

Let L be an elementary abelian p-group of order $q = p^t$, i.e., a finite linear space over $GF(p)$. If $t > 1$ then there exist linear codes over L which are better than linear codes over $GF(q)$. For example, let $B_L(n, 3)$ $(B_q(n, 3))$ be the maximum of the cardinalities of linear n-codes over $(L, +)$ (over $GF(q)$, respectively) with minimal distance 3.

I.3.6 Theorem (Nechaev et al. (1997)) *If*

$$p^{\delta-1}\frac{q^r - 1}{q - 1} < n \leq p^\delta\frac{q^r - 1}{q - 1} - (p^\delta - 1) \text{ for some } k \geq 2 \text{ and } \delta \in \overline{1, t - 1},$$

then $B_L(n, 3) = p^{t-\delta}B_q(n, 3)$.

The attempts to build linear codes over modules which are better than linear codes over linear spaces unexpectedly failed. We say that a n-code $\mathcal{K}$ over M is **majored** by a n-code $\mathcal{L}$ over some alphabet L if $|L| = |M|$, $|\mathcal{L}| \geq |\mathcal{K}|$ and $d(\mathcal{L}) \geq d(\mathcal{K})$.

I.3.7 Theorem (Nechaev (1996a)) *Let $_RM$ be a module over a local ring R, and let $_{\bar{R}}L$ be a linear space of cardinality $|L| = |M|$ over the residue field $\bar{R} = R/\mathfrak{N}$. Then any linear code $\mathcal{K} < _RM^n$ is majored by some linear code $\mathcal{L} < _{\bar{R}}L^n$. If M is a finite abelian group and L is a direct sum of elementary abelian groups of cardinality $|L| = |M|$, then any linear n-code over M is majored by some linear n-code over L.*

The most important modern applications of linear codes over modules are related to linear representations over modules of non-linear codes over rings (e.g., Kerdock, Preparata, Delsarte-Goethals) and to a description of linear recursive MDS-codes (see Kurakin et al. 1999).

I.4 History of Algebra before 1500

by Hans Kaiser in Vienna, Austria

Introduction

Historians of mathematics usually connect the birth of algebra as a discipline of mathematics with the publication of the book "al-jabr wa'l muqabalah" by Muhammed al-Khuwarizmi around 830 A.D. As a matter of fact, the name "algebra" is derived from the word "al-jabr" occurring in the title. Al-jabr, which may be translated as "restoration" (or "completion") means the addition of a term to both sides of an equation to get rid of negative terms, wa'l muqabalah (which can be translated as "reduction") describes the cancelling of terms that occur on both sides of the equation. So the title of the book may be translated as "The art of solving equations".

One characteristic of the early history of algebra (up to 1700 A.D.) is the gradual invention of symbolism and the solving of equations with numerical coefficients. We distinguish three stages: the *rhetorical* or verbal, the *syncopated* (in which abbreviations were used), and the *symbolic*. Symbolism in the notation of equations went through several modifications until it assumed its modern form in the time of François Viète (1540–1603) and was introduced to common use by the Rechenmeister-movement in Europe.

Treatment of equations in ancient civilizations

One source of ancient Egyptian mathematics is the Papyrus Rhind, a collection of problems dealing with matters of everyday life. To each problem a solution is given (without any hints of reasoning). There, linear equations in one unknown quantity are treated by the "method of false assumption".

An example: "A quantity whose fourth part is added to it becomes 15."
Solution: "Take 4, one fourth of it is 1, together 5. Then 15 is divided by 5, and the result is multiplied by 4. Thus the quantity we look for is 12, the fourth part of it is 3, and hence $15 = 12+3$."

In the presentation of the solution, abbreviations are used.

In Babylonian sources we find similar problems which deal with linear equations.

An example: "I found a stone, but I did not weigh it. After I added one-seventh and one-eleventh, I weighed it: 1 ma-an (weight unit). What was the original weight of the stone?"

This problem consists of the treatment of the equation (in modern writing):

$$\left(x + \frac{x}{7}\right) + \frac{1}{11}\left(x + \frac{x}{7}\right) = 1.$$

Babylonians were familiar with the numerical solution of quadratic equations. Their method of solution was essentially our modern algorithm of forming a complete square and extracting the square root.

Also in China we can trace - for example - algebraic activities in the commentary of Liu Hui (around 250) on the book "Nine Sections". We find the "rule of false assumption" and later also treatments of quadratic equations.

The culmination of Chinese algebra is found in the thirteenth century when an ancient method which resembles Horner's algorithm was perfected.

The Greeks of the classical period could solve many algebraic problems of considerable difficulty, but the solutions were all geometric. We even speak of "geometric algebra". Equations were translated into geometric problems and then solved geometrically. Euclid's elements, Book II, Proposition 4, deals with the algebraic identity $(a + b)^2 = a^2 + b^2 + 2ab$. It is formulated the following way:

"If a straight line be divided into any two parts, the square of the whole line is equal to the squares on the two parts, together with twice the rectangle contained by the parts."

The real precursor of algebra in antiquity is Diophantus (around 250). He developed a syncopated style of writing equations in the introduction to his famous "Arithmetica". There Diophantus gave an ingenious treatment of indeterminate equations, usually two or more equations in several variables that have an infinite number of rational solutions.

Hindu writers too developed algebraic methods. For example, in the work of Brahmagupta (7th century) we find satisfactory algorithms for solving special linear or quadratic equations, he even seemed to accept negative and irrational roots. Hindu algebraists also realized that a quadratic equation (with real roots) has two roots.

The birth of "algebra"

As already mentioned, systematic algebraic investigations began with al-Khuwarizmi and the publication of "al-jabr wa'l muqabalah", which took place around 825. This textbook consists of three parts.

The first part consists of a systematic treatment of linear and quadratic equations. All equations are written in words, no symbols occur. Since negative numbers and zero were not in use in al-Khuwarizmi's time, all coefficients of the equations are considered to be positive. First all equations are reduced to the following canonical types (in modern notation):

$$
\begin{aligned}
&(1)\quad ax^2 = bx &\qquad &(4)\quad ax^2 + bx = c \\
&(2)\quad ax^2 = b &\qquad &(5)\quad ax^2 + c = bx \\
&(3)\quad ax = b &\qquad &(6)\quad ax^2 = bx + c
\end{aligned}
$$

Solving such an equation means finding one positive solution. For each case al-Khuwarizmi gives a recipe for the solution and then proves his solution method discussing a typical example in terms of geometric algebra.

For example, his discussion of type (4) runs along the following lines: *"The following is an example for squares and roots equal to numbers: A square and 10 roots are equal to 39 units. The roots in the problem before us are 10. Take one-half of the roots, which is 5. Multiply 5 by itself, which gives 25. This amount you add to 39, giving 64. Having taken then the square root of this, which is 8, subtract from it the half of the roots, 5, leaving 3. The number 3 therefore represents one root of this square, which itself, of course, is 9."*

Written in modern form, the equation is $x^2 + 10x = 39$, the solution is derived in the form $x = \sqrt{\left(\frac{10}{2}\right)^2 + 39} - \frac{10}{2}$. If we replace 10 and 39 by general coefficients b and c we obtain in essence our modern formula. So al-Khuwarizmi solves this case by the method of forming a complete square and extracting the square root. The proof of this algorithm is supplied in the Greek style of geometric algebra:

The proof is that we construct a square of unknown sides, and let this square figure represent the square which together with its roots you wish to find. Let the square then be $a\, b$, of which any side represents one root.

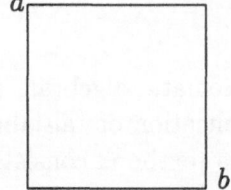

Since then ten roots were proposed with the square, we take a fourth part of the number 10 and apply to each side of the square an area of equidistant sides of which the length should be the same as the length of the square first described and the breadth $2\frac{1}{2}$, which is the fourth part of 10. Therefore, four areas of equidistant sides are applied to the first square, $a\,b$. These are the areas c, d, e, f.

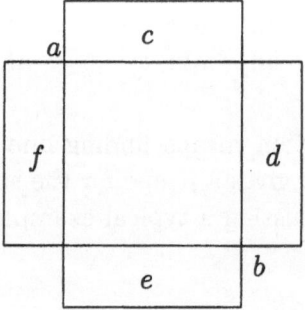

The size of the areas on each of the four corners, which is found by multiplying $2\frac{1}{2}$ by $2\frac{1}{2}$, completes that which is lacking in the larger area. We complete the drawing of the larger area by addition of the four products, the sum of which is 25.

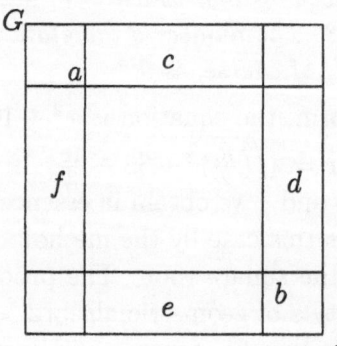

And now it is evident that the first square figure, which represents the square of the unknown and the four surrounding areas make 39. When we add 25 to this, the larger completed square $G\,H$ represents 64, of which 8 is the root, and by this is designated one side of the complete figure. Therefore, when we subtract from 8 twice the fourth part of 10, there will remain but 3 which is equal to one side of the first square *a b*.

In a similar way, al-Khuwarizmi discusses the remaining 5 canonical types of equations.

In the second and third part of the book, the described algorithms are applied to practical problems of everyday life, problems of surveying and problems of inheritance according to Muslim law.

The study of algebra blossomed among Muslim scholars. The peak was reached by 'Umar al-Khayyam (1048–1131). In his treatise we find a systematic classification of equations of degree less or equal to 3. For each of the 25 canonical types of such equations he derives a geometrical solution by using conic sections, but fails to find a numerical solution in the case of cubic equations. He remarks: *"As for the proof in those cases where the subject of the problem is a pure number, we cannot solve it, neither can anyone who knows this art of al-jabr. Perhaps someone else who comes after us may know it ... ".*

This numerical algorithm for cubic equations was found in Renaissance Italy.

For more information, see (Kaiser and Nöbauer 1998), (Karpinski 1919), and (Winter and 'Arafat 1950).

I.5 History of Algebra after 1500

by John Stillwell in Melbourne, Australia

Introduction

The year 1500 is an appropriate starting point for a history of modern algebra because, around that time, algebra took a new direction with the solution of cubic equations. As will be explained below, this led to the theories of groups and fields, and to the complex numbers. When combined with the 17th century development of number theory, which led to the theory of rings, the 16th century theory of equations can be seen as the source of almost all algebra today.

From equations to group theory

Sometime around 1500, Scipione del Ferro of Bologna discovered a formula, involving square roots and cube roots, for solving cubic equations.

The solution was first published in 1545, in Cardano's *Ars Magna*. By this time, it had been joined by a solution of the quartic equation, found by Cardano's student Ferrari.

These successes boosted the confidence of later 16th century algebraists, such as Bombelli and Viète, and led to conceptual and notational advances. Bombelli introduced the algebra of complex numbers in order to explain how Cardano's solution of the cubic equation $x^3 = px + q$, namely

$$x = \sqrt[3]{\frac{q}{2} + \sqrt{\left(\frac{q}{2}\right)^2 - \left(\frac{p}{3}\right)^3}} + \sqrt[3]{\frac{q}{2} - \sqrt{\left(\frac{q}{2}\right)^2 - \left(\frac{p}{3}\right)^3}},$$

could account for the real solution $x = 4$ of the equation $x^3 = 15x + 4$. Viète developed essentially our modern notation for handling polynomials which, when picked up by Fermat and Descartes around 1630, became the starting point for algebraic geometry.

At this stage, it was no doubt expected that further progress in algebra would come through solving equations of degrees 5, 6, and so on. But as it turned out, progress actually came from *getting stuck* on the equation of degree 5, the quintic equation.

Failure to solve the quintic led to Lagrange's theory of equations of 1770, which emphasized permutations of the roots and implicitly contained some ideas of group theory. This was followed by attempts to prove *unsolvability* of the quintic, by Ruffini and Abel, which led to further understanding of permutations and hinted at the theory of fields. Finally, in 1830, a complete understanding of solvability (of equations) was achieved when Galois brought to light the underlying concept of solvability of groups.

Galois isolated the fundamental concepts of group theory, including normal subgroups, quotients and simple groups, and he also drew attention to the concept of field. Indeed, his investigations of equations also involved finite fields and linear groups over them. The depth of Galois' influence can be judged from the fact that the first book on group theory, Jordan's huge *Traité des substitutions* of 1870, is ostensibly an exposition of Galois' ideas.

At about this time, however, geometry began to exert an influence. The theory of invariants developed by Boole, Cayley and Sylvester stimulated group theory when Klein 1872 switched attention from the invariants to the group *under which* they are invariant—in his Erlangen program of classifying geometries by groups. In topology, the fundamental group of Poincaré 1892 took group theory in yet another direction, that of algorithmic problems involving generators and relations. Rather surprisingly, this led to a new kind of unsolvability in algebra—the algorithmic unsolvability of the so-called *word problem*, discovered by Novikov in 1955.

Group theory also took an interesting detour into geometry and back again, in the theory of continuous groups initiated by Sophus Lie in 1873. The *simple* Lie groups were already classified by 1894, through the efforts of Lie, Killing and Cartan. Their classification was the model for the much more difficult classification of finite simple groups, completed around 1980.

For details on the development of group theory, particularly in the 19th century, see Wussing (1984)

From quadratic forms to ring theory

Around 1640, Fermat made claims about the representation of numbers by quadratic forms, such as $x^2 + y^2$, $x^2 + 2y^2$ and $x^2 - ny^2$. Euler and Lagrange proved many of these claims by the introduction of *algebraic integers*, which enabled quadratic forms to be split into linear factors. Despite their success, a general theory of algebraic integers faced severe difficulties—in particular, the failure of unique prime factorization. Kummer overcame the latter problem, in certain cases, with his *ideal numbers* in 1843. However, it remained unclear what ideal numbers were, and where they could be expected to exist.

Gauss, who had perhaps been the first to observe the failure of unique prime factorization, devised an elaborate way around it with his theory of composition of forms in the *Disquisitiones Arithmeticae* of 1801. Among other things, composition of forms exposed the role of finite abelian groups in number theory, and also brought to light their axiomatic properties and structure, but not until much later (Kronecker 1870). And ideal numbers still seemed a better idea, even if their meaning remained elusive.

Clarification was achieved in the general theory of algebraic numbers, developed by Dedekind between 1871 and 1894, and definitively presented in Hilbert's *Zahlbericht* of 1897. By constraining the algebraic integers to lie in a suitable field—one of finite degree over the rationals—Dedekind was able to ensure they had enough properties in common with the ordinary integers. These are the properties we now take to define a *ring*. In 1882, Dedekind and Weber successfully used the same model to develop a theory of algebraic *functions*, and used it to give algebraic proofs of some basic theorems about Riemann surfaces.

Over the next few decades, these examples (and a few others, such as Hensel's *p*-adic numbers) led to both the general theory of fields, presented by Steinitz in 1910, and the general theory of rings, developed by Emmy Noether in the 1920s. In the process, Kummer's "ideal numbers" were finally absorbed into the concept of an *ideal* in a ring. Noether was also the leader in developing concepts suitable for discussing algebraic structures of all kinds, and in pointing out their presence in other parts of mathematics. For example, in 1926 she observed that the "Betti and torsion numbers" of a manifold, introduced by Poincaré, were better captured by a Betti *group*. This launched a "takeover" of topology by algebra whose effects are still strongly felt. A similar takeover of algebraic geometry was launched by Grothendieck in the 1960s.

Today, the position of algebra in geometry does not seem as dictatorial as it did in the 1960s. There has been a fight back by geometric methods in topology and sometimes in algebra itself. Nonetheless, it is clear that abstract algebra has completely transformed mathematics in the 20th century, and we can expect it to remain prominent in the future.

For more on the development of abstraction, with particular emphasis on ring theory, see Corry (1996). An excellent general book on algebra since 1500 is Scholz et al. (1990).

Bibliography

Abe, E.: 1980, *Hopf algebras*. Cambridge University Press, Cambridge, xii+284 pp., translated from the Japanese by Hisae Kinoshita and Hiroko Tanaka.

Abhyankar, S. S.: 1990, *Algebraic geometry for scientists and engineers*. American Mathematical Society, Providence, RI, xiv+295 pp.

— 1998, *Resolution of singularities of embedded algebraic surfaces*. Springer-Verlag, Berlin, second edition, xii+312 pp.

— 2001, *Algebraic Geometry for Scientists and Engineers*. Bulletin of the American Mathematics Society, To Appear.

Abramsky, S., D. M. Gabbay, and T. S. E. Maibaum, eds.: 1994, *Handbook of Logic in Computer Sciencel. 3 Volumes*. The Clarendon Press Oxford University Press, New York, xvi+490 pp.

Adámek, J., H. Herrlich, and G. E. Strecker: 1990, *Abstract and Concrete Categories*. John Wiley & Sons Inc., New York, xiv+482 pp., The joy of cats, A Wiley-Interscience Publication.

Adeleke, S. A. and H. D. Macpherson: 1996, *Classification of infinite primitive Jordan permutation groups*. Proceedings of the London Mathematical Society, **72**, 63–123.

Adian, S. I.: 1967, *Defining relations and algorithmic problems for groups and semigroups*. A.M.S., Providence, RI.

— 1970, *Infinite irreducible systems of group identities (Russian)*. Izv. Akad. Nauk SSSR, Ser. Mat., **34**, 715–734.

— 1979, *The Burnside problem and identities in groups*, volume 99 of *Ergebnisse der Math. und ihrer Grenzgebiete*. Springer-Verlag, Berlin, xi+311 pp.

Adyan, S. I.: 1984, *Studies in the Burnside problem and related questions*. Trudy Mat. Inst. Steklov., **168**, 171–196, algebra, mathematical logic, number theory, topology.

Albert, A. A.: 1961, *Structure of Algebras*. American Mathematical Society, Providence, R.I., xi+210 pp.

Albert, D., R. Baldinger, and J. Rhodes: 1992, *The identity problem for finite semigroups*. J. Symbolic Logic, **57**, 179–192.

Almeida, J.: 1995, *Finite Semigroups and Universal Algebra*. World Scientific, Singapore, english translation.

— 2000, Dynamics of implicit operations and tameness of pseudovarieties of groups. Technical Report CMUP 2000-01, Univ. Porto.

Alonso, J., T. Brady, D. Cooper, V. Ferlini, M. Lustig, M. Mihalik, M. Shapiro, and H. Short: 1991, *Notes on word hyperbolic groups*. Group theory from a geometrical viewpoint (Trieste, 1990), A. H. É. Ghys and A. Verjovsky, eds., World Sci. Publishing Co., Inc., River Edge, NJ, 3–63.

Alperin, R. C. and P. B. Shalen: 1982, *Linear groups of finite cohomological dimension*. Invent. Math., **66**, 89–98.

Amayo, R. K. and I. Stewart: 1974, *Infinite-dimensional Lie algebras*. Noordhoff International Publishing, Leyden, xi+425 pp.

Anderson, D. and D. Dobbs: 1995, *Zero-Dimensional Commutative Rings*. Marcel Dekker.

Anderson, D. D., ed.: 1997, *Factorization in integral domains*, volume 189, Marcel Dekker Inc., New York.

Anderson, D. D. and E. W. Jonson: 1984, *Ideal theory in commutative semigroups*. Semigroup Forum, **30**, 127–158.

Anderson, F. W. and K. R. Fuller: 1992, *Rings and categories of modules*. Springer-Verlag, New York, second edition, x+376 pp.

Anderson, I. and M. Fels: 1997, *Symmetry reduction of variational bicomplexes and the principle of symmetric criticality*. Amer. J. Math., **119**, 609–670.

André, J.: 1961, *Parallelstrukturen, 2: Translationstrukturen*. Math. Z., **76**, 155–163.

Andréka, H., I. Németi, and I. Sain: 2001, *Algebraic Logic*. Handbook of Philosophical Logic, Vol. 2, Kluwer, Dordrecht, second edition, 133–247.

Andréka, H., J. van Benthem, and I. Németi: 1998, *Modal languages and bounded fragments of predicate logic*. Journal of Philosophical Logic, **27**, 217–274.

Andrews, G. E.: 1976, *The theory of partitions*. Addison-Wesley Publishing Co., Reading, Mass.-London-Amsterdam, xiv+255 pp., encyclopedia of Mathematics and its Applications, Vol. 2.

Arnautov, V. I.: 1998, *The theory of radicals of topological rings*. Math. Japon., **47**, 439–544.

Arnautov, V. I., C. I. Beidar, S. T. Glavatsky, and A. V. Mikhalëv: 1992, *Intersection property in the radical theory of topological algebras*. Proceedings of the International Conference on Algebra, Part 2 (Novosibirsk, 1989), Amer. Math. Soc., Providence, RI, 205–225.

Arnautov, V. I., S. T. Glavatsky, and A. V. Mikhalev: 1996, *Introduction to the theory of topological rings and modules*. Marcel Dekker Inc., New York, vi+502 pp.

Arnold, J.: 1973a, *Krull dimension in power series rings*. Trans. Amer. Math. Soc., **177**, 299–304.

— 1973b, *Power series rings over Prüfer domains*. Pacific J. Math., **44**, 1–11.

Arnold, J. and R. Gilmer: 1974, *The dimension sequence of a commutative ring*. Amer. J. Math., **96**, 385–408.

Arshinov, M. and L. Sadovskii: 1972, *Certain lattice-theoretic properties of groups and semigroups (Russian)*. Usp. Mat. Nauk, **27**, 139–180.

Artamonov, V. A.: 1991, *The structure of Hopf algebras*. Algebra. Topology. Geometry, Vol. 29 (Russian), Akad. Nauk SSSR Vsesoyuz. Inst. Nauchn. i Tekhn. Inform., Moscow, 3–63.

Artin, E., C. J. Nesbitt, and R. M. Thrall: 1944, *Rings with Minimum Condition*. University of Michigan Press, Ann Arbor, Mich., x+123 pp.

Artin, E. and O. Schreier: 1927, *Algebraische Konstruktion reeller Körper*. Abh. Math. Sem. Hamburg, **5**, 85–99.

Asano, K.: 1951, *Über kommutative Ringe, in denen jedes Ideal als Produkt von Primidealen darstellbar ist*. J. Math. Soc. Japan, **3**, 82–90.

Aschbacher, M.: 1986, *Finite group theory*. Cambridge University Press, Cambridge, x+274 pp.

Ash, C. J.: 1979, *Uniform labelled semilattices*. J. Austral. Math. Soc., **28**, 385–397.

— 1991, *Inevitable graphs: a proof of the type II conjecture and some related decision procedures*. Int. J. Algebra and Computation, **1**, 127–146.

Atiyah, M. and I. MacDonald: 1969, *Introduction to Commutative Algebra*. Addison-Wesley.

Auinger, K. and M. B. Szendrei: 1999, *On identity bases of epigroup varieties*. J. Algebra, **220**, 437–448.

Bahturin, Y., A. A. Mikhalev, and M. Zaicev: 2000, *Infinite-dimensional Lie superalgebras*. Handbook of algebra, Vol. 2, North-Holland, Amsterdam, 579–614.

Bahturin, Y. A.: 1987, *Identical relations in Lie algebras*. VNU Science Press b.v., Utrecht, x+309 pp., translated from the Russian by Bakhturin.

Bahturin, Y. A., A. A. Mikhalev, V. M. Petrogradsky, and M. V. Zaicev: 1992, *Infinite-dimensional Lie superalgebras*. Walter de Gruyter & Co., Berlin, x+250 pp.

Bak, A.: 1987, *Editorial*. *K*-Theory, **1**, 1–4.

Baker, K. A.: 1977, *Finite equational bases for finite algebras in a congruence-distributive equational class*. Adv. Math, **24**, 207–243.

Bakhturin, Y. A. and A. Y. Ol'shanskiĭ: 1988, *Identities (Russian)*. Itogi Nauki i Tekhniki. Sovremennye Problemy Matematiki. Fundamental'nye napravleniya, Vol. 18, VINITI, Moscow, 117–242.

Balbes, R. and P. Dwinger: 1974, *Distributive Lattices*. University of Missouri Press, Columbia, Mo., xiii+294 pp.

Bartholdi: 2001, Lie algebras and growth in branch groups, pre-print.

Bartholdi, L. and R. I. Grigorchuk: 2000, *Lie methods in growth of groups and groups of finite width*. Computational and geometric aspects of modern algebra (Edinburgh, 1998), Cambridge Univ. Press, Cambridge, 1–27.

Bartholdi, L., R. I. Grigorchuk, and Z. Sunik: 2001, *Branch groups*. Handbook of Algebra, Vol. 3, Elsevier.

Bass, H.: 1960, *Finitistic dimension and a homological generalization of semi-primary rings*. Trans. Amer. Math. Soc., **95**, 466–488.

— 1963, *On the ubiquity of Gorenstein rings*. Math. Z., **82**, 8–28.

— 1964, *K-theory and stable algebra*. Inst. Hautes Études Sci. Publ. Math., 5–60.

— 1968, *Algebraic K-theory*. W. A. Benjamin, Inc., New York-Amsterdam, xx+762 pp.

Bass, H., E. H. Connell, and D. Wright: 1982, *The Jacobian conjecture: reduction of degree and formal expansion of the inverse*. Bull. Amer. Math. Soc. (N.S.), **7**, 287–330.

Becker, B. and V. Weispfenning: 1993, *Gröbner Bases, A Computational Approach to Commutative Algebra*, volume 141 of *Graduate Texts in Mathematics*. Springer, Berlin, Heidelberg, New York, 574 pp.

Behrens, E.-A.: 1972, *Ring theory*. Academic Press, New York, vii+320 pp., translated from the German by Clive Reis, Pure and Applied Mathematics, Vol. 44.

Beidar, K. I. and M. Brešar: 2001, *Extended Jacobson Density Theorem for Rings with Derivations and Automorphisms*. Israel J. Math, **122**, 317–346.

Beidar, K. I. and M. A. Chebotar: 2000, *On functional identities and d-free subsets of rings. I, II*. Comm. Algebra, **28**, 3925–3951, 3953–3972.

Beidar, K. I., S. T. Glavatsky, and A. V. Mikhalev: 1989, *Semisimple classes and lower radicals of topological nonassociative algebras*. Trudy Sem. Petrovsk., 250–261, 268.

Beidar, K. I., W. S. Martindale, III, and A. V. Mikhalev: 1996, *Rings with generalized identities*. Marcel Dekker Inc., New York, xiv+522 pp.

Bell, H. E., A. A. Klein, and L.-C. o. Kappe: 1997, *An Analogue for Rings of a Group Problem of P. Erdős and B. H. Neumann*. Acta Math. Hungar., **77**, 57–67.

Belov, A., V. Borisenko, and V. Latyshev: 1997, *Monomial algebras*. Algebra, 4. J. Math. Sci. (New York), **87**, 3463–3575.

Berberian, S. K.: 1972, *Baer *-rings*. Springer-Verlag, New York, xiii+296 pp., die Grundlehren der mathematischen Wissenschaften, Band 195.

Berezin, F. A.: 1987, *Introduction to superanalysis*. D. Reidel Publishing Co., Dordrecht, xii+424 pp., edited and with a foreword by A. A. Kirillov, With an appendix by V. I. Ogievetsky, Translated from the Russian.

Berglund, J. F., H. D. Junghenn, and P. Milnes: 1989, *Analysis on semigroups*. John Wiley & Sons Inc., New York, xiv+334 pp., function spaces, compactifications, representations, A Wiley-Interscience Publication.

Bergman, C. H., R. D. Maddux, and D. L. Pigozzi, eds.: 1990, *Algebraic Logic and Universal Algebra in Computer Science*, Springer-Verlag, Berlin.

Bergman, G. M.: 1978, *The diamond lemma for ring theory*. Adv. in Math., **29**, 178–218.

Berkovich, Y. G. and E. M. Zhmud': 1998, *Characters of finite groups, Part 1*, volume 172 of *Translations of Mathematical Monographs*. American Math. Society, Providence RI.

— 1999, *Characters of finite groups, Part 2*, volume 181 of *Translations of Mathematical Monographs*. American Math. Society, Providence.

Berman, A. and J. Plemmons: 1979, *Nonnegative matrices in the mathematical sciences*. Academic Press, New York, xviii+316 pp.

Berstel, J. and D. Perrin: 1985, *Theory of Codes*. Academic Press, New York, London.

Besche, H. U., B. Eick, and E. A. O'Brien: 2001, *The groups of order at most 2000*. Electron. Res. Announc. Amer. Math. Soc., **7**, 1–4.

Bhargava, M.: 1998, *Generalized factorials and fixed divisors over subsets of a Dedekind domain*. J. Number Theory, **72**, 67–75.

Binder, F.: 1996, *Fast Computations in the Lattice of Polynomial Rational Function Fields*. ISSAC-96, Y. N. Lakshman, ed., ACM Press, 43–48.

Birkenmeier, G. F.: 1995, *A survey of regularity conditions and the simplicity of prime factor rings*. Vietnam J. Math., **23**, 29–38.

Birkenmeier, G. F. and N. J. Groenewald: 1997, *Prime Ideals in Rings with Involution*. Quaestiones Math., **20**, 591–603.

Birkenmeier, G. F., N. J. Groenewald, and H. E. Heatherly: 1997a, *Minimal and Maximal Ideals in Rings with Involution*. Beiträge Algebra Geom., **38**, 217–225.

Birkenmeier, G. F., H. E. Heatherly, J. Y. Kim, and J. K. Park: 2000, *Triangular matrix representations*. J. Algebra, **230**, 558–595.

Birkenmeier, G. F., H. E. Heatherly, and E. K. Lee: 1993, *Completely prime ideals and associated radicals*. Ring theory (Granville, OH, 1992), World Sci. Publishing, River Edge, NJ, 102–129.

Birkenmeier, G. F., J. Y. Kim, and J. K. Park: 1997b, *Regularity conditions and the simplicity of prime factor rings*. J. Pure Appl. Algebra, **115**, 213–230.

Birkhoff, G.: 1935, *On the structure of abstract algebras*. Proc. Cambridge Philos. Soc., **31**, 433–454.

— 1950, *Hydrodynamics — A Study in Logic, Fact and Similitude*. Princeton Univ. Press, Princeton, N.J.

— 1967, *Lattice theory*. American Mathematical Society, Providence, R.I., vi+418 pp., third edition. American Mathematical Society Colloquium Publications, Vol. XXV.

Birkhoff, G. and J. D. Lipson: 1970, *Heterogeneous algebras*. J. Combinatorial Theory, **8**, 115–133.

Biryukov, A. P.: 1970, *Varieties of idempotent semigroups (Russian)*. Algebra i Logika, **9**, 255–273.

Blum, E. K.: 1965, *A note on free subsemigroups with two generators*. Bull. Amer. Math. Soc., **71**, 678–679.

Bluman, G. and J. Cole: 1969, *The general similarity solution of the heat equation*. J. Math. Mech., **18**, 1025–1042.

Bogdanović, S.: 1985, *Semigroups with a system of subsemigroups*. University of Novy Sad, Novy Sad.

Bokut', L. A.: 1963, *Certain embedding theorems for rings and semigroups (Russian)*. Sib. Mat. J., **4**, 500–518.

— 1976, *Imbeddings into simple associative algebras*. Algebra i Logika, **15**, 117–142, 245.

Bokut', L. A. and D. Collins: 1980, *Mal'cev's problem and groups with a normal form*. Word Problems II, eds: Adian, S.I. and Boon, W.W. and Higman, G., North-Holland Publishing Company, Amsterdam, 29–53.

Book, R. and F. Otto: 1993, *String-rewriting systems*. Springer, Berlin.

Boone, W.: 1959, *The word problem*. Ann. of Math., **70**, 207–265.

Borel, A.: 1999, *Algebraic groups and Galois theory in the work of Ellis R. Kolchin*. Selected Works of Ellis Kolchin with Commentary, American Mathematical Society, Providence, 505–526.

Borho, W. and H. Kraft: 1976, *Über die Gelfand-Kirillov-Dimension*. Math. Ann., **220**, 1–24.

Bourbaki, N.: 1959, *Éléments de mathématique. Première partie: Les structures fondamentales de l'analyse. Livre II: Algèbre. Chapitre 9: Formes sesquilinéaires et formes quadratiques*. Hermann, Paris, 211 pp. (1 insert) pp.

— 1960-1982, *Groupes et Algèbres de Lie*. Hermann (Chap. 1-8) and Masson (Chap. 9), Paris, first edition.

— 1968, *Éléments de mathématique. Fasc. XXXIV. Groupes et algèbres de Lie. Chapitre IV: Groupes de Coxeter et systèmes de Tits. Chapitre V: Groupes engendrés par des réflexions. Chapitre VI: systèmes de racines*. Hermann, Paris, 288 pp. (loose errata) pp., actualités Scientifiques et Industrielles, No. 1337.

— 1971, *Éléments de mathématique. Fasc. XXVI. Groupes et algèbres de Lie. Chapitre I: Algèbres de Lie*. Hermann, Paris, 146 pp. (1 foldout) pp., seconde édition. Actualités Scientifiques et Industrielles, No. 1285.

— 1972a, *Commutative Algebra*. Addison-Wesley.

— 1972b, *Éléments de mathématique. Fasc. XXXVII. Groupes et algèbres de Lie. Chapitre II: Algèbres de Lie libres. Chapitre III: Groupes de Lie*. Hermann, Paris, 320 pp., actualités Scientifiques et Industrielles, No. 1349.

— 1972c, *Elements of mathematics. Commutative algebra*. Hermann, Paris, xxiv+625 pp., translated from the French.

— 1975, *Éléments de mathématique*. Hermann, Paris, 271 pp., fasc. XXXVIII: Groupes et algèbres de Lie. Chapitre VII: Sous-algèbres de Cartan, éléments réguliers. Chapitre VIII: Algèbres de Lie semi-simples déployées, Actualités Scientifiques et Industrielles, No. 1364.

— 1998, *Lie groups and Lie algebras. Chapters 1-3*. Springer-Verlag, Berlin, xviii+450 pp., translated from the French, Reprint of the 1989 English translation.

Box, G. E. P., W. G. Hunter, and J. S. Hunter: 1978, *Statistics for experimenters*. John Wiley & Sons, New York-Chichester-Brisbane, xviii+653 pp., an introduction to design, data analysis, and model building, Wiley Series in Probability and Mathematical Statistics.

Brešar, M.: 1993, *Commuting Traces of Biadditive Mappings, Commutativity-Preserving Mappings and Lie Mappings*. Trans. Amer. Math. Soc., **335**, 525–546.

Brewer, J. W.: 1981, *Power Series over Commutative Rings*. Marcel Dekker Inc., New York, vii+96 pp.

Britton, J.: 1963, *The word problem*. Ann. of Math., **77**, 16–32.

Bröcker, T. and T. tom Dieck: 1985, *Representations of Compact Lie Groups*. Springer, New York, first edition.

Broué, M.: 1995, *Rickard equivalences and block theory*. LMS Lecture Notes Series, **211(1)**, 58–79.

Brown, K. S.: 1982, *Cohomology of groups*. Springer-Verlag, New York, x+306 pp.

Brown, K. S. and R. Geoghegan: 1984, *An infinite-dimensional torsion-free* FP_∞ *group*. Invent. Math., **77**, 367–381.

Brown, T. C.: 1968, *On locally finite semigroups (Russian)*. Ukrain. Math. J., **20**, 732–738.

Bruns, W. and J. Herzog: 1998, *Cohen-Macaulay rings*. Cambridge University Press, Cambridge, revised edition.

Bruns, W. and U. Vetter: 1988, *Determinantal rings*. Springer-Verlag, Berlin, viii+236 pp.

Buchberger, B.: 1965, *Ein Algorithmus zum Auffinden der Basiselemente des Restklassenringes nach einem nulldimensionalen Polynomideal*. Ph.D. thesis, University of Innsbruck, Austria.

— 1970, *An Algorithmic Criterion for the Solvability of Algebraic Systems of Equations (German)*. Aequationes mathematicae, **4/3**, 374–383, published version of the author's PhD thesis, Univ. of Innsbruck (Austria), 1965. English translation in (Buchberger and Winkler (1998)), pp. 535–545.

Buchberger, B. and F. Winkler: 1998, *Gröbner Bases and Applications*, volume 251 of *London Mathematical Society Lecture Notes Series*. Cambridge University Press, 552 pp.

Budkina, L. G. and Al. A. Markov: 1973, *On F-semigroups with three generators. (Russian)*. Mat. Zam., **14**, 267–277.

Burmeister, P.: 1986, *A model theoretic oriented approach to partial algebras*. Number 32 in Mathematical Research, Akademie–Verlag, Berlin.

Burris, S. and S. Nelson: 1971, *Embedding the dual of* Π_∞ *in the lattice of equational classes of semigroups*. Algebra Universalis, **1**, 248–254.

Burris, S. and H. P. Sankappanavar: 1981, *A Course in Universal Algebra*. Springer-Verlag, Berlin–Heidelberg–N.Y.

Byleen, K., J. Meakin, and F. Pastijn: 1978, *The fundamental four-spiral semigroup*. Journal of Algebra, **65**, 6–26.

Cahen, P.-J. and J.-L. Chabert: 1997, *Integer-valued polynomials*. American Mathematical Society, Providence, RI, xx+322 pp.

Cameron, P. J.: 1990, *Oligomorphic Permutation Groups*. London Mathematical Society Lecture Notes **152**, Cambridge University Press, Cambridge.

— 1996, *Cofinitary permutation groups*. Bulletin of the London Mathematical Society, **28**, 113–140.

— 1999, *Permutation Groups*. London Mathematical Society Student Texts **45**, Cambridge University Press, Cambridge.

Camina, R.: 1997, *Subgroups of the Nottingham group*. J. Algebra, **196**, 101–113.

Campbell, C., E. Robertson, N. Ruškuc, and R. Thomas: 1995, *Rewriting a semigroup presentation*. Internat. J. Algebra Comput., **5**, 81–103.

Campbell, L. A.: 1973, *A condition for a polynomial map to be invertible*. Math. Ann., **205**, 243–248.

Camps, R. and W. Dicks: 1993, *On semilocal rings*. Israel J. Math., **81**, 203–211.

Cannon, J. W.: 1984, *The combinatorial structure of cocompact discrete hyperbolic groups*. Geom. Dedicata, **16**, 123–148.

Cantor, D. G. and H. Zassenhaus: 1981, *A new algorithm for factoring polynomials over finite fields*. Math. Comp., **36**, 587–592.

Cartan, E.: 1935, *La Méthode du Repère Mobile, la Théorie des Groupes Continus, et les Espaces Généralisés*. Hermann, Paris.

Cartan, H. and S. Eilenberg: 1956, *Homological Algebra*. Princeton University Press, Princeton, N. J., xv+390 pp.

Ceĭtin, G.: 1958, *An associative calculus with an insoluble problem of equivalence*. Trudy Math. Inst. Steklov, **52**, 172–189.

Chang, C. C. and H. J. Keisler: 1990, *Model theory*. North-Holland Publishing Co., Amsterdam, third edition, xvi+650 pp.

Chapman, S. T. and S. Glaz, eds.: 2000, *Non-Noetherian Commutative Ring Theory*, volume 520 of *Math. Appl.*. Kluwer Acad. Publ., Dordrecht.

Chase, S., D. Harrison, and A. Rosenberg: 1965, *Galois Theory and Cohomology of Commutative Rings*. Mem. Amer. Math. Soc.

Chase, S. U.: 1960, *Direct Products of Modules*. Trans. Amer. Math. Soc., **97**, 457–473.

Chatters, A. W. and C. R. Hajarnavis: 1980, *Rings with chain conditions*. Pitman (Advanced Publishing Program), Boston, Mass., vii+197 pp.

Chevalley, C.: 1943, *On the theory of local rings*. Ann. of Math. (2), **44**, 690–708.

— 1946, *Theory of Lie Groups*. Princeton Univ. Press, Princeton, first edition.

— 1951, *Introduction to the Theory of Algebraic Functions of One Variable*. American Mathematical Society, New York, N. Y., xi+188 pp., mathematical Surveys, No. VI.

— 1954, *The algebraic theory of spinors*. Columbia Univ. Press, New York, reprinted in Collected works II, Springer, 1997.

Choffrut, C. and J. Karhumäki: 1997, *Combinatorics of words*. Handbook of formal languages, Vol. 1, Springer, Berlin, 329–438.

Chrislock, J.: 1969, *A certain class of identites on semigroups*. Proc. Amer. Math. Soc., **21**, 189–190.

Churchill, G. A. and P. G. Trotter: 2000, *A unified approach to biidentities for e-varieties*. Semigroup Forum, **60**, 208–230.

Ćirić, M. and S. Bogdanović: 1995, *Theory of greatest decompositions of semigroups (A survey)*. Filomat (Niš), **9**, 385–426.

Clay, J. R.: 1992, *Nearrings: Geneses and applications*. Oxford University Press, New York.

Clifford, A.: 1954, *Bands of semigroups*. Proc. Amer. Math. Soc., **5**, 499–504.

Clifford, A. H.: 1941, *Semigroups admitting relative inverses*. Ann. of Math. (2), **42**, 1037–1049.

Clifford, A. H. and M. Petrich: 1977, *Some classes of completely regular semigroups*. J. Algebra, **46**, 462–480.

Clifford, A. H. and G. B. Preston: 1967a, *The algebraic theory of semigroups. Vol. I*. Amer. Math. Soc., Providence, R.I., xv+350 pp.

— 1967b, *The Algebraic Theory of Semigroups, Vol. II*. Amer. Math. Soc., Providence, R. I.

Cohen, I.: 1950, *Commutatitve rings with restricted minimum condition*. Duke Math. J., **17**, 27–42.

Cohen, I. S.: 1946, *On the structure and ideal theory of complete local rings*. Trans. Amer. Math. Soc., **59**, 54–106.

Cohn, P. M.: 1966, *On the structure of the GL_2 of a ring*. Inst. Hautes Études Sci. Publ. Math., 5–53.

— 1981, *Universal Algebra*. D. Reidel Publishing Co., Dordrecht, second edition, xv+412 pp.

— 1982, *Algebra*, volume 1. John Wiley & Sons, Chichester–New York–Brisbane–Toronto–Singapore, 2nd edition.

— 1985, *Free rings and their relations*. Academic Press Inc. [Harcourt Brace Jovanovich Publishers], London, second edition, xxii+588 pp.

— 1995, *Skew fields*. Cambridge University Press, Cambridge, xvi+500 pp., theory of general division rings.

Corry, L.: 1996, *Modern algebra and the rise of mathematical structures*. Birkhäuser Verlag, Basel, xiv+460 pp.

Corwin, L. and F. Greenleaf: 1990, *Representations of nilpotent Lie groups and their applications*. Cambridge Univ. Press, Cambridge, first edition.

Cracknell, A. P.: 1968, *Applied group theory*. Pergamon Press, Oxford.

Craig, W.: 1974, *Logic in Algebraic Form*. North-Holland, Amsterdam, viii+204 pp.

Crawley, P. and R. P. Dilworth: 1973, *Algebraic Theory of Lattics*. Prentice Hill, Inglwood Cliffs, N. J.

Curtis, C. and I. Reiner: 1962, *Representation Theory of Finite Groups and Associative Algebras*. Interscience, New York.

Day, A.: 1977, *Splitting lattices generate all lattices*. Algebra Universalis, **7**, 163–170.

Day, A., C. Herrmann, and R. Wille: 1972, *On modular lattices with four generators*. Algebra Universalis, **2**, 317–323.

de Luca, A. and S. Varricchio: 1999, *Finiteness and Regularity in Semigroups and Formal Languages*. Springer-Verlag, Berlin, Heidelberg, New York.

Dedekind, R.: 1877, *Über die Anzahl der Ideal-Klassen in den verschiedenen Ordnungen eines endlichen Körpers*. Gesammelte Mathematische Werke, Chelsea Publishing Co., New York, 1968, volume 1.

— 1897, *Über Zerlegungen von Zahlen durch ihre gröten gemeinsamen Teiler*. Festschrift der Herzogl. Technische Hochschule zur Naturforscher-Versammlung, Braunschweig, reprinted in "Gesammelte mathematische Werke", Vol. 2, pp. 103–148, Chelsea, New York, 1968.

— 1969, *Gesammelte Mathematische Werke, LXIX*. Chelsea.

Dedekind, R. and H. Weber: 1882, *Theorie der Algebraischen Funktionen einer Veränderlichen*. Crelle Journal, **92**, 181–290.

Delsarte, P.: 1973, *An algebraic approach to the association schemes of coding theory*. Philips Res. Rep. Suppl.

Dembowski, P.: 1968, *Finite Geometries*. Springer, Berlin.

DeMeyer, F. and E. Ingraham: 1971, *Separable Algebras over Commutative Rings*. Springer-Verlag, Berlin, iv+157 pp.

Dénes, J. and A. D. Keedwell: 1991, *Latin Squares*. North-Holland Publishing Co., Amsterdam, new developments in the theory and applications.

Diaconis, P.: 1989, *A Generalization of Spectral Analysis With Application to Ranked Data*. The Annals of Statistics, **17**, 949–979.

Diekert, V.: 1986, *Commutative monoids have complete presentations by free (non-commutative) monoids.*. Theoret. Comput. Sci., **46**, 313–318.

Dieudonné, J.: 1944, *Sur les fonctions continues p-adiques*. Bull. Sci. Math. (2), **68**, 79–95.

Dilworth, R. P.: 1954, *Proof of a conjecture on finite modular lattices*. Ann. of Math., **60**, 359–364.

Divinsky, N. J.: 1965, *Rings and radicals*. University of Toronto Press, Toronto, Ont., xii+160 pp., mathematical Expositions No. 14.

Dixmier, J.: 1977, *C*-algebras*. North-Holland Publishing Co., Amsterdam, xiii+492 pp., translated from the French by Francis Jellett, North-Holland Mathematical Library, Vol. 15.

— 1981, *von Neumann algebras*. North-Holland Publishing Co., Amsterdam, xxxviii+437 pp., translated from the second French edition by F. Jellett.

Dixon, J. D., M. P. F. du Sautoy, A. Mann, and D. Segal: 1999, *Analytic pro-p groups*. Cambridge University Press, Cambridge, second edition, xviii+368 pp.

Dixon, J. D. and B. Mortimer: 1996, *Permutation Groups*. Graduate Texts in Mathematics **163**, Springer-Verlag, New York.

do Lago, A. P. and I. Simon: 2001, *Free Burnside semigroups, to appear*. Theoretical Informatics and Applications.

Doran, R. S., ed.: 1994, *C*-algebras: 1943–1993*, American Mathematical Society, Providence, RI, a fifty year celebration.

Doran, R. S. and V. A. Belfi: 1986, *Characterizations of C*-algebras*. Marcel Dekker Inc., New York, xi+426 pp., the Gel'fand-Naĭmark theorems.

Dornhoff, L. and F. E. Hohn: 1978, *Applied Modern Algebra*. Macmillan, New York.

Draxl, P. K.: 1983, *Skew fields*. Cambridge University Press, Cambridge, ix+182 pp.

Drazin, M. P.: 1958, *Pseudo-inverses in associative rings and semigroups*. Amer. Math. Mon., **65**, 506–514.

Drensky, V.: 1998, *Gelfand-Kirillov dimension of PI-algebras*. Methods in ring theory (Levico Terme, 1997), Lecture Notes in Pure and Appl. Math., 198, Dekker, New York, 97–113.

— 2000, *Free algebras and PI-algebras*. Springer-Verlag, Singapore, xii+271 pp., Graduate Course in Algebra.

Drensky, V., J. Gutierrez, and J.-T. Yu: 1999, *Gröbner bases and the Nagata auto-morphism*. J. Pure Appl. Algebra, **135**, 135–153.

Drensky, V. and J.-T. Yu: 2001a, *Automorphisms and Coordinates of Polynomial Algebras*. Combinatorial and Computational Algebra, AMS, Contemporay Math. AMS Series, 264, 179–206.

— 2001b, *Tame and wild coordinates of $K[z][x,y]$*. Trans. Amer. Math. Soc.

du Sautoy, M.: 2000, *Counting p-groups and nilpotent groups*. Inst. Hautes Études Sci. Publ. Math., 63–112 (2001).

Dubreil-Jacotin, M. L.: 1947, *Sur l'immersion d'un demi-groupe dans un groupe*. C. R. Acad. Sci., **225**, 787–788.

Dubrovin, N. I.: 1980, *Chain domains*. Vestnik Moskov. Univ. Ser. I Mat. Mekh., 51–54, 104.

Duncan, A., E. Robertson, and R. N.: 1999, *Automatic monoids and change of generators*. Math. Proc. Cambridge Philos. Soc., **127**, 403–409.

Dvurečenskij, A.: 1993, *Gleason's theorem and its applications*. Kluwer Academic Publishers Group, Dordrecht, xvi+325 pp.

Eilenberg, S.: 1974, *Automata, languages, and machines. Vol. A*. Academic Press, New York, xvi+451 pp., pure and Applied Mathematics, Vol. 58.

— 1976, *Automata, Languages and Machines*, volume B. Academic Press, New York.

Einstein, A.: 1953, *The meaning of relativity*. Princeton University Press, Princeton, N. J., iv+168 pp., 4th ed.

Eisenbud, D.: 1995, *Commutative algebra*. Springer-Verlag, New York, xvi+785 pp., with a view toward algebraic geometry.

Eklof, P. C.: 1977, *Ultraproducts for Algebraists*. Handbook of mathematical logic, Part A, North-Holland, Amsterdam, 3–313. Studies in Logic and the Foundations of Math., Vol. 90.

Epstein, D., D. Holt, and S. Rees: 1991, *The use of Knuth-Bendix methods to solve the word problem in automatic groups*. J. Symbolic Computation, **12**, 397–414.

Epstein, D. B. A., J. W. Cannon, D. F. Holt, S. V. F. Levy, M. S. Paterson, and W. P. Thurston: 1992, *Word Processing in Groups*. Jones and Bartlett, Boston, MA, xii+330 pp.

Evans, T.: 1951a, *On multiplicative systems defined by generators and relations, I*. Proc. Cambridge Philos. Soc., **47**, 637–649.

— 1951b, *The word problem for abstract algebras*. J. London Math. Soc., **26**, 64–71.

Facchini, A.: 1998, *Module theory*. Birkhäuser Verlag, Basel, Endomorphism rings and direct sum decompositions in some classes of modules.

Facchini, A., D. Herbera, L. S. Levy, and P. Vámos: 1995, *Krull-Schmidt fails for Artinian modules*. Proc. Amer. Math. Soc., **123**, 3587–3592.

Faddeev, D. K.: 1964, *On the semigroup of genera in the theory of integral represen-tations*. Izv. Akad. Nauk SSSR Ser. Mat., **28**, 475–478, english transl., Amer. Math. Soc. Transl. (2), 64, 1967, 97–101.

Faith, C.: 1967, *Lectures on injective modules and quotient rings*. Springer-Verlag, Berlin, xv+140 pp., lecture Notes in Mathematics, No. 49.

— 1973, *Algebra: rings, modules and categories. I*. Springer-Verlag, New York, xxiii+565 pp., die Grundlehren der mathematischen Wissenschaften, Band 190.

— 1976, *Algebra. II*. Springer-Verlag, Berlin, xviii+302 pp., Ring theory, Grundlehren der Mathematischen Wissenschaften, No. 191.

— 1982, *Injective modules and injective quotient rings*. Marcel Dekker Inc., New York, viii+105 pp.

— 1999, *Rings and things and a fine array of twentieth century associative algebra*. American Mathematical Society, Providence, RI.

Fässler, A. and E. Stiefel: 1992, *Group Theoretical Methods and Their Applications*. Birkhäuser, Basel, Boston, Berlin.

Fels, M. and P. Olver: 1999, *Moving coframes. II. Regularization and theoretical foundations*. Acta Appl. Math., **55**, 127–208.

Fennemore, C. F.: 1971a, *All varieties of bands I*. Math. Nachr., **48**, 237–252.

— 1971b, *All varieties of bands II*. Math. Nachr., **48**, 253–262.

Fokas, A.: 1980, *A symmetry approach to exactly solvable evolution equations*. J. Math. Phys., **2**, 1318–1325.

Fontana, M., J. A. Huckaba, and I. J. Papick: 1997, *Prüfer Domains*. Marcel Dekker, New York, x+328 pp.

Formanek, E.: 1991, *The polynomial identities and invariants of $n \times n$ matrices*. Conference Board of the Mathematical Sciences, Washington, DC, vi+57 pp.

Fossum, R. M.: 1973, *The divisor class group of a Krull domain*. Springer-Verlag, New York, viii+148 pp., Ergebnisse der Mathematik und ihrer Grenzgebiete, Band 74.

Fountain, J., ed.: 1995, *Semigroups, Formal Languages and Groups*, volume 466 of *Series C: Mathematical and Physical Sciences*, Kluwer, Dordrecht, The Netherlands.

Freese, R.: 1979, *Projective geometries as projective modular lattices*. Trans. Amer. Math. Soc., **251**, 329–342.

— 1980, *Free modular lattices*. Trans. Amer. Math. Soc., **261**, 81–91.

— 1987a, *A decomposition theorem for modular lattices containing an n-frame*. Acta Sci. Math. (Szeged), **51**, 57–71.

— 1987b, *Free lattice algorithms*. Order, **3**, 331–344.

— 1994, *Finitely based modular congruence varieties are distributive*. Algebra Universalis, **32**, 104–114.

Freese, R., J. Ježek, and J. B. Nation: 1993, *Term rewrite systems for lattice theory*. J. Symbolic Computation, **16**, 279–288.

— 1995, *Free Lattices*. Amer. Math. Soc., Providence, Mathematical Surveys and Monographs, vol. 42.

Freese, R. and B. Jónsson: 1976, *Congruence modularity implies the Arguesian identity*. Algebra Universalis, **6**, 225–228.

Freese, R. and R. McKenzie: 1987, *Commutator theory for congruence modular varieties*. Cambridge University Press, Cambridge, iv+227 pp.

Freese, R. and J. B. Nation: 1985, *Covers in free lattices*. Trans. Amer. Math. Soc., **288**, 1–42.

Fu, G.: 1999, *Primary decomposition of ideals*. Marcel Dekker, 351–367 pp.

Fuchs, L.: 1960, *Abelian Groups*. Pergamon Press, New York, 367 pp.

— 1963, *Partially ordered algebraic systems*. Pergamon Press, Oxford, ix+229 pp.

Fujiwara, T.: 1955, *Note on the isomorphism problem for free algebraic systems*. Proc. Japan Acad., **31**, 135–136.

Fulton, W. and J. Harris: 1991, *Representation Theory, A First Course*. Springer, New York.

Ganter, B. and R. Wille: 1999, *Formal Concept Analysis*. Springer-Verlag, Berlin, x+284 pp., Mathematical Foundations, translated from the 1996 German original.

GAP: 2000, *GAP – Groups, Algorithms, and Programming, Version 4.2*. The GAP Group, Aachen, St Andrews, available at http://www-gap.dcs.st-and.ac.uk/~gap.

Gathen, J. von zur and J. Gerhard: 1999, *Modern Computer Algebra*. Cambridge University Press, Cambridge, UK.

Gelfand, I. M. and V. A. Ponomarev: 1970, *Problems of linear algebra and classification of quadruples of subspaces in a finite dimensional vector space*. Hilbert Space Operators, North-Holland Publishing Co., Amsterdam, 163–237, coll. Math. Soc. János Bolyai, vol. 5.

Gel'fand, S. I. and Y. I. Manin: 1996, *Methods of homological algebra*. Springer-Verlag, Berlin, xviii+372 pp., translated from the 1988 Russian original.

Gerhard, J. A.: 1970, *The lattice of equational classes of idempotent semigroups*. J. Algebra, **15**, 195–224.

— 1983, *Free completely regular semigroups. I. Representation*. J. Algebra, **82**, 135–142.

Gersten, S. M. and H. Short: 1991a, *Small cancellation theory and automatic groups, II.*. Invent. Math., **105**, 641–662.

Gersten, S. M. and H. B. Short: 1991b, *Rational subgroups of biautomatic groups*. Ann. of Math. (2), **134**, 125–158.

Ghys, E. and P. de la Harpe, eds.: 1990, *Sur les groupes hyperboliques d'après Mikhael Gromov*, volume 83 of *Progress in Mathematics*, Birkhäuser Boston, Inc., Boston, MA.

Gierz, G., K. H. Hofmann, K. Keimel, J. D. Lawson, M. W. Mislove, and D. S. Scott: 1980, *A compendium of continuous lattices*. Springer-Verlag, Berlin, xx+371 pp., completely revised and updated edition: "Continuous Lattices and Domains", Cambridge University Press, to appear.

Gilmer, R.: 1963, *Rings in which the unique primary decomposition theorem holds*. Proc. Amer. Math. Soc., **14**, 777–781.

— 1972a, *Multiplicative Ideal Theory*. Marcel Dekker, New York, x+609 pp.

— 1972b, *On factorization into prime ideals*. Comment. Mat. Helv., **47**, 70–74.

— 1992, *Multiplicative Ideal Theory*. Queen's Papers Pure Appl. Math., Vol. 90, Kingston, Ontario.

Gilmer, R. and W. Heinzer: 1972, *The Laskerian property, power series rings, and Noetherian spectra*. Proc. Amer. Math. Soc., **34**, 13–16.

Glaz, S.: 1989, *Commutative Coherent Rings*. Springer-Verlag, Berlin, xii+347 pp.

— 1992, *Commutative Coherent Rings: historical perspective and current developments*. Nieuw Arch. Wisk. (4), **10**, 37–56.

— 2001, *Finite Conductor Rings*. Proc. Amer. Math. Soc., **129**, 2833–2843.

Glazek, K.: 2001, *A Short Guide to the Literature on Semirings and their Applications in Mathematics and Computer Science*. Kluwer Academic Publishers, Dordrecht.

Golan, J. S.: 1999a, *Power Algebras over Semirings*. Kluwer Academic Publishers, Dordrecht, x+203 pp.

— 1999b, *Semirings and their Applications*. Kluwer Academic Publishers, Dordrecht, xi+381 pp.

Goldie, A. W.: 1960, *Semi-prime rings with maximum condition*. Proc. London Math. Soc. (3), **10**, 201–220.

Goodearl, K. R.: 1976, *Ring theory*. Marcel Dekker Inc., New York, viii+206 pp., Nonsingular rings and modules, Pure and Applied Mathematics, No. 33.

— 1982, *Notes on real and complex C^*-algebras*. Shiva Publishing Ltd., Nantwich, iv+211 pp.

— 1991, *von Neumann regular rings*. Robert E. Krieger Publishing Co. Inc., Malabar, FL, second edition, xviii+412 pp.

Gorbunov, V. A.: 1976, *On Lattices of Quasivarieties (Russian)*. Algebra i Logika, **15**, 436–457.

— 1983, *Quasiidentities of two-element algebras (Russian)*. Algebra i Logika, **22**, 121–127.

— 1995, *The Structure of Lattices of Varieties and Lattices of Quasivarieties: Their Similarity and Difference. II (Russian)*. Algebra i Logika, **34**, 369–397.

— 1998, *Algebraic theory of quasivarieties*. Kluwer Academic/Plenum Publishers, N.Y., xii+368 pp.

Gorbunov, V. A. and V. I. Tumanov: 1982, *Structure of quasivariety lattices (Russian)*. Trudy Instituta Matematiki SO AN SSSR, Vol. 2, Nauka, Novosibirsk, 12–44.

Gordon, R. and J. C. Robson: 1973, *Krull dimension*. American Mathematical Society, Providence, R.I., ii+78 pp., memoirs of the American Mathematical Society, No. 133.

Gorenstein, D.: 1980, *Finite Groups*. Chelsea, New York, second edition.

— 1982, *Finite simple groups*. Plenum Publishing Corp., New York, x+333 pp., an introduction to their classification.

Gorenstein, D., R. Lyons, and R. Solomon: 1994, *The classification of the finite simple groups*. American Mathematical Society, Providence, RI, xiv+165 pp.

— 1996, *The classification of the finite simple groups. Number 2. Part I. Chapter G*. American Mathematical Society, Providence, RI, xii+218 pp., general group theory.

— 1998, *The classification of the finite simple groups. Number 3. Part I. Chapter A.* American Mathematical Society, Providence, RI, xvi+419 pp., almost simple *K*-groups.

— 1999, *The classification of the finite simple groups. Number 4. Part II. Chapters 1–4.* American Mathematical Society, Providence, RI, xvi+341 pp., uniqueness theorems, With errata: *The classification of the finite simple groups. Number 3. Part I. Chapter A* [Amer. Math. Soc., Providence, RI].

Goto, M. and F. D. Grosshans: 1978, *Semisimple Lie algebras.* Marcel Dekker Inc., New York, vii+480 pp., lecture Notes in Pure and Applied Mathematics, Vol. 38.

Grätzer, G.: 1979, *Universal algebra.* Springer-Verlag, New York, second edition, xviii+581 pp.

— 1998, *General lattice theory.* Birkhäuser Verlag, Basel, second edition, xx+663 pp.

Grätzer, G. and H. Lakser: 1973, *A note on the implicational class generated by a class of structures.* Can. Math. Bull., **16**, 603–605.

Grätzer, G., H. Lakser, and E. T. Schmidt: 1998, *Congruence lattices of finite semimodular lattices.* Canad. Math. Bull., **41**, 290–297.

Grätzer, G., H. Lakser, and F. Wehrung: 2000, *Congruence amalgamation of lattices.* Acta Sci. Math. (Szeged), **66**, 3–22.

Grätzer, G. and E. T. Schmidt: 1962, *On congruence lattices of lattices.* Acta Math. Acad. Sci. Hungar., **13**, 179–185.

— 1999, *Congruence-preserving extensions of finite lattices to sectionally complemented lattices.* Proc. Amer. Math. Soc., **127**, 1903–1915.

Grätzer, G. and E. T. Schmidt: 2000, *Congruence-preserving extensions of finite lattices to semimodular lattices.* Houston J. Math., **26**.

— 2001, *Regular congruence-preserving extensions.* Algebra Universalis, **46**, 119–130.

Grätzer, G. and F. Wehrung: 1999a, *A new lattice construction: the box product.* J. Algebra, **221**, 315–344.

— 1999b, *Proper congruence-preserving extensions of lattices.* Acta Math. Hungar., **85**, 175–185.

— 2000a, *The strong independence theorem for automorphism groups and congruence lattices of arbitrary lattices.* Adv. in Appl. Math., **24**, 181–221.

— 2000b, *Tensor products of semilattices with zero, revisited.* J. Pure Appl. Algebra, **147**, 273–301.

Grayson, D.: 1976, *Higher algebraic K-theory. II (after Daniel Quillen).* Algebraic *K*-theory (Proc. Conf., Northwestern Univ., Evanston, Ill., 1976), Springer, Berlin, 217–240. Lecture Notes in Math., Vol. 551.

Green, J. A.: 1951, *On the structure on semigroups.* Ann. Math., **54**, 163–172.

Green, J. A. and D. Rees: 1952, *On semigroups in which $x^r = x$.* Proc. Camb. Philos. Soc., **48**, 35–40.

Grillet, P. A.: 1995, *Semigroups: an Introduction to the Structure Theory.* Marcel Dekker, New York, Basel, Hong Kong.

Gröbner, W.: 1950, *Elimination Theory (German)*. Monatshefte für Mathematik, **54**, 71–78.

Gross, H., C. Herrmann, and R. Moresi: 1987, *The classification of subspaces in Hermitean vector spaces*. J. Algebra, **105**, 516–541.

Grothendieck, A.: 1968, *Le groupe de Brauer. I,II,III. Algèbres d'Azumaya et interprétations diverses*. Dix Exposés sur la Cohomologie des Schémas, North-Holland, Amsterdam, 46–188.

Gumm, H. P.: 1983, *Geometrical methods in congruence modular algebras*. Mem. Amer. Math. Soc., **45**, viii+79.

Gunarwardena, J.: 1998, *Idempotency*. Cambridge University Press, Cambridge, xii+443 pp.

Gunn, C. and D. Maxwell: 1992, Not knot. Video, ISBN 0-86720-240-8, published by AKPeters, Natick, Mass.

Gupta, N.: 1989, *On groups in which every element has finite order*. Amer. Math. Monthly, **96**, 297–308.

Hagemann, J. and C. Herrmann: 1982, *Arithmetical Locally Equational Classes and Representation of Partial Functions*. Universal algebra (Esztergom, 1977), North-Holland, Amsterdam, 345–360.

Haiman, M.: 2001, *Hilbert schemes, polygraphs, and the Macdonald positivity conjecture*. J. Amer. Math. Soc., **14**, 941–1006.

Hall, M. and R. P. Dilworth: 1944, *The imbedding problem for modular lattices*. Ann. of Math., **45**, 450–456.

Hall, M., Jr. and J. K. Senior: 1964, *The groups of order 2^n ($n \leq 6$)*. The Macmillan Co., New York, 225 pp.

Hall, P.: 1954, *Finiteness conditions for soluble groups*. Proc. London Math. Soc., **4**, 419–436.

— 1962, *Wreath powers and characteristically simple groups*. Proc. Cambridge Philos. Soc., **58**, 170–184.

— 1969, *The Edmonton notes on nilpotent groups*. Queen Mary College Mathematics Notes, Mathematics Department, Queen Mary College, London.

— 1974, *On the embedding of a group in a join of given groups*. J. Austral. Math. Soc., **17**, 434–495, collection of articles dedicated to the memory of Hanna Neumann, VIII.

Hall, T. E.: 1973a, *On regular semigroups*. J. Algebra, **24**, 1–24.

— 1973b, *The partially ordered set of all J-classes of a semigroup*. Semigroup Forum, **6**, 263–269.

— 1989, *Identities for existence varieties of regular semigroups*. Bull. Austral. Math. Soc., **40**, 59–77.

Hall, T. E. and W. D. Munn: 1979, *Semigroups satisfying minimal conditions II*. Glasgow Math. J., **20**, 133–140.

Halmos, P. R.: 1963, *Lectures on Boolean Algebras*. D. Van Nostrand Co., Inc., Princeton, N.J., v+147 pp.

Hammer, P. L. and A. Kogan: 1992, *Horn Functions and their DNFs*. Inform. Process. Lett., **44**, 23–29.

Hawkes, T. O.: 1975, *Two applications of twisted wreath products to finite soluble groups*. Trans. Amer. Math. Soc., **214**, 325–335.

Hazewinkel, M., ed.: 1999–, *Handbook of Algebra, vols. 1–*. North-Holland, Amsterdam.

Hebisch, U. and H. J. Weinert: 1996, *Semirings and semifields*. Handbook of Algebra, Vol. 1, North-Holland, Amsterdam, 435–462.

— 1998, *Semirings - Algebraic Theory and Applications in Computer Science*. World Scientific, Singapore, viii+361 pp.

Heinzer, W. and J. Ohm: 1971, *Locally Noetherian commutative rings*. Trans. Amer. Math. Soc., **158**, 273–284.

— 1972, *On the Noetherian-like rings of E.G. Evans*. Proc. Amer. Math. Soc., **34**, 73–74.

Helgason, S.: 1978, *Differential Geometry, Lie Groups, and Symmetric Spaces*. Acad. Press, San Diego.

Henkin, L., J. D. Monk, and A. Tarski: 1971, 1985, *Cylindric Algebras, Part I*. North-Holland, Amsterdam, second edition, vi+508 pp.

— 1985, *Cylindric Algebras Part II*. North-Holland, Amsterdam, vii+302 pp.

Henkin, L., J. D. Monk, A. Tarski, H. Andréka, and I. Németi: 1981, *Cylindric Set Algebras*. Springer-Verlag, Berlin, v+323 pp., lecture Notes in Mathematics Vol. 883.

Herbera, D. and A. Shamsuddin: 1995, *Modules with semi-local endomorphism ring*. Proc. Amer. Math. Soc., **123**, 3593–3600.

Hereman, W.: 1994, *Review of symbolic software for the computation of Lie symmetries of differential equations*. Euromath Bull., **1**, 45–82.

Herrmann, C.: 1983, *On the word problem for the modular lattice with four free generators*. Math. Ann., **265**, 513–527.

— 1984, *On elementary Arguesian lattices with four generators*. Algebra Universalis, **18**, 225–259.

Herstein, I. N.: 1953, *The Structure of a Certain Class of Rings*. Amer. J. Math., **75**, 864–871.

— 1968, *Noncommutative Rings*. Published by The Mathematical Association of America, xi+199 pp.

— 1969, *Topics in ring theory*. The University of Chicago Press, Chicago, Ill.-London, xi+132 pp.

— 1976, *Rings with involution*. The University of Chicago Press, Chicago, Ill.-London, x+247 pp., chicago Lectures in Mathematics.

Herwig, B. and D. Lascar: 2000, *Extending partial automorphism and the profinite topology on the free groups*. Trans. Amer. Math. Soc., **352**, 1985–2021.

Herzog, M. and J. Schönheim: 1971, *Linear and nonlinear single error-correcting perfect mixed codes*. Information and Control, **18**, 364–368.

Higgins, P.: 1992, *Techniques of semigroup theory*. Oxford University Press, Oxford.

Higman, G.: 1961, *Subgroups of finitely presented groups*. Proc. Royal. Soc. London, **262**, 455–475.

— 1967, *The orders of relatively free groups*. Proc. Internat. Conf. Theory of Groups (Canberra, 1965), Gordon and Breach, New York, 153–165.

Hijikata, H. and K. Nishida: 1992, *Classification of Bass orders*. J. Reine Angew. Math., **431**, 191–220.

— 1997, *Primary orders of finite representation type*. J. Algebra, **192**, 592–640.

Hille, E.: 1976, *Ordinary Differential Equations in the Complex Domain*. John Wiley & Sons, New York.

Hilton, P. J. and U. Stammbach: 1997, *A course in homological algebra*. Springer-Verlag, New York, second edition, xii+364 pp.

Hindman, N. and D. Strauss: 1998, *Algebra in the Stone-Čech compactification*. Walter de Gruyter & Co., Berlin, xiv+485 pp., theory and Applications.

Hironaka, H.: 1964, *Resolution of singularities of an algebraic variety over a field of characteristic zero. I, II*. Ann. of Math. (2) 79 (1964), 109–203; ibid. (2), **79**, 205–326.

Hmelevskii, Y. I.: 1976, *Equations in free semigroups*. Proc. Steklov Inst. Mat., Am. Math. Soc. Transl., **107**.

Ho-Kim, Q. and X.-Y. Pham: 1998, *Elementary particles and their interactions*. Springer, New York.

Hochster, M. and C. Huneke: 1995, *Applications of the existence of big Cohen-Macaulay algebras*. Adv. Math., **113**, 45–117.

Hodges, W.: 1993, *Model theory*. Cambridge University Press, Cambridge, xiv+772 pp.

Hoehnke, H.-J.: 1986, *Fully invariant algebraic closure systems of congruences and quasivarieties of algebras*. Lectures in Universal Algebra (Szeged, 1983), North-Holland, Amsterdam, 189–207.

Hofmann, K. H. and M. W. Mislove: 1996, *Principles underlying the degeneracy of topological models of the untyped lambda calculus*. Semigroup theory and its applications (New Orleans, LA, 1994), Cambridge Univ. Press, Cambridge, 123–155.

Hofmann, K. H. and S. A. Morris: 1998, *The structure of compact groups*. Walter de Gruyter & Co., Berlin, xviii+835 pp., a primer for the student—a handbook for the expert.

Hofmann, K. H. and P. S. Mostert: 1966, *Elements of compact semigroups*. Charles E. Merryll Books, Inc., Columbus, Ohio, xiii+384 pp.

Holland, W. C.: 1969, *The characterization of generalized wreath products*. J. Algebra, **13**, 152–172.

Holt, D. F.: 1995, KBMAG—Knuth-Bendix in monoids and automatic groups. Software package, available by anonymous ftp from ftp.maths.warwick.ac.uk in directory people/dfh/kbmag2.

Holt, D. F. and D. F. Hurt: 1999, *Computing automatic coset systems and subgroup presentations*. J. Symbolic Comput., **27**, 1–19.

Hong, H. and J. Schicho: 1998, *Algorithms for Trigonometric Curves (Simplification, Implicitization, Parameterization)*. J. Symbolic Comput., **26**, 279–300.

Hoogland, E. and M. Marx: 2001, *Interpolation in Guarded Fragments*. Studia Logica.

Hotzel, E.: 1975–1976, *On semigroups with maximal conditions*. Semigroup Forum, **11**, 337–362.

— 1979, *On finiteness conditions in semigroups*. J. Algebra, **60**, 352–370.

Howie, J. M.: 1991, *Automata and languages*. The Clarendon Press Oxford University Press, New York, x+294 pp.

— 1995, *Fundamentals of Semigroup Theory*. The Clarendon Press Oxford University Press, New York, x+351 pp., oxford Science Publications.

Howie, R.: 1992, *A Century of Lie Theory*. Mathematics into the Twenty-first Century, AMS, Providence, 101–320.

Huckaba, J.: 1995, *Commutative Rings with Zero Divisors*. Marcel Dekker.

Humphreys, J.: 1975, *Linear algebraic groups*. Springer-Verlag, New York, xiv+247 pp.

Huppert, B.: 1983, *Endliche Gruppen I*. Springer-Verlag, New York, second edition.

— 1998, *Character Theory of Finite Groups*. Walter de Gruyter, Berlin.

Huppert, B. and N. Blackburn: 1982, *Finite Groups II*. Springer-Verlag, New York, first edition.

Hutchins, H. C. and H. J. Weinert: 1990, *Homomorphisms and kernels of semifields*. Period. Math. Hung., **21**, 113–152.

Hydon, P.: 2000, *Symmetry Methods for Differential Equations*. Cambridge University Press, Cambridge.

Isaacs, I. M.: 1973a, *Characters of solvable and symplectic groups*. Amer. J. of Math., **95**, 594–635.

— 1973b, *Equally partitioned groups*. Pac. J. Math., **49**, 109–116.

— 1994a, *Character Theory of Finite Groups*. Dover, New York.

— 1994b, *The π-character theory of solvable groups*. J. Austral. Math. Soc. (A), **57**, 81–102.

— 1995, *Characters and sets of primes for solvable groups*. Finite and Locally Finite Groups, B. H. et al, ed., Kluwer, Dordrecht, number 471 in NATO ASI Series C, 347–375.

Ivanov, S. V.: 1994, *The free Burnside groups of sufficiently large exponents*. Internat. J. Algebra Comput., **4**, ii+308.

Ivanov, S. V. and A. Y. Ol'shanskiĭ: 1991, *Some applications of graded diagrams in combinatorial group theory*. Groups—St. Andrews 1989, Vol. 2, Cambridge Univ. Press, Cambridge, 258–308.

Iwahori, N. and T. Kondo: 1965, *A criterion for the existence of a nontrivial partition of a finite group with applications to finite reflection groups*. J. Math. Soc. Japan, **17**, 207–215.

Iyama, O.: 1998, *A generalization of rejection lemma of Drozd-Kirichenko*. J. Math. Soc. Japan, **50**, 697–718.

Jacobson, N.: 1962, *Lie algebras*. Interscience Publishers (a division of John Wiley & Sons), New York-London, ix+331 pp., interscience Tracts in Pure and Applied Mathematics, No. 10.

— 1964, *Structure of rings*. American Mathematical Society, Providence, R.I., ix+299 pp.

— 1968, *Structure and representation of Jordan algebras*, volume 39 of *Colloquium Publications*. American Mathematical Society, Providence, RI, x+453 pp.

— 1975, PI-*algebras: An introduction*. Springer-Verlag, Berlin, iv+115 pp., lecture Notes in Mathematics, Vol. 441.

— 1979, *Lie algebras*. Dover Publications Inc., New York, ix+331 pp., republication of the 1962 original.

Jaffard, P.: 1960, *Theorie de la Dimension dans les Anneaux de Polynomes*. Gauthier-Villars.

Jagžev, A. V.: 1980, *Endomorphisms of free algebras*. Sibirsk. Mat. Zh., **21**, 181–192, 238.

James, G. and A. Kerber: 1981, *The representation theory of the symmetric group*. Addison-Wesley Publishing Co., Reading, Mass., xxviii+510 pp.

Jans, J. P.: 1964, *Rings and Homology*. Holt, Rinehart and Winston, New York, vii+88 pp.

Jansen, C. et al.: 1995, *An Atlas of Brauer Characters*. London Mathematical Society, Oxford.

Jensen, C. U. and H. Lenzing: 1989, *Model-theoretic Algebra with Particular Emphasis on Fields, Rings, Modules*. Gordon and Breach Science Publishers, New York, xiv+443 pp.

Ježek, J.: 1976, *Intervals in the lattice of varieties*. Algebra Universalis, **6**, 147–158.

Jipsen, P. and H. Rose: 1992, *Varieties of Lattices*, volume 1533 of *Lecture Notes in Mathematics*. Springer-Verlag, Berlin-New York.

Johnson, J.: 1969, *Differential dimension polynomials and a fundamental theorem on differential modules*. Amer. J. Math., **91**, 239–248.

Jones, P.: 1990, *Inverse semigroups and their lattice of inverse semigroups*. Lattices, Semigroups and Universal Algebra, Edited by J. Almeida et al., Plenum Press, New York, 115–127.

Jónsson, B.: 1966, *The unique factorization problem for finite relational structures*. Colloq. Math., **14**, 1–32.

Jónsson, B.: 1972, *Topics in Universal Algebra*. Lect. Notes Math., Berlin.

Jónsson, B.: 1995, *Congruence distributive varieties*. Math. Japon., **42**, 353–401.

Jónsson, B. and Tarski, A.: 1961, *On two properties of free algebras*. Math. Scand., **9**, 95–101.

Jung, H. W. E.: 1942, *Über ganze birationale Transformationen der Ebene*. J. Reine Angew. Math., **184**, 161–174.

Jura, A.: 1978, *Coset enumeration in a finitely presented semigroup*. Canad. Math. Bull., **21**, 37–46.

Kaarli, K. and A. Pixley: 2001, *Polynomial Completeness in Algebraic Systems*. Chapman & Hall/CRC, Boca Raton - London - New York - Washington, D.C.

Kac, V. G.: 1977a, *Lie superalgebras*. Advances in Math., **26**, 8–96.

— 1977b, *A sketch of Lie superalgebra theory*. Comm. Math. Phys., **53**, 31–64.

Kadison, R. V.: 1958, *Theory of operators. II. Operator algebras*. Bull. Amer. Math. Soc., **64**, 61–85.

Kadourek, J. and M. B. Szendrei: 1990, *A new approach in the theory of orthodox semigroups*. Semigroup Forum, **40**, 257–296.

Kaiser, H. and W. Nöbauer: 1998, *Geschichte der Mathematik*. Oldenbourg, München, second edition.

Kalman, R., P. Falb, and M. Arbib: 1969, *Topics in Mathematical Systems Theory*. McGraw-Hill, New York.

Kalmbach, G.: 1983, *Orthomodular lattices*. Academic Press, London, viii+390 pp.

— 1998, *Quantum measures and spaces*. Kluwer Academic Publishers, Dordrecht, xiv+343 pp.

Kanatani, K.: 1990, *Group-Theoretical Methods in Image Understanding*. Springer-Verlag, Berlin, xii+459 pp.

Kaplansky, I.: 1948, *Locally Compact Rings I*. Amer. J. Math, **70**, 447–452.

— 1950, *The Weierstrass theorem in fields with valuations*. Proc. Amer. Math. Soc., **1**, 356–357.

— 1951, *Locally Compact Rings II*. Amer. J. Math, **73**, 20–24.

— 1968, *Rings of operators*. W. A. Benjamin, Inc., New York-Amsterdam, viii+151 pp.

— 1970, *Commutative Rings*. Allyn and Bacon, Boston, x+180 pp.

— 1976, *An introduction to differential algebra*. Hermann, Paris, 64 pp.

Karpilovsky, G.: 1987, *The algebraic structure of crossed products*. North Holland, Amsterdam, x+348 pp.

— 1989, *Unit groups of group rings*. Longman, Essex, xiv+393 pp.

Karpinski, L. C.: 1919, *Robert of Chester's Latin Translation of the Algebra of al-Khuwarizmi*. Mac Millan, New York.

Karrass, A. and D. Solitar: 1970, *The subgroups of a free product of two groups with an amalgamated subgroup*. Trans. Ameerican Math. Soc., **149**, 237–255.

Karzel, H. and H.-J. Kroll: 1988, *Geschichte der Geometrie seit Hilbert*. Wissenschaftliche Buchgesellschaft, Darmstadt, x+246 pp.

Karzel, H., K. Sörensen, and D. Windelberg: 1973, *Einführung in die Geometrie*. Vandenhoeck & Ruprecht, Göttingen, 219 pp., studia mathematica/Mathematische Lehrbücher, Uni-Taschenbücher, No. 184.

Kasch, F.: 1982, *Modules and rings*. Academic Press, London, xii+372 pp., translated from the German.

Katsov, Y.: 2000, *On diagrams and flatness of functors*. J. Pure Appl. Algebra, **154**, 247–256, category theory and its applications (Montreal, QC, 1997).

Kaďourek, J.: 1992, *On varieties of combinatorial inverse semigroups II*. Semigroup Forum, **44**, 53–78.

Keller, O.: 1939, *Ganze Cremona-Transformationen*. Monatsh. Math. Phys., **47**, 299–306.

Kemer, A. R.: 1991, *Identities of associative algebras*. Algebra and analysis (Kemerovo, 1988), Amer. Math. Soc., Providence, RI, 65–71.

Kerber, A.: 1999, *Applied finite group actions*. Springer-Verlag, Berlin, second edition, xxvi+454 pp.

Kertész, A.: 1987, *Lectures on Artinian rings*. Akadémiai Kiadó, Budapest, 427 pp., translated from the German.

Kharlampovich, O. G.: 1954, *A finitely presented solvable group with unsolvable word problem*. Izv. Akad. Nauk SSSR, **45**, 852–873.

Kharlampovich, O. G. and M. V. Sapir: 1995, *Algorithmic problems in varieties*. Internat. J. Algebra Comput., **5**, 379–602.

Khukhro, E. I.: 1993, *Nilpotent groups and their automorphisms*. Walter de Gruyter & Co., Berlin, xiv+252 pp.

Klaas, G., C. R. Leedham-Green, and W. Plesken: 1997, *Linear pro-p-groups of finite width*. Springer-Verlag, Berlin, viii+115 pp.

Kleiman, E. I.: 1977, *Some properties of the lattice of varieties of inverse semigroups*. Research in Modern Algebra, Sverdlovsk University, Sverdlovsk, 56–72.

Klejman, Y. G.: 1984, *Some questions in the theory of varieties of groups*. Math. USSR, Izv., **22**, 33–65.

Klop, J. W.: 1992, *Term Rewriting Systems*. Handbook of Logic in Computer Science, Volume 2, S. Abramski, D. Gabbay, and T. Maibaum, eds., Oxford Science Publications, 2–117.

Klüners, J.: 1999, *On Polynomial Decompositions*. J. Symbolic Comput., **27**, 261–269.

Knapp, A.: 1986, *Representation Theory of Semisimple Groups*. Princeton Univ. Press, Princeton.

— 1996, *Lie Groups Beyond an Introduction*. Birkhäuser, Boston.

Knus, M.-A.: 1991, *Quadratic and Hermitian Forms over Rings*. Springer-Verlag, Berlin, xii+524 pp., with a foreword by I. Bertuccioni.

Knus, M.-A., A. Merkurjev, M. Rost, and J.-P. Tignol: 1998, *The Book of Involutions*. American Mathematical Society, Providence, RI, xxii+593 pp., with a preface in French by J. Tits.

Knuth, D. E.: 1998, *The Art of Computer Programming, vol.2, Seminumerical Algorithms*. Addison-Wesley, Reading MA, 3rd edition.

Koblitz, N.: 1998, *Algebraic Aspects of Cryptography*. Springer-Verlag, Berlin, x+206 pp.

Kolchin, E. R.: 1964, *The notion of dimension in the theory of algebraic differential equations*. Bull. Amer. Math. Soc., **70**, 570–573.

Kondratieva, M. V., A. B. Levin, A. V. Mikhalev, and E. V. Pankratiev: 1999, *Differential and difference dimension polynomials*. Kluwer Academic Publishers, Dordrecht, xiv+422 pp.

König, S. and A.Zimmermann: 1998, *Derived Equivalences for Group Rings*. Springer, Berlin.

Kontorovich, P.: 1939, *Sur la représentation d'un groupe fini sous la forme d'une somme directe de sous-groupes. 1*. Mat. Sbornik (N.S.), **5**, 289–296.

— 1946, *Sur les groupes à base de partition. 2.*. Mat. Sbornik (N.S.), **19**, 287–308.

Kostrikin, A. I.: 1959, *The Burnside problem*. Izv. Akad. Nauk SSSR, Ser. Mat., **23**, 3–34.

— 1986, *Vokrug Bernsaida*. "Nauka", Moscow, 232 pp.

Kozhukhov, I. B.: 1980, *On semigroups with minimal or maximal condition on left congruences*. Semigroup Forum, **21**, 337–350.

Krasil'nikov, A. N.: 1991, *The identitities of a group with nilpotent commutator subgroup are finitely based*. Math. USSR, Izv., **37**, 539–553.

Krasner, M. and L. Kaloujnine: 1951, *Produit complet des groupes de permutations et problème d'extension de groupes. III*. Acta Sci. Math. Szeged, **14**, 69–82.

Krause, G. R. and T. H. Lenagan: 2000, *Growth of algebras and Gelfand-Kirillov dimension*. American Mathematical Society, Providence, RI, revised edition, x+212 pp., graduate Studies in Mathematics, 22.

Kreuzer, M. and L. Robbiano: 2000, *Computational Commutative Algebra I*. Springer, Berlin, Heidelberg, New York, 321 pp.

Krohn, K. and J. Rhodes: 1965, *Algebraic theory of machines. I. Prime decomposition theorem for finite semigroups and machines*. Trans. Amer. Math. Soc., **116**, 450–464.

Krull, W.: 1929, *Über einen Hauptsatz der allgemeinen Idealtheorie*. S.-B. Heidelberger Akad. Wiss., 11–16.

— 1937, *Beiträge zur Arithmetik kommutativer Integritätsbereiche III*. Math. Zeit., **42**, 745–766.

— 1938, *Dimensionstheorie in Stellenringen*. Crelle Journal, **179**, 204–226.

— 1951, *Jacobsonsche Ringe, Hilbertscher Nullstellensatz, Dimensionstheorie*. Math. Zeit., **54**, 354–387.

Külshammer, B.: 1991, *Lectures on Block Theory*. Cambridge University Press, Cambridge.

Kung, J. P. S.: 1985, *Matchings and Radon transforms in lattices, I. Consistent lattices*. Order, **2**, 105–112.

— 1987, *Matchings and Radon transforms in lattices II. Concordant sets*. Math. Proc. Cambridge Philosophical Society, **101**, 221–231.

Kunz, E.: 1985, *Introduction to Commutative Algebra and Algebraic Geometry*. Birkhäuser.

Kurakin, V. L., M. V. T., A. V. Mikhalev, and A. A. Nechaev: 1999, *Linear and polylinear recurrences over finite rings and modules (review)*. Appl. Alg. and Error Corr. Codes, Springer, volume 1719 of *Lect. Not. Comp. Sci.*.

Kuz'min, E. N. and I. P. Shestakov: 1994, *Nonassociative structures*. Algebra VI, A. I. Kostrikin and I. R. Shafarevich, eds., Springer-Verlag, Berlin, volume 57 of *Encyclopaedia of Mathematical Sciences*, chapter II, 197–276.

Lallement, G.: 1967, *Demi-groupes réguliers*. Ann. Mat. Pura Appl. (4), **77**, 47–129.

— 1979, *Semigroups and Combinatorial Applications*. J. Wiley and Sons, New York.

— 1986, *Some algorithms for semigroups and monoids presented by a single relation*. Semigroup theory and applications, Oberwolfach, Springer, Berlin, 176–182.

Lam, T. Y.: 1980, *The Theory of Ordered Fields*. Ring theory and Algebra, III (Proc. Third Conf., Univ. Oklahoma, Norman, OK, 1979), Dekker, New York, 1–152.

— 1983, *Orderings, Valuations and Quadratic Forms*. Published for the Conference Board of the Mathematical Sciences, Washington, DC, vii+143 pp.

— 1984, *An Introduction to Real Algebra*. Rocky Mountain J. Math., **14**, 767–814, ordered fields and real algebraic geometry (Boulder, Col., 1983).

— 1991, *A first course in noncommutative rings*. Springer-Verlag, New York, xvi+397 pp.

— 1999a, *Bass's work in ring theory and projective modules*. Algebra, K-theory, Groups, and Education (New York, 1997), Amer. Math. Soc., Providence, RI, 83–124.

— 1999b, *Lectures on modules and rings*. Springer-Verlag, New York, xxiv+557 pp.

— 2001, *A first course in noncommutative rings*. Springer-Verlag, New York, second edition.

Lambek, J.: 1976, *Lectures on rings and modules*. Chelsea Publishing Co., New York, second edition, viii+184 pp.

Lampe, W.: 1986, *A property of the lattice of equational theories*. Algebra Universalis, **23**, 61–69.

Lanski, C.: 1997, *An Engel Condition with Derivation for Left Ideals*. Proc. Amer. Math. Soc., **125**, 339–345.

Lasker, E.: 1905, *Zur Theorie der Moduln und Ideale*. Math. Ann., **60**, 20–116.

Lawrence, J.: 1974, *A singular primitive ring*. Proc. Amer. Math. Soc., **45**, 59–62.

Lawson, M. V.: 1998, *Inverse Semigroups: The Theory of Partial Symmetries*. World Scientific, Singapore.

Laywine, C. F. and G. L. Mullen: 1998, *Discrete Mathematics using Latin Squares*. John Wiley & Sons Inc., New York, xviii+305 pp., a Wiley-Interscience Publication.

Le Bruyn, L.: 1997, *Automorphisms and Lie stacks*. Comm. Algebra, **25**, 2211–2226.

Leedham-Green, C. R. and M. F. Newman: 1980, *Space groups and groups of prime-power order. I*. Arch. Math. (Basel), **35**, 193–202.

Leïtes, D. A.: 1984, *Lie superalgebras*. Current problems in mathematics, Vol. 25, Akad. Nauk SSSR Vsesoyuz. Inst. Nauchn. i Tekhn. Inform., Moscow, 3–49, english translation: J. Sov. Math. **30** (1985), 2481–2512.

Lempken, W., P. Tiep, and P. Fleischmann: 1998, *The primitive p-Frobenius Groups*. Proc. Amer. Math. Soc., **126**, 1337–1343.

Lentin, A.: 1972, *Equations dans les Monoïdes Libres*. Gauthier-Villars, Paris.

Levi, F. W.: 1944, *On semigroups*. Bull. Calcutta Math. Soc., **36**, 141–146.

Lidl, R. and H. Niederreiter: 1994, *Introduction to Finite Fields and their Applications*. Cambridge University Press, Cambridge, xii+416 pp.

— 1997, *Finite Fields*. Cambridge University Press, Cambridge, second edition, xiv+755 pp.

Lidl, R. and G. Pilz: 1998, *Applied Abstract Algebra*. Springer-Verlag, New York, second edition, xvi+486 pp.

Lie, S.: 1924, *Gruppenregister*. Gesammelte Abhandlungen, B.G. Teubner, Leipzig, volume 5, 767–773.

Lin, Z. and L. Xu, eds.: 2001, *Applications of Gröbner Bases*, volume 12(3/4) of *Multidimensional Systems and Signal Processing*. Kluwer Academic Publishers, special issue on Applications of Groebner Bases in Multidimensional Systems and Signal Processing.

Lluis-Puebla, E., J.-L. Loday, H. Gillet, C. Soulé, and V. Snaith: 1992, *Higher algebraic K-theory: an overview*. Springer-Verlag, Berlin, x+164 pp.

Loeckx, J., H.-D. Ehrich, and M. Wolf: 1996, *Specification of Abstract Data Types*. Wiley–Teubner, Chichester, xi+260 pp.

— 2000, *Algebraic specification of abstract data types*. Handbook of Logic in Computer Science, Volume 5, S. Abramski, D. Gabbay, and T. Maibaum, eds., Oxford Science Publications, 217–316.

Lothaire, M.: 1983, *Combinatorics on Words. Encyclopedia of Mathematics and its Applications, Vol.17*. Addison-Wesley, Reading, MA.

— 2002, *Algebraic combinatorics on Words*. Cambridge University Press, to appear, Cambridge.

Lyndon, R. C.: 1951, *Identities in two-valued calculi*. Trans. Am. Math. Soc., **71**, 457–465.

— 1952, *Two notes on nilpotent groups*. Proc. Amer. Math. Soc., **3**, 579–583.

Lyndon, R. C. and P. E. Schupp: 1977, *Combinatorial Group Theory*. Springer-Verlag, Berlin, xiv+339 pp., Ergebnisse der Mathematik und ihrer Grenzgebiete, Band 89.

Lysënok, I. G.: 1996, *Infinite Burnside groups of even period*. Izv. Ross. Akad. Nauk Ser. Mat., **60**, 3–224.

Lyubich, Y. I.: 1992, *Mathematical structures in population genetics*. Springer-Verlag, Berlin, ix + 373 pp.

Mac Lane, S.: 1963, *Homology*. Academic Press, New York, x+422 pp., die Grundlehren der mathematischen Wissenschaften, Bd. 114.

— 1998, *Categories for the working mathematician*. Springer-Verlag, New York, second edition, xii+314 pp.

Macdonald, I. G.: 1995, *Symmetric functions and Hall polynomials*. Oxford University Press, New York, second edition, x+475 pp.

MacLane, S. and O. F. G. Schilling: 1939, *Infinite number fields with Noether ideal theories*. Amer. J. Math., **61**, 771–782.

MacWilliams, F. J. and N. J. A. Sloane: 1977, *The Theory of Error-Correcting Codes.*. North-Holland Publishing Co., Amsterdam, 2 Volumes, North-Holland Mathematical Library, Vol. 16.

Magid, A. R.: 1994, *Lectures on Differential Galois Theory*. American Mathematical Society, Providence, xiii+103 pp.

Magill, K. D., Jr.: 1993, *Green's equivalences and related concepts for semigroups of continuous self-maps*. Papers on general topology and applications (Madison, WI, 1991), New York Acad. Sci., New York, 246–268.

Magnus, W., A. Karrass, and D. Solitar: 1966, *Combinatorial Group Theory*. Wiley, New York.

— 1976, *Combinatorial group Theory*. Dover, New York, second edition.

Magri, F.: 1978, *A simple model of the integrable Hamiltonian equation*. J. Math. Phys.; 19, 1156–1162.

Magurn, B. A., ed.: 1985, *Reviews in K-theory, 1940–84*. American Mathematical Society, Providence, R.I., viii+811 pp., reviews reprinted from Mathematical Reviews.

Makanin, G. S.: 1977, *The problem of solvability of equations in a free semigroup*. Mat. USSR Sb., **32**, 129–198.

Mal'cev, A. I.: 1966, *Several remarks about quasivarieties of algebraic systems (Russian)*. Algebra i Logika, **5**, 3–9.

— 1973, *Algebraic systems*. Springer-Verlag, New York, xii+317 pp., posthumous edition, edited by D. Smirnov and M. Taĭclin, Translated from the Russian, Die Grundlehren der mathematischen Wissenschaften, Band 192.

Malle, G. and B. Matzat: 1999, *Inverse Galois Theory*. Springer-Verlag.

Manin, Y. I.: 1988, *Quantum groups and noncommutative geometry*. Université de Montréal Centre de Recherches Mathématiques, Montreal, PQ, vi+91 pp.

Manz, O. and T. R. Wolf: 1993, *Representations of Solvable Groups*. Cambridge University, Cambridge.

Marcinek, W.: 1991, *Generalized Lie algebras and related topics. I, II*. Acta Univ. Bratislav. Mat. Fiz. Astronom., 3–21, 23–52.

Markov, A.: 1947, *On the impossiblity of certain algorithms in the theory of associative systems*. Dokl. Akad. Nauk, **55**, 587–590.

— 1951, *Impossibility of algorithms for recognising some properties of associative systems*. Dokl. Akad. Nauk SSSR, **77**, 953–956.

Markov, Al. A.: 1971, *On finitely generated subsemigroups of a free semigroup*. Semigroup Forum, **3**, 251–258.

Marx, M., L. Pólos, and M. Masuch, eds.: 1996, *Arrow Logic and Multi-Modal Logic*, The European Association for Logic, Language and Information, Stanford University, cSLI Publications.

Matijasevič, J.: 1967, *Simple examples of unsolvable associative calculi*. Dokl. Akad. Nauk SSSR, **173**, 1264–66.

Matsumura, H.: 1989, *Commutative ring theory*. Cambridge University Press, Cambridge, second edition, xiv+320 pp., translated from the Japanese by M. Reid.

Maxson, C. J.: 1985, *Near-rings associated with generalized translation structures.* J. Geometry, **24**, 175–193.

Maxson, C. J. and G. P. Pilz: 1985, *Near-rings determined by fibered groups.* Arch. Math., **44**, 311–318.

— 1989, *Kernels of covered groups, II.* Results Math., **16**, 140–154.

McAlister, D.: 1971, *Representations of semigroups by linear transformations.* Semigroup Forum, 189–263; 283–320.

McAlister, D. B.: 1974a, *Groups, semilattices and inverse semigroups.* Trans. Amer. Math. Soc., **192**, 227–244.

— 1974b, *Groups, semilattices and inverse semigroups, II.* Trans. Amer. Math. Soc., **196**, 351–370.

McConnell, J. C. and J. C. Robson: 1987, *Noncommutative Noetherian rings.* John Wiley & Sons Ltd., Chichester, xvi+596 pp., with the cooperation of L. W. Small, A Wiley-Interscience Publication.

McCoy, N. H.: 1973, *The theory of rings.* Chelsea Publishing Co., Bronx, N.Y., x+161 pp., reprint of the 1964 edition.

McKay, S.: 2000, *Finite p-groups.* University of London, Queen Mary, School of Mathematical Sciences, London, x+102 pp.

McKay, S. and C. R. Leedham-Green: 2002, *The structure of finite p-groups.* Clarendon Press, Oxford.

McKenzie, R.: 1972, *Equational bases and non-modular lattice varieties.* Trans. Amer. Math. Soc., **174**, 1–43.

— 1996, *Tarski's finite basis problem is undecidable.* Int. J. Algebra and Computation, **6**, 49–104.

McKenzie, R., G. McNulty, and W. Taylor: 1987, *Algebras, Lattices, Varieties, Vol. 1.* Wadsworth & Brooks/Cole, Monterey, xii+361 pp.

McKinsey, J. C. C.: 1943, *The decision problem for some classes of sentences without quantifiers.* J. Symbolic Logic, **8**, 61–76.

McNaughton, R. and Y. Zalstein: 1975, *The Burnside problem for semigroups.* J. Algebra, **34**, 292–299.

Meakin, J.: 1980, *The partially ordered set of all J-classes of a semigroup.* J. London Math. Soc., **21**, 244–256.

— 1985, *The Rees construction in regular semigroups.* Semigroups. Structure and universal algebraic problems. Coll. Math. Soc. J. Bolyai, 39, North-Holland, 115–156.

— 1993, *An invitation to inverse semigroup theory.* Ordered Structures and Algebra of Computer Languages (eds. K. P. Shum, P.C. Yeun), World Scientific, Singapore, 91–115.

Mehrtens, H.: 1979, *Die Entstehung der Verbandstheorie.* Gerstenberg Verlag, Hildesheim.

Meldrum, J. D. P.: 1995, *Wreath products of groups and semigroups.* Longman, Harlow, xii+324 pp.

Merkur'ev, A. S. and A. A. Suslin: 1982, *K-cohomology of Severi-Brauer varieties and the norm residue homomorphism*. Izv. Akad. Nauk SSSR Ser. Mat., **46**, 1011–1046, 1135–1136.

Merzljakov, J.: 1986, *Rational groups*. Nauka, Moscow, 464 pp.

Merzljakov, J. I.: 1976, *A Survey of Recent Results on Automorphisms of Classical Groups*. Automorphisms of classical groups (Russian), Izdat. "Mir", Moscow, 250–259.

Micali, A. and P. Revoy: 1979, *Modules Quadratiques*. Bull. Soc. Math. France Mém., 144 pp. (1980).

Mikhalev, A. A. and A. A. Zolotykh: 1995, *Combinatorial aspects of Lie superalgebras*. CRC Press, Boca Raton, FL, viii+260 pp., with 1 IBM-PC floppy disk (3.5 inch; HD).

Mikhalev, A. V. and E. V. Pankratiev: 1980, *Differential dimension polynomial of systems of differential equations*. Algebra (collection of papers), A. I. Kostrikin, ed., Moscow Univ., Moscow, 57–67, (Russian).

Mikhalev, A. V., A. B. Shabat, and V. V. Sokolov: 1991, pp. 115–184, *The symmetry approach to classification of integrable equations*. What is Integrability?, V. Zakharov, ed., Springer Verlag, New York.

Miller, G., H. Blichfeldt, and L. Dickson: 1916 (Dover reprint, 1961), *Theory and Applications of Finite Groups*. Wiley.

Miller, W., Jr.: 1972, *Symmetry Groups and Their Applications*. Academic Press, New York, San Francisco, London.

Milne, J. S.: 1980, *Étale cohomology*. Princeton University Press, Princeton, N.J., xiii+323 pp.

Milnor, J.: 1966, *Whitehead torsion*. Bull. Amer. Math. Soc., **72**, 358–426.

— 1969/1970, *Algebraic K-theory and quadratic forms*. Invent. Math., **9**, 318–344.

— 1971, *Introduction to algebraic K-theory*. Princeton University Press, Princeton, N.J., xiii+184 pp., annals of Mathematics Studies, No. 72.

Mishchenko, S. P.: 1990, *Growth of varieties of Lie algebras*. Uspekhi Mat. Nauk, **45**, 25–45, 189.

Moh, T. T.: 1983, *On the Jacobian conjecture and the configurations of roots*. J. Reine Angew. Math., **340**, 140–212.

Monk, J. D.: 2000, *An introduction to cylindric set algebras*. Logic Journal of IGPL, **8**, 451–506.

Montgomery, S.: 1993, *Hopf algebras and their actions on rings*. Published for the Conference Board of the Mathematical Sciences, Washington, DC, xiv+238 pp.

Morel, F.: 1999, *Suite spectrale d'Adams et invariants cohomologiques des formes quadratiques*. C. R. Acad. Sci. Paris Sér. I Math., **328**, 963–968.

Morita, K.: 1958, *Duality for modules and its applications to the theory of rings with minimum condition*. Sci. Rep. Tokyo Kyoiku Daigaku Sect. A, **6**, 83–142.

Morse, M.: 1921, *Recurrent geodesics on a surface on negative curvature*. Trans. Amer. Math. Soc., **22**, 84–100.

Müller, B. J.: 1971, *Duality theory for linearly topologized modules*. Math. Z., **119**, 63–74.

Munn, W. D.: 1961, *Pseudo-inverses in semigroups*. Proc. Camb. Phil. Soc., **57**, 247–250.

— 1970, *Fundamental inverse semigroups*. Quart. J. Math. Oxford, **21**, 157–170.

— 1974, *Free inverse semigroups*. Proc. London Math. Soc., **29**, 385–404.

Murskiĭ, V. L.: 1964, *The existence in three-valued logic of a closed class with a finite basis, not having a finite complete system of identities*. Soviet Math. Dokl., **6**, 1020–1024.

— 1975, *The existence of a finite basis, and some other properties, for "almost all" finite algebras (Russian)*. Problemy Kibernet., **50**, 43–56.

Nagata, M.: 1962, *Local Rings*. Wiley Interscience.

— 1972, *On automorphism group of k[x, y]*. Kinokuniya Book-Store Co. Ltd., Tokyo, v+53 pp., department of Mathematics, Kyoto University, Lectures in Mathematics, No. 5.

Nambooripad, K. S. S.: 1979, *Structure of regular semigroups. I*. Mem. Amer. Math. Soc., **22**, vii+119.

Narkiewicz, W.: 1990, *Elementary and analytic theory of algebraic numbers*. Springer-Verlag, Berlin, second edition, xiv+746 pp.

Nation, J. B.: 1982, *Finite sublattices of a free lattice*. Trans. Amer. Math. Soc., **269**, 311–337.

Navarro, G.: 1998, *Characters and Blocks of Finite Groups*. Cambridge University Press, Cambridge.

Nechaev, A. A.: 1995, *Finite quasi-Frobenius modules, applications to codes and to linear recurrences*. Fundam. Prikl. Mat., **1**, 229–254.

— 1996a, *Linear codes over modules and over spaces. MacWilliams identity*. Proceedings of the 1996 IEEE Int. Symp. Inf. Theory and Appl., Victoria B. C., Canada, 35–38.

— 1996b, *Polylinear recurring sequences over modules and quasi-Frobenius modules*. First International Tainan-Moscow Algebra Workshop (Tainan, 1994), de Gruyter, Berlin, 283–298.

Nechaev, A. A., A. S. Kuz'min, and V. T. Markov: 1997, *Linear codes over finite rings and modules*. Fundam. Prikl. Mat., **3**, 195–254, functional analysis, differential equations and their applications (Russian) (Puebla, 1995).

Németi, I.: 1991, *Algebraizations of quantifier logics, an introductory overview*. Studia Logica, **50**, 485–569, a longer, updated version can be found in http://www.math-inst.hu/pub/algebraic-logic.

Németi, I. and I. Sain, eds.: 2000, *Logic Journal of IGPL, Vol. 8.*, special issue on Algebraic Logic.

Neukirch, J., A. Schmidt, and K. Wingberg: 2000, *Cohomology of Number Fields*. Springer-Verlag, New York.

Neumann, B. H.: 1937, *Identical relations in groups I*. Math.Ann, **114**, 506–525.

— 1954, *An essay on free products of groups with amalgamations*. Philos. Trans. Roy.Soc.London, **246**, ser. A.

— 1956, *Ascending derived series*. Compositio Math., **13**, 47–64.

— 1963, *Twisted wreath products of groups*. Arch. Math., **14**, 1–6.

— 1967a, *Some remarks on semigroup presentations*. Canad. J. Math., **19**, 1018–1026.

Neumann, B. H. and H. Neumann: 1959, *Embedding theorems for groups*. J. London Math. Soc., **34**, 465–479.

Neumann, B. H., H. Neumann, and P. M. Neumann: 1962, *Wreath products and varieties of groups*. Math. Z., **80**, 44–62.

Neumann, H.: 1967b, *Varieties of Groups*, volume 37 of *Ergebnisse der Math. und ihrer Grenz.*. Springer-Verlag, New York, x+192 pp.

Neumann, P. M.: 1964, *On the structure of standard wreath products of groups*. Math. Z., **84**, 343–373.

Neumann, P. M. and J. Wiegold: 1964, *Schreier varieties of groups*. Math. Z., **85**, 392–400.

Noether, E.: 1918, *Invariante Variationsprobleme*. Nachr. Konig. Gesell. Wissen. Gottingen, Math.–Phys. Kl., 235–257.

— 1921, *Idealtheorie in Ringbereichen*. Math. Ann., **83**, 24–65.

— 1927, *Abstrakter Aufbau der Idealtheorie in algebraischen Zahl- and Funktionenkörpern*. Math. Ann., **96**, 26–61.

Novikov, P.: 1955, *On the algorithmic unsolvability of the word problem in group theory*. Trudy Mat. Inst. Steklov, **44**, 143.

Oates, S. and M. B. Powell: 1964, *Identical relations in finite groups*. J. Algebra, **1**, 11–39.

O'Carroll, L.: 1976, *Embedding theorems for proper inverse semigroups*. J. Algebra, **42**, 26–40.

Okniński, J.: 1991, *Semigroup algebras*. Marcel Dekker, New York, x+357 pp.

— 1998, *Semigroups of matrices*. World Scientific, Singapore, xiv+311 pp.

Ol'shanskiĭ, A. Y.: 1970, *On the finite basis problem for laws in groups*. Izv. Akad. Nauk SSSR, Ser. Mat., **34**, 376–384, russian.

— 1991, *Geometry of defining relations in groups*. Kluwer Academic Publishers Group, Dordrecht, xxvi+505 pp., translated from the 1989 Russian original by Yu. A. Bakhturin.

— 1995, *The geometry of defining relations in groups*. Groups—Korea '94 (Pusan), de Gruyter, Berlin, 263–265.

Olver, P.: 1993, *Applications of Lie Groups to Differential Equations*. Springer-Verlag, New York.

— 1995, *Equivalence, Invariants, and Symmetry*. Cambridge University Press, Cambridge.

Olver, P. and P. Rosenau: 1987, *Group-invariant solutions of differential equations*. SIAM J. Appl. Math., **47**, 263–278.

O'Meara, O. T.: 1974, *Lectures on linear groups*. American Mathematical Society, Providence, R.I., vii+87 pp., expository Lectures from the CBMS Regional Conference, Tempe, Ariz., 1973, Conference Board of the Mathematical Sciences Regional Conference Series in Mathematics, No. 22.

Ore, O.: 1933, *Theory of Noncommutative Polynomials*. Ann. Math., **34**, 480–508.

— 1935, *On the foundations of abstract algebra, I*. Ann. Math., **36**, 406–437.

— 1936, *On the foundations of abstract algebra, II*. Ann. Math., **37**, 265–292.

Osborne, M. S.: 2000, *Basic Homological Algebra*. Springer-Verlag, New York, x+395 pp.

Ovsyannikov, L.: 1982, *Group Analysis of Differential Equations*. Academic Press, New York.

Papasoglu, P.: 1995, *Strongly geodesically automatic groups are hyperbolic*. Invent. Math., **121**, 323–334.

Passi, I. B. S.: 1968, *Polynomial maps on groups*. J. Algebra, **9**, 121–151.

— 1979, *Group rings and their augmentation ideals*. Springer, Berlin, vi+137 pp.

Passman, D. S.: 1968, *Permutation Groups*. Benjamin, New York.

— 1971, *Infinite group rings*. Marcel Dekker Inc., New York, viii+149 pp., pure and Applied Mathematics, 6.

— 1977, *The algebraic structure of group rings*. Wiley-Interscience, New York, xiv+720 pp.

— 1984, *Group rings of polycyclic groups*. Group theory. Essays for Philip Hall, Academic Press, London, 207–256.

— 1989, *Infinite crossed products*. Academic Press, Boston, xii+468 pp.

— 1998, *Semiprimitivity of group algebras: past results and recent progress*. Trends in ring theory (Miskolc, 1996), Amer. Math. Soc., Providence, 127–157.

Pastijn, F.: 1990, *The lattice of completely regular semigroup varieties*. J. Austral. Math. Soc. Ser. A., **49**, 24–42.

Pastijn, F. and M. Petrich: 1986, *Congruences on regular semigroups*. Trans. Amer. Math. Soc., **295**, 607–633.

Pederson, G. K.: 1979, *C*-Algebras and their Automorphism Groups*. Academic Press, London.

Pedicchio, M. C.: 1995, *A categorical approach to commutator theory*. J. Algebra, **177**, 647–657.

Perkins, P.: 1969, *Bases for equational theories of semigroups*. J. Algebra, **11**, 289–314.

Petrich, M.: 1970, *The translation hull in semigroups and rings*. Semigroup Forum, **1**, 283–360.

— 1972, *Regular semigroups satisfying certain conditions on idempotents and ideals*. Trans. Amer. Math. Soc., **170**, 245–267.

— 1973, *Introduction to Semigroups*. Charles E. Merrill, Columbus, Ohio.

— 1974, *The structure of completely regular semigroups*. Trans. Amer. Math. Soc., **189**, 211–236.

— 1977, *Lectures in Semigroups*. Acad. Verlag, Berlin.

— 1984, *Inverse Semigroups*. John Wiley & Sons, New York.

Petrich, M. and N. R. Reilly: 1999, *Completely regular semigroups*. John Wiley & Sons Inc., New York, xii+481 pp., a Wiley-Interscience Publication.

Pierce, R. S.: 1982, *Associative algebras*. Springer-Verlag, New York, xii+436 pp., studies in the History of Modern Science, 9.

Pigozzi, D.: 1988, *Finite basis theorem for relatively congruence-distributive quasi-varieties*. Trans. Am. Math. Soc., **310**, 499–533.

Pilz, G. F.: 1983, *Near-rings*. North-Holland Publishing Co., Amsterdam, second edition, xv+470 pp., the Theory and its Applications.

— 1996, *Near-rings and near-fields*. Handbook of Algebra, Vol. 1, North-Holland, Amsterdam, 463–498.

Pin, J.-E.: 1986, *Varieties of formal languages*. Plenum Publishing Corp., New York, x+138 pp., translated from the French by A. Howie.

Pinus, A. G.: 1986, *Congruence Modular Varieties of Algebras (Russian)*. Irkutsk. Gos. Univ., Irkutsk, 132 pp.

Plotkin, B. I.: 1966, *Groups of Authomorphism auf Algebraic Systems*. Nordoff.

— 2000, *Algebra, categories and databases*. Handbook of Algebra, Vol. 2, North-Holland, Amsterdam, 79–148.

Plotkin, B. I. and S. M. Vovsi: 1983, *Mnogoobraziya predstavlenii grupp*. "Zinatne", Riga, 339 pp., obshchaya teoriya, svyazi i prilozheniya. [General theory, connections and applications].

Pólya, G.: 1919, *Über ganzwertige Polynome in algebraischen Zahlkörpern*. J. reine angew. Math., **149**, 97–116.

Ponizovskii, J.: 1975, *On matrix semigroups (Russian)*. Tez. Dokl. 13 Vsesouz. alg. simpos. Part 2, Gomel, 233.

— 1982, *On irreducible matrix semigroups*. Semigroup Forum, 117–148.

Pontrjagin, L.: 1973, *Continuous Groups*. Nauka, Moscow.

Porteous, I. R.: 1995, *Clifford Algebras and the Classical Groups*. Cambridge University Press, Cambridge, x+295 pp.

Post, E.: 1947, *Recursive unsolvability of a problem by Thue*. J. Symb. Logic, **12**, 1–11.

Prestel, A.: 1984, *Lectures on Formally Real Fields*. Springer-Verlag, Berlin, xi+125 pp.

Preston, G. B.: 1954, *Inverse semi-groups*. J. London Math. Soc., **29**, 396–403.

Pride, S.: 1995, *Geometric methods in combinatorial semigroup theory*. Semigroups, formal languages and groups, Kluwer, Dordrecht, 215–232.

Priestley, H. A.: 1972, *Ordered Topological Spaces and the Representation of Distributive Lattices*. Proc. London Math. Soc. (3), **24**, 507–530.

Procesi, C.: 1973, *Rings with polynomial identities*. Marcel Dekker Inc., New York, viii+190 pp., pure and Applied Mathematics, 17.

Puig, L.: 1999, *On the Local Structure of Morita and Rickard Equialences between Brauer Blocks*. Birkhäuser, Basel.

Putcha, M.: 1988, *Linear algebraic monoids*. Cambridge University Press, Cambridge, x+171 pp.

— 1995, *Monoids of Lie type*. Semigroups, Formal Languages and Groups, Kluwer Academic Publishing, Dordrecht, 353–367.

Quillen, D.: 1973, *Higher algebraic K-theory, I*. Algebraic K-theory, I: Higher K-theories (Proc. Conf., Battelle Memorial Inst., Seattle, Wash., 1972), Springer, Berlin, 85–147. Lecture Notes in Math., Vol. 341.

Rasiowa, H.: 1974, *An Algebraic Approach to Non-Classical Logics*. North-Holland Publishing Co., Amsterdam, xv+403 pp., studies in Logic and the Foundations of Mathematics , Vol. 78.

Rasiowa, H. and R. Sikorski: 1970, *The Mathematics of Metamathematics*. PWN—Polish Scientific Publishers, Warsaw, third edition, 519 pp., monografie Matematyczne, Tom 41.

Rautenberg, W.: 1981, *2-element matrices*. Studia Logica, **40**, 315–353.

Razmyslov, Y. P.: 1989, *Identities of algebras and their representations*. "Nauka", Moscow, 432 pp., russian, with an English summary.

— 1994, *Identities of algebras and their representations*. American Mathematical Society, Providence, RI, xiv+318 pp., translated from the 1989 Russian original.

Rees, D.: 1940, *On semi-groups*. Proc. Camb. Philos. Soc., **36**, 387–400.

Rees, S.: 1998, *Hairdressing in groups: a survey of combings and formal languages*. The Epstein birthday schrift, I. Rivin, C. Rourke, and C. Series, eds., Geometry and Topology, Coventry, volume 1 of *Geom. Topol. Monogr.*, 493–509.

Reilly, N. R.: 1980, *Varieties of completely semisimple inverse semigroups*. J. Algebra, **65**, 427–444.

Reilly, N. R. and P. G. Trotter: 1986, *Properties of relatively free inverse semigroups*. Trans. Amer. Math. Soc., **294**, 243–262.

Reiner, I.: 1970, *A survey of integral representation theory*. Bull. Amer. Math. Soc., **76**, 159–227.

— 1975, *Maximal orders*. Academic Press, London-New York, xii+395 pp., london Mathematical Society Monographs, No. 5.

Reusch, B.: 1975, *Generation of Prime Implicants from Subfunctions and a Unifying Approach to the Covering Problem*. IEEE Trans. Computers, **C-24**, 924–930.

Reuter, K.: 1987, *Matchings for linearly indecomposable modular lattices*. Discrete Math., **63**, 245–249.

Rhodes, J.: 1999, *Undecidability, automata and pseudovarieties of finite semigroups*. Int. J. Algebra and Computation, **9**, 455–473.

Ribes, L. and P. A. Zalesskiĭ: 1993, *On the profinite topology on a free group*. Bull. London Math. Soc., **25**, 37–43.

Robinson, G.: 2000, *Dade's Projective Conjecture for p - Solvable Groups*. J. of Algebra, **229**, 234–248.

Roggenkamp, K. W.: 1990, *Indecomposable representations of orders*. Topics in algebra, Part 1 (Warsaw, 1988), PWN, Warsaw, 449–491.

Roggenkamp, K. W. and M. Ştefănescu, eds.: 1996, *Representation theory of groups, algebras, and orders*, "Ovidius" University Press, Constanţa, an. Ştiinţ. Univ. Ovidius Constanţa Ser. Mat. **4** (1996), no. 1.

Rognes, J. and C. Weibel: 2000, *Two-primary algebraic K-theory of rings of integers in number fields*. J. Amer. Math. Soc., **13**, 1–54.

Roseblade, J. E.: 1978, *Prime ideals in group rings of polycyclic groups*. Proc. London Math. Soc. (3), **36**, 385–447.

Rosenberg, I.: 1970, *Über die funktionale Vollständigkeit in den mehrwertigen Logiken. Struktur der Funktionen von mehreren Veränderlichen auf endlichen Mengen*. Rozpravy Československé Akad. Věd Řada Mat. Přírod. Věd, **80**, 3–93.

Rosenberg, J.: 1994, *Algebraic K-theory and its applications*. Springer-Verlag, New York, x+392 pp.

Rotman, J. J.: 1995, *An introduction to the theory of groups*, volume 148 of *Graduate Texts in Mathematics*. Springer-Verlag, New York, fourth edition, xvi+513 pp.

Rowen, L. H.: 1980, *Polynomial identities in ring theory*. Academic Press, New York, xx+365 pp.

— 1988a, *Ring theory. Vol. I*. Academic Press Inc., Boston, MA, xxiv+538 pp.

— 1988b, *Ring theory. Vol. II*. Academic Press Inc., Boston, MA, xiv+462 pp.

Ruiz, J. M.: 1993, *The Basic Theory of Power Series*. Friedr. Vieweg & Sohn, Braunschweig, x+134 pp.

Ruppert, W.: 1984, *Compact semitopological semigroups: an intrinsic theory*. Springer-Verlag, Berlin, v+260 pp.

Saliĭ, V. N.: 1969, *Equationally normal varieties of semigroups (Russian)*. Izv. Vyssh. Uchebn. Zaved. Mat, 61–68.

— 1988, *Lattices with unique complements*. American Mathematical Society, Providence, RI, x+113 pp., translated from the Russian.

Salomaa, A.: 1981, *Jewels of Formal Language Theory*. Pitman, London.

Saltman, D. J.: 1999, *Lectures on Division Algebras*. Published by American Mathematical Society, Providence, RI, viii+120 pp.

Sanders, J. and J. Wang: 1998, *On the integrability of homogeneous scalar evolution equations*. J. Diff. Eq., **147**, 410–434.

Sapir, M. V.: 1980, *On the quasivarieties generated by finite semigroups*. Semigroup Forum, **20**, 73–88.

— 1988, *Problems of Burnside type and the finite basis property in varieties of semigroups*. Math. USSR–Izv., **30**, 295–314.

Sautoy, M. d., D. Segal, and A. Shalev, eds.: 2000, *New Horizons in pro-p Groups*. Birkhäuser, Boston.

Schafer, R. D.: 1995, *An introduction to nonassociative algebras*. Dover Publications, New York, x + 166 pp.

Scheiblich, H. E.: 1973, *Free inverse semigroups*. Proc. Amer. Math. Soc., **38**, 1–7.

— 1974, *Kernels of inverse semigroup homomorphisms.* J. Austral. Math. Soc., **18**, 289–292.

Schenzel, P.: 1982, *Dualisierende Komplexe in der lokalen Algebra und Buchsbaum-Ringe.* Springer-Verlag, Berlin, vii+161 pp., with an English summary.

Scheunert, M.: 1979, *The theory of Lie superalgebras.* Springer, Berlin, x+271 pp., an introduction.

Schmidt, E. T.: 1974, *Every finite distributive lattice is the congruence lattice of some modular lattice.* Algebra Universalis, **4**, 49–57.

Schmidt, R.: 1994, *Subgroup lattices of groups.* Walter de Gruyter & Co., Berlin, xvi+572 pp.

Scholz, E., K. Andersen, H. J. M. Bos, and et al.: 1990, *Geschichte der Algebra.* Bibliographisches Institut, Mannheim, viii+506 pp., eine Einführung. [An introduction].

Schreier, O.: 1927, *Die Untergruppen der freien Gruppen.* Abh. Math. Sem. Univ. Hamburg, **5**, 161–183.

Schubert, H.: 1972, *Categories.* Springer-Verlag, New York, xi+385 pp., translated from the German by Eva Gray.

Schützenberger, M. P.: 1956, *Une théorie algébrique du codage.* C. R. Acad. Sci., **242**, 862–864.

Scott, D. S.: 1993, *A Type-Theoretical Alternative to ISWIM, CUCH, OWHY.* Theoret. Comput. Sci., **121**, 411–440, a collection of contributions in honour of Corrado Böhm on the occasion of his 70th birthday.

Sehgal, S. K.: 1978, *Topics in group rings.* Marcel Dekker, New York, vi+251 pp.

— 1993, *Units in integral group rings.* Longman, Essex, xii+357 pp.

Seidenberg, A.: 1953, *A note on the dimension theory of rings.* Pacific J. Math., **3**, 505–512.

Serre, J.-P.: 1965, *Algèbre locale. Multiplicités.* Springer-Verlag, Berlin, vii+188 pp. (not consecutively paged) pp., cours au Collège de France, 1957–1958, rédigé par Pierre Gabriel. Seconde édition, 1965. Lecture Notes in Mathematics, 11.

— 1987, *Complex semisimple Lie algebras.* Springer-Verlag, New York, x+74 pp., translated from the French by G. A. Jones.

— 1992a, *Lie Algebras and Lie Groups.* Springer, New York, second edition.

— 1992b, *Topics in Galois Theory.* Jones and Bartlet, Boston.

Sharpe, D. W. and P. Vámos: 1972, *Injective modules.* Cambridge University Press, London, xii+190 pp., cambridge Tracts in Mathematics and Mathematical Physics, No. 62.

Shevrin, L. N.: 1960, *On subsemigroups of free semigroups.* Sov. Math. Dokl., **1**, 892–894.

— 1965, *On locally finite semigroups (Russian).* Dokl. AN SSSR. Ser. Mat., **162**, 770–773.

— 1974, *A general theorem on semigroups with some finiteness conditions (Russian).* Mat. Zametki, **15**, 925–935.

— 1991, *Semigroups*. Handbook "General Algebra", ed. Skornyakov, L.A., Vol.II (Russian), "Nauka", Moscow, 11–191.

— 1992, *On two longstanding problems concerning nilsemigroups*. "Semigroups with applications" (Proc. of the Conference in Oberwolfach), ed. by J. M. Howie, W. D. Munn, H. J. Weinert, World Scientific, 222–235.

— 1995, *On the theory of epigroups. I, II*. Russian Acad. Sci. Sb. Math., **82,83**, 485–512,133–154.

Shevrin, L. N. and A. J. Ovsyannikov: 1983, *Semigroups and their subsemigroup lattices*. Semigroup Forum, **27**, 1–154.

— 1996, *Semigroups and their subsemigroup lattices*. Kluwer Academic Publishers, Dordrecht–Boston–London, 390 pp.

Shevrin, L. N. and E. V. Sukhanov: 1989, *Structural aspects of the theory of semigroup varieties*. Soviet Math. Izv. VUZ, **33**, 1–34.

Shevrin, L. N. and M. V. Volkov: 1985, *Identities of semigroups*. Soviet Math. Izv. VUZ, **29**, 1–64.

Shield, D.: 1977, *The class of a nilpotent wreath product*. Bull. Austral. Math. Soc., **17**, 53–89.

Shirshov, A. I.: 1962, *Some algorithmic problems for Lie algebras*. Sibirsk. Mat. Z., 292–296.

Shmel'kin, A. L.: 1964, *Wreath products and varieties of groups, (Russian)*. Dokl. Akad. Nauk SSSR, Soviet Mat., **157**, 1063–1065.

Shmel'kin, D. A.: 1963, *The semigroup of varieties (Russian)*. Dokl. Akad. Nauk SSSR, Soviet Mat., **149**, 543–545, transl: The semigroup of group manifolds, Soviet Math. Dokl. 4(1963), 449–451.

Shneperman, L. B.: 1975, *On local finiteness of matrix periodic semigroups (Russian)*. Tez. Dokl. 4 Respubik. konf. Belorus. Part 2, Minsk, 71.

— 1982, *The Shur theorem for periodic semigroups of linear relations*. Semigroup Forum, **25**, 203–211.

Shparlinski, I. E.: 1999, *Finite Fields: Theory and Computation*. Kluwer Academic Publishers, Dordrecht, xiv+528 pp.

Shpilrain, V. and J.-T. Yu: 2000, *Polynomial retracts and the Jacobian conjecture*. Trans. Amer. Math. Soc., **352**, 477–484.

Shutov, E. G.: 1963, *Embeddings of semigroups in simple and complete semigroups (Russian)*. Mat. Sbornik, **62**, 496–511.

Shyr, H. J.: 1991, *Free Monoids and Languages*. Hon Min Book Company, Taichung, Taiwan.

Silcock, H. L.: 1977, *Generalized wreath products and the lattice of normal subgroups of a group*. Algebra Universalis, **7**, 361–372.

Sims, C.: 1994, *Computation with finitely presented groups*. Cambridge University Press, Cambridge.

Simson, D.: 1992, *Linear representations of partially ordered sets and vector space categories*. Gordon and Breach Science Publishers, Montreux, xvi+499 pp.

Sit, W. Y.: 1975, *Well-ordering of certain numerical polynomials*. Trans. Amer. Math. Soc., **212**, 37–45.

Skolem, T.: 1920, *Logisch-kombinatorische Untersuchungen über die Erfüllbarkeit und Beweisbarkeit mathematischer Sätze nebst einem Theorem über dichte Mengen*. Videnskapsselskapets skrifter I, Matematisk-naturvidenskabelig klasse, Videnskabsakademiet i Kristiania, **4**, 1–36.

— 1936, *Ein Satz über ganzwertige Polynome*. Det Kongelige Norske Videnskabers Selskab (Trondheim), **9**, 111–113.

— 1970, *Select Works in Logic*. Scandinavian University Books, Oslo.

Skornjakov, L. A.: 1966, *Locally bicompact biregular rings*. Mat. Sb. (N.S.), **69 (111)**, 663.

Skornyakov, L. A.: 1964, *Complemented modular lattices and regular rings*. Oliver & Boyd, Edinburgh, viii+182 pp.

Skorobogatov, A. N.: 1999, *Beyond the Manin obstruction*. Invent. Math., **135**, 399–424.

Smale, S.: 1998, *Mathematical problems for the next century*. Math. Intelligencer, **20**, 7–15.

Smirnov, D. M.: 1989, *Varieties of Algebras (Russian)*. Nauka, Novosibirsk.

Smith, M. K.: 1989, *Stably tame automorphisms*. J. Pure Appl. Algebra, **58**, 209–212.

Smoktunowicz, A.: 2000, *Polynomial rings over nil rings need not be nil*. J. Algebra, **233**, 427–436.

Solomon, L.: 1995, *An introduction to reductive monoids*. Semigroups, Formal Languages and Groups, Kluwer Academic Publishing, Dordrecht, 295–352.

Spangenberg, N.: 1990, Familienkonflikte eßgestörter Patientinnen. Eine empirische Untersuchung mit der Repertory Grid Technik. Habilitationsschrift, Universität Gießen.

Spanier, E. H.: 1966, *Algebraic Topology*. McGraw-Hill, New York.

Srinivas, V.: 1996, *Algebraic K-theory*. Birkhäuser Boston Inc., Boston, MA, second edition, xviii+341 pp.

Stanley, R. P.: 1986, *Enumerative combinatorics*, volume 1. Wadsworth & Brooks / Cole, Monterey.

— 1996, *Combinatorics and commutative algebra*. Birkhäuser Boston Inc., Boston, MA, second edition, x+164 pp.

Steinitz, E.: 1910, *Algebraische Theorie der Körper*. Crelle Journal, **137**, 163–308, reprinted by Chelsea Publishing Company in New York in 1950.

Stenström, B.: 1975, *Rings of quotients*. Springer-Verlag, New York, viii+309 pp., die Grundlehren der Mathematischen Wissenschaften, Band 217.

Stewart, I. and D. Tall: 1987, *Algebraic number theory*. Chapman & Hall, London, second edition, xx+262 pp.

Stückrad, J. and W. Vogel: 1986, *Buchsbaum rings and applications*. Springer-Verlag, Berlin, 286 pp.

Stumme, G. and R. Wille, eds.: 2000, *Begriffliche Wissensverarbeitung: Methoden und Anwendungen*. Springer, Heidelberg.

Suschkewitsch, A. K.: 1928, *Über die endlichen Gruppen ohne das Gesetz der eindeutigen Umkehrbarkeit*. Math. Ann., **99**, 30–50.

Suzuki, M.: 1982, *Group theory, I*. Springer-Verlag, Berlin, xiv+434 pp., translated from the Japanese.

— 1986, *Group theory, II*. Springer-Verlag, New York, x+621 pp., translated from the Japanese.

Swan, R. G.: 1970, *K-theory of finite groups and orders*. Springer-Verlag, Berlin, iv+237 pp., lecture Notes in Mathematics, Vol. 149.

Sweedler, M. E.: 1969, *Hopf algebras*. W. A. Benjamin, Inc., New York, vii+336 pp., mathematics Lecture Note Series.

Świerczkowski, S.: 1960, *On isomorphic free algebras*. Fund. Math., **50**, 35–44.

Szász, G.: 1963, *Introduction to Lattice Theory*. Academic Press, New York, 229 pp.

Szidarovszky, F. and A. T. Bahill: 1992, *Linear systems theory*. CRC Press, Boca Raton, FL, xx+425 pp.

Tachikawa, H.: 1973, *Quasi-Frobenius rings and generalizations. QF − 3 and QF − 1 rings*. Springer-Verlag, Berlin, xi+172 pp., lecture Notes in Mathematics, Vol. 351.

Takesaki, M.: 1979, *Theory of operator algebras. I*. Springer-Verlag, New York, vii+415 pp.

Tamura, T.: 1956, *Indecomposable completely simple semigroups except groups*. Osaka Math. J., **8**, 35–42.

Tarski, A. and S. Givant: 1987, *A Formalization of Set Theory without Variables*. American Mathematical Society, Providence, Rhode Island, xxi+318 pp., colloquium Publications Vol. 41.

Taylor, W.: 1977, *Varieties of topological algebras*. J. Austral. Math. Soc. Ser. A, **23**, 207–241.

— 1979, *Equational Logic*. Houston J. Math., Survey issue, i–iii, 1–69.

— 1988, *Cursillo on Varieties of Algebras*. Notes on a series of lessons given at the 9th Escuela Latinoamericana de Matemática, Santiago.

Thue, A.: 1906, *Über unendliche Zeichenreihen*. Norske Vidensk. Selsk. Skrifter. I. Mat.-Nat. K1, Christiania, **7**, 1–22.

— 1912, *Über die gegenseitige Lage gleicher Teile gewisser Zeichenreihen*. Norske Vidensk. Selsk. Skrifter. I. Mat.-Nat. K1, Christiania, **10**, 1–67.

Tignol, J.-P.: 1988, *Galois's Theory of Algebraic Equations*. Wiley.

Todd, J. and H. Coxeter: 1936, *A practical method for enumerating cosets of a finite abstract group*. Proc. Edinburgh Math. Soc., **5**, 26–34.

Trahtman, A. N.: 1974, *On covering elements in the lattice of varieties of algebras (Russian)*. Matem. Zametki, **15**, 307–312.

— 1983, *The finite basis problem for semigroups of order less than six*. Semigroup Forum, **27**, 387–389.

— 1991, *Finiteness of identity bases of 5-element semigroups (Russian)*. Semigroups and Their Homomorphisms, Russian State Pedagogical Univ., Leningrad, 76–97.

Trotter, P. G.: 1984, *Free completely regular semigroups*. Glasgow Math. J., **25**, 241–254.

— 1996, *E-varieties of regular semigroups*. Semigroups, Automata and Languages, World Scientific, Singapore, 247–262.

Tschantz, S. T.: 1990, *Infinite intervals in free lattices*. Order, **6**, 367–388.

Tuganbaev, A. A.: 1998, *Semidistributive modules and rings*. Kluwer Academic Publishers, Dordrecht, x+352 pp.

— 1999, *Distributive Modules and Related Topics*. Gordon and Breach Science Publishers, Amsterdam, xvi+258 pp.

Ufnarovskij, V.: 1995, *Combinatorial and asymptotic methods in algebra*. Algebra, VI, Encyclopaedia Math. Sci., 57, Springer, Berlin, 1–196.

Utumi, Y.: 1956, *On quotient rings*. Osaka Math. J., **8**, 1–18.

van Benthem, J.: 1996, *Exploring Logical Dynamics*. The European Association for Logic, Language and Information, Stanford University, xi+329 pp., cSLI Publications.

van den Essen, A.: 1997, *Polynomial automorphisms and the Jacobian conjecture*. Algèbre non commutative, groupes quantiques et invariants (Reims, 1995), Soc. Math. France, Paris, 55–81.

van der Kulk, W.: 1953, *On polynomial rings in two variables*. Nieuw Arch. Wiskunde (3), **1**, 33–41.

van der Waerden, B.: 1950, *Modern Algebra, Vol. II*. Ungar.

van Lint, J. H.: 1992, *Introduction to Coding Theory*. Springer-Verlag, Berlin, second edition, xii+183 pp.

van Lint, J. H. and R. M. Wilson: 1992, *A Course in Combinatorics*. Cambridge University Press, Cambridge, xii+530 pp.

Van Oystaeyen, F. and M. Saorin, eds.: 2000, *Interactions between ring theory and representations of algebras*, Marcel Dekker Inc., New York.

Vasconcelos, W. V.: 1976, *The Rings of Dimension Two*. Marcel Dekker Inc., New York, x+101 pp., lecture Notes in Pure and Applied Mathematics, Vol. 22.

Vaughan-Lee, M. R.: 1970, *Uncountably many varieties of groups*. Bull. London Math. Soc., **2**, 280–286.

— 1990, *The restricted Burnside problem*. The Clarendon Press Oxford University Press, New York, xiv+209 pp., oxford Science Publications.

Vernikov, B. M. and M. V. Volkov: 2000, *The structure of lattices of nilsemigroup varieties (Russian)*. Proc. Ural State Univ. Math. and Mech., 34–52.

Voevodsky, V.: 1999, *Voevodsky's Seattle lectures: K-theory and motivic cohomology*. Algebraic K-theory (Proc. AMS-IMS-SIAM Joint Summer Reserach Conf., Seattle, Wash., 1997), American Mathematical Society, Providence, R.I., 283–303. Proc. Symp. Pure Math., Vol. 67.

Vogt, F.: 1995, *Subgroup lattices of finite Abelian groups: structure and cardinality*. Lattice theory and its applications, K. A. Baker and R. Wille, eds., Heldermann-Verlag, Berlin, 241–259.

Volkov, M. V.: 1994, *Young diagrams and the structure of the lattice of overcommutative semigroup varieties*. Transformation Semigroups, Univ. of Essex, Colchester, 99–110.

— 2001, *The finite basis problem for finite semigroups*. Math. Japonica, **53**, 171–199.

von Neumann, J.: 1960, *Continuous Geometry*. Princeton University Press, Princeton, N. J.

Wagner, V. V.: 1952, *On the theory of partial transformations*. Doklady Akad. Nauk. SSSR, **84**, 653–656.

Wallach, N.: 1988,1992, *Real Redactive Groups I,II*. Acad. Press, Boston, first edition.

Wang, S. S. S.: 1980, *A Jacobian criterion for separability*. J. Algebra, **65**, 453–494.

Warner, S.: 1989, *Topological fields*. North-Holland Publishing Co., Amsterdam, xiv+563 pp., notas de Matemática [Mathematical Notes], 126.

— 1993, *Topological rings*. North-Holland Publishing Co., Amsterdam, x+498 pp.

Watier, G.: 1997, *On the word problem for single relation monoids with an unbordered relator*. Internat. J. Algebra Comput., **7**, 749–770.

Wehrfritz, B.: 1973, *Infinite linear groups*. Springer-Verlag, Berlin, xiv+229 pp.

Weibel, C. A.: 1994, *An introduction to homological algebra*. Cambridge University Press, Cambridge, xiv+450 pp.

— 1999a, The development of algebraic K-theory before 1980, preprint, available at http://www.math.rutgers.edu/~weibel/khistory.dvi.

— 1999b, *An introduction to algebraic K-theory*. Work in progress, available at http://www.math.rutgers.edu/~weibel/Kbook.html.

Weinstein, A.: 1996, *Groupoids: unifying internal and external symmetry. A tour through some examples*. Notices Amer. Math. Soc., **43**, 744–752.

Weir, A. J.: 1955, *The Sylow subgroups of the symmetric groups*. Proc. Amer. Math. Soc., **6**, 534–541.

Whitman, P. M.: 1941, *Free lattices*. Annals of Math., **42**, 325–330.

— 1942, *Free lattices II*. Ann. of Math, **43**, 104–115.

Wielandt, H.: 1964, *Finite Permutation Groups*. Academic Press, London.

Wieslaw, W.: 1988, *Topological fields*. Marcel Dekker, New York, Basel, Hong Kong, monographs and Textbooks in Pure and Applied Mathematics], 119.

Wille, R.: 1982, *Restructuring lattice theory: An approach based on hierarchies of concepts.*. Ordered sets, I. Rival, ed., Reidel, Dordrecht–Boston, 445–470.

— 1985, *Complete tolerance relations of concept lattices*. Contributions to general algebra, G. Eigenthaler, H. K. Kaiser, W. B. Müller, and W. Nöbauer, eds., Hölder–Pichler–Tempsky, Wien, volume 3, 397–415.

— 1999, *Conceptual Landscapes of Knowledge: A Pragmatic Paradigm for Knowledge Processing*. Classification in the Information Age, W. Gaul and H. Locarek-Junge, eds., Springer, Heidelberg, 344–356.

— 2000, *Contextual Logic Summary*. Working with Conceptual Structures. Contributions to ICCS 2000, G. Stumme, ed., Shaker-Verlag, Aachen, 265–276.

Wille, U.: 1991, *Eine Axiomatisierung bilinearer Kontexte*. Mitt. Math. Sem. Gießen, **200**, 72–112.

Winkler, F.: 1996, *Polynomial Algorithms in Computer Algebra*. Springer-Verlag, Wien New York, Wien.

Winter, H. J. J. and W. 'Arafat: 1950, *The algebra of 'Umar Khayyām*. J. Roy. Asiatic Soc. Bengal. Sci., **16**, 27–77.

Wisbauer, R.: 1991, *Foundations of module and ring theory*. Gordon and Breach Science Publishers, Philadelphia, PA, german edition, xii+606 pp., a handbook for study and research.

Wood, J. A.: 1999, *Duality for modules over finite rings and applications to coding theory*. Amer. J. Math., **121**, 555–575.

Wussing, H.: 1984, *The genesis of the abstract group concept*. MIT Press, Cambridge, Mass., 331 pp., a contribution to the history of the origin of abstract group theory, Translated from the German by Abe Shenitzer and Hardy Grant.

Xue, W.: 1992, *Rings with Morita duality*. Springer-Verlag, Berlin, x+167 pp.

Young, J.: 1927, *On the partitions of a group and the resulting classifications*. Bull. Amer. Math. Sci., **33**.

Yu, J. T.: 1995, *On the Jacobian conjecture: reduction of coefficients*. J. Algebra, **171**, 515–523.

Zaicev, M. V.: 1993, *Special Lie algebras*. Uspekhi Mat. Nauk, **48**, 103–140.

Zalesskiĭ, A. E.: 1995, *Group rings of simple locally finite groups*. Finite and locally finite groups (Istanbul, 1994), Kluwer, Dordrecht, 219–246.

Zariski, O. and P. Samuel: 1958, *Commutative Algebra, Vol. I*. D. Van Nostrand Inc., Princeton, N.J., xi+329 pp.

— 1960, *Commutative Algebra, Vol. II*. Van Nostrand.

— 1975, *Commutative Algebra*. Graduate Texts in Mathematics, vols. 28, 29, Springer-Verlag, New York, reprint of the 1958/1960 edition.

Zassenhaus, H.: 1939, *Über endliche Fastkörper*. Abh. Math. Sem. Univ. Hamburg II, 187–220.

Zelmanov, E. I.: 1989, *On the restricted Burnside problem*. Sibirsk. Mat. Zh., **30**, 68–74.

— 1990a, *On the restricted Burnside problem*. Siberian Math. J., **30**, 885–891.

— 1990b, *Solution of the restricted Burnside problem for groups of odd exponent*. Izv. Akad. Nauk SSSR Ser. Mat., **54**, 42–59, 221.

— 1991, *Solution of the restricted Burnside problem for 2-groups*. Mat. Sb., **182**, 568–592.

Zelmanowitz, J.: 1976, *An extension of the Jacobson density theorem*. Bull. Amer. Math. Soc., **82**, 551–553.

Zhevlakov, K. A., A. M. Slinko, I. P. Shestakov, and A. I. Shirshov: 1982, *Rings that are nearly associative*, volume 104 of *Pure and Applied Mathematics*. Academic Press, New York, xi+371 pp.

Zhil'tsov, I. Y.: 2000, *On identities of epigroups (Russian)*. Dokl. RAN, **375**, 10–13.

Zinov'ev, V. A. and T. Èrikson: 1996, *On Fourier-invariant partitions of finite abelian gro ups and on the MacWilliams identity for group codes*. Problemy Peredachi Informatsii, **32**, 137–143.

About the Authors

Without photos:

H. Jerome KEISLER, Dept. of Mathematics, Univ. of Wisconsin, Madison, WI53706-1313, USA
keisler@math.wisc.edu

A. G. PINUS, Dept. of Mathematics, Russian Acad. Sci., Universitetskii pr 4, 630090 Novosibirsk, Russia
algebra@nstu.nsk.su

David SALTMAN, Dept. of Mathematics, Univ. of Texas, Austin, Tx 78712-1082, USA
saltman@mail.ma.utexas.edu

Mikhail V. ZAICEV, Dept. of Mech. & Math., Moscow State Univ., 119899 Moscow, Russia
root@gmzmvg.math.msu.su

Shreeram S. ABHYANKAR
Dept. of Mathematics and Comp. Sci.
Purdue Univ. West Lafayette
IN 47907-1968, USA
ram@cs.purdue.edu

Jorge ALMEIDA
Centro de Matematica, Univ. of Porto
P.Gomes Teixeira
4099-002 Porto, Portugal
jalmeida@fc.up.pt

David ANDERSON
Dept. of Mathematics, Univ. of
Tennessee, Ayres Hall
Knoxville, TN 37996-0001, USA
anderson@math.utk.edu

Hajnal ANDRÉKA
Renyi Institute for Mathematics
P.O.Box 127
1364 Budapest, Hungary
andreka@renyi.hu

Vyacheslav A. ARTAMONOV
Dept. of Mech. & Math.
Moscow State Univ.
119899 Moscow, Russia
artamon@mech.math.msu.su

Gilbert BAUMSLAG
Dept. of Mathematics
City College of New York
New York, NY 10031, USA
gilbert@rio.sci.ccny.cuny.edu

Kostia BEIDAR
Dept. of Mathematics
National Cheng-Kung Univ.
700 Tainan, Taiwan
beidar@mail.ncku.edu.tw

Howard E. BELL
Dept. of Mathematics
Brock University
St.Catharines, Ont. L2S 3A1, Canada
hbell@spartan.ac.BrockU.CA

Alexei J. BELOV
Dept. of Mech. & Math.
Moscow State University
119899 Moscow, Russia
kanel@dnttm.ru

Franz BINDER
Inst. Algebra
Johannes Kepler Univ. Linz
4040 Linz, Austria
Franz.Binder@algebra.uni-linz.ac.at

Gary F. BIRKENMEIER
Dept. of Mathematics
Univ. Louisiana
Lafayette, La 70504-1010, USA
gfb1127@louisiana.edu

Leonid BOKUT
Dept. of Mathematics, Russian Acad. Sci.
Universitetskii pr 4,
630090 Novosibirsk, Russia
bokut@math.nsc.ru

Bruno BUCHBERGER
Inst. Symb. Comp.
Johannes Kepler Univ. Linz
4040 Linz, Austria
Bruno.Buchberger@risc.uni-linz.ac.at

Peter BURMEISTER
FB 4, AG 1., TU Darmstadt
Schlossgartenstr.7, 64289 Darmstadt
Germany
burmeister@mathematik.tu-darmstadt.de

Paul-Jean CAHEN
Dept. Math. Com A. Fac. St.Jerome
13397 Marseille, Cedex 20
France
paul-jean.cahen@VMESA12.u-3mrs.fr

Peter J. CAMERON
School Math. Sc. Queen Mary
Univ. of London
London E1 3JT, UK
p.j.cameron@qmw.ac.uk

Jean-Luc CHABERT
Fac. de Math., Univ. de Picardie
33 rue Saint Leu
80039 Amiens Cedex 01, France
jlchabert@worldnet.fr

Paul M. COHN
Dept. Math.
Univ. College London
Gower Street, London WC1 6BT, UK
pmc@math.ucl.ac.uk

Vesselin DRENSKY
Dept. of Math. and Inf., Bulg. Acad.Sci.
Akad. Georgy BonchevStr. Blk.8
1113 Sofia, Bulgaria
drensky@math.bas.bg

Hans-Dieter EHRICH
Inst. f. Software, TU Braunschweig
Postfach 3329, 38023 Braunschweig
Germany
HD.Ehrich@tu-bs.de

Albert FÄSSLER
Hochschule für Technik und
Architektur Biel, P.O.Box 1180
2501 Biel, Switzerland
Albert.Faessler@hta-bi.bfh.ch

Peter FLEISCHMANN
Chair of Pure Math., Dept. of Math.
Univ. of Kent
Canterbury CT2 7NF, UK
P.Fleischmann@ukc.ac.uk

Marco FONTANA
Dip. di Mat., Univ. Roma Tre
Largo San Leonardo Murialdo, 1
00146 Roma, Italy
fontana@mat.uniroma3.it

Ralph FREESE
Dept. of Math., Univ. of Hawaii
2526 The Mall
Honolulu, HI 96822, USA
ralph@math.hawaii.edu

Laszlo FUCHS
Dept. of Mathematics
Tulane Univ.
New Orleans, LA 70118-5698, USA
fuchs@tulane.edu

Joachim von zur GATHEN
Fachb. Mathematik/Informatik
Univ. Paderborn
33095 Paderborn, Germany
gathen@math.uni-paderborn.de

Robert GILMER
Dept. of Mathematics
Florida State Univ.
Tallahassee, Fl. 32306-4510, USA
gilmer@zeno.math.fsu.edu

Sergei T. GLAVATSKY
Dept. of Mech. & Math.
Moscow State Univ.
119899 Moscow, Russia
glav@mech.math.msu.su

Sarah GLAZ
Dept. of Mathematics
Univ. of Connecticut
Storrs, CT 06269, USA
glaz@uconnvm.uconn.edu

George GRATZER
Dept. of Mathematics
Univ. of Manitoba
Winnipeg MB R3T 2N2, Canada
gratzer@cc.umanitoba.ca

Rostislav I. GRIGORCHUK
Inst. Mech.
Steklov Institute
117966 Moscow, Russia
grigorch@mi.ras.ru

Henry E. HEATHERLY
Dept. of Mathematics
Univ. Louisiana
Lafayette, La 70504-1010, USA
gfb1127@louisiana.edu

Udo HEBISCH
Fak. Math. u. Inf., TU Bergakad. Freiberg
Bernhard-von-Cotta-Str. 2
09596 Freiberg, Germany
hebisch@math.tu-freiberg.de

Jaques HELMSTETTER
Inst. Fourier, Univ. de Grenoble
BP 74, 38402, St. Martin d´Hères
France
Jacques.Helmstetter@ujf-grenoble.fr

Jürgen HERZOG
FB 6, Univ. Essen
PF 103764, 45117 Essen
Germany
juergen.herzog@uni-essen.de

Peter M. HIGGINS
Dept. of Mathematics, Univ. of Essex
Wivenhoe Park
Colchester CO4 3SQ, UK
Peteh@essex.ac.uk

Joachim HILGERT
Inst. Math., TU Clausthal
Erzstr. 1, 38678 Clausthal-Zellerfeld
Germany
majhi@math.tu-clausthal.de

Peter HILTON
Math. Sci. Dept., SUNY
Binghampton
NY 13902-6000, USA
marge@math.binghampton.edu

Karl H. HOFMANN
FB Math., Techn. Hochschule
Schloßgartenstr.7, 64289 Darmstadt
Germany
hofmann@mathematik.tu-darmstadt.de

Derek F. HOLT
Dept. of Mathematics
Univ. of Warwick
Coventry CV4 7AL, UK
dfh@maths.warwick.ac.uk

John M. HOWIE
Dept. of Mathematics
Univ. of St. Andrews
St. Andrews KY16 9SS, UK
jmh@st-and.ac.uk

I. Martin ISAACS
Dept. of Mathematics, Univ. of Wisconsin
480 Lincoln Dr., Madison
WI 53706-1388, USA
isaacs@math.wisc.edu

Awad A. ISKANDER
425 Dover Dr., Lafayette
La 70503, USA
awadiskander@hotmail.com

Kalle KAARLI
Dept. of Mathematics
Univ. of Tartu
Tartu, EF 2400, Estonia
kaarli@math.ut.ee

Hans KAISER
Inst. Algebra, TU Wien
Wiedner Hauptstr. 8-10
1040 Wien, Austria
hans.kaiser@tuwien.ac.at

Gudrun KALMBACH, H.E.
Inst. Math., Univ. Ulm
89069 Ulm
Germany
073162193-0001@t-online.de

Kenichi KANATANI
Dept. Inform. Technology
Okayama Univ.
Okayama 700-8530   Japan
kanatani@suri.it.okayama-u.ac.jp

Helmut KARZEL
Lehrst. f. Geom., TU München
Postf. 202420, 80290 München
Germany
karzel@mathematik.tu-muenchen.de

Yefim KATSOV
Dept. of Mathematics
Hanover Coll.
Hanover, IN 47243, USA
katsov@hanover.edu

Vladimir KIRICHENKO
Dept. of Mech. & Math.
Univ. Kiev
Ukraine
vkir@mechmat.univ.kiev.ua

Tsit-Yuen LAM
Dept. of Mathematics
Univ. of Calif. Berkeley
Berkeley, CA 94720-0001, USA
lam@math.berkeley.edu

Hans LAUSCH
Dept. of Mathematics
Monash Univ.
Clayton, Vic 3168, Australia
Lausch@sci.monash.edu.au

Charles R. LEEDHAM-GREEN
School Math. Sc.
Queen Mary & Westfield Coll.
Mile End Rd., London E1 4NS, UK
C.R.Leedham-Green@qmw.ac.uk

Rudolf LIDL
Dept. Vice Chancellor
Univ. of Tasmania
POB 1214, Launceston, Australia
Rudi.Lidl@utas.edu.au

I. G. LYSENOK
Inst. Mech.
Steklov Institute
117966 Moscow, Russia
lysenok@mi.ras.ru

Judit X. MADARÁSZ
Renyi Institute for Mathematics,
P.O.Box 127, 1364 Budapest
Hungary
madarasz@renyi.hu

Andy MAGID
Dept. of Mathematics
Univ. of Oklahoma
Norman, OK 73019-0001, USA
amagid@AFTERMATH.math.ou.edu

Kenneth D. MAGILL, Jr
Dept. of Mathematics
State Univ. of New York, 106 Dief.Hall
Buffalo, NY 14214-3093, USA
kdmagill@acsu.buffalo.edu

Victor T. MARKOV
Dept. of Mech. & Math.
Moscow State Univ.
119899 Moscow, Russia
markov@mech.math.msu.su

Wallace S. MARTINDALE, 3rd
8640 Montgomery Ave.
Glenside PA 19038
USA
jmartind@chapline.net

Carlton J. MAXSON
Dept. of Mathematics
Texas A&M Univ.
College Station, Tx 77843-3368, USA
cjmaxson@math.tamu.edu

John D.P. MELDRUM
Dept. of Mathematics
Univ. of Edinburgh
Mayfield Rd., Edinburgh EH9 3JZ, UK
meldrum@maths.ed.ac.uk

Alexander A. MIKHALEV
Dept. of Mech. & Math.
Moscow State Univ.
119899 Moscow, Russia
mikhalev@shade.msu.su

Alexander V. MIKHALEV
Dept. of Mech. & Math.
Moscow State Univ.
119899 Moscow, Russia
aamikh@cnit.math.msu.su

Alexander A. NECHAEV
Dept. of Mech. & Math.
Moscow State Univ.
119899 Moscow, Russia
nechaev@cnit.msu.ru

Istvan NÉMETI
Renyi Institute for Mathematics
P.O.Box 127, H-1364 Budapest
Hungary
nemeti@renyi.hu

Harald NIEDERREITER
Dept. Math. National Univ. of Singapore
2 Science Drive 2
Singapore 117543
nied@math.nus.edu.sg

Jan OKNINSKI
Dept. of Mathematics
Warsaw Univ.
Banacha 2, Warsaw 02-97, Poland
okninski@mimuw.edu.pl

Peter J. OLVER
Dept. of Mathematics, Univ. of Minn.
206 Church Street, Minneapolis
MN 55455-0488, USA
olver@ima.umn.edu

A. J. OVSYANNIKOV
Dept. Algebra, Ural State Univ.
Lenina 51, 620083 Ekaterinburg
Russia
Alexander.Ovsyannikov@usu.ru

E. V. PANKRATIEV
Dept. of Mech. & Math.
Moscow State Univ.
119899 Moscow, Russia
pankrat@shade.msu.ru

Ira J. PAPICK
Dept. of Mathematics
Univ. of Missouri
Columbia, Missouri 65211, USA
papicki@missouri.edu

Donald S. PASSMAN
Dept. of Mathematics, Univ. of Wisconsin
480 Lincoln Dr., Madison
WI 53706-1388, USA
passman@math.wisc.edu

Peter PAULE
Inst. Symb. Comp.
Johannes Kepler Univ. Linz
4040 Linz, Austria
Peter.Paule@risc.uni-linz.ac.at

Luiz A. PERESI
Dept. of Mathematics, Univ. of Sao Paulo
Rua do Matao 1010
Sao Paulo 05508-900, Brasil
peresi@ime.usp.br

Günter F. PILZ
Inst. Algebra
Johannes Kepler Univ. Linz
4040 Linz, Austria
gun@jku.at

Boris I. PLOTKIN
Dept. of Mathematics
Hebrew University of Jerusalem
91904, Givat Ram, Jerusalem, Israel
plotkin@macs.biu.ac.il

Jonathan M. ROSENBERG
Dept. of Mathematics
Univ. of Maryland
College Park, MD 20742-4015, USA
jmr@math.umd.edu

Joseph J. ROTMAN
Dept. of Mathematics
Univ. of Illinois at Urbana-Champaign
Urbana, Ill 61801-2975, USA
rotman@math.uiuc.edu

Louis H. ROWEN
Dept. of Mathematics
Bar Ilan Univ.
Ramat Gan, Israel
rowen@macs.biu.ac.il

Nikola RUSKUC
Dept. of Mathematics
Univ. of St. Andrews
St. Andrews KY16 9SS, UK
nik@dcs.st-and.ac.uk

Vlacheslav N. SALII
ul Shekhurdina 10 A, Kv 63,
410020 Saratov
Russia
SaliiVN@info.sgu.ru

E. Tamás SCHMIDT
Math. Inst., Budapest Univ. of Technology
Müegyetem rkp 3-9
H-1111 Budapest, Hungary
schmidt@math.bme.hu

Lev N. SHEVRIN
Ural State Univ.
Lenina 51, Ekaterinburg 620083
Russia
Lev.Shevrin@usu.ru

Lance W. SMALL
Dept. of Math., Univ. of Calif.@San Diego
9500 Gilman Drive, La Jolla
CA 92093-0112, USA
lwsmall@ucsd.edu

Ronald SOLOMON
Dept. of Mathematics, Ohio State Univ.
231 W. 18th Str., Columbus
OH 43210-1101, USA
solomon@MATH.ohio-state.edu

John STILLWELL
Dept. of Mathematics
Monash Univ.
Clayton, Vic 3168, Australia
john.stillwell@sci.monash.edu.au

Evgeny V. SUKHANOV
Dept. Algebra, Ural State Univ.
Lenina 51, 620083 Ekaterinburg
Russia
evgeny.sukhanov@usu.ru

Peter G. TROTTER
School of Math. a. Phys.
Univ. of Tasmania, Box 252 C
Hobart, Tasm. 7001, Australia
trotter@hilbert.maths.utas.edu.au

Askar A. TUGANBAEV
pereulok Prechistenskii 18-9,
Moscow, 119034
Russia
askar@tuganbaev.mccme.ru

Frank J. VOGT
Rohlederstr. 13
60435 Frankfurt/Main
Germany
frank.j.vogt@t-online.de

Mikhail V. VOLKOV
Dept. Algebra, Fac. Math. and Mech.
Ural State Univ.
620083 Ekaterinburg, Russia
Mikhail.Volkov@usu.ru

Hanns J. WEINERT
Inst. Math., TU Clausthal
Erzstr. 1, 38678 Clausthal-Zellerfeld
Germany
mauh@ibm.tu-clausthal.de

Richard WIEGANDT
A. Renyi Inst. of Mathematics
POB 127, 1364 Budapest
Hungary
wiegandt@renyi.hu

Rudolf WILLE
FB Math., TU Darmstadt
Schlossgartenstr.7, 64289 Darmstadt
Germany
wille@mathematik.tu-darmstadt.de

Franz WINKLER
Inst. Symb. Comp.
Johannes Kepler Univ. Linz
4040 Linz, Austria
Franz.Winkler@risc.uni-linz.ac.at

Jietai YU
Dept. pof Mathematics, Hong Kong Univ.
Pokfulam Rd, HK Island
Hong Kong
yujt@hkusua.hku.hk

Efim ZELMANOV
Dept. of Mathematics, Yale Univ.
POB 208283, New Haven
Ct 06520-8283, USA
zelmanov@math.yale.edu

Andrej ZOLOTYKH
Chair of Higher Algebra
Moscow State Univ.
119899 Moscow, Russia
zolotykh@lsili.ru

Index

*-ring, 293
*-algebra, 297
*-ring, 297
0−1-order, 279
0-disjoint union, 50
2-cocycle, 310
2-primal, 252

abelian, 71
abelian algebra, 472
abelian color Lie superalgebra, 340
abelian congruence, 472
abelian group, 71
abelian Lie algebra, 331
abelian regular, 255
abelian variety , 472
absolute geometry, 144
absolutely flat, 230, 253, 509
absolutely free, 102
absorbing zero, 319
abstract data type, 486
acceleration, 149
action, 86
     of a group, 120
action on a tree, 106
action on an algebra, 311
acyclic sequence, 498
Adian, 143
adjoint, 297
adjoint representation, 331
admissible, 263
admissible representation, 407
admits quantifier elimination, 525
Ado's Theorem, 92
affine algebra, 286
affine coordinate ring, 356
affine model, 357
affine over an abelian group, 459
affine semigroup, 181
affine-input systems, 354

AI-systems, 354
al-jabr, 535
Albert algebra, 346
Albert's Problem, 344
algebra
     absolutely free, 462
     affine complete, 460
     affine over abelian group, 459
     central simple, 194
     cyclic, 194
     finitely based, 468
     finitely q-based, 468
     free (in a class), 461
     freely generated by a set, 461
     functionally complete, 458
     Lie, 89
     locally affine complete, 460
     locally functionally complete, 458
     locally primal, 458
     many-sorted, 482
     partial, 482
     primal, 458
     relatively free, 462
     strictly locally affine complete, 460
     total, 482
algebra in a category, 306
algebra of differential operators, 327
algebra of subrelations, 480
algebra with polynomial identity, 284
algebraic closure operation, 438
algebraic group, 92
algebraic lattice, 438
algebraic path problem, 322
algebraic semigroup, 52
algebraically independent, 356
algorithm
     Dehn, 118
     Knuth-Bendix, 16
     Todd-Coxeter, 16
algorithmic canonical simplifier, 266

almost factorial, 163
almost strong Skolem property, 190
α-derivation, 240
alphabet, 8
alternative algebra, 344
amalgam, 454
amalgamated product, 105, 454
Amalgamation Property, 446
amply f-supplemented, 255
angular momentum, 149
annihilators, 530
anti-isomorphism, 56
antichain, 25
anticommutativity, 330
antipode, 308
archimedean, 312
arithmetical, 459
arity, 451
Artinian, 258
Artinian module, 244, 511
Artinian ring, 244
associated group, 51
associated prime ideal, 154
asymptotically stable, 352
asynchronous automatic group, 117
atom, 235, 241, 468
atomic, 163, 468
atomic semifir, 235
attribute concept, 447
attributes, 447
augmentation ideal, 384
automatic group, 115
automatic monoid, 118
automatic semigroup, 118
automaton, 67
    complete deterministic, 67
    finite state, 115
automorphic, 64
automorphism
    regular, 129
    tame, 235
automorphism group, 74
automorphisms, 455
AW*-algebra, 299
axiomatized classes, 455

B-Theorem, 101
Bézout cofactors, 347
Baer *-algebra, 299

Baer lower radical, 250
Baer radical, 215
Baker-Campbell-Hausdorff identity, 90
balanced incomplete block design, 325
band, 5, 41, 61
    left of semigroups, 5
    normal, 41
    of semigroups
        of a class $\mathcal{K}$, 5
    of subsemigroups, 5
    rectangular, 5, 41
    right of semigroups, 5
barycenter, 149
baryon, 149
base group, 124
basis, 219
    regular, 189
Baumslag-Solitar group, 117
BCH-codes, 529
Bergman Gap Theorem, 302
Bezout, 280
Bezout domain, 168
Bezout formula, 241
Bezout module, 228, 254
bialgebra, 307
biautomatic group, 116
BIB-design, 325
bicharacter, 340
bicyclic semigroup, 42
birationally equivalent, 357
biregular, 184, 256, 317
Birkhoff, 141, 456
Birman, 128
block
    of a tolerance relation, 138
block in a lattice, 430
block-structure, 430
blocks, 325
blocks in a group algebra, 400
blow-up, 358
Bol condition, 145
Boolean algebra, 427, 432, 478
Boolean isomorphic, 433
Boolean lattice, 425, 427
Boolean polynomials, 433
Boolean ring, 433
Boolean semifield, 319
Boolean semiring, 319
Boolean space, 435

Borel subgroup, 52
boundaries, 498
bounded lattice, 426
bracket
    Lie, 89
Brandt groupoid, 276
Brauer, 388, 401
Brauer character, 399
Brauer group, 195, 519
Brauer's reciprocity theorem, 399
Brauer-Cartan-Hua theorem, 295
Brauer-Hasse-Noether theorem, 195
breadth of a lattice, 54
Brown-McCoy radical, 215, 250
Bruck loop, 145
Bruhat decomposition, 52
Burnside Lemma, 120
Burnside problems, 28, 103, 111
Burnside variety, 103, 142
Burnside's $p^a q^b$ Theorem, 83, 99
Burnside, W., 387

C^*-algebra, 297
cancellation rule, 424
canonical form, 440
canonical module, 180
canonical word, 477
Capelli polynomial, 285
Cartan decomposition
    of a Lie group, 91
Cartan matrix, 398
Cartan subalgebra, 335
Cauchy-Frobenius Lemma, 120
Cayley graph, 115
Cayley's Theorem, 86
CD-lemma, 270
center, 193, 344, 429, 430
central closure, 204
central congruence, 472
central extension, 517
    universal, 517
central simple algebras, 194
centralizer condition, 148
centralizer near-ring, 323
centralizes, 471
chain complex, 498
chain homotopic, 499
chain homotopy, 499
chain homotopy equivalences, 499

chain homotopy equivalent, 499
chain map, 498
chain of semigroups, 6
character, 386
character group, 74
character table, 387
characteristic closure, 287
characteristic of a free module, 233
characteristic term algebra, 490
characterization of characters, 388
charge, color, 150
check matrix, 528
Chinese Lantern, 428
Chinese remainder algorithm, 349
Chinese remainder problem, 347
Chinese Remainder Theorem for ideals,
    167
Chow ring, 522
Church-Rosser property, 477
class
    Fitting, 84
    lattice-characterized, 55
    lattice-elementary, 55
    lattice-universal, 56
    normal, 85
    pseudo-primitive class, 486
    Schunck, 84
class function, 386
class group, 515
class number, 515
class of groups, 83
classical absolute geometry, 145
classical quotient ring, 243
classical ring of right quotients, 202
classically semisimple, 509
classification theorem, 99
Clifford algebra, 273, 289
Clifford group, 292
closed ideal, 437
closed submodule, 207
closed subset, 438
closed subset of a partial algebra, 482
closed under essential extensions, 217
closed under extensions, 216
closure, 319
    completely 0-simple, 50
closure operation, 438, 481
CM, 178
coaction, 311

coadjoint action, 408
coalgebra, 306
cochain complex, 500
cocharacter, 339
coclass, 78
cocommutative, 306
cocycle map, 194
code, 68, 526
    biprefix, 69
    homogeneous, 68
    prefix, 69
    suffix, 69
    uniform, 68
codimension, 339
    nth, 287
coequalizer, 495
Cohen-Macaulay, 178
coherent, 510
Cohn, 128
coinvariant, 311
colength, 339
colimit, 494
collinear, 146
color charge, 150
color Lie p-superalgebra, 341
color Lie superalgebra, 340
combable group, 118
commutant, 297
commutation factor, 340
commutator, 149, 471
commutator of congruences, 471
commute, 429
compact element, 439
compactly generated lattice, 439
compatible valuation, 374
compensator, 352
complement, 426
    Frobenius, 131
complemented, 426
complete deterministic finite state automaton, 67
complete homomorphism, 437
complete lattice, 436
complete space, 314
complete sublattice, 437
complete weight enumerators, 532
completely 0-simple closure, 50
completely decomposable, 73
completely normal basis, 361

completely prime, 249
completely prime ideal, 249
completely prime ring, 249
completely semiprime, 249
completely semiprime ideal, 249
completely semiprime ring, 249
completely simple semigroups, 63
Completion by cuts, 437
complexification, 300
component
    archimedean, 7
    of a band, 5
composition, 161
composition algebra, 344
composition of f, g, 269, 271
Composition-Diamond lemma, 270
comultiplication, 306
concept lattice, 448
concrete lattice, 429
conditional equations, 490
confluence, 477
confluence condition, 270, 477
confluent, 489
conformal algebra, 327
congruence
    $\mathcal{H}$-congruence, 467
    fully invariant, 467
congruence distributive, 469
congruence lattice, 437
congruence modular, 469
congruence partition, 134
congruence permutable, 469
congruence regularity, 472
congruence relation, 488
congruence uniformity, 472
congruence-preserving extension, 419
conjugacy class, 120, 386
connected group, 52
consensus, 434
consequence, 143
conservation law, 148
constant functor, 494
constituent, 387
constructive specifications, 490
content
    of a map, 122
continuous geometry, 427
continuous lattice, 439
continuous lattices, 66

contractions
 linear, 62
controller, 352
convolution, 307
coordinated, 456
coproduct, 454, 494
corestriction, 503
counit, 306
countably based, 80
countably injective, 183, 230
cover, 134, 441
covering map, 365
covering property, 430
critical, 144, 279
crossed product, 195, 310, 383
 Grothendieck group, 385
 Noetherian, 384
crown product, 127
cryptogroup, 41
cycle, 498
cycle index, 122
cycle indicator polynomial, 122
cyclic code, 529
cylinder, 479
cylindric algebra
 full, 479
 representable, 479
cylindric-relativized set algebras, 480
Czerniakiewicz, 129

data type, 486
Davenport constant, 164
decomposition
 greatest
 band, 6
 left, 6
 matrix, 6
 right, 6
 semilattice, 6
 left, 6
 matrix, 6
 rectangular, 6
 right, 6
 semilattice, 6
decomposition problem, 161
Dedekind domain, 155, 162, 165, 251
Dedekind ring, 515, 522
defect, 401
defect group, 401

defining relation, 14, 72
deg-lex order, 269
degree function, 232, 347
degree of a character, 386
degree of a representation, 390
degree of a variety, 356
deGroot, 456
Dehn algorithm, 118
Dehn function, 108
Dehn problem, 475
Delta-method, 384
dense, 203
derivation, 193, 236, 331, 366
 α-, 240
derivation operators, 447
derived length, 388, 389
derived series, 331
derived subalgebra, 331
determinant, 516
determinantal sum, 200
diagonal sum, 200
diagram
 van Kampen, 107
diamond, 423
diamond condition, 270, 477
Dickson near-field, 323
differential
 extension, 366
 Galois group, 366
 Galois theory, 366
 strong isomorphism, 368
differential dimension, 369
differential dimension polynomial, 369
differential field, 366
 constants, 366
 finitely generated, 367
 operator, 366
 ordinary, 369
 partial, 369
 Picard-Vessiot, 367
 strongly normal, 368
 universal, 367
differential ideal, 369
differential polynomial, 369
differential polynomial ring, 236
differential type, 370
dihedral group, 76
dimension, 356
dimension of a projective closure, 357

dimension subgroup, 385
direct limit, 453
direct product
    of partial algebras, 482
directly representable, 473
discrete category, 494
discrete logarithm problem, 377
discrete series representation, 407
disjunctive normal form, 433
distinct-degree factorization, 160
distinguishing character, 532
distributive, 208, 221, 228, 280
distributive lattice, 423
distributively generated, 281
divisible, 72
divisible hull, 72
divisible module, 223
divisible part, 72
division algebra, 51, 193, 198
division ring, 193, 198
divisor, 25
    of a semigroup, 19
    Rees, 25
divisors of zero, 19
    proper, 19
domain, 202, 249
    Bezout, 168
    Dedekind, 165
    Euclidean, 347
    Prüfer, 167
    principal ideal, 239
    principal right ideal, 240
    unique factorization, 241
dominates, 358
domination problem, 358
don't-care, 434
Dong Lemma, 327
Dorninger, D., 460
dual algebra, 307
dual bialgebra, 308
dual code, 529, 531
dual module, 318
dual ring, 258

ε-anti-commutativity, 340
ε-commutative, 340
E-equation, 483
    validity, 483
ε-Heisenberg algebra, 341

ε-Jacobi identity, 340
ECE-equation, 483
egg-box picture, 4, 51, 63
Eigenthaler, D., 460
elasticity, 164
elementarily equivalent, 524
elementary embedding, 524
elementary extension, 524
elementary implication, 483
elimination of the leading word, 269
ELW, 269
empty category, 495
encoding, 526
endodistributive, 282
endomorphism, 455
endomorphism group, 74
endomorphism ring, 74, 205
energy, 149
Engel's problem, 338
enough injectives, 492, 500
enough projectives, 492, 500
epic R-field, 198
epidivisor, 25
epigroup, 23
epimorphism
    between partial algebras, 484
    extendable, 485
    universal epimorphism, 485
equal partition, 134
equal-degree factorization, 160
equalizer, 495
equation
    E-equation, 483
    existence equation, 483
    existentially conditioned, 483
    QE-, 483
    quasi-existence, 483
equational theory, 467
equivalence sequence, 489
equivalent homomorphisms, 199
equivalent representations, 390
equivalent theories, 523
essential, 224, 260
essential extension, 224, 260
essential right ideal, 204
Euclidean domain, 347
Euclidean ring, 240
exact sequence, 498
exceptional algebra, 346

exchange, 221
exchange property, 207, 430
exchange ring, 207
existence equation, 483
existence equation
    existentially conditioned, 483
existentially closed, 525
existentially conditioned existence equation, 483
exponent, 196
exponential growth, 301
exponential map, 90
extendable
    epimorphism, 485
extended center, 204
extended commutator, 193
extended Euclidean algorithm, 348
extended module, 168
extension
    central, 517
    no new constants, 366
    universal central, 517
extensionality, 438
extent, 447
exterior algebra, 273, 290

$\mathscr{F}$-groups, 83
F-injector$\mathscr{F}$-injector, 84
$\mathscr{F}$-projector, 83
f. c. center, 383
factor, 10, 19, 25
    left, 10
    proper, 10
    right, 10
factorization
    distinct-degree, 160
    equal-degree, 160
factorization structure, 485
faithful, 332
faithful representation, 390
Fano plane, 427
feedback, 352
Feit-Thompson-Theorem, 83
fellow-traveler property, 115
fibration, 134
field
    differential, 366
    ordered, 373
field coproduct, 201, 235

field degree, 356
field extension, 355
field of fractions, 198
field spectrum, 200
filter, 430, 435
filter product, 452
filtered diagram, 497
finite, 407
finite conductor ring, 507
finite conjugate center, 383
finite exchange property, 207, 220
finite field, 359
finite obliquity, 80
finite prime field, 359
finite rank, 80
finite state automaton, 115
finite type, 329
finite width, 80
finitely axiomatizable approximation, 480
finitely based, 102, 143
finitely injective, 230
finitely presented, 230, 509
finiteness condition, 25
fir, 231
first order formula
    for partial algebras, 483
Fitting class, 84
fixed field, 363
fixed point set, 120
flat, 221, 227, 254, 509
follows, 337
force, 149
form
    quadratic, 520
formal concept, 447
formal context, 447
formal distribution, 326
formal language, 67, 321
formal power series, 321
formally real, 312, 373
formation
    amalgamated, 446
formula
    first order
        for partial algebras, 483
    preservation of a, 484
    reflection, 484
fractional subset, 189
frame of a semigroup, 4

frames, 425
free, 219, 227
free abelian group, 72
free associative algebra, 231
free base, 461
free color Lie superalgebra, 343
free constructors, 490
free functors, 495
free generating set, 102
free ideal ring, 231
free Lie algebra, 270, 337
free module, 509
free product, 105, 426, 454
free resolution, 491
free right ideal ring, 231
free semigroup, 8
free set of generators, 72
free Σ-algebra, 487
free topological module, 315
freely generated, 102
Frobenius, 131
Frobenius algebra, 258
Frobenius automorphism, 360
Frobenius complement, 131
Frobenius Formula, 404
Frobenius group
　　p-local, 134
Frobenius kernel, 131
Frobenius reciprocity, 388
Frobenius, G., 387
full, 428
　　homomorphism, 484
full R-lattice, 273
full cylindric algebra, 479
full inverse subsemigroup, 57
full matrix, 199, 234
full relation algebra, 478
full transformation semigroup, 46
fully atomic, 235
fully invariant, 141
fully ordered ring, 312
function
　　Dehn, 108
function field, 356, 357
function ring, 313
functionally complete, 458
fuzzy set, 322

G-GCD ring, 507

G-graded Lie algebra, 340
G-set, 120
Galois
　　differential, 366
Galois extension, 202, 364
Galois field, 360
Galois group, 363
Galois theory
　　differential, 367, 368
γ-radical, 216
Gaussian polynomial, 140
Gel'fand-Kirillov dimension, 301
Gellmann matrix, 151
general linear color Lie superalgebra, 341
general ZPI-ring, 156
generalized decomposition numbers, 402
Generalized GCD ring, 507
generalized nil radical, 250
generalized translation structure, 135
generated field, 355
generated ring, 355
generator polynomial, 529
generators, 393
generic algebra, 286
generic division algebra, 196
generic zero, 369
genus, 276
geometric cover, 134
geometric quantization, 408
geometry, continuous, 427
geometry, projective, 427
Glauberman's Z^*-Theorem, 99
Glauberman, G., 389
Glauberman-Isaacs correspondence, 389
Gleason's theorem, 428
global dimension, 503
gluon, 150
Golay-codes, 528
Goldie ring, 247
Goldie's Theorem, 243
Grätzer, G., 460
graded central simple algebra, 291
graph
　　automorphism, 110
　　Cayley, 115
Grassman algebra, 273
greatest common divisor, 347
Green's relations, 3
Grothendieck construction, 495

Grothendieck group, 385, 513, 515
ground terms, 487
group
    absolutely free, 102
    acting on a graph, 110
    action, 120
    almost nilpotent, 51
    associated, 51
    asynchronous automatic, 117
    automatic, 115
    Baumslag-Solitar, 117
    biautomatic, 116
    binary icosahedral, 518
    combable, 118
    critical, 144
    free, 102
    Frobenius, 131
    Grothendieck, 513, 515
    Heineken, 116, 119
    hyperbolic, 118
    left [right], 21
    Lie, 89
    locally finite, 384
    metabelian, 104
    nilpotent, 75
    orbitally sound, 385
    perfect, 518
    permutation, 86
    polycyclic-by-finite, 384
    rectangular, 41
    relatively free, 102
    soluble, 82
    stable general linear, 515
    Steinberg, 518
    Suzuki, 133
    symmetric, 86
    topological, 62
    von Dyck, 118
    Weyl, 52
    word problem, 107
    word-hyperbolic, 118
    Zassenhaus, 133
group $\mathcal{H}$-class, 4
group algebra, 383
group completion, 513
group element, 23
group of extensions, 74
group of units, 64
group part, 23

group ring, 245, 383
    almost simple, 384
    Artinian, 384
    augmentation ideal, 384
    catenary, 385
    dimension subgroup, 385
    domain, 385
    idempotent, 385
    Jacobson radical of, 384
    nilpotent element, 385
    Noetherian, 384
    p. i., 384
    prime, 384
    prime length, 385
    prime spectrum, 385
    primitive, 384
    primitive length, 385
    semiprime, 384
    semiprimitive, 384
    trace, 385
group with zero, 48
group, rotation, 149
group, special unitary, 150
group, transformation, 148
group, translation, 149
group-graded ring, 383
group-like element, 307
growth function of algebra, 300
growth of algebra, 301
    exponential, 301
    intermediate, 301
    polynomial, 301
    sub-exponential, 301
growth of function, 301
Gröbner basis, 264, 269
Gröbner-Shirshov basis, 269
GS-basis, 269, 272
guaranteed rank, 532
Gumm, H. P., 459

Hadamard matrix, 379
Hagemann, J., 461
Hahn embedding, 374
Hahn valuation, 374
half-factorial domain, 163
half-spin representations, 293
Hall π-subgroup, 83
Hall subgroup, 388, 389
Hall's Theorem, 83

Hamming bound, 527
Hamming code, 528
Hamming distance, 526, 530
Harish-Chandra module, 408
Harrison topology, 376
Hasse invariant, 275
Hattori torsion-free, 227
Hausdorff space, 314
HCLF, 241
height of an ideal, 158
height-one, 162
Heineken group, 116, 119
Hensel Lemma, 350
Hensel lemma, 350
Hensel lifting, 160
hereditarily just infinite, 80
hereditary, 214, 284, 511
hereditary ring, 506
Hermitian form, 432
Herrmann, C., 461
Heyting lattice, 425
HFD, 163
higher K-theory, 520–522
    of finite field, 521
    of ring of integers, 521
    of scheme, 521
highest common left factor, 241
Higman variety, 476
Hilbert Basis Theorem, 243
Hilbert lattice, 430
Hilbert series, 288
Hilbert space, 428
Hilbert symbol, 519
Hilbert-Samuel polynomial, 171
HNN-extension, 105
homogeneous, 289
homogeneous element, 341
homology module, 498
homomorphically closed, 216
homomorphism, 332, 455
    between partial algebras, 484
    full, 484
hook length formula, 404
Hopf algebra, 308
Hopf condition, 338
horizontal sum, 429
HSP theorem, 466
Hu's theorem, 435, 459
Hu, T. K., 459

hyperbolic group, 118
hyperbolic loops, 148
hypersurface, 356

I-adic topology, 315
ideal, 319, 430
    completely prime, 249
    completely semiprime, 249
    height-one prime, 162
    maximal, 355
    of strong finite type, 158
    primary, 153
    prime, 162, 249, 355
    primitive, 213
    semiprime, 249
    strongly primary, 153
    T-nilpotent, 248
    unitary, 190
ideal class group, 162
ideal completion, 438
ideal extension of a semigroup, 2
ideal lattice, 424
ideal of a lattice, 438
ideal of a semigroup, 1
    0-minimal, 3
    left, 1
    minimal, 2
    principal, 2
    proper, 1
    right, 1
    two-sided, 1
idealizer, 30
    condition, 30
ideals of values, 190
idempotency, 438
idempotent
    primitive, 17
idempotent matrix, 514
identical relation, 102
identity, 284, 319, 465
    Baker-Campbell-Hausdorff, 90
    heterotypical, 58
    homotypical, 58
    satisfied by an algebra, 465
    semigroup, 51
ight skew power series ring, 283
Image parameters, 409
implication, 430
    elementary, 483

incidence loop, 147
incidence space, 146
independent
    set of laws, 143
index of an element, 23
induced block, 401
induced character, 388
induced permutation representation, 120
induced representation, 407
induction theorem, 388
inductive property, 216
inertial degree, 275
inertial system, 149
infimum-dense, 448
infinite distributive law, 425
initial abstract data type specification, 489
initial algebra, 487
initial object, 494
initial specification, 489
injective, 220, 223, 227, 260, 282, 509
injective dimension, 171
injective hull, 224, 511
injective object, 500
injective resolution, 501
injective with respect to N, 223, 259
inner derivation, 236
inner quasidirect product, 146
input function, 352
integer-valued polynomial, 168, 189
integral dependence, 157
integrally closed, 166
intent, 447
interaction, 149
interaction, strong, 150
interaction, weak, 149
intermediate growth, 301
interpolation problem, 348
interpretation
    partial interpretation, 483
interval algebras, 435
invariant, 228, 311
invariant factors, 241
invariant reflection structure, 145
inverse, 35
inverse limit, 454
inverse monoid, 52
inverse semigroup, 42
    T_E, 44

congruence, 43
E-unitary, 44
free inverse semigroup, 45
fundamental, 44
greatest idempotent separating congruence, 43
least group congruence, 43
symmetric, 43
variety, 45
involuted monoid, 481
involution, 293
irreducible, 332, 334
irreducible character, 386
irreducible element, 241
irreducible ideal, 358
irreducible representation, 390, 406
irreducible variety, 356
irredundant
    join, 441
    meet, 441
irredundant polynomial, 434
isomorphism, 455
    between partial algebras, 484
    stable, 514
isotoncity, 438
isotypic components, 391
iterated monoidal transform, 359
Iwasawa decomposition
    of a Lie group, 91

Jónsson's lemma, 443, 470
Jacobi identity, 149
Jacobian identity, 330
Jacobson, 346
Jacobson radical, 213
Jacobson radical class, 213
Jacobson radical ring, 213
Jacobson's Conjecture, 215
Jauch-Piron state, 429
join, 436
join irredundant, 441
join refines, 440
join-irreducible, 424
Jordan algebra, 345
Jordan group, 89
Jordan isomorphism, 237
Jordan product, 236, 345
Jordan-Wielandt Theorem, 88
Jordan-Zassenhaus Theorem, 277

just infinite, 80

k-closure, 319
$\mathcal{K}$-congruence, 6
K-finite, 407
k-ideal, 319
K-loop, 146
K-semigroup, 54
K-spectrum
    of a Lie group representation, 408
K-theory
    higher, 520–522
    Localization Theorem, 522
    lower, 517
    Milnor, 519, 521, 522
    of finite field, 521
    of ring of integers, 521
    of scheme, 521
K_0, 513
    reduced, 514
K_1, 515
    of PID, 516
K_2, 518
    connection with Brauer group, 519
    of $\mathbb{Q}$, 519
    of finite field, 518
König's Lemma, 454
Köthe radical, 214, 215
Köthe's Problem, 215
Kaplansky's Theorem, 170
kernel, 36, 43, 424
    Frobenius, 131
kernel of a semigroup, 2
kernel of character, 387
Killing form, 332, 336
Kleene's Theorem, 68
Kleinfeld, 345
Knuth-Bendix algorithm, 16
Knuth-Bendix completion algorithm, 489
K_p-series, 126
Krull dimension, 156, 170, 185, 248, 251
Krull domain, 162
Krull's Dimension Theorem, 170
Krull's Intersection Theorem, 170
Krull's principal ideal theorem, 158

Λ-lattices, 278
lambda-group, 475
Lang's Homomorphism Theorem, 375
Langlands parameters, 408

Langlands program, 409
language
    variety, 31
Laskerian ring, 154
Latin square, 379
lattice
    continuous, 66
    left Λ, 276
    modular, 420
    orthomodular, 429
    representable, 418
    weakly atomic, 441
lattice characteristic, 55
lattice isomorphism, 56
lattice, bounded, 426
lattice, complemented, 426
lattice, complemented modular, 427
lattice, ortho-, 428
lattice, relatively complemented, 427
lattice-ordered, 312
Laurant series
    right skew, 184
Laurent polynomials, 327
law, 102
    orthomodular, 429
law, conservation, 148
leading power product, 263
leading word, 269
left adjoint, 311, 496
left coherent, 230
left derived functor, 501
left exact, 501
left H-comodule, 311
left H-module, 311
left hereditary, 214, 218
left integral element, 309
left invariant, 228
left Krull dimension, 248
left order, 247, 273
left perfect, 176, 259
left resolution, 500
left semiperfect, 176
left skew Laurent series ring, 283
left skew power series ring, 231, 283
left strong, 214, 218
length of a word, 9
lepton, 149
letter, 8
Levi part

of a Lie group, 91
Levi subalgebra, 332
Levitzki's Theorem, 244
Lie algebra, 89, 330
Lie bracket, 89, 331
Lie composition, 271
Lie ELW, 271
Lie group, 89
Lie isomorphism, 237
Lie product, 236
Lifting Theorem, 350
limit, 494
linear character, 386
linear code, 528, 530
linear continuous system, 352
linear contractions, 62
linear identity, 383
linear syzygies, 267
linearly compact, 226, 317
linearly independent, 219
linearly reductive, 181
linearly topologized, 317
lines, 146
link of F, 181
Lipschitz group, 292
local, 182, 281
local cohomology, 179
local homomorphism, 199
local polynomial operations, 458
local ring, 169, 356
    regular, 357
locality, 328
localization, 170, 200
Localization Theorem, 522
locally finite, 142
locally finite group, 384
locally functionally complete, 458
locally primal, 458
locally solvable, 93
Loop Lemma, 430
loose specification, 490
Lorentz transformation, 149
Lorentz boost, 147
lower K-theory, 517
lower central series, 331
LS-word, 271
LSA-word, 270
Lyndon, 143
Lyndon-Shirshov associative word, 270

Lyndon-Shirshov word, 271

M-group, 388
M-regular sequence, 178
m-system, 249
MacWilliams Identity, 529, 532
Makar-Limanov, 129
Mal'cev, 456
Maltsev condition, 469
Maltsev term, 469
many-sorted algebra, 482
map
    exponential, 90
matched, 38
matric extensible, 215
matrix
    elementary, 515
    idempotent, 514
matrix ideal, 200
matrix semigroup, 49
    nilpotent component of, 51
    uniform component of, 51
matrix, Gellmann, 151
Matsumoto's Theorem, 518
maximal chain of prime ideals, 157
maximal class, 77
maximal ideal, 249
maximal order, 273
maximal right ring of quotients, 204
maximal torus, 406
maximum likelihood decoding, 527
Maxwell's equations, 372
McKenzie, R., 459
measures, 310
median law, 423
meet, 436
meet irredundant, 441
Merkur'ev-Suslin Theorem, 197
meson, 149
metabelian, 104, 141, 331
Milnor K-theory, 519, 521, 522
Milnor Conjecture, 519, 521
minimal distance, 526
Minkowski metric, 149
mod-p-reduction, 399
model, 488, 523
model completion, 525
modular, 214, 345
modular character, 399

modular law, 420
modular representation, 398
modular representations, 390
module, 258, 331
    Artinian, 244
    continuous, 207
    semi-Artinian, 247
module topology, 315
momentum, 149
momentum, angular, 149
monadic algebras, 481
monic word, 269
monoid, 62
    automatic, 118
    bicyclic, 14
    equidivisible, 11
    free, 10
    involuted, 481
    symmetric inverse, 52
    transformation, 67
monoidal transform, 359
monomial group, 388
monomorphic, 486
Morita-dual, 530
Mostowski-Ehrenfeucht, 455
motion group, 144
multiplication module, 283
multiplicatively cancellative, 319
multiplicity, 171
multiplier, 115
mutual commutator, 331
mutually local, 326

$[n, k, d]$-code, 533
n-ary operation, 451
n-ary terms, 458
n-fir, 233
N-injective, 259
N-projective, 220, 260
n-system, 249
Nakayama's Lemma, 170
near-field, 323
near-ring, 322
Neumanns-Šmel'kin, 143
neutral algebra, 471
neutral variety, 471
Nielsen, 128
nil radical, 214, 215
nil-element, 24

nil-extension, 24
nilalgebra, 344
nilpotent, 344
    of class m, 331
nilpotent algebra, 472
nilpotent blocks, 402
nilpotent component, 51
nilpotent of class c, 75
nilpotent radical, 332
nilradical, 344
nilsemigroup, 24
Noether Axioms, 166
Noetherian, 242, 258, 356, 511
Noetherian reduction system, 489
Noetherian ring, 245
non-singular projective model, 357
nonlinear syzygies, 267
nonsingular, 225, 254
nonsingular ideal, 358
norm residue symbol, 519
normal form, 489
normal ideal, 276
normal semigroup, 181
normalization lemma, 157
normalizer of isolated orbitals, 385
Not Knot, 116
Nottingham group, 81
nucleon, 149

O'Nan-Scott Theorem, 87
Oates and Powell, 143
object concept, 447
object parameter, 409
objects, 447
Odd Order Theorem, 100
Ω-algebras, 455
Ω-groups, 452
OML, 429
one-sidedly topological semigroups, 62
operation
    n-ary, 451
    affine over abelian group, 459
    local polynomial, 458
    local term, 458
    partial term operation, 483
    polynomial, 458
    projection, 458
    term, 458
operation symbols, 487

orbit, 87, 120
orbit method, 408
orbitally sound group, 385
order, 247, 278
    left, 273
    maximal, 273
    R-, 273
    right, 273
order of a finite field, 359
ordered field, 373
ordinal sum, 8
ordinary representations, 390
Ore domain, 198
Ore right multiple condition, 198
orthochronous, 147
orthocomplementation, 428
orthogonal Lie algebra, 336
orthogroup, 41
    normal, 41
ortholattice, 428
orthomodular lattice, 429
orthomodular law, 429
orthomodular space, 432
outer automorphism, 129
outer quasidirect product, 146
output function, 352
overring, 167

p-adic analytic, 80
p-decomposition matrix, 399
p-filter, 430
p-group
    p-group
        maximal, 82
    p-group
        of maximal class, 77
    p-group
        powerful, 75
p-group, 75
p-ideal, 430
p-local Frobenius group, 134
p-weight, 401
Pólya substitution, 123
Pólya's theorem, 122
parabolic subgroup, 407
paragroups, 63
parameters of a design, 325
parity-check code, 528
parity-check matrix, 532

partial algebra, 482
    (closed) subalgebra of a, 482
    closed subset of a, 482
    direct product of, 482
    E-equation, 483
    ECE-equation, 483
    epimorphism, 484
    equation
        ECE-, 483
        existence equation, 483
    existence equation, 483
    existentially conditioned existence
        equation, 483
    first order formula, 483
    full homomorphism, 484
    generation of a, 482
    homomorphism, 484
    isomorphism, 484
    many-sorted, 482
    partial interpretation, 483
    partial term operation, 483
    preservation of a formula, 484
    QE-equation, 483
    quasy-existence-equation, 483
    reduced product, 485
    reflection of a formula, 484
    satisfaction
        of an E-equation, 483
    term, 482
    term existence statement, 483
    theory, 482
    valuation, 483
partial interpretation
    in a partial algebra, 483
partial term operation, 483
partially ordered group, 312
partially ordered ring, 312
particle, 150
partition, 134
    congruence, 134
    equal, 134
partition lattice, 436
Payley construction, 379
pentagon, 423
perfect code, 527
perfect ideal, 369
peripheral, 64
permutation group, 86
    cofinitary, 89

finitary, 88
multiply transitive, 87
oligomorphic, 88
primitive, 87
transitive, 87
Pfister's Local-Global Principle, 376
phase transformation, 149
photon, 149
PI-algebra, 284
π-injective, 226, 261, 509
π-projective, 223, 261, 509
π-regular, 256
Picard-Vessiot, 367
PID, 239
Pierce stalk, 255
Pixley, A., 459, 461
planar near-ring, 324
Plotkin, 457
pointed Hopf algebra, 309
points at infinity, 357
pole assignment property, 353
polycyclic-by-finite group, 384
polymorphic, 486
polynilpotent variety, 142
polynomial, 458
    integer valued, 189
polynomial growth, 301
polynomially complete, 458
polynomially equivalent, 189
positive chain complex, 500
positive cochain complex, 500
positive cone, 312, 373
positive definite, 293
positive formula, 524
positive type, 135
Post algebra, 426
power series
    right skew, 182
power-associative, 344
powerful, 75
PP ring, 506
Prüfer domain, 157, 167
Prüfer theorems, 72
prefix, 10
preordering, 376
presentation, 14
    semigroup, 14
preservation
    of a formula, 484

preserve colimits and limits, 497
preserves, 336
PRID, 240
Priestley duality, 424
primal, 458
primary ideal, 153
prime, 243, 249
prime ideal, 249, 424
prime matrix ideal, 200
prime radical, 215, 250
prime ring, 204, 243, 249
prime structure, 525
primitive, 213, 345
primitive element, 307, 360
primitive polynomial, 360
principal, 239
principal factor of a semigroup, 19
principal ideal domain, 239
principal ideal theorem, 158
principal indecomposable modules, 398
principal projective ring, 506
principal series representation, 407
pro-p-groups, 79
probability, 428
product
    direct
        of partial algebras, 482
    reduced product, 485
product of subgroups, 104
profinite group, 364
projection, 294, 299
projection operation, 458
projection theorem, 432
projective, 219, 227, 281, 509
projective cover, 176, 220, 227, 248, 510
projective dimension, 171, 502
projective functors, 496
projective geometry, 427
projective model, 357
projective object, 500
projective resolution, 500
projective with respect to, 220, 260
projectivity, 56
prolonged action, 93
pseudo-primitive class, 486
pseudoboolean lattice, 425
pseudocomplement, 425
pseudoinverse, 27
pseudovariety, 31

σ-reducible, 33
    decidable, 32
    implicit operation, 33
    tame, 33
pure, 207
Pure transcendental differential extensions, 371
pure-injective, 207
Pólya action, 121

QE-equation, 483
QF-module, 530
QF-ring, 258, 512
quadratic form, 520
quadratic ideal, 346
quadratic space, 291
quantum structures, 429
quasi-Baer, 253
quasi-equational theory, 467
quasi-existence-equation, 483
quasi-Frobenius, 512, 530
quasi-Frobenius ring, 258
quasi-identity, 465
    satisfied by an algebra, 465
quasi-injective, 225, 261, 512
quasi-invertible, 213
quasi-isomorphic, 73
quasi-polyadic algebras, 481
quasi-projective, 222, 261, 512
quasi-regular, 213, 345
quasicyclic groups, 71
quasidirect product
    inner, 146
    outer, 146
quasigroup, 475
quasiprimal, 460
quasivariety, 465
    finitely based, 468
    generated by a class, 466
    non-trivial, 466
quaternion algebra symbol, 519
quaternion group, 76
quotient, 347
quotient lattice, 138
quotient term algebra, 488

R-closed, 532
R-field, 198
R-order$R$-order, 273
radical, 216, 345

Jacobson, 246
    of a Lie group, 91
ramification index, 275
rank, 73, 335
    of a free algebra, 231, 461
    of a free group, 140
    of a free semigroup, 9
    of a group, 73
    of a root system, 334
    of an BN-pair, 98
rank of a Lie algebra, 335
ranking of shape λ, 413
rational language, 67
real-closed field, 374
reciprocal, 530
recognized, 31
recovery equation, 409
rectangular band
    of semigroups, 5
rectangular bands, 41
rectangular group, 41
recursively presented, 476
reduce, 263
reduced, 72, 182, 228, 269
    product, 485
reduced K_0, 514
reduced group, 72
reduced primary decomposition of an ideal, 154
reduced ring, 249
reduced Witt ring, 376
reduction sequence, 489
reduction system, 489
reductive group, 52
Reed-Muller code, 528
Reed-Solomon code, 528
Rees congruence, 2
Rees factor, 51
Rees quotient semigroup, 2
reflection
    of a formula, 484
reflection groups, 92
reflexive, 258
regular, 184, 230, 243, 253, 254, 280
    von Neumann, 509
regular $\mathscr{D}$-class, 4
regular class of rings, 216
regular element, 202, 243
regular local ring, 171

regular representation, 406
regular right ideal, 203
regular ring, 221, 505
regular semigroup, 35
regular semigroups
    E-solid, 37
    e-variety, 38
    bi-free, 38
    biordered set, 36
    fundamental, 36
    sandwich set, 36
    trace- kernel, 37
regulator, 352
relation algebra
    full, 478
    representable, 478
relation semigroups, 69
relative pseudocomplement, 425
relatively complemented, 427
relatively free, 102, 286
relatively pseudocomplemented lattice, 425
remainder, 347
remainder function, 263
Renner monoid, 52
repetition code, 528
representable, 338, 495
representable cylindric algebra, 479
representable relation algebra, 478
representation, 331, 386, 390, 405, 455, 495
    admissible, 407
    completely reducible, 406
    discrete series, 407
    faithful, 332, 455
    induced, 407
    irreducible, 332, 406
    of a Lie group, 405
    principal series, 407
    regular, 406
    strongly continuous, 407
    unitarizable, 408
representation modules, 278
representations
    similar, 457
represented, 495
residually finite, 86, 476
residually small, 472, 473
residue field, 169

resolution of singularities, 358
restricted Burnside problem, 103
restricted universal enveloping algebra, 341
retract, 25
reversion, 290
rewriting system, 477
Rickart *-algebra, 299
Rickart *-ring, 294
right V-ring, 255, 512
right adjoint, 496
right Bezout, 182
right d-dependent, 232
right derived functor, 502
right distributive, 183
right exact, 501
right fir, 231
right Ikeda-Nakayama, 259
right integral element, 309
right invariant, 183, 221, 226, 228, 281
right Kasch, 259
right localization of A with respect to B, 282
right Noetherian, 242
right order, 273
right Ore condition, 202
right perfect, 220, 510
right PF-ring, 259
right pseudo-Frobenius, 259
right QF-3-ring, 261
right quasi-invariant, 281
right resolution, 500
right Rickartian, 221
right self-injective, 222, 225, 260, 509
right semihereditary, 221
right skew Laurent series ring, 184
right skew power series ring, 182
right t-nilpotent, 510
right uniserial, 183
right weakly regular, 256
right order, 202
rigid tree algebra, 436
ring
    Artinian, 244
    biregular, 184
    coherent, 505
    Dedekind, 515, 522
    finite conductor, 507
    general ZPI, 156

generalized GCD, 507
Goldie, 247
hereditary, 506
Laskerian, 154
Noetherian, 245
of strong finite type, 158
perfect, 248
principal projective, 506
reduced, 182
regular, 184, 505
right Bezout, 182
right distributive, 183
right invariant, 183
right uniserial, 183
semi-Artinian, 248
semilocal, 174
semiperfect, 248
semiprimary, 174
semiprimitive, 213
semisimple, 246
special principal ideal, 156
strongly Laskerian, 154
strongly regular, 183
zero-dimensional, 156
ring extension, 355
ring of algebraic integers, 168
ring of differential polynomials, 369
ring of entire functions, 168
ring of finite representation type, 278
ring of integers of F, 162
ring of stable range 1, 254
ring topology, 314
root, 336
root system, 333
Rosenberg, 459
Rosenberg classes, 459
rotation group, 149

Sabidussi, 456
sandwich matrix, 21
    regular, 22
satisfaction
    of an E-equation
        in a partial algebra, 483
satisfies the identity, 337
saturated, 457, 525
scalar product, 290
Schreier, 143
Schreier varieties, 144

Schreier variety, 337
Schunck class, 84
semi-Artinian, 226
semi-dihedral group, 76
Semi-Thue system, 477
semiautomorphism, 130
semidirect product, 44
semidistributive, 281
semidistributive lattice, 440
semifield, 319
    Boolean, 319
semifir, 233
semigroup, 53, 54
    0-bisimple, 18
    0-simple, 18
    $\mathcal{J}$-trivial, 32
    $\mathcal{K}$-indecomposable, 6
    almost nilpotent, 30
    aperiodic, 31
    archimedean, 2, 58
    automatic, 118
    bicyclic, 20, 47
    bisimple, 17
    Clifford, 40
    combinatorial, 61
    compact, 62
    completely 0-simple, 18
    completely regular, 23, 39, 61
    completely semisimple, 19
    completely simple, 17, 39, 63
    congruence-free, 18
    congruence-simple, 18
    crumbly, 8
    decomposable into a band, 5
    division, 34
    finitely assembled, 26
    finitely based, 59
    finitely presented, 14
    four-spiral, 20
    free, 51
    free Burnside, 29
    fundamental regular, 36
    group-bound, 24
    idempotent-free, 18
    inverse, 39
    Krohn-Rhodes complexity, 34
    lattice-determined, 57
    left 0-simple, 18
    left simple, 17

left singular, 5
left zero, 5
local semilattice, 32
locally finite, 28
locally nilpotent, 30
matrix, 49
nilpotent, 30
normal, 181
null, 18
of finite breadth, 54
periodic, 23
profinite, 32
pseudoinvertible, 24
quasi-completely regular, 24
quasiperiodic, 24
Rees matrix
    over a group with zero, 22
    over a semigroup, 21
regular, 35, 52
right 0-simple, 18
right simple, 17
right singular, 5
right zero, 5
semisimple, 19
semitopological, 62
separated, 47
separative, 7
simple, 17
singular, 5
topological, 62
torsion, 23
unary, 27
uniform, 50
unipotent, 24
unipotently partitionable, 24
with left [right] division, 21
semigroup identity, 51
semihereditary, 511
semihereditary ring, 506
semilattice, 5, 42
    of semigroups, 5
        strong, 7
semilattice of completely simple semi-
        groups, 40
semilocal, 174, 182, 230, 260, 281, 510
semiperfect, 510
semiprimary, 221, 262
semiprimary ring, 174
semiprime, 243, 249

semiprime ideal, 249
semiprime ring, 249
semiprimitive, 213
semiregular, 221, 255
semiring, 319, 321
    Boolean, 319
    zero-sum-free, 319
semisimple, 230, 253, 332, 509
    Lie group, 90
semisimple class, 213, 216
semisimple Lie superalgebra, 343
semisimple module, 224, 260
semitopological semigroup, 62
separable K-algebra, 274
separable extension, 363
serial, 511
SFT-ideal, 158
SFT-ring, 158
Σ-algebra, 487
Σ-algebra morphism, 487
sigma-notation, 306
Σ-semiring, 321
signature, 455, 487
Silver-Pohlig-Hellman algorithm, 378
similar elements, 235, 241
simple, 332, 509
simple points, 357
simple ring, 235, 249
simplest Boolean polynomial, 434
simplex code, 529
Sims' Conjecture, 88
singular point, 357
SK_1, 516
skeleton, 138
skew elements, 236
skew field, 198
skew group ring, 383
skew polynomial ring, 240
skew product, 310
skew trace, 293
skew-injective, 226, 261, 512
skew-projective, 222, 261, 512
skew-symmetric, 293
slim triangles, 118
small, 220, 227, 510
small category, 494
smooth vectors
    of a Lie group representation, 407
socle, 87, 260

soluble, 82
soluble series, 82
solution
    universal solution, 485
solvable, 344
    of length m, 331
solvable radical, 332
sorts, 487
space, Hilbert, 428
Specht problem, 338
special, 338
special algebra, 346
special class, 217
special linear Lie algebra, 336
special linear Lie superalgebra, 341
special principal ideal ring, 156
special radical, 214, 217
special unitary group, 150
specialisation, 199
specification, 486
spectral analysis, 413
spectrum, 356, 408
Sperner Space, 136
spin representations, 292
spinorial group, 292
SPP$_U$ theorem, 466
squarefree part, 159
stabilizer, 120
stable system, 352
stably isomorphic, 514
standard basis, 269
standard identity, 284
Stanley-Reisner ring, 181
state, 428
state of a system, 352
statement
    term existence statement, 483
Steinberg group, 518
Steinberg symbols, 518
Stone's Theorem, 433
strictly linear compact, 317
strong, 368
strong interaction, 150
strong semilattice structure, 40
strong Skolem property, 190
strongly π-regular, 256
strongly continuous representation, 407
strongly indecomposable, 177
strongly Laskerian ring, 154

strongly normal, 368
strongly primary ideal, 153
strongly regular, 183, 255, 280, 317
strongly right d-dependent, 232
strongly two generated, 191
structure
    factorization structure, 485
structures, quantum, 429
SU(2), 150
SU(3), 150
sub-exponential growth, 301
subalgebra
    of a partial algebra, 482
subalgebra lattice, 437
subconcept-superconcept-relation, 447
subdirect product, 452
subepigroup, 24
subgroup
    fully invariant, 141
    verbal, 141
submodule
    closed, 207
    pure, 207
subsemigroup
    inverse full, 57
substitution-cylindrification algebras, 481
subvariety, 356
suffix, 10
$\perp$-sum, 375
supernilpotent radicals, 217
supersolvable, 388
supremum-dense, 448
Suslin-Merkurjev Theorem, 519
Suzuki, 133
Swan's Theorem, 515
Sylow's Theorem, 100
symbol, 519
    Hilbert, 519
    norm residue, 519
    quaternion algebra, 519
    Steinberg, 518
symmetric, 293
symmetric element, 237, 299
symmetric group, 86
symmetric ring of quotients, 204
symmetry, 148
symmetry class, 121
symmetry group, 93
symplectic Lie algebra, 336

syntactic monoid, 68
system
    linear continuous, 352
    regular, 93
system of parameters, 170
system of simple roots, 333

T-ideal, 287
T-isomorphic, 46
T-isomorphism, 46
T-retract, 46
tame automorphism, 128
tame case, 161
Tarski's Transfer Principle, 375
tensor product, 74, 375, 419
term, 454
    n-ary, 458
    for partial algebras, 482
term algebra, 463
term existence statement, 483
term operation
    partial term operation, 483
term operations, 458
term rewriting system, 489
terminal object, 494
ternary discriminator, 460
Theorem
    Łos, 453
    Ado, 92
    Artin-Procesi, 285
    Baer-Kaplansky, 74
    Birkhoff, 452
    Brauer-Hasse-Noether, 195
    Burnside, 99
    Burnside's $p^a q^b$, 83
    classification, 99
    Feit-Thompson, 83, 100
    Frobenius, 99, 131
    Glaubermann, 99
    Gleason, 428
    Goldie, 247
    Hall, 83
    Hilbert Basis, 243
    Hu, 435, 459
    Kaplansky, 285
    Leptin-Liebert, 74
    Levitzki, 244
    Levitzki-Amitsur, 285
    Litoff-Ánh, 239

Merkur'ev-Suslin, 197
Neumann-Wiegold, 144
O'Nan-Scott, 87
of Adian, 16
of Brauer-Cartan-Hua, 295
of Dieckert, 16
of Goldie, 243
of Ito-Michler, 389
of Magnus, 16
of Mal'cev-Schmidt, 86
of Novikov-Boone, 107
of Redei, 16
Pólya, 122
Prüfer, 72
projection, 432
Ritt-Raudenbush, 369
stongly embedded, 101
Sylow, 100
Thompson, 132
Wedderburn-Artin, 246
theory, 523
    of partial algebras, 482
Thompson, 132
Thompson, J., 389
Thue problem, 475
Thue-Morse words, 13
Tietze problem, 475
tiled order, 279
time, 352
Tits alternative, 51
Todd-Coxeter algorithm, 16
tolerance relation, 137
top group, 124
topological group, 62
topological homomorphism, 314
topological isomorphism, 314
topological ring, 314
topological semigroup, 62
torsion class, 23
torsion group, 71
torsion module, 242
torsion part, 71
torsion product, 74
torsion submodule, 242
torsion theory, 219
torsion-free, 71, 227, 242
torsion-free group, 71
torsionless modules, 278
torus

maximal, 406
total algebra, 482
total divisor, 241
totally ordered ring, 312
totally transcendental, 201
trace, 43, 150, 385
trace iddentity, 288
trace ring, 287
transcendence basis, 356
transcendence degree, 356
transfer function, 354
transformation group, 148
transformation monoid, 67
transformation semigroup, 46
transformation, Lorentz, 149
transformation, phase, 149
translation, 147
translation group, 149
translation structure, 136
    generalized, 135
tree algebra, 436
trivial polynomial mod S, 269
trivial relation, 233
trivial ring, 231
trivialisable, 233
twisted group ring, 383
type, 451
typical differential dimension, 370

U(1), 149
U-band, 8
U-semigroup, 53
ultraproduct, 453
uncertainty relation, 150
uniform component, 51
uniform module, 225
uniform semigroup, 50
uniformizer, 275
unipotency class, 23
unique factorisation domain, 241
uniquely divisible by, 145
uniserial, 223, 280, 511
unit, 62
unit-regular, 254
unitarizable representation, 408
unitary representation, 406
    of a Lie group, 406
universal
    epimorphism, 485

solution, 485
universal R-field, 199
universal algebra, 451
universal element, 495
universal enveloping algebra, 272, 333
    of a Lie algebra, 406
universal formula, 524
universal mapping property, 461
universal property, 453
universal structure, 525
universal theory, 524
universally constructed, 495
upper radical, 216

V-formation, 446
validity
    of an E-equation
        in a partial algebra, 483
valuation
    in a partial algebra, 483
valuation ring, 169
van Kampen diagram, 107
vanish at infinity, 298
variable, 483
variety, 102, 356, 465
    Burnside, 142
    equationally complete, 468
    finitely based, 102, 445, 468
    generated by a class, 466
    non-trivial, 466
    of semigroups, 58
        overcommutative, 60
        periodic, 60
    ofgroups, 140
    polynilpotent, 142
    Schreier, 144
variety-product, 142
vector bundle, 515
verbal ideal, 337
verbal subgroup, 141
Veronese subring, 178
vertex algebra, 329
Virasoro algebra, 327
von Dyck groups, 118
von Neumann algebra, 297
von Neumann regular, 230, 253, 509
von Neumann regular radical, 215
von Neumann regular ring, 506

W-boson, 150

Wall finiteness obstruction, 515
Wave equation, 372
weak interaction, 149
weak action, 310
weak algorithm, 232
weak inverse, 33
weakly Krull, 165
weakly special class, 217
Wedderburn's theorem, 51
weight enumerator polynomial, 529
weight of a word, 528
well below, 439
Werner, H., 460
Weyl algebra, 236
Weyl group, 52, 92
Whitehead torsion, 516
Whitman's condition, 440
Wielandt, 132
Witt ring, 375, 520
Wolf, T., 389
word, 140
    cube-free, 12
    empty, 10
    over an alphabet, 8
    primitive, 9
    square-free, 12
word problem, 474
    decidable, 474
    solvable, 474
word-acceptor, 115
word-hyperbolic group, 118
wreath product
    complete, 124
    complete permutational, 123
    permutational, 123
    restricted, 124
    restricted permutational, 123
    standard, 124
    twisted, 127
    verbal, 127

Yoneda Lemma, 495

Z-order, 278
Z-ring, 278
Zariski closure, 51
Zassenhaus, 132
Zassenhaus group, 133
Zelmanov, 346
zero, 319

zero differentiation, 498
zero-dimensional ring, 156
zero-divisor, 319